Operations Research Proceedings 2005

Selected Papers
of the Annual International Conference
of the German Operations Research Society (GOR)

Bremen, September 7–9, 2005

Hans-Dietrich Haasis · Herbert Kopfer
Jörn Schönberger (Editors)

Operations Research Proceedings 2005

Selected Papers
of the Annual International Conference
of the German Operations Research Society (GOR),

Bremen, September 7–9, 2005

With 165 Figures and 109 Tables

Springer

Prof. Dr. Hans-Dietrich Haasis
University of Bremen
Institute of Shipping Economics and Logistics
Dept. Logistics Systems
Universitätsallee GW1 Block A
28359 Bremen
Germany
e-mail: haasis@isl.org

Prof. Dr.-Ing. Herbert Kopfer
University of Bremen
Faculty of Business Studies and Economics
Chair of Logistics
PO Box 33 04 40
28334 Bremen
Germany
e-mail: kopfer@logistik.uni-bremen.de

Dr. Jörn Schönberger
University of Bremen
Faculty of Business Studies and Economics
Chair of Logistics
PO Box 33 04 40
28334 Bremen
Germany
e-mail: sberger@logistik.uni-bremen.de

ISSN 0721-5924
ISBN-10 3-540-32537-9 Springer Berlin Heidelberg New York
ISBN-13 978-3-540-32537-6 Springer Berlin Heidelberg New York

Springer is a part of Springer Science+Business Media

springer.com

Production: LE-TEX Jelonek, Schmidt & Vöckler GbR, Leipzig
Cover design: eStudio Calamar S.L., F. Steinen-Broo, Pau/Girona, Spain

SPIN 11675679 42/3100/YL 5 4 3 2 1 0 Printed on acid-free paper

Preface of the Conference Chairs

From September 7-9, 2005 the international scientific annual conference Operations Research 2005 of the Gesellschaft für Operations Research (GOR) took place at the Bremen University for the first time. More than 600 participants from Germany and from all over the world participated in this scientific meeting. The program consisted of 2 plenary, 15 semi-plenary, and more than 400 contributed presentations, which were organized in 20 sections. The two special interest sections on "logistics" and "new maritime businesses" defined the major topics of the conference.

From the overwhelming amount of talks held during this conference, 128 long versions were accepted for publication. The accepted contributions represent a wide portfolio chosen from the comprehensive spectrum of Operations Research in theoretical research and practical experience. The great response to the conference proved the high significance of Operations Research in economics and science.

The plenary talks were held by Professor Gilbert Laporte, Université de Montréal, Canada, to "Overview of Recent Metaheuristics for the Vehicle Routing Problem" as well as by Professor Bernhard Gläser, Social Science Research Center Berlin, to "Integriertes Küstenzonenmanagement und neue maritime Wertschöpfung".

During the conference the GOR Dissertation Prizes as well as the GOR Diploma Prizes were awarded. We congratulate all winners, especially Professor Rudolf Möhring from the Mathematical Institute of the Technical University Berlin on receiving the GOR Scientific Prize award.

Various persons and organizations contributed to the great success of the conference. We would like to thank the GOR-board as well as the members of the program and organisation committees of the congress. Furthermore, we are grateful to all speakers from Germany and from all over the world for their active participation and to the section leaders as well as to the session chairs for their professional moderation of the interesting talks and discussions. Our thanks are offered to the sponsors, whose financial contributions meant a lot for the realization of this conference.

We express our special thanks to our staff members Dr. Jörn Schönberger and Dipl.-Wirtschaftsing. Hendrik Wildebrand for their excellent job during the congress and their great engagement in the tasks before, during and after the congress. We would also like to thank Bärbel Niedzwetzki from GOR administrative office as well as Dr. Werner A. Müller and Barbara Feß from Springer-Verlag for their strong support in publishing this proceedings volume.

Bremen,
December 2005

Univ.-Prof. Hans-Dietrich Haasis
Univ.-Prof. Herbert Kopfer

Committees

Program

Prof. Dr. Hans-Dietrich Haasis (Bremen, chair)
Prof. Dr. Herbert Kopfer (Bremen, chair)
Prof. Dr. Horst W. Hamacher (Kaiserslautern)
Prof. Dr. Michael Jünger (Cologne)
Prof. Dr. Josef Kallrath (Weisenheim)
Prof. Dr. Martin G. Möhrle (Bremen)
Prof. Dr. Thorsten Poddig (Bremen)
Prof. Dr. Heinrich Rommelfanger (Frankfurt/M.)
Prof. Dr. Thomas Spengler (Braunschweig)

Organization

Prof. Dr. Hans-Dietrich Haasis
Prof. Dr. Herbert Kopfer
Prof. Dr. Martin G. Möhrle
Prof. Dr. Thorsten Poddig
Katrin Dorow
Dr. Jörn Schönberger
Nadja Shigo
Hendrik Wildebrand

Scientific Sections and Section Leaders

Logistics
H. Kopfer (Bremen), J. Dethloff (Bremen)

New Maritime Businesses
H.-D. Haasis (Bremen)

Production & Supply Chain Management
H.-O. Günther (Berlin)

Finance, Banking and Insurance
H. Locarek-Junge (Dresden), T. Poddig (Bremen)

Artificial Intelligence and Fuzzy Logic
H. Rommelfanger (Frankfurt)

Discrete & Combinatorial Optimization
J. Kallrath (Weisenheim), A. Martin (TU Darmstadt)

Routing and Networks
D. Mattfeld (Braunschweig), J. Schönberger (Bremen)

OR Applications in Health and Life-Sciences
S. Fleßa (Greifswald), S. Pickl (Cologne)

Continuous Optimization
V. Schulz (Trier)

Econometrics, Game Theory and Mathematical Economics
M. Lehmann-Waffenschmidt (Dresden), C. Puppe (Karlsruhe)

e-Business and Computer Sciences
U. Hasenkamp (Marburg), S. Voß (Hamburg)

Sustainable Systems
A. Tuma (Augsburg)

Revenue Management
A. Kimms (Freiberg), R. Klein (Darmstadt)

Marketing
D. Baier (Cottbus), U. Wagner (Vienna)

Managerial Accounting
H.-U. Küpper (Munich), C. Hofmann (Hannover)

Tourism, Entertainment and Sports
A. Drexl (Kiel), S. Knust (Osnabrück)

Scheduling and Project Management
E. Pesch (Siegen), R. Kolisch (Munich)

Technology and Innovation
M. G. Möhrle (Bremen), C. Stummer (Vienna)

Decision Theory
W. Habenicht (Hohenheim), C. Tammer (Halle)

Applied Probability
K.-H. Waldmann (Karlsruhe)

Contents

Part XII Econometrics, Game Theory and Mathematical Economics

Part XIII e-Business and Computer Sciences

Part XIV Sustainable Systems

Part XV Revenue Management

Part XVI Marketing

Part XVII Managerial Accounting

Part XVIII Tourism, Entertainment and Sports

Part XIX Scheduling and Project Management

Part XXII Applied Probability

Part I

Dissertation Award of the GOR

Zeitkontinuität in zeitdiskreten Modellen – Neue Ansätze für die Produktionsplanung in der Prozessindustrie

Christopher Sürie

Feldstraße 45, 63225 Langen, csurie@web.de
Dissertation erstellt am Fachgebiet Produktion & Supply Chain Management, Technische Universität Darmstadt, Hochschulstraße 1, 64289 Darmstadt

Zeitdiskrete Modelle sind dadurch gekennzeichnet, dass sie Zustände an bestimmten (diskreten) Zeitpunkten beschreiben sowie die Summe der Ereignisse zwischen diesen Zeitpunkten. Dementsprechend führt die Abbildung von zeitkontinuierlichen Prozessen – wie sie häufig in der Prozessindustrie anzutreffen sind – zu Abbildungsdefekten an diesen Diskretisierungspunkten.

Wünschenswert ist es jedoch, auch zeitkontinuierliche Prozesse in zeitdiskreten Modellen adäquat abbilden zu können. Daher werden die hieraus resultierenden Abbildungsdefekte zunächst ausführlich analysiert, und es wird dargestellt, wie sie durch geschickte Modellierung weitgehend verhindert werden können.

Anschließend werden die entwickelten gemischt-ganzzahligen Modellformulierungen mit aus der Literatur bekannten Modellen verglichen sowie die Ergebnisse von Rechentests präsentiert, welche ihre Leistungsfähigkeit bestätigen. Als Lösungsverfahren dienen Standardsolver für gemischt-ganzzahlige Optimierungsprobleme in Verbindung mit Schnittebenen und einer zeitlichen Dekompositionsheuristik.

Inhalt der Dissertation

Das Ziel der Dissertation (Suerie 2005) besteht darin, einen Modellierungsansatz zu entwickeln, der es ermöglicht, zeitkontinuierliche Prozesse – wie sie häufig in der Prozessindustrie anzutreffen sind – in zeitdiskreten Modellen abzubilden.

Ausgangspunkt hierfür ist eine Klassifikation von Losgrößenmodellen. Als Hauptunterscheidungsmerkmal wird die relative zeitliche Länge der Perioden gewählt. Findet in jeder Periode nur maximal ein Rüstvorgang statt, so spricht man von „small-bucket“ Modellen, können in jeder Periode hingegen mehrere Rüstvorgänge eingeplant werden, so spricht man von „big-bucket“ Modellen. Neben diesen beiden Klassen wird noch die Klasse der „hybriden“ Modelle eingeführt, um Modelle einzuordnen, die Charakteristika beider Klassen vereinen. Verschiedene Losgrößenmodelle werden in diese drei Klassen eingeordnet, und ihre Vor- und Nachteile werden herausgearbeitet. Um Zeitkontinuität in zeitdiskreten Modellen abbilden zu können, müssen aufeinander folgende Perioden verknüpft werden, das heißt, der Zustand an den Periodengrenzen muss eindeutig definiert sein. Dies ist nur bei den vorgestellten „small-bucket“ Modellen sowie den „hybriden“ Modellen der Fall. Entsprechend werden das *P*roportional *L*ot-Sizing and *S*cheduling *P*roblem (PLSP, vgl. Drexl und Haase 1995) als ausdrucksstärkstes „small-bucket” Modell sowie das *C*apacitated *L*ot-*S*izing *P*roblem with *L*inked Lot Sizes (CLSPL, vgl. Suerie und Stadtler 2003) aus der Klasse der „hybriden“ Modelle als Basismodelle zur weiteren Betrachtung ausgewählt.

Im dritten Kapitel der Dissertation werden die verschiedenen Erweiterungen für zeitdiskrete Losgrößenmodelle diskutiert, die notwendig sind, um Aspekte von Zeitkontinuität abzubilden. Als erster wichtiger Eckpunkt wird die Modellierung von Rüstzuständen an den Periodengrenzen identifiziert (vgl. Fig. 1). Periodengrenzen resultieren aus der Einteilung des Planungszeitraums in Perioden. Von der korrekten Modellierung dieser Punkte hängt wesentlich ab, ob der resultierende Plan an diesen Stellen Unstetigkeiten aufweist. Mit Hilfe von Beispielen, die der Literatur entnommen sind, kann gezeigt werden, dass sich Pläne, in denen der Erhalt von Rüstzuständen über Periodengrenzen hinweg abgebildet ist, fundamental von solchen Plänen unterscheiden, in denen dies nicht der Fall ist. Dieser Unterschied wird auch mit Ergebnissen eigener Rechentests belegt.

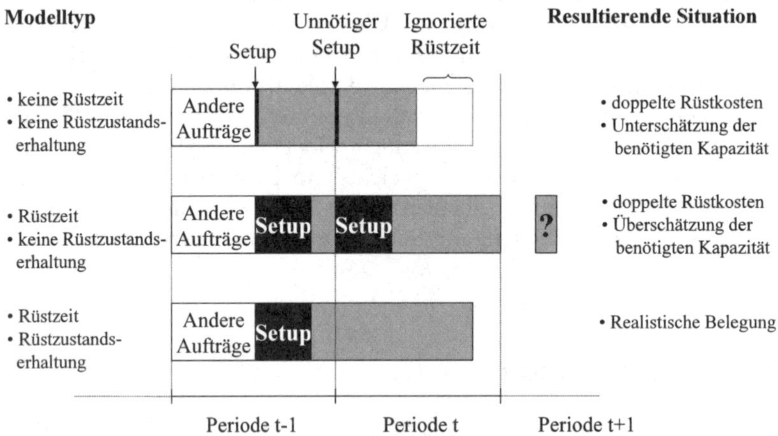

Fig. 1. Modellierung der Periodengrenzen (Suerie 2005, S. 34).

Als zweiter wesentlicher Eckpunkt zur Modellierung von Zeitkontinuität in zeitdiskreten Losgrößenmodellen wird die Modellierung von Losgrößen identifiziert, die sich über zwei oder mehrere Perioden erstrecken. Speziell in der Prozessindustrie tritt häufig die Problematik auf, dass Losgrößen bestimmte Mindestgrößen nicht unterschreiten oder bestimmte Höchstgrenzen nicht überschreiten dürfen. Zudem muss die Losgröße häufig als ein ganzzahliges Vielfaches der Größe eines „Batch“ darstellbar sein (vgl. Kallrath 2002). Diese Beschränkungen sind meist durch die Produktionsprozesse oder Produktionsanlagen bedingt. Mindestgrößen können sich beispielsweise durch das Erreichen einer zu einer chemischen Reaktion notwendigen kritischen Masse ergeben. Höchstgrenzen sind meist dadurch bedingt, dass nach Erreichen einer maximalen Produktionsmenge die verwendeten Anlagen gereinigt werden müssen. Die Beschränkung auf ganzzahlige Vielfache einer „Batch“-Größe ergibt sich hingegen durch vorgegebene Behältergrößen (Reaktoren, Tanks). Zeitdiskrete Losgrößenmodelle weisen in diesem Zusammenhang die Schwäche auf, dass sich durch sie geplante Mengen auf Perioden beziehen. Hier ist es jedoch entscheidend, die jeweiligen Restriktionen nicht auf die in den einzelnen Perioden zu produzierenden Mengen zu beziehen, sondern auf das komplette Los, dessen Produktion sich über zwei oder mehr Perioden hinzieht. Fig. 2 zeigt beispielhaft, wie sich die optimale Lösung eines Problems verändert, sobald diese zusätzlichen Restriktionen berücksichtigt werden.

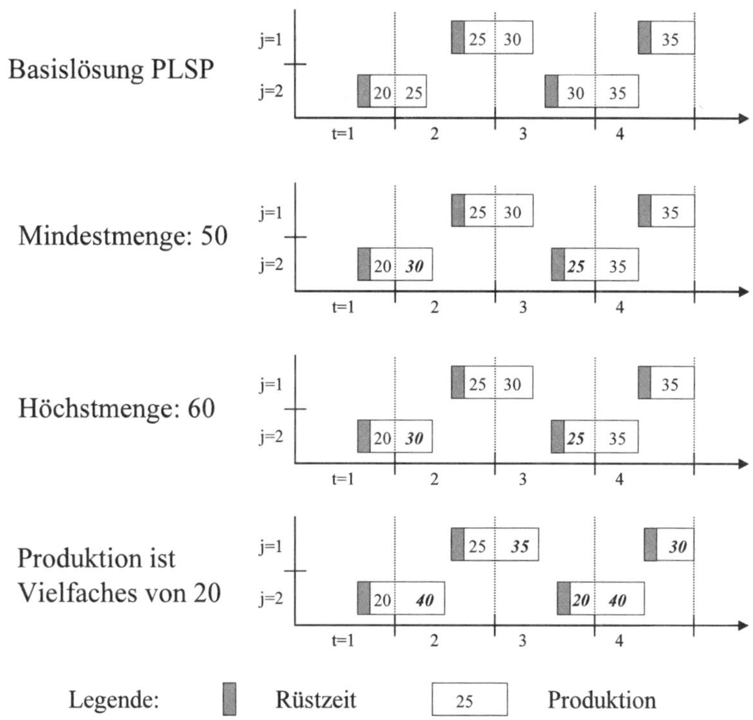

Fig. 2. Veränderung der Basislösung bei Berücksichtigung verschiedener Restriktionen (Änderungen der Produktionsmengen jeweils kursiv; Suerie 2005, S. 38)

Die exakte Abbildung von Rüstvorgängen wird als dritter Eckpunkt identifiziert, da auch Rüstvorgänge eine zeitliche Dauer aufweisen, die auf eine Periodengrenze fallen kann. Schließlich rundet die Analyse des Einflusses dieser drei Eckpunkte auf die Auslastung der Produktionsanlagen die Diskussion der notwendigen Erweiterungen von zeitdiskreten Modellen zur Abbildung von Zeitkontinuität ab.

Im vierten Kapitel der Dissertation wird ein Überblick der relevanten Literatur gegeben. In diesem Literaturüberblick werden zwei Ziele verfolgt. Zum einen werden die im zweiten Kapitel vorgestellten Losgrößenmodelle daraufhin untersucht, inwieweit die im dritten Kapitel erarbeiteten Erweiterungen zur Abbildung von Zeitkontinuität in zeitdiskreten Modellen bereits in ihnen enthalten sind. Zum anderen werden Modelle aus der Prozessindustrie als dem vorrangigen Anwendungsgebiet vorgestellt. Obwohl diese oft zusätzliche Nebenbedingungen aufweisen, stellt sich ihre Analyse als wertvoll heraus, da sie sowohl eine große Vielfalt an Modellierungsideen bereitstellen als auch ein Hauptanwendungsgebiet der im Weiteren entwickelten Modelle darstellen. Um die Erweiterungsmöglichkeit des vorgestellten Modellierungs- und Lösungsansatzes zu unterstreichen, wird der vorgestellte Ansatz drei aus diesem Teil der Literatur entnommenen Modellformulierungen im Rahmen der durchgeführten Rechentests gegenübergestellt.

Das fünfte Kapitel der Dissertation gibt eine kurze Einführung in das Konzept der Advanced Planning Systeme. Die im Weiteren vorgestellten Modellformulierungen und der gewählte Lösungsansatz sind sehr gut dafür geeignet, um in einem solchen System eingesetzt zu werden. Des Weiteren werden die Grundlagen der verwendeten Lösungstechniken (gemischt-ganzzahlige Optimierung, Dekomposition) vorgestellt. In diesem Zusammenhang wird auch ein Überblick über die derzeitigen Fähigkeiten von Standardsoftware zur gemischt-ganzzahligen Optimierung gegeben.

Der Hauptbeitrag der Dissertation liegt in der Modellierung von Aspekten von Zeitkontinuität in zeitdiskreten Losgrößenmodellen. Im sechsten Kapitel der Dissertation werden daher gemischt-ganzzahlige Modellformulierungen für die im dritten Kapitel identifizierten Eckpunkte entwickelt. Die Modellformulierungen werden für beide in Kapitel zwei als geeignet identifizierten Basismodelle (PLSP und CLSPL) vorgestellt. Dies ist notwendig, um die unterschiedlichen Charakteristika dieser beiden Modelle zu berücksichtigen. Gleichzeitig ist es aber auch möglich, diese beiden Basismodelle zu kombinieren.

Die entwickelten Erweiterungen hinsichtlich (a) der Modellierung von Rüstzuständen an Periodengrenzen, (b) periodenübergreifender Losgrößen, (c) periodenübergreifender Rüstzeiten sowie (d) der verschiedenen Restriktionen bezüglich der Auslastung der Produktionsanlagen werden nach dem Baukastenprinzip entworfen und sind beliebig kombinierbar. Damit ist es möglich, durch Kombination aller genannten Erweiterungen beliebige Pläne, die auf einem kontinuierlichen Zeitstrahl abbildbar sind, auch in einem zeitdiskreten Raster darzustellen.

Um die (rechentechnische) Leistungsfähigkeit der Modellformulierungen zu verbessern, werden zudem Schnittebenen entwickelt. Da jede Erweiterung eigene (zusätzliche) Variablen sowie Nebenbedingungen erfordert, empfiehlt es sich, nicht immer ein komplettes Modell zu verwenden, sondern sich des Baukastenprinzips zu bedienen und nur die in der vorliegenden Planungssituation tatsächlich notwendigen Erweiterungen zu berücksichtigen. Trotzdem kann es vorkommen, dass die gemischt-ganzzahlige Optimierung mit Hilfe von Standardsoftware an ihre Grenzen stößt. Aus diesem Grund wird eine Dekompositionsheuristik (Stadtler 2003) auf die entwickelten Modellformulierungen adaptiert.

Schließlich werden die vorgestellten Modellformulierungen sowie die Dekompositionsheuristik im siebten Kapitel ausführlichen Rechentests unterzogen. Eine Analyse optimaler Lösungen belegt den fundamentalen Unterschied zwischen Modellen, die eine Erhaltung von Rüstzuständen an Periodengrenzen erlauben, und solchen, die dies nicht tun. Hinsichtlich der Restriktionen, die sich aus Sicht der Prozessindustrie für Losgrößen ergeben, kann gezeigt werden, dass die Einhaltung von Mindest- beziehungsweise Höchstmengen deutlich weniger Rechenaufwand erfordert als die Beachtung der Beschränkung der Losgröße auf ein ganzzahliges Vielfaches einer „Batch"-Größe. Allerdings ist der Fortschritt hinsichtlich der Lösbarkeit von Modellen mit diesen Restriktionen, der sich aus dem Vergleich des entwickelten Modellierungsansatzes mit einem Benchmark aus der Literatur ergibt, insbesondere hier sehr groß. Sämtliche verwendeten Benchmarks aus der Literatur werden sowohl hinsichtlich der Lösungsgüte als auch hinsichtlich der benötigten Rechenzeit deutlich übertroffen.

Die Erweiterbarkeit der vorgestellten Modellformulierungen hinsichtlich weiterer Nebenbedingungen wird dadurch demonstriert, dass das Modell mit drei aus der prozessindustriespezifischen Literatur stammenden Modellen auf Basis eines von diesen verwendeten Testdatensatzes verglichen wird. Hier zeigt sich, dass die entwickelte Modellformulierung nicht nur hinsichtlich der benötigten Rechenzeit konkurrenzfähig ist. Vielmehr kann auch ein neue, bisher unbekannte, optimale Lösung erzeugt werden. Da alle Rechentests mit der Standardsoftware XpressMP (Release 2003G) durchgeführt worden sind, werden zudem Testergebnisse präsentiert, die mit verschiedener Standardsoftware zur gemischt-ganzzahligen Optimierung generiert worden sind. Es zeigt sich, dass die Ergebnisse unabhängig von der verwendeten Optimierungssoftware sind.

Zusammenfassend lässt sich festhalten, dass im Rahmen der Dissertation neue Modellformulierungen entwickelt wurden, die es erlauben, alle Pläne, die auf einer zeitkontinuierlichen Achse darstellbar sind, auch in einem zeitdiskreten Planungsumfeld abzubilden. Es verbleiben lediglich kleine Einschränkungen, nämlich dass *in jeder Periode* maximal ein Rüstvorgang stattfinden darf (wenn als Basismodell das PLSP gewählt wird) beziehungsweise dass *in jeder Periode für jedes Produkt* nur ein Rüstvorgang stattfinden darf (wenn das CLSPL als Basismodell gewählt wird). Auf der anderen Seite empfiehlt es sich, mit kurzen Perioden zu planen, da anderenfalls die korrekte Modellierung der Periodengrenzen an Relevanz verliert, so dass die genannte Einschränkung vernachlässigbar bleibt. Die Integration der entwickelten Modellbausteine in ein Planungssystem führt zu Plänen, die keine Unstetigkeiten mehr durch die Diskretisierung der Zeit in Perioden aufweisen. Zudem haben sich die vorgestellten Modellformulierungen als rechentechnisch effizient erwiesen.

Literatur

Drexl A, Haase K (1995) Proportional Lotsizing and Scheduling. International Journal of Production Economics 40:73-87

Kallrath J (2002) Planning and Scheduling in the Process Industry. OR Spectrum 24:219-250

Stadtler H (2003) Multilevel Lot-Sizing with Setup Times and Multiple Constrained Resources: Internally Rolling Schedules with Lot-Sizing Windows. Operations Research 51: 487-502

Suerie C (2005) Time Continuity in Discrete Time Models. New Approaches for Production Planning in Process Industries. Lecture Notes in Economics and Mathematical Systems No. 552. Springer, Berlin Heidelberg New York

Suerie C, Stadtler H (2003) The Capacitated Lot-Sizing Problem with Linked Lot-Sizes. Management Science 49:1039-1054

A Hierarchical Production Planning Approach for Multiprocessor Flow Shops

Daniel Quadt

Department of Production, Logistics and Operations Management, Catholic University of Eichstätt-Ingolstadt, Auf der Schanz 49, 85049 Ingolstadt, Germany
daniel.quadt@ku-eichstaett.de

Summary. We consider the lot-sizing and scheduling problem for multiprocessor flow shops. In this environment, lot-sizing and scheduling are interdependent. A hierarchical planning approach is presented together with an overview of the developed solution procedures. The objective is to minimizes setup, inventory holding and back-order costs as well as the mean flow time. Computational results are illustrated.

1 Introduction

A multiprocessor flow shop is a flow shop with several parallel machines on some or all production stages. Figure 1 shows an example. Multiprocessor flow shops are ubiquitously found in various industries, for example in the automotive, the chemical, the electronics, the packaging and the textile industries (see e.g. [1, 2, 15, 18]).

We consider the short-term production planning problem for multiprocessor flow shops, which comprises the lot-sizing and the scheduling problem. The

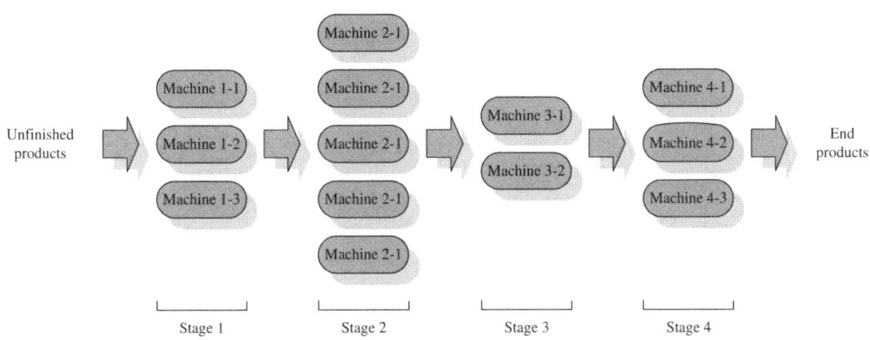

Fig. 1. Example of a multiprocessor flowshop

lot-sizing problem is to calculate production orders (lot-sizes) for an upcoming, short-term planning horizon, which typically covers one to several weeks. The production orders have to meet certain demand volumes for a number of products. If a demand cannot be met, the volume is back-ordered and has to be produced after its due date. The scheduling problem is to assign one of the parallel machines on each stage and to define precise production start- and end-times for all jobs and thus the job sequence on each machine.

Several jobs have to be scheduled for a number of products. Setup times and costs are incurred when changing a machine from one product to another. For this reason, a batching procedure is usually embedded in the scheduling phase. The batching procedure determines how many units of a product to consecutively produce on a specific machine. In the industrial practice, products can often be aggregated to (logistical) product families. A solution procedure should exploit this fact, as typically, setup times and costs are lower when changing a machine between products of the same family.

The machines of a stage are assumed to be identical, meaning that all machines of a stage can produce all products, and their process times are the same. However, the process times may vary for different products and production stages. Hence, to reach similar production rates among stages, the number of machines per stage may be different as well. A solution procedure has to consider the parallel machines individually, as typically, a setup has to be performed on all machines that produce a product. In contrast, using aggregated machines and production volumes usually implies that these setups cannot be modeled accurately.

Lot-sizing and scheduling decisions are interdependent: It is not possible to determine a schedule without knowing the lot-sizes that have to be produced. On the other hand, one cannot calculate optimal lot-sizes without knowing the machine assignments and the product sequences. These interdependencies make it impossible to effectively solve the problems separately [10].

This paper is organized as follows: Sect. 2 presents a short literature review. An overview of a new, integrative solution procedure is illustrated in Sect. 3. Computational results are presented in Sect. 4. We give a summary and draw conclusions in Sect. 5.

2 Literature Review

So far, the lot-sizing and the scheduling problem for multiprocessor flow shops have mainly been considered separately. Such dichotomic approaches cannot coordinate the interdependencies of the two problems. Nevertheless, both the lot-sizing and the scheduling problem are known to be NP-hard, even when solved separately.

There is little research on the lot-sizing problem for multiprocessor flow shops. Some authors consider similar systems, such as parallel machines or job shops with alternative routings (e.g. [4, 5, 8, 9]).

There are numerous studies covering the stand-alone scheduling component of the problem, e.g. [3, 7, 12, 16, 17]. A recent survey on optimal solution procedures is given in [6], a survey focussing on heuristics in [14].

There is a lack of solution procedures that consider the combined lot-sizing and scheduling problem. Such approaches are needed to incorporate the interdependencies between the two problems.

3 An Integrative Solution Approach

We illustrate an integrative lot-sizing and scheduling approach for multiprocessor flow shops. The objective is to minimize setup, inventory holding and back-order costs as well as the mean flow time through the flow shop.

The solution approach consists of the phases 'Bottleneck planning', 'Schedule roll-out' and 'Product-to-slot assignment'. Figure 2 shows an overview and depicts the developed procedures. Although the first phase is bottleneck oriented, a special model formulation ensures that the other stages are taken into account implicitly also in this phase.

The first phase of the solution approach solves a single stage lot-sizing and scheduling problem on the bottleneck stage. Products are aggregated to product families. A novelty of the procedure is that it is based on a new 'heuristic' model formulation that uses integer variables in contrast to binary variables as employed by standard formulations. This makes it possible to solve the model to optimality or near-optimality using standard algorithms that are embedded in commercial software packages (e.g. CPLEX). However, the heuristic model cannot capture all capacity restrictions as imposed by the original problem. For this reason, it is embedded in a period-by-period heuristic, which iteratively solves instances of the model using standard algorithms (CPLEX).

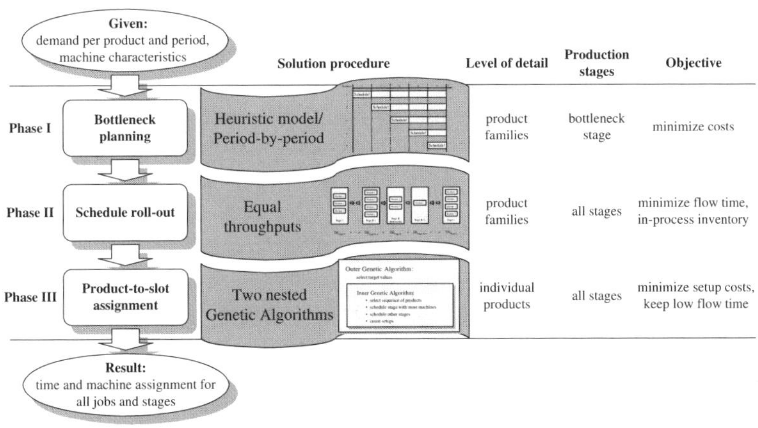

Fig. 2. Summary of the solution approach and the developed procedures

The second phase generates a product family plan on all production stages. Thus, it keeps the aggregation level of product families, but explicitly includes the non-bottlenecks stages. It assigns a production time and a machine to each product family unit, which means that it solves a multi-stage scheduling and batching problem. However, the number of machines per product family and the batch-sizes are pre-determined by the first phase. The objective is to minimize the mean flow time. This implies a minimization of waiting times between stages, and hence of in-process inventory. As a side effect, the usage of intermediate buffers is also minimized. The basic idea of the solution procedure is to establish the same product family throughput on all stages. In this way, each preceding stage supplies the intermediate products just in time for the subsequent stage. Thus, the products do not have to wait between stages and the flow time as well as the inventory volumes of intermediate products are minimized. The result of Phase II are so-called 'machine/time slots', signalling at what time and on which machine a product family unit is to be produced.

The third phase considers the individual products and calculates the final schedule, which consists of exact production times and machine assignments for every product unit on all stages. The problem is to determine how many and which machines to set up and when to produce the individual product units. Hence, it is a scheduling and batching problem for all stages on product level. The solution is calculated using the machine/time slots from the second phase: We assign an individual product unit of the respective family to each of the slots under the objective of minimizing intra-family setup costs, while keeping the low flow time of the second phase. There is a trade-off between the two objectives. Two nested Genetic Algorithms are employed to solve the problem. A novelty of the approach is the representation scheme: An outer Genetic Algorithm sets a target number of setups and a target batch-size for each product of the current family. With these target values, the problem resembles a two-dimensional packing problem: On a Gantt-Chart, the available slots for the product family form a pallet and the products form rectangular boxes that have to be placed on the pallet. An inner Genetic Algorithm tries to find an optimal packing pattern. The packing pattern may in turn be interpreted as a schedule.

A detailed description of the phases and the developed solution procedures can be found in [10], a more detailed overview in [13].

4 Computational Results

Several variants of the algorithm for the first phase have been implemented and compared with a direct implementation of a mixed integer programming model of the problem in CPLEX. For 960 small test instances with 5 products families, 4 periods and 4 machines each, the best heuristic leads to solutions that are 23% worse than the optimal solutions computed by CPLEX. The average computation time of the heuristic is 11 seconds, compared with 45 seconds required by CPLEX to prove optimality. In addition, 288 large test instances have been generated with 20 product families, 6 periods and 10 machines each. Using the direct implementation, CPLEX is able to solve only 97 of these instances within a time-limit of 1 hour. None of these solutions is proven to be optimal. The best heuristic is able to solve 285 of the instances, with an average computation time of approximately 25 minutes. On average, the heuristic solutions are approximately 24% better than the time-truncated CPLEX solutions [11].

The heuristics of the second and the third phase have been evaluated jointly [10]. 384 instances have been generated covering 3 production stages, 20 product families, up to 10 products per family and up to 80 machines on a stage. The findings show that—besides the number of products and the number of machines—also the demand pattern has a significant influence on the number of setups: The number of setups is lower if there are high and low volume products within a family compared with an evenly distributed volume among the products of a family. The computation time depends on the problem size. On average, it is 11 minutes. For large problems with 10 products per family, it is approximately 21 minutes.

5 Summary and Conclusions

We have presented an integrative solution approach for the combined lot-sizing and scheduling problem for multiprocessor flow shops. Computational tests have shown that the procedures are able to solve practically sized problems in a moderate amount of time. Thus, they can be used in real business applications. The procedures presented for the second and third phase are embedded in a scheduling system for a semiconductor assembly facility. In a comparative study with the performance of the real shop floor that has been planned manually, the system has led to a flow time reduction of 55%, while at the same time, only 80% of the capacity on the bottleneck stage has been utilized. The system has also been used for a capacity planning study [10].

There are some aspects open for further research: Alternative solution approaches should be developed to compare the heuristic. Especially optimal procedures would be beneficial. At the same time, the presented planning approach is flexible in the sense that it does not depend on the heuristics developed for each of the phases. Alternative procedures could be developed for one or more phases to take different scenarios into account, such as non-stable bottlenecks or different objectives.

References

1. Leonard Adler, Nelson Fraiman, Edward Kobacker, Michael Pinedo, Juan Carlos Plotnicoff, and Tso Pang Wu. BPSS: A scheduling support system for the packaging industry. *Operations Research*, 41(4):641–648, 1993.

2. A. Agnetis, A. Pacifici, F. Rossi, M. Lucertini, S. Nicoletti, F. Nicolo, G. Oriolo, D. Pacciarelli, and E. Pesaro. Scheduling of flexible flow lines in an automobile assembly plant. *European Journal of Operational Research*, 97:348–362, 1997.
3. Meral Azizoglu, Ergin Cakmak, and Suna Kondakci. A flexible flowshop problem with total flow time minimization. *European Journal of Operational Research*, 132(3):528–538, 2001.
4. Matthias Carl Derstroff. *Mehrstufige Losgrößenplanung mit Kapazitätsbeschränkungen*. Physica, Heidelberg, 1995.
5. K. Haase and A. Kimms. Lot sizing and scheduling with sequence dependent setup costs and times and efficient rescheduling opportunities. *International Journal of Production Economics*, 66:159–169, 2000.
6. Tamás Kis and Erwin Pesch. A review of exact solution methods for the non-preemptive multiprocessor flowshop problem. *European Journal of Operational Research*, 164(3):592–608, 2005.
7. Mary E. Kurz and Ronald G. Askin. Scheduling flexible flow lines with sequence-dependent setup times. *European Journal of Operational Research*, 159(1):66–82, 2004.
8. Linet Özdamar and Gülay Barbarosoglu. Hybrid heuristics for the multi-stage capacitated lot sizing and loading problem. *Journal of the Operational Research Society*, 50:810–825, 1999.
9. Linet Özdamar and Mehmet Ali Bozyel. Simultaneous lot sizing and loading of product families on parallel facilities of different classes. *International Journal of Production Research*, 36:1305–1324, 1998.
10. Daniel Quadt. *Lot-Sizing and Scheduling for Flexible Flow Lines*. Springer, 2004.
11. Daniel Quadt and Heinrich Kuhn. Capacitated lot-sizing and scheduling with parallel machines, back-orders and setup carry-over. Working Paper, Catholic University of Eichstätt-Ingolstadt, Germany, submitted, 2004.
12. Daniel Quadt and Heinrich Kuhn. Batch scheduling of jobs with identical process times on flexible flow lines. *International Journal of Production Economics*, 2005. Forthcoming.
13. Daniel Quadt and Heinrich Kuhn. A conceptual framework for lot-sizing and scheduling of flexible flow lines. *International Journal of Production Research*, 43(11):2291–2308, 2005.
14. Daniel Quadt and Heinrich Kuhn. A taxonomy of flexible flow line scheduling procedures. Working Paper, Catholic University of Eichstätt-Ingolstadt, Germany, submitted, 2005.
15. Fouad Riane. *Scheduling Hybrid Flowshops: Algorithms and Applications*. Ph.D. Thesis, Facultés Universitaires Catholiques de Mons, 1998.
16. F. Sivrikaya Serifoglu and G. Ulusoy. Multiprocessor task scheduling in multi-stage hybrid flow-shops: A genetic algorithm approach. *Journal of the Operational Research Society*, 55:504–512, 2004.
17. Bagas Wardono and Yahya Fathi. A tabu search algorithm for the multi-stage parallel machine problem with limited buffer capacities. *European Journal of Operational Research*, 155(2):380–401, 2004.
18. R. J. Wittrock. An adaptable scheduling algorithm for flexible flow lines. *Operations Research*, 36(4):445–453, 1988.

Representing Labor Demands in Airport Ground Staff Scheduling

Jörg Herbers

INFORM Institut für Operations Research und Management GmbH
Joerg.Herbers@inform-ac.com

Summary. Airport ground staff scheduling gives rise to a number of challenging optimization problems. We give an overview of airport shift scheduling, focusing on suitable representations for labor demands. We argue that especially in short-term shift planning, traditional models using a demand curve representation of workloads are not always appropriate in ground handling. We therefore formulate task-level shift planning as the problem of creating a cost-minimal set of shift duties which cover the given set of work tasks. The resulting model integrates aspects from shift scheduling and vehicle routing. Depending on the number of additional constraints imposed, different solution techniques seem appropriate. We outline a branch-and-price algorithm which is able to solve an important class of task-level shift planning problems from the practice of airlines and ground handling companies to proven optimality within reasonable run-times.

1 Introduction

Within its ground time at the airport, an aircraft requires different services, e.g. the unloading and loading of baggage, the cleaning of the aircraft cabin, fuel and water supply etc. Additionally to the aforementioned *ramp services*, *passenger services* e.g. comprise check-in and boarding. Many of these services relate to single flight events. As an example, the unloading of baggage for a medium-size aircraft may require four workers for a duration of 30 minutes after landing. The same flight may additionally require several agents for dedicated check-in (referring to this flight only) for a given time period before departure. Such different services can then be represented in a unifying framework.

Reflecting agreements of handling contracts, *engagement standards* (task generation rules) define how many workers are required in which time intervals for given sets of flight events, including inbound and outbound flights. Matching the engagement standards to the set of relevant flight events results in a set of *work tasks* which define the workload for the given company or department. Note that the definition of work tasks for other services (like common check-in for several flights) may ask for the use of queuing models; alternatively, work tasks may be independent from any flight events (e.g. ticket counters, aircraft maintenance) [1].

Each work task can be characterized by a number of attributes. While for most ground handling services, tasks have fixed start times (e.g. baggage loading and unloading have to start/end at fixed times), tasks for other services may be movable within given boundaries. As an example, cabin cleaning can normally be carried out within the ground time of an aircraft. Additionally, we are given a fixed duration and a location (e.g. a gate or check-in counter) for each task. Some tasks may ask for special qualifications like language skills for check-in agents.

Planners at airports often try to gain insight into the temporal evolution and magnitude of workloads by analyzing a *demand curve* which gives the number of required staff for each interval (e.g. on a 15-minutes discretization level) of the planning horizon (e.g. one day, one week). If all tasks have fixed start times and if travel times between different locations can be neglected, the demand curve can be built by superposing all work tasks. If, however, start time intervals or travel times are involved, the process of building a demand curve may require leveling procedures [5].

Staff scheduling models traditionally build upon a demand curve representation of workloads, see [4] for an overview. The *shift scheduling problem* consists in covering the workload for a given service by a cost-minimal set of shift duties [6]. These shift duties must adhere to a set of *shift types* which are due to legal and union regulations or company policies. Each shift type defines a start and an end time for a shift duty. Additionally, a shift of a given type must usually contain one or several lunch or relief breaks within given intervals.

In the following, we focus on suitable representations for workloads in airport shift planning. The following section contrasts models using a demand curve representation with procedures which are based on single work tasks. We characterize the importance of task-level shift planning in airport ground staff scheduling. Finally, we outline a branch-and-price solution methodology for the solution of real-world task-level shift planning problems and give some experimental results. It should be noted that while our focus is on airport staff scheduling, the general ideas may apply to other sectors as well.

2 Representing Labor Demands

As mentioned above, nearly all models in the workforce scheduling literature are based on a demand curve representation of workloads [4]. Dantzig [2] was the first to propose a generalized set covering formulation for the shift scheduling problem:

$$\min \sum_{k \in K} c^k x^k \tag{1}$$

$$\text{s.th.} \sum_{k \in K} a_t^k x^k \geq d_t \qquad \forall t \in T \tag{2}$$

$$x^k \geq 0 \text{ and integer } \forall k \in K \tag{3}$$

with K denoting the set of shift types, x^k the number of shifts and c^k the cost of shift type $k \in K$. T is the discretized time horizon and d_t the labor demand (number of required workers) for period d_t. The coefficient a_t^k is set to 1 if period t is covered by shift type k and to 0 otherwise[1].

In airport shift planning, the demand curve only represents an aggregated view on workloads which are originally defined by work tasks. In the following set partitioning formulation for task-level shift planning[2], we enumerate all shifts each of which has a fixed shift type and contains a sequence of work tasks:

$$\min \sum_{k \in K} c^k \sum_{p \in \Omega^k} \theta_p^k \tag{4}$$

$$\text{s.th.} \sum_{k \in K} \sum_{p \in \Omega^k} a_{ip}^k \theta_p^k = 1 \qquad \forall i \in I \tag{5}$$

$$\theta_p^k \in \{0,1\} \; \forall k \in K, \forall p \in \Omega^k \tag{6}$$

with K the set of shift types, Ω^k the index set for different shifts of type $k \in K$, $\theta_p^k = 1$ if and only if shift p of type k is part of the solution, I the set of all work tasks and $a_{ip}^k = 1$ if shift p of type k covers work task $i \in I$ ($a_{ip}^k = 0$ otherwise).

In formulation (4)-(6), each index $p \in \Omega^k$ describes one valid shift, containing a sequence of tasks and one or several lunch and relief breaks such that all relevant constraints are met (e.g. time window constraints for tasks and breaks). Since work tasks may be placed at different locations, each worker on a shift must move from one location to the other, carrying out a sequence of tasks. The gaps between the tasks must allow for the necessary travel times.

Task-level shift planning can be interpreted as an intermediate between the classical shift scheduling problem and vehicle routing models. Shifts represent tours of "customers" (tasks) which have to be served at given times (in the *vehicle scheduling* case) or within given time windows (analogously to the *vehicle routing problem with time windows*). Shift start and end times as well as lunch and relief breaks limit the extent of tours as defined by the set of valid shift types. Constraints with regard to the aggregated qualification requirements of each shift can be interpreted as special kinds of capacity restrictions. Task-level shift planning also bears similarities to crew scheduling problems, see e.g. [4]; however, the latter usually do not make reference to shift types or travel distances.

---

[1] This assumes that each shift types defines only fixed breaks or no breaks at all.

[2] We use the term *shift planning* in order to avoid confusion with the traditional shift scheduling problem which is based on a demand curve representation.

3 Task-Level Shift Planning for Airport Operations

Despite its relevance to ground staff scheduling, it is interesting to note that there are virtually no publications on the aforementioned task-level approach to airport shift planning. One reason for this is that in other application areas, a demand curve is a natural representation of workloads, e.g. when labor demands result from customer arrivals (in call centers, at toll booths). Furthermore, a demand curve representation may give a sufficient approximation to task-level workloads if the workforce is homogeneous or if we face considerable uncertainty with regard to the exact workloads (e.g. in long-term planning). As an example, Dowling et al. [3] describe an airport setting in which long-term shift planning is based on a demand curve formulation while the tasks are allocated to shifts only shortly before the day of operations. Finally, the potential complexity of solving task-level models is an important reason for the widespread use of demand-level models[3] [4].

Especially in short-term and operational ground staff planning, demand-level models are not always appropriate. However, task-level optimization problems can become quite complex. As an example, it may be forbidden to combine certain types of tasks in shifts (e.g. work tasks for water and lavatory services). In order to reflect the mix of workers at hand, we may impose constraints on the combination of qualification requirements in shifts. Furthermore, we may have to consider task splitting (preemption). A task thus does not have to be entirely covered by a single shift, but can be handed over from one shift to another at a certain point in time (e.g. if tasks represent service counter occupations). Team work may require several tasks and shifts to be placed in parallel.

It should be noted that not all of the above constraints are equally important in the practice of airport ground handling. However, even basic task-level shift planning is NP-hard in the strong sense [5]. Nevertheless, it turns out that moderately-sized planning scenarios are amenable to exact solution approaches.

4 A Branch-and-Price Solution Methodology

Based on formulation (4)-(6), we have conceived a branch-and-price algorithm for basic task-level shift planning. In our setting, we are given a set of work tasks which are fixed in time. Tasks can be placed at different locations with travel times between all locations given. Each shift starts and ends at a single depot. Shift types can define one or several breaks which have to be placed within given time windows. There can be minimum and maximum restrictions on the number of shifts of given shift types or groups of shift types.

[3] We use the term *demand-level models* (in contrast to task-level models) even if this might be misleading since work tasks represent labor demands as well.

Model (4)-(6) lends itself well for column generation solution techniques. It turns out that (4)-(6) can be interpreted as the result of applying Dantzig-Wolfe decomposition to a multi-commodity flow formulation of the shift planning problem [5]. We start by solving a restricted master program which results from omitting most of the shifts $\bigcup_{k \in K} \Omega^k$ from (4)-(6). In each iteration, a shortest path problem is solved in order to obtain one or several new shifts which may improve the objective function. We obtain integer solutions by a branch-and-price solution method with branches applied to the original flow variables. Details of the solution approach can be found in [5].

The algorithm has been implemented on the basis of the COIN-OR libraries BCP and CLP. It has been compiled using gcc 3.3.4 and run on an AMD Athlon A3000+ computer (1 GB main memory, Linux 2.6.8). The first six columns of Table 1 characterize 17 real-world airport planning scenarios used for the experimental evaluation. For each test case, the number of days of the scheduling horizon and the numbers of work tasks and shift type realizations (shift types unfolded over the scenario days) are given. The test cases often naturally decompose by day. We have used an automatic decomposition technique; the row *components* indicates the resulting numbers of components which have been separately submitted to the branch-and-price algorithm. On each component, the algorithm was stopped when a solution within 0.5% of the LP relaxation was found or 4000 secs. runtime were reached. The columns *LP relaxation*, *solution value* and *running time* represent sums over all optimization components.

Table 1. Branch-and-price results

No.	days	tasks	shift type realizations	quantitative restrictions	components	LP relaxation	solution values	run-time (secs.)
1	8	2934	284	0	9	252380.0	252764.0	995.03
2	8	2813	1584	0	8	56327.3	56568.0	12440.82
3	8	2128	593	0	9	22841.5	22873.0	1503.71
4	8	1429	536	0	8	22121.0	22126.0	208.51
5	8	1027	520	0	8	16330.0	16333.0	83.64
6	8	1598	592	0	8	26329.0	26373.0	324.43
7	4	2588	48	20	4	52320.0	52320.0	832.64
8	4	1816	48	16	4	40960.0	40960.0	219.51
9	3	1718	51	24	3	40133.4	40160.0	621.12
10	3	860	33	0	3	24800.0	24800.0	11.76
11	3	327	51	0	3	7040.0	7040.0	0.96
12	3	572	42	0	4	12400.0	12400.0	6.46
13	8	3517	72	49	8	94915.0	94980.0	712.69
14	8	1297	191	0	1	12550.0	12550.0	3627.97
15	8	149	785	0	1	3090.0	3100.0	0.31
16	8	255	750	0	9	7020.0	7020.0	0.52
17	7	2256	36	0	1	44329.5	44329.5	938.24

It can be seen that the LP relaxation provides very tight bounds with an average gap of only 0.08% to the integer solution values. On 8 out of 17 test cases, the solution value equals the LP bound, meaning that the results are proven optimal. It is remarkable that even on scenarios 14 and 17 which cannot be decomposed at all, optimal solutions are found. The run-times are very moderate with an average of little more than 22 minutes.

5 Summary

We have given an overview of real-world airport shift planning, focusing on the representation of labor demands. We have contrasted models based on a demand curve representation and task-level formulations. We have argued that demand-level models are not always appropriate in ground staff scheduling. Using branch-and-price and automatic problem decomposition, we have shown that task-level planning problems from the practice of airlines, airports and ground handling companies can often be solved to proven optimality within moderate run-times.

An interesting subject of future research would be a detailed study on the relationship between demand-level and task-level models. For the solution of more complex task-level shift planning problems, combined CP/OR techniques like constraint-based column generation seem to be promising.

References

1. M. J. Brusco, L. W. Jacobs, R. J. Bongiorno, D. V. Lyons, and B. Tang. Improving personnel scheduling at airline stations. *Operations Research*, 43:741–751, 1995.
2. G. B. Dantzig. A comment on Edie's "Traffic delays at toll booths". *Journal of the Operations Research Society of America*, 2:339–341, 1954.
3. D. Dowling, M. Krishnamoorthy, H. Mackenzie, and D. Sier. Staff rostering at a large international airport. *Annals of Operations Research*, 72:125–147, 1997.
4. A. T. Ernst, H. Jiang, M. Krishnamoorthy, and D. Sier. Staff scheduling and rostering: A review of applications, methods and models. *European Journal of Operational Research*, 153:3–27, 2004.
5. J. Herbers. *Models and Algorithms for Ground Staff Scheduling on Airports*. PhD thesis, Rheinisch-Westfälische Technische Hochschule Aachen, 2005.
6. J. M. Tien and A. Kamiyama. On manpower scheduling algorithms. *SIAM Review*, 24:275–287, 1982.

Rapid Mathematical Programming or How to Solve Sudoku Puzzles in a Few Seconds

Thorsten Koch[1]

Zuse Institute Berlin, Takustr. 7, 14195 Berlin, koch@zib.de

1 Introduction

Using the popular puzzle game of Sudoku, this article highlights some of the ideas and topics covered in the author's PhD thesis [8]. The thesis deals with the implementation and application of out-of-the-box tools in linear and mixed integer programming. It documents the lessons learned and conclusions drawn from five years of implementing, maintaining, extending, and using several computer codes to model and solve real-world industrial problems.

By means of several examples it is demonstrated how to apply a modeling language to rapidly devise mathematical models of real-world problems. It is shown that today's MIP-solvers are often capable of solving the resulting mixed integer programs, leading to an approach that delivers results very quickly, even on problems that required the implementation of specialized branch-and-cut algorithms a few years ago.

2 The modeling language ZIMPL

The presentation is centered around the newly developed algebraic modeling language ZIMPL [8], which is similar in concept to well known languages like GAMS [2] or AMPL [6]. Algebraic modeling languages allow to describe a mathematical model in terms of sets depending on parameters. This description is translated automatically into a mixed integer program which can be fed into any out-of-the-box MIP-solver.

If AMPL could do this in 1989 why would one bother writing a new program to do the same in 1999? One reason is that all major modeling languages for linear and mixed integer programs are commercial products [7]. None of these languages is available as source code. None can be given to colleagues or used in classes for free, apart from very limited "student editions". Usually, only a limited number of operating systems and architectures are supported. The situation has improved somewhat since 1999 when the development of ZIMPL started. Today at least one other open source modeling system is available; the GNU MATHPROG language [13].

What ZIMPL distinguishes from other modelling languages is the use of rational arithmetic. With a few exceptions, all computations in ZIMPL are done with infinite precision rational arithmetic. This ensures that no rounding errors can occur. One might think that the use of rational arithmetic results in a huge increase of computation time and memory. But experience shows that this seems not to be relevant with current hardware. ZIMPL has been successfully used to generate integer programs with more than 30 million non-zero coefficients.

An introduction into modeling with Zimpl together with a complete description of the language can be found in [8]. Also details of the implementation are described. Both theoretical and practical considerations are discussed. Aspects of software engineering, error prevention, and detection are addressed. ZIMPL is still under active development and available from the author's website at www.zib.de/koch/zimpl.

3 Real-world projects

In the second part of the thesis, several real-world projects that employed the methodology and the tools developed in the first part of the thesis are examined. Figure 1 shows a typical solution process cycle. In our experience customers, from industry share a few attributes. They

- do not know exactly what they want,
- need it next week,
- have not yet collected the data necessary or do not even have all the data,
- often need only one shot studies,
- are convinced *"our problem is unique"*.

This mandates an approach that is fast and flexible. And this is what general tools are all about: Rapid prototyping of mathematical models, quick integration of data, and a fast way to check whether the approach is getting to be feasible. Due to the ongoing advances in hardware and software, the number of problems that can be successfully tackled with this approach is steadily increasing.

While most of the research is aimed at improving solution techniques, we focus on mathematical model building and on how to conveniently translate the model and the problem data into solver input.

The benefits of this approach are demonstrated in [8] by a detailed presentation of four projects from telecommunication industry dealing with facility location and UMTS planning problems. Problems, models, and solutions are discussed. Special emphasis is put on the dependency between the precision of the input data and the results. Possible reasons for unexpected and undesirable solutions are explained. Furthermore, the Steiner tree packing problem

Fig. 1. Modeling cycle according to [10]

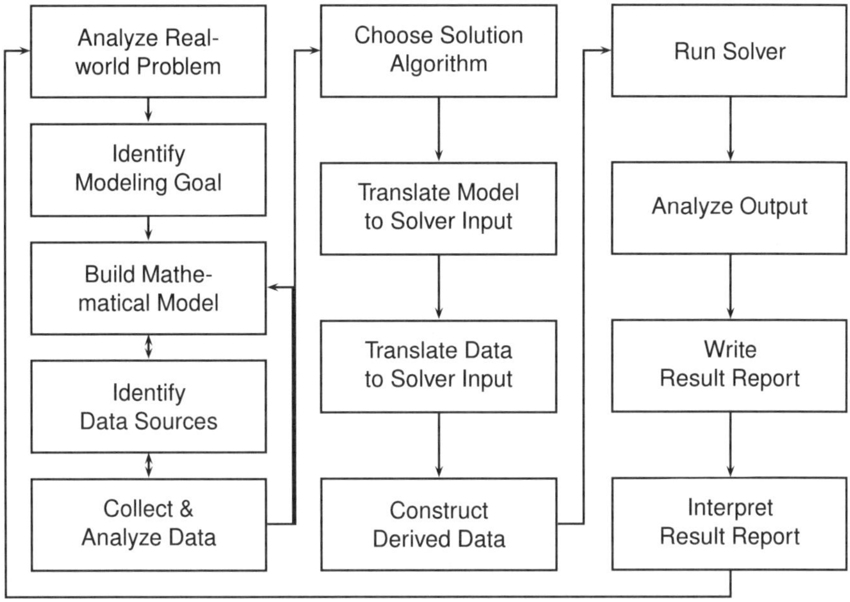

in graphs, a well-known hard combinatorial problem, is revisited. A formerly known, but not yet used model is applied to combine switchbox wire routing and via minimization in VLSI design. All instances known from the literature are solved by this approach, as well as some newly generated bigger problem instances. The results show that the improvements in solver technology, as claimed in [4, 3], allow our rapid prototyping strategy to succeed even on difficult problems, provided a suitable model is chosen.

4 Sudoku

To give an impression of how to use ZIMPL, we demonstrate the approach on the popular puzzle game Sudoku [5]. The aim of the puzzle is to enter a numeral from 1 through 9 in each cell of a 9×9 grid made up of 3×3 subgrids. At the beginning several cells are already given preset numerals. At the end, each row, column and subgrid must contain each numeral exactly once. Figure 3(a) shows an example (for details see, e. g., en.wikipedia.org/wiki/Sudoku). Obviously, the problem can be stated using constraint programming as a collection of *alldifferent* constraints [11]. But how to formulate this as an integer program? ZIMPL can automatically generate IP's for certain constructs such as the absolute value of the difference of two variables (`vabs`). Using 81 inte-

Listing 1. A ZIMPL model to solve Sudoku using integer variables

```
param p            := 3;
set J              := { 0 .. p*p-1 };
set KK             := { 0 .. p-1} * { 0 .. p-1};
set F              := { read "fixed.dat" as "<1n,2n>" };
param fixed[F]     := read "fixed.dat" as "<1n,2n>3n";
var x              [J * J] integer >= 1 <= 9;

subto rows: forall <i,j,k> in J*J*J with j < k do
   vabs(x[i,j]-x[i,k]) >= 1;
subto cols: forall <i,j,k> in J*J*J with j < k do
   vabs(x[j,i]-x[k,i]) >= 1;
subto squares: forall <m,n> in KK do
   forall <i,j,k,l> in KK*KK with p*i+j < p*k+l do
      vabs(x[m*p+i,n*p+j] - x[m*p+k,n*p+l]) >= 1;
subto fixed: forall <i,j> in F do x[i,j] == fixed[i,j];
```

ger variables in the range $\{1..9\}$ the *alldifferent* constraint can be formulated by demanding that the absolute difference of all pairs of relevant variables is greater than or equal to one. This leads to the ZIMPL program shown in Listing 1.

Fig. 2. Sudoku puzzle and solution

						8	9	
		1	3		9			
	5					2		4
					2	6	8	
9			7		6			5
	3	6	5					
1		3					5	
			8		7	1		
	4	5						

(a) Sudoku puzzle

3	6	7	4	2	5	8	9	1
4	2	1	3	8	9	5	7	6
8	5	9	6	7	1	2	3	4
5	7	4	1	3	2	6	8	9
9	1	8	7	4	6	3	2	5
2	3	6	5	9	8	4	1	7
1	8	3	9	6	4	7	5	2
6	9	2	8	5	7	1	4	3
7	4	5	2	1	3	9	6	8

(b) Solution

How does the `vabs` construct work? Given a bounded integer variable $l_x \leq x \leq u_x$, where $l_x, x, u_x \in \mathbb{Z}$, two additional binary variables b^+ and b^- are introduced as indicators for whether x is positive or negative, i.e., $b^+ = 1$ if and only if $x > 0$ and $b^- = 1$ if and only if $x < 0$. In case of $x = 0$, both b^+ and b^- are zero. Two additional non-negative variables x^+ and x^- are introduced to hold the positive and negative part of x. This can be formulated as an integer program as follows:

$$\begin{array}{rcl} & x^+ - x^- & = x \\ b^+ \leq & x^+ & \leq \max(0, u_x) b^+ \\ b^- \leq & x^- & \leq |\min(0, l_x)| b^- \\ & b^+ + b^- & \leq 1 \\ & b^+, b^- & \in \{0, 1\} \end{array} \tag{1}$$

Note that the polyhedron described by the linear relaxation of System (1) has only integral vertices (see [8] for details).

Using System (1), the following functions and relations can be expressed using $x = v - w$, where $v, w \in \mathbb{Z}$ with $l_x = l_v - u_w$ and $u_x = u_v - l_w$ whenever two operands are involved:

$$\begin{array}{ll} \text{abs}(x) = x^+ + x^- & v \neq w \Leftrightarrow b^+ + b^- = 1 \\ \text{sgn}(x) = b^+ - b^- & v = w \Leftrightarrow b^+ + b^- = 0 \\ \min(v, w) = w - x^- & v \leq w \Leftrightarrow b^+ = 0 \\ \max(v, w) = x^+ + w & v < w \Leftrightarrow b^- = 1 \\ & v \geq w \Leftrightarrow b^- = 0 \\ & v > w \Leftrightarrow b^+ = 1 \end{array}$$

More information on this topic can be found, for example, in [12, 9] or at the GAMS website http://www.gams.com/modlib/libhtml/absmip.htm.

Unfortunately, the IP resulting from Listing 1 is hard to solve. CPLEX 9.03 was not able to find a feasible solution after more than six hours and a million branch-and-cut nodes. As an alternative we modeled the problem using binary variables as shown in Listing 2. With this formulation all Sudoku puzzles we have tried so far were solved either by preprocessing or at the root node of the branch-and-bound tree.

Choosing the right formulation is often more important than having the best solver algorithm. Especially with real-world problems, having the ability to experiment swiftly with different formulations is essential.

5 Conclusion and outlook

It turned out that regarding real-world problems understanding the problem itself and the limitations presented by the available data are often a bigger obstacle than building and solving the resulting mathematical model. The ability to easily experiment with formulations is a key factor for success.

The use of automatically generated constructs in modeling languages makes it even easier to turn complex problems into models. It seems likely that future solvers will "understand" these extended functions directly and either convert them into whatever suits them best or handle them directly [1].

Listing 2. A ZIMPL model to solve Sudoku using binary variables

```
param p := 3;
set J   := { 0 .. p*p-1 };
set KK  := { 0 .. p-1} * { 0 .. p-1 };
set F   := { read "fixed.dat" as "<1n,2n,3n>" };
var x    [J*J*J] binary;

subto rows: forall <i,j> in J*J do sum <k> in J:x[i,j,k]==1;
subto cols: forall <j,k> in J*J do sum <i> in J:x[i,j,k]==1;
subto nums: forall <i,k> in J*J do sum <j> in J:x[i,j,k]==1;
subto fixed:forall <i,j,k> in F do                x[i,j,k]==1;
subto squares: forall <m,n,k> in KK*J do
                        sum <i,j> in KK:x[m*p+i,n*p+j,k]==1;
```

The free availability and the simplicity of use make ZIMPL a well suited tool for teaching modeling linear and mixed integer programs.

References

1. Tobias Achterberg. SCIP – a framework to integrate constraint and mixed integer programming. Technical Report 04-19, Zuse Institute Berlin, 2004. See scip.zib.de.
2. J. Bisschop and A. Meeraus. On the development of a general algebraic modeling system in a strategic planning environment. *Mathematical Programming Study*, 20:1–29, 1982.
3. Robert E. Bixby. Solving real-world linear programs: A decade and more of progress. *Operations Research*, 50(1):3–15, 2002.
4. Robert E. Bixby, Marc Fenelon, Zonghao Gu, Ed Rothberg, and Roland Wunderling. MIP: Theory and practice – closing the gap. In M. J. D. Powell and S. Scholtes, editors, *System Modelling and Optimization: Methods, Theory and Applications*. Kluwer, 2000.
5. David Eppstein. Nonrepetitive paths and cycles in graphs with application to Sudoku. ACM Computing Research Repository, 2005.
6. R. Fourer, D. M. Gay, and B. W. Kernighan. *AMPL: A Modelling Language for Mathematical Programming*. Brooks/Cole, 2nd edition, 2003.
7. Josef Kallrath, editor. *Modeling Languages in Mathematical Optimization*. Kluwer, 2004.
8. Thorsten Koch. *Rapid Mathematical Programming*. PhD thesis, Technische Universität Berlin, 2004. Corrected version available as ZIB technical report 04-58. Regarding software see www.zib.de/koch/zimpl.
9. Frank Plastria. Formulating logical implications in combinatorial optimization. *European Journal of Operational Research*, 140:338–353, 2002.
10. Hermann Schichl. Models and the history of modeling. In Josef Kallrath, editor, *Modeling Languages in Mathematical Optimization*, pages 25–36. Kluwer, 2004.
11. W.J. van Hoeve. The alldifferent constraint: A survey. In *6th Annual Workshop of the ERCIM Working Group on Constraints*. Prague, June 2001.
12. H. Paul Williams and Sally C. Brailsford. Computational logic and integer programming. In J. E. Beasley, editor, *Advances in Linear and Integer Programming*, pages 249–281. Oxford University Press, 1996.
13. GNU linear programming toolkit glpsol version 4.7. www.gnu.org/software/glpk.

Part II

Diploma Award of the GOR

Standortplanung von Einsatzkräften bei Großereignissen

Julia Drechsel

Technische Universität Bergakademie Freiberg, Fakultät für Wirtschaftswissenschaften, Lehrstuhl für Allgemeine Betriebswirtschaftslehre, insbesondere Industriebetriebslehre/Produktionswirtschaft und Logistik, Lessingstraße 45, 09599 Freiberg, Germany, E-Mail: julia.drechsel@bwl.tu-freiberg.de

Abstract

Großveranstaltungen sind auf Grund der enormen Menschenansammlungen eng mit einem erhöhten Gefahrenpotenzial verbunden. Zur Absicherung müssen deswegen immer Einsatzkräfte von Rettungsdiensten, Polizei, Feuerwehr etc. vor Ort sein. Für die verantwortlichen Entscheidungsträger ist es jedoch ein schwieriges Problem, die Anzahl sowie die Standorte der benötigten Einsatztrupps festzulegen. Mittels einer Analyse der praktischen Einsatzplanung werden konkrete Fragestellungen definiert, die die Basis für eine Modellierung bilden. An Hand verschiedener mathematischer Modelle und deren Lösung wird deutlich, dass Instrumente des Operations Research die Entscheidungsfindung der Verantwortlichen unterstützen können. Die vorgestellten Modelle konnten auf einen praktischen Fall einer Großveranstaltung in Dresden mit 50.000 Besuchern erfolgreich angewendet werden.

1 Einführung

Um den sicheren Ablauf von Großveranstaltungen zu gewährleisten, müssen Kräfte von Polizei, Feuerwehr und Rettungsdiensten eingesetzt werden, die in Notfallsituationen an Ort und Stelle sind und sofort eingreifen können.

In der Praxis ist es ein schwieriges Problem, die Anzahl der nötigen Einsatzkräfte sowie deren Standorte auf der Veranstaltungsfläche zu ermitteln. Da es an allgemeinen gesetzlichen Vorschriften mangelt, gibt es kein einheitliches Vorgehen zur Bestimmung der gesuchten Größen. Stattdessen werden in der Praxis in einem gedanklichen Prozess alle Informationen, die im Vorlauf der Veranstaltung bekannt sind, zusammengetragen (z. B. erwartete Besucherzahl, Größe und Beschaffenheit der Veranstaltungsfläche, Anwesenheit wichtiger Persönlichkeiten, polizeiliche Erkenntnisse über das Gewaltpotenzial, den erwarteten Alkoholkonsum etc.). Basierend auf ihren Erfahrungswerten aus vergangenen Veranstaltungen bestimmen die Einsatzplaner dann die Anzahl der nötigen Einsatzkräfte und die Standorte der einzelnen Einsatztrupps.

In der Literatur finden sich nur wenige Quellen zu diesem speziellen Problem (Maurer 2001). Allerdings gibt es verwandte Probleme, die sich mit ähnlichen Fragestellungen beschäftigen: z. B. die Standortplanung von Rettungs- oder Feuerwachen in Städten (Badri et al. 1998, Pirkul und Schilling 1988, Toregas et al. 1971, Werners et al. 2001, 2002, 2003).

Für eine detaillierte Betrachtung konnten drei Zielstellungen identifiziert werden, die in den folgenden Abschnitten modelliert werden:

1. Wie viele Einsatzkräfte sind nötig und wo sind diese zu positionieren, um die gesamte Veranstaltungsfläche abzudecken? Aus Kostengründen sollte diese Anzahl möglichst gering sein. Allerdings muss sichergestellt werden, dass jeder Punkt der Veranstaltungsfläche, an dem es zu einem Unfall oder ähnlichem kommen könnte (Einsatzorte), innerhalb einer bestimmten Zeit (Hilfsfrist) vom Standort der Einsatzkräfte erreichbar ist.
2. Ist auf Grund von Kapazitätsengpässen die Anzahl der Einsatzkräfte begrenzt, müssen Standorte gefunden werden, so dass jeder Einsatzort innerhalb einer möglichst kurzen Hilfsfrist erreicht werden kann.
3. Sind sowohl die Anzahl der Einsatzkräfte als auch die einzuhaltende Hilfsfrist vorgegeben, erhält man eine dritte mögliche Fragestellung: Gesucht werden nun Standorte, so dass eine maximale Anzahl von Einsatzorten abgedeckt ist, d. h. innerhalb der vorgegebenen Hilfsfrist erreichbar ist.

2 Minimierung der Anzahl der Einsatzkräfte

Für die folgende Modellierung wird die Veranstaltungsfläche als Graph dargestellt. Zur Bestimmung der Anzahl der benötigten Einsatzkräfte sind folgende Parameter nötig: erstens die Menge der Knoten $V = \{1,\ldots,n\}$ zur Abbildung der Einsatzorte, zweitens die kürzesten Entfernungen $d_{ij} \geq 0$ zwischen jeder Kombination aus zwei Knoten $(i,j) \in V \times V$ und drittens der Parameter r, der die einzuhaltende Hilfsfrist angibt. Die binäre Entscheidungsvariable x_{ij} ist gleich eins, wenn der Einsatzort j vom Einsatztrupp im Standort i abgedeckt wird und sonst null. z_i ist ebenfalls eine binäre Entscheidungsvariable und gibt an, ob ein Standort im Knoten i eingerichtet wird oder nicht. Die Zielfunktion lautet:

$$\min \sum_{i \in V} z_i \qquad (1)$$

unter den Nebenbedingungen:

$$d_{ij} \times x_{ij} \leq r \qquad (i,j) \in V \times V \qquad (2)$$

$$\sum_{i \in V} x_{ij} \geq 1 \qquad j \in V \qquad (3)$$

$$x_{ij} \leq z_i \qquad (i,j) \in V \times V \qquad (4)$$

$$x_{ij} \in \{0,1\} \qquad (i,j) \in V \times V \qquad (5)$$

$$z_i \geq 0 \qquad i \in V \qquad (6)$$

Für jeden eingerichteten Standort wird das entsprechende z_i gleich eins. Die Summe über alle z_i gibt somit die Anzahl der benötigten Einsatztrupps an (1). Nebenbedingung (2) sorgt dafür, dass der Abstand zwischen einem Einsatzort und einem zugeordneten Standort nicht größer als die vorgegebene Hilfsfrist r ist. Damit jeder Einsatzort von mindestens einem Einsatztrupp innerhalb der Hilfsfrist erreichbar ist, wurde Nebenbedingung (3) formuliert. Ein Einsatzort kann nur von einem Standort abgedeckt werden, wenn in diesem Knoten auch ein Einsatztrupp stationiert wurde (4). Wegen den Nebenbedingungen (4) und (5) genügt es, $z_i \geq 0$ anstatt $z_i \in \{0,1\}$ festzulegen.
Es ist denkbar, dass eine Abdeckung der gesamten Fläche zu aufwendig ist und dass es genügt, D% aller Einsatzorte innerhalb der Hilfsfrist zu erreichen. Für diese Betrachtung muss das Modell (1), (2), (4), (5), (6) um folgende Nebenbedingungen erweitert werden:

$$\sum_{j \in V} y_j \geq \frac{D \times n}{100} \qquad (7)$$

$$\sum_{i \in V} x_{ij} \geq y_j \qquad j \in V \qquad (8)$$

$$0 \leq y_j \leq 1 \qquad j \in V \qquad (9)$$

Die zusätzliche Entscheidungsvariable y_j gibt an, ob der Knoten j von mindestens einem Einsatztrupp abgedeckt wird oder nicht. Wie für das z_i ist die Formulierung $0 \leq y_j \leq 1$ ausreichend. Die Summe aller y_j zählt die abgedeckten Knoten. Wird diese Anzahl ins Verhältnis zur Gesamtknotenzahl n gesetzt, ergibt sich der Parameter D (7). Nebenbedingung (8) stellt den Zusammenhang zwischen x_{ij} und y_j her, so dass ein Einsatzort j als abgedeckt gilt, wenn er einem Standort i zugeordnet wurde. Zu beachten ist, dass y_j nicht größer als eins wird, selbst wenn der

Einsatzort von mehreren Standorten abgedeckt wird, um Doppelzählungen mehrfach abgedeckter Einsatzorte zu vermeiden.

3 Minimierung der Hilfsfrist

Bei der zweiten Fragestellung wird eine bestimmte Anzahl k von Einsatztrupps vorgegeben und eine möglichst kleine Hilfsfrist r gesucht, innerhalb der alle Einsatzorte erreichbar sind. Dafür ergibt sich folgendes Modell:

$$\min r \tag{10}$$

unter den Nebenbedingungen (2), (4), (5), (6), (7), (8), (9) sowie:

$$\sum_{i \in V} z_i \leq k \tag{11}$$

$$r \geq 0 \tag{12}$$

Weil die Hilfsfrist r in diesem Fall ermittelt werden soll, ist sie nicht mehr Parameter sondern Entscheidungsvariable (12). Neu im Modell ist lediglich Nebenbedingung (11), die die Anzahl der Einsatztrupps nach oben hin beschränkt. Das Modell beinhaltet bereits den Parameter D, so dass wie im vorangegangenen Abschnitt Abdeckungsgrade kleiner als 100% vorgegeben werden können.

4 Maximierung der Abdeckung

Als dritte Möglichkeit sollen nun möglichst viele Einsatzorte abgedeckt werden unter der Maßgabe, dass sowohl die Anzahl der Einsatztrupps als auch die Hilfsfrist begrenzt sind:

$$\max \frac{100}{n} \sum_{j \in V} y_j \tag{13}$$

unter den Nebenbedingungen (2), (4), (5), (6), (8), (9) sowie (11).

In der Zielfunktion wird jetzt der Wert für den Abdeckungsgrad maximiert. Dafür wird die Anzahl der abgedeckten Einsatzorte (Knoten) gezählt und ins Verhältnis zur Gesamtknotenzahl gesetzt. Der konstante Faktor 100/n spielt für die Optimierung keine Rolle, er dient nur einer einfacheren Interpretation des Zielfunktionswertes.

5 Abschließende Bemerkungen

Die hier dargestellten Modelle zur Standortplanung von Einsatzkräften bei Großveranstaltungen sind nur eine mögliche Modellierungsvariante. In der zu Grunde liegenden Diplomarbeit werden des weiteren Modelle zur diskreten und kontinuierlichen Standortplanung in der Ebene vorgestellt. Außerdem findet sich dort ein Modell zur Umpositionierung der Einsatzkräfte, das zur Anwendung kommt, wenn sich im Laufe einer Veranstaltung die Parameter ändern oder es sich um eine mobile Veranstaltung handelt.

In der Diplomarbeit ist ausführlich ausgeführt, wie die Modelle mit Standardsoftware zur mathematischen Optimierung gelöst werden können. Verschiedene Rechenstudien machen deutlich, dass z. B. die hier vorgestellten Probleme mit bis zu 50 Knoten in Sekundenbruchteilen mit Hilfe der Standardsoftware AMPL/CPLEX auf einem PC mit Intel Pentium 4 Prozessor (3,1 GHz) optimal gelöst werden können.

Die Methoden wurden erfahrenen Einsatzplanern vorgestellt und stießen auf großes Interesse. An Hand eines praktischen Beispiels (Großveranstaltung in Dresden mit 50.000 Besuchern) konnte gezeigt werden, dass die Verfahren als Planungsgrundlage und Kontrollinstrument in der Praxis einsetzbar sind.

5.1.1 Danksagung

In erster Linie möchte ich der GOR für die Auszeichnung meiner Diplomarbeit danken und dass sie mir damit die Möglichkeit gegeben hat, die Ergebnisse meiner Arbeit einem großen Kreis zu präsentieren. Natürlich geht mein Dank ebenfalls an Herrn Prof. Dr. Kimms, der mich so hervorragend betreut hat.

Außerdem bedanke ich mich recht herzlich für die inhaltliche Unterstützung bei Michael Bonert (Deutsches Zentrum für Luft- und Raumfahrt), Jörg Kästner (Brand- und Katastrophenschutzamt Dresden), Georg Förster (Deutsches Rotes Kreuz), Harald Lewin (Malteser Hilfsdienst), Wolfgang Herold (Johanniter-Unfall-Hilfe) und Peter Brunke (Polizei Köln).

5.1.2 Literatur

Ahuja RK, Magnanti TL, Orlin JB (1993) Network Flows – Theory, Algorithms, and Applications. Prentice Hall, Upper Saddle River

Badri MA, Mortagy AK, Alsayed CA (1998) A Multi-Objective Model for Locating Fire Stations. European Journal of Operational Research 110:243-260

Domschke W, Drexl A (1985) Location and Layout Planning – An International Bibliography. Springer, Berlin

Drechsel J (2005) Standortplanung von Einsatzkräften bei temporären und mobilen Großereignissen. diploma thesis, Technical University of Freiberg

Landeshauptstadt Dresden Brand- und Katastrophenschutzamt Abteilung Rettungsdienst und Abteilung Einsatz (eds.) (2004) Richtlinie zur sanitäts- und rettungsdienstlichen Absicherung von Veranstaltungen, Konzerten, Demonstrationen und anderen Ereignissen, bei welchen mit einer hohen Konzentration von Menschen zu rechnen ist.

Malteser Hilfsdienst (2000) Leitfaden Planung und Durchführung von Sanitätseinsätzen.

Maurer K (2001) Einsatzplanung bei Großveranstaltungen. In: Mitschke T, Peter H (eds.) Handbuch für Schnell-Einsatz-Gruppen. 3rd edn. Stumpf und Kossendey, Edewecht, pp 271-295

Pirkul H, Schilling D (1988) The Siting of Emergency Service Facilities with Workload Capacities and Backup Service. Management Science 34: 896-908

Toregas C, Swain R, ReVelle C, Bergman L (1971) The Location of Emergency Service Facilities. Operations Research 19:1363-1373

Werners B, Drawe M, Thorn J (2001) Standortplanung für das Rettungswesen. Wirtschaftswissenschaftliches Studium 12:653-658

Werners B, Meyer D, Thorn J (2002) Unterstützung der Standortplanung für die Feuerwehr Bochum. Working Paper 2002/02, University of Bochum

Werners B, Thorn J, Hagebölling D (2003) Standortplanung in Bochum – Wohin mit den Rettungswachen?. Rettungsdienst 26:237-241

Part III

Logistics

Customer Selection and Profit Maximization in Vehicle Routing Problems

Deniz Aksen[1], Necati Aras[2]

[1] Koç University, College of Administrative Sciences and Economics, Rumelifeneri Yolu, Sarıyer, 34450 İstanbul, Turkey

[2] Boğaziçi University, Department of Industrial Engineering, Bebek, 34342 İstanbul, Turkey

1 Introduction

The capacitated vehicle routing problem (CVRP or simply VRP) is one of the most studied combinatorial optimization problems in the literature of operations research. The main reason for this much attention is the abundance of its real-life applications in distribution logistics and transportation. In this study we focus on the single-depot capacitated VRP with profits and time deadlines (VRPP-TD).VRPP-TD is a generalization of the VRP where visiting each customer incurs a fixed revenue, and it is not necessary to visit all customers. The objective is to find the number and routes of vehicles under time deadline restrictions so as to maximize the total profit, which is equal to the total revenue collected from the visited customers less the traveling cost. For this problem we propose an efficient iterative marginal profit analysis method called iMPA applied in a two-phase framework. The first phase involves solving a time-deadline constrained vehicle routing problem (VRP-TD) using simulated annealing given a set of customers. The second phase is the implementation of iMPA where each customer's marginal profit value is calculated with respect to the set of routes found in the first phase. It is decided upon which customers to retain and which ones to discard. With the remaining set of customers determined in phase 2, phase 1 is repeated. A final correction is performed on the final solution in order to make sure that all routes have positive total profit values. In our numerical experiments we report the results obtained with this framework and assess the solution quality of our approach.

2 Literature Review

A recent paper by Feillet et al. (2005) elaborates on the traveling salesman problem with profits (TSPP) which is a generalization of the traveling salesman problem (TSP) where it is not necessary to visit all vertices of the given graph. With each customer is associated a profit that is known a priori. TSPP can be formulated as a discrete bicriteria optimization problem where the two goals are maximizing the profit and minimizing the traveling cost. It is also possible to define one of the goals as the objective function and the other as a satisfiability constraint. In one version, which is known as orienteering problem (OP), selective TSP (STSP), or maximum collection problem (MCP) in the literature, the objective is the maximization of the collected profit such that the total traveling cost (distance) does not exceed an upper bound. The other version, named as the prize collecting TSP (PCTSP), is concerned with determining a tour with minimum total traveling cost where the collected profit is greater than a lower bound. Feillet et al. (2005) provide an excellent survey of the existing literature on TSPP. Their survey lists various modeling approaches to TSSP and exact as well as heuristic solution techniques.

The extension of TSPP to multiple vehicles is referred to as the VRP with profits (VRPP). The multi-vehicle version of the OP is called the team orienteering problem (TOP) which is studied by Chao et al. (1996). The authors propose a 5-step metaheuristic based on deterministic annealing for its solution. A very recent paper on TOP is due Tang and Miller-Hooks (2005) who develop a tabu search heuristic for the problem. Butt and Cavalier (1994) address the multiple tour maximum collection problem (MTMCP) in the context of recruiting football players from high schools. They propose a greedy tour construction heuristic to solve this problem. Later on, Butt and Ryan (1999) develop an exact algorithm for the MTMCP based on branch and price solution procedure. Gueguen (1999) also proposes branch and price solution procedures for the so-called selective VRP and for the prize collecting VRP both of which are further constrained by time windows.

In this paper we focus on the VRPP-TD and propose a new heuristic called iMPA for its solution. Given a complete graph $G=(N, E)$, where $N=\{0,1,2,\ldots,n\}$ is the set of $(n+1)$ vertices (customers and one depot) and $E=\{(i,j): i,j\in N \wedge i\neq j\}$ is the edge set, the objective of the VRPP-TD is to find the best routes for vehicles which depart from the depot (vertex 0), visit a set of customers, and then return to the depot so as to maximize the total profit. Profit equals the total revenue collected from the visited customers less the traveling cost. We assume that demand and profit of each customer, customer locations and the location of the depot are known with certainty. In our formulation we take into account customer-specific temporal constraints referred to as time deadlines, maximum route duration/length constraints and a uniform (homogeneous) vehicle capacity. Capacity and time deadline constraints in addition to arbitrary customer demands differentiate our problem from the TOP studied by Chao et al. (1996) and Tang and Miller-Hooks (2005).

3 Model Formulation and Solution Methodology

In the mixed-integer linear program (*MIP*) below, p_i is the revenue collected by visiting customer i; q_i is the demand of customer i; d_{ij} is distance between customers i and j; β denotes the unit traveling cost, and Q is the vehicle capacity. Decision variable y_i is 1 if customer i is visited by some vehicle, 0 otherwise; x_{ij} is 1 if customer j is visited after customer i by some vehicle, 0 otherwise; u_i is a weight associated with each customer i bounded between the demand of the customer and the vehicle capacity.

$$\text{maximize} \quad \sum_{i\in N\setminus\{0\}} p_i y_i - \beta \sum_{i\in N} \sum_{\substack{j\in N\\ i\neq j}} d_{ij} x_{ij} \tag{1}$$

$$\text{s.t.} \quad \sum_{\substack{j\in N\\ j\neq i}} x_{ij} = \sum_{\substack{j\in N\\ j\neq i}} x_{ji} \qquad i\in N \tag{2}$$

$$\sum_{\substack{j\in N\\ j\neq i}} x_{ij} = y_i \qquad i\in N\setminus\{0\} \tag{3}$$

$$Q \sum_{i\in N\setminus\{0\}} x_{0i} \geq \sum_{i\in N\setminus\{0\}} q_i y_i \tag{4}$$

$$u_i - u_j + Qx_{ij} + (Q - q_i - q_j)x_{ji} \leq Q - q_j \qquad i,j\in N\setminus\{0\} \wedge i\neq j \tag{5}$$

$$q_i \leq u_i \leq Q \qquad i\in N\setminus\{0\} \tag{6}$$

$$y_i, x_{ij} \in \{0,1\} \qquad i,j\in N \tag{7}$$

In this formulation, the first constraint is a degree balance constraint for all vertices in N. The second constraint ensures that a customer has no incoming and outgoing arcs unless it is visited. The third constraint imposes a minimum number of vehicle routes according to visits to customers. The next two constraints are lifted Miller-Tucker-Zemlin subtour elimination constraints for the VRP first proposed by Desrochers and Laporte (1991), corrected later by Kara et al. (2005). Finally, the last constraint pertains to the integrality of y_i and x_{ij}'s. Note that Q must be chosen larger than the maximum customer demand such that every customer is eligible to be visited by a vehicle.

The *MIP* formulation in (1)-(7) needs to be supplemented by time deadline constraints and consequently arrival time variables for each customer node if we want to solve a VRPP-TD instance with it. However, after the incorporation of those extra constraints and variables, a commercial solver such as CPLEX cannot solve problems with more than 20 customers. Therefore, we need efficient heuristics to tackle large size instances of VRPP-TD. To this end, we propose a new heuristic referred to as iterative marginal profit analysis (iMPA) which is based on the idea of the marginal profit of a customer. We employ iMPA in the following solution framework.

1. Solve the given problem instance as a VRP-TD with the current set of visited customers using simulated annealing (SA).

2. For the current set of routes apply iMPA until the marginal profit of each and every remaining customer is positive.
3. If iMPA does not modify the set of visited customers (i.e., no customers are dropped from the current routes), then go to step 4. Otherwise (i.e., some customers are dropped), go to step 1.
4. Check the profit of all routes. If they are positive, then stop. Otherwise, for each route with a nonpositive total profit drop the customer with the lowest marginal profit until the total profit of the route becomes positive *and* the removal of this customer does not improve the total profit of the route.

Step 4 of the above solution framework is necessary since it is possible that the total profit associated with a route can be negative even if the marginal profits of customers on that route are all positive. Therefore, we make a final check to detect such routes, and try to modify them by removing some customers so as to make the profits of these routes positive. It is clear that at the very beginning of step 1 the initial set of visited customers is taken as the set of all customers $N\setminus\{0\}$.

In step 1 of the above procedure we solve a VRP-TD using SA. Given a combinatorial optimization problem with a finite set of solutions and an objective function, the SA algorithm is characterized by a rule to randomly generate a new feasible solution in the neighborhood of the current solution. The new solution is accepted if there is an improvement in the objective value. In order to escape local minima, new solutions with worse objective values are also accepted with a certain probability that depends both on the magnitude of the deterioration Δ and the annealing temperature T. The acceptance probability is taken as $e^{-\Delta/T}$. A certain number of iterations (L_k) are performed at a fixed temperature (T_k), then the temperature is reduced every L_k iterations by the cooling rate α. The most important characteristic of an SA-based heuristic is the definition of the neighborhood structure that determines how new solutions are generated from the current solution. We use three different neighborhood structures, which are 1-0 move, 1-1 exchange, and 2-Opt. In 1-0 move, a customer selected arbitrarily is removed from its current position and inserted in another position on the same or a different route. 1-1 exchange swaps the positions of two customers that are either on the same route or on two different routes. Finally, 2-Opt removes two arcs, which are either in the same route or in two different routes, and replaces them with two new arcs. For a detailed explanation and pictorial description of these local improvement heuristics we refer the reader to Tarantilis et al. (2005).

In step 1 of the proposed procedure where VRP-TD is solved by SA with the current set of visited customers, SA has to be provided with an initial solution. For this purpose we use the parallel savings heuristic of Clarke and Wright (1964). At each iteration of the SA heuristic we randomly choose one of the local improvement heuristics to generate a new feasible solution from the current solution. Note that it is possible that one move of the selected local improvement heuristic may result in an infeasible solution to the VRP-TD because this solution may violate vehicle capacity or time deadline constraints. If this happens, we try all possible moves that can be performed by the local improvement heuristic until a feasible solution is found. This implies that one of the feasible solutions in the neighborhood of the current solution is found with certainty if there exists one. In case there are no feasible solutions with respect to the selected local improvement heuristic, then we randomly choose another one. If we cannot generate a feasible solution, the procedure is stopped, and we report the best feasible solution found so far. In step 2 of our solution framework, we apply iMPA for the current set of routes found in step 1. Given a customer *i*, its marginal profit is defined as $\pi_i = p_i - \beta(d_{ki} + d_{il} + d_{kl})$ where *k* and *l* are those nodes that, respectively, precede and succeed customer *i*. If $\pi_i \leq 0$, then customer *i* is not worth visiting. It can be dropped from its route. Otherwise, it is profitable to visit customer *i*; thus it is kept between nodes *k* and *l*. The formal definition of iMPA is given below.

1. For the current set of routes compute each customer's marginal profit π_i. Sort $\pi_i, i \in N \setminus \{0\}$ values in nondecreasing order and obtain a sorted stack $\pi_{[i]}$.
2. If $\pi_{[1]}$ of customer $i_{[1]}$ (i.e, customer with the lowest marginal profit) is positive, then exit iMPA. Otherwise, go to step 3.
3. Let $succ_{i[1]}$ and $pred_{i[1]}$ be the successor and predecessor nodes, respectively, of customer $i_{[1]}$ on its current route *r*. Delete $i_{[1]}$ from the route. Update the π values of $succ_{i[1]}$ and $pred_{i[1]}$ if they are customer nodes. Cancel route *r* if $i_{[1]}$ was the only customer on it.
4. Set $\pi_{[1]}$ to infinity such that it is put at the end of the stack. Restore the nondecreasing order of π_i values in the stack and go to step 2.

4 Computational Results

In order to test the proposed heuristic we generate random VRPP-TD instances with up to 20 customers. These instances are solved by the proposed heuristic involving iMPA, and the objective values are compared to those found optimally by the solver CPLEX 8.1 operating under GAMS. The heuristic is coded and compiled in Visual C++ 6.0 and run on a 3.20GHz Pentium 4/HT PC with 1 GB RAM.

In the implementation of SA, we adopt the following choices. The number of iterations, L_k, which have to be performed at temperature T_k is set to $250n^2$ where *n* is the number of customers. The parameter α which determines the cooling rate is assigned a value of 0.95. SA algorithm is terminated when the improvement in the objective value of the best solution during the last five temperature updates is less than 1%. The routes of the best solution found by the SA are an input to our heuristic involving iMPA. Results are shown in Table 1 where we report the percent gaps from the best objective values found by CPLEX. In some instances, as can be seen in the table, the maximum allowable CPU time (3 hours) is not sufficient for CPLEX to arrive at a proven optimal solution.

One possible improvement to the method proposed in this paper is in step 2 where iMPA is carried out on a given solution with a set of visited customers. On the basis of marginal profits, iMPA identifies which customers should be dropped from this set. It does not take into account the possibility of adding any currently unvisited customer to this set. We are currently working on the extension of our method that will account also for the repatriation of discarded customers.

Table 4.1. Accuracy of the results obtained by iMPA

Prob. No.	No. of Customers.	Best Value	Proven Optimal	Heuristic	Gap %
1	10	312.81	yes	312.81	0.00
2	10	100.41	yes	91.81	8.56
3	10	108.53	yes	97.99	9.71
4	10	191.06	yes	181.92	4.78
5	10	213.14	yes	194.55	8.72
6	10	194.67	no	252.07	-29.39
7	10	25.54	yes	25.54	0.00
8	15	26.97	no	24.92	7.60
9	15	17.23	yes	11.56	32.91
10	20	90.72	no	90.72	0.00
11	20	247.23	no	230.45	6.78
12	20	186.61	no	156.71	19.08

References

Butt SE and Cavalier T (1994) A heuristic for the multiple tour maximum collection problem. Computers & Operations Research 21: 101–111.

Butt SE and Ryan DM (1999) An optimal solution procedure for the multiple tour maximum collection problem using column generation. Computers & Operations Research 26: 427–441.

Chao I, Golden B and Wasil E (1996) The team orienteering problem. European Journal of Operational Research 88: 464–474.

Clarke G, Wright JW (1964) Scheduling of vehicles from a central depot to a number of delivery points. Operations Research 12: 568-581.

Desrochers M and Laporte G (1991) Improvements and extensions to the Miller-Tucker-Zemlin subtour elimination constraints. Operation Research Letters 10: 27–36.

Feillet D, Dejax P and Gendreau M (2005) Traveling salesman problems with profits. Transportation Science 39:188–205.

Gueguen C (1999) Méthodes de résolution exacte pour les problèmes de tournées de véhicules. Ph.D. thesis, Laboratoire Productique Logistique, Ecole Centrale Paris.

Kara İ, Laporte G and Bektaş T (2004) A note on the lifted Miller–Tucker–Zemlin subtour elimination constraints for the capacitated vehicle routing problem. European Journal of Operational Research 158: 793–795.

Tang H and Miller-Hooks E (2005) A tabu search heuristic for the team orienteering problem. Computers & Operations Research 32: 1379–1407.

Tarantilis CD, Ioannou G, Kiranoudis CT and Prastacos GP (2005) Solving the open vehicle routeing problem via a single parameter metaheuristic algorithm. Journal of the Operational Research Society 56: 588–596.

A Decision Support System for Strategic and Operational Planning of Road Feeder Services

Paul Bartodziej, Ulrich Derigs, Boris Grein

Department of Information Systems and Operations Research (WINFORS), University of Cologne, Pohligstr. 1, 50969 Cologne, Germany
derigs@winfors.uni-koeln.de, paul.bartodziej@uni-koeln.de

1 Introduction

Generally, all major airlines operate so-called hub and spoke networks. That is, passengers as well as cargo are transported from many different origins to a small number of so-called hubs where these units are consolidated and then transported to other hubs using high-capacity airplanes. There the entities are separated again and transported to their respective destinations.

For cargo the airline operates a multi-modal network, i.e. transportation to and from the hubs is mostly done on the ground using specifically equipped trucks. Moreover, within Europe the majority of air cargo is generally transported via trucks due to time and cost reasons and thus only longer intercontinental transportation is done by airplane. Road Feeder Service (RFS) is the name of this transportation mode which today covers an essential part of the air cargo transport process and has obtained a significant role within the multi-modal air cargo business.

Obviously, as with every multi-modal transportation, there are complex organizational and physical handling problems to be tackled at the interface between the modes: air transportation and ground transportation. Yet, moreover there are significant different concepts and regulations which rule the operation of a cargo airline and a trucking company, respectively, with the different duty regulations for pilots and truckers being an obvious and significant example. Also, different views of the basic services to be operated may lead to significant inefficiencies for both partners. In this paper we survey several decision problems within RFS, and outline a model and a heuristic for supporting planning as well as operation of RFS. The model and the associated heuristic have been implemented in a decision support system (DSS) with the focus to support the trucking company as well as the cargo airline.

2 RFS: Planning problems

In cargo airlines planning is done comparably to passenger airlines: They establish and publish a weekly timetable where the time of service for all relations - the so-called origin and destination pairs (ODs) - is stated. This timetable is constructed every six months on the basis of the estimated market demand. A typical task submitted to the RFS-trucking company like "from Frankfurt 16:30 to Helsinki 11:00 (+2 days) on Tue,Wed,Thu,Sat,Sun" is called a "line" and has to be served regularly every week during the planning period. For the trucking company such a line breaks down to an eventually large set of equal atomic tasks, so-called trips during the weeks of the planning period.

Now, in an announcement or auction, bundles of lines are submitted to the trucking companies who are asked to price the single lines as well as the whole bundle, and to decide on an offer to the airline. The trucking company serving this line has to set up a plan for the entire set of tasks. During operation of the timetable airlines will eventually ask for additional transportation tasks on the spot - so-called "ad-hocs" - and thus they require pricing information on potential ad-hocs in the announcement, already. Note, that there are specific regulations for the required lead time and the cost of a cancellation of a trip of a line on a specific day. Thus depending on this cost, the price for an ad-hoc and the probability of the demand the airline may decide not to create and announce a line but to request ad-hocs on demand instead.

Given a line or a bundle of lines for bidding, the trucking company has to check how this (set of) line(s) can be operated. A suitable type of truck, as for instance with a specific conture or cooling device etc., has to be selected and it may be necessary to operate the associated trips with two drivers due to distance and time requirements. As in the example above there may be involved a train and/or a ferry which is operating on a fixed schedule and thus determines the routing and the timing of such trips. Based on these and other side-constraints which are higly influencing the feasibility with respect to legal driving times etc., the trucking company has to decide on the cost/bid-price of a line.

Now, the trucking company should not calculate every line by its own but thoroughly investigate the potential of combining trips of different lines (on different days) to feasible round-trips for the vehicles and drivers. That is, the problem of the trucking company is to optimally allocate its resources, trucks and drivers, to the set of trips obtained from the bundle of lines such that a number of constraints as for instance the legal regulations for drivers and the on-time requirements of the airline are fulfilled and that contribution to profit is maximized. The result of resource allocation is a set of so-called "blocks", i.e. sequences of trips which can be operated sequentially by one vehicle/driver(s)-pair without violation of any constraint. To support this complex problem of "lines to blocks transformation" is the focus of our model, heuristic and decision support system.

In fact, our model and DSS can be used to tackle several different problems in connection with planning and operational control of RFS:

1. On a strategic level, the DSS can support the trucking company in pricing a bundle of lines and deciding on what lines to bid for.
2. On the other side, the DSS can advice the airline company on how to optimally combine lines into sets of blocks for the announcement/bidding process.
3. On a tactical level, the trucking company is supported when searching for a transportation plan for the next period (e.g. a month) given a set of tasks which results from lines and other transportation services.
4. On the operational level, the dispatch of the trucking company can be supported when solving the problem on how to incorporate an ad-hoc task into the existing plan and on deciding whether to accept such an order and how to calculate and set the associated bid-price for the airline.

In the following we will discuss the tactical block planning problem (BPP) of the trucking company. BPP is very close to variants of the vehicle and crew scheduling problem which have been studied in airline as well as in public transportation.

3 The block planning problem (BPP)

Abstracting from the specific requirements in the RFS-domain the block planning problem can be formulated as follows: Given a fixed planning period $1, ..., T$ with a set F of transportation tasks/trips with some domain-specific requirements and a non-homogenous set V of vehicle/driver-combinations which are available for serving the trips, find an assignment of the vehicles to the trips forming blocks of trips which can be operated sequentially by a vehicle such that within blocks certain constraints are fulfilled. Here a block b is a sequence $(f_0, f_1, ..., f_n)$, $f_i \in F$, of trips ordered by increasing departure time with the sequence of trips starting and ending in a fixed location, the depot or hub.

Given a block $b \in B$ we can calculate

c_b^v = contribution to profit of block $b \in B$, if it is served by vehicle $v \in V$

BT_b^v = departure time from the depot of block $b \in B$, if served by vehicle $v \in V$; $BT_b^v \in T$

ET_b^v = arrival time at the depot of block $b \in B$, if served by vehicle $v \in V$; $ET_b^v \in T$

$$\gamma_b^v = \begin{cases} 1 \text{ if block } b \in B \text{ can be served by vehicle } v \in V \\ 0 \text{ else} \end{cases}$$

$$\delta_b^f = \begin{cases} 1 \text{ if block } b \in B \text{ contains trip } f \in F \\ 0 \text{ else} \end{cases}$$

Assume that we have constructed the set B of all possible blocks. Introducing the decision variable

$$x_b^v = \begin{cases} 1 \text{ if block } b \in B \text{ is served by vehicle } v \in V \\ 0 \text{ else} \end{cases}$$

we can model the BPP as a "set partition like" mathematical program:

$$\text{Maximize} \quad \sum_{v \in V} \sum_{b \in B} c_b^v x_b^v \tag{1}$$

$$\text{Subject to} \quad \sum_{v \in V} \sum_{b \in B} x_b^v \delta_b^f = 1 \qquad \forall f \in F \tag{2}$$

$$\sum_{\{b \in B | BT_v^b \leq t \leq ET_v^b\}} x_b^v \leq 1 \qquad \forall t \in T, \forall v \in V \tag{3}$$

$$x_b^v \leq \gamma_b^v \qquad \forall b \in B, \forall v \in V \tag{4}$$

$$x_b^v \in \{0, 1\} \quad \forall b \in B, \forall v \in V \tag{5}$$

The objective function (1) maximizes the sum of all contribution to profits obtained from the single blocks which are determined by trip-revenues, vehicle cost and driver pay. Here we assume that for every trip revenue is fix and thus the objective is equivalent to minimizing operational cost. The set of constraints (2) guarantees that every trip is contained in exactly one selected block and thus implicitly we guarantee that each selected block is assigned exactly one vehicle. The set of constraints (3) assures that at any time there is at most one block assigned to a vehicle. The set of constraints (4) comprises the condition that only blocks are selected and served which do not violate any problem-specific criteria.

Here we have modelled all domain-specific constraints through the function γ_b^v. Thus this function contains all the the knowledge about time constraints, technical restrictions, and, especially the necessary requirements stemming from legal regulations with respect to driving and rest periods for the drivers which have to be obeyed by the RFS-trucking company. A thorough requirements analysis and the precise definition of the "rule-base" for the DSS-model has been a major task within the DSS-development process. A description of the entire RFS-BPP with the precise rule-base will be given in a forthcoming paper.

4 The Heuristic

Here, we outline the basics of the heuristic search procedure for generating solutions for RFS-BPP. The heuristic is based on a specific representation and consists of two phases: a *construction phase*, where an initial feasible solution is generated through path decomposition of the compatibility graph and a *local search improvement phase* where blocks are split and joined.

Representation: The compatibility graph The concept of the compatibility graph is well known and widely used in scheduling mass transit crews and vehicles (cf. [1]), and in airline crew scheduling (cf. [2], [3]). Here the tasks which have to be combined, i.e. the trips, flight services, etc. build the node-set V of a directed graph $G = (V, E)$, and an arc (i, j) connecting two nodes i and j is introduced into E if the associated tasks can be combined, i.e. if after having performed task i the driver, the crew, etc. is able to perform j. Also two dummy nodes representing the "start" and "end" of the planning period are introduced into V and connected with all other nodes. Since the tasks (and the transfers from task i to task j) are time-consuming the resulting graph is acyclic.

In many applications paths in G correspond to feasible combinations and vice versa, and the problem reduces to finding an optimal partitioning of G into disjoint paths. For this property to hold, the compatibility of two tasks must depend solely on the properties of these two tasks. Yet, in our application the compatibility of two trips, i and j say, does not depend on i and j only, but also on the trips performed prior to i. Nevertheless, the initial definition of the compatibility graph which we use is also based on "local constraints" on two consecutive trips only. Then the path-condition has to be relaxed to the relation that every feasible block corresponds to a path in G, but not vice versa. Thus the path/block-feasibility captured in constraint (4) of the BPP-model and the rule-base of the DSS, respectively, has to be guaranteed through the application of a rather complex partitioning/path finding strategy.

Construction: Block extension through optimal matching Our construction procedure is an immediate application of the matching-based piece-construction heuristic proposed in [1] for scheduling mass transit crews and vehicles.

In a first step, the nodes of the compatibility graph G, are partitioned into levels, where the level of a node i equals the length of the shortest path from *start* to i. Here the length of a path is defined as the number of edges the path contains. Since G is acyclic, the levels can be computed rather efficiently, using simple depth first search.

Let l_{max} be the highest level of a node in G. Then in a second step l_{max} iterations are performed to generate paths which represent feasible blocks. After k iterations the nodes in level 1 to level k are partitioned into paths representing feasible blocks, and every path/block is assigned a vehicle/driver-combination which can perform the block. Then in iteration $k + 1$ the paths are eventually extended by nodes/tasks from level $k + 1$ solving a perfect matching problem on a weighted bipartite auxiliary graph. After performing the last iteration l_{max} we obtain a partitioning of the graph/the tasks into paths/feasible blocks.

Improvement: Splitting and joining blocks The improvement heuristic is a local search procedure which is based on two operations/moves: split and join, which, given a feasible solution i.e. a set of blocks, combine two blocks to one block (join) or partition a block in two blocks (split) and assign

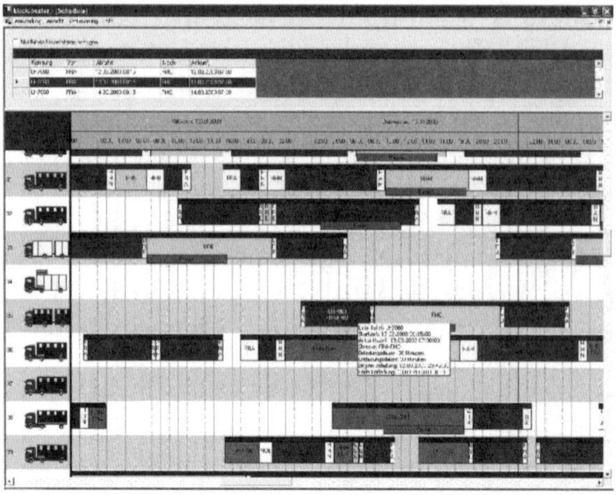

Fig. 1. Screenshot of the DSS-User Interface

compatible vehicle/crew-combinations to the new block(s). This set of moves has already been proposed in [1]. A detailed description of the entire heuristic will be given in a forthcoming paper.

5 Concluding remarks

We have implemented the model and the heuristic within a decision support system which offers basically two functions:

1. Given a set of lines the the system generates a "good" feasible solution/block plan which is displayed in a user-friendly graphical interface (Figure 1 shows a screenshot displaying historical data).
2. Using the drag and drop-functionality of the user-interface the user can modify a given block plan and have the system check feasibility and calculate cost/contribution to profit. Through interactive planning the user can incorporate additional knowledge like additional rules and/or relax constraints from the rule-base.

References

1. Ball M., Bodin L., Dial R. (1983) *A Matching Based Heuristic for Scheduling Mass Transit Crews and Vehicles*, Transportation Science 17:4–31
2. Ball M., Roberts A. (1985) *A Graph Partitioning Approach to Airline Crew Scheduling*, Transportation Science 19:107–126
3. Lavoie S., Minoux M., Odier E. (1988) *A new approach for crew pairing problems by column generation with an application to air transportation*, European Journal of Operational Research 35:45–58

Mehrdepot-Umlaufplanung: Berücksichtigung von Verschiebeintervallen für Fahrten in einem Time-Space-Netzwerk-basierten Modell

Stefan Bunte, Natalia Kliewer, Leena Suhl

Decision Support & Operations Research Lab, Universität Paderborn, Warburger Straße 100, 33100 Paderborn {stbunte|kliewer|suhl}@uni-paderborn.de

1 Einleitung

Bei dem Mehrdepot-Umlaufplanungsproblem handelt es sich um eine im Planungsprozess der Busunternehmen auftretende Entscheidungssituation. Dabei sollen die Tagesumläufe der Fahrzeuge festgelegt und den Depots sowie Fahrzeugtypen zugewiesen werden, so dass alle Beförderungsfahrten eines Betriebstages mit minimalen Gesamtkosten bedient werden.

Diese Aufgabe wird unter anderem in [2], [3] und [4] als ein Mehrgüterflussproblem modelliert. Dabei verwenden [2] und [4] eine explizite Abbildung aller möglichen Fahrtenverknüpfungen in einem Connection-Netzwerk und [3] - ein Time-Space-Netzwerk mit stark reduzierter Anzahl der zu berücksichtigenden Fahrtenverknüpfungen. In dieser Arbeit erweitern wir den Time-Space-Netzwerk-basierten Ansatz aus [3] um die Betrachtung von zeitlichen Verschiebeintervallen für Fahrten.

Durch geringfügiges zeitliches Verschieben von Fahrten soll dabei eine möglichst hohe Einsparung im Fahrzeugbedarf erzielt werden. Das Potenzial der zeitlichen Fahrtenverschiebungen liegt in den besseren Anschlussmöglichkeiten, so dass Fahrtenverknüpfungen zustande kommen können, die ohne Verschiebung nicht möglich wären. Das zulässige Verschiebeintervall kann unterschiedlich groß für verschiedene Fahrten sein. Zum Beispiel dürfen Schulfahrten tendenziell weiter verschoben werden als normale Taktfahrten.

Da die Beförderungsfahrten im öffentlichen Personennahverkehr in der Regel minutengenau geplant werden, sind die Werte der Ankunfts- und Abfahrtszeiten diskret. Bei der Berücksichtigung zeitlicher Verschiebeintervalle für Fahrten können die Fahrten somit immer um ein vielfaches der entsprechenden Zeiteinheit (hier Minuten) verschoben werden. Die Betrachtung diskreter Verschiebeintervalle wurde zum Beispiel auch in [1] vorgenommen.

2 Berücksichtigung von Zeitfenstern

Im Time-Space-Netzwerkmodell aus [3] werden die potentiellen Abfahrts- und Ankunftsereignisse in Ereignis-Zeitlinien (Timelines) der Haltestellen und Depots durch Kanten für mögliche Fahrzeugaktivitäten verbunden. Dies sind z.B. Kanten für Beförderungsfahrten, Umsetzfahrten zu anderen Haltestellen/Depots und für Standzeiten der Fahrzeuge in den Haltestellen.

Um die möglichen Fahrtenverschiebungen im Netzwerkmodell abzubilden, wird ein neuer Typ von Kanten für das Netzwerk definiert: die **Zeitfenster-Kanten**. Wenn eine Beförderungsfahrt-Kante in einem vorgegebenen Zeitfenster verschiebbar sein soll, werden mehrere Zeitfenster-Kanten in die Umgebung der Beförderungsfahrt-Kante eingefügt und die Abfahrts- und Ankunftszeiten entsprechend angepasst. Bei der Optimierung kann zwischen einer Beförderungsfahrt-Kante und einer der dazugehörigen Zeitfenster-Kanten gewählt werden.

Das Hinzufügen der Zeitfenster-Kanten macht das Netzwerk und somit auch das mathematische Modell komplexer und das ohnehin NP-harte Problem noch schwieriger lösbar. Um die Modellgröße nicht unnötig anwachsen zu lassen, werden nur die Zeitfenster-Kanten in das Netzwerk eingefügt, die tatsächlich einen neuen Anschluss ermöglichen können und nicht-redundant sind, was aus der Netzwerkstruktur ablesbar ist. Bild 1 zeigt ein Beispiel, in dem von vier möglichen Zeitfenster-Kanten für eine Beförderungsfahrt von der Haltestelle 2 zu der Halstestelle 1 nur zwei sinnvoll sind, weil sie neue Anschlüsse in einer der beiden Haltestellen ermöglichen.

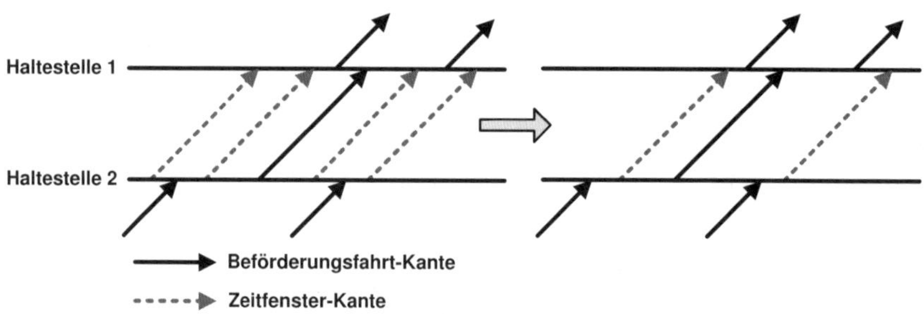

Abbildung 1. Reduktion überflüssiger Zeitfenster-Kanten

Das mathematische Modell für die Mehrdepot-Umlaufplanung entspricht einem *minimum cost flow problem* mit zusätzlichen Restriktionen (sog. *Cover Constraints*), die die Bedienung einer Fahrt von genau einem Depot und Fahrzeugtyp sicherstellen. Für die Berücksichtigung der zeitlichen Fahrtenverschiebungen ist eine entsprechende Anpassung des mathematischen Modells notwendig. Die Zielfunktion und die Fluss-Restriktionen bleiben dabei unverändert. Die Cover Constraints müssen nun zusätzlich zu den regulären

Beförderungsfahrt-Kanten auch die eingefügten Zeitfenster-Kanten berücksichtigen. Somit wird sichergestellt, dass bei der Lösung des Modells zwischen den Beförderungsfahrt- und Zeitfenster-Kanten alternativ gewählt werden kann, so dass bei kostengünstigeren Konstellationen eine zeitliche Verschiebung von Fahrten zustande kommt.

3 Ermittlung verschiebbarer Fahrten

Die Menge der verschiebbaren Fahrten kann auf unterschiedliche Weise vorgegeben werden. Möglich ist die Vorgabe eines **globalen Zeitfensters**, das für sämtliche Beförderungsfahrten des Fahrplans gilt. Diese Art der Vorgabe verschiebbarer Fahrten hat einen entscheidenden Nachteil: die Problemgröße explodiert mit wachsender Länge des Verschiebeintervalls. Um dies zu vermeiden werden nachfolgend einige Möglichkeiten vorgestellt, die Menge der verschiebbaren Fahrten zu begrenzen, um die Größe des zu lösenden Problems beherrschbar zu halten.

Eine **benutzerdefinierte Definition** der Menge verschiebbarer Fahrten erfolgt durch eine explizite Angabe der Zeitfenster für jede Fahrt durch den Endanwender und ermöglicht somit die Freigabe von einer Teilmenge der Fahrplanfahrten für die Verschiebung. Die Modellgröße wird dadurch gegenüber der Variante mit globalen Zeitfenstern verkleinert, da nur Kanten für die explizit als verschiebbar ausgezeichneten Fahrten in das Netzwerk eingefügt werden müssen. Dies soll für praxisrelevante Instanzen zu einer Lösungszeit führen, die im Vergleich zu dem Fall ohne Berücksichtigung von Fahrtenverschiebungen nicht viel größer ist.

Häufig besteht das Problem, dass die Menge der zur Verschiebung freizugebenden Fahrten nicht vom Endanwender (Planer im Busunternehmen) festgelegt werden kann. In solchen Fällen wird ein Verfahren benötigt, das eine Menge von verschiebbaren Fahrten vorgibt, da ein pauschales Versehen aller Fahrplanfahrten mit Verschiebeintervallen (s.o. - globale Zeitfenster) tendenziell zu zu großen Modellen führen würde. Diese Menge soll Fahrten enthalten, die ein hohes Potential für die Einsparung von Umläufen versprechen. Sie wird nachfolgend als die **Menge der kritischen Fahrten** bezeichnet. Für die Ermittlung dieser werden nachfolgend zwei heuristische Methoden vorgeschlagen.

Verkürzungs-Heuristik

In der Verkürzungs-Heuristik werden zunächst probeweise kürzere Fahrtzeiten für alle Fahrten des Fahrplans angenommen. Dafür werden alle Ankunftsereignisse künstlich um eine bestimmte (durch einen Parameter vorgegebene) Zeit verfrüht. Für einen auf diese Weise veränderten Fahrplan wird ein optimaler Umlaufplan berechnet. Eine anschließende Untersuchung dieses Umlaufplans führt zu der Ermittlung kritischer Stellen - dies sind die Fahrtenverknüpfungen, die ohne die vorgenommene künstliche Verkürzung von Fahrten nicht zustandekommen könnten. Die beiden an einem solchen 'Konflikt' beteiligten Fahrten werden entsprechend der zeitlichen Überschneidung für die nachfolgende 'echte' Umlaufplanung verschiebbar gemacht, also in die Menge der kritischen Fahrten aufgenommen.

Schnitt-Heuristik

Die Schnitt-Heuristik zur Bestimmung verschiebbarer Fahrten basiert auf der Beobachtung, dass Fahrpläne in der Regel deutliche Auslastungsspitzen aufweisen. Ziel ist es, solche Spitzen 'abzutragen' und die Auslastung auf umliegende Gebiete zu verteilen. Grafisch äußert es sich in einer Glättung der Auslastungskurve. Im Gegensatz zu der oben beschriebenen Schnitt-Heuristik ist dafür kein vorgelagerter Optimierungsvorgang notwendig, sondern die Informationen werden ausschließlich aus dem gegebenen Fahrplan gewonnen.

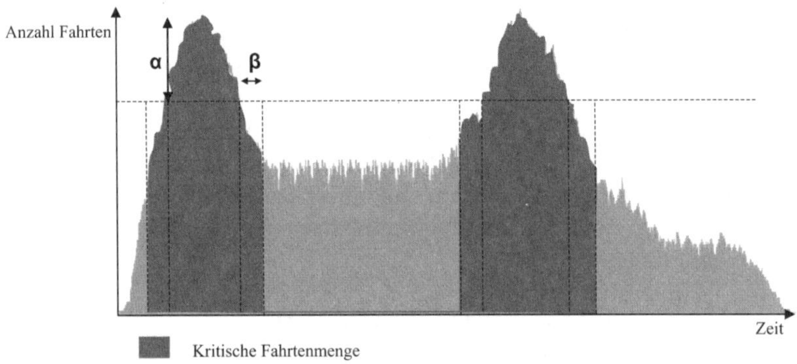

Abbildung 2. Ermittlung kritischer Fahrten in der Schnitt-Heuristik

In Abbildung 2 ist die Auslastung durch einen typischen Fahrplan durch eine Tagesganglinie grafisch dargestellt. Die horizontale Achse beschreibt die Zeit und die vertikale stellt die Anzahl der zu einem Zeitpunkt gleichzeitig stattfindenden Beförderungsfahrten dar. Das Maximum dieser Linie stellt eine untere Grenze für die Anzahl der insgesamt benötigten Fahrzeuge dar. Die Fahrten, die sich in den Spitzen befinden, sollen verschiebbar gemacht werden, so dass die Auslastungskurve geglättet werden kann. Für diese Art der Bestimmung kritischer Fahrten wird ein horizontaler Schnitt in der Fahrtenmenge vorgenommen, der um einen Wert α von der Spitze entfernt ist. Die kritische Menge von Fahrten befindet sich in der Säule, die sich aus der Projektion des Schnittes auf die Zeitachse ergibt.

Zusätzlich zu diesem Schnitt kann eine β-Umgebung angegeben werden. Diese erweitert die Menge der kritischen Fahrten über die Grenzpunkte des durch α definierten Schnittes hinaus. Statt absoluter Werte für die Höhe des Schnittes kann auch eine relative Höhe vorgegeben werden, ab der die Fahrten in die kritische Fahrtenmenge aufgenommen werden sollen.

4 Testergebnisse

Nachfolgend werden Testergebnisse für drei unterschiedlich große reale Probleminstanzen vorgestellt. Die Anzahl der Beförderungsfahrten sowie die durchschnittliche Gruppengröße (durchschnittliche Anzahl der erlaubten Fahrzeugtyp-Depot-Kombinationen pro Fahrt) für getestete Instanzen sind in der ersten Spalte der Tabelle 2 angegeben.

Für jede Instanz wurden Testläufe mit globaler Zeitfensterangabe, und mit den im Abschnitt 3 beschriebenen Heuristiken zur Ermittlung verschiebbarer Fahrten durchgeführt. Für die Schnitt-Heuristik wurden dabei sowohl absolute (durch α und β festgelegte) als auch prozentuale Schnitte getestet. Für alle Kombinationen variierte das maximal zulässige Intervall für die zeitliche Verschiebung einer Fahrt von 1 bis 6 Minuten.

Die Kosten für eine verschobene Fahrt wurden auf das Dreifache der normalen Kosten gesetzt, so dass Fahrten möglichst nur dann verschoben wurden, wenn dadurch die Anzahl der Fahrzeuge reduziert wird und nicht etwa die operativen Kosten.

In Tabelle 1 sind die prozentualen Fahrzeugeinsparungen und die dazu notwendigen Verschiebungen von Beförderungsfahrten dargestellt. Sie ergeben sich aus der jeweiligen Strategie zur Ermittlung der kritischen Fahrtenmenge sowie der Größe des jeweiligen Verschiebeintervalls.

Die in der Tabelle 2 angegebenen Laufzeiten gelten für eine Optimierung mit dem Branch&Cut-Solver ILOG CPLEX 9.0 mit dem dualen Simplex-Algorithmus zur Berechnung der LP-Relaxationen. Alle Testläufe wurden auf einem XEON 2,20 GHz Prozessor mit 2 GB RAM unter Microsoft Windows XP Professional ausgeführt. Die mit einem '–' gekennzeichneten Testläufe wurden nach einem Zeitlimit von 24 Stunden abgebrochen.

Folgende Schlussfolgerungen können aus den Tests gewonnen werden:

- Eine Berücksichtigung von Verschiebeintervallen führt bereits bei geringer Verschiebung der Fahrten zu einer Einsparung von Fahrzeugen.
- Globale Zeitfenster liefern durch Berücksichtigung aller Fahrten die größten Einsparungen, benötigen allerdings eine sehr große Laufzeit zur Lösung.
- Für eine vergleichbare Fahrzeugeinsparung bietet die Verkürzungs-Heuristik in vielen Fällen eine schnelle Lösung, benötigt allerdings wesentlich mehr Fahrtenverschiebungen.
- Die Schnitt-Heuristik liefert sowohl mit absoluten als auch mit relativen Parametern sehr gute Lösungen in Bezug auf Qualität und Laufzeit.

Tabelle 1. Einsparung von Fahrzeugen in % (Gesamtverschiebung in Minuten)

Verschiebeintervall (in Min.):	1	2	3	4	5	6
Saarbrücken						
– Global	0(0)	2,0(16)	4,1(81)	8,2(300)	12,2(770)	14,3(1087)
– Verkürzung	0(0)	0(0)	0(0)	2,0(22)	2,0(39)	4,1(145)
– Schnitt(10/10)	0(0)	2,0(14)	4,1(76)	8,2(291)	12,2(749)	14,3(1062)
– Schnitt(50%)	0(0)	2,0(14)	2,0(16)	8,2(299)	8,2(279)	8,2(310)
Halle						
– Global	4,4(50)	7,0(127)	7,8(225)	8,7(291)	9,6(372)	11,3(569)
– Verkürzung	2,6(18)	6,1(372)	7,0(279)	7,8(376)	7,8(413)	9,6(1012)
– Schnitt(10/10)	3,5(26)	7,0(125)	7,8(231)	8,7(283)	9,6(344)	9,6(378)
– Schnitt(50%)	4,4(50)	7,0(127)	7,8(226)	8,7(300)	9,6(362)	11,3(598)
München						
– Verkürzung	0(0)	–	–	–	–	–
– Schnitt(50%)	1,8(7)	–	–	–	–	–

Tabelle 2. Laufzeit (in Sekunden)

Instanz	Verschiebeintervall	1	2	3	4	5	6
Saarbrücken	– Global	14	39	106	310	1648	2563
(1296 Fahrten,	– Verkürzung	174	175	172	184	191	195
$\oslash G = 2,0$)	– Schnitt(10/10)	10	30	70	246	1302	2602
	– Schnitt(50%)	8	28	28	277	327	475
Halle	– Global	1600	3600	7847	11688	17109	21075
(2047 Fahrten,	– Verkürzung	190	703	621	791	1751	2620
$\oslash G = 3,2$)	– Schnitt(10/10)	268	483	644	1873	1522	2145
	– Schnitt(50%)	313	592	1132	2421	3115	3883
München	– Verkürzung	50114	–	–	–	–	–
(1808 Fahrten, $\oslash G = 13,0$)	– Schnitt(50%)	80908	–	–	–	–	–

Literatur

1. Daduna J, Völker M (1997) Fahrzeugumlaufbildung im ÖPNV mit unscharfen Abfahrtszeiten. Der Nahverkehr 11:39–43
2. Forbes M, Holt J, Watts A (1994) An exakt algorithm for multiple depot bus scheduling. European Journal of Operations Research 72(1):115–124
3. Kliewer N, Mellouli T, Suhl L (2005) A time-space network based exact optimization model for multiple-depot bus scheduling. European Journal of Operations Research (erscheint 2005)
4. Löbel A (1997) Optimal Vehicle Scheduling in Public Transit. PhD Thesis, TU Berlin

Adaptive Dienst- und Umlaufplanung im ÖPNV

Vitali Gintner[1], Stefan Kramkowski[1], Ingmar Steinzen[1] und Leena Suhl[1]

Decision Support & OR Lab und International Graduate School of Dynamic Intelligent Systems, Universität Paderborn, Warburger Str. 100, 33100 Paderborn
{gintner,ivan80,steinzen,suhl}@upb.de

Zusammenfassung. Die Probleme der Umlauf- und Dienstplanung im Öffentlichen Personennahverkehr (ÖPNV) werden traditionell nacheinander gelöst. Somit liegen für die Generierung möglicher Dienste feste Umläufe zugrunde. Das führt oft dazu, dass die Dienstpläne unzulässig bzw. nicht kostenoptimal sind. Wir präsentieren ein adaptives Verfahren, dass der Dienstplanung eine Rückkopplung zu der Umlaufplanung erlaubt. Dabei können bei der Dienstplanung die gegebenen Umläufe ggf. in gewissem Maße verändert werden, so dass ihre Gesamtkosten aber optimal bleiben. Die durchgeführten Tests zeigen, dass durch diese Rückkopplung eine enorme Verbesserung der Dienstpläne erreicht werden kann.

1 Einleitung

Umlauf- und Dienstplanung sind zwei Hauptaufgaben des operativen Planungsprozesses eines Verkehrsbetriebs im ÖPNV. Sie werden in der Literatur und Praxis sehr häufig als Aufgaben betrachtet, die wegen ihrer Komplexität und Größe nacheinander abgearbeitet werden. Es ist aber bekannt, dass eine gleichzeitige Betrachtung zu Effizienzgewinnen führen kann. Die integrierte Behandlung wurde in den letzten Jahren in der Literatur untersucht (z.B. [3], [4], [1] und [2]) und zur Lösung von Problemen kleiner bis mittlerer Größe erfolgreich eingesetzt.

Durch die *Umlaufplanung* werden Fahrzeuge einer Menge von vorgegebenen Fahrten zugewiesen, sodass die Fahrzeugkosten minimal sind. Die Fahrzeugkosten setzen sich aus Fixkosten pro eingesetzte Fahrzeug und variablen Kosten pro Zeiteinheit außerhalb des Depots sowie gefahrenen Kilometern zusammen. Der daraus resultierende Umlaufplan ist genau dann zulässig, wenn alle Fahrten genau einem Fahrzeug zugewiesen sind, jedes Fahrzeug in einem Depot startet und nach einer Abfolge von Fahrten dorthin wieder zurückkehrt. Wir definieren zwei Umlaufpläne als gleichwertig, wenn die gesamten Kosten der Umläufe gleich sind.

Für die *Dienstplanung* werden die Umläufe an gegebenen *Ablösepunkten* geteilt und die so entstandenen *Dienstelemente* Diensten kostenminimal zugeordnet. Eine Lösung des Dienstplanungsproblems ist genau dann zulässig, wenn jedes Dienstelement ausgeführt wird und jeder Dienst die gesetzlichen sowie betrieblichen Regelungen erfüllt. Ein Dienst besteht aus einem oder mehreren solcher *Dienststücken* unterteilt durch anrechenbare Pausen. Ein Dienststück ist eine Abfolge von Dienstelementen, die ein Fahrer an einem Fahrzeug ohne Pausenunterbrechung bedient. Das Dienstplanungsproblem wird meist als *Set Partitioning* (bzw. *Covering*) Problem formuliert und mit Hilfe des *Column Generation*-Verfahrens gelöst.

Die traditionelle feste Kopplung der Dienstplanung an die Umlaufplanung schränkt die Dienstvielfalt stark dadurch ein, dass die Umläufe und somit alle Leerfahrten schon fest vorgegeben sind. Dies führt sehr oft dazu, dass die Dienstpläne nicht kostenoptimal bzw. sogar unzulässig sind.

Im Abschnitt 2 präsentieren wir ein alternatives Vorgehen, bei dem während der Dienstplanung anstatt eines festen Umlaufplans implizit mehrere gleichwertige Umlaufpläne gleichzeitig betrachtet werden. Die so geschaffenen zusätzlichen Freiheitsgrade ermöglichen uns, einen besseren Dienstplan zu finden, für den der vorgegebene Umlaufplan durch Umschichtung der Umläufe adaptiert wird. In Abschnitt 3 werden die durchgeführten Testläufe und Ergebnisse präsentiert.

2 Adaptive Umlauf- und Dienstplanung

Zur Veranschaulichung unseres Verfahrens betrachten wir zunächst ein Beispiel. Sei ein Umlaufplan UP mit drei Umläufe U_1, U_2 sowie U_3 und Fahrten wie in Tabelle 1 gegeben. Wir nehmen an, dass sich in einem Bus außerhalb des Depots immer ein Fahrer befinden muss, d.h. die Wartezeit im Bus zwischen zwei Fahrten gilt nicht unbedingt als Arbeitszeitunterbrechung. Die maximale Dauer eines Dienststückes sei auf vier Stunden beschränkt. Somit können die Umläufe U_2 und U_3 jeweils komplett durch einen Fahrer bedient werden. Die Gesamtdauer des Umlaufs U_1 beträgt 4:45 Stunden. Durch die Überschreitung des Limits von vier Stunden sind zwei Fahrer zur Bedienung dieses Umlaufs notwendig.

UmlaufID	FahrtID	Fahrttyp	von	nach	Abfahrt	Ankunft
U_1	f_1^D	Ausrückfahrt	Depot	C	8:15	8:30
	f_2	Fahrgastfahrt	C	A	8:30	9:30
	f_3	Fahrgastfahrt	A	B	11:10	12:00
	f_4^L	Leerfahrt	B	C	12:00	12:15
	f_5	Fahrgastfahrt	C	B	12:30	12:45
	f_6^D	Einrückfahrt	B	Depot	12:45	13:00
U_2	f_7^D	Ausrückfahrt	Depot	B	9:10	9:25
	f_8	Fahrgastfahrt	B	A	9:25	10:00
	f_9	Fahrgastfahrt	A	C	10:15	11:00
	f_{10}^D	Einrückfahrt	C	Depot	11:00	11:15
U_3	f_{11}^D	Ausrückfahrt	Depot	B	10:00	10:15
	f_{12}	Fahrgastfahrt	B	A	10:15	10:50
	f_{13}	Fahrgastfahrt	A	B	11:35	12:15
	f_{14}	Fahrgastfahrt	B	A	12:15	13:00
	f_{15}^D	Einrückfahrt	A	Depot	13:00	13:20

Tabelle 1. Umlaufplan UP mit drei Umläufen

Betrachtet man die Umläufe genauer, wird man feststellen, dass die Umläufe U_1 und U_2 beide zwischen 10:00 und 10:15 Uhr an Haltestelle A warten. Die Gesamtkosten der Umläufe ändern sich nicht, wenn man ihre Teile ab diesem Zeitpunkt (grau markierte Flächen in der Tabelle) vertauscht. Wir erhalten somit einen alternativen, gleichwertigen Umlaufplan UP' mit Umläufen $U_1' = \{f_1^D, f_2, f_9, f_{10}^D\}$, $U_2' = \{f_7^D, f_8, f_3, f_4^L, f_5, f_6^D\}$ und $U_3' = \{f_{11}^D, f_{12}, f_{13}, f_{14}, f_{15}^D\}$ mit Umlaufdauern von 3:00, 3:50 und 3:20 Stunden. Legt man UP' anstelle von UP in der Dienstplanung zu Grunde, sind nur drei Dienste notwendig: ein Dienst für jeden Umlauf.

Im Folgenden wird ein Verfahren für Dienstgenerierung vorgestellt, das implizit mehrere gleichwertige Umlaufpläne gleichzeitig betrachtet. Dies geschieht mit Hilfe eines *Dienststücksplanungsgraphes.*

2.1 Dienstgenerierung

Die Menge aller zulässigen Dienste wird während der Dienstplanung in einem zweistufigen Prozess erzeugt. Zuerst werden alle zulässigen Dienststücke generiert und erst daraus Dienste als zulässigen Kombination mehrerer Dienststücke zusammengefügt. Für die erste Phase generieren wir einen Dienststücksplanungsgraph (DSPG) als Hilfsnetzwerk. Dieser Graph hat die Struktur eines Zeit-Ort-Netzwerks (*engl.: Time-Space Network*), in dem Knoten Ort-Zeit-Allokationen und Kanten Aktivitäten dazwischen darstellen.

Für jede Aktivität (Fahrgast-, Depot- oder Leerfahrt) aus dem gegebenen Umlaufplan werden zwei Knoten sowie eine Kante erzeugt. Die Knoten repräsentieren Fahrtanfang und Fahrtende mit den dazugehörigen Abfahrts- und Ankunftszeit sowie Start- und Endhaltestellen. Sind zwei aufeinander folgenden Aktivitäten direkt nacheinander, ohne Wartezeit dazwischen auszuführen, wird der Endknoten der ersten Aktivität mit dem Startknoten der Zweite zusammengefasst, ansonsten werden sie durch eine Wartekante verbunden. Alle Knoten sind nach Haltestellen gruppiert und nach Zeit sortiert. Die Wartekanten repräsentieren wartende Fahrzeuge, wobei zu jedem Zeitpunkt an jeder Haltestelle maximal eine Wartekante existiert. Somit werden die Wartekanten mehrerer gleichzeitig wartender Fahrzeuge zusammengefasst. *Die Flussgröße* jeder Kante gibt an, wie viele Fahrzeuge sie repräsentiert. Nur die zusammengefassten Kanten haben eine Flussgröße größer eins. Existiert mehr als ein Depot oder/und Fahrzeugtyp, wird für jede Kombination *Depot*× *Fahrzeugtyp* ein eigener DSPG mit den relevanten Umläufen erzeugt.

Abbildung 1 zeigt einen DSPG des erläuterten Beispiels aus Tabelle 1. Die Flussgröße z_i ist gleich zwei für die beiden Kanten e_2 und e_4 und eins sonst (nicht abgebildet). Die Abbildung zeigt, dass die Pfade, die Umläufe darstellen, durch die Zusammenfassung der Wartekanten an Haltestelle 1 nicht mehr disjunkt sind.

Der DSPG für den alternativen Umlaufplan UP' aus dem obigen Beispiel ergibt den gleichen Graphen wie Abbildung 1. Somit repräsentiert der DSPG für einen Umlaufplan implizit mehrere gleichwertige Umlaufpläne, die aus unterschiedlicher Zerlegung von DSPG in disjunkte Pfade gewonnen werden können.

Wir nehmen an, dass die Zulässigkeit eines Dienststückes nur durch seine Dauer beschränkt ist und jeder Fahrtanfang und Fahrtende gültiger Ablösepunkt ist. Eine Erweiterung für den Fall mit beliebigen Ablösepunkten ist dabei möglich. Ein Pfad mit zulässiger Dauer zwischen je zwei Knoten in DSPG stellt ein Dienststück dar. Da die Pfade nicht mehr disjunkt sind, lassen sich auch Dienststücke finden, die aus

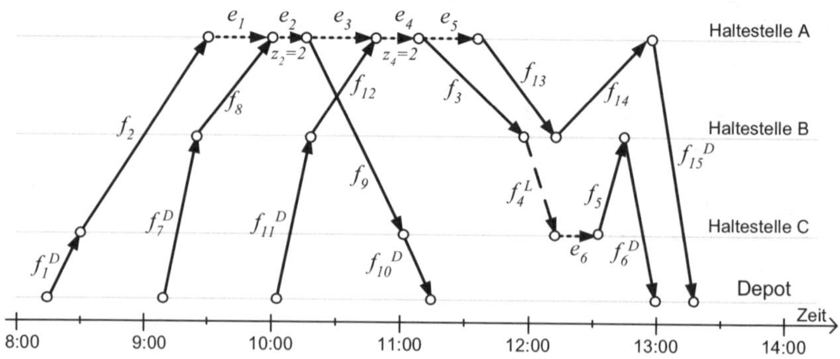

Abb. 1. Dienststückplanungsgraph für Umlaufplan UP

Kanten unterschiedlicher Umläufe bestehen (auch wenn diese Umläufe keinen gemeinsamen Schnittpunkt haben, wie z.B. das Dienststück $\{f_8,\ e_2,\ e_3,\ e_4,\ e_5,\ f_{13}\}$). Werden solche Dienststücke in die Lösung ausgewählt, ist DSPG so in disjunkte Pfade/Umläufe zu zerlegen, dass jedes Dienststück nur Kanten eines Umlaufs beinhaltet (vgl. Definition eines Dienststückes). Im Prinzip werden die Umläufe aus den Dienststücken also nachträglich erzeugt.

Werden beispielsweise folgende Dienststücke aus Abb.1 in die Lösung des Dienstplanungsproblems ausgewählt: $D_1 = \{f_1^D,\ f_2,\ e_1,\ e_2,\ f_9,\ f_{10}^D\}$, $D_2 = \{f_7^D,\ f_8,\ e_2,\ e_3,\ e_4,\ f_3,\ f_4^L,\ e_6,\ f_5,\ f_6^D\}$ und $D_3 = \{f_{11}^D,\ f_{12},\ e_4,\ e_5, f_{13},\ f_{14},\ f_{15}^D\}$, lassen sie DSPG in den alternativen Umlaufplan UP' aus dem obigen Beispiel zerfallen.

Da für jede Kombination *Depot*×*Fahrzeugtyp* ein eigener Graph mit nur relevanten Umläufen existiert und außerdem keine zusätzlichen Warte- bzw. Leerfahrtzeiten betrachtet werden, ist sichergestellt, dass jede Zerlegung einen zulässigen Umlaufplan mit gleichen Kosten darstellt.

2.2 Das mathematische Modell

Sei D die Menge aller zulässigen Dienste, die durch zulässige Kombinationen mehrere Dienststücke unter Erfüllung der Dienstregeln erzeugt sind. Seien weiterhin E die Menge aller Kanten im DSPG und $D(e)$ die Menge aller Dienste, die die Kante $e \in E$ beinhalten. Für jeden Dienst $d \in D$ definieren wir eine binäre Entscheidungsvariable x_d, die angibt, ob der Dienst d für die Lösung ausgewählt wurde oder nicht. Seien c_d die operativen Kosten eines Dienstes $d \in D$ und z_e die Flussgröße einer Kante $e \in E$. Das Dienstplanungsproblem kann folgendermaßen als ein MIP formuliert werden (DPP):

$$min \quad \sum_{d \in D} c_d x_d \tag{1}$$

$$s.t. \quad \sum_{d \in D(e)} x_d = z_e \qquad \forall\, e \in E \tag{2}$$

$$x_d \in \{0, 1\} \qquad \forall\, d \in D \tag{3}$$

In (1) werden die gesamten Dienstkosten minimiert. Restriktionen (2) stellen sicher, dass jede Kanten genau so oft mit Diensten überdeckt ist, wie ihre Flussgröße ist oder, mit anderen Worten, wie viele Fahrzeuge sie repräsentiert.

Die Formulierung ist ein *Generalized Set Partitioning Problem.* Ersetzt man Gleichungen (2) durch $\geq$-Ungleichungen, erhält man ein *Generalized Set Covering Problem*, das generell einfacher zu lösen ist. Dabei repräsentiert die Mehrfachüberdeckung von Kanten Fahrer, die als Fahrgäste mitfahren.

2.3 Lösungsverfahren

Wir betrachten den Fall ohne globale Einschränkung der Dienstanzahl. Somit sind die Probleme der Dienstplanung für jedes Depot unabhängig voneinander und werden separat gelöst. Aufgrund der großen Anzahl zulässiger Dienste verwenden wir zum Lösen von DPP ein *Column Generation*-Verfahren.

Zuerst wird mit Hilfe einer Greedy-Methode eine Menge der Dienste/Spalten bestimmt, die alle Kanten/Zeilen überdecken. Dann wird die LP-Relaxation von DPP mit diesen Spalten mit Hilfe des Optimierungssoftware CPLEX gelöst. Die Lösung liefert eine untere Schranke für das Gesamtproblem und duale Werte für Kanten. Nun werden neue Dienste/Spalten mit negativen reduzierten Kosten erzeugt und dem Modell hinzugefügt. Dienste werden in einem zweistufigen Verfahren erzeugt. In der ersten Phase werden die zulässigen Dienststücke als Pfade zwischen zulässigen Knotenpaaren in DSPG berechnet wobei nicht alle möglichen Pfade erzeugt werden müssen. Wir verwenden ein alternatives Vorgehen, das für ein ähnliches Unterproblem in [3] vorgeschlagen wurde. Für jede Kante $e \in E$ setzen wir als Distanz ihre reduzierten Kosten $\bar{c}_e = c_e - d_e$, als Differenz zwischen operativen Personalkosten c_e der Kante und ihrem dualen Wert d_e aus der letzten Lösung der LP-Relaxation. Dann wird für jedes zulässiges Knotenpaar jeweils nur den kürzesten Weg bestimmt und daraus ein Dienststück erzeugt. In der zweiten Phase erzeugen wir aus diesen Dienststücken durch Permutation eine Menge neuer Dienste, die jeweils alle Dienstregeln erfüllen und negative reduzierten Kosten haben müssen. Findet man keine Dienste mit negativen reduzierten Kosten aus solchen Dienststücken, dann existieren überhaupt keine Dienste mit negativen reduzierten Kosten mehr, auch wenn man alle möglichen Dienststücken betrachten würde (Beweis in [3]).

Wurden Dienste mit negativen reduzierten Kosten gefunden, werden sie den aktuellen Diensten hinzugefügt. Die LP-Relaxation wird mit der aktualisierten Spaltenmenge wieder gelöst, es werden neue duale Werte für Kanten ermittelt und neue Dienste generiert. Das Verfahren wiederholt sich solange, bis es keine neue Dienste mit negativen reduzierten Kosten gefunden werden können oder ein anderes Abbruchkriterium erfüllt ist.

Existiert mehr als ein Diensttyp, wird die zweite Phase für jeden Diensttyp durchgeführt. Für den Fall mit mehr als drei erlaubten Dienststücken pro Dienst werden in der Literatur oft Dienststücke und Dienste mit Hilfe eines Ressourcen-begrenzten-kürzesten-Wege Problems (*engl: Resource-Constrained-Shortest-Path*) anstelle von Permutation erzeugt.

Im letzten Schritt wird eine zulässige ganzzahlige Lösung mit Hilfe der Branch& Bound-Prozedur von CPLEX berechnet. Die Lösung repräsentiert Dienste des gesuchten Dienstplans. Schließlich werden aus den Dienststücken dieser Dienste neue Umläufe des alternativen Umlaufplans erzeugt.

3 Ergebnisse

Wir haben unsere Methode an realen Daten regionaler Verkehrsbetriebe zweier deutschen Städte getestet. Als Ablösepunkte wurden Anfang und Ende jeder Fahrt gewählt. In jeder Iteration des *Column Generation* wurden maximal 20000 neue Spalten erzeugt. Hatte die Problemgröße eine kritische Grenze erreicht, wurden Spalten mit hohen reduzierten Kosten wieder gelöscht. Alle Berechnungen wurden auf einem Windows-PC mit einer Pentium IV 2,2 GHz CPU und 1 GB RAM ausgeführt.

Alle drei Instanzen wurden erst rein sequenziell in traditioneller Weise und dann mit unserer Methode gelöst. Die resultierende Anzahl der Dienste sowie Laufzeit für beiden Methoden sind in der Tabelle 2 dargestellt, die außerdem Instanzkennzahlen - wie Anzahl der Fahrten, Depots und Umläufe - beinhaltet.

Instanz	DP-Verfahren	CPU (sek)	#Dienste	
city1	traditionell	15	72	
(423 Fahrten, 1 Depot, 26 Umläufe)	adaptiv	345	66	
			-6	(8,3%)
city2a	traditionell	119	225	
(1077 Fahrten, 2 Depots, 67 Umläufe)	adaptiv	293	198	
			-27	(9,2%)
city2b	traditionell	511	419	
(2047 Fahrten, 2 Depots, 114 Umläufe)	adaptiv	3605	316	
			-103	(24,5%)

Tabelle 2. Testergebnisse

Wie aus der Tabelle zu sehen ist, kann die vorgeschlagene Methode bei der Dienstplanung zu enormer Reduktion der Dienstanzahl führen. Dabei steigt die prozentuale Einsparung mit der Problemgröße, da sich mehr Möglichkeiten für alternative Dienststücke ergeben. Der Laufzeitanstieg ist begrenzt (etwas über 1 Stunden für die größte Instanz) und kann, angesichts der Kosteneinsparung, in Kauf genommen werden.

Literaturverzeichnis

1. Borndörfer, Löbel, and Weider. A bundle method for integrated multi-depot vehicle and duty scheduling in public transit. Technical Report ZR-04-14, ZIB - Zuse Institute Berlin, Berlin, Germany, 2004.
2. Desaulniers, Cordeau, Desrosiers, and Villeneuve. Simultaneous multi-depot bus and driver scheduling. In *TRISTAN IV preprints.* 2001.
3. Freling. *Models and Techniques for Integrating Vehicle and Crew Scheduling.* PhD thesis, Tinbergen Institute, Erasmus University Rotterdam, 1997.
4. Huisman. *Integrated and Dynamic Vehicle and Crew Scheduling.* PhD thesis, Tinbergen Institute, Erasmus University Rotterdam, 2004.

Timber Transport Vehicle Routing Problems: Formulation and Heuristic Solution

Manfred Gronalt and Patrick Hirsch

BOKU – University of Natural Resources and Applied Life Sciences, Institute of Production Economics and Logistics, Feistmantelstrasse 4, 1180 Wien, Austria

Abstract. We present a model formulation and a Tabu Search based solution method for the Timber Transport Vehicle Routing Problem (TTVRP). A fleet of m trucks which are situated at the respective homes of the truck drivers has to fulfil n transports of round timber between different wood storage locations and industrial sites. All transports are carried out as full truck loads. Since the full truck movements are predetermined our objective is to minimize the overall distance of empty truck movements. In addition to the standard VRP we have to consider weight constraints at the network, multi-depots, and time windows. The optimum solution of this problem with common solver software is only possible for very small instances. Therefore we develop a customized heuristic to solve real-life problems. We modify the traditional Tabu Search by delimiting the neighborhood in some iterations and verify our heuristic with extensive numerical studies.

1 Introduction and Problem Description

An important challenge in wood flow planning is the optimization of transports between harvest areas and industrial sites. This problem is described in the literature as Timber Transport Vehicle Routing Problem (TTVRP) (see e.g. [5]) and Log-Truck Scheduling Problem (LTSP) (see e.g. [6]). Our work can be characterized as follows: a fleet of m trucks which are situated at the respective homes of truck drivers has to fulfill n transports of round timber between different wood storage locations and industrial sites, like pulp mills and sawmills, during a given time period. All transports are carried out as full truck loads; the vehicle is loaded at the wood storage location and unloaded at the industrial site. Each tour starts at the home of the truck driver who leaves with an empty truck to his first wood storage location to load round timber. After this he drives to the belonging industrial site and completes the transport. The truck driver can now finish his tour and return back home or start a new transport. Since the full truckload movements are predetermined they are considered as tasks in the following.

The following constraints must be taken into consideration: Some parts of the forest road networks are not suitable for bigger trucks due to weight restrictions. Therefore some wood storage locations can only be reached by trucks with a certain capacity. We denote this as route weight constraint. There are also time windows at the industrial sites which have predefined opening hours at their intakes. Time windows also occur at the truck starting points since truck drivers are only on duty at certain times. Additionally we have to observe tour length constraints and capacity constraints. Our aim is to minimize the overall distance of empty truck movements.

The TTVRP is related to the Multi Depot Vehicle Routing Problem with Pickup and Delivery and Time Windows; supplementary one has to deal with specific route weight constraints and full truck loads. Gronalt et al. [4] describe the TTVRP as a special application of full truck load scheduling problems. An overview of Vehicle Routing Problems can be found for example in Toth and Vigo [7].

Our modeling approach is based on the perception of the TTVRP as a special case of a Stacker Crane Problem (SCP). The SCP is a sequencing problem that consists of finding the minimum cost cycle on a mixed graph with arcs and edges $G = (V, E, A)$. The predefined arcs correspond to our tasks. A feasible solution must include all tasks. Since empty truck movements are also arcs all nodes are endpoint of at least one arc. Coja-Oghlan et al. [1] give an example of a SCP which describes the scheduling of a delivery truck.

To make things clearer we present a small example of the TTVRP. Each truck is assigned to a specific tour r. It starts from the truck starting point and returns to this point at the end of the tour. An artificial task r is introduced to link the endpoint to the starting point. The set of these tasks is denoted by $R = \{1, \ldots, \omega\}$. The number of elements of set R is equal to the number of available trucks m. The elements of R are also used to label the tours. Wood storage locations and industrial sites can be visited more than once during a planning period. This leads to the following formulation: $W = \{1, \ldots, w\}$ is the set of tasks, $M = \{m_1, \ldots, m_w\}$ is the set of all wood storage locations, and $N = \{n_1, \ldots, n_w\}$ represents all industrial sites at which the wood is unloaded. In our example we have to complete 8 tasks and two artificial tasks. The sets are defined in the following way: $R = \{A, B\}$, $M = \{P1, P1, P2, P3, P4, P5, P6, P5\}$, $N = \{I1, I1, I1, I2, I3, I2, I2, I3\}$, and $W = \{1, 2, 3, 4, 5, 6, 7, 8\}$. The first element in M is allocated to the first element in N, the second to the second, and so on. In Figure 1a) the given tasks and in 1b) the obtained optimal solution for given durations and transport data is shown. A1 to A7 and B1 to B11 are the trips on tour A and B respectively. In Figure 1 the number of rectangles at an industrial site is equal to the number of visits at this location during the planning horizon. The same is true for the number of triangles at the wood storage locations.

In section 2 we present the model formulation of the TTVRP. The model is validated for small instances, using the Xpress-MP software. For real-life problems we developed a customized heuristic based on Tabu Search. The variants of our heuristic are discussed in section 3. We have made extensive numerical studies to test our algorithm and its variants; the results and an outlook on future research are presented in section 4.

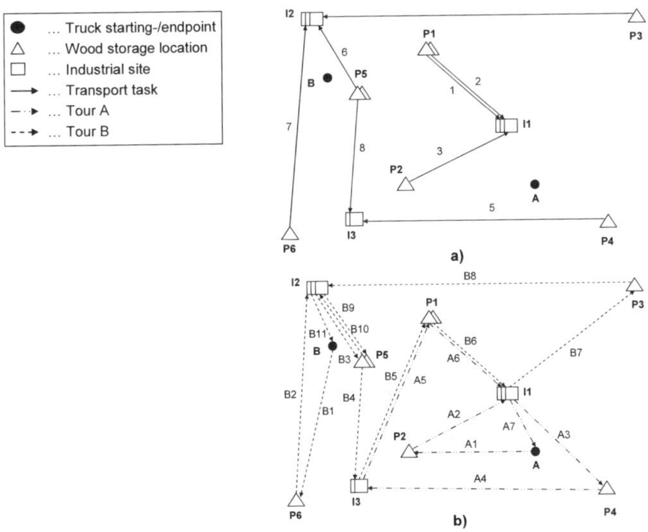

Fig. 1. Example of a TTVRP

2 Model formulation

First we give the notation for the used parameters and variables. A tour r has a maximum capacity Q_r and a maximum tour length T_r. The tour is finished at time v_r. We also have to observe a time window $[e_r, l_r]$ and a service time s_r for each tour. The time window represents the working hours of the truck drivers, the service time is needed for preparing activities. Each task i has the following attributes: a starting point m_i, an endpoint n_i, loading time a_i at the wood storage location, route weight constraint k_i given in units of weight, order quantity q_i, unloading time s_i at the industrial site, time window $[e_i, l_i]$ at the industrial site, and traveling time u_i. Each truck is allowed to arrive at an industrial site at a time $0 \leq b_i \leq l_i$; if the truck arrives at a time $b_i < e_i$ it has to wait for the time period $w_i = e_i - b_i$. t_{ij} represents the time that is needed to get from task i to task j; these are the empty truck movements. We use the following binary decision variables:

- $x_{ijr} = 1$, if task j is visited directly after task i on route r; 0 otherwise.
- $y_{ir} = 1$, if task i is on route r; 0 otherwise.

The set presented in equation (1) includes all tasks.

$$\widetilde{W} = W \cup R \tag{1}$$

The objective function (2) minimizes the duration of empty truck movements.

$$\min \sum_{r \in R} \sum_{i \in \widetilde{W}} \sum_{j \in \widetilde{W}} t_{ij} \cdot x_{ijr} \tag{2}$$

The following constraints have to be fulfilled:

$$\sum_{i \in \widetilde{W}} x_{ihr} - \sum_{j \in \widetilde{W}} x_{hjr} = 0 \quad \ldots \forall h \in \widetilde{W}, r \in R \tag{3}$$

$$\sum_{r \in R} \sum_{j \in \widetilde{W}} x_{ijr} = 1 \quad \ldots \forall i \in \widetilde{W} \tag{4}$$

$$\sum_{j \in W \cup \{r\}} x_{rjr} = 1 \quad \ldots \forall r \in R \tag{5}$$

$$\sum_{i \in W \cup \{r\}} x_{irr} = 1 \quad \ldots \forall r \in R \tag{6}$$

$$y_{ir} = \sum_{j \in \widetilde{W}} x_{ijr} \quad \ldots \forall i \in \widetilde{W}, r \in R \tag{7}$$

$$q_i \cdot y_{ir} \le Q_r \quad \ldots \forall i \in W, r \in R \tag{8}$$

$$k_i \cdot y_{ir} \ge Q_r \quad \ldots \forall i \in W, r \in R \tag{9}$$

$$\sum_{i \in \widetilde{W}} \sum_{j \in \widetilde{W}} t_{ij} \cdot x_{ijr} + \sum_{i \in W} u_i \cdot y_{ir} \le T_r \quad \ldots \forall r \in R \tag{10}$$

$$b_i + w_i + s_i + t_{ij} + a_j + u_j - M \cdot (1 - x_{ijr}) \le b_j \quad \ldots \forall i \in \widetilde{W}, j \in W, r \in R \tag{11}$$

$$b_i + w_i \ge e_i \quad \ldots \forall i \in \widetilde{W} \tag{12}$$

$$b_i \le l_i \quad \ldots \forall i \in \widetilde{W} \tag{13}$$

$$b_i + w_i + s_i + t_{ir} - M \cdot (1 - x_{irr}) \le v_r \quad \ldots \forall i \in \widetilde{W}, r \in R \tag{14}$$

$$v_r \le l_r \quad \ldots \forall r \in R \tag{15}$$

$$x_{ijr} \in \{0,1\} \quad \ldots \forall i \in \widetilde{W}, j \in \widetilde{W}, r \in R \qquad y_{ir} \in \{0,1\} \quad \ldots \forall i \in \widetilde{W}, r \in R \tag{16}$$

$$w_i \ge 0 \quad \ldots \forall i \in \widetilde{W} \qquad b_i \ge 0 \quad \ldots \forall i \in \widetilde{W} \tag{17}$$

Constraints (3) guarantee a tour, (4) to (6) are defining predecessor and successor relationships, and (7) links the binary variables. Constraints (8) to (10) ensure the observance of truck capacity, route weight constraints, and maximum travel times. (11) to (15) deal with the time windows at the industrial sites and truck starting points. Constraints (16) define the binary variables and (17) the nonnegativity constraints. Additionally we need to apply anti-subcycle constraints to guarantee that every tour starts at the starting point and ends at the endpoint of an artificial task r.

3 Heuristic Solution

We use Tabu Search (see e.g. Glover and Laguna [3]) as a solution method for the TTVRP. Cordeau et al. [2] developed a Tabu Search method which combines simple vertex moves with local reoptimization. It uses fixed but problem size dependent tabu durations, intermediate infeasible solutions, aspiration criteria which are attribute related, and penalization of worsening candidate solutions by adding costs which are dependent on how often an attribute was used in a solution. We have adapted this method for our specific problem. To generate an initial solution we use a myopic heuristic in which we calculate a regret-value that gives the additional costs if the second best option is taken; we use this value to decide which tasks should be inserted in a tour.

The Standard Tabu Search explores the whole non-tabu neighborhood of a solution in every iteration step. This procedure may make sense if the constraints of the problem are very tight and feasible solutions are hard to find; but it is also very time-consuming. Toth and Vigo [8] proposed the Granular Tabu Search to restrict the neighborhood of solutions drastically and reduce computing times. They try to limit moves that insert “long” arcs in the current solution. Our approach concentrates on a certain fraction of empty truck movements in the current solution; only these links are chosen to be removed in neighbor solutions. Other links can only be modified if a task from a removed link is inserted. To do that we first sort the links according to their duration in a descending order. Then we choose a predefined number of links starting from the one with the longest duration. The number of used links is calculated as a fraction of all existing empty truck loads; the divider is set as a parameter. This strategy seems to be myopic since “shorter” links are not affected directly. To overcome this problem we merge full and restricted neighborhood search in our algorithm; after a predefined number of iteration steps with a restricted neighborhood we set an iteration step with full neighborhood search. We call this alternating strategy.

A post-optimization strategy based on 2opt is also applied to improve single tours. Extensive numerical studies were made to test our algorithm and its variants.

4 Results and Conclusion

Our introductory example with 2 trucks and 8 tasks can be solved optimally with a standard MIP package in negligible computation time. However for real life cases we need to apply heuristics to gain reasonable results. We have implemented our algorithm and its variants in Visual Basic. Our algorithm is able to solve the introductory example within a few iteration steps. In order to learn how the algorithm will perform two sets of test cases are developed. Each set consists of 20 instances with 30 tasks and 10 trucks which is a typical real life problem size. The first set of instances has weaker constraints than the second one in terms of the average task duration and the traveling times between the tasks. We have tested our algorithm in the following variants: Tabu Search with a full neighborhood (TSFN), Tabu Search with a restricted neighborhood and no alternating strategy (TSRN), Tabu Search with an alternating strategy (TSAS), and Tabu Search with an alternating strategy and post-optimization (TSASP). We also varied the following parameters in our test runs: Tabu Search specific parameters, number of iteration steps until the restricted neighborhood is followed by a full neighborhood in our alternating strategy, and the divider which specifies the fraction of links used for neighborhood search. We can clearly state that the TSRN, even though it offers the shortest computation times, is not recommendable with respect to solution quality. The TSAS offers an excellent solution quality in much shorter computing times than TSFN. But it is advantageous to use a full neighborhood search in more iteration steps if the constraints are tight; therefore the TSFN seems to be a good method if feasible solutions are very hard to find. The TSASP is recommendable for smaller problem instances since the post-optimization strategy improves single tours much quicker than Tabu Search. For larger problem sizes it may make sense to save the computing time used for post-optimization since the tours are rebuilt very often.

Future research will concentrate on an extension of our model to a multi-period problem. In this case we know the tasks for one week and we should find an optimal allocation of these tasks to every day of this week. An evenly distributed workload and optimized daily trips are the objective.

References

1. Coja-Oghlan, A., Krumke, S. O., Nierhoff, T. (2004): A Heuristic for the Stacker Crane Problem on Trees which is Almost Surely Exact. Journal of Algorithms, In Press.
2. Cordeau, J. F., Laporte, G., Mercier, A. (2001): A unified tabu search heuristic for vehicle routing problems with time windows. Journal of the Operational Research Society **52**, 928-936.
3. Glover, F. and Laguna, M. (1997): Tabu Search. Kluwer Academic Publishers, Boston, USA.
4. Gronalt, M., Hartl, R., Reimann, M. (2003): New savings based algorithms for time constrained pickup and delivery of full truckloads. European Journal of Operational Research **151**, No. 3, 520-535.
5. Karanta, I., Jokinen, O., Mikkola, T., Savola, J., Bounsaythip, C. (2000): Requirements for a Vehicle Routing and Scheduling System in Timber Transport. Proceedings of the IUFRO International Conference: Logistics in the forest sector, 235-250.
6. Palmgren, M., Rönnqvist, M., Värbrand, P. (2003): A near-exact method for solving the log-truck scheduling problem. International Transactions in Operational Research **10**, 433-447.
7. Toth, P. and Vigo, D. (eds.) (2002): The Vehicle Routing Problem. SIAM, Philadelphia, USA.
8. Toth, P. and Vigo, D. (2003): The Granular Tabu Search and Its Application to the Vehicle-Routing Problem. INFORMS Journal on Computing **15**, No. 4, 333-346.

Robustness in the Context of Autonomous Cooperating Logistic Processes: A Sustainability Perspective

Lars Arndt and Georg Müller-Christ

Chair of Sustainable Management, Department of Economics and Business Studies, University of Bremen, Wilhelm-Herbst-Str. 12, 28359 Bremen, Germany, {larndt,gmc}@uni-bremen.de

Abstract. Autonomous cooperating logistic processes seem to be a promising approach to increase the robustness of logistics systems. Searching for the necessary organizational prerequisites for the successful implementation and sustainment of autonomous cooperating logistic processes we pick up the concept of robustness and use the New Systems Theory to outline a notion of organizational robustness, which can be regarded as an important factor enabling businesses to adopt innovations in spite of related uncertainties.

1. Introduction

In recent times, researchers in the field of logistics and supply chain management have increasingly directed their attention to concepts like robustness, resilience or risk management (cf. e.g. Christopher/Peck 2000; Norman/Lindroth 2004). All these concepts refer to the question how logistics systems or supply chains can function effectively while being confronted with complex and dynamic environmental conditions and demands.

A particular approach to deal with this challenge is adressed by the German Collaborative Research Center (CRC) 637 „Autonomous Cooperating Logistic Processes". The CRC focuses on a paradigm shift in logistics based upon fundamental changes in decision-making processes within logistics systems due to the fact that the dynamic and structural complexity of logistics networks makes it more and more impossible to provide a central planning and control unit with all decision-relevant information and thus requires adaptive logistic processes. The notion of autonomous cooperating logistic processes refers to the decentralised coordination of autonomous logistic objects in a heterarchical organizational structure. Precondition as well as driving force of autonomous logistic processes are developments in information and communication technologies, like RFID (Radio Frequency Identification) technology or wireless communication networks.

Against the background of increasing dynamic and structural complexity within logistics networks – caused by changing conditions in the markets like the shift from seller to buyer markets and the increasing importance of customer orientation and individualisation – autonomous cooperating logistic processes are intended to provide an improved capability to react to unanticipated events maintaining a high level of efficiency and effectiveness and thereby increasing the robustness of the logistics system.

While this effect seems to be favorable from the perspective of businesses, autonomous cooperating logistic processes also confront them with new challenges. An increase in the level of autonomous control of intra-organizational and inter-organizational logistic processes has considerable consequences on the management level. These consequences can be inferred from the fundamental dilemma in regard to autonomous control: Decentral information processing and decision-making increases flexibility but at the same time makes it more difficult to ensure that the local decisions are in the best interest of the business. Correspondingly, the gain in robustness on the level of a single logistic process may not be a sufficient incentive for businesses to invest in the infrastructure necessary for autonomous cooperating logistic processes.

Against this background we suggest to distinguish between the notion of robustness on the process level and a concept of organizational robustness which refers to the organizational structures in which autonomous logistics systems are embedded. The purpose of this paper is to outline this concept of organizational robustness and to examine how organizational robustness can help businesses to deal with the fundamental dilemma of autonomous control.

2. The Concept of Robustness

While the concept of robustness is popular within natural and engineering sciences, its application in the context of social systems stands only at its beginning. Therefore we cannot offer a general, widely accepted definition of robustness. Referring to Jen (2003), we define robustness as a measure of feature persistence of a system under changing environmental conditions, which provide the system with unforeseen perturbations. This very general definition, however, requires further specification. At first, it is necessary to specify the features, whose persistence is to be investigated. Here, we want to draw upon the New Systems Theory, which sets aside any ontological notion of systems and replaces it by an observer-relative understanding. Organizations as social systems are then considered as self-referential, operationally closed unities. This means that by establishing a boundary between themselves and the environment organizations continually create themselves and the whole bandwidth of features, which can be attributed to them.

Corresponding to the notion of self-referential closure an organization arranges all its operations such that they contribute to the reproduction of the system/environment-distinction and thus to the self-reproduction of the organization.

If, as suggested above, we connect the idea of feature persistence with the system/environment-distinction and therefore with the self-reproduction of the organization, the link between robustness and the way the organization creates its own boundaries becomes obvious. By sustaining the system/environment-distinction the system creates a difference between internal and external complexity that marks the boundary between system and environment. It should be emphasized that the New Systems Theory considers social systems, including organizations, as sense-systems. Accordingly, the system/environment-distinction does not refer to some kind of spatial boundaries but to a difference based on sense. Correspondingly, organizational boundaries have to be considered as based on sense as well (cf. Ortmann/Sydow 1999). Luhmann (1997; 2000) specifies the New Systems Theory's notion of organization by characterizing it as a recursive unity of decisions. Decisions are perceived as a specific form of communications which constitute the emergence of organizations as social systems. If an organization succeeds in continuing this self-referential circle of decisions, which marks the inside of the system, its self-reproduction is successful. The notion that organizations act to some purpose or have to achieve specified functions plays an important role in sustaining the organization as a recursive unity of decisions.

In the following we will address the question under which circumstances an organization's specific way of establishing and maintaining its boundaries contributes to its robustness. In order to find an answer to this question we have to direct our attention to the system/environment-distinction again. Referring to the inside of this distinction, an organization is able to develop a certain identity, which is largely based upon specific functions the organization strives to achieve. This organizational identity can be considered as a self-description that serves as a basis for the future operations of the system. It is exactly this self-description and its capacity to provide the organization with a simplified notion of the relation between system and environment that enables an organization to act under complex environmental conditions without possessing the variety to confront the whole environmental complexity within the system. Recalling that an organization can be considered as a recursive unity of decisions, it is obvious that past decisions play an important role in the creation of an organization's identity and thus enable and restrain future decisions at the same time.

The organization's environment, which is at the outside of the system/environment-distinction, functions as a negative correlative of the organization's unity. System and environment thus simply mark different sides of the same form. Against this background it seems only reasonable that all organizations exhibit a certain tendency to protect their identity from changes and to maintain rigid boundaries. This form of closure is a constitutive characteristic of self-referential systems and intended to ensure the self-reproduction of the organization.

What can we infer from these considerations regarding the robustness of an organization? Is rigidity of an organization's identity and its boundaries sufficient to qualify an organization as robust? In order to negate this question we only have to take into account that the establishment of sense-based boundaries does not cut through causal relationships between system and environment. Even if – as the New Systems Theory suggests – the system/environment-distinction is the result of an internal activity of the system, this distinction cannot be maintained against causal relations. In other words, there must be a fit between the system and the environment, which does not only depend on factors internal to the system. Especially against the background of changing environmental conditions an organization tends to compromise its fit with the environment by rigorously clinging to a given identity and boundaries.

Referring to the notion of an organization as a recursive unity of decisions, we can use the terms redundancy and variety to address the problem mentioned above. According to Luhmann (1988) redundancy is a measure of the structural rigidity of an organization as a unity of decisions. If the scope of decisions possible in the future is narrowed to a relevant extent by previous decisions and resulting organizational structures, we can speak of redundancy. The bias of organizations towards redundancy can be considered as another expression for the aforementioned tendency of organizations to protect their identity from changes and to maintain rigid boundaries in order to ensure their self-reproduction. What we stated above about the perils of rigidity of identity and boundaries holds true for redundancy as well. Thus, redundancy can compromise the fit between the organization and its environment, especially if the system operates in a rapidly changing environment. Redundancy alone does not make organizations robust. Instead we have to turn our attention to variety, which, according to Luhmann, describes the dissimilitude of decisions within an organization. Variety enables organizations to continue their decision-making on the basis of a wide range of options. Variety gives organizations the opportunity to decide about previous decisions, a property which Baecker (2003) characterizes as re-entry of uncertainty into the organization, which is originally intended to absorb uncertainty. At the same time this property is an important prerequisite for reflexivity, which increases the likelihood of the fit between organization and environment.

Referring to the terms redundancy and variety, it can be argued that the robustness of an organization marks the optimal level of redundancy and variety. This optimal level ensures the maintenance of the system/environment-distinction and thus the continuation of the self-referential closure of the organization. At the same time it enables the organization to process perturbations which result from changes in the environmental conditions and to which the system must react in order to ensure its fit with the environment. A robust organization neither clings rigidly to its given identity by imposing non-fitting boundaries on its environment nor opens its boundaries to an extent that endangers its self-reproduction.

It has to be emphasized that there are no blueprints for organizational robustness. The optimal level of redundancy and variety varies from organization to organization. The question which conditions lead to the emergence of robustness and how the creation of robustness can be addressed within the management process must be the subject of further research.

3. Organizational Robustness and Autonomous Cooperating Logistic Processes

Finally we want to discuss the meaning of the outlined concept of organizational robustness in the context of autonomous cooperating logistic processes and the fundamental dilemma of autonomous control. Robustness plays an important role in regard to the management and organization level of autonomous logistics systems. It is likely that already the decision to participate and to invest in an autonomous logistics system is affected by an organization's robustness, as a robust organization will not consider the uncertainties associated with autonomous control of logistic processes as perils to the organization's identity and thus to the sustainment of the organization itself. A robust organization is able to deal with these uncertainties without compromising the basis of its future operations. In addition, robustness increases the sensitivity of the organization towards changes in its environment which require the system to act in order to maintain the fit between organization and environment. This increases the probability that the organization will be able to deal with the fundamental dilemma of autonomous control in a constructive manner. A robust organization is likely to welcome innovations like autonomous logistic processes. Instead of perceiving them as impediments to achieving certain functions, they will be considered as necessary preconditions in order to be able to achieve functions at all. Robustness thus implies the ability to restrain an organization's effort to achieve its function in the short run in the interest of the sustainment of the organization's ability to continue its existence and to achieve its function in the long run. Thus, it becomes obvious that the outlined concept of organizational robustness is strongly linked to the concept of sustainability. An organization is sustainable if it succeeds to secure its continued existence in the long run. As can be inferred from the considerations above, a robust organization is more likely to be sustainable than a less robust one.

4. Summary

In this paper we argued that in the context of autonomous cooperating logistic processes it makes sense to distinguish an original technical notion of robustness, referring to the logistic processes itself, from a concept of organizational robustness which has a positive influence on the successful implementation of autonomous logistic processes. The concept of organizational robustness was outlined on the basis of the New Systems Theory and focussing on the relation between redundancy and variety. Furthermore, it was shown that organizational robustness is strongly linked to a systemic notion of sustainability.

Considering the importance of organizational robustness in the context of autonomous logistic processes, it seems indicated to undertake further research effort in order to examine the emergence of organizational robustness.

Acknowledgement

This research was supported by the German Research Foundation (DFG) as part of the Collaborative Research Centre 637 "Autonomous Cooperating Logistic Processes".

References

Baecker, D. (2003): Organisation und Management. Suhrkamp, Frankfurt a.M.

Christopher, M./Peck, H. (2004): Special Feature – Building the Resilient Supply Chain. In: The International Journal of Logistics Management 15 (2): 1-14

Jen, E. (2003): Stable or robust? What's the difference? In: Complexity 8 (3): 12-18

Luhmann, N. (1988): Organisation. In: Küpper, W./Ortmann, G. (eds.): Mikropolitik: Rationalität, Macht und Spiele in Organisationen. Westdeutscher Verlag, Opladen, pp. 165-186

Luhmann, N. (1997): Die Gesellschaft der Gesellschaft. Suhrkamp, Frankfurt a.M.

Luhmann, N. (2000): Organisation und Entscheidung. Westdeutscher Verlag, Opladen.

Norman, A./Lindroth, R. (2004): Categorization of Supply Chain Risk and Risk Management. In: Brindley, C. (ed.): Supply Chain Risk. Ashgate Publishing, Aldershot, pp. 14-27

Ortmann, G./Sydow, J. (1999): Grenzmanagement in Unternehmensnetzwerken: Theoretische Zugänge. In: Die Betriebswirtschaft 59 (2): 205-220

Open Vehicle Routing Problem with Time Deadlines: Solution Methods and an Application

Zeynep Özyurt[1], Deniz Aksen[2], Necati Aras[3]

[1,2] Koç University, College of Administrative Sciences and Economics, Rumelifeneri Yolu, Sarıyer, 34450 İstanbul, Turkey
[3] Boğaziçi University, Industrial Engineering Dept., Bebek, 34342 İstanbul, Turkey

Abstract. In the open route version of the well-known vehicle routing problem, vehicles are not required to return to the depot; or if they are required, then they return by traveling the same route back. In this study, we present a modified Clarke-Wright parallel savings algorithm, a nearest insertion algorithm and a tabu search heuristic for the open vehicle routing problem with time deadlines. Some random test problems and a real-life school bus routing problem are solved by these heuristics, and results are compared.

1 Introduction

Capacitated vehicle routing problem (VRP) can be defined as determining a set of routes for a fleet of vehicles based at one or several depots. The objective of the VRP is to deliver a set of geographically dispersed sites or customers with known demands on minimum-cost vehicle tours originating and terminating at a depot. Open vehicle routing problem (OVRP) is a variant of VRP where vehicles are not required to return to the depot, or if they are required, then they return by traveling the same route back. Although OVRP received little attention from researchers until recent years, it has been commonly occurring in the transportation business.

In this study, OVRP with time deadlines (OVRP-TD) is solved with a modified Clarke-Wright parallel savings algorithm, with a greedy nearest insertion algorithm, and with a tabu search heuristic. In OVRP-TD, each customer must be visited before a certain time deadline. The timing of service delivery which arises as vehicle arrival time in routing problems is an important Quality of Service (QoS) guarantee given to meet customer expectations in service systems.

2 Literature Review and Problem Analysis

OVRP studies reported in the literature are not as abundant as the studies on VRP. First, Schrage (1981) mentions OVRP in an article mentioning real-life routing problems. Sariklis and Powell (2000) solve symmetric OVRP by a two phase algorithm which uses cluster-first-route-second mechanism.

Tarantilis and Kiranoudis (2002) solve a real-life instance of multi-depot OVRP for fresh meet distribution by a meta-heuristic they called "list based threshold accepting algorithm" (LBTA). A spatial decision support system (SDSS) is proposed by Tarantilis et al. (2004). A genetic solution procedure called BoneRoute is used for the OVRP. Tarantilis et al. (2004) propose a single parameter simulated annealing-based algorithm for the same problem. Brandão (2004) proposes a tabu search algorithm (TS) for OVRP with maximum route length constraint. Another TS algorithm is due Fu et al. (2005) again subject to maximum route length constraint. These two TS algorithms differ in their initial solutions, neighborhood structures, objective function and tabu definitions. Both algorithms seem to outperform Sariklis and Powell's solutions; however, CPU times of Sariklis and Powell are considerably better. Although Fu et al. improve the solutions for several of the problems in Brandão's paper, for some others they find worse solutions in terms of total traveling distance and the number of used vehicles.

Eliminating the constraint that all vehicles have to return to the depot does not make OVRP a simpler problem. Also, a good solution for VRP cannot be converted to a good OVRP solution by simply dropping incoming arcs of the depot. Thus, OVRP is to be studied separately. Our problem differs from the current OVRP literature in two points: First is the incorporation of time deadlines. Each customer must be visited before his time deadline. The second difference is the constraint that each route terminates at one of the driver nodes which are specified beforehand. Driver nodes practically correspond to parking lots or homes of drivers. The presence of such fixed driver nodes suits especially those situations in which deliveries to customers are outsourced to a shipping company, or drivers use the same vehicles also to commute between home and depot.

2.1 Clarke-Wright Parallel Savings Algorithm Modified for OVRP

This method (CW) is proposed by Clarke and Wright (1964) for the single depot VRP. Since the algorithm is efficient and simple to implement, it still remains popular to date. In order to adapt CW to OVRP-TD, we modify the distances between customers and depot, and drivers and depot. The modified distances are assigned as follows. Customer-Depot distance is set to infinity, because a vehicle is not to return to the depot directly from a customer node. Driver-Customer distance is also infinity, because a vehicle is not allowed to proceed from a driver to a customer. The same is true for the Driver-Driver distance as well. Finally, Driver-Depot distance is taken as zero to assure that a vehicle will return to the depot from a driver node without increasing the objective value.

As a result of these modified distances, a route is guaranteed to start from the depot, visit one or more customers and end at a driver node.

2.2 Tabu Search Algorithm

Tabu search (TS) is a meta-heuristic algorithm that guides a local search to prevent it from being trapped in premature local optima by prohibiting those moves that cause to return to previous solutions and cycling. TS starts with an initial solution. At each iteration, a neighborhood of solutions is generated, and the best one from this neighborhood is selected as the new solution. Certain attributes of previous solutions are kept in a tabu list which is updated at the end of each iteration. The selection of the best solution in the neighborhood is done such that it does not adopt any of the tabu attributes. Best feasible solution so far (incumbent) is updated if the new current solution is better and feasible. The procedure continues until any of two stopping criteria is met, which are maximum number of iterations performed and maximum number of non-improving iterations during which the incumbent does not improve. Characteristics of TS heuristic proposed for OVRP-TD can be stated as follows.

Initial Solution

Two different methods of initial solution generation are used at the beginning of TS. First one is the well-known Clarke-Wright parallel savings algorithm (CW). If CW fails to create a feasible initial solution due to time deadline stringency, we try to correct or at least minimize this infeasibility by shifting customer nodes from their current positions to new positions on the same or on a different route.

Second one is a greedy constructive heuristic called nearest insertion method (NI). In this method, we start with as many routes as the number of drivers. Each route initially consists of the depot and a driver node. Customers are then inserted into these routes one by one. All feasible insertion positions are examined for all customers awaiting insertion. Each time, that particular customer is selected which has the least expensive insertion position. The procedure is repeated until all customers have been inserted. For n customers, NI creates the initial solution in $O(n^3)$ time. When a feasible insertion point cannot be found at all, then the least infeasible position with respect to vehicle capacity and time deadlines is chosen.

In both methods described above, when a feasible initial solution cannot be generated, it is hoped that feasibility will be restored during the TS iterations.

Evaluation of Solutions

In our TS heuristic, we apply strategic oscillation by admitting infeasible solutions into the procedure. The evaluation of such solutions is different from that of feasible solutions in that a penalty cost for violating capacity and time constraints will be added to their objective value. This penalty is added to prevent the algorithm from exploring the infeasible regions of the search space in excess.

Penalty costs rise and fall according to the number of feasible and infeasible solutions visited. Every 10 iterations, the numbers of visited feasible and infeasible solutions are compared. If more feasible solutions are visited than infeasible ones, penalty terms are divided by 1.5; otherwise penalty terms are multiplied by 1.5. The objective value for a solution is obtained by $\sum_{r=1}^{K}\left[D(r)+p_c V_c(r)+p_t V_t(r)\right]$ where $D(r)$ denotes total distance traveled on route r, K denotes the total number of routes, V_c denotes overcapacity (total demand of customers in route r – vehicle capacity), V_t denotes total tardiness in route r, p_c and p_t denote penalty coefficients for overcapacity and for total tardiness in a route, respectively.

Neighborhood Structure

Three move operators are used to generate a neighborhood to the current solution. For each move two customers are selected randomly as pilot nodes:

i. 1-0 move: One of the selected nodes is taken from its current position and inserted after the other node.
ii. 1-1 exchange: Two selected nodes are swapped by preserving their original positions.
iii. 2-Opt move: For two pilot nodes in the same route, visiting order between these is reverted. If the pilot nodes are in different routes, then the route segments following them are swapped preserving the order of nodes on each segment.

Besides the neighborhood generation, local search with these moves is incorporated into TS as a tool of local post optimization (LPO). In the application of LPO, all customers are set one by one as the first pilot node. For a certain customer set as the first pilot node, second pilot node of the move is selected such that the move yields the highest improvement in total distance traveled without causing any infeasibility. At the end of every 100 iterations as well as when the incumbent solution is updated, a series of LPO is applied to the current solution. This LPO comprises 1-1 exchange, 2-Opt move, 1-0 move and one more time 2-Opt move.

Tabu Attributes and Tabu Tenure

Tabu attribute definitions for three move operators are as follows:

1. 1-0 move: If customer i is inserted after customer j, the position of customer i cannot be changed by the same move while it is tabu-active.
2. 1-1 exchange: If customers i and j are swapped, i and j cannot be swapped again while they are tabu-active.
3. 2-Opt move: If it is applied to customers i and j, it cannot be applied again to the same customers while they are tabu-active.

At each iteration tabu tenure is selected randomly between 5 and 15 iterations. In some cases, namely if aspiration criterion is satisfied, a move can be executed although its attributes are tabu-active. Aspiration criterion is satisfied if the total distance resulting from the move is better than the incumbent's objective value.

3 Computational Results

All codes in the study are written in ANSI C language, compiled and executed in Visual C++ 6.0 on a 3.40 GHz Pentium 4/HT PC with 2 GB RAM. Five random OVRP-TD instances and a real-life OVRP are solved with TS as well as CW and NI followed by rigorous LPO. The real-life instance is taken from a company that carries students of an elementary school in the metropolitan city of İstanbul. 22 vehicles pick 434 students from home in the morning and carry them back home in the afternoon.

In Tables 1 and 2, TS-NI denotes TS whose initial solution is generated by the nearest insertion method (NI), and TS-CW denotes TS whose initial solution is generated by CW. Their results are compared to see the initial solution effect. Best, average, and worst total distances and CPU seconds of 20 random runs are reported. Table 1 shows the results for five random test problems, and Table 2 shows the results for the school bus problem. TS is also compared against the pure CW and pure NI both of which are followed by LPO. CW+LPO denotes CW with a series of LPO consisting of 1-1 exchange, 2-Opt move, 1-0 move and one more time 1-1 exchange. Similarly, NI+LPO denotes NI with the same LPO series.

Compared with the LPO-enhanced classical heuristics, TS finds better solutions with both initial solution generation methods. The only exception to this is the problem with 75 customers. Here, CW+LPO finds a slightly better total distance value than the best total distance of TS-NI. The test runs with the TS are inconclusive about the effect of the initial solution method as can be seen in the tables. Finally, the total Euclidean distance traveled by the school bus company in the real-life example amounts to 1,192,230 m according to their current routing plan. This distance in the best solution found by CW+LPO is 351,809 m, while average distance is computed as 354,987 m. TS-CW provides 70.5% improvement over the company's routing plan and 2.3% improvement over the CW+LPO heuristic.

4 Conclusion

In this paper, OVRP-TD is presented with the constraint that routes terminate at one of the driver nodes. The problem is solved with two classical heuristics enhanced by local post optimization, and with a tabu search meta-heuristic. In the latter, infeasible solutions are penalized dynamically, which distinguishes it from the previous meta-heuristics proposed in the OVRP literature. In test problems which range from 25 to 100 customers in size, tabu search with the Clarke-Wright initial solution performs better than the classical heuristics with local post optimization. Limited empirical evidence shows that the initial solution's effect on the quality of the final tabu search solution is problem-dependent. For the real-life school bus routing problem tabu search with Clarke-Wright initial solution improves the company's current routing plan by 70.5%.

5 References

Brandão J, (2004). A tabu search heuristic algorithm for open vehicle routing problem. European Journal of Operational Research 157: 552–564.

Clarke G, Wright JW (1964). Scheduling of vehicles from a central depot to a number of delivery points. Operations Research 12: 568–581.

Fu Z, Eglese R, Li LYO (2005). A new tabu search heuristic for the open vehicle routing problem. Journal of the Operational Research Society 56: 267–274.

Sariklis D, Powell S (2000). A heuristic method for the open vehicle routing problem. Journal of the Operational Research Society 51: 564–573.

Schrage L (1981). Formulation and structure of more complex/realistic routing and scheduling problems. Networks 11: 229–232.

Tarantilis CD, Diakoulaki D, Kiranoudis CT (2004). Combination of geographical information system and efficient routing algorithms for real life distribution operations. European Journal of Operational Research 152: 437–453.

Tarantilis CD, Ioannou G, Kiranoudis CT, Prasdacos GP (2004). Solving the open vehicle routing problem via single parameter meta-heuristic algorithm. Journal of the Operational Research Society: 1–9.

Tarantilis CD, Kiranoudis CT (2002). Distribution of fresh meat. Journal of Food Engineering 51: 85–91.

Table 5.1. Total distances and CPU times for random test problems

Problem Size	*TS-CW*			*TS-NI*			*Classical Heuristics*	
	minimum	**average**	**maximum**	**minimum**	**average**	**maximum**	**CW+LPO**	**NI+LPO**
25 customers *5 drivers*	2,030.1[a] 5.09[b]	2,080.0 6.12	2,193.9 6.42	1,782.9 4.67	2,071.9 5.82	2,193.8 6.61	2,030.2 0.00	2,033.2 0.00
50 customers *10 drivers*	1,104.9 1.59	1,116.8 13.19	1,140.6 22.13	1,081.1 1.75	1,108.5 11.05	1,157.6 19.70	1,148.2 0.0	1,142.5 0.00
75 customers *10 drivers*	1,199.8 6.83	1,223.1 24.43	1,274.0 31.41	1,211.6 15.56	1,228.6 24.97	1,275.1 29.92	1,210.7 0.02	1,231.2 0.00
80 customers *8 drivers*	1,467.8 6.91	1,501.9 29.18	1,530.6 50.09	1,476.3 6.94	1,521.7 22.80	1,558.2 43.42	1,562.9 0.00	1,622.1 0.00
100 customers *15 drivers*	2,382.9 4.64	2,445.7 10.12	2,516.3 25.80	2,362.1 6.42	2,443.2 11.33	2,527.7 21.75	2,431.5 0.02	2,411.4 0.02

[a,b] In each cell, the first figure shows the total distance value while the figure below indicates the CPU time in seconds

Table 5.2. Total distances and CPU times for the real-life school bus routing problem

	TS-CW			*TS-NI*			*Classical Heuristics*	
	minimum	**average**	**maximum**	**minimum**	**average**	**maximum**	**CW+LPO**	**NI+LPO**
Total Distance [m]	351,809.4	354,986.7	358,381.6	355,695.4	366,647.7	386,925.8	359,917.4	397,202.0
CPU time [sec]	48.3	105.45	166.77	50.19	117.70	244.22	0.64	1.16

An Optimal Control Policy for Crossdocking Terminals

Matthias Stickel[1] and Kai Furmans[2]

[1] Institut für Fördertechnik und Logistiksysteme, Universität Karlsruhe (TH)
stickel@ifl.uni-karlsruhe.de
[2] Institut für Fördertechnik und Logistiksysteme, Universität Karlsruhe (TH)
furmans@ifl.uni-karlsruhe.de

Summary. Crossdocking terminals are transhipment facilities without stock, for the rapid consolidation and shipment of products. The difference to traditional distribution centers is the complete elimination of all storage functions. In consequence of this elimination the incoming and outgoing shipments have to be exactly coordinated to achieve a transshipment operation at minimum cost. This paper describes a mixed-integer model for a centralized optimal control policy for crossdocking facilities, which takes into account the shipments of goods to the crossdocking terminal, the transfers inside the terminal as well as the shipments to the customers. A Branch-and-Bound algorithm has been applied to obtain an exact solution for the optimization model which is tested on different data sets. The results show that an exact solution can be obtained for small instances. Finally, the results for a decomposed model are presented. The decomposition yields faster results for larger instances and therefore is more applicable in practice.

1 Crossdocking

Following the success of the Just-In-Time-Concept (JIT) in the manufacturing industry, the consumer goods industry introduced the concept of Efficient Consumer Response (ECR) in the early nineties. This concept follows the same idea as JIT by trying to have the right goods in the right quantity at the right location and time, in order to eliminate inventories, reduce cycle times and hence react more effectively to changes in consumer demand (van der Heydt (1998)). Along with the demand for flexible supply chains, new distribution techniques were necessary to reduce inventories in traditional warehouse systems. Within this context the crossdocking concept was developed which describes the rapid consolidation and shipment of products with the focus on inventory elimination. The corresponding locations for this activity are crossdocking terminals. There are numerous classification possibilities for crossdocking implementations. The most common is the differentiation into single- and two-stage crossdocking. In single-stage crossdocking the supplier is performing the task of customer specific order picking (Gue (2001)). At the crossdocking terminal the deliveries are then simply transferred to a waiting truck at a "dock across", hence the name "Crossdocking". In the case of two-stage crossdocking the inbound trucks deliver single-product pallets. These products are then sorted and order-picked within the crossdocking terminal (Gue and Kang (2001)). Figure 1 shows the concept of (single-stage) crossdocking. It is obvious, that the in- and outbound trucks have to be scheduled very precisely, since the delay of a single truck can disturb the complete transshipment process.

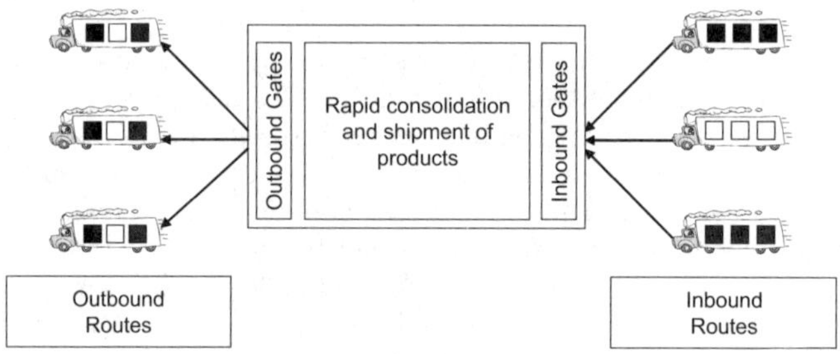

Fig. 1. Crossdocking Concept

2 A centralized model for a crossdocking terminal

This paper considers a crossdocking terminal for the consolidation of goods deliveries. The proposed model focuses on retail distribution, however it is not limited to that. The model reflects the situation that a 3rd party logistics provider (3PL) is responsible for picking up goods at the suppliers' manufacturing or distribution sites, the transshipment at the crossdocking terminal and the deliveries to retail stores in e.g. urban areas. It is assumed that demand of these retail stores is constant over the planning period.

Figure 2 illustrates the logistic problems that need to be solved by the 3PL in order to operate at minimum costs. The resulting problem consists of three different optimization problems which have to be looked at simultaneously. First a vehicle routing problem appears for the pick-up of goods from the suppliers. Since suppliers might easily be spread over a large geographical area, the separate tours can start from more than one depot. An analog problem appears when distributing products from the crossdocking terminal to retail stores. When trucks arrive or depart from the crossdocking terminal, an assignment of trucks to strip (inbound) or stack (outbound) doors has to be done. Depending on this dock door assignment the resource scheduling inside the terminal has to be performed. In this context resource scheduling is

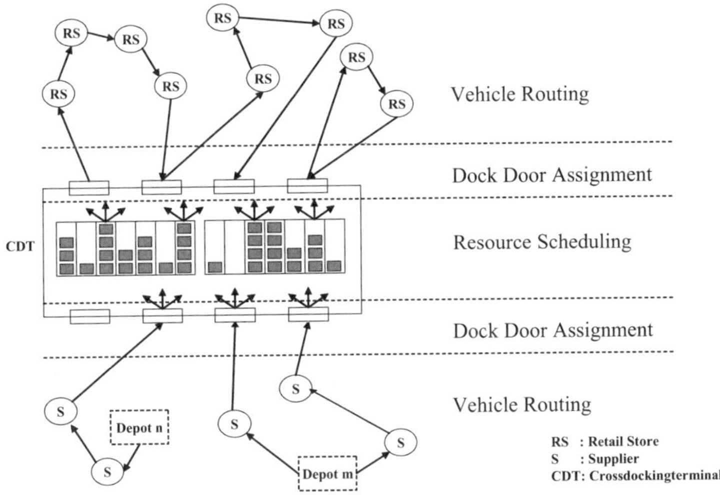

Fig. 2. To control the complete crossdocking process optimally the separated logistic problems have to be coordinated

mainly focused on personnel planning since most of the work is done manually. To these problems a large amount of sophisticated publications are available. However, all of them consider each problem individually and do not try to solve the problems simultaneously. For a good overview on Vehicle Routing Problems refer to Savelsberg and Sol (1995) and Toth and Vigo (2002). A solution to the dock door assignment problem is presented by Tsui and Chang (1992) and Bermudez and Cole (2001). Gue (1999) shows, that the dock door assignment in crossdocking terminals influences the labor cost for internal material handling up to 20 %. The internal processes have been modelled by Li et al. (2004), who applied a modified job-shop-scheduling problem and solved it with a genetic algorithm.

Describing the complete optimization model would be beyond the scope of this paper. For a detailed description please refer to Stickel and Metzelaers (2005). The resulting problem formulation is denoted as *Crossdocking Scheduling Problem (CDSP)*. The goal function consists of the components and decision variables described in tables 1 and 2.

Table 1. Components of the CDSP

a/b	fixed cost for delivery/pick-up vehicles
c/d	variable cost for delivery/pick-up tours
e/f	waiting cost for outbound/inbound vehicles at the terminal
g	penal cost for lateness at retail stores
i/j	handling cost for outbound/inbound vehicles
k	inventory cost for buffering of products
m	transport cost inside the terminal

Table 2. Decision variables (binary) of the CDSP

$X_{i,j}^{k}$	route i,j is traveled in outbound tour k
$Y_{i,j}^{k}$	route i,j is traveled in inbound tour k
u_{i}^{v}	timewindow v is used at retail store i
$u_{n}^{v,h}$	pick-up route h reaches the supplier in timewindow v
$u_{q}^{v,k}$	delivery route k starts in timeslot v from stack door q
$\delta_{k,k'}$	delivery route k and k' are consecutive

$$(CDSD)\min \underbrace{\left(\sum_{k\in K}\sum_{j\in F} X_{d,j}^{k} - \sum_{k\in K}\sum_{k'\in K}\delta_{k,k'}\right) c_{fix}}_{a} + \underbrace{\sum_{h\in H}\sum_{n\in N} c_{fix}^{h} Y_{\overline{d},n}^{h}}_{b} +$$

$$\underbrace{\sum_{k\in K}\sum_{i\in\overline{F}}\sum_{j\in\overline{F}} c_{i,j}\, X_{i,j}^{k}}_{c} + \underbrace{\sum_{h\in H}\sum_{m\in\overline{N}}\sum_{n\in N} c_{m,n} Y_{m,n}^{h}}_{d} + \underbrace{\sum_{j\in F} c_{wait}^{j}\, t_j}_{e} + \underbrace{\sum_{h\in H}\sum_{n\in N} c_{wait}^{n}\, t_n^{h}}_{f} +$$

$$\underbrace{\sum_{v\neq 0}\sum_{h\in H} c_{hand}^{v}\sum_{q\in Q} u_q^{v,h}}_{i} + \underbrace{\sum_{v\neq 0}\sum_{k\in K} c_{hand}^{v}\sum_{q\in Q} u_q^{v,k}}_{j} + \underbrace{\sum_{i\in F} c_{late}^{i}\, u_{late}^{i}}_{g} +$$

$$\underbrace{\sum_{n\in N}\sum_{v'\in V} c_{stag}^{n}\left(\sum_{v=1}^{v'}\sum_{h\in H} l_n^{v,h} - \sum_{v=1}^{v'}\sum_{k\in K} d_n^{v,k}\right)}_{k} + \underbrace{\sum_{i\in N}\sum_{v\in V}\sum_{p\in P}\sum_{q\in Q} c_{p,q}\, l_{i,v}^{p,q}}_{m}$$

3 Implementation and Results

The model has been implemented on a P4, 2,4 Ghz-System in ILOG OPL Studio, which uses CPLEX 8.0 to solve the problem, applying a Branch-and-Bound algorithm. To test the implementation, a data set has been generated, which relies on the retail store locations of a large german supermarket chain. The suppliers are located in distances ranging from 20 to 100 km from the retail stores. The demand for the retail stores have been chosen to range between 25 % - 50 % of a delivery truck's capacity. The observed crossdock had 3 strip and 3 stack doors with a planning horizon of 6 hours. The model has been tested in different configurations, regarding the number of suppliers and retailers. The results are shown in table 3. It is obvious that the proposed model is only sufficient for small instances. However, instances for a cross-docking terminal with up to 5 suppliers and 5 retailers could be solved in acceptable time.

During the investigation a decomposed model has also been implemented. In this modification the model has been separated into two components which

Table 3. Runtimes for different problemsizes

Num. of	**Number of suppliers**			
retail stores	**2**	**3**	**4**	**5**
2	1 s	5,4 s	70,2 s	3,4 min
3	1,3 s	11,03 s	87,1 s	4,5 min
4	3,3 s	12,38 s	2,3 min	74,3 min
5	10,2 min	59,7 min	28,4 h	8 d
6	3,2 min	6,3 h	18,8 h	$\geq$ 10 d
7	26,4 min	2,9 d	19,9 h	$\geq$ 10 d

have been solved sequentially. With the abdication of optimality the solution was of course obtained much faster. First, vehicle routings were optimized, without the assignment to a specific dock door, only to a specific time window. Second, the internal processes were optimized with the given time windows from step one, by assigning the arriving and departing trucks within each time window to a specific dock. Since only a small crossdocking terminal has been investigated up to this point the deviation from the optimum could be neglected. However, for large crossdocking terminals with a large number of doors this simplification might not be made anymore. Table 4 shows the improved runtime results. One can see that the runtimes decrease significantly and that larger instances are solvable.

Table 4. Runtimes for different problemsizes for the decomposed model. The values represent only the runtime for the vehicle routing, since the runtimes for dock door assignment were always in seconds range

Num. of	**Number of Suppliers**			
retail stores	**2**	**3**	**4**	**5**
2	0,3 s	1,5 s	2,5 s	13,1 s
3	0,3 s	2,3 s	4,6 s	27,3 s
4	0,8 s	6,2 s	31,7 s	3,7 min
5	9,3 s	13,3 s	9,2 min	44,9 min
6	64 s	2,5 min	14,2 min	1,5 h
7	6,4 min	13,7 min	1,1 h	2,7 h

4 Outlook

The proposed model is part of broader research concerned with planning & control policies of transshipment facilities. On one hand a centralized planning & control approach (as presented) is pursued. In this approach it is assumed that all decision relevant information is available and that a central instance (e.g. a 3PL) has the necessary decision power. In this context the presented model is under further development, especially with heuristic solution methods being investigated.

On the other hand, a decentralized approach is under consideration, in which the mentioned assumptions do not hold anymore. Here, the planning and decision power is not owned by a single entity. Hence decentral obtained planning solutions have to be coordinated in order to promote cost effective logistic processes. The dock door assignment is interpreted as an interface between separate logistic planning processes and independent partners have different valuations for arriving in specific time windows, therefore these time windows are allocated among the cooperating partners by means of combinatorial auctions.

References

Bermudez, R. and M. H. Cole (2001). A genetic algorithm approach to door assignments in breakbulk terminals. Technical Report MBTC-1102, Mack-Blackwell Transportation Center, University of Arkansas, Fayetteville, Arkansas.

Gue, K. R. (1999, 11). The effects of trailer scheduling on the layout of freight terminals. *Transportation Science 33*(4), 419–443.

Gue, K. R. (2001). Crossdocking: Just-in-time for distribution. Working paper, Graduate School of Business & Public Policy Naval Postgraduate School, Monterey, USA.

Gue, K. R. and K. Kang (2001). Staging queues in material handling and transportation systems. In *Proceedings of the 2001 Winter Simulation Conference.*

Li, Y., A. Lim, and B. Rodrigues (2004). Crossdocking - JIT scheduling with time windows. *Journal of Operational Research Society 10*(1057), 1–10.

Savelsberg, M. and M. Sol (1995). The general pickup and delivery problem. *Transportation Science 29*, 17–29.

Stickel, M. and M. Metzelaers (2005). Ein zentrales Steuerungsmodell für bestandslose Umschlagzentren. Working paper, Universität Karlsruhe (TH), Institut für Fördertechnik und Logistiksysteme.

Toth, P. and D. Vigo (2002). *The Vehicle Routing Problem.* SIAM Monographs on Discrete Mathematics and Applications. Philadelphia: SIAM.

Tsui, L. Y. and C.-H. Chang (1992). An optimal solution to a dock door assignment problem. *Computers and Industrial Engineering 23*(1-4), 283–286.

van der Heydt, A. (1998). *Efficient Consumer Response (ECR)* (3 ed.). Frankfurt: Peter Lang Verlag.

An Enumerative Approach to Rail-Truck Intermodal Transportation of Mixed Shipments

Manish Verma[1], Vedat Verter[2]

[1] Faculty of Business Administration, Memorial University, St. John's, Canada

[2] Faculty of Management, McGill University, Montreal, Canada.

The global chemical industry, fuelled by the ever increasing demand in China, is growing steadily. In 2002 chemicals had the highest global trade growth rate of 10%, which in turn translated into US \$ 660 billion export value. While pipelines and bulk tankers are used primarily as modes to transport large quantities of a single product, parcel tankers are used to simultaneously transport multiple cargoes. The tank containers, also referred to as intermodal (ISO or IMO) tanks, are designed for intermodal transportation i.e., movement of freight by more than one mode of transportation.

In Europe, a high percentage of chemical industry products are carried by intermodal (IM) transport. Just like for rail and trucks, intermodal dangerous goods regulation is developed by United Nations, which is then implemented at the national level. The U.N. approach, being mode oriented, develops regulation for each mode, which has to be complied within that link of the intermodal chain. In the United States, roughly 9% of the multiple mode shipments are hazardous in nature while 1.5% of the total hazardous shipments are moved by the combination of more than one mode. The 2002 Transportation–Commodity Flow Survey further adds that these numbers may have been underestimated since the shipper does not always know the modal combinations used to transport the goods, and some shipments moving by more than one mode may be reported as a single mode shipment (U.S. DOT 2004). If the impressive growth in intermodal traffic is any indication, the volume of chemicals and hazardous materials moved via intermodal channels can only increase in the future, and hence this emerging area warrants increased attention from academics and practitioners.

Macharis and Bontekoning (2003), and Bontekoning et al. (2004) provide excellent review of research in the intermodal transportation domain, but despite the volume of chemicals and other hazardous cargo moved in intermodal tanks there is no risk assessment work to date. This work is motivated by the desire to contribute in the realm of rail-truck intermodal transportation of hazardous materials, wherein efficiencies of trucks and economies of railroads are combined to move shipments. Unfortunately studies comparing the safety of railroads and trucks for transporting hazardous materials are conditional on a range of factors (like volume, distance, shipment-frequency, accident rates, etc.), and hence are not conclusive. In this work an intelligent approach that ensures routing of rail-truck intermodal shipments based on both *risk* and *cost* and not just cost is presented.

In a rail-truck intermodal transportation system a shipment that needs to be transported from a shipper to a receiver is first transported by truck to a rail intermodal terminal. There it is transshipped from truck to a train, which undertakes the rail-haul part of the journey to the destination terminal. Intermodal trains are distinct from traditional freight trains, operate on a fixed-schedule, are usually quite punctual and offer multiple types of services. At the other end of the transport chain the shipment is transshipped from train to truck and delivered by truck to the receiver.

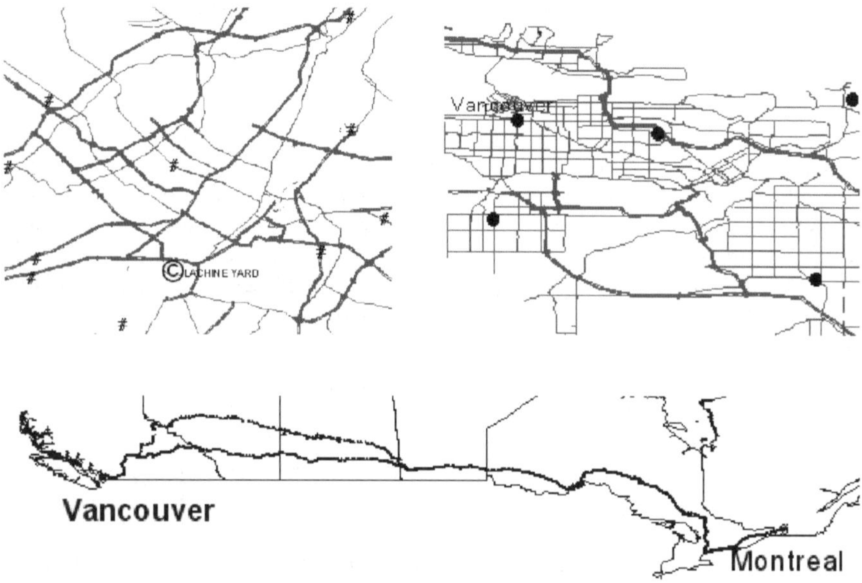

Fig. 1. Sections of a Rail-Truck Intermodal Chain

For this illustrative example there are ten shippers in Quebec spread around the Island of Montreal and ten receivers in British Columbia, and each is linked to the intermodal terminal via a number of paths. Each of the ten receivers has demanded shipments from each of the ten shippers, and hence there are 100 supply-demand pairs with specified delivery *times*. Two types of intermodal train service namely *regular* and *premium* is available between the two terminals. Figure 1 depicts the shippers, receivers, drayage paths and IM train service between Montreal and Vancouver.

Demand, generated hypothetically, is in terms of number of intermodal containers/trailers, and is categorized into hazardous and non-hazardous cargo. Attached to each demand is a delivery time, and the shipper has to ensure that the said demand is met by then. It is essential to make the delivery time assumption, otherwise there is no need to use the expensive intermodal trains when the same shipments could be moved by a normal freight train at a lower cost, and reach their destinations at a later time. For an intermodal chain to be feasible, the sum of times taken by each activity (drayage, transshipment, rail-haul, etc.) forming the chain should not exceed the "*time*" specified by the receiver. For the purpose of illustration we will track the shipment from the first shipper (i.e. *Repentigny*) to the first receiver (i.e. *Kelowna*), but other shipments can be evaluated similarly. Receiver at *Kelowna* has specified a period of 5 days or 120 hours from the time an order has been placed to receive shipments.

Table 1. Attributes of the routes to Repentigny

Paths	**1**	**2**	**3**	**4**
Time (hrs.)	2.08	2.28	2.12	3.00
Distance (kms.)	83	91	85	120
Cost ($)	254	264	256	300
Population Exposure (people)	2959	2787	2557	1266

On receiving request from the shipper, the driver(s) leaves the Lachine yard with the empty truck-trailer(s) and takes the shortest path to the shipper location. The moment the driver-trailer leaves the IM yard, the time-counter starts for the shipper. It is reasonable to assume that the shipper has been apprised of the road and rail travel times, departure and cut-off time for the IM train, and hence needs to prepare the shipments accordingly. Three things are happening at this location: the container (trailer) is being loaded; driver-truck is waiting and adding to the cost; and, a portion of the specified "*time*" is elapsing. There are four paths from the shipper at Repentigny to the *Lachine* yard. It is expected that the driver will take the shortest path, since it is the cheapest. The shortest path from the yard to this shipper is 41.5 km, and hence path 1 indicates using the shortest path for both segments of inbound drayage (41.5 * 2 =83). Path 2 through path 4, imply taking the shortest path to the shipper, and then a different one to get back to the IM yard. Associated with each path is a *cost* and a *risk* attribute. All the cost numbers were estimated using the publicly available data, while population exposure risk was calculated in ArcView GIS. The effectiveness of *population exposure* as an aggregate measure of risk in Verma and Verter (2004) and in Verma et al. (2005) motivates its usage for risk assessment in this work. Risk accrues in two states: stationary at the different sites; and, during transportation. In the former instance a danger circle centered at the point of handling will be generated, while in the latter case exposure bands along the transportation corridors would be carved out due to the movement of hazardous shipments.

At *Repentigny* 1,557 people were exposed due to the handling of a single container with hazardous cargo, and a total of 82,521 people due to the handling of 53 containers with hazardous cargo. It is evident from Table 1 that path 1 has the highest population exposure while path 4 the lowest, and that the latter is the longest of the four paths available to this shipper. It is not prudent to take path 1 if hazardous material is being transported since it goes through downtown Montreal and exposes more than double the number than does path 4, which bypasses downtown Montreal.

In fact a *risk-cost* tradeoff analysis on "***time***" dimension should be conducted, while being aware of the following: the departure times of the IM trains are fixed as they are *schedule-based*; and, the specified "***time***" at the receivers' site is a hard constraint. By spending an extra $46 and taking path 4, population exposure can be brought down by 1,693 people. Clearly this would mean that the shipper should have the shipments ready an hour earlier than when taking path 1, in order to make the same IM train even after traveling on the longest route i.e. path 4. In here the increased cost and/or lower risk is acceptable only if the specified "*time*" element is not violated. Each of the four paths in table 1 for the shipper at *Repentigny* is feasible if the shipments are readied at appropriate times. If the 53 trucks (containers) with hazardous cargo demanded from this shipper take path 4 to the IM terminal, instead of the shortest route path 1, then 89,729 fewer people would be exposed. On the other hand, since path 4 is the longest it will mean having all the containers ready an hour early to be able to make to that IM train, and incurring an extra $2,438 ($15,900-$13,462) for the entire shipment. As alluded to earlier *time-dimension* will drive the determination and consequent selection of feasible routes, while a JIT approach wherein a tractor-trailer does not have to wait in line to be loaded/transferred/unloaded is being incorporated for evaluation. It has been assumed that enough flat-cars and IM trains for all the incoming intermodal units are available, and the yard-master makes the operations decision about the number of trains required on a particular route.

Once the containers reach the IM yard, they are placed on rail flatcars using a gantry or overhead crane. While the *cost* will be incurred for each container, *risk* will accrue only from containers with hazardous cargo due to additional handling at the yard. *Lachine* yard receives the containers from other nine shippers as well, and hence the cumulative exposure from the hazardous cargo from all the ten shippers is 837,666 people. This is the risk stemming from handling 538 containers with hazardous cargo. Canadian Pacific Railroad's (CPR) intermodal network was recreated in *ArcView* GIS between Montreal and Vancouver, and the route through Edmonton measured 2,920 miles, while the one through Calgary 2,713 miles. IM train speed was calculated using information provided on CPR's website. The regular train service ***(R-IM)*** stops in Edmonton for traffic swaps, which is estimated to take 6 hours on average, for a total yard-to-yard time of 103 hours. On the other hand the premium train service ***(P-IM)***, being both non-stop and faster, will cover the same distance in 73 hours. The yard-to-yard time on the route through Calgary is 96 and 68 hours for the two train services.

The time elapsed (rounded to decimal places) at each stage of the movement, using ***R-IM***, of intermodal units from the shipper at *Repentigny* till they reach *Kelowna* was calculated. It was noticed that some intermodal combinations were infeasible. For example, path 1 of inbound drayage ***R-IM***, and path 1 of outbound drayage is not a viable option since it will need 5.03 days to complete the journey while the specified time is 5 days. It is the onus of the shipper (and the intermodal company) to deliver shipments before the scheduled delivery time. One way to remove infeasibility is by allocating more efficient resource at the bottleneck of the intermodal chain, i.e., by using faster intermodal trains. When ***P-IM*** is employed to move traffic between the IM yards, shipments can arrive at their destination before the cut-off time. Clearly there is no need to load all the traffic on ***P-IM*** since it is more expensive than ***R-IM***, but shipments on the infeasible route combination should be considered. On reaching the Vancouver yard, one would expect the driver to take any path except path 1 (is not just the longest, but also infeasible when used in conjunction with ***R-IM***), when going to *Kelowna*. If ***P-IM*** is used to move all the shipments destined for *Kelowna*, then path 1 will be feasible. It is important to note if Calgary route is chosen to move shipments, no ***P-IM*** would be required since all intermodal combinations become feasible.

It is assumed that the IM service provider has the capacity and equipment to move shipments as per demand, and hence these containers will not be stranded at the yards waiting to be westbound. Although it is assumed that each shipment will arrive at their destination on schedule, one still needs to ascertain the loading of each IM train in order to calculate the exact *population exposure* due to a particular train (Verma and Verter, 2004), (Verma et al., 2005). For train loading any assignment heuristic could be used, although an interesting makeup motivated by our earlier work is to run IM train equivalent of hazmat *unit-train*. By employing two unit trains and two mixed trains, as opposed to four mixed trains, the population exposure risk was reduced by 24%. In should be noted that the cost remains the same since the number of railcars to be moved is unchanged, but there is a possibility to reduce risk by implementing an intelligent loading strategy.

Table 2. Outbound Drayage to Kelowna

Paths	1	2	3	4	5
Time (hrs.)	21.30	18.38	18.35	18.40	18.50
Distance (kms.)	852	735	734	736	740
Cost ($)	1215	1068	1068	1070	1075
Population Exposure (people)	37	87	86	88	89

Table 2 presents the five possible ways to reach *Kelowna* from the Vancouver yard, and path 1 is the most expensive. The other four paths, while being very similar to each other from both cost and risk standpoint, expose roughly 2.5 times the number of people exposed if path 1 is used to move the hazardous cargo. The receiver at *Kelowna* needs 53 containers with hazardous cargo and the population exposure difference between path 1 and the best among the other four paths (namely, path 3) is 49 people per container. 2,597 more people are exposed when all 53 containers move on path 3, as opposed to when they move on path 1. Of course not taking the shortest path will imply more travel cost. An additional $147/container will be incurred if the truck driver takes path 1 instead of the shortest path (path 3). Hence, by spending an additional $7,791, the population exposure risk could be reduced by 2,597 people.

Once the shipments have been delivered the dollar cost incurrence and the specified "***time***" counter stops, although the same cannot be said of population exposure. At the receiver's site, population exposure stems from the number of people within the danger circle centered at the point of container handling. This will be incurred for all the hazardous containers. At *Kelowna* 32 people are exposed due to the handling of a single hazardous container, which results in a cumulative total of 1,696 people for 53 containers. On reaching the location, the containers (trailers) are unloaded and the driver returns to the yard for future movement or waits in the region & awaits order for another inbound drayage.

In conclusion, this work was motivated by the desire to combine the advantages of two modes namely trucks and railroads. An intelligent enumeration method was employed to solve a 100 supply-demand pair problem, and societal risk reduction, given flexible time-schedule, was demonstrated. It can be said that it is possible to reduce population exposure risk by spending more money and/or readying the shipments at an earlier time in the event of inbound trucking, and by taking the *premium* IM train service for the rail-haul in order to enable the outbound trucking to take a *risk-cost* weighted path to the receiver's site. In addition, *population exposure* stemming from IM trains can be further reduced by implementing a train make-up scheme on the lines of (hazardous) *unit-trains*. In general, risk reduction via trade-off analysis is possible only when all the IM parties are concerned about safety and not driven just by the desire to minimize cost. Our immediate future research direction is the development of a mathematical formulation and a solution procedure for the general case of the problem presented in this paper, which involves multiple IM terminals.

Reference:

Bontekoning YM, Macharis C, Trip JJ (2004) Is a new applied transporation research field emerging? –A review of intermodal rail-truck freight transport literature. *Transportation Research –A*, 38:1-34.

Macharis C, Bontekoning YM (2003) Opportunities for OR in intermodal freight transport research: A review. *European Journal of Operational Research,* Article in Press. 2003.

U.S. Department of Transportation (2004) 2002 Economic Census: Transportation and Commodity Flow Survey.

Verma M, Verter V (2004) Railroad Transportation of Dangerous Goods: Population Exposure to Airborne Toxins. Accepted December 2004, forthcoming in *Computers & Operations Research.*

Verma M, Verter V, Gendreau M (2005) A Tactical Planning Model for the Railroad Transportation of Mixed Freight. Working Paper, Faculty of Business Administration, Memorial University.

Some Remarks on the Stability of Production Networks

Bernd Scholz-Reiter[1], Fabian Wirth[2], Michael Freitag[1], Sergey Dashkovskiy[2], Thomas Jagalski[1], Christoph de Beer[1], and Björn Rüffer[2]

[1] University of Bremen, Department of Planning and Control of Production Systems, Germany {bsr,fmt,jag,ber}@biba.uni-bremen.de

[2] University of Bremen, Zentrum für Technomathematik, Germany {fabian,dsn,rueffer}@math.uni-bremen.de

Summary. The increasing complexity of production and logistics networks and the requirement of higher flexibility lead to a change of paradigm in control: Autonomously controlled systems where decisions are taken by parts or goods themselves become more attractive. The question of stability is an important issue for the dynamics of such systems. In this paper we are going to touch this question for a special production network with autonomous control. The stability region for a corresponding fluid model is found empirically. We point out that further mathematical investigations have to be undertaken to develop some stability criteria for autonomous systems.

1 Introduction

In view of modern information and communication technologies and the dynamics and complexity of production and logistics networks, the idea to employ decentralized autonomous control, i.e., to design a network as interconnected autonomous units able to make decisions themselves, seems to be a new paradigm in logistics due to its flexibility and robustness [1], [2], [3], [4], [5]. Networks with centralized control commonly used in past decades are well studied in the sense that there are different models like queuing, fluid and discrete models proposed. These models allow to predict the behavior of a system, its efficiency and provide the designer with criteria of stability cf. [6], [7]. In this paper we concentrate on the stability properties of production networks with autonomous control. We consider the same production scenario as in [3] and [4] and state a fluid model for it. With help of simulation we find the stability area of parameters. In Section 2 we briefly describe the model. In the Section 3 we introduce the notion of stability and quote some known results. Section 4 contains simulation results. We collect some conclusions in Section 5.

2 Model Description

The considered production network is a dynamic flow-line manufacturing system. It consists of n parallel production lines each with m machines M_{ij} and an input buffer B_{ij} in front of each machine (see Fig. 1). Every line processes a certain kind of product $1, 2, \ldots, K$ by m job steps. The raw materials for each product enter the system via sources; the final products leave the system via drains. The production lines are coupled at every stage and every line is able to process every type of product within a certain stage. The decision about changing the line is made as an autonomous decision by the part itself on the basis of local information about buffer levels and expected waiting times until processing.

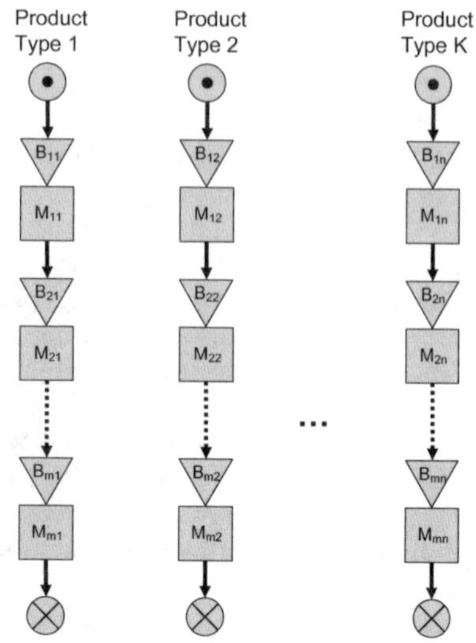

Fig. 1. $m \times n$ machines production network

To handle the complexity of the production network, the described scenario is reduced to 3×3 machines, i.e., three production lines each with three stages. The parts are autonomous in their decision which machine to choose. They take into account the fact that the processing times are higher on foreign lines than on their associated production line. At each production stage the parts compare the processing times of the parts in the buffers and their own processing time on the respective machine and choose the machine with the minimal time for being processed. Table 1 shows the processing times and the resulting processing rates for the three different product types on the three production lines.

Table 1. Processing times and resulting processing rates of the 3x3 machine model

	Processing times [h:min]/ Processing rates [parts/hour] at production line j		
	1	2	3
Part Type 1	2:00 / 0.5	2:30 / 0.4	3:00 / 0.33
Part Type 2	3:00 / 0.33	2:00 / 0.5	2:30 / 0.4
Part Type 3	2:30 / 0.4	3:00 / 0.33	2:00 / 0.5

To analyse the system's behaviour at varying demand and workload fluctuations, an arrival function $\lambda(t)$ is defined and set as a sine function:

$$\lambda(t) = \lambda_m + \alpha \times \sin(t + \varphi) \tag{1}$$

Here, λ_m is the mean arrival rate, α is the amplitude of the sine function, and φ indicates a phase shift. The arrival functions for the three product types 1, 2 and 3 are identical except for the phase shift of 1/3 period. This phase shift is chosen to simulate a seasonal varying demand for the three different products.

3 The Notion of Stability

There are several notions of stability for different models in the literature, see [7] for Harris recurrence of queuing networks, [6] for weak and strong stability of fluid models, [5] for Input-to-State stability in control systems. Roughly speaking this properties mean that the state of the system (or length of the queues) remains bounded if the external input to the system is bounded.

Let us consider a queuing network with K classes of customers processed with service times m_k, $k = 1, \ldots, K$; routing matrix P, external arrival rates $\lambda = (\lambda_1, \ldots, \lambda_K)'$ and let C_i be the set of classes processed on the server i, $i = 1, \ldots, I$. The effective arrival rate $\lambda_{\text{eff},k}$ to the class k is given by $\lambda_{\text{eff}} = (I - P')^{-1}\lambda$. Then for some special networks and some special service disciplines the condition

$$\rho_i := \sum_{k \in C_i} \lambda_{\text{eff},k} m_k < 1 \tag{2}$$

was found to be a sufficient for stability of the corresponding network. However changing the discipline may cause instability, see [8].

It is known that stability of fluid limit guarantees the stability of the corresponding queuing network. However if the fluid limit is unstable one can conclude nothing about the stability of queuing network. In the following we describe some simulation results concerning stability of the described 3×3 machines model. The known criteria are not applicable for this model because of the autonomous routing by the parts themselves.

4 Simulation Results

4.1 Stability Region Using Fluid Models

The fluid model for the 3×3 machines model is given by the following set of equations:

$$Q_{ijk}(t) = Q_{ijk}(0) + \int_0^t \lambda_{ijk}(s) - \mu_{ijk}\dot{T}_{ijk}(s)ds, \tag{3}$$

$$\lambda_{ik} \equiv \sum_{j=1}^{3} \lambda_{ijk}, \tag{4}$$

$$\lambda_{ijk}(t) = \lambda_{ik}(t) \times \mathbb{1}_{\left\{W_{ij}(t)+\frac{\lambda_{ik}(t)\times dt}{\mu_{ijk}} \leq \min_{l\neq j}\left\{W_{il}(t)+\frac{\lambda_{ik}(t)\times dt}{\mu_{ilk}}\right\}\right\}}, \tag{5}$$

$$\lambda_{i+1,k}(t) = \sum_{j=1}^{3} \mu_{ijk}\dot{T}_{ijk}(t), \text{ for } i = 1, 2, \tag{6}$$

$$\dot{T}_{ijk}(t) = \begin{cases} 0, & \text{if } Q_{ijk}(t) = 0, \\ \frac{Q_{ijk}(t)}{\sum_{l=1}^{3} Q_{ijl}(t)}, & \text{else,} \end{cases} \tag{7}$$

$$W_{ij}(t) = \sum_{k=1}^{k} \frac{Q_{ijk}(t)}{\mu_{ijk}}. \tag{8}$$

Here i and j denote the row and column of each machine in the network, k refers to the product type, Q is the queue length in fractions of products, W the amount of work in the queue, μ is constant and describes the maximal possible service rates, T is the cumulative allocation time per machine and product type, it increases whenever the respective server spends time serving the corresponding type of part. λ_{ik} denotes the arrival rate of part k in row i, which then is directed to exactly one machine.

Fluid limits do not capture individual parts, they can be seen as a macroscopic view of the process, such that all external arrival rates become their averages. The autonomous routing is captured in (5) under the restriction of (4), that is, the input rate can only turn to one machine at the time. By the macroscopic perspective we may assume (7), since in the FIFO queues of each server and by the intelligent routing algorithm, the parts are assumed to be equally distributed, so that each server spends service time for each class of products as it relates to the total queue length in front of the machine.

Using standard discrete time numerical methods, we calculate estimates on the stability region of the fluid limit, which in turn is a subset of the stability region of the discrete event system, see Fig. 2 lower curve (*solid line*). Here we used an exemplary arrival rate of 0.4 parts / hour (as in [3, 4]) for product type 3 and varied the arrival rates for types 1 and 2.

4.2 Stability Region Using Discrete Event Simulation

To analyse the stable parameter region using the discrete event simulation analogue to the continuous method the arrival rate for part type 3 is set as constant, in this case $\lambda_3 = 0.4$ parts / hour. The other two arrival rates are still sine functions with an amplitude of $\alpha = 0.15$ parts / hour. The mean of the sine curves are independently varied, i.e., one of the mean arrival rates is

Fig. 2. Subset of the stability region of the 3×3 machine model with a given arrival rate of 0.4 parts of type 3 per hour

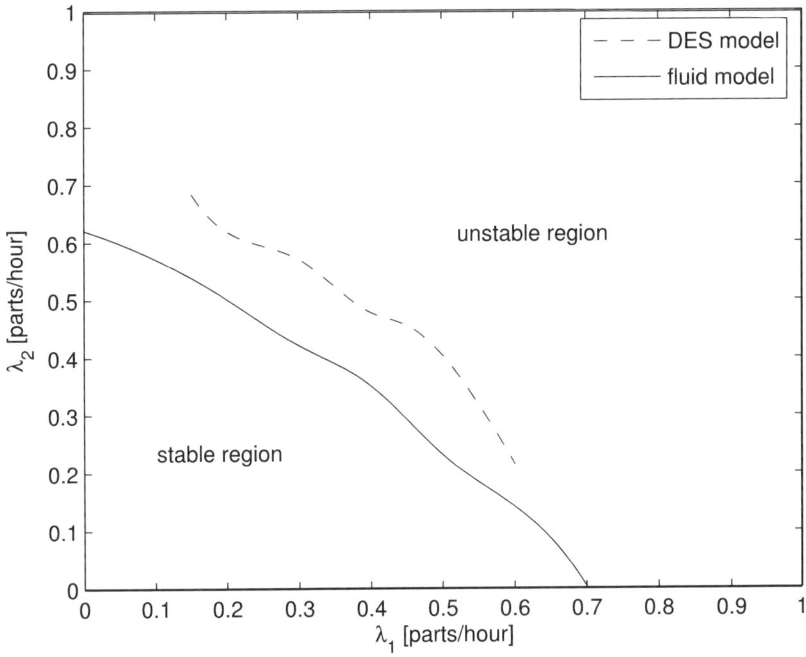

held constant while the other is increased unless the buffer levels begin to rise to infinity. The maximum mean arrival rate before the buffer levels begin to rise to infinity is called the critical rate. The result of this stability analysis is shown in Fig. 2 upper curve (*dashed line*).

The minimal mean arrival rates λ_1 and λ_2 are 0.15 parts / hour because the amplitude is $\alpha = 0.15$ parts / hour and no negative arrival rates are allowed.

5 Conclusions

Using the fluid model, a general statement about stability of the 3×3 machines production network could be derived, i.e., a stability margin (cf. lower curve (*solid line*) in Fig. 2) independent from a specific arrival function $\lambda(t)$. The DES model, on the other hand, provided a stability margin (cf. upper curve (*dashed line*) in Fig. 2) for a specific arrival function $\lambda(t)$. Due to the particular parameter settings, the stability margin of the DES model is above the margin of the fluid model.

The stability of the network depends on the specific parameters for the arrival rates and service rates, but a definitely stable parameter region could be found by the fluid model.

6 Acknowledgments

This research is funded by the German Research Foundation (DFG) as part of the Collaborative Research Centre 637 "Autonomous Cooperating Logistic Processes: A Paradigm Shift and its Limitations" (SFB 637).

References

1. Scholz-Reiter B, Freitag M, Windt K (2004) Autonomous logistic processes. In: Proceedings of the 37th CIRP International Seminar on Manufacturing Systems 357–362
2. Dashkovskiy S, Wirth F, Jagalski T (2004) Autonomous control of Shop Floor Logistics: Analytic models. In: Proceedings of the IFAC Conference on Manufacturing, Modelling, Management and Control. On CD-ROM
3. Scholz-Reiter B, Freitag M, de Beer C, Jagalski T (2005) Modelling dynamics of autonomous logistic processes: Discrete-event versus continuous approaches. Annals of the CIRP 55:413–416
4. Scholz-Reiter B, Freitag M, de Beer C, Jagalski T (2005) Modelling and analysis of autonomous shop floor control. In: Proceedings of the 38th CIRP International Seminar on Manufacturing Systems. On CD-ROM
5. Dashkovskiy S, Rüffer B, Wirth F (2005) An ISS Small-Gain Theorem for General Networks. In: Proceedings of the Joint 44th IEEE Conference on Decision and Control and European Control Conference (CDC-ECC'05). To appear.
6. Chen H (1995) Fluid approximations and stability of multiclass queueing networks: work-conserving disciplines. In: Annals of Applied Probability 5:637–665
7. Dai JG (1995) On positive Harris recurrence of multiclass queueing networks: a unified approach via fluid limit models. In: Annals of Applied Probability 5:49–77
8. Bramson M (1994) Instability of FIFO queueing networks. Annals of Applied Probability 4:414–431

Simulating Dispatching Strategies for Automated Container Terminals

Dirk Briskorn[1] and Sönke Hartmann[2]

[1] Lehrstuhl für Produktion & Logistik, Christian-Albrechts-Universität zu Kiel, Germany, briskorn@bwl.uni-kiel.de

[2] HPC Hamburg Port Consulting and HHLA Container Terminal Altenwerder, Hamburg, Germany, s.hartmann@hpc-hamburg.de

1 Introduction

The practical relevance of container terminal logistics has led to an enormous scientific interest in this field. Publications cover not only optimization algorithms (e.g., for equipment scheduling and vehicle routing) and decision problems (e.g., for finding grounding positions for containers in the yard) but also simulation models to study such approaches in a realistic dynamic environment. Recent literature surveys have been given by Steenken et al. [3] and Vis and Koster [4]. In this paper, we consider a highly automated terminal with quai cranes, automated guided vehicles, and automated stacking cranes. Such a terminal configuration is sketched in Figure 1.

Arriving containers are stored in the terminal for a certain period of time until they are picked up either on the landside by trucks or trains or on the seaside by vessels. The stacking area is divided into blocks. Each of them is served by one or more rail mounted gantry cranes (RMG). The RMGs put arriving containers into the stack, remove containers from the stack when they are picked up, and carry out so-called shuffle moves, i.e., move containers to another position within the stack if they stand on top of containers that have to be moved out of the stack.

Vessels are served by quai cranes (QC). They either discharge containers from vessels or load them onto vessels. This has to be done according to a given stowage plan. The latter imposes precedence relations on some of the containers. This leads to a partial order in which containers to be loaded have to arrive at a QC.

The transportation of containers between RMGs and QCs is done by automated guided vehicles (AGV). AGVs are unmanned vehicles that are unable to load and unload themselves, thus quai cranes and stacking cranes have to load and unload them. The AGV area can be divided into lanes alongside the quay (which are also used as handover lanes at the quai cranes), lanes alongside the stacking area, lanes connecting quai and stacking area, and handover lanes at each RMG stack. At each QC, some lanes are used as a buffer area where AGVs wait either because another AGV with a predecessor container has to pass or because the handover position is still occupied by another AGV.

Terminals with this type of configuration exist in Rotterdam, Netherlands (Delta Terminal of Europe Container Terminals, ect) and Hamburg, Germany (Container Terminal Altenwerder, CTA). This study has been carried out in cooperation with CTA. The focus of this paper is on the simulation of scheduling strategies for AGVs. First, we describe a simulation model for testing those strategies. Subsequently, we propose two specific strategies, employ them in our model and present first results.

2 Simulation Model

2.1 Material Flow

In what follows, we describe the material flow components of the model which cover the behavior of the equipment. The underlying approach is to model the equipment behavior by means of time distributions for their actions (instead of modeling the behavior in full detail).

Quai Cranes. The QCs repeatedly either handle containers or wait for AGVs. When working, we have a cyclic process of serving vessels and AGVs. While the duration of QC waiting phases results from the AGVs, the duration of the container handling cycles of the QCs must be defined. We divide a cycle into two parts: The first one is the interaction with an AGV, namely releasing/picking the container and lifting the spreader high enough such that

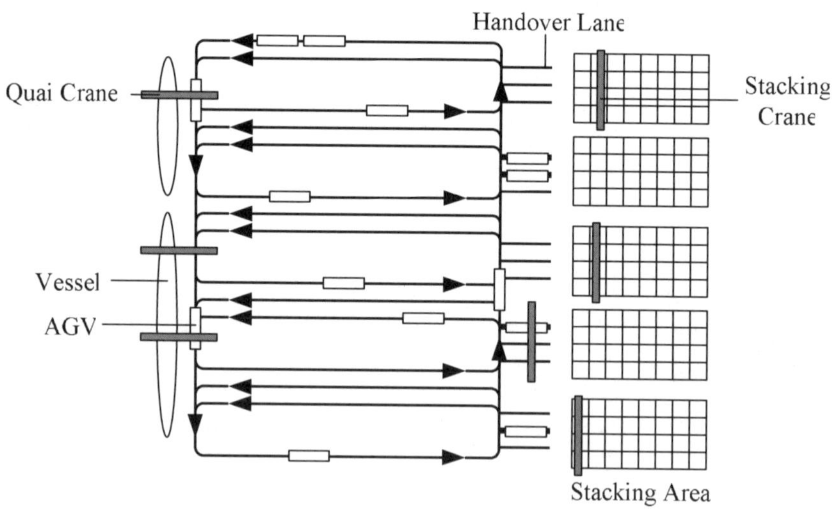

Fig. 1. Layout of the container terminal

the AGV can leave. This part represents the handover time of an AGV at the QC. We define a distribution describing the duration of this part. The second part covers the remaining time of the QC cycle, that is, the time for transporting a container between AGV and vessel and getting ready for the next AGV. For the duration of this part of the cycle, we again define a distribution. This part is the time that has to elapse until the next AGV-related handover time can start. After an AGV has been served, the next AGV which does not have to wait for a predecessor is allowed to move from the buffer to the QC.

Automated Guided Vehicles. The behavior of the AGVs consists of either carrying out transportation jobs or waiting for further jobs. When an AGV handles a job the following process can be divided into four parts. First, it drives to the container's pick-up location (if it is not equal to the AGV's current position). Afterwards, it waits for service either by QC (if the container is a discharged one) or by RMG (if the container is to be loaded). When the AGV has received the container, it drives to the container's destination. Finally, it has to wait for service again to be unloaded. Note that if the container's destination is a QC this final part consists of waiting in the buffer (if necessary), driving to the QC and waiting for service.

Waiting times in handover lanes do not depend on AGVs themselves, thus, they are determined by QCs and RMGs. However, driving times have to be defined to describe the AGV behavior accurately. To do so, we calculate an ideal driving time based on the distance and the number of curves etc. Subsequently, this ideal time is multiplied by a factor that leads to the actual driving time. This factor reflects AGV interferences on the layout such as congestions. For this factor, a distribution is employed.

Rail Mounted Gantry Cranes. Besides serving AGVs, RMGs carry out many tasks such as shuffling containers, serving the landside interface of the stacking area, etc. Because these activities are beyond the scope of our simulation purposes, we do not consider them explicitly. Therefore, instead of modeling the RMG behavior itself, we define a distribution for each RMG representing the time span an AGV is tied up in the handover lane including waiting for service and lifting or releasing the container.

2.2 Information Flow

In addition to the behavior of the equipment, the simulation model contains the decision logic of the terminal control system. We obtain the information flow components described in the following.

Jobs. We consider only AGV jobs. Each job is associated with a container and hence with a pick-up and a delivery location. For each QC, we generate a list of related AGV jobs. In case of a discharging QC, the delivery location (stack) is selected using a distribution which assumes that the containers from one QC are brought to the k_D nearest blocks. Analogously, for a loading QC, the pick-up location (stack) is determined with a distribution that assumes that containers to be loaded were stacked into the k_L nearest blocks. This way, we do not have to consider a stacking strategy or positions inside the blocks. Instead, we use distributions to reflect the impact of the strategy, namely that the AGV should have a short way to drive and that the nearest block cannot always be selected because this would lead to excessive workloads of individual RMG blocks.

The overall number of jobs is large (virtually unlimited), such that the simulation model will lead to the maximum waterside productivity (jobs completed per hour) that the AGV dispatching strategy is able to achieve. Note that this is realistic since a real terminal also tries to discharge and load containers as fast as possible.

AGV Assignment Procedure. The core of the strategies to be simulated is the decision which AGV shall carry out which job at which time. Whenever an AGV finishes its current job or a new job enters the system the assignment procedure is executed. Only the assignments of currently available AGVs are fixed, the remaining AGVs are considered again when the assignment procedure is executed the next time. The basic idea of the strategy is presented in Section 3. The strategy makes use of estimates for the availability time of an AGV, that is, the completion time of its current job. This estimate is generated a certain time before the job is actually completed (i.e., before the AGV is unloaded by a QC or an RMG). Both the time the estimate is generated in advance and its deviation from the actual availability time is controlled by corresponding distributions.

RMG Assignment Procedure. Since we do not model the behavior of the stacking cranes explicitly (see Section 2.1), we did not have to consider an RMG scheduling procedure that selects the next job for a crane. The impact of the stacking cranes on the AGVs has been captured by explicitly incorporating the distribution of the AGV waiting times at the stack. Moreover, the RMG scheduling selects the job for a crane and hence determines the container to be loaded onto an empty AGV (note that we propose that the AGV scheduling plans for a container when sending an empty AGV to a block but the container to be actually loaded onto the AGV is determined by the RMG scheduling). Here, we assume that the RMG assignment procedure selects the most urgent container. The urgency of a container is defined analogously to the AGV strategy under consideration. Note that we only consider container urgency here and not other criteria of the RMG scheduling such as emtpy travel time minimization.

3 AGV Dispatching Strategy based on Inventory Levels

AGVs either supply QCs with containers or receive them from QCs. Therefore, the main objective for AGV scheduling strategies is to guarantee a constant inflow of AGVs to the QCs. This is a substantial condition for a high productivity of the terminal measured by discharged or loaded containers. In order to obtain a balanced inflow of AGVs, we count the AGVs currently on the way to a QC q. We interpret this number as an "inventory level" belonging to the QC. Consequently, the QC having the smallest inventory level and hence the smallest number of AGVs should be most likely to receive the next AGV that becomes available.

We obtain a standard linear assignment problem with AGVs and transportation jobs (which is solved to optimality using the Hungarian method, see Kuhn [1]). The cost c_{ja} of assigning job j to AGV a represents the time which will pass until a can pick up the container related to j as well as the urgency of job j which is measured by the inventory level of the related QC. Then we determine the number of AGVs to be considered, denoted as n, and select the n most urgent jobs with respect to the inventory levels. We compared two variants of this assignment approach which differ in the way the AGVs to be considered are determined:

- **Few AGVs.** This variant considers only the AGVs which are currently free and could start the next job immediately. Note that in case of a high workload this will usually be a single AGV which will be assigned to the most urgent container (which is similar to a greedy heuristic employing a priority rule).
- **More AGVs.** Here, we consider those AGVs which are currently free and those which will finish their current job within a certain time horizon (we assume that we have an estimate of the availability times of these AGVs).

4 Results

We implemented the simulation model and the two dispatching strategies in the Java-based simulation framework Desmo-J (cf. Page et al. [2]). The model includes 10 QCs, 20 RMG blocks and 40 AGVs. We created five scenarios which differ in the structure of the precedence relations between containers (see Section 1). For each scenario and each strategy, 100 simulation runs were carried out. The distributions for modelling the equipment behavior were derived from original statistics of CTA.

In order to evaluate the two strategies, we consider the terminal's waterside productivity given as the average number of discharged and loaded containers per QC and hour. The results are given in Table 1 where the scenarios are listed in decreasing order of the density of the precedence relations. For reasons of confidentiality, we cannot give absolute values. Instead, we selected the variant with few AGVs as a base (having a value of 1.0 in each column) and give relative results for the variant with more AGVs. We can see that the case with more AGVs is clearly superior since it leads to a productivity that is about 10 % higher in every scenario.

Table 1. Simulation results for inventory approach (relative productivity)

Approach	scenario 1	scenario 2	scenario 3	scenario 4	scenario 5
Few AGVs	1.000	1.000	1.000	1.000	1.000
More AGVs	1.090	1.097	1.099	1.095	1.107

5 Conclusions

We proposed a simulation model for the seaside processes at automated container terminals. The model contains the equipment behavior as well as the strategies of the terminal control system. Rather than considering every detail, the model employs distributions that cover processing times of the equipment as well as stacking decisions. The main advantage of this simulation model is that it can easily be adapted to a specific real-world terminal. This can be done by using statistics of the terminal for the distributions defined in the model. Moreover, one can focus on the strategies to be examined—in our case, we employed strategies for AGV dispatching, but instead of also considering RMG dispatching strategies that were not within focus, we simply incorporated the AGV waiting time distribution at the RMG blocks. Hence, we modelled the impact of the RMGs on the AGVs rather than the RMGs themselves. Summing up, we obtain a simple model with short run times which can easily be configured and produces realistic results.

The simulation model was used to test two variants of a strategy for AGV dispatching. The variant with a longer horizon (which also considers AGVs which are not available yet) proved to be superior in terms of terminal productivity. The reason for this is that this variant includes more degrees of freedom for optimization. This seems to be more important than the potential drawback of uncertainty of the time estimates for AGV availability. The next step will be a more in-depth analysis of the inventory-based strategy and a comparison with a conventional time-based scheduling strategy.

References

1. H. W. Kuhn. The hungarian method for the assignment problem. *Naval Research Logistics Quarterly*, 2:83–97, 1955.
2. B. Page, T. Lechler, and S. Claassen, editors. *Objektorientierte Simulation in Java mit dem Framework DESMO-J.* BoD GmbH, Norderstedt, 2000.
3. D. Steenken, S. Voß, and R. Stahlbock. Container terminal operations and operations research – a classification and literature review. *OR Spectrum*, 26:3–49, 2004.
4. I. F. A. Vis and R. de Koster. Transshipment of containers at a container terminal: An overview. *European Journal of Operational Research*, 147:1–16, 2003.

Part IV

New Maritime Businesses

Integration of Berth Allocation and Crane Assignment to Improve the Resource Utilization at a Seaport Container Terminal

Frank Meisel[1] and Christian Bierwirth[2]

[1] Martin-Luther-University Halle-Wittenberg
frank.meisel@wiwi.uni-halle.de
[2] Martin-Luther-University Halle-Wittenberg
christian.bierwirth@wiwi.uni-halle.de

Abstract. This talk deals with the combination of two decision problems, which occur consecutively while planning the charge and discharge operations of container ships in container terminals. The Berth Allocation Problem (BAP) considers the allocation of ships to berths in the course of time. The Crane Assignment Problem (CAP) addresses the assignment of quay cranes to ships. We provide a heuristic approach for the integrated solution of these problems and present computational results based on real world data.

1 Introduction

As seaport terminals are often a bottleneck in the transport chain, the organization and control of container handling processes receives increasing attention. Terminal operations planning involves several tasks on the tactical as well as on the operational level [7, 8]. In this paper we concentrate on the quay side tasks in a container terminal (CT) by an investigation of the integration of the BAP and the CAP. It is organized as follows. In Section 2 we introduce the optimization problems under consideration, their integration and the related objective function. Section 3 presents a solution method which has been adopted from heuristics for the resource constrained project scheduling problem (RCPSP). Finally, some computational results are presented.

2 Problem Description

2.1 Berth Allocation Problem

A berth plan determines the quay positions and the berthing times for all container vessels which have to be served within a certain period. The BAP usually aims at finding a berth plan which minimizes the total stay or delay times of vessels at a port. If the quay is partitioned into several berths with predetermined lengths it is only allowed to moor one vessel per berth at one time. Otherwise, if no such partition is given, vessels can be moored wherever enough space (including clearance) is available. In the first case the problem is referred to as the discrete BAP and as the continuous BAP in the other case [3]. This paper deals with the continuous type of BAP which has been previously investigated, cf. [2, 4].

2.2 Crane Operations Planning

The charge and discharge operations at a container vessel are performed by so called quay cranes (QCs). Several optimization problems have to be solved while planning the operations of QCs. First, in the Crane Assignment Problem (CAP) cranes must be assigned to the vessels over time. Second, in the Crane Split bay areas are assigned to QCs and the sequences in which cranes process the bays must be determined. Finally, in the Crane Scheduling Problem a detailed schedule for the charge and discharge operations at each bay has to be built. We consider only the CAP, i.e. decide how many QCs must work on each vessel at a certain point in time. Again the port stay times or the delay times are minimized.

2.3 Integration of BAP and CAP

The BAP and the CAP strongly interact. The CAP determines the vessel's port stay time which, at the same time, is an input for the BAP. Moreover, the BAP determines the vessel's time to berth which again is an input for the CAP . Therefore, the integration of both problems, which we refer to as the Berth Allocation & Crane Assignment Problem (BACAP), is particularly focused in the literature, cf. [1, 6].

Fig. 1 shows a feasible solution of an exemplary BACAP instance on the left hand side. In the space time diagram vessels are represented by rectangles with horizontal dimension equal to their length and vertical dimension expressing their port stay time. A gray shaded box within a vessel's rectangle indicates a QC being assigned to the vessel at an associated time t. It can be seen that the crane assignment influences the port stay times and therefore may render the berth plan infeasible, e.g. V_1 is served three additional hours if only four QCs are assigned and thus conflicts with V_2.

2.4 A Resource Oriented Objective Function

The most widespread objectives of terminal service-providers aim at minimizing the port stay times or the delays of vessels. These goals are important to fulfill customer expectations but they do not take the cost of operations into

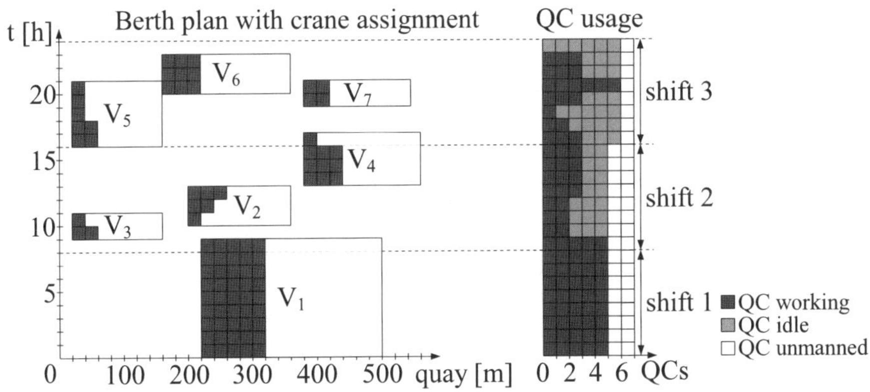

Fig. 1. Berth plan with crane assignment and resulting QC usage

account. As the regional competition of terminal providers grows, however, cost orientation receives increasing importance. For this reason we introduce a new objective function that aims at a reduction of QC idle times. Idle times occur if a QC has been manned by a gang (seven or more workers) for a shift, but is not assigned to a vessel. This can e.g. happen due to the workload fluctuation. Especially difficult are large vessels which are served in parallel by many QCs at the beginning or at the end of a shift.

The occurrence of crane idle time is demonstrated for three consecutive working shifts (0-8, 8-16, 16-24) by the QC usage depicted at the right hand side of Fig. 1. We obtain a demand of five QCs for the first two shifts. However, in the shift 2 these QCs are only required from 8-9 to finish serving vessel V_1. Afterwards three and later two of the manned cranes get idle for the rest of the shift. A similar situation occurs in shift 3 where we observe a maximum demand of six QCs for only one hour.

To achieve a quantitative formulation of the objective function we introduce the following index variables:

t	working hours (enumerated), $t = 1, \dots T$
s	working shift $s = 1, \dots S$ with 8 hours per shift, $s(t) = \lceil \frac{t}{8} \rceil$
i	index of vessel V_i, $i = 1, \dots n$

Every solution of the BACAP provides the number of QCs assigned to V_i at time t denoted as r_{it}. The required number of manned QCs in shift s and the corresponding utilization rate are given by

$$c_s = \max \left\{ \sum_{i=1}^{n} r_{it} \mid t = 8(s-1)+1, 8(s-1)+2, \dots 8(s-1)+8 \right\} \quad (1)$$

$$u = \frac{\text{demanded QC capacity}}{\text{provided QC capacity}} = \frac{\sum_{t=1}^{T} \sum_{i=1}^{n} r_{it}}{\sum_{s=1}^{S} c_s \cdot 8} \quad (2)$$

Since the demanded capacity of QCs is preset, the utilization rate can only be improved by reducing the amount of capacity provided. This leads to the following objective function, which attempts to reduce the provided but unused QC capacity for a finite number of consecutive working shifts.

$$\min \rightarrow c = \sum_{s=1}^{S} \sum_{t=1}^{8} \left(c_s - \sum_{i=1}^{n} r_{i,8(s-1)+t} \right) \tag{3}$$

3 Scheduling Algorithm

This section outlines a heuristic scheduling algorithm for the BACAP which is based on priority-rule methods for the RCPSP. In our approach each vessel is represented by an activity. The required amount of the resource QC is allocated to the vessel activity over its duration. As there are several ways to allocate QCs, different modes of a vessel activity are created. An example is illustrated in Fig. 2 showing four modes of QC allocations for vessel V_4. This vessel requires a total of 11 QC-hours of service. The maximum number of parallel working QCs is three (*dashed line*). Using mode (a) the vessel is served from its berthing time b_4 by this maximum number of QCs which leads to $r_{4,b_4} = r_{4,b_4+1} = r_{4,b_4+2} = 3$ and $r_{4,b_4+3} = 2$. In mode (b) the vessel is served by at most two QCs and thus requires a single one during the last hour. Note that the vessel's port stay time increases from 4 to 6 hours if (b) is used instead of (a). Modes (c) and (d) show patterns with multiple changes of the crane assignment during a serving process. This can be a useful option if e.g. QCs become available during a shift or if a new vessel arrives which has to be served urgently. Many other modes are possible. However, modes with a lot of fluctuation are not welcome because they enforce frequent set-ups for the QCs.

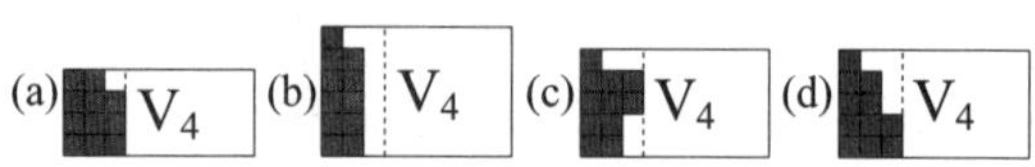

Fig. 2. QC allocation modes for a vessel

In order to take the arrival times of vessels into consideration an additional activity must be included into the project for each vessel. Its processing time represents the lead time (head) for the arrival. This activity requires no resource and is a predecessor activity of the vessel's service activity. Predecessor relationships can also arise among the service activities of vessels, e.g. if a feeder vessel has to be delayed until a deep sea vessel discharged its containers.

To solve such problems heuristically, we apply a simple priority-rule based method. First, for every service activity we create eight QC allocation modes. Next, activities are introduced to ensure the arrival times. All these activities are inserted into the *open set*, which contains all so far unscheduled activities. Further required activity sets are the *active set* (actually processing activities), the *decision set* (activities with all predecessor activities already scheduled) and the *done set* (finished activities).

The procedure repeats the following steps until either the vessels have been scheduled completely or the allowed project duration T is exceeded, where T is set to the latest allowed departure time of all vessels. The project time t is incremented by discrete time steps (e.g. hours). If it turns out that an activity in the *active set* is completed in step t, it is inserted into the *done set*. The *decision set* is updated by inserting those activities of the *open set* which are ready to be scheduled but not yet scheduled. For all modes of these activities the increase of the objective function value is computed by assuming that the vessel is scheduled at time t. Then the activity modes are sorted in ascending order of their contribution to the objective function value increase. The first activity in the sorted set for which appropriate QC capacity and quay space are available is scheduled. The vessel's berthing time is set to t and its crane allocation vector r_{it} is set to the corresponding activity mode. Finally, the activity is deleted from the *decision set* and added to the *active set* and the associated data is updated.

Notice that the above procedure makes decisions regarding the berthing times and positions of vessels as well as the particular service modes to be used. Of course, more sophisticated RCPSP-techniques can be involved, cf. [5], but this is beyond the scope of this paper.

4 Computational Results

We applied the solution method to a real world problem provided by a major CT operator in Germany. The data reflects a period of one week in which 52 vessels had to be served. From this data six instances are generated which respectively include all vessels to be served at two consecutive days. These instances are denoted as I_{d_1/d_2} with d_1 and d_2 as consecutive days.

The scheduling algorithm was implemented in Java. All tests were performed on a PC Pentium 4 with 3.06 GHz. Table 1 compares the schedules which have been manually generated in practice with the solutions found by the proposed algorithm. The results include the idle times of QCs c and the resulting utilization rate u, compare Equations (3) and (2). Furthermore, the average departure time d is listed to observe the impact of the used objective function on the port stay times. The computational time lies below one second for all six instances and therefore, it is not shown in the table.

It can be seen that for each instance c decreases and therewith u increases. For $I_{1/2}$ and $I_{6/7}$ even the average departure time is shortened. At first glance the proposed approach appears promising although a careful analysis is necessary to really understand the interaction of the potentially conflicting ob-

Table 1. Computational results

	n	Manually found solution			Scheduling algorithm			Relative deviation		
		c [QC-hrs.]	u [%]	d [hrs.]	c [QC-hrs.]	u [%]	d [hrs.]	c [%]	u [%]	d [%]
$I_{1/2}$	13	79	71	31.7	47	80	30.8	−41	+13	−3
$I_{2/3}$	12	49	69	28.8	41	73	29.0	−16	+ 6	+1
$I_{3/4}$	16	65	77	28.7	41	84	29.3	−37	+ 9	+2
$I_{4/5}$	14	103	75	29.4	39	87	30.8	−62	+16	+5
$I_{5/6}$	11	81	76	38.0	57	82	38.4	−30	+ 8	+1
$I_{6/7}$	18	96	68	32.7	49	81	32.5	−49	+19	−1

jectives. If successful, further research will concentrate on the incorporation of powerful improvement heuristics to solve larger problem instances.

5 Conclusions

The paper introduced a new objective function for the integrated BACAP occurring at seaport CTs. It considers the terminal operator's labor cost by minimizing the idle time of QCs. The problem was solved heuristically by a priority-rule based method. First computational tests came along with good results encouraging further research in the field.

References

1. Daganzo C F (1989) The crane scheduling problem, Transportation Research B 23/3: 159–175
2. Guan Y, Cheung R K (2004) The berth allocation problem: models and solution methods, OR Spectrum 26: 75–92
3. Imai A, Sun X, Nishimura E, Papadimitriou S (2005) Berth allocation in a container port: using a continuous location space approach, Transportation Research B 39: 199–221
4. Kim K H, Moon K C (2003) Berth scheduling by simulated annealing, Transportation Research B 37: 541–560
5. Neumann K, Zimmermann J (2000) Procedures for resource leveling and net present value problems in project scheduling with general temporal and resource constraints, European Journal of Operational Research 127: 425–443
6. Park Y M, Kim K H (2003) A scheduling method for berth and quay cranes, OR Spectrum 25: 1–23
7. Steenken D, Voß S, Stahlbock R (2004) Container terminal operation and operations research - a classification and literature review, OR Spectrum 26: 3–49
8. Vis I F A, de Koster R (2003) Transshipment of containers at a container terminal: an overview, European Journal of Operational Research 147: 1–16

Simulation der Supply Chain für Offshore-Wind-Energie-Anlagen

Sebastian Gabriel, Carsten Boll

Institut für Seeverkehrswirtschaft und Logistik, Abt. Planungs- und Simulationssysteme, Stresemannstr. 46, 27570 Bremerhaven, Tel.: +49(471)140-440, E-Mail: gabriel@isl.org, boll@isl.org

1 Ausgangslage

Die deutsche Bundesregierung verfolgt in ihrer Offshore-Strategie das Ziel, bis zum Jahr 2030 25.000 Megawatt Windkraftleistung im Meer zu realisieren [1]. Unter anderem auf Grund von Naturschutzbestimmungen liegen die potenziellen Standorte für Windparks überwiegend in der Ausschließlichen Wirtschaftszone (AWZ) Deutschlands, so dass bedingt durch den relativ großen Abstand zur Küste in einer Wassertiefe von über 25 Metern gegründet werden muss. Derzeit liegen dem Bundesamt für Seeschifffahrt und Hydrographie (BSH) 27 Anträge für die Errichtung von Offshore-Windparks in der AWZ vor. In Zusammenhang mit der großen Wassertiefe und dem Küstenabstand sind aus Wirtschaftlichkeitsgründen Wind-Energie-Anlagen (WEA) der 5 Megawatt-Klasse für den Betrieb in den Windparks vorgesehen, die sich zur Zeit jedoch noch im Probebetrieb befinden.

Die Errichtung eines Offshore-Windparks, bestehend aus den Grundbestandteilen Gründungskörper, Transformator, Seekabel und WEA, stellt die Logistik insbesondere unter den genannten Bedingungen vor große Herausforderungen. Allein die Massen der Komponenten einer WEA der 5 Megawatt-Klasse (Turmsegment ca. 150 t, Rotornabe ca. 60 t, Rotorblatt ca. 20 t) verdeutlichen die Anforderungen an die Logistik. Bei den in der Supply Chain fließenden Gütern bzw. Logistikobjekten handelt es sich somit nicht um standardisierte Ladeeinheiten, sondern um extrem großvolumige und schwere Güter, die außerdem eine hohe Kapitalbindung bedeuten. Einen weiteren wichtigen Aspekt stellt der Einfluss des Wetters auf die Logistikkette bzw. Supply Chain dar. Der Transport und insbesondere die Montage auf See können nur bis zu einer bestimmten Wellenhöhe und Windstärke durchgeführt werden. Neben der Restriktion, dass die Errichtung eines

Windparks lediglich in einem beschränkten Zeitfenster innerhalb eines Jahres vorgenommen werden kann, besteht eine stetige Unsicherheit über Auftritt und Länge von wetterbedingten Unterbrechungen der Prozesse innerhalb der Supply Chain. Diesen Aspekten muss in einem schlüssigen Logistikkonzept Rechnung getragen werden. Auf Grund der Untersuchungsrelevanz des zeitlichen Verlaufs der Supply Chain-Zustände und des Einflusses stochastischer Elemente bietet sich für die Untersuchung der möglichen Supply Chains von den Produzenten der Zulieferteile bis hin zur Montage auf See die Anwendung der ereignisdiskreten, stochastischen Simulation an [2].

2 Das Simulationsmodell

Das Simulationsmodell „Wind-Energie-Anlagen Logistiksimulation" (WEA-LogSim) wurde im Rahmen eines Kooperationsprojektes des Instituts für Seeverkehrswirtschaft und Logistik (ISL) und der Logistik-Service-Agentur (LSA) entwickelt. Als Entwicklungsumgebung wurde das objektorientierte Simulationswerkzeug eM-Plant 7.0 von Tecnomatix verwendet. Zielsetzung des Projektes war die Erstellung einer Simulationsanwendung zur Modellierung und Überprüfung möglicher Supply Chains für die Errichtung von WEA in Offshore-Windparks, mit der die Projektbeteiligten bei der Planung und Gestaltung ihrer logistischen Netzwerke sowie bei der Konfiguration ihrer Prozesse und Standortstrukturen unterstützt werden können. Besonderes Augenmerk wurde dabei auf die Umschlag-, Produktions- und Montageprozesse gelegt, die detailliert über Teilprozesse mit ihrem Zeit- und Ressourcenbedarf (Inputfaktoren) beschrieben werden können [3].

2.1 Modellparameter

Das Simulationsmodell ermöglicht dem Benutzer die Modellierung unterschiedlich konfigurierter Supply Chains bzgl. der:

- Definition und Parametrisierung der WEA-Komponenten und Zulieferteile
- Standortwahl und -parametrisierung
- Definition und Parametrisierung der Ressourcen
- Definition der logistik- und fertigungsbezogenen Prozesse
- Auftragsvergabe

Definition und Parametrisierung der WEA-Komponenten und Zulieferteile

Je nach den zu untersuchenden WEA-Typen und der zu modellierenden Fertigungstiefe können (als die in dem Simulationsmodell fließenden Logistikobjekte)

WEA-Komponenten und Zulieferteile definiert und über ihren Wert parametrisiert werden.

Standortwahl und -parametrisierung

Bei der Erstellung einer Supply Chain müssen den Supply Chain Leistungsstellen (im Weiteren Bausteine genannt) Zulieferer, Hersteller, Umschlagplatz, Lager, Offshore-Basishafen sowie Windpark entsprechende Standorte in dem in der Simulation hinterlegten Wege- bzw. Standortnetzwerk zugeordnet werden. Zudem muss jeder neu eingerichtete Baustein parametrisiert werden. Je nach Baustein sind u.a. die Kapazitäten und Kosten des Eingangs- und des Ausgangslagers sowie die Ressourcenausstattung zu bestimmen. Abhängig von der Funktion des Bausteins können zuvor baustein- und komponentenspezifische Prozesse angewählt werden, die in dem Baustein ausgeführt werden.

Definition und Parametrisierung der Ressourcen

Die Ressourcen bzw. Potenzialfaktoren teilen sich in die Gruppen Personal und Arbeitsmittel auf. Ihnen können Schichtpläne sowie fixe und variable Kosten zugeordnet werden. Falls eine Ressource der Gruppe Arbeitsmittel im Bedarfsfall angemietet werden soll, müssen fixe und variable Leihkosten festgelegt werden.

Bei den Transportmitteln, die für den Logistikprozess Transport eingesetzt werden und somit nicht einem Baustein zugeordnet sind, sind die durchschnittliche Geschwindigkeit, die Verteilungsparameter der zeitlichen Verspätung ihrer Bereitstellung und gegebenenfalls die maximal zulässige Windstärke und Wellenhöhe zu bestimmen, bei denen sie eingesetzt werden können.

Definition der logistik- und fertigungsbezogenen Prozesse

Die logistik- und fertigungsbezogenen Prozesse, die in den Bausteinen ausgeführt werden, werden über ihre Teilprozesse und deren jeweiligen Ressourcen- und Zeitbedarf definiert. Der jeweilige Zeitbedarf eines Teilprozesses kann durch eine stochastische Verteilung und deren Parameter beschrieben werden. Zu den bausteinbezogenen Prozessen gehören die Umschlagprozesse sowie die Produktions- bzw. Montageprozesse. Die Montage- und Umschlagprozesse im Windpark können wie die Verkehrsmittel Wellenhöhen- und Windstärkenrestriktionen unterworfen werden.

Auftragsvergabe

Die Auftragsvergabe erfolgt über einen zentralen Auftragsmanager, in dem sämtliche Produktions- bzw. Montageaufträge sowie Transportaufträge terminiert und von dort ausgelöst werden.

Der Einfluss des Wetters wird durch ein separates Wettermodul determiniert. Auf Grundlage von über einen Zeitraum von einem Jahr von der Forschungsstation Fi-no I aufgezeichneten Wetterdaten bzw. der Wellenhöhe (Seegang) und Windstärke erzeugt das Wettermodul in halbstündigen Abständen zufällig generierte Ausprägungen der aktuellen Wellenhöhe und der Windstärke, die ggfs. die Ausführung von Transport- und Montageprozessen beeinflussen.

Die Simulationszeit kann dabei in beliebige Zeitintervalle aufgeteilt werden, in denen die jeweils verwendeten Parameter im Wettermodul angepasst werden. Die Parameter beinhalten die durchschnittliche Wellenhöhe und Windstärke, die Varianz der Wellenhöhen- und der Windstärkedifferenz zwischen zwei aufeinander folgenden Werten sowie die jeweils maximale und minimale Wellenhöhe bzw. Windstärke. Die jeweilige Wetteränderung ist dabei eine gemäß den Parametern normalverteilte Zufallsvariable, um die die vorherige Wetterausprägung unter bestimmten Nebenbedingungen erhöht bzw. erniedrigt wird. Hierbei wird gewährleistet, dass die Differenz zwischen zwei aufeinander folgenden Ausprägungen der Wellenhöhe und der Windstärke eine bestimmte Toleranzbreite nicht überschreitet und somit unrealistisch hohe kurzfristige Schwankungen vermieden werden.

2.2 Aufbau und Ablauf der Simulation

Das Supply Chain Modell teilt sich in zwei bzw. drei Ebenen auf (vgl. Fig. 1).

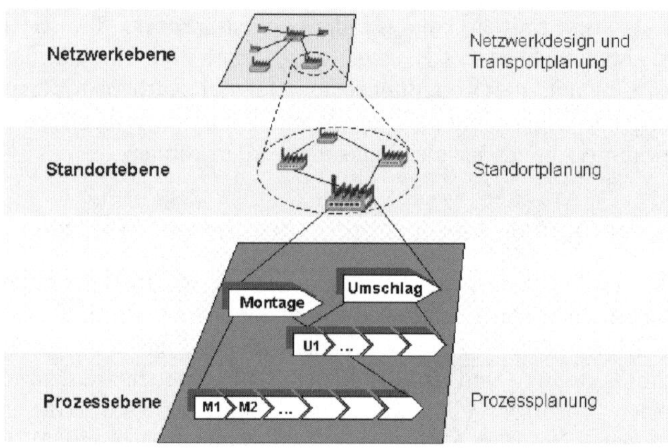

Fig. 1. Ebenen des Simulationsmodells

Die Netzwerkebene beschreibt die räumliche Zuordnung der Bausteine bzw. ihre Position innerhalb des hinterlegten Wegenetzwerkes und damit die Struktur des zu modellierenden Systems Supply Chain (vgl. Fig. 2) [4].

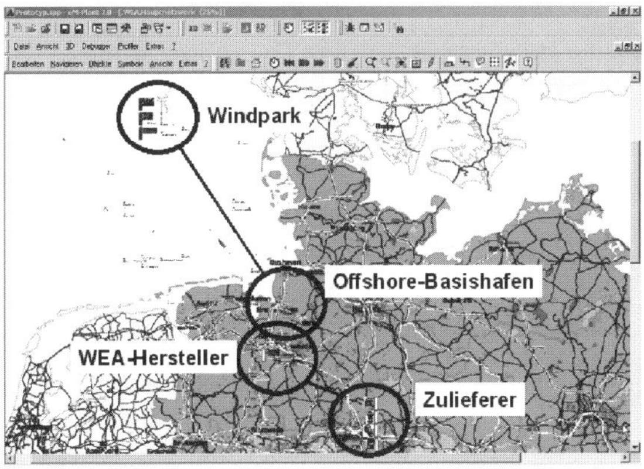

Fig. 2. Struktur einer Supply Chain für Offshore-WEA

Dementsprechend findet in dieser Ebene das räumliche Netzwerkdesign bzw. die Transportplanung statt. Durch die Unterebene Standortebene besteht die Möglichkeit innerhalb eines Offshore-Basishafens ein eigenes Unternetzwerk aus Bausteinen zu erstellen. Analog zu der (Haupt-) Netzwerkebene können auch hier Bausteine innerhalb eines Wegenetzwerkes räumlich angeordnet werden. Die Prozessebene beschreibt die Planung der bausteinspezifischen Prozesse.

Für die Erstellung einer neuen Supply Chain erfolgt ein Großteil der Definitionen und Parametrisierungen in dem eigenständigen, selbstentwickelten Eingabe- und Prozessbeschreibungstool Prozessingenieur. Dies beinhaltet die Definition und Parametrisierung der WEA-Komponenten und Zulieferteile, die Definition und Parametrisierung der Ressourcen sowie die Definition der Prozesse. Die Standortwahl und –parametrisierung erfolgt in der Simulationsanwendung selbst. Das Programm Prozessingenieur ist zudem für die Auswertung eines Simulationslaufs verantwortlich. Wie in Fig. 3 dargestellt erfolgt der Datentransfer zwischen dem Prozessingenieur und der Simulationsanwendung über die datenbanksystemunabhängige Datenbankschnittstelle Open Database Connectivity (ODBC).

2.3 Auswertung der Simulation

Zur Bewertung eines Simulationslaufs einer Supply Chain können die entstandenen Logistikkosten (Transport-, Umschlag- und Lagerkosten), die Wartezeiten der Produktions-, Montage- und Transportaufträge sowie die Auslastung der Ressourcen berechnet werden. Entsprechend können unterschiedliche Supply Chain Konfigurationen verglichen und Engpässe u.a. bzgl. der Lagerkapazitäten sowie der Ressourcenausstattung identifiziert werden. Des Weiteren können unterschiedliche Wetterszenarien gestestet und ihre Auswirkungen auf die Performance der Supply Chain bewertet werden. Damit bietet sich das Modell insbesondere für die simulative Risikoanalyse möglicher Supply Chains an.

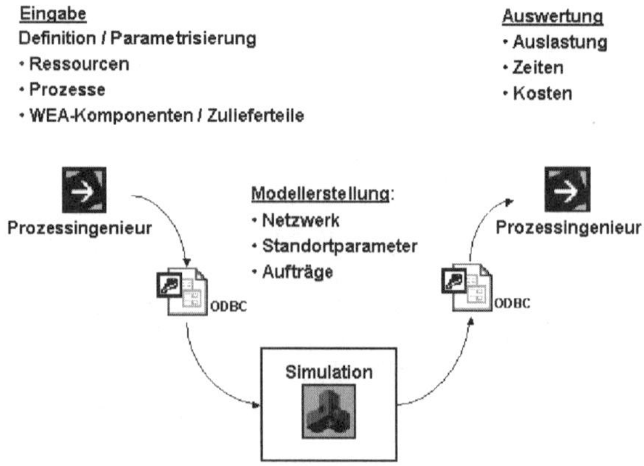

Fig. 3. Ablauf eines Simulationslaufs

3 Ausblick

Die Simulation kann sowohl von Projektplanungsgesellschaften, Herstellern und Standortentwicklern, als auch von Logistikdienstleistern oder Versicherungen für die Überprüfung und Anpassung ihrer Planungen eingesetzt werden. Das Simulationsmodell unterstützt dabei eine reibungslose Projektsteuerung und -abwicklung, die Erarbeitung branchenorientierter Standortangebote, die Angebotserarbeitung von Projektbeteiligten, die Organisation der Hafenprozesse sowie die Identifizierung und Bewertung von Projektrisiken.

Literatur

[1] o.V. (2002) Strategie der Bundesregierung zur Windenergienutzung auf See im Rahmen der Nachhaltigkeitsstrategie der Bundesregierung
[2] Law, Averill M., Kelton, W. David (2000) Simulation and Modeling Analysis. McGraw-Hill, Boston
[3] Delfmann, Werner (1997) Analyse und Gestaltung integrierter Logistiksysteme auf der Basis von Prozessmodellierung und Simulation. In: Wildemann, Horst (Hrsg.) Geschäftsprozessorganisation. TCW-Verlag, München
[4] Gudehus, Timm (2004) Logistik. Springer, Berlin

Modeling and Solution for Yard Truck Dispatch Planning at Container Terminal

Hua-An Lu and Jing-Yi Jeng

Dept. of Shipping & Transportation Management, National Taiwan Ocean University. #2, Pei-Ning Road, Keelung 202, Taiwan

Abstract

This research addresses the dispatch planning of yard trucks to support transferring tasks of containers between shipside and yard areas. With considering the fundamental operations of fixed number of trucks serving one quay crane, we discuss the possible serving status of yard trucks in the connection of task by task. Upon these limited connection patterns, we formulate a min-max nonlinear integer programming model for dispatching yard trucks to efficiently complete the system works. A heuristic approach with two phases of selecting locations for stored or picked containers then assigning served trucks is suggested. We apply it with four proposed principles of location assignment on the analysis of the real-world cases from a dedicated terminal. The results show that the closest position assignment principle is better in the larger scale problems. However, more trucks than the critical number can not increase the performance.

Keywords: yard trucks, quay crane, yard area, min-max nonlinear integer programming model

1. Introduction

Except the efficiency of quay cranes, the performance of container terminal at port still has something to do with the logistical operations of transferring containers. Dispatching transit vehicles between shipside and yard areas plays one of the most important roles in this process system. Since there are different characteristics with the deployed equipments for container moves, we focus on yard truck (YT) operation which is one of the popular transit vehicle systems besides automatic guide vehicle (AGV) and straddle carrier (SC).

Some studies have contributed on the overview of container terminal operations (Vis and Koster 2003; Steenken et al. 2004). As to the YT, Bish et al. (2001) have proved the problem of vehicle dispatching between shipside and yard is NP-hard. Bish (2003) furthermore discussed how to optimize scheduling multiple vehicles to support multiple vessels. Nishimura and Imai (2000) formulated a mathematical model to deal with the joint dispatches of YTs in the public terminal. They proposed a genetic algorithm to solve the supporting pattern in simulation of the real-world cases, but the flexibility of truck assignment was not discussed.

However, little research has dealt with the fundamental dispatching problem of YTs by the systematic analysis method. In this paper, we try to formulate the mathematical model from the basic loading status of YTs and connecting patterns of shipside tasks. The solving heuristics and some tests are also discussed.

2. Problem Description

A truck is either in loaded or empty status during the outbound or inbound transitions in stevedoring operations. There are four kinds of connected patterns of the consecutive tasks to a yard truck.

1. Discharging followed by discharging
   YT loads the container moving from shipside to yard area with loaded status then comes back to the shipside yard in empty status. It successively accepts another container from the gantry crane to repeat the movement of last task.
2. Discharging followed by loading
   YT loads the container moving from shipside to yard area with the loaded status, but directly goes to another storage location with empty status for another outbound container. It is loaded at turning to the shipside.
3. Loading followed by discharging
   YT goes to the shipside loaded with the export container then accepts inbound container to move to the yard area for storage. This connected pattern keeps consecutive tasks in loaded status without empty tours.
4. Loading followed by loading
   This is a revised cycle of discharging by discharging. YT loads a container to shipside but comes back to the yard area in empty status for next task.

The connected assemblies of consecutive tasks depend on the sequences of loading and discharging containers. The task assignment of each truck can have many choices according to the number of supporting trucks. Meanwhile, multiple YTs will be grouped up to exclusively served the operations of a specific quay crane until finishing its loading and discharging tasks. In this paper, we deal with the planning of dispatching multiple YTs to support stevedoring operations of one quay crane with known and consistent moving speed between ship bays.

3. Model Formulation and Solving Algorithm

To refer each task of crane loading or discharging as a served job, the set of jobs is denoted as $\boldsymbol{J}$. The number of $\boldsymbol{J}$ is $|\boldsymbol{J}|$ with the index of j, $j = 1, 2, 3, \ldots, |\boldsymbol{J}|$. $j = 0$ specially means the initial status. We also denote the set of YTs as $\boldsymbol{K}$ and the set of yard bays available for storage or pick-up as $\boldsymbol{L}$. Meanwhile, the set $\boldsymbol{J}_l$ represents all suitable jobs to be stored in bay l, while $\boldsymbol{L}_j$ is the set of all bays which can store the container of job j. For distinguishing the attributes of each job, we let the set of discharging jobs as $\boldsymbol{D}$ and the set of loading as $\boldsymbol{V}$. Decision variables are included:

$e_{jj'}$: time expenditure in empty status of truck executing job j by j',

f_j : time expenditure in full-load status of truck executing job j,

$s^k_{j'j}$: truck k executing job j after finishing job j' or not, 1 for yes, 0 otherwise,

t^k_j : accumulated working time of truck k after finishing job j, but $t^k_0 = 0, \ \forall k$,

x^k_j : job j assigned to truck k or not, 1 for yes, 0 otherwise, and

y_{jl} : job j stored in yard l to load or discharge or not, 1 for yes, 0 otherwise.

Other parameters are:

w_j : waiting time of truck staying under quay crane in executing job j, normally it is randomly changed with the cooperation between crane and truck,

ψ_j : waiting time of truck operating at yard bay in executing job j, normally it is randomly changed with the cooperation between yard handling equipment and truck,

U_l : stored capacity of yard l for discharging jobs,

g_l : time expenditure of truck moving from shipside to yard l,

r_l : time expenditure of truck moving from yard l to shipside, and

$\tau_{ll'}$: time expenditure of truck moving from yard l to yard l'.

The mathematical model is formulated as follows.

$$Min. \ \underset{k, j=|\boldsymbol{J}|}{Max.} \ t^k_j \tag{1}$$

$$s.t. \ \sum_k x^k_j = 1 \quad \forall j = 1,2,\ldots,|\boldsymbol{J}| \tag{2}$$

$$\sum_{l \in L_j} y_{jl} = 1 \quad \forall j = 1,2,\ldots,|\boldsymbol{J}| \tag{3}$$

$$\sum_{j \in J_l} y_{jl} \le U_l \quad \forall l \tag{4}$$

$$\sum_{j'=0}^{j-1} s^k_{j'j} = x^k_j \quad \forall j = 1,2,\ldots,|\boldsymbol{J}|, \ k \tag{5}$$

$$2 s^k_{j'j} \le x^k_{j'} + x^k_j \quad \forall j' = 1,2,\ldots,|\boldsymbol{J}|-1, \ \ j = j'+1,\ldots|\boldsymbol{J}|, \ k \tag{6}$$

$$f_j = \begin{cases} \sum_l g_l \ y_{jl} & \text{if } j \in \boldsymbol{D} \\ \sum_l r_l \ y_{jl} & \text{if } j \in \boldsymbol{V} \end{cases} \qquad \forall j = 1,2,...,|\boldsymbol{J}| \tag{7}$$

$$e_{jj'} = \begin{cases} \sum_l g_l \ y_{jl} & \text{if } j \in \boldsymbol{V}, \ j' \in \boldsymbol{V} \\ \sum_l r_l \ y_{jl} & \text{if } j \in \boldsymbol{D}, \ j' \in \boldsymbol{D} \\ 0 & \text{if } j \in \boldsymbol{V}, \ j' \in \boldsymbol{D} \\ \sum_{(l,l')} \tau_{ll'} y_{jl} y_{j'l'} & \text{if } j \in \boldsymbol{D}, \ j' \in \boldsymbol{V} \end{cases} \qquad \forall j = 1,2,...,|\boldsymbol{J}|-1, \quad j' = j+1,...,|\boldsymbol{J}| \tag{8}$$

$$t_j^k = t_{j-1}^k + \sum_{j'=0}^{j-1} s_{j'j}^k \ e_{jj'} + f_j \ x_j^k + \psi_j \ x_j^k + w_j x_j^k \qquad \forall k, j = 1,...,|\boldsymbol{J}| \tag{9}$$

$$x_j^k \in \{0, \ 1\} \qquad \forall j = 1,2,...,|\boldsymbol{J}|, \ k \tag{10}$$

$$y_{jl} \in \{0, \ 1\} \qquad \forall j = 1,2,...,|\boldsymbol{J}|, \ l \tag{11}$$

$$s_{j'j}^k \in \{0, \ 1\} \qquad \forall j' = 0,...,|\boldsymbol{J}|-1, \quad j = j'+1,,...,|\boldsymbol{J}|, \ k \tag{12}$$

$$f_j \geq 0 \qquad \forall j = 1,2,...,|\boldsymbol{J}| \tag{13}$$

$$e_{jj'} \geq 0 \qquad \forall j = 1,2,...,|\boldsymbol{J}|-1, \quad j' = j+1,...,|\boldsymbol{J}| \tag{14}$$

$$t_j^k \geq 0 \qquad \forall j = 1,2,...,|\boldsymbol{J}|, \ k \tag{15}$$

Objective function (1) minimizes the last finished time of trucks after jobs assignment to every YTs. Equation (2) limits each job can only assign to a truck. Equation (3) allows each job can store in or extract from the one of the available yards. Equation (4) is the capacity limitation for the number stored in each yard. Equations (5) and (6) keep the exact connection of sequential served jobs for each truck. Equations (7) and (8) describe the moving time of assigned truck in full-load and empty status. Equation (9) records the accumulative finished time of assigned jobs before each one completion for each truck. Equations (10) to (15) are the variable constraints.

A two-phase heuristics is proposed to solve this problem. We first decide the locations for all of jobs then assign the served trucks with following the sequential jobs. In the first part, four principles judged by the traveling time of trucks between shipside and stored bays are suggested to assign.

1. Closest position (P1): This principle selects the nearest possible bay with available capacity for each job.
2. Farthest position (P2): This principle selects the farthest possible bay with available capacity for each job.
3. Closest by Farthest (P3): This principle alternately selects the nearest and farthest possible bays with available capacities for all of jobs.
4. Random selection (P4): This principle randomly selects the possible bay with available capacity for each job.

In the second part, we assign the truck that can arrive earliest the initial position of the considered job after finishing the last one to serve it. While more than one truck can be considered, the truck with the less accumulative service time takes a prior assignment. We pile up the whole algorithm as follows.

Step 0 Input all of initial data. If there are n containers, let job $j = 1$.

Step 1 Assign the bay position of job j according to the applied principle. Reduce one capacity of assigned bay, $j = j + 1$.

Step 2 If $j > n$, go to *Step* 3. Otherwise, go to *Step* 1.

Step 3 Let $j = 1$.

Step 4 If $j > n$, stop and output the assigned results. Otherwise, go to *Step* 5.

Step 5 Let truck $k = 1$, earliest time $t = \infty$, assigned truck $d = 0$, the least accumulative service time $q = \infty$.

Step 6 Let the accumulative service time of truck $k = s$ and the time of truck k to arrive the initial position of the current job $j = r$. Calculate $s + r$.

Step 7 If $s + r < t$ or $s + r = t$, $s < q$, then $d = k$, $t = s + r$, $q = s$.

Step 8 If all of trucks are compared, assign truck d to job j then go to *Step* 9. Otherwise, $k = k + 1$, go to *Step* 6.

Step 9 Update the accumulative service time of truck d included the waiting time in shipside or yard area operations, $j = j + 1$, go to *Step* 4.

4. Numerical Examples

We applied the algorithm to some cases provided from the studied company operating at the wharf 70 in port of Kaoshiung. Table 1 displays the data of these cases. Basically, the company deployed 6 trucks to support one crane operations.

Table 1. Data for test cases

Case	Total Moves	Discharged Moves	Loaded Moves	Remarks
1	117	67	50	Outbound containers almost loaded after discharging
2	189	21	168	Outbound containers almost loaded after discharging
3	190	71	119	Outbound containers almost loaded after discharging

The real working times provided from the studied company include the interruption from the short meal duration, unexpected breakdown of the crane and the irregular waiting of crane operations in shipside and yard. For adjusting to the same basis of comparison in continuous handling, we deduct the longer waiting hours and the extra handing times beyond the maximum range of randomly handling times that provided by the studied company as the estimated plan interval. We can find from Table 2 that our truck dispatches are equivalent in four principles of location assignment, and are better than the real and estimated.

The consistent solving results imply that the size of truck fleet is too large to absorb the effect of various assignment principles. We further made the sensitivity analysis to the size of trucks. The results shown from Fig. 1 can conclude that the first principle, i.e. to assign the closest position, has a better truck dispatch than others. The working time will converge at different levels of the fleet size, while truck fleet is more than 6 vehicles.

Table 2. Solving results for the real-world cases

Unit: Minute

Case	Location assigned Principles				Real working time	Estimated planned time
	P1	P2	P3	P4		
1	188	188	188	188	279	188
2	309	309	309	309	380	311
3	309	309	309	309	443	347

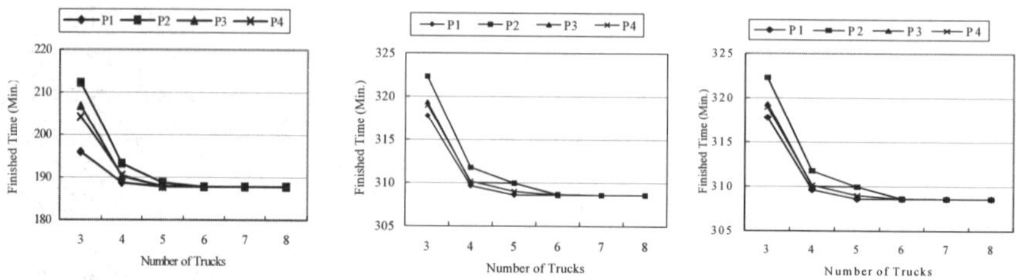

Fig. 1. Fleet analysis for case 1, 2, and 3

5. Conclusions and Suggestions

This research has focused on the basic problem of dispatching YTs supporting one quay crane to formulate the mathematical model and to develop a heuristics. We divided the problem into location assignment of each task and truck sequential assignment. Four principles of location assignment were suggested. From tests of the studied company cases, to assign the closest position had a better effects in general size of trucks. However, their dispatches were indifferent while the fleet of trucks is more than a critical number of vehicles. These results can provide the reference to the further studies on the complex dispatches of YT operations.

References

Bish EK, Leong TY, Li CL, Ng JWC, Simchi-Levi D (2001) Analysis of a new vehicle scheduling and location problem. Naval Research Logistics 48:363-385

Bish EK (2003) A multiple-crane-constrained scheduling problem in a container terminal. European Journal of Operational Research 144:83-107

Nishimura E, Imai A (2000) Optimal vehicle routing for container handling in a multi-user container terminal. In: Proceedings of the international seminar on global transportation network, Kobe, pp 123-132.

Steenken D, Voβ S, Stahlbock R (2004) Container terminal operation and operations research – a classification and literature review. OR Spectrum 26:3-49

Vis IFA, Koster R (2003) Transshipment of containers at a container terminal: an overview. European Journal of Operational Research 147:1-16

Strategic Tools for the Sustainable Development of Maritime Regions

Hans-Dietrich Haasis and Oliver Möllenstädt

Abstract

In line with the sustainable development of maritime regions and the design of new maritime businesses strategic management tools can be applied. These tools take into account regional characteristics as well as new value added processes within supply chains. Moreover they are based on the concept of knowledge regions. A knowledge region means a better communication and knowledge management between enterprises and decision makers in a region. Knowledge management is used to improve the personal, business process orientated and informational interfaces between decision makers. In this paper these tools are presented; their application within regional decision processes will be demonstrated.

1. Requirements of sustainable development in maritime regions

Maritime regions nowadays face the strong global interregional competition. The competition between ports and terminal operateurs in the different port ranges is only one specific aspect of this development. A more and more differentiated demand for logistics services shows changing requirements of customers in industry and retail business. On the supply side of logistics services a broad range of new business models of service providers is evolving. On the demand side the structures of logistics systems in industry and retail business are changing:

- E-business normally has to distribute large volumes of small loading units to regional distributed customers. These systems work with central logistics centres.
- Other than these businesses, large retailers often use systems of regional distributed warehouses (Kujath 2003).

From a business point of view, warehouse and logistics facilities have to bundle regional and temporal distributed flows of goods. These logistics nodes guarantee an efficient utilisation of the different modes of transportation and infrastructures. The development of such nodes in the international logistics systems is a necessary precondition for

- the appliance of mass-efficient modes of transport,
- the modal shift,
- the use of effects of net and scale and
- a high delivery service.

In the past, seaports had the function of bundling and coordinating transregional freight transportation as logistical nodes. Nowadays seaports stay in the role of centres of cumulation and handling in intermodal global transport chains. In the last years a phase of deep restructuring of freight transport on the European continent has started. This development is motivated in new logistics and supply chain concepts and has been supported by the further integration of the Single European Market. Industry and retailers build up their own logistics systems – more and more independently from traditional or public planned transport structures.

Continuing, Seaports have an interface function in global transport chains. The largest part of global transport will further on be carried though the large ports. But seaports do not automatically offer the best business conditions for the cumulation, packing, further processing and dispersion of goods and products at the same time. The ongoing containerisation makes further dislocation of such value added activities in other regions possible. Logistics processes like containerisation and further processing of goods often can better be carried out efficiently in large consolidation centres in the hinterland than in the urbanized port areas in retaining small time slots (Kujath 2003). New nodes of international logistics systems arise in suburban areas or in periperhies of densely populated areas. So the share of the value added in the port region on the overall value added could decrease in the future. That is the reason why the scientific perspective on seaports and port regions has changed during the last years towards a new paradigm considering ports as elements in value-driven chain systems (Robinson 2002).

For a sustainable development of maritime regions it is essential to ensure that an adequate amount of value adding within the overall value chains, beyond the transport and container handling operations, is carried out directly in the maritime region (Haezendonck/Coeck/Verbeke 2000). Otherwise the economic, ecologic and social disadvantages of transport flows through the region like congestion of transport systems, environmental pollution and space requirements will overbalance the economic benefits (Hesse 1998). For regional development that means not to “sell the region as a hub” or, staining in metaphorical language, seeing its main function as a “container sluice”.

2. The knowledge regions approach

Maritime Regions can stabilise their position in the European logistics systems under conditions of changing logistics and supply chain concepts if they align their logistics compentences with customer requirements.[1] Maritime regions have to drive their development towards the vision of a learning and knowlegde-based economy, which is intensively promoted by the European Union since its Lissabon agenda. In modern regional economics different approaches have been developed to describe the influence of knowledge, learning and innovation on regional economic development. These approaches have been illustrated with titles like 'innovative milieus', 'learning regions', 'regional clusters' or 'regional innovation systems' (Fürst 2002). Cooke states that it is necessary to adapt the regional innovation system for stimulating innovative interaction between global value chains and regional clusters to take account of the strong emphasis of learning that is central to the upgrading process (Cooke 2003). The theoretical approaches describe different facets of the same phenomenon: Regional knowledge networks of businesses, science and regional adminstration enchance the learning and innovation capabilities of regional actors. Further on we will subsume these approaches under the term 'knowledge regions'. A knowledge region means a better communication and knowledge management between enterprises and decision makers in a region. Learning capabilities lead the region to upgrade and adapt its knowledge base, and by this its competences, to the value chain's requirements. Regional knowledge networks and regional knowledge management can make it easier for businesses to face challenges, question traditional routines and to identify new action and decision opportunities. Furthermore regional networks can be an instrument for the necessary harmonisation of economic, social and ecologic interests of the different regional actors (Hesse 1998).

3. Management of knowledge regions

The theoretical description and case studies of knowledge regions lead to the question, how these regions can be designed, how learning processes can be initiated or supported, which governance models are seen to be suitable for the according regional initiatives, in general how successful knowledge regions have to be 'managed' by public and private actors (regional government, business firms etc.). Former approaches of strategic management of martime regions like the framework for functional analysis of port performance presented by Teurelincx are not sufficient for the management of the interaction of the actors in the regional knowledge networks (Teurelincx 2000). These tools concentrate on the strategic analysis of ports as value chain elements and the creation of ideas for

---

[1] Lawson introduces the competence-based theory to regional economics and describes regional competences as a key determinant of the interregional competition (Lawson 1997).

strategic actions but lose sight of other management aspects like strategy implementation and controlling.

The quality of regional management and with it the management instruments differ from the management of business firms and should not be confused with hierarchical decision making. Because of the independency of the autonomous acting regional actors (business firms, scientific institutions, politicians/administration) conventional approaches of strategic planning are not considered to be suitable for the management of a knowledge region. Regions seem to be self-organised and self-directed complex systems. Regional development is a specific kind of evolutionary process. According to systems theory on the one hand, this high system complexity ensures the systems survival. On the other hand it complicates management actions for a goal-oriented development. Evolutionary management approaches estimate that complex systems can be lead by a kind of controlled evolution, which takes into account the self-directing forces of the system (Bleicher 1999). Additional strategic tools supporting this management philosophy, beyond portfolio analysis approaches, are needed.

4. Strategic tools

In the context of project activities of the ISL a phase model for the identification and assessment of measures has been developed. It includes six phases:

- Impulse and kick-off,
- Stakeholder identification and logistics team building,
- Assessment of location performance,
- Identification of measures,
- Estimation of effects and prioritisation and
- Documentation.

Tools for the strategic management of knowledge regions should support feedback processes for strategic controlling. Tools which can be used in different phases of the development of a knowledge region are:

- *Stakeholder analysis:* Regions consist of a system of stakeholder groups. It is necessary to take into account the existence of each external and internal stakeholder group. The stakeholder analysis, developed by Probst/Gomez (1989), identifies all stakeholders of a region and analyses the influenceability of each group. Internal stakeholders of a region are the partners of the knowledge region initiative and are located in the region. This are e.g. regional businesses of the logistics and transport sector. External stakeholders can be businesses, groups of interest, politicians etc. which affect a region from outside. It is helpful to analyse the regional stakeholders at the beginning of a knowledge region development project (Endres 2003).

- *Balanced Scorecard:* The Balanced Scorecard (BSC) is a management approach developed by Kaplan/Norton. In its focus are a common vision and strategy more than hierarchical control. Objectives and indicators from different perspectives are connected. Successful knowledge regions requires clearly definded visions and objectives. The Balanced Scorecard delivers a system of key indicators and clearly defined strategic actions to reach the objectives. The Balanced Scorecard with its system of objectives and the system of key indicators takes into account different perspectives on the knowledge region and its environment. The Balanced Scorecard guarantees a 'balanced' formulated strategy. The Balanced Scorecard is an instrument for strategy development and strategy controlling which is ideal especially for networks. It activates stakeholders and reduces specific problems of the controlling in membership organisations (Endres 2003, Gmür/Brandl 2000).

- *Knowledge Mapping:* Knowledge maps visualise the knowledge of an organisation or a region within all its network relationships. Knowledge Mapping is a tool for exploration of the knowledge base of a single business firm or a region. Knowledge maps are graphical representations of knowledge owners, knowledge assets, knowledge structures and knowledge applications. Knowledge maps can help to estimate the development of knowledge assets or the selection of expert teams. Knowledge maps can help to reduce the cost of acquistion of specific knowledge. Knowledge maps can be developed for the knowledge region and can be an important tool for the implementation of the knowledge exchange processes. The interaction of firms and institutions in the region will be supported by this tool during all phases of a knowlegde regions project.

5. Summary

Maritime regions face the challenges of globalisation. They have to react flexible on changing requirements of global transport and logistics chains. Furthermore ecological and social aspects requirements have to be taken into account in the context of sustainable development in maritime regions. Regional logistics competences can be adjusted by regional learning and innovation processes which are characterised by an intensive interaction of regional actors from economy, administration and science. This interaction can also ensure a harmonisation of different interests of the various groups of interest. Network-based knowledge regions are suitable environments for this interaction. For stimulating these learning processes and managing the knowledge region, new strategic tools are needed. Stakeholder analysis, Balanced Scorecard and Knowledge Mapping techniques support regional learning and innovation processes.

Literature

Bleicher, K. (1999): Das Konzept integriertes Management. Visionen – Missionen – Programme. Frankfurt a.M./New York.

Cooke, P. (2003): Regional Innovation and Learning Systems, Clusters, and Local and Global Value Chains. In: Bröcker, J./Dohse, D./Soltwedel, R. (2003): Innovation Clusters and Interregional Competition. Berlin et al. P. 28-51.

Endres, E. (2003): Qualitätsmanagement beim Aufbau von Lernenden Regionen. In: Elsner, W./Biesecker, A. (Hrsg.): Neuartige Netzwerke und nachhaltige Entwicklung. Komplexität und Koordination in Industrie, Stadt und Region. Frankfurt a. M.

Fürst, D. (2002): Region und Netzwerke. Aktuelle Aspekte zu einem Spannungsverhältnis. In: DIE Zeitschrift für Erwachsenenbildung I/2002. P. 22-24.

Haezendonck, E./Coeck, C./Verbeke, A. (2000): The Competitive Position of Seaports: Introduction of the Value Added Concept. In: International Journal of Martime Economics. Vol. II, number 2. P. 107-118.

Hesse, M. (1998): Raumentwicklung und Logistik. In: Raumordnung und Raumforschung H. 2/3 1998. P. 125-135.

Kujath, H. J. (2003): Logistik und Raum – Neue regionale Netzwerke der Güterverteilung und Logistik. Working Paper. Erkner.

Lawson, C. (1997): Towards a competence theory of the region. ESRC Centre for Business Research, University of Cambridge. Working paper No. 81.

Lehner, F. (2000): Organisational Memory. Konzepte und Systeme für das organisatorische Lernen und das Wissensmanagement. München.

Probst, G./Gomez, P. (1989): Vernetztes Denken. Wiesbaden.

Robinson, R. (2002): Ports as elements in value-driven chain systems: the new paradigm. In: Maritime Policy and Management, Vol. 29, No. 3. P. 241-255.

Teurelincx, D. (2000): Functional Analysis of Port Performance as a Strategic Tool for Strenghtening a Port's Competitive and Economic Potential. In: International Journal of Martitime Economics. Vol. II, number 2. P. 119-140.

Part V

Production & Supply Chain Management

A Two-echelon Model for Inventory and Returns

Allen H. Tai[1] and Wai-Ki Ching[2]

Department of Mathematics, The University of Hong Kong, Pokfulam Road, Hong Kong, China. h0150695@graduate.hku.hk, wkc@maths.hku.hk

Summary. We consider an Markovian model for a two-echelon inventory/return system. The system consists of a supply plant with infinite capacity and a central warehouse for inventory and returns. There is also a number of local warehouses which are also able to re-manufacture products. To obtain a high service level of handling inventory and returns, lateral transshipment of demands is allowed among the local warehouses.

1 Introduction

The efficiency of product/service delivery is one of the major concerns in many industries including re-manufacturing industry. Since customers are usually scattered over a large regional area, a network of locations (local warehouses) for inventory of products and handling of returns is necessary to maintain a high service level. In our study, Lateral Transshipment (LT) is allowed among the local warehouses to enhance the service level. LTs are also very practical in many organizations having multiple locations linked by computers. Substantial savings can be realized by the sharing of inventory in the local warehouses [15]. A number of research publications have been appeared in this area. Kukreja et. al [9] developed a single-echelon and multi-location inventory model for slow moving and consumable products. Moinzadeh and Schmidt [12] studied the emergency replenishment for a single-echelon model with deterministic lead times. Aggarwal and Moinzadeh [2] then extended the idea to a two-echelon model. Ching [5] considered a multi-location inventory system where the process of LT is modelled by Markov-modulated Poisson Process (MMPP). Both numerical algorithm and analytic approximation have been developed to solve the steady-state probability distribution of the system [5, 8]. Lee [10] and Axsäter [3] considered a two-echelon system in which the local warehouses are grouped together. Within the group, the warehouses were assumed to be identical. Simulation study of a two-echelon system can also be found in [14]. Alfredsson and Verrijdt [1] considered a two-echelon inventory system for service parts with emergency supply options in terms of LT and direct delivery. In this paper, we consider an inventory/returns model based on the framework and analysis discussed in [1, 10]. The model of the system consists of a supply plant with infinite capacity, a central warehouse and a number of local warehouses with re-manufacturing capacity. Here we consider a queueing model for a two-echelon inventory system. Queueing model is a useful tool for many inventory models and manufacturing systems [4, 5, 7, 8].

2 The Two-echelon System

In this section, we present a two-echelon system based on the framework discussed in [1, 10]. The system consists of a supply plant with infinite capacity, a central warehouse (maximum capacity C) and n identical local warehouses (each has a capacity of L), see Fig. 1. The arrival processes of both the demands and the returns at each local warehouse are assumed to follow independent Poisson processes with mean rates λ and σ respectively. If the local warehouse is full, the returns will be disposed. Otherwise the returns will be inspected and the inspection time is assumed to be negligible. If the returns are repairable, they will be repaired and ready to be delivered again. If the returns are not repairable, they will be disposed [6]. The probability that a return is repairable is ρ $(0 < \rho < 1)$. Moreover, the demands are satisfied in the following manner:
(a1) An arrived demand is first filled by the stock at the local warehouse and at the same time a replenishment order is issued to the central warehouse. The replenishment time from the central warehouse to a local warehouse is assumed to be exponentially distributed with mean μ^{-1}. The demand is first-come-first-served and the replenishment is one-for-one. In many situations, this replenishment policy has been shown to be optimal [13]. If a repaired return arrives before the replenishment order is satisfied, the replenishment order will be cancelled.
(a2) If a local warehouse is out of stock, an arrived demand will then be satisfied by a LT from another randomly chosen local warehouse and at the same time the local warehouse that makes the LT will issue a replenishment order to the central warehouse. Here in this situation, we assume that the LT time is negligible when compare with the time for a replenishment order from the central warehouse.
(a3) If unfortunately all the local warehouses are also out of stock, then an arrived demand will be satisfied by a direct delivery from the central warehouse and at the same time a replenishment order is issued to the plant from the central warehouse. The replenishment time from the plant to the central warehouse is assumed to be exponentially distributed with mean γ^{-1}.
(a4) Finally, if all the warehouses (including the central warehouse) are out of stock, the demand is satisfied by direct delivery from the plant.

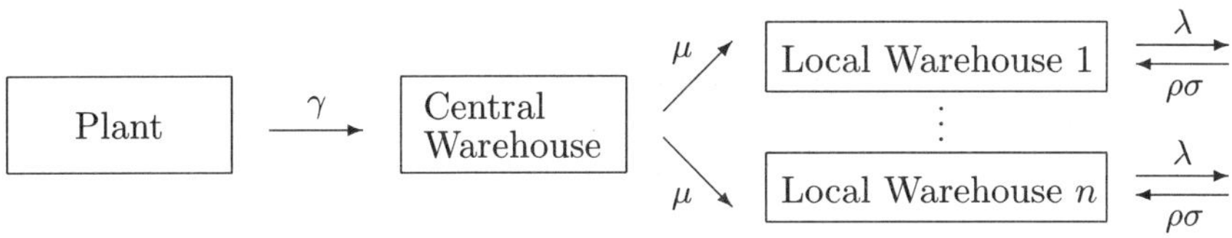

Fig. 1. The Two-echelon Model.

3 The Local Warehouses

In this section, we discuss the service level of each local warehouse. We define the following probabilities:

p, the common starving probability of the local warehouses (the probability that a warehouse has no stock on hand).

β, the common probability that a demand is met from stock on-hand at the local warehouse at which it occurs.

α, the common probability that a demand is met by a LT.

θ, the common probability that a demand cannot be met either from stock at the location at which it occurs or by a LT.

The probability θ can be calculated by grouping all the local warehouses. The aggregated demand arrival rate is $n\lambda$. Assuming low rejection rate of the repairable returns the total repairable return arrival rate can be approximated by $n\rho\sigma$. The maximum inventory capacity is nL. As long as any location has stock, demand can be met and therefore we have (see [4])

$$\theta = \frac{(n\lambda)^{nL}}{\prod_{k=1}^{nL}(k\mu + n\rho\sigma)} \left[1 + \sum_{j=0}^{nL-1} \frac{(n\lambda)^{nL-j}}{\prod_{l=1}^{nL-j}(l\mu + n\rho\sigma)} \right]^{-1}. \tag{1}$$

Obviously we have

$$\alpha + \beta + \theta = 1 \quad \text{and} \quad p = 1 - \beta. \tag{2}$$

To determine all the probabilities, one needs an extra equation. The additional equation can be determined by analyzing the inventory process at a local warehouse. We first determine the rate of demand that a local warehouse experiences while it has stock. Suppose x is the demand that occurs at a location that is met by LT from another location during time t, then one has $nx = n\alpha\lambda t$ and the rate that demand occurs during the time a location has stock due to transshipment to other location is given by $\frac{x}{\beta t} = \frac{\alpha\lambda t}{\beta t} = \frac{\alpha\lambda}{\beta}$. The total demand rate τ at a location conditioned on having stock is given by

$$\tau = \lambda + \frac{\alpha\lambda}{\beta} = \frac{\lambda(\alpha + \beta)}{\beta} = \frac{\lambda(1 - \theta)}{\beta}. \tag{3}$$

The maximum inventory in a local warehouse is L. According to the flow of stock (order) described in (a1)-(a4) of Section 2, the replenishment rate is

$(L-i)\mu+\rho\sigma$ when the inventory level is i. We let $\mathbf{p}=(p_0,p_1,\ldots,p_L)^t$ be the steady-state probability vector for a local warehouse system. Here p_i is the steady-state probability that the inventory level is i. We order the states from 0 to L and obtain the following generator matrix A for the system [5]:

$$A=\begin{pmatrix} * & -\tau & & & 0 \\ -L\mu-\rho\sigma & * & -\tau & & \\ & \ddots & \ddots & \ddots & \\ & & -2\mu-\rho\sigma & * & -\tau \\ 0 & & & -\mu-\rho\sigma & * \end{pmatrix}. \quad (4)$$

Here "$*$" is such that each column sum is zero. The meaning of the generator matrix is that the steady state-probability vector $\mathbf{p}$ satisfies

$$A\mathbf{p}=\mathbf{0} \quad \text{and} \quad \sum_{i=0}^{L} p_i = 1. \quad (5)$$

It can be shown that

$$p_i=\begin{cases} \dfrac{\tau^{L-i}}{\prod_{k=1}^{L-i}(k\mu+\rho\sigma)}\left[1+\displaystyle\sum_{j=0}^{L-1}\frac{\tau^{L-j}}{\prod_{l=1}^{L-j}(l\mu+\rho\sigma)}\right]^{-1} & i=0,\ldots,L-1 \\ \left[1+\displaystyle\sum_{j=0}^{L-1}\frac{\tau^{L-j}}{\prod_{l=1}^{L-j}(l\mu+\rho\sigma)}\right]^{-1} & i=L, \end{cases} \quad (6)$$

see for instance [4]. Hence, we have

$$\beta=\sum_{i=1}^{L}p_i=\left(1+\sum_{i=1}^{L-1}\frac{\tau^{L-i}}{\prod_{k=1}^{L-i}(k\mu+\rho\sigma)}\right)\left[1+\sum_{j=0}^{L-1}\frac{\tau^{L-j}}{\prod_{l=1}^{L-j}(l\mu+\rho\sigma)}\right]^{-1}. \quad (7)$$

Since $\tau=\lambda(1-\theta)/\beta$, if we let $\omega(i)=\lambda^i(1-\theta)^i\left(\prod_{k=1}^{i}(k\mu+\rho\sigma)\right)^{-1}$, then we have

$$f(\beta)=\beta\left(1+\sum_{i=1}^{L}\frac{\omega(i)}{\beta^i}\right)-\left(1+\sum_{i=1}^{L-1}\frac{\omega(i)}{\beta^i}\right)=0. \quad (8)$$

The following proposition gives the condition for the probability β to be unique.

Proposition 1. *The function $f(\beta)$ has exactly one zero in $(0,1)$ if $\mu+\rho\sigma\geq\lambda$.*

By solving (1), (2) and (8), one can obtain all the probabilities β,α,θ and p. Unfortunately there is no analytic solution for the probability β. However, one may apply the bisection method [11] to obtain an approximated solution. By using the bisection method, we compute the probability β, and hence the probabilities θ and α for different values of L,n,λ,μ and $\rho\sigma$ in Table 1.

Table 1. Probabilities β, θ, α of the system for $\lambda = 5$ and different μ, n and L

	$\mu = 4, \quad \rho\sigma = 1$			$\mu = 1, \quad \rho\sigma = 4$		
	$L = 2$	$L = 4$	$L = 8$	$L = 2$	$L = 4$	$L = 8$
	$\beta = 0.7263$	$\beta = 0.9763$	$\beta = 0.9999$	$\beta = 0.5940$	$\beta = 0.8749$	$\beta = 0.9952$
$n = 2$	$\theta = 0.1069$	$\theta = 0.0017$	$\theta = 0.0000$	$\theta = 0.1659$	$\theta = 0.0288$	$\theta = 0.0001$
	$\alpha = 0.1667$	$\alpha = 0.0220$	$\alpha = 0.0000$	$\alpha = 0.2401$	$\alpha = 0.0962$	$\alpha = 0.0047$
	$\beta = 0.6883$	$\beta = 0.9762$	$\beta = 0.9999$	$\beta = 0.5063$	$\beta = 0.8639$	$\beta = 0.9952$
$n = 4$	$\theta = 0.0377$	$\theta = 0.0000$	$\theta = 0.0000$	$\theta = 0.0812$	$\theta = 0.0040$	$\theta = 0.0000$
	$\alpha = 0.2739$	$\alpha = 0.0238$	$\alpha = 0.0000$	$\alpha = 0.4125$	$\alpha = 0.1321$	$\alpha = 0.0048$
	$\beta = 0.6708$	$\beta = 0.9762$	$\beta = 0.9999$	$\beta = 0.4501$	$\beta = 0.8621$	$\beta = 0.9952$
$n = 8$	$\theta = 0.0071$	$\theta = 0.0000$	$\theta = 0.0000$	$\theta = 0.0330$	$\theta = 0.0001$	$\theta = 0.0000$
	$\alpha = 0.3220$	$\alpha = 0.0238$	$\alpha = 0.0000$	$\alpha = 0.5169$	$\alpha = 0.1378$	$\alpha = 0.0048$

4 The Aggregated Model for the Central Warehouse

In this section, we propose an aggregated inventory model for the central warehouse and the local warehouses by aggregating the demand and inventory of the local warehouses, see Fig. 2. The aggregated demand for the central warehouse is the superposition of all the Poissonian demand from the local warehouses which is still a Poisson process with mean rate $n\lambda$. Again we assume a low rejection rate of repairable returns, therefore the total repairable return arrival rate can be approximated by $n\rho\sigma$. In the model, we let $x(t)$ be the inventory at the central warehouse and $y(t)$ be the total inventory level of the sum of all the local warehouses at time t.

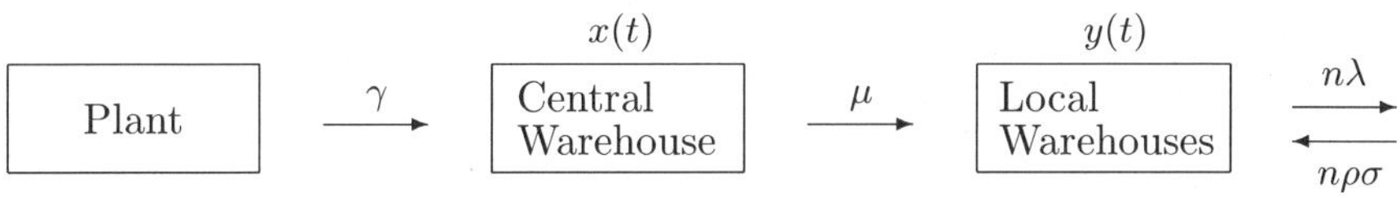

Fig. 2. The Aggregated Model.

According to the flow of stock (order) described in (a1)-(a4) of Section 2, the states of the joint inventory process $\{(x(t), y(t)), t \geq 0\}$ is a continuous time Markov chain taking values in the state-space

$$S = \{(x(t), y(t)) : x(t) = 0, \ldots, C;\ y(t) = 0, \ldots, nL.\}. \tag{9}$$

Each time when visiting a state, the process stays there for a random period of time that follows the exponential distribution and is independent of the past behavior of the process. We remark that the states of the inventory system forms an irreducible Markov chain and therefore the steady-state probability distribution exists, see for instance [4].

5 Concluding Remarks

In this paper, we present a two-echelon inventory/return system. The system consists of a supply plant with infinite capacity and a central warehouse for inventory. There is a number of local warehouses having capacities for re-manufacturing of products. Lateral transshipment of demands is allowed among the local warehouses. The followings are some of our future research issues: (i) To consider the minimization of the average running cost of the system by including costs associated with the normal replenishment, lateral transshipment and inventory costs at all the warehouses. (ii) To develop fast numerical method for solving the large Markov chain problem in Section 4.

References

1. P. Alfredsson and J. Verrijdt (1999) *Modeling Emergency Supply Flexibility in a Two-Echelon Inventory System*, Management Sci. 45, 1416–1431.
2. P. Aggarwal and K. Moinzadeh (1994) *Order Expedition in Multi-echelon Production/Distribution Systems*, IIE Trans. 26, 86–96.
3. S. Axsäter (1990) *Modelling Emergency Lateral Transshipments in Inventory Systems*, Management Sci. 36, 1329–1338.
4. J. Buzacott and J. Shanthikumar (1993) *Stochastic Models of Manufacturing Systems*, Prentice-Hall, New Jersey.
5. W. Ching (1997) *Markov Modulated Poisson Processes for Multi-location Inventory Problems*, Int. J. Production Economics 53, 217–223.
6. W. Ching, W. Yuen and A. Loh (2003) *An Inventory Model with Returns and Lateral Transshipments*, Journal of Operational Research Society, 54, 636–641.
7. W. Ching and X. Zhou (2000) *Circulant Approximation for Preconditioning in Stochastic Automata Networks*, Comput. Math. Appl., 39, 147–160.
8. W. Ching, R. Chan and X. Zhou (1997) *Circulant Preconditioners for Markov Modulated Poisson Processes and Their Applications to Manufacturing Systems*, SIAM J. Matrix Anal. Appl. 18, 464–481.
9. A. Kukreja, C. Schmidt and D. Miller (2001) *Stocking Decisions for Low Usage Items in a Multi-location Inventory System*, Management. Sci. 47, 1371–1383.
10. H. Lee (1987) *A Multi-echelon Inventory Model for Repairable Items with Emergency Lateral Transshipments*, Management Sci. 33, 1302–1316.
11. M. Maron (1987) *Numerical Analysis : A Practical Approach*, 2nd Edition, Macmillan, New York.
12. K. Moinzadeh and C. Schmidt (1991) *An $(S-1, S)$ Inventory System with Emergency Orders*, Oper. Res. 39, 308–321.
13. J. Muckstadt (2005) *Analysis and Algorithms for Service Parts Supply Chains*, Springer Series in Operations Research and Financial Engineering, Springer, N.Y.
14. C. Pyke (1990) *Priority Repair and Dispatch Policies for Repairable Item Logistic Systems*, Naval Res. Logist. 37, 1–30.
15. L. Robinson (1990) *Optimal and Approximate Policies in Multi-period, Multi-location Inventory Models with Transshipments*, Oper. Res. 38, 278–295.

Bestimmung von Losgrößen, Transportzyklen und Sicherheitsbeständen in unternehmensübergreifenden Wertschöpfungsketten

Heinrich Kuhn[1] and Fabian J. Sting[2]

[1] Katholische Universität Eichstätt-Ingolstadt, Lehrstuhl für Produktionswirtschaft, Auf der Schanz 49, D-85049 Ingolstadt, Tel: 0841-937-1820, Fax: 0841-937-1955, heinrich.kuhn@ku-eichstaett.de

[2] WHU - Otto Beisheim School of Management, Lehrstuhl für Produktionsmanagement, Burgplatz 2, 56179 Vallendar, Tel: 0261-6509-387, Fax: 0261-6509-389, fabian.sting@whu.edu

Abstract: In dem Beitrag wird ein zweistufiges Planungskonzept vorgestellt, um im Zuge der Gestaltung einer Supply Chain die Losauflagen, Transportzyklen und Sicherheitsbestände für die Produkte der Supply Chain festzulegen. In der ersten Planungsstufe werden die Transport- und Losgrößen festgelegt. Die zweite Stufe bestimmt unter Einhaltung eines vorgegebenen Serviceniveaus am Ende der Supply Chain die notwendigen Sicherheitsbestände je Lagerstufe.

1 Problemstellung

Die Gestaltung unternehmensübergreifender Wertschöpfungsketten bzw. von Supply Chains erfolgt in der Regel im Zuge des Entstehungsprozesses eines Produktes und beginnt aus Sicht eines Original Equipment Manufacturer (OEM) mit der Auswahl der 1st-Tier Lieferanten. In einigen Industriebranchen findet dieser Auswahlprozess bereits ein bis zwei Jahre vor dem Anlauf der Serienfertigung (Start of Production, SOP) statt, so dass bereits in dieser frühen Phase potentielle Supply Chains gestaltet und bewertet werden müssen. Vor allem die Abschätzung der in einer Supply Chain entstehenden Transport-, Lager- und Handlingkosten setzt voraus, dass der Entscheidungsträger eine Vorstellung darüber hat, mit welchen Losauflagen, Transportzyklen und Sicherheitsbeständen die einzelnen Mitglieder der Supply Chain (Lieferanten und Sublieferanten) in der Zukunft operieren werden bzw. wie diese Größen im Gesamtverbund günstig dimensioniert sein sollten. Um diese Kenngrößen einer Supply Chain hinreichend genau vorherzusagen bzw. in einem ersten Ansatz festzulegen, wird in dem Beitrag ein Modell und Planungskonzept zur Bestimmung der optimalen Losgrößen, Transportzyklen und Sicherheitsbestände in einer mehrstufigen Produktions- und Lieferkette (Supply Chain) vorgestellt. Jeder Knoten der Supply Chain besteht dabei aus einem oder mehreren Eingangslagern, einer Produktionsstufe und einem Ausgangslager. Lieferbeziehungen bestehen zwischen einem Ausgangslager (Lieferant) und genau einem Eingangslager der unmittelbar nachgelagerten Stufe (Abnehmer). Es wird somit von einer konvergierenden Lieferanten- und Erzeugnisstruktur ausgegangen. Abbildung 1 verdeutlicht das Modell der zu Grunde gelegten Supply Chain-Struktur.

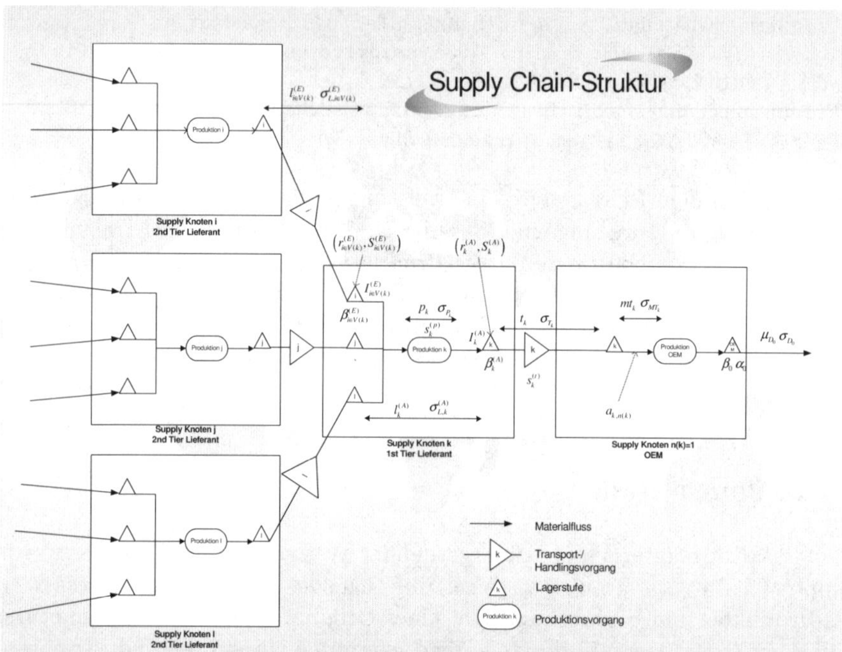

Abbildung 1. Modell der Struktur der Supply Chain

2 Zweistufiges Planungskonzept

Die Bestimmung der optimalen Losgrößen, Transportzyklen und Sicherheitsbestände erfolgt mit Hilfe eines zweistufigen Planungskonzepts. Ausgangspunkt der Planung ist die Kenntnis einer bestimmten Supply Chain Topologie, die durch das Angebot eines Lieferanten definiert wird. Anschließend werden im Rahmen der ersten Planungsstufe die Transport- und Losgrößenplanung für die gesamte Supply Chain simultan gelöst. Ziel ist es, die Summe aus Supply Chain übergreifenden Transport-, Handling- und Lagerkosten zu minimieren. Entscheidungsvariablen sind die Transport-, Handling- und Losauflagefrequenzen, die Auswahl bestimmter Transportmittel bzw. -modi. Als Nebenbedingungen werden unter anderem Restriktionen des Lager- und Transportvolumens berücksichtigt.

Die zweite Planungsstufe bestimmt auf der Grundlage der Ergebnisse der ersten Planungsstufe für jede Lagerstufe die notwendigen Sicherheitsbestände, so dass ein vorgegebenes Serviceniveau am Ende der Supply Chain eingehalten wird. Berücksichtigt werden dabei die drei wesentlichen Einflussgrößen auf das jeweilige Serviceniveau einer Lagerstufe (vgl. Tempelmeier 2003): Inputunsicherheit, Transformationsunsicherheit, Outputunsicherheit. Zielgröße der zweiten Planungsphase ist die Minimierung der Kosten für die jeweiligen Sicherheitsbestände unter Einhaltung eines vorgegebenen Servicegrads am Ende der Supply Chain.

2.1 Planung von Losgrößen und Transportzyklen

In jedem Knoten der Supply Chain sind zwei grundsätzliche Entscheidungen zu treffen. Zum einen ist der Produktionszyklus (Produktionslosgröße) und zum anderen ist der Transportzyklus (Transportlosgröße) zu bestimmen, mit der die Endprodukte vom Ausgangslager in das Eingangslager der nächsten Stufe zu transportieren ist.

Grundlage zur Festlegung dieser beiden Gruppen von Entscheidungsvariablen ist das folgende Modell:

$$\min \quad Z = \sum_{k=1}^{K} \left[\frac{s_k^{(p)}}{r_k^{(A)}} + e_k \cdot \frac{D_k \cdot r_k^{(A)}}{2} + \sum_{i \in \mathcal{V}(k)} \left(\sum_{z=1}^{Z} s_{ikz}^{(t)} \cdot D_{ik} \cdot x_{ikz} + e_{ik} \cdot \frac{D_{ik} \cdot r_{ik}^{(E)}}{2} \right) \right] \tag{1}$$

u.B.d.R.

$$\frac{r_i^{(A)}}{r_{ik}^{(E)}} \geq 1 \text{ und ganzzahlig} \qquad i = 2, 3, ..., K; k \in \mathcal{N}(i) \tag{2}$$

$$\frac{r_{ik}^{(E)}}{r_k^{(A)}} \geq 1 \text{ und ganzzahlig} \qquad k = 1, 2, ..., K; i \in \mathcal{V}(k) \tag{3}$$

$$D_{ik} \cdot r_{ik}^{(E)} \geq q_{ik}^{(min)} \qquad k = 1, 2, ..., K; i \in \mathcal{V}(k) \tag{4}$$

$$D_{ik} \cdot r_{ik}^{(E)} \leq q_{ik}^{(max)} \qquad k = 1, 2, ..., K; i \in \mathcal{V}(k) \tag{5}$$

$$\left(D_{ik} \cdot r_{ik}^{(E)} - u_{ikz} \right) \cdot x_{ikz} \geq 0 \qquad k = 1, 2, ..., K; i \in \mathcal{V}(k); z = 1, 2, ..., Z \tag{6}$$

$$\sum_{z=1}^{Z} x_{ikz} = 1 \qquad k = 1, 2, ..., K; i \in \mathcal{V}(k) \tag{7}$$

$$x_{ikz} \in \{0, 1\} \qquad k = 1, 2, ..., K; i \in \mathcal{V}(k); z = 1, 2, ..., Z \tag{8}$$

Dabei bedeuten:

Indizes:

k	Index der Knoten der Supply Chain ($k = 1, 2, ..., K$)
z	Index der Transporttarife ($z = 1, 2, ..., Z$)

Daten:

D_k	Bedarf für Erzeugnis k im Ausgangslager von Knoten k
D_{ik}	Bedarf für Erzeugnis i im Eingangslager von Knoten k
e_k	Marg. Lagerkostensatz für Erzeugnis k im Ausgangslager von Knoten k
e_{ik}	Marg. Lagerkostensatz für Erzeugnis i im Eingangslager von Knoten k
$q_{ik}^{(max)}$	Maximale Transportmenge zwischen Knoten i und k
$q_{ik}^{(min)}$	Mindesttransportmenge zwischen Knoten i und k
$s_k^{(p)}$	Rüstkostensatz für die Produktion in Knoten k
$s_{ikz}^{(t)}$	Transportkostensatz zwischen Knoten i und k in der Traifstufe z
u_{ikz}	Mindestmenge für den Transportkostensatz z zwischen Knoten i und k
$\mathcal{V}(k)$	Menge der direkten Vorgänger von Knoten k
$\mathcal{N}(i)$	Menge der direkten Nachfolger von Knoten i

Variablen:

$r_k^{(A)}$	Produktionszyklus für das Erzeugnis k im Knoten k
$r_{ik}^{(E)}$	Transportzyklus für das Erzeugnis i in das Eingangslager von Knoten k
x_{ikz}	1, wenn Transportkostensatz z zwischen Knoten i und k gilt, ansonsten 0

Zur Lösung des beschriebenen Modells wurde eine Verfahren entwickelt, das auf dem von Heinrich (1987) vorgeschlagenen Verfahren zur Lösung des dynamischen Mehrprodukt-Losgrößenproblems aufbaut. Hier wird jedoch von einem konstanten Bedarfsverlauf (statische Problemstellung) ausgegangen (vgl. Blackburn/Millen, 1982). Das Verfahren von Heinrich (1987) wurde derart modifiziert, dass die Restriktionen der Mindest- und Maximaltransportmengen und die Wahl unterschiedlicher Transportmodi berücksichtigt werden.

2.2 Planung lokaler Sicherheitsbestände

Die Planung der lokalen Sicherheitsbestände basiert auf der stochastischen Lagerhaltungstheorie einstufiger Systeme. Die (r, S)-Lagerhaltungspolitik bildet die praktischen Lagerprozesse in der untersuchten Supply Chain sehr gut ab, zudem können die Resultate aus dem ersten Planungsschritt weiterverarbeitet werden.

Ausgehend von einem einstufigen Lagerhaltungssystem wird das Konzept der stochastischen Lagerhaltung auf die gesamte Supply Chain ausgeweitet, d.h., dass jeder einzelne Supply Chain Knoten durch mehrere Eingangslager für alle eingehenden Vorprodukte sowie ein Ausgangslager für das jeweilige Erzeugnis abgebildet wird. Zwischen den Lagerstufen gibt es Wechselwirkungen, die insbesondere durch Input-, Transformations- und Outputunsicherheiten bedingt sind. Ein instabiler Fertigungsprozess (Transformationsunsicherheit) eines Supply Chain Knotens, dessen Ausgangslager zudem einem stark schwankenden Nachfrageprozess (Outputunsicherheit) gegenübersteht, erhöht das Beschaffungsrisiko des stromabwärts gelagerten Supply Chain Knoten (Inputunsicherheit). Aggregiert werden Input- und Transformationsunsicherheit für eine Lagerstufe zunächst in der Zufallsvariable L, der Wiederbeschaffungszeit. Für das Ausgangslager k bestimmt sich die Zufallsvariable der Wiederbeschaffungszeit L_k wie folgt:

$$L_k = P_k + \max_{i \in \mathcal{A}} \{J_i\} \tag{9}$$

Deutlich wird, dass die Wiederbeschaffungszeit von k nicht nur von der Fertigungszeit P_k abhängt, sondern ebenso von den Verzögerungszeiten J_i im Falle eines Fehlmengenereignisses eines der Eingangslager. Die Menge $\mathcal{A}$ bezeichnet dabei die vorgelagerten Eingangslager, bei denen zum Zeitpunkt der Wiederbeschaffung eine Lieferunfähigkeit vorliegt. Für die vereinfachende Annahme, dass in einer Periode bei maximal einem Eingangslager ein Fehlmengenereignis auftritt, ergibt sich als Erwartungswert der Wiederbeschaffungszeit:

$$E\{L_k\} = E\{P_k\} + \sum_{i \in \mathcal{V}(k)} (1 - \alpha_i) E\{J_i\} \tag{10}$$

Dabei kennzeichnet α_i den ereignisbezogenen Servicegrad der Eingangslagerstufe i und $\mathcal{V}(k)$ die Menge aller Eingangslager, die Ausgangslager k vorgeordnet sind. Weiterhin kann die Varianz der Wiederbeschaffungszeit analytisch bestimmt werden. Anschließend werden unter Beachtung des stufenspezifischen stochastischen Nachfrageprozesses und der Wiederbeschaffungszeit die relevanten Momente der Nachfrage in der Wiederbeschaffungszeit Y_k für alle Eingangs- und Ausgangslager ermittelt. Auf diesem Weg wird die Outputunsicherheit einer Lagerstufe in die Betrachtung einbezogen. Die Größe Y_k ist diejenige Zufallsvariable, die alle modellierten stochastischen Einflussgrößen einer Lagerstufe miteinander kombiniert.

Sicherheitsbestände in den einzelnen Lagerstufen dienen dazu, die zufällig schwankende Nachfrage in der Wiederbeschaffungszeit auszugleichen. Die Wahrscheinlichkeit, dass die Nachfrage am Ausgangslager k in der Wiederbeschaffungszeit vollständig bedient werden kann, ist wiederum durch den Servicegrad α_k gegeben.

Das nun vorliegende Optimierungsproblem kann wie folgt charakterisiert werden: Die *Entscheidungsvariablen* sind die lagerstufenspezifischen Servicegrade, die unter Einhaltung der *Nebenbedingung* einer vorbestimmten Service-Restriktion am OEM optimiert werden. Die *Zielfunktion* ist die Summe der Lagerkosten für die Supply Chain übergreifenden Sicherheitsbestände.

Ein Algorithmus zur heuristischen Lösung wurde entwickelt, der stromabwärts rekursierend auf Basis des Algorithmus von Nelder und Mead (1965) eine Optimierung der Servicegrade durchführt. Stromabwärts rekursiv bedeutet dabei, dass die relevanten Unsicherheitsparameter und Servicegrade zunächst für die vom OEM am weitesten entfernten Tiers (Rohstofflieferanten) ermittelt werden. Anschließend werden die nachfolgenden Knoten betrachtet bis der OEM erreicht wird.

3 Ergebnisse und Ausblick

Erste numerische Analysen zeigen für den vorgeschlagenen Lösungsansatz nachvollziehbare Ergebnisse. Zum einen kann beobachtet werden, dass die unterschiedlichen Eingangslager eines bestimmten Knotens dem entsprechenden Ausgangslager tendenziell gleichgerichtet Serviceniveaus bieten. Es scheint nicht effizient zu sein, unterschiedliche Servicegrade von seinen Lieferanten zu fordern, dies ist vor allem bei abgestimmtem Materialfluß verständlich.

Ferner zeigt sich, dass Stufen, die signifikanten Sicherheitsbestand bevorraten, sich zumeist in einem zusammenhängenden Pfad von Lagerstufen mit hohem Sicherheitsbestand befinden. Der Aufbau von Sicherheisbeständen ist offenbar nur dann sinnvoll, wenn die Kette der Service gewährenden Supply Chain Stufen bis zum OEM nicht abreißt.

Bei einem Beispiel mit identischen Lagerkostensätzen für alle Stufen der Supply Chain zeigt sich, dass die optimale Strategie eine alleinige Bevorratung von Sicherheitsbestand auf der letzten Stufe ist. Durch dieses „Risikopooling“ werden alle Input-, Transformations- und Outputunsicherheiten der gesamten Supply Chain bestmöglich aufgefangen.

Literatur

1. Blackburn, J.; und Millen, R. Improved heuristics for multi-stage requirements planning systems. *Management Science 28* (1982), 44–56.
2. Heinrich, C. *Mehrstufige Losgrößenplanung in hierarchisch strukturierten Produktionsplanungssystemen.* Berlin et al., 1987.
3. Lee, H.L.; und Billington, C. Material management in decentralized supply chains. *Operations Research 41* (1993), 835–847.
4. Nelder, J.A.; und Mead, R. A simplex method for function minimization. *Computer Journal 7* (1965), 308–313.
5. Tempelmeier, H. *Material-Logistik: Modelle und Algorithmen für die Produktionsplanung und -steuerung und das Supply Chain Management*, 5. ed. Berlin et al., 2003.

A Group Setup Strategy for PCB Assembly on a Single Automated Placement Machine

Ihsan Onur Yilmaz and Hans-Otto Günther

Dept. of Production Management, TU Berlin
Wilmersdorfer Str. 148, D-10585 Berlin, Germany
E-mail: ihsan.o.yilmaz@tu-berlin.de

Balancing the savings between component setup and unit assembly times is an important issue in medium-variety, medium volume printed circuit board (PCB) production. We present a group setup strategy for a single automated placement machine, which focuses on minimizing the total makespan for a given number of batches of different PCB types. Thus, agglomerative clustering techniques are applied using the component similarities and considering the limited feeder capacity of the component magazine. In each grouping step, a magazine setup is generated and the component feeders are re-arranged with regard to the batch sizes of the individual PCB types. Using this magazine setup, the sequence of placement operations for each PCB type and the total assembly time for the group of PCBs are determined. Grouping of PCBs is only allowed if the total makespan can be reduced.

1. Introduction

Utilization of the assembly machines and thus reducing the assembly costs is an important issue in PCB assembly. Because of the increasing variety of products in the market, medium-variety medium-volume PCB assembly can be seen in many production systems. Hence, grouping similar PCBs into setup families in order to reduce setup times is a common application. In this paper, a group setup strategy is proposed for a single collect-and-place machine, which is considered as the bottleneck of the assembly system.

The offline setup procedure has become popular in production environments due to the development of machinery with interchangeable feeder trolleys. In offline setup, a feeder trolley can be detached and replaced with another trolley in a short period of time. Hence, the feeder trolley for the next PCB family can be prepared while the machine continues assembling other PCBs. The opportunity of this offline setup procedure makes the group setup strategy preferable against other strategies from the literature.

In a group setup, the feeder-slot assignment in the magazine cannot be tailored to the individual PCB. Hence, by enlarging the groups, setup time is saved but at the expense of increased assembly time per PCB. The trade-off between the savings achieved through reduced setup effort and the increase in single PCB assembly times must be considered by taking the makespan minimization as the main objective.

Most of the papers on group setup strategies in PCB assembly focus on the analysis of component similarity between PCB types. Shtub and Maimon (1992) were among the first to develop clustering procedures for PCB grouping problems based on similarity measures of the PCB types. Other authors use integer linear programming models, e.g. Maimon and Shtub (1991) and Daskin et al. (1997), or graph-theoretic approaches, e.g. Bhaskar and Narendran (1996), to generate groups of PCB types. Knuutila et al. (2001) show how real-life problems of job grouping can be solved by efficient heuristics, binary programming, and constraint programming techniques. An issue of practical importance for certain types of assembly machinery is the consideration of different feeder types. Knuutila et al. (2004) propose several efficient heuristics by considering different feeder types.

In addition, the medium-variety, medium-volume production environment has received attention in various research papers, e.g. Peters and Subramanian (1996) and Leon and Peters (1996) and (1998). They focus on partial setup strategies and compare their results with other setup strategies, including group setup. A performance evaluation of minimum and group setup strategies in a high-variety, low-volume environment can also be found in Smed et al. (2003).

Common to all of the above mentioned papers is that actual PCB assembly times are either based on rough estimates or assumed to remain constant irrespective of the composition of the PCB families. This assumption, however, is not rectified in most industrial applications (cf. Laakso et al. 2002). In this paper, we present a methodology which adjusts the setup of the component magazine and determines the actual assembly time per board in each step of the grouping procedure. This is achieved by integrating algorithms for the optimization of the assembly machine operations. The underlying objective is to minimize the total makespan for a given number of batches of different PCB types. In addition, batch-size considerations are taken into account in order to achieve the best magazine layout for the whole group of PCBs.

2. Generation of setup families using inclusion trees

Our approach employs the so-called "inclusion measure" as a similarity coefficient, which is more appropriate for PCB assembly compared with the other conventional similarity measures. Conventional similarity measures do not provide information about which PCB type best comprises or includes another one. However, this information might be quite valuable in scheduling PCB assembly.

For instance, a so-called perfect subset situation exists if one type of PCB uses a subset of component types from another PCB. Combining these two PCB types does not require any additional slots in the component magazine (i.e. no additional setup of a component feeder is required) and thus may prove to be also advantageous in terms of makespan. This analysis can be performed by calculating so-called asymmetric inclusion measures (cf. Raz and Yaung 1994), which can be defined as:

$$IM(i,j) = \frac{n(i,j)}{n(j)} \tag{1}$$

where

IM (i,j)	inclusion measure (extent to which PCB *j* is included in PCB *i*)
n(i,j)	number of common components of PCB *i* and PCB *j*
n(j)	number of components of PCB *j*

For grouping PCB types, we apply a modified hierarchical clustering procedure proposed by Raz and Yaung (1994). Their approach is to group entities using a tree structure. Arcs connecting entities (PCB types) are assigned a weight – the inclusion measure of the two entities. After the complete tree has been constructed, it is cut at "weak" branches into sub-trees corresponding to individual clusters.

In the first step, inclusion measures are calculated for all pairs of PCBs. Because a PCB can not be included in a smaller PCB, i.e. one with a smaller number of component types, PCBs are sorted in descending order of size, i.e. their number of different component types. The largest PCB, i.e. the one with the largest number of different component types, is assigned to the root of the inclusion tree. Other PCBs, in descending order of their size, are assigned as a descendant node of the node in the tree to which they show the highest value of the inclusion measure. (In case of ties, the PCB type is assigned as a descendant of the node with the smallest size among those in the tie group.) This procedure is repeated until all PCB types are assigned in the tree.

In the original algorithm, clusters are formed by cutting arcs in the tree connecting nodes with an inclusion measure less than a user defined threshold value. In our specific case, however, this procedure could frequently lead to groups of PCBs, which violate the magazine capacity constraints and which do not promise a reduction in total makespan. Therefore, the original procedure, which uses threshold values for cutting the arcs, is not applicable here. Instead, we modified the original algorithm in order to explicitly consider the constraints arising from group setup strategies in PCB assembly.

After the complete inclusion tree has been constructed, the arc with the maximum weight in the tree is identified. (In case of ties, the arc at the lowest level in the tree is chosen, as this would require a smaller number of component types, and thus, occupy less component magazine capacity.) For the selected arc, two conditions must be satisfied: (1) the magazine capacity constraint must not be violated through the common component type requirements and (2) grouping is only allowed if the total makespan is reduced. If both conditions are met, the two corresponding PCB types are merged into a group. If due to the magazine capacity constraint no other PCB types can be added to that group, it is excluded from the inclusion tree. Otherwise, the newly created group constitutes a new node in the inclusion tree and the inclusion measures of all other PCB types with respect to this group of PCBs are updated. In the subsequent steps of the clustering procedure, this group may be again joined with PCB types represented by neighboring nodes in the tree. Thus, groups consisting of more than two PCB types may be created. This procedure is repeated until no further groups can be formed. Again, it should be noted that the calculation of makespan requires specific algorithms for scheduling the machine operations. These algorithms are explained in section 3.

3. Optimization of machine operations

As mentioned before, our approach attempts to model the placement times of automated assembly machines more realistically than other approaches known from the academic literature. Thus machine-specific algorithms for the optimization of the machine operations have to be integrated into the solution procedure. In particular, the assignment of component feeders to slots in the component magazine and the placement sequence of the components must be determined at each step in the grouping procedure considering the individual operation mode of the assembly machine. In the sequel, the integration of machine scheduling algorithms into the proposed hierarchical clustering approach is exemplified for the case of a single collect-and-place machine. Nevertheless, our approach can be applied to any other type of machinery.

3.1. Magazine setup

In the heuristic solution procedure of Grunow et al. (2004), the assignment of component feeders to slots in the magazine of the placement machine is determined first. For the solution to this sub-problem, a heuristic approach is suggested which analyses the neighborhood relations between the different types of components and the corresponding placement locations on the board. In this paper, we extend this basic single-PCB approach for the case of determining the magazine setup for a group of different PCB types.

In our approach, we take the size of the batches to be produced of each PCB type into account. This aspect has been neglected previously in the literature. Instead of using a composite (super) PCB approach, the defined neighborhood relations are weighed by the batch size of the corresponding PCB and a neighborhood matrix is generated. By summing up all weighted PCB-specific neighborhood matrices, the aggregate neighborhood matrix for the entire group of PCBs is obtained. Based on this aggregate matrix, the component setup is determined much more realistically compared to the conventional composite PCB approaches.

3.2. Component placement sequence

After the magazine setup for the group of PCBs is determined using the aggregate neighborhood matrix, the next step is to derive the component placement sequence and to calculate the total assembly time per board. For the collect-and-place machine considered, we apply the sequencing heuristic from Grunow et al. (2004) for each type of PCB considering the magazine layout for the group as input.

Given the assignment of component feeders to magazine positions, the problem of sequencing the placement operations for a collect-and-place machine is similar to the well-known vehicle routing problem with the placement head corresponding to the (single) vehicle with a limited loading capacity. Based on this correspondence, Grunow et al. (2004) adapted standard methods for vehicle routing problems and developed efficient heuristic algorithms. Due to their small computational effort they can easily be integrated into the clustering procedures of the group setup approach.

3.3. Improvement of magazine setup and component placement sequence

In our approach, an extended version of the 2-opt-exchange procedure proposed by Grunow et al. (2004) is applied. Once the groups of PCBs have been identified, the 2-opt procedure is called up to improve the magazine setup obtained for the groups and the placement sequence for each individual type of PCB. In order to reduce the computational effort, the extended 2-opt-exchange procedure is applied only as a final step to improve the feeder assignment and placement sequence obtained for the various groups of PCBs.

3.4. Determination of the total makespan

The heuristic procedures indicated in sections 3.1 and 3.2 are integrated into the clustering approaches presented in section 2 in order to consider the group magazine setup and the assembly time per PCB more realistically. Given the assembly time and the corresponding batch sizes of all PCBs $i \in I$, which are elements of group j ($i \in I_j$), as well as the time required for setting up the magazine for group j, the total assembly time for the entire PCB group j can easily be derived. The total makespan is determined by adding up the total assembly times for all groups j:

$$\textit{total makespan} = \sum_{j \in J} \left(\textit{setup time}_j + \sum_{i \in I_j} \left(\textit{assembly time}_i \times \textit{batch size}_i \right) \right) \tag{2}$$

This way, it is possible to examine if a reduction of the total makespan is achieved in the grouping procedure. In case of offline setup, the setup time corresponds to the time required for exchanging the feeder trolley.

4. Conclusions

In this paper, a novel approach for group setup strategies has been presented fot the case of a single collect-and-place assembly machine. Minimizing the makespan for a given number of batches of different PCB types is regarded as the objective function. The proposed methodology applies machine-specific algorithms for optimizing magazine setups for each PCB family and generating placement sequences for each PCB type. Thus, realistic assembly times and optimized magazine setups are used in contrast to other conventional approaches for grouping PCBs. Additionally, batch sizes are also integrated into the group formation procedure. Development of other group formation methodologies based on realistic assembly times is considered as a future research topic.

References

Bhaskar G, Narendran TT (1996) Grouping PCBs for set-up reduction: a maximum spanning tree approach. International Journal of Production Research: 34,621–632

Daskin MS, Maimon O, Shtub A, Braha D (1997) Grouping components in printed circuit board assembly with limited component staging capacity and single card setup. International Journal of Production Research 35:1617–1638

Grunow M, Günther HO, Schleusener M, Yilmaz IO (2004) Operations planning for collect-and-place machines in PCB assembly. Computers and Industrial Engineering 47:409–429

Knuutila T, Hirvikorpi M, Johnsson M, Nevalainen O (2004) Grouping PCB assembly jobs with feeders of several types. The International Journal of Flexible Manufacturing Systems 16:151–167

Knuutila T, Puranen M, Johnsson M, Nevalainen O (2003) Three perspectives for solving the job grouping problem. International Journal of Production Research 39:4261–4280

Laakso T, Johnsson M, Johtela T, Smed J, Nevalainen O (2002) Estimating the production times in PCB assembly. Journal of Electronics Manufacturing 11:161–170

Leon VJ, Peters BA (1996) Replanning and analysis of partial setup strategies in printed circuit board assembly systems. The International Journal of Flexible Manufacturing Systems 8:389–412

Leon VJ, Peters BA (1998) A comparison of setup strategies for printed circuit board assembly. Computers and Industrial Engineering 34:219–234

Maimon O, Shtub A (1991) Grouping methods for printed circuit board assembly. International Journal of Production Research 29:1379–1390

Peters BA, Subramanian GS (1996) Analysis of partial setup strategies for solving the operational planning problem in parallel machine electronic assembly systems. International Journal of Production Research 34:999–1021

Raz T, Yaung T (1994) Heuristic clustering based on a measure of inclusion. International Journal of Industrial Engineering 1:57–65

Shtub A, Maimon O (1992) Role of similarity measures in PCB grouping procedures. International Journal of Production Research 30:973–983

Smed J, Salonen K, Johnsson M, Johtela T, Nevalainen O (2003) A comparison of group and minimum setup strategies in PCB assembly. The International Journal of Flexible Manufacturing Systems 15:19–35

Optionsbündelung und Fertigungsablauf in der Automobilindustrie

Nils Boysen und Christian M. Ringle

Universität Hamburg, Institut für Industriebetriebslehre und Organisation
Von-Melle-Park 5, 20146 Hamburg; {boysen, cringle}@econ.uni-hamburg.de

Zusammenfassung. Bei der Optionsbündelung werden einzelne Ausstattungsmerkmale (Optionen) zu Paketen (Bündeln) zusammengefasst. Im Mittelpunkt dieses Beitrags steht die Beantwortung der Frage, ob eine solche Optionsbündelung den Produktionsablauf positiv beeinflusst. Dabei richtet sich das Hauptaugenmerk auf eine Variantenfließfertigung, die etwa in der Endmontage der Automobilindustrie eine dominierende Stellung einnimmt. Eine umfangreiche Simulationsstudie soll den positiven Effekt der Optionsbündelung nachweisen. Ferner wird untersucht, welche Bündelungsart im Rahmen einer Variantenfließfertigung besonders viel versprechend erscheint.

Einführung

Die Zusammenfassung mindestens zweier ebenso getrennt anbietbarer Güter oder Dienstleistungen zu einem einheitlichen Angebot mit einem einzigen Preis nennt man „Preisbündelung“ [7]. Beispiele für die Bündelung reichen von Fast-Food-Menüs, über Saisonkarten bei Sportveranstaltungen bis zu Office-Paketen in der Bürokommunikation. Eine Bündelung ist aber nicht auf Endprodukte beschränkt [4] – ebenso ist auch eine „Optionsbündelung“ von Bedeutung [6]. So können Automobilhersteller einzelne Optionen der Fahrzeugausstattung wie etwa Zentralverriegelung, Diebstahlsicherung und Airbags zu einem Sicherheitspaket bündeln. Ein Zweck der Optionszusammenfassung besteht darin, den Kunden zum Kauf höherwertiger Bündel zu bewegen [6]. Ein weiterer Effekt liegt in der Komplexitätsreduktion für den Kunden. Statt aus allen möglichen Optionskombinationen wählen zu müssen (so bietet BMW theoretisch 10^{32} Varianten an [10]), kann die Bündelung dem Kunden die Zusammenstellung seiner Variante erleichtern [13].

In diesem Beitrag soll aber weder der Einfluss der Bündelung auf den Absatz [8] noch die Wahrnehmung des Kunden [9] untersucht werden, sondern die Wirkung auf den Ablauf der Fertigung. Durch weniger Produktvarianten verringert sich die Varietät in der Fertigung und erleichtert somit die Koordination innerhalb der Wertschöpfungskette. Der Zusammenhang zwischen der Anzahl der Produktvarianten und den Fertigungskosten wird in der Literatur häufig mit einer Zunahme der Umrüstvorgänge bei steigender Variantenanzahl begründet [14].

Im Bereich der Endmontage trifft man aber häufig auf eine so genannte Variantenfließfertigung. In einem solchen Produktionssegment sind die Umrüstvorgänge soweit reduziert, dass die einzelnen Varianten in der Losgröße Eins gefertigt werden können [1]. Dadurch besteht nicht länger ein Zusammenhang zwischen der Anzahl der gefertigten Produktvarianten und den Umrüstkosten. Trotzdem verspricht eine mit der Optionsbündelung verbundene Reduktion der Produktvarianten eine vereinfachte Koordination in der Endmontage. Dieser Zusammenhang ist aber nicht so offensichtlich wie hinsichtlich der Umrüstkosten einer Sorten- oder Serienfertigung, weshalb zur genaueren Analyse eine Simulationsanalyse durchgeführt wird. Die bis dato einzige Untersuchung in dieser Richtung stammt von Fisher und Ittner [5]. Allerdings beschränkt sich jene Untersuchung auf Just-in-Time-Belange (Level-Scheduling) und eine einzige Art der Optionsbündelung („alles-oder-nichts"). Dementsprechend sollen die unterschiedlichen Arten der Optionsbündelung erhoben werden, um eine differenzierte Beurteilung – auch im Hinblick auf die Auslastung der Ressourcen (Car-Sequencing) – zu leisten.

Arten der Optionsbündelung

Die traditionelle Systematisierung der Bündelung unterscheidet in: "unbundling" (Angebot einzeln), "pure bundling" (Angebot im Bündel) und "mixed bundling" (Angebot im Bündel und einzeln) [7]. Aus Sicht der Optionsbündelung ist diese Einteilung aufgrund der Vielzahl an Optionen zu verfeinern (Abbildung 1).

Abb. 1. Arten der Optionsbündelung

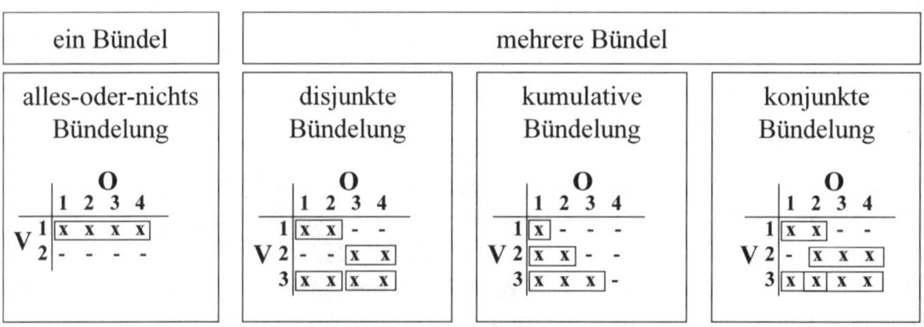

Die einfachste Art der Bündelung entsteht, wenn lediglich ein einziges Bündel angeboten wird. Dem Kunden verbleibt entweder die Auswahl aller Optionen, oder der Verzicht auf jegliche Zusatzausstattung. Damit ist die Anzahl der angebotener Varianten V bei der „alles-oder-nichts" Bündelung auf genau zwei beschränkt (V=2). Als „disjunkte" Bündelung wird der Fall bezeichnet, in dem einzelne Bündel B bezüglich der enthaltenen Optionen überschneidungsfrei sind: eine Option kann immer nur in genau einem Bündel vorkommen: $V = 2^B$. Eine

„kumulative“ Bündelung („trade-up bundles“ [4]). liegt vor, wenn die einzelnen Bündel sich sukzessive erweitern, mithin ein umfangreicheres Bündel alle Optionen untergeordneter Bündel enthält und diese zusätzlich um eine oder mehrere Optionen erweitert. Wenn alle Optionen als mögliche Bündelgrenzen zugelassen sind, richtet sich die Anzahl der theoretisch möglichen Varianten V nach der Anzahl der Optionen O ($V = O+1$). Bei der „konjunkten Bündelung“ können lediglich einzelne Optionen in mehreren Bündeln vorkommen. Sind alle Bündelkombinationen erlaubt, wovon im Weiteren ausgegangen wird, so ergibt sich die theoretische Variantenanzahl V wie im disjunkten Fall aus: $V = 2^B$. Als Vergleichsmaßstab ist zusätzlich der Fall zu betrachten, in dem überhaupt keine Bündelung vorliegt und die Kunden frei aus allen Optionen wählen können (unbundling): $V = 2^O$.

Gibt man nun für die Anzahl an Optionen und Bündel eine bestimmte Größe vor, so lassen sich die Bündelungsarten nach Maßgabe der theoretisch möglichen Variantenanzahl in eine Rangfolge bringen. Folgt man der Ansicht, dass der positive Einfluss auf den Fertigungsablauf größer ausfällt, je weniger Varianten aus dem Bündelungskonzept hervorgehen, so gilt diese Rangfolge auch für den Einfluss auf den Fertigungsablauf. Diese These gilt es im Weiteren, durch eine Simulationsuntersuchung zu stützen.

Reihenfolgeplanung bei der Variantenfließfertigung

In einer Variantenfließfertigung können die Varianten in wahlfreier Reihenfolge auf einem Fließsystem gefertigt werden. An den einzelnen Stationen werden die einzelnen Optionen entsprechend der Spezifikation der jeweiligen Produktvariante eingebaut.

Tabelle 1. Verwendete Symbole für die Planungsansätze der Reihenfolgeplanung

T	Anzahl der Fertigungstakte (mit $t = 1,\ldots,T$)
O	Menge der Optionen (mit $o \in O$)
V	Menge der Varianten (mit $v \in V$)
a_{vo}	Bedarfsfaktor (1, Variante v enthält Option o; 0, sonst)
d_v	Anzahl an zu fertigenden Einheiten der Variante v im Planungshorizont
x_{vt}	Binärvariable (1, Variante v wird in Takt t gefertigt; 0, sonst)

Die Versorgung der Stationen mit den dafür benötigten Bauteilen erfolgt “just-in-time“ von vorgelagerten Produktionsstufen. Dementsprechend zielt das sog. Level-Scheduling darauf, die Fertigungsfolge in der Endmontage so anzuordnen, dass der Bedarf an eingehenden Optionen möglichst gleichmäßig auf die Fertigungsfolge verteilt wird [11][12]. Formal lässt sich diese Planungsaufgabe unter Verwendung der in Tabelle 1 zusammengefassten Symbole wie folgt als Optimierungsmodell (1)-(4) darstellen:

$$\text{Minimiere } z = \sum_{t=1}^{T} \sum_{o \in O} \left(\sum_{v \in V} \sum_{t'=1}^{t} x_{vt'} \cdot a_{vo} - t \cdot \frac{\sum_{v' \in V} d_{v'} \cdot a_{v'o}}{T} \right)^2 \quad (1)$$

$$\sum_{v \in V} x_{vt} = 1 \quad \forall t = 1,...,T \quad (2)$$

$$\sum_{t=1}^{T} x_{vt} = d_v \quad \forall v \in V \quad (3)$$

$$x_{vt} \in \{0,1\} \quad \forall v \in V;\ t = 1,...,T \quad (4)$$

Dieser Ansatz beachtet allein die Just-in-time-Belange vorgelagerter Produktionsstufen. Trotz der prinzipiellen Möglichkeit einer wahlfreien Reihenfolge der Varianten, nimmt die Fertigungsfolge aber auch Einfluss auf die Ressourcen der Endmontage selber. In Abhängigkeit von den optionalen Bauteilen, die in der Variante aufgehen, brauchen einzelne Varianten eine unterschiedlich lange Zeit, bis die Montage beendet ist. Folgen zwei oder mehr für eine Station sehr arbeitsintensive Varianten hintereinander, so kommt es an dieser Station zu Überlastungen. In der Praxis kommen daher H_o:N_o-Regeln zum Einsatz. Danach dürfen von N_o aufeinander folgenden Varianten maximal H_o die Option o enthalten, da sonst Überlastungen entstehen. Somit verfolgt die Zielfunktion (5) in Verbindung mit den Gleichungen (2)-(4) das Ziel, diese Überlastungen zu minimieren [1][2]:

$$\text{Minimiere } z = \sum_{t=1}^{T} \sum_{o=1}^{O} \max \left\{ \sum_{t'=t}^{\min\{t+N_o-1;T\}} \sum_{v \in V} a_{vo} \cdot x_{vt'} - H_o ; 0 \right\} \quad (5)$$

Simulationsuntersuchung

Insgesamt sollen acht Testfälle mit insgesamt 192 Testinstanzen betrachtet werden. Dabei erfolgt für jede der vier Bündelungsarten ein Vergleich mit dem Fall ungebündelter Optionen. Um einen objektiven Vergleich sicherzustellen, darf innerhalb einer einzelnen Testinstanz pro Option nicht die Anzahl der einzubauenden Optionen zwischen gebündeltem (b) und ungebündeltem (u) Fall divergieren, sondern lediglich die Verteilung der Optionen auf die vorab fixierte Anzahl an Produktvarianten. Es gilt:

$$\sum_{v \in V} a_{vo}^{b} \cdot d_v = \sum_{v \in V} a_{vo}^{u} \cdot d_v \forall o \in O .$$

Zusätzlich sollen kleine und große Testinstanzen untersucht werden. Die kleinen Testinstanzen sind so gewählt, dass sie in vertretbarer Zeit exakt gelöst werden können (Xpress^{MP} Release 2003). Hingegen erfolgt die Lösung der großen Testinstanzen mittels eines heuristischen Threshold Accepting (TA)-Ansatzes [3].

Die Lösungsrepräsentation innerhalb des TA erfolgt durch einen Reihenfolgevektor π, wobei jede Position π_t (t=1,...,T) die Nummer v der zu fertigenden Variante im Takt t aufnimmt. Eine Startlösung für jeden der zwei Ansätze kommt dadurch zustande, dass die Varianten entsprechend des jeweiligen Bedarfs d_v zufällig auf den Reihenfolgevektor verteilt werden. Für das Level-Scheduling und das Car-Sequencing kann anschließend die Bewertung der Reihenfolge mit den Zielfunktionen (1) bzw. (5) erfolgen. Ausgehend von dieser Startlösung werden Nachbarschaftslösungen erzeugt, indem zufallsgestützt zwei beliebige Varianten der Fertigungsreihenfolge vertauscht werden. Diese veränderte Reihenfolge wird daraufhin – sofern die Differenz aus neuem und altem Zielfunktionswert einen Schwellenwert nicht überschreitet – als Ausgangspunkt für die weitere Nachbarschaftssuche akzeptiert und der Schwellenwert abgesenkt.

Tabelle 2. Erhobene Kriterien zur Beurteilung

Krit.	Beschreibung
V(diff)	Differenz der Variantenanzahl zwischen ungebündeltem und gebündeltem Fall.
B	Belegung der Bedarfsmatrizen a_{vo} mit Optionen: $B = \sum_{v \in V} \sum_{o \in O} a_{vo} \cdot d_v \Big/ T \cdot \lvert O \rvert$
Abs(X)	absolute Abweichung der Zielfunktionswerte zwischen ungebündeltem und gebündeltem Fall beim Car-Sequencing (X=Car) bzw. Level Scheduling (X=Lev)
Rel(X)	relative Abweichung der Zielfunktionswerte z zwischen ungebündeltem u und gebündeltem Fall g ($X \in \{Car, Lev\}$): Rel(X) = (z(u) – z(g)) / z(u)
Rel(TA)	relative Abweichung der Zielfunktionswerte z des Threshold Accepting TA von der optimalen Lösung opt: Rel(TA) = (z(TA) – z(opt)) / z(opt)
N	Anzahl Testinstanzen pro Testfall

Als ein wichtiges Ergebnis der kleinen Testfälle lässt sich festhalten, dass der TA-Ansatz sehr gute Ergebnisse für alle vier Bündelungsarten liefert (62 von 64 optimal). Damit dürfte auch für die großen Testfälle die Vergleichbarkeit der Bündelungsarten anhand des heuristischen TA-Verfahrens gegeben sein. Die Ergebnisse der großen Testinstanzen sind in Tabelle 3 zusammengefasst.

Tabelle 3. Ergebnisse der großen Testfälle

	alles oder nichts		kumulativ		konjunkt		disjunkt	
	μ	σ	μ	σ	μ	σ	μ	σ
V(diff)*	143,38	48,79	115,66	44,78	104,25	57,73	107,44	56,35
B	0,52	0,12	0,48	0,08	0,51	0,07	0,53	0,07
Abs(Car)*	119,84	57,61	56,75	34,96	38,53	25,38	31,53	22,6
Abs(Lev)*	5839	3443	2873	1668	860	912	932	708
Rel(Car)*	11,9%	4,5%	7,9%	3,5%	4,3%	2,9%	3,4%	2,9%
Rel(Lev)*	20,8%	9,3%	13,6%	7,1%	3,3%	2,75%	3,3%	3,6%
N	32	32	32	32				

* Signifikant unterschiedliche Mittelwerte im Vergleich zum Mittelwert der Gesamtverteilung (Irrtumswahrscheinlichkeit 10%), einfaktorielle Varianzanalyse (ANOVA)

Die inhaltliche Interpretation der Ergebnisse erfolgt, indem die theoretisch erwartete Rangfolge der Bündelungsarten bezüglich ihres positiven Einflusses auf die Reihenfolgeplanung den über die Zielfunktionswerte tatsächlich ermittelten Rangfolgen gegenübergestellt wird:

Theoretisch:	alles-oder-nichts > kumulativ > disjunkt > konjunkt
Car-Sequencing:	alles-oder-nichts > kumulativ > konjunkt > disjunkt
Level-Scheduling:	alles-oder-nichts > kumulativ > konjunkt = disjunkt

Aus dieser Gegenüberstellung wird ersichtlich, dass die theoretische Überlegung bestätigt werden kann, dass der Einfluss auf die Güte der Fertigungsfolgen allein durch die Anzahl der resultierenden Varianten bestimmt wird. Für die Praxis können daraus zwei Schlussfolgerungen abgeleitet werden: (i) Aus Produktionssicht sind die Optionsbündel so zu wählen, dass die Anzahl der resultierenden Varianten soweit als möglich begrenzt wird. (ii) Da die alles-oder-nichts Bündelung speziell in der Automobilindustrie die Freiheitsgrade der Variantenkonfiguration zu sehr einschränkt, erscheint vor allem die kumulative Bündelung als gute Wahl, um einen effizienteren Fertigungsablauf in der Endmontage zu ermöglichen.

Literatur

[1] Boysen N (2005) Reihenfolgeplanung bei Variantenfließfertigung: ein integrativer Ansatz. Zeitschrift für Betriebswirtschaft 75: 135-156

[2] Drexl A, Kimms A (2001) Sequencing JIT mixed-model assembly lines under station-load and part-usage constraints. Management Science 47: 480-491

[3] Dueck G, Scheuer T (1990) Threshold accepting: A general purpose optimization algorithm appearing superior to simulated annealing. Journal of Computational Physics 90: 161-175

[4] Eppen GD, Hanson WA, Martin RK (1991) Bundling: new products, new markets, low risk. Sloan Management Review 32: 7-14

[5] Fisher ML, Ittner CD (1999) The impact of product variety on automobile assembly operations: empirical evidence and simulation analysis. Management Science 45: 771-786

[6] Fürderer R, Huchzermeier A (1997) Optimale Preisbündelung unter Unsicherheit. Zeitschrift für Betriebswirtschaft Ergänzungsheft 1: 117-133

[7] Guiltinan JP (1987) The price bundling of services: a normative framework. Journal of Marketing 51: 74-85

[8] Hanson WA, Martin RK (1990) Optimal bundle pricing. Management Science 36: 155-174

[9] Jedidi K, Jagpal S, Manchanda P (2003) Measuring heterogeneous reservation prices for product bundles. Marketing Science 22: 107-130

[10] Meyr H (2004) Supply chain planning in the German automotive industry. OR Spectrum 26: 447-470

[11] Miltenburg J (1989) Level schedules for mixed-model assembly lines in just-in-time production systems. Management Science 35: 192-207

[12] Monden Y (1998) Toyota production system: an integrated approach to just-in-time, 3. edn. Engineering and Management Press, Norcross

[13] Pil FK, Holweg M (2004) Linking product variety to order-fulfilment strategies. Interfaces 34: 394-403

[14] Thonemann UW, Bradley JR (2002) The effect of product variety on supply-chain performance. European Journal of Operational Research 143: 548-569

A Heuristic Method for Large-Scale Batch Scheduling in the Process Industries

Norbert Trautmann[1] and Christoph Schwindt[2]

[1] Departement für Betriebswirtschaftslehre, Universität Bern, 3012 Bern, Switzerland, norbert.trautmann@ifm.uni-bern.de

[2] Institut für Wirtschaftswissenschaft, TU Clausthal, 38678 Clausthal-Zellerfeld, Germany, christoph.schwindt@tu-clausthal.de

Summary. In the process industries, final products arise from chemical and physical transformations of materials on processing units. In batch production mode, the total requirements for intermediate and final products are divided into individual batches. Storage facilities of limited capacity are available for stocking raw materials, intermediates, and final products. We present a novel approach to solving large-scale instances of the minimum-makespan production scheduling problem. The basic idea consists in constructing a production schedule by concatenating copies of a cyclic subschedule. For generating an appropriate subschedule we formulate a mixed-integer nonlinear program providing the set of batches of one cycle and the number of cycles needed to satisfy the primary requirements. The subschedule is then obtained by allocating the processing units, intermediates, and storage facilities over time to the batches executed in the cycle.

1 Introduction

We deal with short-term planning of batch production in the process industries. To produce a batch, at first the inputs are loaded into a processing unit. Then a transformation process, called a task, is performed, and finally the batch is unloaded from the processing unit. We consider multi-purpose processing units, which can operate different tasks. Symmetrically, a task may be executed on different processing units. The duration of a task depends on the processing unit used. The minimum and the maximum filling level of a processing unit give rise to a lower and an upper bound on the batch size. Between consecutive executions of different tasks on a processing unit, a changeover with sequence-dependent duration is necessary. In general, storage facilities of limited capacity are available for stocking raw materials, intermediates, and final products. Some products are perishable and must be consumed immediately after production. Material flows can be linear, divergent, convergent, or general. In the latter case, the product structure may also contain cycles. The input and the output proportions of the products consumed or produced, respectively, by a task are either fixed or variable within prescribed bounds. For a practical example of a batch production we refer to the case study presented by Kallrath in [8].

Typically, a plant is operated in batch production mode when a large number of different products are processed on multi-purpose equipment. In this case, the plant is configured according to (a subset of) the required final products. Before processing the next set of final products, the plant has to be reconfigured, which requires the completion of all operations. In order to ensure high resource utilization and short customer lead times, the objective of makespan minimization is particularly important. That is why we consider the short-term planning problem which for given primary requirements consists in computing a feasible schedule with minimum makespan.

Various solution methods for this problem are known from literature. Most of them follow a monolithic approach, which addresses the problem as a whole, starting from a mixed-integer linear programming formulation of the problem. In those models, the time horizon is divided into a given number of time periods, the period length being either fixed (time-indexed formulations, cf. e.g. [9]) or variable (continuous-time formulations, see e.g. [4] or [7]). The main disadvantage of the monolithic approaches is that the CPU time requirements for solving real-world problems tend to be prohibitively high (cf. [10]). To overcome this difficulty, heuristics reducing the number of variables have been developed (cf. e.g. [2]).

A promising alternative approach is based on a decomposition of the short-term production planning problem into interdependent subproblems, as it e.g. has been proposed in [3], [10], and [11]. The solution approach developed in what follows is based on the hierarchical decomposition into a batching and a batch-scheduling problem presented in [11]. Batching provides a set of batches for the intermediate and final products needed to satisfy the primary requirements. Batch scheduling allocates the processing units, intermediates, and storage facilities over time to the processing of all batches. The batching problem can be formulated as a MINLP of moderate size, which can be solved using standard software. For the solution of the batch-scheduling problem, a truncated branch-and-bound method and a priority-rule-based method have been proposed in [11] and [12], respectively. Within a reasonable amount of CPU time, good feasible solutions to problem instances with up to 100 batches can be computed with both methods. Recently, Gentner et al. ([6]) have proposed to decompose the batch-scheduling problem into a set of subproblems that are solved iteratively. The assignment of the batches to the individual subproblems is determined by solving a binary linear problem for each subproblem. This decomposition method is able to approximatively solve batch-scheduling instances with up to 3065 batches (cf. [5] and [6]) in the space of several hours of CPU time.

In this paper, we present a cyclic scheduling approach for the short-term planning problem. A preliminary version of this approach can be found in [13].

The basic idea consists in reducing the size of the batch-scheduling problem by computing a cyclic subschedule, which is executed several times. The set of batches belonging to one cycle is determined by solving a mixed-integer nonlinear program, which also provides the number of cycles needed to satisfy the primary requirements (cyclic batching problem), where we impose an upper bound on the number of batches in one cycle. The subschedule is then obtained by scheduling the batches on the processing units subject to material-availability and storage capacity constraints (cyclic batch-scheduling problem). The latter problem can be solved using one of the methods proposed in [11] and [12].

The remainder of this paper is organized as follows. In Section 2 we formulate the cyclic batching problem as a MINLP. In Section 3 we show how the copies of the subschedule can be efficiently concatenated to obtain the complete production schedule. In Section 4 we report on results of an experimental performance analysis.

2 Cyclic batching

Let T be the set of all tasks and let β_τ and ε_τ be the batch size and number of batches for task $\tau \in T$. By Π_τ^- and Π_τ^+ we denote the sets of input and output products, respectively, of task $\tau \in T$. $\Pi_\tau := \Pi_\tau^- \cup \Pi_\tau^+$ is the set of all input and output products of task τ, and $\Pi := \cup_{\tau \in T} \Pi_\tau$ is the set of all products considered. In addition to β_τ and ε_τ, the (negative) proportions $\alpha_{\tau\pi} < 0$ of all input products $\pi \in \Pi_\tau^-$ and the proportions $\alpha_{\tau\pi} > 0$ of all output products $\pi \in \Pi_\tau^+$ have to be determined for all tasks $\tau \in T$ such that

$$\sum_{\pi \in \Pi_\tau^+} \alpha_{\tau\pi} = - \sum_{\pi \in \Pi_\tau^-} \alpha_{\tau\pi} = 1 \quad (\tau \in T) \tag{1}$$

Batch sizes β_τ and proportions $\alpha_{\tau\pi}$ have to be chosen within prescribed intervals $[\underline{\beta}_\tau, \overline{\beta}_\tau]$ and $[\underline{\alpha}_{\tau\pi}, \overline{\alpha}_{\tau\pi}]$, i.e.,

$$\underline{\alpha}_{\tau\pi} \leq \alpha \leq \overline{\alpha}_{\tau\pi} \quad (\tau \in T,\ \pi \in \Pi_\tau) \tag{2}$$

$$\underline{\beta}_\tau \leq \beta \leq \overline{\beta}_\tau \quad (\tau \in T) \tag{3}$$

Let T_π^- and T_π^+ be the sets of all tasks consuming and producing, respectively, product $\pi \in \Pi$ and let $\Pi^p \subset \Pi$ be the set of perishable products. Then equations

$$\alpha_{\tau\pi}\beta_\tau = -\alpha_{\tau'\pi}\beta_{\tau'} \quad (\pi \in \Pi^p, (\tau, \tau') \in T_\pi^+ \times T_\pi^-) \tag{4}$$

ensure that the amount of a perishable product π produced by one batch of some task $\tau \in T_\pi^+$ can immediately be consumed by any task $\tau' \in T_\pi^-$ consuming π.

Let $\Pi^i \subset \Pi$ be the set of intermediate products. In order to obtain a cyclic solution allowing for executing the same subschedule an arbitrary number of times, the amount of π produced within one cycle must be equal to the amount of π consumed, i.e.,

$$\sum_{\tau \in T} \alpha_{\tau\pi}\beta_\tau\varepsilon_\tau = 0 \quad (\pi \in \Pi^i) \tag{5}$$

Note that constraint (5) also ensures that we avoid any excess of the storage capacities for intermediates.

Proportions $\alpha_{\tau\pi}$, batch sizes β_τ, and the numbers of batches ε_τ define the set of batches belonging to one cycle. The number $\xi \in \mathbb{Z}_{\geq 0}$ of cycles is a decision variable, whose value depends on the given primary requirements. By ϱ_π we denote the primary requirement less the initial stock of product π. We assume that there are no primary requirements for intermediates. The final inventory of product π then equals $\xi \sum_{\tau \in T} \alpha_{\tau\pi}\beta_\tau\varepsilon_\tau$. This amount must be sufficiently large to match the requirements ϱ_π for π, i.e.,

$$\xi \sum_{\tau \in T} \alpha_{\tau\pi}\beta_\tau\varepsilon_\tau \geq \varrho_\pi \quad (\pi \in \Pi \setminus \Pi^i) \tag{6}$$

In addition, the number of batches within one cycle must not exceed the prescribed upper bound $\overline{\varepsilon}$, i.e.,

$$\sum_{\tau \in T} \varepsilon_\tau \leq \overline{\varepsilon} \tag{7}$$

Finally, let p_τ be the mean processing time of task τ on the alternative processing units. To minimize the workload to be scheduled in the batch-scheduling step, the objective function is chosen to be the total mean processing time $\xi \sum_{\tau \in T} p_\tau\varepsilon_\tau$. In sum, the cyclic batching problem reads

$$\left\{\begin{array}{ll} \text{Minimize} & \xi \sum_{\tau \in T} p_\tau\varepsilon_\tau \\ \text{subject to (1) to (7)} & \\ & \varepsilon_\tau \in \mathbb{Z}_{\geq 0} \quad (\tau \in T) \\ & \xi \in \mathbb{Z}_{\geq 0} \end{array}\right.$$

3 Cyclic batch scheduling and concatenation

For generating the complete production schedule, we proceed as follows. At first, we compute a feasible subschedule S for executing the $\sum_{\tau \in T} \varepsilon_\tau$ batches of one cycle. S defines a partial ordering among those batches. We represent this relation by precedence relationships between the batches. Moreover, the completion time of the last batch that is processed on a processing unit defines a release date for the changeover to the first batch on that unit in the next execution of the subschedule. Analogously, the last change in the inventory level of an intermediate gives rise to a release date for the first batch that subsequently produces or consumes that intermediate.

The start and completion times for the processing of the batches in the first cycle equal those of subschedule S. For computing the start and completion times of the batches in the next cycle, we solve a temporal scheduling problem which consists in computing an earliest schedule for those batches subject to the precedence relationships between and the release dates for the batches. This temporal scheduling problem represents a longest path problem and can be solved efficiently by standard network flow algorithms (see [1]). Thus, the concatenation of the subschedule copies providing the sought production schedule can be performed in polynomial time.

4 Experimental performance analysis

We have compared the new heuristic to the decomposition approach by Gentner et al. [6]. We have used a test set introduced in [5], which has been constructed by varying the primary requirements for final products in the case study from [8]. For each instance, we have computed a locally optimal solution to the cyclic batching problem with Frontline Systems' Solver package. The subschedules have been computed by using the truncated branch-and-bound method proposed in [11]. The tests have been performed on an 800 MHz Pentium III PC. The results for the decomposition approach have been reported in [5], where a 1400 MHz Pentium IV PC has been used.

The results obtained for the 13 problem instances are displayed in Table 1. For each problem instance the new method is able to find a markedly better solution. Especially for larger problem instances, the required CPU time is significantly smaller than for the decomposition approach. Having prescribed an upper bound of $\overline{\varepsilon} = 100$ batches, about 75 seconds are required for solving

Table 1. Computational results

Instance	This paper			Decomposition [6]	
	# of batches	Makespan	CPU time [s]	Makespan	CPU time [s]
WeKa0_1	176	272	89	352	38
WeKa0_2	264	402	89	474	53
WeKa0_3	352	532	89	612	120
WeKa0_4	440	662	89	738	209
WeKa0_5	528	792	89	906	178
WeKa0_6	616	922	90	1046	215
WeKa0_7	704	1052	91	1199	323
WeKa0_8	792	1182	91	1334	281
WeKa0_9	880	1312	91	1548	399
WeKa0_10	968	1442	91	1740	431
WeKa0_15	1408	2092	91	2123	644
WeKa0_20	1848	2742	91	2899	1500
WeKa0_30	2728	4042	92	4416	5235

the cyclic batching problem. The truncated branch-and-bound method has been stopped after 15 seconds of CPU time. The concatenation has always required less than 1 second of CPU time.

Acknowledgements. This research has been supported by the Deutsche Forschungsgemeinschaft under Grant Schw1178/1.

References

1. Ahuja RK, Magnanti TL, Orlin JB (1993): Network Flows. Prentice Hall, Englewood Cliffs
2. Blömer F, Günther HO (1998): Scheduling of a multi-product batch process in the chemical industry. Computers in Industry 36:245–259
3. Brucker P, Hurink J (2000): Solving a chemical batch scheduling problem by local search. Annals of Operations Research 96:17–36
4. Castro P, Barbosa-Póvoa AP, Matos H (2001): An improved RTN continuous-time formulation for the short-term scheduling of multipurpose batch plants. Industrial & Engineering Chemistry Research 40:2059–2068
5. Gentner K (2005): Dekompositionsverfahren für die ressourcenbeschränkte Projektplanung. Shaker Verlag, Aachen
6. Gentner K, Neumann K, Schwindt C, Trautmann N (2004): Batch production scheduling in the process industries. In: Leung JYT (ed.) Handbook of Scheduling: Algorithms, Models, and Performance Analysis. CRC Press, Boca Raton
7. Ierapetritou MG, Floudas CA (1998): Effective continuous-time formulation for short-term scheduling. 1. Multipurpose batch processes, Industrial & Engineering Chemistry Research 37:4341–4359
8. Kallrath J (2002): Planning and scheduling in the process industry. OR Spectrum 24:219–250
9. Kondili E, Pantelides CC, Sargent RWH (1993): A general algorithm for short-term scheduling of batch operations: I. MILP Formulation. Computers & Chemical Engineering 17:211–227
10. Maravelias CT, Grossmann IE (2004): A hybrid MILP/CP decomposition approach for the continuous time scheduling of multipurpose batch plants. Computers & Chemical Engineering 28:1921–1949
11. Neumann K, Schwindt C, Trautmann N (2002): Advanced production scheduling for batch plants in process industries. OR Spectrum 24:251–279
12. Schwindt C, Trautmann N (2004): A priority-rule based method for batch production scheduling in the process industries. In: Ahr D, Fahrion R, Oswald M, Reinelt G (eds) Operations Research Proceedings 2003. Springer, Berlin
13. Trautmann N (2005): Operative Planung der Chargenproduktion. Deutscher Universitäts-Verlag, Wiesbaden

Planning Disassembly for Remanufacturing Under a Rolling Schedule Environment

Tobias Schulz and Ian M. Langella

Faculty of Economics and Management, Otto-von-Guericke University Magdeburg, Postfach 4120, 39016 Magdeburg, Germany.
{tobias.schulz;ian.langella}@ww.uni-magdeburg.de

Abstract. This contribution examines the performance of both exact as well as heuristic solution methods for the disassemble-to-order problem under a rolling schedule planning environment. The results indicate that the problem-specific heuristic exhibits good average performance, and low dispersion and low maximum penalty values. Even so, the cost penalties remain higher than that of applying exact methods to the rolling schedule problem.

1 Introduction

Product recovery management, where firms take back products following their use by consumers, has resulted in positive steps toward a more sustainable economy by decreasing both virgin material consumption as well as the need for scarce landfill space [2]. Within this field, remanufacturing (when firms disassemble the returned products, inspect the parts, and reassemble them into "good as new" remanufactured products) has been shown to be a particularly advantageous option, enabling the recovery of not only the material contained in the returned products (as would be the case in recycling), but also a portion of the embedded economic value [6].

In order to remanufacture products, the requisite components must be harvested from returned products through disassembly. Facing a given demand for the remanufactured product, the amount of components we need can be calculated using a bill of materials explosion. Determining how many of each type of returned product to disassemble in order to fulfill this demand is the task of so-called disassemble-to-order problems. Both exact (viz. mixed integer linear programming) as well as heuristic methods exist to solve this problem, the latter being motivated by the fact that obtaining an exact solution is not always possible for realistic problem sizes.

Planning disassembly in practice is typically done considering a rolling time horizon consisting of several periods, as is similar to production planning in traditional systems. Data on demand is available and specified for the first N periods (called a *forecast window*), and resulting decisions are only implemented for the first k periods, after which the schedule is rolled and the next N-period problem is solved, and so on. When this is done, the decisions resulting from the exact solution need not be optimal and the performance might be surpassed by using simpler heuristics, as the case in [1] which deals with lot-sizing. This contribution seeks to examine the loss of optimality which is incurred when using exact models for planning over a rolling schedule, and furthermore how a problem-specific heuristic performs in this environment.

In the following section, we will introduce the reader to the problem area, as well as the exact and heuristic solution methods. In the third section, we will describe the performance study and elaborate on its results. Concluding remarks and an outlook on further research goals are afforded in the final section.

2 Problem setting and solution methods

We can start by introducing the terminology which is used in this research. Old products which are returned and disassembled are referred to as *cores* (we use the set I to refer to the set of cores, using $i \in I$ as the index). Parts which are gained from disassembly (which are the inputs to remanufacturing) are called *leaves* (denoted by the set L, with $l \in L$ for the index). Each core type contains a specific amount of each leaf, which is called a *yield*, with $n_{i,l}$ representing the amount of leaf l contained in core i. Some leaves are only contained in one core (*unique*), while others (*common*) might be contained in several or even all of the core types.

Each core type has a given acquisition and separation cost (called a *core cost*) which represents the cost to compensate the customer for the core, transport it to the disassembly facility, and completely disassemble it down to the leaves (we denote this c_i^s). Other decision relevant costs include those related to the leaves. We can procure leaves (as an alternative to harvesting them through disassembly) at a given per unit cost of c_l^p, and leaves resulting from disassembly for which there is no immediate demand can be disposed of (which costs c_l^d per unit) or held to satisfy demand in future periods (costing c_l^h per part-period). We further denote the planning horizon T, using $t = 1, ..., T$ for the time index. The demand for the leaves which must be satisfied in each period is represented by $D_{l,t}$. Starting inventory for the leaves would be specified as $y_{l,0}$. The decisions sought include the core procurement and separation ($x_{i,t}^s$), leaf procurement ($x_{l,t}^p$), disposal ($x_{l,t}^d$), and holding ($y_{l,t}$) decisions. Generally, the problem's difficulty stems from the fact that disassembling a certain core to obtain leaves which are immediately demanded might result in obtaining leaves for which there is no immediate demand, which must be either held for future demand or disposed of.

An exact solution to this problem can be obtained from a mixed integer linear programming (MILP) model formulation, as follows:

$$Min \quad C = \sum_{t=1}^{T} \left[\sum_{i \in I} c_i^s \cdot x_{i,t}^s + \sum_{l \in L} \left(c_l^p \cdot x_{l,t}^p + c_l^d \cdot x_{l,t}^d + c_l^h \cdot y_{l,t} \right) \right] \quad (1)$$

$$y_{l,t} = y_{l,t-1} + \sum_{i \in I} x_{i,t}^s \cdot n_{i,l} + x_{l,t}^p - x_{l,t}^d - D_{l,t} \quad l \in L \quad t = 1, ..., T \quad (2)$$

$$y_{l,T} = 0 \quad l \in L \quad (3)$$

$$x_{i,t}^s, x_{l,t}^p, x_{l,t}^d, y_{l,t} \geq 0 \text{ and integer} \quad \forall i, l \quad t = 1, ..., T$$

The objective function (1) minimizes the relevant costs over the planning horizon. The constraints maintain the inventory balance for the leaves (2), ensure that leaves are not held to the end of the planning horizon in lieu of disposal (3), and force the decisions to take on non-negative integer values.

The fact that planning disassembly in practice requires the consideration of a large number of different core types over a time horizon consisting of many periods results in problem sizes too large to be reliably solved within a reasonable amount of time. This has motivated the development of problem-specific heuristics, among them the *two-phase heuristic* which is described in detail in [4]. The heuristic works by subjecting each period in the planning horizon to the two phases of the heuristic. In the first phase, demand in each period is satisfied by the disassembly of cores selected based on the core's $profitability$. Here, profitability refers to the difference between the 'revenue' of the core (in other words, the procurement cost of the resulting leaves for which there is demand in period t, costs which are saved by obtaining the leaves through disassembly) and its 'cost' (which is the sum of the core costs plus disposal costs for contained leaves for which there is no current period demand). Once the demand in the considered period has been fulfilled (and as such the solution is feasible), the second phase attempts to improve this solution by reversing disassembly decisions made in the first phase, and obtaining the leaves by either holding leaves which would have been disposed of in previous periods or through external leaf procurement, and terminates when no further improvements are possible.

3 Experimental design and results

In order to glean insight into the performance of the exact and heuristic methods in a rolling schedule environment, we start by generating large $M = 69$ period problems. The MILP and heuristic methods are then given the first $N = 10$ periods ($t = 1, ..., 10$) of demand information and solved, but only the decisions of the first $k = 1$ period are implemented. The schedule is then rolled, and the procedure repeated for $t = 2, ..., 11$, and so on until in the last roll the time periods $t = 60, ..., 69$ and the decisions for $t = 60$ are implemented. The decisions and costs resulting from both the MILP (which we will refer to as the *optimal rolling solution*) as well as the heuristics for these 60 periods are relevant for the comparison. This is compared with the MILP solution to the 69-period problem, whereas only the first 60 periods are used for the comparison. We refer to this as the *true optimal solution*. The commercial solver XPRESS-MP was used to provide solutions for all of the MILP instances.

In order to generate the parameters needed for the problem, we followed a specific experimental design based on our experience with several automobile engine remanufacturers which now will be detailed. Our study used problems with 12 core types, each containing 12 leaves. One aspect which we particularly wanted to examine was differences in the level of *commonality*, in other words the extent to which the leaves were common to different core types. To analyze this factor, we constructed problems with *high* commonality (where 8 of the 12 leaves were common to all of the cores) and *low* commonality (where 4 of the 12 leaves were common to all the cores). Leaves which were not common to all the cores were assumed to be unique to only one core. While this rules out the case that certain leaves belong to *some* but not *all* of the cores, we believe it will be sufficient to examine the role of commonality on the performance. We further assumed that at most one of each leaf was contained in the cores, i.e. $n_{i,l} \in \{0,1\} \forall i, l$. While in realistic bills of materials it is often the case that $n_{i,l} > 1$ for at least one leaf l (an aspect referred to in the disassembly literature as *multiplicity*, see e.g. [3]), in our experience most of the leaves are contained only once in the cores.

The relevant costs were drawn at random from given probability distributions. Starting with the procurement cost for the leaves, the cost was drawn from the uniform distribution with a minimum and maximum of 1 and 10 respectively. Leaf disposal costs were also taken from the uniform distribution between 0 and 2. For holding cost, we assumed it to be proportional to the procurement cost, but were also interested in ascertaining its effects on performance. As such, we constructed two levels for this factor; a *high* holding cost scenario where it was assumed to be 4% of the leaf procurement cost, and a *low* holding cost scenario where it was 2%. Core cost was drawn from the normal distribution with the mean of $0.5 \cdot \sum_{l \in L} c_l^p \cdot n_{i,l}$, which centers the core cost on half of the summed procurement costs of the leaves contained therein. Demands were also drawn from the normal distribution, with a mean of $50 \cdot \sum_{i \in I} n_{i,l}$, making the average demand proportional to the number of cores which contained the leaf.

As we have two experimental factors (commonality and holding cost) each with two levels, this results in 4 combinations in this full factorial study. For each factor combination, data for 10 instances of a 69 period problem were generated and used for the comparison. The generation and use of 10 randomly generated instances allows us to gain information on the average performance and also the dispersion, while keeping the computational effort within reasonable limits for this pilot study. As a performance measure, we use the *percent penalty* (see e.g. [5]) of the heuristic solution when compared with the true optimal solution.

Table 1. Percent penalty from optimal rolling solution

	high holding cost	**low holding cost**
high commonality	0.14 / 0.21 / 0.25	0.28 / 0.37 / 0.44
low commonality	0.27 / 0.34 / 0.39	0.51 / 0.65 / 0.79

We can start by examining the percent penalty of the optimal rolling solution shown in Table 1 where the minimum, average, and maximum penalties over the 10 problems of each scenario are given. As can be seen, there is a loss of optimality, but this loss is relatively small never exceeding 1%. This is in contrast to the loss of optimality exhibited by the Wagner-Whithin algorithm for lot-sizing in [1]. One can also see that the penalty increases both with decreasing commonality and holding costs. With respect to holding cost, we can see that a lower holding cost allows us to take more advantage of holding leaves for future demand, an aspect which shows up in the true optimal solution. As such, the optimal rolling solution (which only has access to N periods of demand data) cannot take advantage of leaf holding as well as the true optimal solution can. Regarding commonality, we opine that in problems where commonality is low, information on future demand for leaves is also of value. Due to the fact that within these scenarios, each core consists of less common leaves than in the high commonality scenarios, the value of such information seems to be higher. This might explain why the percent penalty from the optimal rolling solution when compared to the true optimal solution appears to be larger for the low commonality cases.

Table 2. Percent penalty from two-phase heuristic

	high holding cost	**low holding cost**
high commonality	1.6 / 1.7 / 1.9	1.7 / 1.9 / 2.4
low commonality	2.0 / 2.3 / 2.6	2.6 / 2.8 / 3.2

We can next examine the performance of the two-phase heuristic, using the (minimum, average, and maximum) results shown in Table 2. As can be seen, the performance is quite good both with respect to average performance, as well as maximum penalties. Specifically, in none of the instances did the heuristic deviate from the true optimal by 4%. Also supportive is that the dispersion of the cost penalty between the 10 instances in each scenario was fairly small. This suggests that the heuristic will deliver good results reliably, a desirable characteristic for a heuristic (see e.g. [5]).

4 Conclusion and outlook

We have seen that industrial sized disassemble-to-order problems consist of a large enough amount of integer decision variables as so to render the exact solution unattainable within consistently acceptable computation times. In practice, this planning is done under a rolling schedule, which causes the exact solution to lose optimality, albeit with relatively small penalties. Our results indicate that the two-phase heuristic works well in this environment, with low average penalties and a low dispersion suggesting reliably good results. Further research can increase the scope of the performance study, in order to ascertain if the results obtained are generally valid. Another interesting aspect might be to incorporate the aspect of *planning nervousness*, where demands in periods which lie at the end of the forecast horizon suffer from uncertainty. When this is the case, the exact solution performance might deteriorate faster than that given by myopic heuristic. This will be relegated, though, to future research efforts.

References

[1] J. D. Blackburn and R. A. Millen. Heuristic lot-sizing performance in a rolling-schedule environment. *Journal of Operations Management*, 11: 691–701, 1980.

[2] M. P. de Brito and R. Dekker. A framework for reverse logistics. In R. Dekker, M. Fleischmann, K. Inderfurth, and L. N. Van Wassenhove, editors, *Reverse Logistics: Quantitative Models for Closed-Loop Supply Chains*, pages 3–27. Springer, Berlin, 2004.

[3] A. J. D. Lambert and S. M. Gupta. *Disassembly Modeling for Assembly, Maintenance, Reuse, and Recycling.* CRC Press, Boca Raton, 2005.

[4] I. M. Langella and T. Schulz. Effects of a rolling schedule environment on the performance of disassemble-to-order heuristics. Working paper, Faculty of Economics and Management, Otto-von-Guericke University Magdeburg, Germany, 2005.

[5] E. A. Silver. An overview of heuristic solution methods. *Journal of the Operational Research Society*, 55(9):936–954, 2004.

[6] M. Thierry, M. Salomon, J. Van Nunen, and L. Van Wassenhove. Strategic issues in product recovery managment. *California Management Review*, 37(2):114–135, 1995.

An LP-based Heuristic Approach for Strategic Supply Chain Design

Rafael Velásquez[1,2], M.Teresa Melo[1,3], and Stefan Nickel[1,4]

[1] Fraunhofer Institute for Industrial Mathematics (ITWM), D 67663 Kaiserslautern, Germany
[2] velasque@itwm.fhg.de
[3] melo@itwm.fhg.de
[4] Chair of Operations Research and Logistics, Saarland University, D 66041 Saarbrücken, Germany, s.nickel@orl.uni-saarland.de

Summary. A novel heuristic approach is proposed for solving a large scale facility location problem arising in supply chain design. The problem formulation includes many practical aspects such as a dynamic planning horizon, generic supply chain structure, inventory and distribution of goods, budget constraints, and storage limitations. Moreover, facility location decisions are modelled through the gradual relocation of existing facilities to new sites over a given planning horizon. The heuristic approach explores the solution of the linear relaxation of the problem. It successively rounds the fractional variables corresponding to the 0/1 decisions of changing the facilities' status (i.e., open new / close existing facilities), and it is also used to roughly estimate the total number of facility configuration changes over the planning horizon. The proposed heuristic performs very well on a large set of randomly generated problems, producing feasible solutions that on average only deviate 1.4% from the optimum.

1 Introduction

The design of supply chain networks is a complex decision making process. The typical inputs comprise a set of products to be manufactured and distributed to customer zones with known demands. Based on these inputs as well as on information about future conditions and costs, companies have to decide on the location, size and number of new facilities (e.g. warehouses) to operate, as well as the flow of goods through the supply chain network to satisfy known demands. Although facility location and configuration of supply chains have been studied for many years, a number of important practical issues have not received adequate attention in the literature or have been treated individually by several authors as shown in [1]. However, network design is strongly affected by the simultaneous consideration of issues such as the external supply of commodities, inventory opportunities for goods, storage limitations, and availability of capital for investing in facility (re)location. Neglecting these aspects results in formulations with limited applicability to real world supply chains. The main objective of this paper is to fill this gap by proposing a realistic model for strategic supply chain design and developing an efficient heuristic to solve it. In the next section, the problem is presented and a mixed integer linear formulation is briefly described. Sect. 3 is devoted to a heuristic method for solving the problem. Sect. 4 reports on the computational experience while Sect. 5 presents some conclusions.

2 Problem Description and Formulation

It is assumed that a company is considering to gradually relocate part or all of the capacity of some of its existing facilities to new locations during a certain time horizon. Prior to the planning project, a set of candidate sites has been identified where new facilities can be established. Furthermore, in each time period a given budget is available for investing in capacity transfers, in the setup of new facilities and in the shutdown of existing facilities. Any capital available in a period, but not invested then, is subject to an interest rate and the returned value can be used in subsequent periods. Relocation costs incurred by capacity shifts are assumed to depend on the amount moved from an existing facility to a new site, and account, e.g. for workforce and equipment transfers. Other facility costs (i.e., opening, closing and operating) are fixed and independent of the size of the facility. Since the setup of a new facility is usually a time-consuming process, it is assumed that its investment takes place in the period immediately preceding the start-up of operations. Hence, if a new facility starts operating in some period t, fixed costs are charged with respect to its setup in period $t-1$. On the other hand, when an existing facility ceases operating at the end of some period t, shutdown costs are charged in the following period. Next, the required notation is introduced.

Index sets L = set of facilities; S = set of *selectable* facilities with $S \subset L$ and $S = S^c \cup S^o$; S^c = set of *existing* facilities that can be closed; S^o = set of potential sites for establishing *new* facilities; P = set of product types; T = set of periods with $\mid T \mid = n$.

Parameters $\overline{K}_\ell^t$ = maximum capacity of facility $\ell \in L$ in period $t \in T$; $\underline{K}_\ell^t$ = minimum required throughput at the selectable facility $\ell \in S$ in period $t \in T$; $\mu_{\ell,p}$ = unit capacity consumption factor of product $p \in P$ at facility $\ell \in L$; $H_{\ell,p}$ = stock of product $p \in P$ at facility $\ell \in L$ at the beginning of the planning horizon (observe that $H_{\ell,p} = 0$ for every $\ell \in S^o$); $D_{\ell,p}^t$ = external demand of product $p \in P$ at facility $\ell \in L$ in period $t \in T$; B^t = available budget in period $t \in T$; α^t = unit return factor on capital not invested in period $t \in T \setminus \{n\}$, that is, $\alpha^t = 1 + \beta^t/100$ with β^t denoting the interest rate in period t.

Costs $PC^t_{\ell,p}$ = variable cost of producing or purchasing (from an external supplier) one unit of product $p \in P$ by facility $\ell \in L$ in period $t \in T$; $TC^t_{\ell,\ell',p}$ = variable cost of shipping one unit of product $p \in P$ from facility $\ell \in L$ to facility $\ell' \in L \setminus \{\ell\}$ in period $t \in T$; $IC^t_{\ell,p}$ = variable inventory carrying cost per unit on hand of product $p \in P$ in facility $\ell \in L$ at the end of period $t \in T$; $MC^t_{i,j}$ = unit variable cost of moving capacity at the beginning of period $t \in T \setminus \{1\}$ from the existing facility $i \in S^c$ to a new facility established at site $j \in S^o$; OC^t_ℓ = fixed cost of operating facility $\ell \in S$ in period $t \in T$; SC^t_i = fixed cost charged in period $t \in T \setminus \{1\}$ for closing existing facility $i \in S^c$ at the end of period $t-1$; FC^t_j = fixed setup cost charged in period $t \in T \setminus \{n\}$ when a new facility established at site $j \in S^o$ starts its operation at the beginning of period $t+1$.

Decision variables $b^t_{\ell,p}$ = amount of product $p \in P$ produced or purchased by facility $\ell \in L$ in period $t \in T$; $x^t_{\ell,\ell',p}$ = amount of product $p \in P$ shipped from facility $\ell \in L$ to facility $\ell' \in L \setminus \{\ell\}$ in period $t \in T$; $y^t_{\ell,p}$ = amount of product $p \in P$ held in stock at facility $\ell \in L$ at the end of period $t \in T \cup \{0\}$ (observe that $y^0_{\ell,p} = H_{\ell,p}$); $z^t_{i,j}$ = amount of capacity shifted from the existing facility $i \in S^c$ to a newly established facility at site $j \in S^o$, at the beginning of period $t \in T$; ξ^t = capital not invested in period $t \in T$; $\eta^t_\ell = 1$ if selectable facility $\ell \in S$ changes its status in period $t \in T$, and 0 otherwise. Observe that if an existing facility $i \in S^c$ is closed at the end of period t then $\eta^t_i = 1$. Similarly, if a new facility is opened in site $j \in S^o$ at the beginning of period t then $\eta^t_j = 1$.

At the beginning of the planning horizon, all existing facilities in the set S^c are operating. Afterwards, capacity can be shifted from these facilities to one or more new facilities selected from sites in S^o. In view of the assumptions made on the time points for paying fixed facility costs, it follows that a new facility can never operate in the first period since that would force the company to invest in its setup before the beginning of the planning horizon ($\eta^1_j = 0$, $j \in S^o$). Analogously, an existing facility cannot be closed at the end of the last period since the corresponding shutdown costs would be charged in a period beyond the planning horizon ($\eta^n_i = 0$, $i \in S^c$). A mixed integer linear formulation of the problem, denoted by **P**, was first introduced in [1] and consists of the following five groups of constraints: (*i*) general product flow conservation equations for each facility and time period, with product flow being triggered by customer demands $D^t_{\ell,p}$ that have to be served; (*ii*) capacity relocation to a new site can only be executed from operating existing facilities and can only occur after a new facility has been opened at the new site. Moreover, when an existing facility has transferred all of its capacity, its shutdown takes place; (*iii*) total product flow in each facility and period cannot exceed the maximum capacity of the facility. Also, it is only worth operating a selectable facility $\ell \in S$ in period $t \in T$ if its total product flow is at least $\underline{K}^t_\ell$; (*iv*) budget B^t is available in every period t to cover capacity transfers as well as closing and setup costs of existing and new facilities, respectively. Capital ξ^{t-1} not invested in period $t-1$ gains interest at a rate of α^{t-1} and can be used in period t; (v) a facility may change its status at most once, meaning that once an existing (new) facility is closed (opened), it is not allowed to start operating (to be closed) in a later period. Finally, integrality and non-negativity conditions on the decision variables are imposed. The objective function of **P** minimizes the overall costs which include variable production/purchase, transportation and inventory holding costs for all facilities and products during the entire planning horizon. In addition, fixed facility operating costs are also charged. As shown in [1], **P** generalizes many well-known dynamic facility location models including those restricted to opening new facilities and closing existing facilities (no relocation opportunities). Since the facility status variables η_ℓ^t are the only binary decision variables (all other are non-negative continuous variables), this feature is exploited in the solution approach as described in the next section.

3 The Heuristic Approach

The heuristic developed to solve **P** is based on assigning different combinations of 0/1 values to the facility status variables η_ℓ^t, $\ell \in S$, $t \in T$. Observe that for a fixed 0/1 combination, the original problem reduces to a linear program **P-LP** which can be solved in polynomial time and whose optimal solution is also feasible for **P**.

A 0/1 status matrix Ψ of size $|S| \times |T|$, which groups the facility status variables η_ℓ^t, states for each selectable facility $\ell \in S$ and period $t \in T$ whether a status change occurs or not. In the first case, the corresponding entry has value "1", otherwise it is "0". Since each row refers to a facility, its entries can be altered as long as at most one value "1" appears in the row (recall that a facility can have its status changed at most once).

To determine a good combination of 0/1 values, **P** is first relaxed by allowing the η_ℓ^t variables to take any value in the interval $[0, 1]$. The optimal solution of **P-LP** usually contains some variables with a tendency towards one of the two binary values (0 or 1). Furthermore, those variables whose optimal values do not show a tendency towards zero are saved in a set of so-called *CDD-pairs*. Each CDD-pair corresponds to a facility ℓ and a period t for which the optimal value of η_ℓ^t in **P-LP** is larger than a given threshold.

The heuristic starts by using integer rounding: those η_ℓ^t variables with a strong tendency towards zero ($\eta_\ell^t < \underline{\eta}$) or one ($\eta_\ell^t > \overline{\eta}$) are fixed at zero or one respectively, with $\overline{\eta}$ and $\underline{\eta}$ positive pre-defined thresholds. All other η_ℓ^t variables with values in $[\underline{\eta}, \overline{\eta}]$ are left as continuous variables. Solving the resulting linear program (LP) leads to further η_ℓ^t variables with a strong tendency towards zero or one, which are fixed accordingly. The procedure is repeated until no more fixing of variables is possible. If there are remaining fractional variables, these are ordered in decreasing value. Denoting by h the sum of the fractional values, the $\lfloor h \rfloor$ first η_ℓ^t variables are set to one, while the remaining variables are set to zero. The resulting integer status matrix Ψ can lead to a first feasible solution for **P**.

If this is not the case, and assuming a binding budget which limits the total number of facility configuration changes to at most, say $\overline{a}$, the next step aims at estimating $\overline{a}$. Initially, all η_ℓ^t variables not included in the set of CDD-pairs are set to zero, while the remaining variables can take any value in the interval $[0, 1]$. The latter are then successively selected at random and set to one if the resulting LP is still feasible, or set to zero otherwise. The process is repeated until all variables have an integer value. In the case that the resulting 0/1 combination does not yield a feasible solution for **P**, the sum $\overline{a}$ of those variables currently with value 1 is built. Next, $\overline{a}$ facilities are randomly selected from the CDD-pairs and set to one, while all other variables are set to zero. This procedure is applied a pre-defined number of times. In case no feasible solution is found, it is again repeated for $\overline{a} := \overline{a} - 1$ and $\overline{a} := \overline{a} + 1$.

The last step of the heuristic attempts to improve a previously found feasible solution. It is also applied when feasibility was not attained before. It consists of searching for a new solution by altering the status matrix Ψ of the best solution found so far or the resulting status matrix obtained after the integer rounding approach. Let Q^1 be the set of η_ℓ^t variables fixed at one in Ψ. In each iteration, m different η_ℓ^t variables are selected at random and fixed at "0", while at the same time, m other η_ℓ^t variables are randomly chosen and fixed at "1" ($m \in \{1, 2, \ldots, \omega\}$ and $\omega < |Q^1|$). This procedure is repeated a pre-defined number of times for every m, always keeping the best solution. For further information on the parameters used and a detailed description of the heuristic the interested reader is referred to [2].

4 Computational Experience

The heuristic was implemented in C++, while the LP subproblems were solved with ILOG Cplex 7.5 on a Pentium III PC with a 2.6 GHz processor and 2 GB RAM. Eighteen problem types were randomly generated corresponding to three different problem classes. Details of the problem characteristics can be found in [1]. For each problem type, five different instances were randomly generated yielding in total 90 test problems. Each instance was solved six times with the heuristic. In addition, to evaluate the quality of the solutions generated by the heuristic, all test problems were solved to optimality with the MIP solver of ILOG Cplex 7.5. Table 1 summarizes the results obtained. Columns 2 and 3 report the average computational time required by the heuristic and the MIP solver, while column 4 indicates the number of times the heuristic is faster than the general purpose solver. Column 5 is a performance indicator stating the average deviation of the objective function value of the heuristic w.r.t. the optimal solution. Finally, column 6 lists the number of test runs that lead to a feasible solution.

Table 1. Avg. running time and performance of the heuristic

Prob. type	Average CPU time (s) Heuristic	OPT	Heur. vs. OPT (times faster)	Obj. Funct. Dev. (%)	# feas. sol. found (max = 30)
P1	20.0	33.8	1.7	1.34	30
P2	22.4	177.8	7.9	0.64	30
P3	43.4	1329.4	30.6	2.65	30
P4	65.4	5742.6	87.8	1.83	30
P5	25.0	97.0	3.9	1.34	30
P6	36.6	156.0	4.3	11.64	30
P7	2.8	49.6	17.7	0.01	30
P8	37.8	412.2	10.9	0.88	30
P9	12.4	265.2	21.4	0.02	30
P10	12.6	138.4	11.0	0.31	30
P11	10.4	59.2	5.7	0.14	30
P12	6.8	63.4	9.3	0.00	30
P13	95.6	595.8	6.2	2.54	30
P14	24.4	395.6	16.2	0.01	30
P15	9.6	31.8	3.3	0.20	30
P16	85.0	190.2	2.2	0.54	20
P17	152.2	1689.2	11.1	0.49	24
P18	22.4	64.2	2.9	0.28	29
Avg.	38.0	638.4	14.1	1.38	29.1

As shown in Table 1, in 523 out of 540 runs (i.e. 96.9%), at least one feasible solution was found by the heuristic. Furthermore, the solutions obtained deviated on average only 1.38% from the optimum and were generated, on average, at a rate 14.1 faster than the computational time required by ILOG Cplex.

5 Conclusions and Outlook

As described in the previous section, the proposed heuristic performs very well with respect to solution quality and computational time. Regarding the latter, the longest the heuristic needed to solve a problem instance was a little under 9 minutes, while the commercial solver required almost four hours. This leads to the belief that the heuristic can solve much larger problems for which an optimal solution would be too time consuming or even impossible to obtain due to memory restrictions. Further studies will be directed towards reducing the computational effort and finding a feasible solution for every instance.

References

1. Melo MT, Nickel S, Saldanha da Gama F (2006) Dynamic multi-commodity capacitated facility location: A mathematical modeling framework for strategic supply chain planning. C&OR 33(1):181–208
2. Velásquez R (2005) On solving a large-scale dynamic facility location problem with variable neighborhood and token ring search. MA Thesis, Technical University Kaiserslautern, Kaiserslautern, Germany

Der Einfluss von alternativen Bezugsquellen auf die optimale Beschaffungsstrategie

Ivo Neidlein

Fakultät für Wirtschaftswissenschaft, Otto-von-Guericke-Universität, Postfach 4120, 39016 Magdeburg ivo.neidlein@ww.uni-magdeburg.de

Zusammenfassung. In dieser Arbeit wird mithilfe eines Newsboy-Ansatzes der Einfluss von (kurzfristigen) Spotmärkten auf langfristige Kontrakte zwischen Supply Chain Partnern betrachtet.

1 Einleitung

In der klassischen Theorie des Supply Chain Management wird oftmals davon ausgegangen, dass ein (Groß-)Händler seinen Bedarf nur bei einem einzigen Lieferanten deckt. Diese Annahme ist sinnvoll, da üblicherweise bei mehreren potentiellen Lieferanten der günstigste gewählt wird und somit eine einfache Lieferanten-Abnehmer-Beziehung entsteht.

Allerdings kann es für den Händler auch vorteilhaft sein, sein Material von mehrere Lieferanten zu beziehen. Ein Grund dafür ist, dass der Lieferant nicht immer lieferfähig sein muss; dieses Risiko kann durch die Beschaffung bei mehreren Lieferanten abgefedert werden. Als weiterer Grund lassen sich strategische Überlegungen anführen: Falls die Zahl potentieller Anbieter gering ist, besteht die Gefahr, dass eine Monopolsituation entsteht und auf den Abnehmer in Zukunft höhere Kosten zukommen. (vgl. Klotz und Chatterjee (1995) [2])

In der Realität unterscheiden sich Lieferanten meist nicht nur durch den Preis, sondern auch durch die Art der Lieferbeziehung. So werden beispielsweise Rohstoffe sowohl über längerfristige Kontrakte als auch kurzfristig über Spotmärkte gehandelt. Dies ist z.B. der Fall bei Aluminium, hier werden 90% der gehandelten Menge über Kontrakte bezogen, 10% über Spotmärkte (vgl. Peleg et al. (2005) [3] und Wu und Kleindorfer (2005) [5]).

Als kurzfristige Marktplätze existieren jedoch nicht nur Spotmärkte, sondern auch Auktionen. Den Einsatz von Auktionen bei der Beschaffung behandeln beispielsweise Seshadri et al. (1991) [4] Ein Beispiel: In Holland werden Zierpflanzen und Blumen nicht nur über langfristige Kontrakte gehandelt, sondern schon seit langem auch auf so genannten „Holländischen Blumenauktionen“. Während sich die Preise bei längerfristigen Kontrakten über die Zeit betrachtet nur wenig verändern, unterliegen die Spotmarktpreise beziehungsweise die auf Auktionen erzielten Preise starken Schwankungen. In dieser Arbeit soll nun ein Modell aufgestellt werden, welches die zu unterschiedlichen Zeitpunkten vorhandenen Informationen über die kurz- und langfristigen Beschaffungsalternativen bei unsicherer Nachfrage abbildet und somit eine optimale Kombination der Beschaffungsalternativen liefert. Zunächst wird nun ein Basismodell für langfristige Kontrakte entwickelt, das im weiteren auf zwei Arten erweitert wird. Einerseits wird der Einfluss eines Spotmarkts untersucht, andererseits der Einfluss einer Auktion. Für beide Erweiterungen wird eine eine optimale Beschaffungsentscheidung analysiert.

2 Basismodell

Wir gehen hier von einem einperiodigen Newsboymodell aus. Grundlage ist eine stochastische Endkundennachfrage r mit Dichtefunktion $\varphi(r)$ und Verteilungsfunktion $\Phi(r)$, die zwischen den Grenzen r^u und r^o schwankt. Der Händler muss bereits vor der Realisation der Nachfrage festlegen, welche Menge B er beschafft. Ist B kleiner als r, so kann er nur die Menge B zum Preis v verkaufen. Die restliche Nachfrage kann er nicht bedienen, und es entgeht ihm Gewinn. Hat er zu viel beschafft, kann er den Überschuss nur noch zum Schrottwert s verkaufen. Wenn es nur eine Materialquelle gibt, bei der der Händler zum Preis p einkaufen kann, so ist seine optimale Bestellmenge

$$B^* = \Phi^{-1}\left(\frac{v-p}{v-s}\right).$$

3 Spotmarktfall

Erweitern wir nun dieses Modell auf den Fall, dass der Händler zwei alternative Beschaffungsquellen hat. Er kann, um die Endkundennachfrage zu bedienen, sein Material entweder über einen langfristigen Kontrakt beschaffen, oder seinen Bedarf kurzfristig über einen Spotmarkt decken. Um die Lang- und Kurzfristigkeit der verschiedenen Alternativen darzustellen, wird von folgender zeitlicher Abfolge der Entscheidungen ausgegangen.

Der Händler bestellt zum Zeitpunkt $t = 1$ über einen langfristigen Kontrakt die Menge B_K und zahlt dabei pro Stück den Kontraktpreis p_k. Zum Zeitpunkt $t = 2$ kauft er die Menge B_M auf dem Spotmarkt und zahlt dafür den Spotmarktpreis p_M. Schließlich realisiert sich zum Zeitpunkt $t = 3$ die Marktnachfrage, die er mit dem zuvor beschafften Material bedient. Der Spotmarktpreis p_M ist zum Zeitpunkt des Kontraktabschlusses unsicher. Wir nehmen an, dass in $t = 1$ nur Dichte- und Verteilungsfunktion $\psi(p_M)$ und $\Psi(p_M)$ des Spotmarktpreises bekannt sind.

Wenn der erwartete Spotmarktpreis μ_{p_M} kleiner oder gleich dem Kontraktpreis p_K ist, so wird der Händler, der sich nur am erwarteten Gewinn orientiert, seinen Bedarf nur über den Spotmarkt decken. Der Händler wird also nur dann einen längerfristigen Kontrakt eingehen, wenn der Kontraktpreis kleiner als der erwartete Spotmarktpreis ist. In diesem Fall ist die Kontraktmenge also größer gleich Null, und der Händler steht vor dem Optimierungsproblem, wieviel er über den Kontrakt und wieviel er über den Spotmarkt beschaffen soll. Die Gewinnfunktion ist dann

$$\begin{aligned} G_1(B_K, B_M) = v \int_{r^u}^{B_K+B_M} r\varphi(r)dr + v \int_{B_K+B_M}^{r^o} (B_K + B_M)\varphi(r)dr \\ +s \int_{r^u}^{B_K+B_M} (B_K + B_M - r)\varphi(r)dr - p_K B_K - \int_{p_M^u}^{p_M^o} p_M B_M \psi(p_M). \end{aligned} \tag{1}$$

Diese Gewinnfunktion ist von den beiden Variablen B_M und B_K abhängig. Betrachten wir zunächst die optimale Menge B_M für gegebenes B_K. Wenn B_K fest ist, dann realisiert sich als nächstes in $t = 2$ der Marktpreis p_M. Für nun gegebenes p_M und B_K sieht die Gewinnfunktion wie folgt aus:

$$\begin{aligned} G_2(B_M) = v \int_{r^u}^{B_K+B_M} r\varphi(r)dr + v \int_{B_K+B_M}^{r^o} (B_K + B_M)\varphi(r)dr \\ +s \int_{r^u}^{B_K+B_M} (B_K + B_M - r)\varphi(r)dr - p_M B_M, \end{aligned} \tag{2}$$

und es lässt sich analog dem Newsboy-Ansatz die optimale Menge B_M bestimmen:

$$B_M^* = \max\left[\Phi^{-1}\left(\frac{v - p_M}{v - s}\right) - B_K, 0\right]. \tag{3}$$

In Abhängigkeit von der Kontraktmenge existiert also ein maximaler Preis, zu dem der Händler noch bereit ist etwas auf dem Spotmarkt zu kaufen, nämlich $p_M^{\max} = v - (v - s)\Phi(B_K)$. Aus Gleichung (1) und Gleichung (3) erhält man nun die Gewinnfunktion $G_1(B_K, B_M^*)$in Abhängigkeit von p_M und B_K

$$\begin{aligned} G_1(B_K) = &\int_{p_M^u}^{v-\Phi(B_K)(v-s)} \int_{r^u}^{\Phi^{-1}(\frac{v-p_M}{v-s})} vr\varphi(r)dr\psi(p_M)dp_M \\ &+ \int_{v-\Phi(B_K)(v-s)}^{p_M^o} \int_{r^u}^{B_K} vr\varphi(r)dr\psi(p_M)dp_M \\ &+ \int_{p_M^u}^{v-\Phi(B_K)(v-s)} \int_{\Phi^{-1}(\frac{v-p_M}{v-s})}^{r^o} v\Phi^{-1}\left(\frac{v-p_M}{v-s}\right)\varphi(r)dr\psi(p_M)dp_M \\ &+ \int_{v-\Phi(B_K)(v-s)}^{p_M^o} \int_{B_K}^{r^o} vB_K\varphi(r)dr\psi(p_M)dp_M \\ &+ \int_{p_M^u}^{v-\Phi(B_K)(v-s)} \int_{r^u}^{\Phi^{-1}(\frac{v-p_M}{v-s})} s(\Phi^{-1}\left(\frac{v-p_M}{v-s}\right) - r)\varphi(r)dr\psi(p_M)dp_M \\ &+ \int_{v-\Phi(B_K)(v-s)}^{p_M^o} \int_{r^u}^{B_K} s(B_K - r)\varphi(r)dr\psi(p_M)dp_M \\ &- p_K B_K - \int_{p_M^u}^{v-\Phi(B_K)(v-s)} p_M(\Phi^{-1}(\tfrac{v-p_M}{v-s}) - B_K)\psi(p_M)dp_M. \end{aligned} \tag{4}$$

Wenn man nun diese Gewinnfunktion nach B_K ableitet, erhält man folgende Optimalitätsbedingung:

$$p_K = \mu_{p_M} - \int_{p_M^{\max}(B_K)}^{p_M^o} (p_M - p_M^{\max}(B_K))\psi(p_M)dp_M.$$

Die Grenzkosten des Kontrakts entsprechen im Optimum den Grenzkosten des Spotmarktes. Letztere bestehen aus dem erwarteten Spotmarktpreis μ_{p_M} und einem Korrekturterm, welcher den Fall berücksichtigt, dass auf dem Spotmarkt wegen eines zu hohen Preises nichts gekauft wird.

Für gegebene Verteilungsfunktionen von Φ und Ψ lässt sich die optimale Kontraktmenge B_K explizit angeben. Für allgemeine Verteilungsfunktionen lassen sich folgende Aussagen bezüglich der Parameter treffen:

Der Anteil der Kontraktmenge an der Gesamtbeschaffungsmenge ist zum einen vom Kontraktpreis p_K abhängig und zum anderen von der Verteilung des Marktpreises. Wenn der Kontraktpreis größer ist als der erwartete Marktpreis, ist die Kontraktmenge null. Ansonsten gilt, ass mit steigendem Kontraktpreis die Kontraktmenge fällt und mit steigendem erwarteten Spotmarktpreis die Kontraktmenge steigt.

Die Varianz des Spotmarktpreises hat ebenfalls Einfluss darauf, wieviel über den Kontrakt und wieviel über den Spotmarkt gekauft wird. Allerdings spielt hierbei die Art der Dichtefunktion eine Rolle. Man kann sagen, dass eine größere Varianz bei einer großen Klasse von Verteilungsfunktionen (der beispielsweise die Normalverteilung und die Gleichverteilung angehören) mit einer geringeren Kontraktmenge einhergeht. Es lassen sich aber auch Dichtefunktionen konstruieren, bei denen für bestimmte Parameterkonstellationen eine größere Varianz eine größere Kontraktmenge bedeutet.

4 Auktion

Wir betrachten nun den Fall, dass die zweite Beschaffungsquelle nicht ein Spotmarkt, sondern eine Auktion ist. Es wird zum Zeitpunkt $t = 2$ eine feste Menge B_A versteigert.

In unserem Fall betrachten wir eine verdeckte Zweitpreisauktion. Das bedeutet, dass alle Gebote verdeckt abgegeben werden und am Ende der Bieter mit dem höchsten Gebot den Zuschlag erhält; der zu bezahlende Preis ist jedoch nur das zweithöchste Gebot. Der Händler steht nun vor der Fage, wieviel er für die Menge B_A bieten soll. Aus der Auktionstheorie ist bekannt, dass es in diesem Fall für ihn optimal ist, seine maximale Zahlungsbereitschaft zu bieten (vgl. Klemperer (1999)[1]). Diese wird mit $\pi_A^{\max}$ bezeichnet. Diese entspricht gerade dem erwarteten zusätzlichen Erlös, den er mit der Menge B_A erzielen kann, bei gegebenem B_K.

Mit seinem Gebot kann der Händler nun also den Preis, den er bezahlt, sowie die Wahrscheinlichkeit, mit der er den Zuschlag erhält, beeinflussen, nicht aber die (feste) Menge B_A.

Um sein Verhalten zu optimieren, benötigt der Händler noch eine Annahme über das Gebotsverhalten der anderen Auktionsteilnehmer, genauer: über die Verteilung des Höchstgebots der anderen. Diese habe die Dichtefunktion $f(\pi)$ und Verteilungsfunktion $F(\pi)$. Dann ist $F(\pi_A^{\max}(B_K))$ die Wahrscheinlichkeit, dass der Händler bei der Auktion den Zuschlag erhält, und $\int_{\pi_u}^{\pi_A^{\max}(B_K)} \pi f(\pi) d\pi$ ist der erwartete Preis, den er für B_A bezahlen muss.

Zur Veranschaulichung gehen wir zunächst davon aus, dass keine Unsicherheit in der Nachfrage existiert, die Nachfrage also ein festes R ist. Die Gewinnfunktion ist dann nicht vollständig, sondern nur noch abschnittsweise differenzierbar.

Wenn die verauktionierte Menge kleiner als die Nachfrage R und und $F(\pi_A^{\max}(B_K = R - B_A)) > \frac{v-p_K}{v-s}$ ist, existieren zwei lokale Optima. Diese bezeichnen wir im folgenden als auktionsdominiertes bzw. kontraktdominiertes Optimum. Das kontraktdominierte Optimum zeichnet sich dadurch aus, dass die Kontraktmenge B_K genau der Nachfrage R entspricht. Beim auktionsdominierten Optimum geht der Händler davon aus, dass er die Auktion gewinnen wird, und beschafft nur so viel über den Kontrakt, wie notwendig ist, um auch die Nachfrage zu decken, die über die versteigerte Menge B_A hinausgeht. Folglich ist die Kontraktmenge im auktionsdominierten Optimum gleich $R - B_A$ und im kontraktdominierten Optimum gleich R. Wenn B_A größer als R ist, so ist die Randlösung $B_K = 0$ ein Lösungskandidat, da negative B_K nicht zulässig sind. Ist $F(\pi_A^{\max}(B_K = R - B_A)) \leq \frac{v-p_K}{v-s}$ so existiert nur ein Optimum, nämlich $B_K = R$.

Im weiteren gehen wir nun davon aus, dass die Nachfrage unsicher ist, wieder mit der Nachfrageverteilung $\Phi(r)$. Außerdem wird vorausgesetzt, dass $\Phi(r)$ über den gesamten relevanten Bereich differenzierbar ist (somit ist auch die Gewinnfunktion differenzierbar). Der erwartete Nutzen von B_A, also die maximale Zahlungsbereitschaft des Händlers, beträgt dann

$$\pi_A^{\max}(B_K) = vB_A - (v-s)\int_{r^u}^{B_K} B_A \varphi(r) dr - (v-s)\int_{B_K}^{B_K+B_A} (B_K + B_A - r)\varphi(r) dr.$$

Die Gewinnfunktion des Händlers ist dann

$$\begin{aligned} G(B_K) = {} & v \int_{r^u}^{B_K} r\varphi(r) dr + v \int_{B_K}^{r^o} B_K \varphi(r) dr + s \int_{r^u}^{B_K} (B_K - r)\varphi(r) dr \\ & + F(\pi_A^{\max}(B_K))\pi_A^{\max}(B_K) - pB_K - \int_{\pi_u}^{\pi_A^{\max}(B_K)} \pi f(\pi) d\pi \end{aligned}$$

mit der Optimalitätsbedingung

$$\frac{dG(B_K)}{dB_K} = v - p_K - (v-s)\Phi(B_K) + F(\pi_A^{\max}(B_K))\frac{d\pi_A^{\max}(B_K)}{dB_K} = 0.$$

In diesem stochastischen Fall bleiben einige Eigenschaften der Problemstruktur aus dem deterministischen Fall erhalten.

Diese Gewinnfunktion ist wie vorher nicht zwingend konkav in B_K, es können also wiederum mehrere lokale Optima existieren. Wenn B_A sehr groß ist, kann das auktionsdominierte Optimum eine Randlösung ($B_K = 0$) sein. Die unsichere Nachfrage bewirkt aber auch, dass die Auktion Einfluss auf das kontraktdominierte Optimum hat. Dieses ist nun kleiner gleich dem Optimum des (stochastischen) Basismodells aus Abschnitt 2. Der Händler versucht also sowohl den langfristigen Kontrakt als auch die Auktion zu nutzen. Ebenso beeinflusst der Kontrakt das auktionsdominierte Optimum. Es ist außerdem möglich, dass kontrakt- und auktionsdominiertes Optimum zusammenfallen und somit die Gewinnfunktion konkav ist.

5 Zusammenfassung und Ausblick

Wir haben ein Modell zur Optimierung der Beschaffungsmenge bei unsicherer Nachfrage, unsicherem Bezugsbedingungen und einer Kombination von kurz- und langfristigen Beschaffungsalternativen entwickelt. Dabei haben wir festgestellt, ob und wie bei Vorhandensein einer weiteren Quelle diese auch genutzt wird. Ein Teil des Bedarfes wird in der Regel über die kurzfristige Beschaffungsalternative gedeckt. Das bedeutet, dass weniger über den langfristigen Kontrakt beschafft wird als im Basismodell. Im Spotmarktfall existiert eine optimale Kombination aus Kontrakt- und Spotmarktbeschaffungsmenge, wobei der Anteil, der über den Spotmarkt beschafft wird, anwächst, wenn der Kontraktpreis steigt oder der erwartete Spotmarktpreis fällt. Außerdem wurde gezeigt, dass im Auktionsfall lokale Optima existieren können.

Als nächster Schritt ist denkbar, den Einfluss einer Mehrgüterauktion zu untersuchen, d.h. wie das Bietverhalten und die Abnahmemenge aussehen, wenn der Händler nicht nur den Bietpreis, sondern auch die Abnahmemenge beeinflussen kann. Ferner bleibt zu untersuchen, welchen Einfluss komplexere Kontraktformen zur Supply Chain Koordination, wie Mengenrabattkontrakte oder Kontrakte mit Rückkaufoption, auf die vom Händler über den Kontrakt bezogene Menge haben.

Literaturverzeichnis

1. Klemperer P (1999) Auction Theory: A Guide to the Literature. Journal of Economic Surveys 13: 227-286
2. Klotz D E, Chatterjee K (1995) Dual Sourcing in Repeated Procurement Competitions. Management Science 41: 1317-1327
3. Peleg B, Lee H L, Hausmann W H (2002) Short-Term E-Procurement Strategies versus Long-Term Contracts. Production and Operations Management 11: 458-479
4. Seshadri S, Chatterjee K, Lilien G L (1991) Multiple Source Procurement Competitions. Marketing Science 10: 246-263
5. Wu D J, Kleindorfer P R (2005) Competitive Options, Supply Contracting and Electronic Markets. Management Science 51: 452-466

Distributed Planning in Product Recovery Networks

Eberhard Schmid, Grit Walther, and Thomas Spengler*

Technische Universität Braunschweig, Inst. f. Wirtschaftswissenschaften, Abt. BWL, insbes. Produktion und Logistik
{e.schmid|g.walther|t.spengler}@tu-bs.de

Summary. We consider a scenario, where a network of disassembly companies treats waste electronic equipment due to environmental regulation. Legal requirements like collection and recycling targets are formulated as network wide constraints of a network flow model. With regard to a distributed decision situation where a focal company with bounded knowledge coordinates activities of the disassembly network, fulfilment of these common constraints requires coordination. Assuming that the focal company's aim is to maximise network wide profit, a decentralised problem formulation using Lagrangean relaxation is presented.

1 Introduction

Due to legal requirements and the insight that recycling, remanufacturing and reuse of products may provide economic benefits, the field of reverse logistics and product recovery rapidly gains attention. In this paper, we focus on the recycling of waste electric and electronic equipment (WEEE). According to the european WEEE-directive (EC 2003), original equipment manufacturers (OEMs) are obliged to take back their equipment and to pay for the proper treatment. Further, they have to fulfill collection, recycling and recovery targets. It is expected that OEMs will outsource the treatment of WEEE to specialised companies. One part of the recovery process is the disassembly of scrap. In Germany, disassembly is mainly performed by networks of small and medium sized independent companies (Walther 2005). In this paper, we develope a negotiation oriented planning approach to support decentralised planning in such a disassembly network. In the next Section, we outline the planning problem and present a central optimisation approach. We then develope a decentralised negotiation model and give a small example.

* This work has been promoted by the German "Deutsche Forschungsgemeinschaft" (DFG). The authors would like to thank for the support

2 Framework and centralised model

We consider a network of independent disassembly companies which are coordinated by a focal company called "network centre". The flows in the network are shown in figure 1. y_{iqu}^{AQ} denotes the mass of product $i \in P$ that is delivered from source $q \in Q$ to dissassembly company $u \in U$. x_{ju} denote the number of executions of disassembly operation $j \in J$ in company $u \in U$. The choice of disassembly operations determines the disassembly depth. Disassembly depths differ regarding the disassembly time and cost needed but also regarding the material fractions generated. y_{iur}^{DR} is the mass of product / disassembly fraction $i \in P \cup F$ (F denotes the set of material fractions) that are delivered from disassembly company $u \in U$ to sink $r \in R \cup L$, where R denotes the set of reycling companies and L denotes the set of other disposal facilities.

We assume that the network centre is mandated by an OEM to collect a specific mass of electronic scrap and to fulfil the legal recycling targets. The disassembly companies' cost strucures and capacities are private knowledge. We therefore want to construct a mechanism that allows the network centre to negotiate contracts with disassembly companies with regard to quantities, transfer prices for discarded products and targets to be fulfilled by each company. Thereby, we assume that the network centre aims at maximising network wide profit. The framework is outlined in figure 2. We concentrate on the part that is marked by the frame.

To develope a negotiation mechanism, we first present a centralised planing model which can be used to calculate the first best solution assuming perfect hindsight of the central decision maker. The allocation of scrap to the companies and the indivual choices of disassembly depths which determine the material fractions and thus the recycling targets, can be determined using linear optimisation (Walther and Spengler 2005).

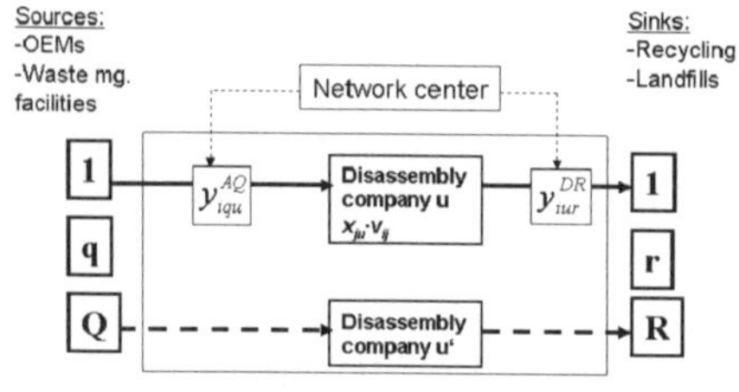

Fig. 1. Material flows

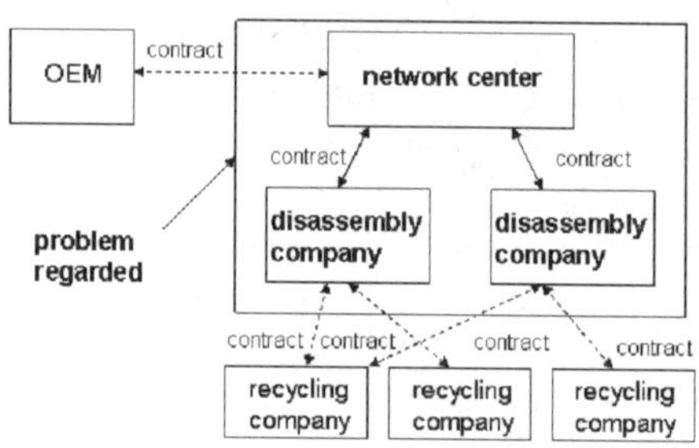

Fig. 2. Planning framework

The objective function maximises the total contribution margin:

$$v^* = \max_{y_{iqu}^{AQ}, y_{iur}^{DR}, x_{ju}} \sum_{u \in U} \sum_{i \in P} \sum_{q \in Q} e_i^A y_{iqu}^{AQ} + \sum_{u \in U} \sum_{r \in R \cup L} \sum_{i \in P \cup F} e_{ir}^V y_{iur}^{DR} - \sum_{u \in U} \sum_{j \in J} c_{ju}^Z x_{ju} \quad (1)$$

where e_i^A denotes the acceptance fee that the network gets per mass unit of product i and c_{ju}^Z is the cost of executing disassembly activity j in company u. e_{ir}^V denotes the selling price (+) / disposal cost (-) of delivering fraction i to facilitiy r. The mass balance equation is given as follows:

$$\sum_{q \in Q} y_{iqu}^{AQ} + \sum_{j \in J} x_{ju} \cdot v_{ij} = \sum_{r \in R \cup L} y_{iur}^{DR} \qquad \forall i \in P \cup F; u \in U. \tag{2}$$

where v_{ij} denotes the mass that is consumed (-) / generated (+) of product / fraction i by the execution of disassembly operation j. The achievement of the collection and recycling targets can be written as:

$$\sum_{u \in U} y_{iqu}^{AQ} = AD_{iq} \qquad \forall i \in P, q \in Q \tag{3}$$

$$\sum_{i \in P} \sum_{u \in U} \sum_{q \in Q} y_{iqu}^{AQ} \cdot \alpha \leq \sum_{u \in U} y_u^{OR}, \qquad y_u^{OR} = \sum_{i \in F} \sum_{r \in R} y_{iur}^{DR} \tag{4}$$

where AD_{iq} denotes the mass that is to be collected from product i at source q. The recycling target, i.e. the percentage of the input mass that is to be recycled, is denoted by α. y_u^{OR} is the recycled mass. To keep the analysis simple, we renounce the seperate modelling of recycling and recovery targets. We assume that material fractions are deemed to be recycled if they are delivered to specialised recycling companies ($r \in R$). More sophisticated concepts use recycling coefficients, accepting only a percentage of the mass entering a recycling company as recycled. This method, as well as the complete model containing several constraints regarding local capacities and the NNCs can be found in Walther and Spengler (2005).

3 Decentralised approach

3.1 Basic mechanism

When independent companies are involved in such a network, coordination cannot be achieved by a central plan because the planner does not posses all relevant information. Therefore, we propose a negotiation oriented approach with sparse information exchange to achieve coordination. Such a negotiation oriented approach can be implemented as a multi agent system (Jennings 2001), where the agents represent the disassembly companies respectively the network centre. To this end, we relax the common constraints ((3) and (4)), which are the collection and recycling targets, and integrate them into the objective function via Lagrangean multipliers. The problem can then be decomposed into independent subproblems, one for each disassembly company, which can be solved independently with respect to the Lagrangean multipliers. For each disassembly company, the following problem can be formulated:

$$v_u^{*l}(\lambda,\pi) = \max_{y_{iqu}^{AQ}, y_{iur}^{DR}, x_{ju}} \lambda y_u^{OR} + \sum_{i\in P}\sum_{q\in Q}(e_i^A - \lambda\alpha - \pi_{iq})y_{iqu}^{AQ} + \sum_{r\in R\cup L}\sum_{i\in P\cup F} e_{ir}^V y_{iur}^{DR} - \sum_{j\in J} c_{ju}^Z x_{ju} \quad (5)$$

where $y_u^{OR} = \sum_{i\in F}\sum_{r\in R} y_{iur}^{DR}$. The Lagrangean multipliers, π_{iq} and λ can be interpeted as contract parameters between the network centre and a disassembly company. An economic interpretation would be an additional charge or bonus for accepted scrap and additional bonus payments for recycled masses. The determination of the optimal Lagrangean multipliers can then be performed by a subgradient procedure (e.g. Holmberg 1995). Thereby, the network centre proposes a set of Lagrangean multipliers and in iteration k the disassembly companies reply $y_{iqu}^{AQ,k}$ and $y_u^{OR,k}$ as tentative contractual parameters. The network centre then decides whether to continue the negotiation. The negotiation terminates if the common constraints are satisfied. Else, the multipliers are updated as follows. First, the actual exceedance of each constraint is determined, which is a subgradient of the Lagrangean dual of (1):

$$\Delta\lambda^k = \sum_{i\in P}\sum_{u\in U}\sum_{q\in Q} y_{iqu}^{AQ,k}\cdot\alpha - \sum_{u\in U} y_u^{OR,k}, \qquad \Delta\pi_{iq}^k = \sum_{u\in U} y_{iqu}^{AQ,k} - AD_{iq} \quad (6)$$

Then, the Lagrangean multipliers can be updated:

$$\lambda^{k+1} = \max(0, \lambda^k + t_k\cdot\Delta\lambda^k), \qquad \pi_{iq}^{k+1} = \pi_{iq}^k + t_k\cdot\Delta\pi_{iq}^k \quad (7)$$

where t_k denotes a step-specific parameter.

3.2 Obtaining feasible solutions

For linear problems, such a price directed mechanism in general may not generate feasible resp. optimal solutions since the solutions reported by the disassembly companies are not necessarily optimal respectively feasible for the overall problem (e.g. Sherali and Choi 1996). Therefore, our algorithm works in two phases. In the first phase, the standard subgradient mechanism proceeds. In the second phase, the algorithm aims at constructing feasible solutions in a negotiation oriented style. With respect to the recycling constraint (4) feasible (but not optimal) solutions may be found. However, the treatment constraints (3) are to be fulfilled exactly. To make the disassembly companies report solutions that result in an exact fulfillment of the overall constraints (3), we propose the following scheme: If the companies are willing to treat less than is available at the sources, the standard adjustment procedure continues. If in iteration k they want to treat more scrap than available at the sources ($\sum_{u\in U} y_{iqu}^{AQ,k} > AD_{iq}$), the network centre provides each company a maximum treatment quantity (MTQ), MTQ_{iqu}^k for every constraint that is exceeded. This MTQ can be interpreted as an additional contract parameter besides the prices. This is repeated until a feasible solution is found.

If the network centre knew the optimal values of y_{iqu}^{AQ} (y_{iqu}^{AQ*}), it could provide these values as MTQ for the disassembly companies and for some specific values of the Lagrangean multipliers the disassembly companies would choose these values. However, this is not the case. One intuitive idea is to define a fraction fr_{iqu}^{k} of the mass that is to be treated at the sources as MTQ for each company. The MTQ could then be calculated as $fr_{iqu}^{k} \cdot AD_{iq}$. Since this may restrict the solution space of the companies too much, no feasible solution might be found. We therefore choose the MTQ as a convex combination of a fraction of the mass that is available at the sources and the quantity every company wanted to treat in the last iteration:

$$MTQ_{iqu}^{k} = \epsilon \cdot fr_{iqu}^{k} \cdot AD_{iq} + (1-\epsilon) \cdot y_{iqu}^{AQ,k-1} \quad \forall i,u,q \quad 0 < \epsilon \leq 1, \tag{8}$$

with $\sum_{u \in U} fr_{iqu}^{k} = 1 \quad \forall i,q$. To determine fr_{iqu}^{k} we use the following: From Sherali and Choi (1996) we know that if the step size and some weights μ are choosen properly, $\hat{y}_{iqu}^{AQ,k} = \sum_{\tau=1}^{k} \mu_{\tau}^{k} y_{iqu}^{AQ,\tau}$ converges to the optimal values y_{iqu}^{AQ*} as $k \to \infty$. Further, a step size $t_k = a/(b+dk)$ $(a > 0, b \geq 0, d > 0)$ and weights $\mu_{\tau}^{k} = 1/k \quad \forall \tau = 1,\ldots,k$ satisfy theese properties. We use this to construct fr_{iqu}^{k} in the following manner:

$$fr_{iqu}^{k+1} = \hat{y}_{iqu}^{AQ,k} / \sum_{u \in U} \hat{y}_{iqu}^{AQ,k} \tag{9}$$

4 Example

In this section, we provide a small numerical example. In our example, there is one product available at three sources. The following quantities are to be collected from the sources by the disassembly network: $AD_{11} = 135,000$, $AD_{12} = 50,000$ and $AD_{13} = 55,000$. There exists one network centre, coordinating collection activities of three disassembly companies which have limited disassembly capacities. The product ($i = 1$) may be disassembled by disassembly activity 1 into fractions 2 and 3. Alternatively, a complete disassembly ($j = 2$) of the product yields fractions 2, 4 and 5. Let us assume that company 1 may only execute disassembly operation 1. The other companies are able to execute both operations. Fractions 2 and 3 may be disposed of at landfills at a price of 0.1 per kilogramme scrap. Fractions 4 and 5 may either be sold to recycling companies at a price of 1 per kilogramme or be disposed of at landfills. The legal recycling target is 40%. The following table shows the results for different cost structures of the companies and for different ϵ. For the step size we choose $a = 0.05$, $b = 0$ and $d = 1$.

The first line in every scenario shows the optimal costs and the cost deviations of the decentralised model. The second line shows the optimal and the decentrally achieved recycling targets. It can be seen that with an increasing number of iterations, the quality of the negotiated contract improves.

$u=1$	$u=2$		$u=3$		v^*	1000 iterations			5000 iterations			9000 iterations		
c_{11}^Z	c_{12}^Z	c_{22}^Z	c_{13}^Z	c_{23}^Z		ϵ=1	ϵ=0.5	ϵ=0.1	ϵ=1	ϵ=0.5	ϵ=0.1	ϵ=1	ϵ=0.5	ϵ=0.5
1	1	2	1	2	350,000	3.6%	3.6%	3.5%	2.3%	2.3%	2.3%	1.8%	1.8%	1.8%
					40.0%	43.7%	45.8%	45.6%	43.7%	43.7%	43.6%	43.0%	43.0%	43.0%
1	1	2	2	3	422,400	11.4%	11.4%	11.2%	5.7%	5.7%	5.6%	3.4%	3.4%	3.4%
					40.0%	45.8%	45.8%	45.6%	42.6%	42.6%	42.6%	41.0%	41.1%	41.4%
1	1	2	3	4	494,400	16.7%	16.9%	16.7%	7.6%	7.6%	7.6%	4.2%	4.2%	4.2%
					40.0%	45.8%	45.8%	45.7%	42.4%	42.4%	42.4%	41.4%	41.4%	41.4%

Table 1. numeric results for different nubers of iterations

5 Conclusions

Starting with a centralised planning approach, we developed a decentralised model in order to negotiate contracts between decision makers in a recycling network. This mechanism requires only sparse information exchange between the decision makers. Further research has to be done improving the quality of the algorithm, in particular the covergence speed. Moreover, we only examined a decision situation where the goal of the network centre is to optimise global network profit. If the goals are conflicting, further aspects like "double marginalisation" may arise and more generic concepts of distributed decision making (e.g. Schneeweiss 2003) have to be applied. Besides advances in methodolgy, applications to real world decision situations are a challenging goal to evaluate the practicability of such approaches.

References

EC. Directive 2002/96/ec of the european parliament and of the council of 27 january 2003 on waste electrical and electronic equipment (WEEE). Technical report, 2003.

K. Holmberg. Primal and dual decomposition as organizational design: Price and/or resource directive decomposition. In R. M. Burton and B. Obel, editors, *Design Models for Hierarchical Organization*, pages 61–92. Kluwer Academic Publishers, Boston, 1995.

N. R. Jennings. An agent based approach for building complex software systems. *Communications of the ACM*, 44(4):35–41, 2001.

C. Schneeweiss. *Distributed Decision Making*. Springer, Berlin, 2003.

H. D. Sherali and G. Choi. Recovery of primal solutions when using subgradient optimization methods to solve Langrangian duals of linear programs. *Operations Research Letters*, 19:105–113, 1996.

G. Walther. *Recycling von Elektro- und Elektronik-Altgeräten: Strategische Planung von Stoffstrom-Netzwerken für kleine und mittelständische Unternehmen*. Deutscher Universitätsverlag, Wiebaden, 2005.

G. Walther and T. Spengler. Impact of WEEE-directive on reverse logistics in Germany. *International Journal of Physical Distribution and Logistics Management*, 35(5):337–361, 2005.

Valuing Product Portfolios Under Uncertainty and Limited Capacity

Philippe Schiltknecht and Marc Reimann

ETH Zurich
Institute for Operations Research
Clausiustrasse 47, CH-8092 Zurich
{schiltknecht,reimann}@ifor.math.ethz.ch

Summary. This paper deals with the investigation of product portfolios in a make-to-order manufacturing setting characterized by demand uncertainty and limited production capacity. Using a simple two-period model we address the general question of planning under uncertainty and show the profit/cost implications of an individual contract in the portfolio and their dependence on capacity tightness.

1 Introduction

The research presented in this paper is motivated by an industrial project with a company from the chemical industry serving the market of exclusive fine chemicals. On this market highly specialized companies are entrusted with the manufacturing of inter-mediates and active ingredients for the life science industry. The business of exclusive fine chemicals is characterized by a small density of customers and traditionally a make-to-order approach is applied, based on customer-supplier contracts. Usually these contracts feature high customer flexibility in terms of demand quantity and delivery date. Undoubtedly this increases customers' loyalty and satisfaction but exposes the manufacturer to uncertainty and risk.

In order to fulfill the flexibilities granted and to deal with changing customer requirements, suppliers can basically react in two ways: either they try to increase their *operational flexibility*, that is their capability to adapt planning and production or they try to change *contractual flexibility*. Changing contractual flexibility thereby means to introduce e.g. (real) options, such as the possibility to renegotiate contract details as due-dates.

Clearly, contractual and operational aspects are closely interlinked, as the profit/cost effects of contractual settings are influenced by the operational planning process. Thus, the quality of a specific contract will not only depend on its own revenue and cost structure but also on the available capacity and on the other products it competes with. The value of a certain product/contract can thus not be fully determined by a 'static' analysis of a contract's specifications, rather it is also necessary to take into account the operational interaction of all products. In this paper we present a simple model that links the operational planning with contractual issues and we shed some light on the appropriate valuation of contracts. Especially we would like to focus on the following questions

- What is the influence of a particular product/contract within a product portfolio under uncertain demand?
- How does this value depend on operational characteristics, such as cost structures, capacity tightness or level of uncertainty?

The remainder of the paper is organized as follows. In the next section we introduce and discuss our model and briefly review some related approaches. Section 3 provides some first computational results. Finally, we conclude this paper with a short summary and an outlook on future research.

2 Related Work and Model Formulation

Related to contractual questions various contributions can be found in academic literature (e.g. the valuation of contracts and (real) options as addressed in Barnes-Schuster et al. [1] or Tsay [7]). Unfortunately most articles thereby neglect operational aspects as known from classical production planning. With respect to operational processes, especially in the field of planning and scheduling, numerous articles have been published in OR literature (see e.g. Clark [2] or Gupta et al. [6]). Compared to that handling of uncertainty in operational processes is relatively new, but attention has constantly increased over the last few years and many different approaches have been proposed, e.g. stochastic programming models as mentioned in Clay and Grossmann [5], simulative methods as proposed in Xie et al. [9] or scenario-based approaches as cited in Gatica et al. [4]. Current models for solving stochastic (planning) problems are practically limited to single- or two-stage problems.

The *2-stage model* presented in this work is structured as follows: every product i is characterized by a contract, which specifies a demand distribution D_i as well as a due date t_i^{due} and a prior reveal date t_i^{reveal}, at which the true demand becomes known ($t_i^{\text{reveal}} \leq t_i^{\text{due}}$). The uncertainty stemming from the demand distributions D_i is modeled by J different scenarios. Thereby for every scenario and each product a realization is sampled from the according distribution. In the following we assume that all N products have the same due date and the same reveal date. Denoting by Cap_1 the capacity before the reveal date, and by Cap_2 the capacity after the reveal date, the setting of the model can be summarized as shown in figure 1.

Basically we have to solve the following decision problem: (i) which quantities Q_i shall be produced in the first planning section before the revelation of demand (one decision for each product *independent* of any scenario) and (ii) which quantities Q_{ij} shall be produced in the second planning section, when the exact demands are known (one decision for each product and scenario, i.e. $N \cdot J$ decisions).

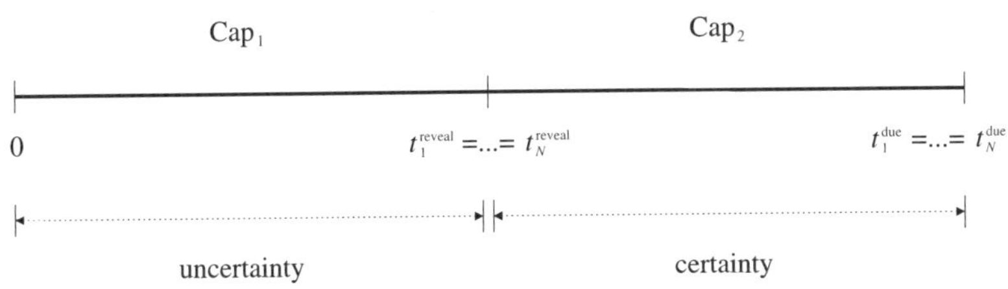

Fig. 1. Graphical illustration of the *2-stage model*

From the operational point of view especially two aspects have to be considered. First, the available capacity is limited, i.e. the production decisions have to respect the according limitations induced by Cap_1 and Cap_2. Secondly every product needs a foregoing setup. These setups are *sequence independent* but occupy capacity and cause additional costs.

With respect to the objective function we want to maximize the expected profit, given as the mean over all scenarios. The profit for the scenarios is thereby calculated as the sum over the product-wise shares, whereby given a scenario j and a product i, the following profit function is assumed:

$$\begin{aligned} \Pi(Q_{ij}^{\text{tot}}) = \; & \min(D_{ij}, Q_{ij}^{\text{tot}}) \cdot \text{price}_i - Q_{ij}^{\text{tot}} \cdot \text{productioncosts}_i \\ & - \max(D_{ij} - Q_{ij}^{\text{tot}}, 0) \cdot \text{penalty}_i - \text{setupcosts}_i \end{aligned}$$

Therein D_{ij} denotes the realized demand of product i in scenario j and $Q_{ij}^{\text{tot}} = Q_i + Q_{ij}$ the according total production quantity. We see that the income is truncated by the realized demand, and that under- and overproduction is penalized. Furthermore if a product is produced setup costs are charged[1].

Even though our model describes a very special instance, the optimization problem turns out to be very complex and difficult to solve. That stems mainly from the setups, which are modeled as binary variables resulting in a mixed integer formulation. It is clear, that there are thus computational limits for large instances (w.r.t. number of products and scenarios). Nevertheless even for small instances, interesting results can be observed.

[1] The following special case has to be observed: No setup is needed in the second planning section, if the according product (or setup) has been scheduled at the end of planning section 1.

3 Preliminary Results

In the following illustrative example we assume a portfolio of 5 products while uncertainty is modeled by 100 demand scenarios. For the products 1, 3, 4 and 5 the underlying distributions are left-truncated normal distribution with the following parameters (mean μ, standard deviation σ): Prod. 1 (21,3), Prod. 3 (27,4), Prod. 4 (5,0.8), Prod. 5 (17, 2.5).

For product 2 we will assume a distribution consisting of two steps: (i) Given a probability of rejection (default $\mathbb{P}[\text{rejection}] = 0.5$) it is determined whether or not the according customer will place an order or not. (ii) If *no* order is placed the according demand will be set to zero. If an order is placed a second sampling takes place and the demand is set according to an uniform distribution on the interval [12,15]. In the following we will refer to this product as the 0/1-product[2]. The operational parameters are given in table 1.

Table 1. Operational parameters of the different products

	setup time [days]	setup cost [$]	prod.-cost [$/day]	price [$/day]	penalty [$/day]
Prod. 1	18	500'000	125'000	165'000	40'000
Prod. 2	14	500'000	200'000	280'000	80'000
Prod. 3	18	420'000	125'000	155'000	35'000
Prod. 4	10	100'000	60'000	80'000	20'000
Prod. 5	10	150'000	80'000	100'000	20'000

The capacity of the first planning section Cap_1 (before revelation of the demand), is assumed to be infinitely large. By placing high production quantities in the first planning section it would be possible to satisfy the customers' demands in all scenarios. However such a wasteful strategy does not pay off, as in most of the scenarios the production cannot be fully sold. Hence the question arises how much of the different products is optimally produced under uncertainty. As seen in table 2 the according decisions strongly depend on the capacity after the reveal date[3]:

Table 2. Production quantities Q_i and profits for different values of Cap_2.

	Cap_2 – Capacity after reveal date (in [days])							
	0	25	50	75	100	125	∞	[days]
Prod. 1	20.46	20.42	20.32	0.00	0.00	0.00	0.00	
Prod. 2	0.00	0.00	0.00	0.00	0.00	0.00	0.00	
Prod. 3	**25.94**	**23.02**	**13.18**	**21.59**	**21.13**	**16.24**	**16.24**	[days]
Prod. 4	4.73	4.75	4.75	4.75	4.75	0.00	0.00	
Prod. 5	15.67	15.41	15.41	15.38	0.00	0.00	0.00	
$\mathbb{E}$[Profit]	-309'982	420'336	839'492	1'018'218	1'117'675	1'209'597	1'219'764	[$]

[2] Products within clinical trial processes can be modeled as 0/1-products

[3] The bold quantities denote the last product of planning section 1. If planning section 2 continues with the same product no setup is needed.

The results are quite intuitive. We can see that given more capacity after the reveal dates, the production under uncertainty is reduced and shifted to the second planning section, leading to a substantial increase in profit. Focusing on the 0/1-product we observe that no production takes place under uncertainty. Instead it seems more profitable to bring forward other products as an implicit capacity reservation for a possible production of product 2 in the second planning section.

In the following we will focus on the 0/1-product and investigate how, for different capacity situations, changes of $\mathbb{P}$[rejection] affect the planning strategy, and the expected profit of the whole portfolio. In table 3, the results are given for second-stage capacities of 25 and 50 and $\mathbb{P}[\text{rejection}] = 0.25, 0.5, 0.75$ and 1:

Table 3. Production Quantities Q_i and profits, given $\text{Cap}_2 = 25$ and 50, and different rejection probabilities.

Capacity after reveal date			$\mathbb{P}$[rejection]				
			0.25	0.5	0.75	1	
	25	Prod. 1	20.42	20.42	20.42	20.46	
		Prod. 2	**0.00**	0.00	0.00	0.00	
		Prod. 3	25.94	**23.02**	**22.31**	**12.99**	[days]
		Prod. 4	4.75	4.75	4.75	4.75	
		Prod. 5	15.55	15.41	15.38	15.55	
		$\mathbb{E}$[Profit]	530'039	420'336	471'155	542'867	[$]
	50	Prod. 1	20.36	20.32	20.36	0.00	
		Prod. 2	0.00	0.00	0.00	0.00	
		Prod. 3	**13.18**	**13.18**	**16.30**	**21.59**	[days]
		Prod. 4	4.75	4.75	0.00	4.75	
		Prod. 5	15.55	15.41	15.38	15.67	
		$\mathbb{E}$[Profit]	978'111	839'492	717'859	736'373	[$]

For $\text{Cap}_2 = 25$ the minimal profit is achieved for $\mathbb{P}[\text{rejection}] = 0.5$. Given $\text{Cap}_2 = 50$ the according minimum depends more on the absence of the 0/1-product and is achieved for $\mathbb{P}[\text{rejection}] = 0.75$. Comparing the profits for $\mathbb{P}[\text{rejection}] = 0.5$ and $\mathbb{P}[\text{rejection}] = 0.75$, the information that in the latter case only in 25% of the scenarios an order will be placed, reduces penalty-costs and allows a better planning. Planning adaptations and penalty reductions thus outperform the loss of expected income, caused by the fall in demand of 25%. Interesting also to see, that for a second-stage capacity of 25 and $\mathbb{P}[\text{rejection}] = 0.25$, it pays off to perform a setup for the 0/1 -product while no production occurs. No similar decision can be observed for any of the other cases. We also observe, that under uncertainty production of products 1,3,4,5 is reduced with increasing rejection probabilities. This might be explained by the fact, that for higher rejection probabilities (implicit) capacity reservation for the 0/1-product becomes obsolete.

4 Conclusion and Future Research

In this paper we investigated the relation between individual contracts and the final profit & loss of the whole portfolio while taking into account production and planning processes. Even for the very stylized production setting given in this paper, we were able to exemplify the influence of the operational environment. Moreover we have shown that changing the available capacity in combination with contractual adaptations may lead to different planning strategies as well as changes in profit & loss.

Future work will now deal with a more extensive study of different parameters characterizing the problem environment. Additionally we also plan to incorporate a risk oriented optimization view similar to the ideas given in Eppen et al. [3] and based on the concepts presented in Rockafellar et al. [8]. Further, the simple model has to be embedded into a more realistic rolling-horizon based setting. Finally we would like to draw implications for designing contracts in order to support sales managers in dealing with the customers.

References

1. Barnes-Schuster D., Bassok Y., Anupindi R. (2002). Coordination and Flexibility in Supply Contracts with Options. Journal of Manufacturing & Service Operations Management **4**(3):171-207.
2. Clark A.R. (2000). Rolling-Horizon Lot-Sizing when Set-up Times are Sequence-Dependent. International Journal of Production Research **38**(10):2287-2307.
3. Eppen G.D., Martin R.K., Schrage L. (1989). A Scenario Approach to Capacity Planning. Operations Research **37**(4):517-527.
4. Gatica G., Papageorgiou L.G., Shah N. (2003). Capacity Planning under Uncertainty for the Pharmaceutical Industry. Transactions of the Institutions of Chemical Engineers **81**(Part A):665–678.
5. Clay R.L., Grossmann I.E. (1997). A disaggregation algorithm for the optimization of stochastic planning models. Computers and Chemical Engineering **21**(7):751–774.
6. Gupta D., Magnusson Th. (2003). The Capacitated Lot-Sizing and Scheduling Problem with Sequence-Dependent Setup Costs and Setup Times. Computers & Operations Research **41**:727–747.
7. Tsay A. (1999). The Quantity Flexibility Contract and Supplier-Customer Incentives. Management Science **45**(10):1139–1358.
8. Rockafellar R.T., Uryasev S. (2000). Optimization of conditional Value-at-Risk. Journal of Risk **2**(3):21–41.
9. Xie J., Zhao X., Lee T.S. (2002). Freezing the Master Production Schedule under Single Resource Constraint and Demand Uncertainty. International Journal of Production Economics **83**:65–84.

Entwicklung eines reaktiven Schedulingsystems für die Prozessindustrie

Ulf Neuhaus, Hans Otto Günther

Fachgebiet Produktionsmanagement, Technische Universität Berlin, Wilmersdorfer Str. 148 10585 Berlin, Germany, Ulf.Neuhaus@tu-berlin.de

Da insbesondere in der chemischen Industrie die Anlagenbelegungsplanung einer Vielzahl von Nebenbedingungen unterliegt, können diese durch einen menschlichen Planer nur schwer gleichzeitig berücksichtigt werden. Treten darüber hinaus während des Produktionsablaufes unvorhergesehene Ereignisse ein, wie z.B. Anlagenausfälle oder das Eintreffen von Eilaufträgen, ist eine manuelle Anpassung des bestehenden Produktionsplans nur mit großem Aufwand möglich. Zur Unterstützung dieser Problemstellung schlagen wir eine reaktive Planungsmethodik vor, die den speziellen Anforderungen der chemischen Sortenproduktion entspricht.

1. Einführung

Die chemische Sortenproduktion ist durch den Einsatz von Mehrzweckanlagen gekennzeichnet, die nach dem Prinzip der Batchproduktion arbeiten. Auf den Mehrzweckanlagen können durch die Variation der Prozessbedingungen und durch flexible Verrohrungen der einzelnen Anlagenteile eine Vielzahl von unterschiedlichen Prozessschritten ausgeführt werden. Aufgrund der erhöhten Prozessflexibilität dieses Anlagentyps sind kurzfristige Reaktionen auf Änderungen der Nachfrage von Endprodukten möglich. Im Rahmen der Anlagenbelegungsplanung führt dies jedoch auch zu einer stark erhöhten Problemkomplexität, da nicht nur die Art und Anzahl der auszuführenden Prozessschritte geplant werden müssen, sondern auch deren genaue zeitliche Zuordnung zu den einzelnen Anlagenteilen. Weiterhin können bei dem Wechsel zwischen verschiedenen Produktvarianten z.B. infolge von Reinigungs- und Umkonfigurationsarbeiten reihenfolgeabhängige Rüstzeiten und -kosten auftreten, die im Hinblick auf eine effiziente Anlagenauslastung zu vermeiden sind. Da diese Planungsaufgabe eine besondere Herausforderung darstellt, wurden in den letzten Jahrzehnten zahlreiche wissenschaftliche Beiträge zu dieser Thematik veröffentlicht (vgl. z.B. Pinedo, 1999, Baptise 2001).

Die Mehrzahl dieser Ansätze geht dabei von einer statischen Planungsumgebung aus. Dabei wird angenommen, dass alle planungsrelevanten Informationen im Voraus bekannt sind und dass in dem betrachteten Produktionssystem bisher keine Aufträge ausgeführt werden.

In realen Produktionssystemen kann jedoch bei der Produktionsausführung eine Vielzahl von unvorhersehbaren Ereignissen auftreten, die eine Fortführung des bestehenden Plans verhindern und somit eine Plananpassung erfordern. Beispiele für solche Störereignisse sind der Ausfall von einzelnen Teilanlagen, das Eintreffen von Eilaufträgen oder Abweichungen von Prozesszeiten und Produktspezifikationen. Aus diesem Grund sind für den praktischen Einsatz von Schedulingverfahren erweiterte Methoden zur Plananpassung von sehr hoher Relevanz. Um den Entscheidungsprozess des Planers sachgerecht zu unterstützen, ist eine zweckmäßige Reschedulinglogik anzuwenden, die den Erfordernissen des jeweiligen Produktionsumfeldes Rechnung trägt. Einen umfassenden Überblick solcher Reschedulingstrategien geben Viera et al. (2003). Die überwiegende Anzahl der dort beschriebenen Ansätze ist jedoch für den Bereich der diskreten Stückgutfertigung vorgesehen. Für die chemische Sortenproduktion existieren bisher jedoch nur vereinzelte Ansätze. (vgl. z.B. Lee et al. 2001, Méndez et al. 2003) Aus diesem Grund ist es das Ziel des vorliegenden Beitrages, eine reaktive Planungsmethodik für diese Anwendungsumgebung zu präsentieren.

2. Architektur des reaktiven Schedulingsystems

Das von uns vorgestellte reaktive Planungssystem basiert auf einer prädiktiv-reaktiven Reschedulingstrategie und besteht im Wesentlichen aus vier Einzelmodulen, deren Interaktion im folgenden kurz skizziert werden soll. Das *Scheduler-Modul* dient zur Erstellung eines prädiktiven Ausgangsplans. Dieser Plan wird anschließend einem *Simulationsmodell* übergeben, welches das Verhalten des realen Produktionsumfeldes repräsentiert. Basierend auf dem Ausgangsplan wird die Simulation gestartet und bis zum Auftreten eines zufällig auftretenden Störereignisses ausgeführt. Bei der diesem Beitrag zu Grunde liegenden Version des Systems werden Anlagenausfälle als Störereignisse berücksichtigt. Beim Auftreten einer Störung wird zunächst der *Analyser* aktiviert, der zunächst entscheidet, ob eine Plananpassung erforderlich ist. Eine Plananpassung ist dann erforderlich, wenn während der gesamten Störungsdauer auf der gestörten Anlage mindestens ein Auftrag eingeplant ist. Ist dies der Fall, erfolgt die Aufbereitung der Ausgangsdaten für den *Rescheduler*, der einen aktualisierten Produktionsplan unter Berücksichtigung der aktuellen Planungssituation erstellt. Dieser Plan wird dann wieder der Simulation übergeben, die mit dessen Abarbeitung fortfährt bis zum Auftreten der nächsten Störung oder bis zum Ende des Planungshorizontes. Da der Analyser und Rescheduler die für die Umsetzung der vorgestellten Reschedulinglogik zentralen Module Module sind, wird in den folgenden Abschnitten deren Funktionsweise ausführlicher erläutert. Dabei werden die folgenden Symbole verwendet:

Indizes und Indexmengen:

$t \in \{0,...,H\}$ Perioden

$u \in U$ Anlageneinheiten

$i \in I$ chemische Prozessschritte

$s \in S$ Produkte, z.B. Rohstoffe, End- und Zwischenprodukte

$f \in F$ Menge aller relevanten Störungen zum aktuellen Zeitpunkt t^{act}

$u \in U^{End}$ Anlageneinheiten die Endprodukte produzieren

$u \in U_i$ Anlagenreinheiten die Prozessschritt i ausführen können

$u \in U_f$ Anlageneinheiten, die von Störung f betroffen sind

$i \in I_u$ Prozessschritte, die auf Anlageneinheit u ausgeführt werden können

$i \in I_s^{in}$ Menge der Prozessschritte i, die Produkt s verwenden

$i \in I_s^{out}$ Menge der Prozessschritte, die Produkt s produzieren

$f \in F^{act}$ Menge von Störungen, die zum aktuellen Zeitpunkt t^{act} auftreten

$f \in F^{old}$ Menge der relevanten Störungen im letzten Umplanungszeitpunkt

$(u,i,t) \in X^{fix}$ Menge der fixierten Aufträge, die zum aktuellen Zeitpunkt auf nicht gestörten Anlagen ausgeführt werden

Parameter:

k Letzter Umplanungszeitpunkt

t^{act} aktueller Zeitpunkt

e_f Endzeitpunkt von Störung f

$d_{s,t}$ Endproduktbedarf von State s am Ende von Periode t

$\tau_{u,i}$ Ausführungsdauer von Prozessschritt i auf Anlageneinheit u

$b_{u,i}$ Batchgröße von Prozessschritt i auf Anlageneinheit u

$P_{s,t^{act}}$ Lagerbestand von State s zum aktuellen Zeitpunkt t^{act}

$P_{s,k}$ Lagerbestand von State s am Ende von Periode k

$x_{u,i,t}^{old}$ = 1, wenn auf Anlageneinheit u zum Beginn von Periode t Prozessschritt i im bisherigen Anlagenbelegungsplan gestartet wurde

$S_{u,i}$ Reinigungszeit von Prozessschritt i auf Anlageneinheit u

$SC_{u,i}$ Reinigungskosten von Prozessschritt i auf Anlageneinheit u

RC_i Umplanungskosten für Prozessschritt i

Entscheidungsvariablen:

$x_{u,i,t}$ = 1, wenn auf Anlageneinheit u zum Beginn von Periode t Prozessschritt i gestartet wird (= 0, sonst)

$p_{s,t}$ Lagerbestand von State s am Ende von Periode t

$s_{u,i,t}$ = 1, wenn auf Anlageneinheit u zum Beginn von Periode t ein Reinigungsvorgang nach Prozessschritt i gestartet wird (= 0, sonst)

$e_{u,i,t}$ = 1, wenn zum Beginn von Periode t obwohl im bisherigen als auch im neuen Schedule Prozessschritt i auf Anlageneinheit u startet (= 0, sonst)

Im Fall einer planungsrelevanten Störung, wird mit der Aufbereitung der Daten zur aktuellen Planungssituation fortgefahren. Wie in Abbildung 1 dargestellt, wird zunächst die Menge der aktiven Störungen ermittelt, wobei eine Störung der Menge der aktiven Störungen zugewiesen wird, wenn sie zum aktuellen Zeitpunkt t^{act} noch nicht abgeschlossen ist. Außerdem wird für jeden Auftrag aus dem alten Schedule, der nicht auf der gestörten Anlage eingeplant war, geprüft, ob er bereits gestartet wurde und zum aktuellen Zeitpunkt t^{act} noch in Bearbeitung ist. Sind diese Bedingungen erfüllt, wird er der Menge der fixierten Aufträge zugewiesen. Abschließend werden für alle End- und Zwischenprodukte die Lagerbestände zum aktuellen Zeitpunkt berechnet.

```
// Bestimmung der zum aktuellen Zeitpunkt relevanten Störungen
for f ∈ F^old ∪ F^act do
    if e_f > t^act then
        F := F ∪ {f};
    endif;
endfor;
// Bestimmung der Menge der fixierten Aufträge
for f ∈ F and u ∈ U\U_f and i ∈ I_u and t ∈ {k,...,t^act} do
    if x^old_{u,i,t} = 1 and t + τ_{u,i} > t^act then
        (u,i,t) ∈ X^fix;
    endif;
endfor;
// Berechnung der Lagerbestände zum aktuellen Zietpunkt
for s ∈ S do
```

$$P_{s,t^{act}} = P_{s,k} - \sum_{t\in\{k..t^{act}\}} d_{s,t} - \sum_{i\in I_s^{in}} \sum_{u\in U_i} \sum_{t\in\{k..t^{act}\}} b_{u,i} \cdot x_{u,i,t}^{old} + \sum_{i\in I_s^{out}} \sum_{u\in U_i} \sum_{t\in\{k..t^{act}\}|t\geq \tau_{u,i}+1} b_{u,i} \cdot x_{u,i,t-\tau_{u,i}}^{old};$$

```
endfor;
```

Abb. 1. Pseudocode zur Aufbereitung der aktuellen Planungssituation

Das *Rescheduling-Modul* besteht aus einem gemischt-ganzzahligen Optimierungsmodell, das auf dem Modellierungsprinzip eines State-Task-Netzwerk basiert und in der wissenschaftlichen Literatur schon häufig zur Anwendung kam (vgl. Kondilli et al. 1993). Im Hinblick auf die speziellen Erfordernisse des Rescheduling, wurden einige Modellerweiterungen eingeführt. Dabei musste neben der Berücksichtigung von fixierten Aufträgen auch sichergestellt werden, dass auf den gestörten Anlagen während der Störungsdauer keine Aufträge eingeplant werden können. Darüber hinaus sollte als Zielfunktion nicht nur die Minimierung der Rüstkosten angestrebt werden, sondern auch eine Vermeidung unnötiger Planänderungen. Die konkrete Modellformulierung ergibt sich wie folgt:

Zielfunktion:

$$\min \sum_{u\in U} \sum_{i\in I_u} \sum_{t\in\{t_{act}..H\}} \left[s_{u,i,t} \cdot SC_{u,i} + (x_{u,i,t} - e_{u,i,t}) \cdot RC_i \right] \tag{1}$$

Unter Berücksichtigung der Nebenbedingungen:

$$p_{s,t^{act}+1} = P_{s,t^{act}} - d_{s,t^{act}+1} - \sum_{i\in T_s^{in}} \sum_{u\in U_i} b_{u,i} \cdot x_{u,i,t^{act}+1} + \sum_{i\in I_s^{out}} \sum_{u\in U_i|t\geq\tau_{u,i}+1} b_{u,i} \cdot x_{u,i,t-\tau_{u,i}+1} \quad \forall\, s\in S \tag{2}$$

$$p_{s,t} = p_{s,t-1} - d_{s,t} - \sum_{i\in T_s^{in}} \sum_{u\in U_i} b_{u,i} \cdot x_{u,i,t} + \sum_{i\in I_s^{out}} \sum_{u\in U_i|t\geq\tau_{u,i}+1} b_{u,i} \cdot x_{u,i,t-\tau_{u,i}} \quad \forall\, s\in S \quad \forall t \in \{t^{act}+2,...,H\} \tag{3}$$

$$\sum_{i\in I_u}\left[\sum_{z\in\{t-\tau_{u,i},\cdots,t|t\geq\tau_{u,i}\}} x_{u,i,z} + \sum_{z\in\{t-S_{u,i},\cdots,t|t\geq S_{u,i}\}} s_{u,i,z}\right]\leq 1 \;\forall\, u\in U, t\in\{t^{act}+1,\ldots,H\} \tag{4}$$

$$x_{u,i,t} + s_{u,i,t} \geq x_{u,i,t-\tau_{u,i}} \quad \forall\, u\in U, i\in I_u, t\in\{t^{act},\ldots,H\} \tag{5}$$

$$x_{u,i,t} = 0 \;\; \forall f\in F, u\in U_f, i\in I_u, t=\{t^{act}+1,\ldots,e_f\} \qquad 6)$$

$$x_{u,i,t} = 1 \;\; \forall\, (u,i,t)\in X^{fix} \tag{7}$$

$$x_{u,i,t} \leq e_{u,i,t} \;\; \forall\, (u,i,t)\in X^{old} \tag{8}$$

$$x_{u,i,t}, s_{u,i,t}, e_{u,i,t} \in \{0,1\} \;\; \forall\, u\in U, i\in T_u, t\in\{t^{act}+1,\ldots,H\} \tag{9}$$

$$p_{s,t} \in Z_+^0 \;\; \forall\, s\in S, t\in\{t^{act}+1,\ldots,H\} \tag{10}$$

Die Zielfunktion (1) minimiert die Summe der Rüst- und Umplanungskosten. Umplanungskosten entstehen immer dann, wenn zu einem Auftrag im aktualisierten Schedule kein Auftrag mit gleichem Startzeitpunkt im alten Schedule existiert, der auf der gleichen Anlage den gleichen Prozessschritt ausführt. Die Nebenbedingung (2) initialisiert die Lagerbestände, die durch die Gleichung (3) fortgeschrieben werden. Gleichung (4) bewirkt, dass auf einer Anlage entweder nur ein Prozessschritt oder ein Reinigungsvorgang ausgeführt werden kann und stellt somit die Einhaltung der Anlagenkapazität sicher. Falls eine Anlage nach Abschluss eines Prozessschrittes nicht durch einen direkt nachfolgenden Prozessschritt gleichen Typs belegt wird, leitet Gleichung (5) einen Reinigungsvorgang ein. Gleichung 6) verhindert die Verwendung der gestörten Anlagen während der gesamten Störungsdauer. Durch Gleichung (7) wird der Startzeitpunkt und die Anlagenzuordnung der bereits gestarteten Aufträge aus dem alten Schedule übernommen. Falls zum gleichen Zeitpunkt und auf der gleichen Anlage der gleiche Prozessschritt sowohl im alten als auch im neuen Schedule startet, nimmt durch Gleichung (8) der Wert der Variablen $e_{u,i,t}$ den Wert eins an und indiziert somit, dass für diesen Auftrag keine Umplanung erfolgt ist. Die Ausdrücke (9) und (10) definieren den Wertebereich der Entscheidungsvariablen.

3. Numerisches Beispiel

Die Funktionsweise des reaktiven Schedulingsystems soll anhand einer erweiterten Version des von Shah et al. (1993) vorgestellten Fallbeispiel demonstriert werden. Basierend auf den Auftragsdaten aus Tabelle 1 wurde mit Hilfe des Scheduler-Moduls ein Ausgangsplan erstellt, der anschließend an das entsprechende Simulationsmodell zur Ausführung übergeben wurde. Während der Ausführung des Ausgangsplanes traten auf den Anlagen U3 und U31 jeweils ein Anlagenausfall auf. Die resultierenden Planungsdaten der entsprechenden Reschedulingläufe wurden in der Tabelle 2 zusammengefasst.

Table 1. Auftragsdaten

Order	State	Order size	Due date	Order	State	Order size	Due date
1	S9	6000	150	5	S9	1700	70
2	S7	2500	140	6	S7	2000	100
3	S5	5500	150	7	S4	2700	70
4	S4	2300	130	8	S5	2000	90

Table 2. Störungsdaten und Ergebnisse der Umplanungsläufe

Störung	Start	Ende	Anlage	Variablen.	Nebenbedingungen.	Rechenzeit [sec]
1	45	55	U31	2719	4454	13.656
2	60	65	U3	2355	3840	124.634

4. Zusammenfassung

In dem vorliegenden Beitrag wurde ein reaktives Schedulingsystem vorgestellt, das den Entscheidungsprozess bei Umplanungsaktivitäten unterstützen soll. Als Störungsursachen wurden Anlagenausfälle berücksichtigt. Anhand eines numerischen Beispiels wurde die Funktionsweise des Systems erläutert. Als Erweiterungen für das System sind die Berücksichtung zusätzlicher Störereignisse und die Implementierung alternativer Schedulingmethoden vorgesehen.

Literatur

Baptiste, P., Le Pape, C. and Nuijten, W. (2001), Constraint-Based Scheduling, Kluwer Academic Publishers, Boston, Dordrecht, London.

Kondili, E., Pantelides, C. C. and Sargent, R. W. H. (1993), A general algorithm for short-term scheduling of batch operation - I. MILP formulation, Computers & Chemical Engineering 17: 211-227.

Lee, Y. G. and Malone, M. F. (2001), Batch process schedule optimization under parameter volatility, International Journal of Production Research 39: 603-623.

Mendez, C. A. and Cerda, J. (2003), Dynamic scheduling in mulitproduct batch plants, Computers & Chemical Engineering.27: 1247-1259.

Pinedo, M. and Chao, X. (1999), Operations Scheduling, McGraw - Hill, Boston.

Shah, N., Pantelides, C. C. and Sergent, R. W. H. (1993), A General algorithm for Short-Therm Scheduling of Batch Operations - II. Computational Issues, Computers & Chemical Engineering 17: 229-244.

Vieira, G. E., Herrmann, J. W. and Lin, E. (2003), Rescheduling manufacturing systems: A framework of strategies, policies and methods, Journal of Scheduling 6: 39-62.

Recovery Knowledge Acquisition in Medium and Long Term Planning of a Joint Manufacturing / Remanufacturing System

Rainer Kleber

Faculty of Economics and Management, Otto-von-Guericke University Magdeburg, POB 4120, 39016 Magdeburg, Germany
rainer.kleber@ww.uni-magdeburg.de

Summary. In this work, the impact of knowledge acquisition in product recovery on optimal stock-keeping decisions as well as on the question on whether at all and when to start remanufacturing is discussed.

1 Introduction

Within the field of product recovery management, remanufacturing has been shown to be a particularly advantageous option, since a large part of the embedded economic value is retained. This option has achieved special interest both from research and practice since remanufactured products can sometimes be regarded to be as 'good as new', and are used to replace production of new products. This leads to a new managerial challenge, because decision making in such *integrated* product recovery systems must be coordinated. Most quantitative approaches in product recovery management (see [1] for an overview) necessitate that all processes keep an initial performance and constant cost rates. However, ongoing competition and the search for profit maximization provide incentives for productivity improvements.

This paper relates to an empirical phenomenon that can be found quite often in practical applications dating back to the 1920's and intends to explore effects of acquiring knowledge in product recovery. The learning curve, introduced by Wright [9], shows a (potential) relationship between cumulative output and labor hours per unit produced. Because of the shortening of manufacturing times or more generally, reduction of input quantities, learning leads to a decrease of direct production costs.

Cost reductions due to learning are not restricted to production alone. It can also be presumed to occur for remanufacturing processes (which often are labor intensive, see [2]), if there are repeated operations performed on a large number of similar items as for instance is the case for single-use cameras

[6]. It must also be stated, however, that if there exists a large diversity of remanufactured products with only little information on how to deal with them correctly, sufficient experience might be more difficult to obtain.

Although the integration of learning effects into production planning has attracted some attention in research (for a survey see [3]), quantitative approaches devoted to effects of knowledge accumulation in product recovery can hardly be found. Exceptions are offered in the context of Total Quality Management [7]. A possible explanation might be that in *pure* remanufacturing systems similar effects occur as in (pure) production systems. New results are to be expected when dealing with *integrated* product recovery systems.

This paper extends earlier works on dynamic product recovery (see [4] for an overview) to allow for remanufacturing knowledge accumulation. Possible research questions are how stock-keeping policies change and what would be the effect on optimal remanufacturing and production policies. For instance, as a strategic implication of productivity improvements, remanufacturing can be profitable in the long run, even if there is no immediate cost advantage over producing new items, because subsequent unit remanufacturing costs are lowered. The remainder of this paper is organized as follows. In Section 2 an optimal control model is introduced and main results are presented in Section 3. Final conclusions are given in Section 4.

2 A model with learning in the remanufacturing shop

In this section a model is presented where current decisions on remanufacturing have an additional impact on future cash flow, because they change a knowledge stock which in turn lowers unit remanufacturing costs. The model rests upon the basic formulation introduced in [8] (also referred to as the *basic model*) which has been adapted in order to account for knowledge acquisition. We consider a single product and single stage product recovery system under deterministic but dynamic conditions. An external demand $d(t)$ has to be immediately satisfied either by producing new items at a rate $p(t) \geq 0$ or by remanufacturing used products with rate $r(t) \geq 0$. Used products returning with given rate $u(t)$ can also be disposed of ($w(t) \geq 0$) or stored in a recoverables inventory $y_u(t)$. Initial and final recoverables inventory is zero $y_u(0) = y_u(T) = 0$ and the marginal increase of recoverables inventory is given by return rate minus the sum of remanufacturing and disposal rates, i.e. $\dot{y}_u(t) = u(t) - r(t) - w(t)$. Since all processes are instantaneously implemented and supposed to be unrestricted, a serviceables inventory is not considered. Production quantity $p(t)$ immediately follows when setting a corresponding remanufacturing rate $r(t)$ from the necessity to immediately satisfy all demand. Therefore, we have $p(t) = d(t) - r(t)$. Non-negativity of the production rate requires the return rate not to exceed the demand rate ($r(t) \leq d(t)$).

The objective of the model is to satisfy customer demands with minimal discounted or undiscounted cash outflow for production, remanufacturing, disposal, and holding recoverables inventory within a finite planning horizon $[0, T]$. Payments for production and disposal depend linearly on the respective decision, i.e. there are time independent per unit payments for production $c_p > 0$ and disposing of items c_w. The out-of-pocket holding cost rate per item and unit time for recoverables is $h_u > 0$. Moreover, it should be not advantageous to hold unneeded returned products, i.e. $h_u > \alpha c_w$, with α denoting the discount rate.

Learning in the remanufacturing shop is approximated by the cumulative remanufacturing volume R which is derived, given an initial stock of knowledge R_0, by using $\dot{R}(t) = r(t) \geq 0$ and $R(0) = R_0 \geq 0$. Remanufacturing unit costs are given by a generic exogenous function $c_r(R)$ measuring the learning curve effect realized so far. For this function we assume

$$c_r'(R) < 0,\ c_r''(R) > 0, \text{ as well as } \lim_{R \to \infty} c_r(R) \geq 0, \tag{1}$$

i.e. unit remanufacturing costs decrease with a decreasing rate. Initial remanufacturing costs $c_r(R_0)$ can be high enough that there is a situation with a negative initial remanufacturing cost advantage ($c_p + c_w - c_r(R_0) < 0$), that would not allow for remanufacturing in the basic model.

An optimal control problem with two states (R and y_u) and two control variables (r and w) has to be solved. The problem is subject to the state equations, a pure state constraint, initial and (partly) terminal conditions for the state variables as well as non-negativity constraints for control variables and an upper limit for the remanufacturing rate

$$\begin{aligned}
\min NPV = & \int_0^T e^{-\alpha t} \left(c_p(d(t) - r(t)) + c_r(R(t))r(t) + c_w w(t) + h_u y_u(t) \right) dt \\
s.t. \quad & \dot{R}(t) = r(t) \geq 0 \text{ and } R(0) = R_0 > 0, \qquad\qquad (2) \\
& \dot{y}_u(t) = u(t) - r(t) - w(t),\ y_u(t) \geq 0\ \ y_u(0) = 0,\ y_u(T) = 0, \\
& d(t) - r(t) \geq 0, r(t) \geq 0, w(t) \geq 0.
\end{aligned}$$

Here, benefits are derived from using an optimal control framework because, as shown in more detail in [5], also the indirect effect of current decisions on future expenses is valued. This is applied by so-called co-state variables which correspond to each state. For instance, we can derive a *value of acquiring knowledge* which rates the impact of current remanufacturing decisions on future costs.

3 Main results

This section deals with main qualitative additions of remanufacturing knowledge acquisition to results already known from the basic model. One can distinguish between (a) operational and (b) strategic effects. Regarding the first (a), we concentrate on one of the main results of dynamic product recovery, namely the maximal amount of time that returns should be held in stock (see [8]) which sometimes is called Maximal Holding Time (MHT). Strategic effects (b) refer to questions like whether to remanufacture at all or at which time to start remanufacturing. In this context we further distinguish between low (**L**), moderate (**M**), and high (**H**) initial remanufacturing expenses $c_r(R_0)$ being defined as follows

- **L**ow: $c_r(R_0) \leq c_p + c_w$ (and consequently, $\alpha(c_r(R_0) - c_p) < h_u$), There immediately exists a positive recovery cost advantage.
- **M**oderate: $c_r(R_0) > c_p + c_w$ and $\alpha(c_r(R_0) - c_p) \leq h_u$, Remanufacturing does not immediately pay off. Out of pocket holding costs are higher than savings from deferring to remanufacture an item.
- **H**igh: $h_u < \alpha(c_r(R_0) - c_p)$ (including $c_r(R_0) > c_p + c_w$), Remanufacturing is so expensive, that holding an item is cheaper than the interests being saved when postponing remanufacturing of that item.

Under (**L**) conditions remanufacturing always starts at the beginning of the planning period. A common property of (**M**) and (**H**) conditions is that once started remanufacturing is performed until the end of the planning period.

Since the effects differ considerably depending on whether discounting can be neglected or not, in the next two subsections results for the case of a zero and nonzero interest rate are sketched.

3.1 Optimal policy with a zero interest rate

Operational effects. Optimal stock-keeping intervals, i.e. periods where $y_u > 0$, require attributes like a Location Property and Inventory Conditions which do not differ from those given in the basic model. Regarding the Maximal Length Property it is to be said that the MHT immediately anticipates all later acquired experience. It is given by $\tau := \max\left\{0, \frac{c_p + c_w - c_r(R^*(T))}{h_u}\right\}$ and it incorporates remanufacturing costs valid at the end of the planning period.

Strategic effects. The case of negligible discounting is characterized by an equal valuation of all payments independently of their respective timing. A postponement of remanufacturing decisions does therefore not make sense and the solution takes on a simple structure. Either remanufacturing takes place starting at the begin of the planning period or it does not take place at all.

If we have a situation with moderate initial remanufacturing costs where $c_r(R_0) > c_p + c_w$ (**M**, while **H** conditions are excluded by definition for $\alpha = 0$) it is questionable whether the investments spent for 'riding down the experience curve' can later be paid back. This point can be answered by using a break-even like analysis. Let $\tilde{R}$ be the quantity at which total remanufacturing costs equal total costs of disposing of this number of returns and producing new items, i.e. $\int_{R_0}^{R_0+\tilde{R}} c_r(x)\, dx = (c_p + c_w)\tilde{R}$. Then, remanufacturing takes place if the total remanufacturing quantity would surpass a break-even total remanufacturing quantity, i.e. $R^*(T) \geq \tilde{R}$.

Since total remanufacturing quantity $R^*(T)$ plays a role both in determining τ and in the question whether to remanufacture or not, the choice of the planning horizon T becomes a critical decisive factor. A longer planning period would lead to an increased MHT and thus, to more stock-keeping. But also the pay off condition would more likely be satisfied because total remanufacturing rises. This result concurs with the strategic focus of the learning curve approach.

3.2 Optimal policy with a positive interest rate

Operational effects. In the case of a positive discount rate, not all later acquired experience is anticipated because respective cost savings are valued less than current expenses. Since the differences in time values decrease as time advances, the MHT rises with time, i.e. a later period where recoverables are stored is allowed to be longer than an earlier one.

Strategic effects. If initial remanufacturing expenses exceed the sum of direct cost of producing a new item and disposing of the old one (**M**) it is possible to start remanufacturing later than at time zero. For such a time point, a trade-off is struck between early remanufacturing that leads to higher later direct cost savings, and a lower discounted value of initial expenses if it is started later. This leads to a break-even like condition for a nonzero start time of remanufacturing. At that time, the value of acquiring knowledge (co-state) must exactly outweigh the initial recovery cost disadvantage.

Under high remanufacturing cost conditions (**H**), a strategic inventory might be used in order to further postpone the start of remanufacturing. The stored returns are used to bundle initial (expensive) remanufacturing while still being able to use these returns to lower remanufacturing unit costs. Here, a trade-off between decreasing the time value of remanufacturing payments and holding costs must be solved.

Similarly to the zero interest rate case, having a time point where remanufacturing would start does not necessarily mean that investments in knowledge acquisition pay off. But in contrast to that case it is not possible to formulate a simple condition as given above.

4 Conclusions

In this paper the effects of introducing a 'learning' remanufacturing process into the optimal control framework introduced by [8] were investigated. The anticipation of later knowledge acquisition led to the possibility to remanufacture returned products even if there exists no immediate cost advantage. In case of a positive interest rate remanufacturing might start later than at the beginning of the planning period. This situation is further characterized by two main results with respect to stock-keeping. Firstly, the maximal length of return collection intervals increases with time and secondly, we discovered yet another motivation for stock-keeping.

Results are sensitive to planning horizon changes. This especially holds for small planning periods where there still exists a considerable potential for cost-improvements after the end of the planning period. Since the learning curve concept is a strategic one, the planning horizon should therefore be chosen sufficiently large. But used together with the product life cycle concept, this analysis can be a helpful tool for deciding on whether to engage in remanufacturing at all. For a more detailed treatment of the planning problems discussed in this paper see [5].

References

[1] Dekker R, Fleischmann M, Inderfurth K, Van Wassenhove LN (2004) Quantitative models for reverse logistics decision management. In: Dekker R, Fleischmann M, Inderfurth K, Van Wassenhove LN (eds) Reverse Logistics: Quantitative Models for Closed-Loop Supply Chains, Springer, Berlin Heidelberg New York, pp 25-41

[2] Guide Jr. VDR, Jayaraman V, Srivastava R, Benton WC (2000) Supply-chain management for recoverable manufacturing systems. Interfaces 30(3):125–142

[3] Gulledge Jr. TR, Khoshnevis B (1987) Production rate, learning, and program costs: Survey and bibliography. Engineering Costs and Production Economics 11:223–236

[4] Kiesmüller GP, Minner S, Kleber R (2004) Managing dynamic product recovery: An optimal control perspective. In: Dekker R, Fleischmann M, Inderfurth K, Van Wassenhove LN (eds) Reverse Logistics: Quantitative Models for Closed-Loop Supply Chains, Springer, Berlin Heidelberg New York, pp 221-247

[5] Kleber R (2005) Dynamic inventory management in reverse logistics. Ph.D. thesis, Otto-von-Guericke University, Magdeburg, Germany

[6] Kodak (1999) Annual Report. The Kodak Corporation, Rochester, NY

[7] Lapré MA, Mukherjee AS, Van Wassenhove LN (2000) Behind the learning curve: Linking learning activities to waste reduction. Management Science 46:597–611

[8] Minner S, Kleber R (2001) Optimal control of production and remanufacturing in a simple recovery model with linear cost functions. OR Spektrum 23:3–24

[9] Wright TP (1936) Factors affecting the cost of airplanes. Journal of Aeronautical Sciences 3:122–128

Part VI

Finance, Banking and Insurance

Performance Measurement of Hedge Fund Indices - Does the Measure Matter?

Martin Eling and Frank Schuhmacher

Dipl.-Kfm. Martin Eling, University of St. Gallen, Institute of Insurance Economics, Kirchlistrasse 2, 9010 St. Gallen, Switzerland, Tel.: +41 71 243 40 93, E-Mail: martin.eling@unisg.ch.

Prof. Dr. Frank Schuhmacher, University of Leipzig, Department of Finance, Jahnallee 59, 04009 Leipzig, Germany, Tel.: +49 341 973 36 70, E-Mail: schuhmacher@wifa.uni-leipzig.de.

1. Introduction

Performance measurement is an integral part of investment analysis and risk management. The goal of performance measurement is to build a ranking of different investments on the basis of risk-adjusted returns in order to evaluate the relative success of these investments. The Sharpe Ratio is the best-known measure of this type. It considers the first two moments of the return distribution (expected value and standard deviation) and is an adequate performance measure if the returns of the investment fund are normally distributed and the investor wishes to place all his risky assets in just one investment fund. Nevertheless, many other performance measures in addition to the Sharpe ratio exist in theory and practice. One can identify two levels of argumentation for the application of alternative performance measures.

First, the choice of the "correct" performance measures depends on the concrete decision making situation of the investor. This means that a different performance measure is adequate for an investor who invests all his risky assets in just one investment fund than for an investor that splits his risky assets for example in a market index and an investment fund (See Scholz/Wilkens (2003) for example). The Sharpe ratio is sufficient in the first case, while in the second case, a performance measure that also takes account of the correlation between the market index and the respective investment fund is adequate. Such measures are the Jensen, Treynor and Treynor-Black measures (See Treynor/Black (1973), Jensen (1968) and Treynor (1965)).

Second, the choice of an adequate performance measure depends on how the returns of an investment fund are distributed. In the case of normally distributed returns, performance measures that rely on the first two moments of the return distribution (expected value, standard deviation) such as the Sharpe ratio are sufficient. In the case of non-normally distributed returns, the standard deviation of the returns fails to adequately display risk and the Sharpe ratio inadequately displays performance. In the analysis of hedge funds, it is frequently stated that the returns they generate have a non-normal distribution and thus hedge funds cannot be adequately evaluated using the classic Sharpe ratio (See for example Mahdavi (2004), p. 47, Brooks/Kat (2002), p. 37, Bacmann/Scholz (2003)). For example, in the case of hedge funds the use of derivative instruments results in an asymmetric return distribution as well as fat tails. For this reason, there is a danger that the use of standard risk and performance measures will result in an underestimation of the risk and an overestimation of the performance (see Lo (2002), Kat (2003)). Consideration of these issues has led to the development of some new performance measures that are currently debated in hedge fund literature.

In Eling/Schuhmacher (2005) we analyse and compare the performance measures that have been developed with regard to these latter issues. In the following sections we present some results of this study. We do not, however, seek to provide a justification for individual performance measures on a theoretical level but set out to answer a very practical question: Does it really matter which performance measure one chooses to evaluate the performance of hedge funds? To answer this question, we consider the criticised Sharpe Ratio and ten alternative approaches to measure the performance of hedge funds. The key idea is to compare the rankings provided by the different performance measures with Spearman's rank correlation coefficient. This investigation is motivated by papers that in other contexts or on the basis of other data have concluded that the choice of a particular risk measure or performance measure has no significant influence on the evaluation of an investment.

For example, Pfingsten/Wagner/Wolferink (2004) compared the rank correlations for various risk measures on the basis of an investment bank's 1999 trading book. In doing so, they found that different measures result in largely identical evaluations of risk. Pedersen/Rudholm-Alfvin (2003) compared risk-adjusted performance measures for various asset classes over the period from 1998 to 2003. The authors found that there was a high rank correlation between the performance measures. The questions that were treated only as a side issue by Pedersen/Rudholm-Alfvin (2003) are the focus of our study. In addition, we concentrate on hedge funds as an asset class, the performance measures that have been proposed for measuring the performance of hedge funds, and the related debate concerning the suitability of classic and newer performance measures for evaluation of hedge funds.

The remainder of the paper consists of three sections. First, we present our data and methodology (Section 2). Then eleven performance measures are employed for determining the performance of ten hedge fund indices as well as for five equity and bond indices (Section 3). Finally, the results of the study are summarised (Section 4).

2. Data and Methodology

In hedge fund literature it is argued that it is inappropriate to assess hedge funds using the classic Sharpe Ratio because hedge fund returns display atypical skewness and excess values, values the Sharpe Ratio does not reflect (see Mahdavi (2004), Brooks/Kat (2002)). Instead, measures that take into account the risk of loss should be used to measure hedge fund performance, in particular the lower partial moments, the drawdown, and the value at risk.

We explore this argument by comparing the classic Sharpe Ratio and ten newer approaches to performance measurement based on the risk of loss: Omega, the Sortino Ratio, Kappa 3, the Upside Potential Ratio, the Calmar Ratio, the Sterling Ratio, the Burke Ratio, Excess Return on Value at Risk, the Conditional Sharpe Ratio, and the Modified Sharpe Ratio (for a full description of each measure, see Eling/Schuhmacher (2005)).

We measure performance using monthly returns of ten Credit Suisse First Boston/Tremont (CSFB) hedge fund indices over the period from January 1994 to December 2003. The hedge fund indices are also compared with five market indices; two of them measure equity performance, two measure bond performance and the remaining index measures an equally weighted investment in the equity and bond indices.

Our methodology takes the following three steps. First, we measure the distributional characteristics and the performance of various hedge fund indices and traditional investments (such as stocks and bonds). Next, we calculate the correlation between the rankings based on different performance measures using Spearman's rank correlation coefficient and also test for significance using Hotelling Pabst statistics. Finally, we check the robustness of our results by varying several parameters. However, in this contribution, we report only the Spearman's rank correlation coefficient results. For a detailed survey of the performance measures, a full description of our measurement results, and several robustness checks, we refer to Eling/Schuhmacher (2005).

3. Findings

After measuring the distributional characteristics and the performance of the hedge fund indices and the traditional investments, we calculate the correlation between rankings based on different performance measures using Spearman's rank correlation coefficient (See Table 1).

Table 1: Rank Correlation of Rankings Based on Different Performance Measures

Performance measure	Sharpe Ratio	Omega	Sortino Ratio	Kappa 3	Upside Potential Ratio	Calmar Ratio	Sterling Ratio	Burke Ratio	Excess Return on Value at Risk	Conditional Sharpe Ratio	Modified Sharpe Ratio
Sharpe Ratio											
Omega	1.00										
Sortino Ratio	0.99	0.98									
Kappa 3	0.95	0.94	0.97								
Upside Potential Ratio	0.83	0.80	0.88	0.95							
Calmar Ratio	0.88	0.85	0.85	0.83	0.73						
Sterling Ratio	0.98	0.97	0.95	0.92	0.80	0.95					
Burke Ratio	0.95	0.92	0.92	0.90	0.82	0.98	0.98				
Excess Return on VaR	0.99	0.99	0.99	0.96	0.86	0.86	0.97	0.93			
Conditional Sharpe Ratio	0.94	0.93	0.97	0.98	0.93	0.82	0.91	0.90	0.96		
Modified Sharpe Ratio	1.00	0.99	0.99	0.96	0.84	0.87	0.98	0.94	1.00	0.95	
Average	0.95	0.94	0.95	0.94	0.84	0.86	0.94	0.92	0.95	0.93	0.95

All performance measures produce almost similar rankings and thus result in a widely identical evaluation of the investments. Our three main findings are as follows.

1. Spearman's rank correlation coefficient between the Sharpe Ratio and the alternative approaches exhibits very high values for almost all measures and is on average 0.95. The picture is the same when comparing the new performance measures with each other (on average 0.92). Hence, all performance measures deliver nearly the same rankings.

2. Two performance measures (the Upside Potential Ratio and the Calmar Ratio) result in slight changes in the evaluation of hedge funds as compared to all the other methods. However, following the Hotelling Pabst statistics, none of the new performance measures (including the Upside Potential Ratio and the Calmar Ratio) show a significantly different ranking than the Sharpe Ratio (see Eling/Schuhmacher (2005)).

3. These findings prove to be very robust when varying both the subject of study and the period of time under consideration, as well as with respect to any change in further given parameters, such as the minimal acceptable return or the confidence level (also see Eling/Schuhmacher (2005)).

In conclusion, on the basis of our data, none of the new performance measures result in significant changes in the evaluation of hedge funds. Nine of the eleven new performance measures produce rankings that are largely the same as those resulting from the Sharpe ratio. Only two of the performance measures, the upside potential ratio and the Calmar ratio, produce a slight change in the rankings. However, after testing using the Hotelling-Pabst statistic, no ranking was significantly different from that produced by the Sharpe ratio, not even the ranking resulting from the upside potential ratio or from the Calmar ratio.

4. Conclusion

It does not matter too much which performance measure one chooses to evaluate hedge funds. Because the newer performance measurement approaches result in rankings that are the same and thus result in the same assessments of hedge funds, use of the classic Sharpe ratio (even if it displays some undesirable features) is justified, at least from a practical perspective.

The results of this study will be helpful for investing in hedge funds that are constructed like indices. However, an important question is whether the choice of a specific performance measure matters when evaluating single hedge funds. The findings of the current study have motivated us to procure return data for single hedge funds in order to analyse this question in the future. However, the indices we examined in this paper appear to be representative for single hedge funds insofar as the return distributions deviate significantly from a normal distribution.

Another question not answered in this study is the relevance of performance measures that are evolved on the basis of correlations, such as the Jensen, Treynor and Treynor-Black measures. This question should also be examined in future studies.

References

Bacmann, Jean-Francois/Scholz, Stefan (2003): Alternative Performance Measures for Hedge Funds, in: AIMA Journal, June 2003.

Brooks, Chris/Kat, Harry M. (2002): The Statistical Properties of Hedge Fund Index Returns and Their Implications for Investors, in: The Journal of Alternative Investments, Vol. 5, No. 2, Fall, pp. 26-44.

Eling, Martin/Schuhmacher, Frank (2005): Hat die Wahl des Performancemaßes einen Einfluss auf die Beurteilung von Hedgefonds-Indizes?, Working Paper, University of St. Gallen/University of Leipzig, to appear in: Kredit und Kapital, Vol. 39.

Kat, Harry M. (2003): 10 Things that Investors Should Know about Hedge Funds, in: The Journal of Wealth Management, Vol. 5, No. 4, pp. 72–81.

Jensen, Michael (1968): The Performance of Mutual Funds in the Period 1945-1968, in: The Journal of Finance, Vol. 23, No. 2, pp. 389-416.

Lo, Andrew W. (2002): The Statistics of Sharpe Ratios, in: Financial Analysts Journal, Vol. 58, No. 4, pp. 36–52.

Mahdavi, Mahnaz (2004): Risk-Adjusted Return When Returns Are Not Normally Distributed: Adjusted Sharpe Ratio, in: The Journal of Alternative Investments, Vol. 6, No. 4, Spring, pp. 47-57.

Pedersen, Christian S./Rudholm-Alfvin, Ted (2003): Selecting a Risk-Adjusted Shareholder Performance Measure, in: Journal of Asset Management, Vol. 4, No. 3, pp. 152–172.

Pfingsten, Andreas/Wagner, Peter/Wolferink, Carsten (2004): An Empirical Investigation of the Rank Correlation Between Different Risk Measures, in: Journal of Risk, Vol. 6, No. 4, pp. 55–74.

Scholz, Hendrik/Wilkens, Marco (2003): Zur Relevanz von Sharpe Ratio und Treynor Ratio: Ein investorspezifisches Performancemaß, in: Zeitschrift für Bankrecht und Bankwirtschaft, Vol. 15, No. 1, pp. 1-8.

Treynor, Jack L. (1965): How to Rate Management of Investment Funds, in: Harvard Business Review, Vol. 43, No. 1, pp. 63-75.

Treynor, Jack L./Black, Fisher (1973): How to Use Security Analysis to Improve Portfolio Selection, in: Journal of Business, Vol. 46, No. 1, pp. 66-88.

On the Applicability of a Fourier Based Approach to Integrated Market and Credit Portfolio Models

Peter Grundke[1]

Abstract:
Based on a version of the well-known credit portfolio model CreditMetrics extended by correlated interest rate and credit spread risk the application of a Fourier based method for calculating credit risk measures is demonstrated. The accuracy and speed of this method is compared with standard Monte Carlo simulation by means of numerical experiments.

I. Introduction

For calculating risk measures of credit portfolios, such as Value-at-Risk or expected shortfall, a range of models have been developed. Most of them rely on Monte Carlo simulations for calculating the probability distribution of the future credit portfolio value, which can be quite computer time consuming.

Moreover, a typical shortcoming of most credit portfolio models is that relevant risk factors, such as interest rates or credit spreads, are not modeled as stochastic variables and hence are ignored during the revaluation at the risk horizon. For example, fixed income instruments, such as bonds or loans, are revalued using the current forward rates and (rating class specific) forward credit spreads for discounting future cash flows. Thus, the stochastic nature of the instrument's value in the future which results from changes in factors other than credit quality is neglected. Various studies (see Kijima and Muromachi 2000, Barnhill and Maxwell 2002, Kiesel et al. 2003 and Grundke 2004, 2005a) show that the missing stochastic modeling of risk factors can cause a severe underestimation of economic capital, especially for high grade credit portfolios with a low stochastic dependence between the obligors' credit quality changes.

However, adding these risk factors as additional ingredients of a credit portfolio model, the computational burden of full Monte Carlo simulations increases and the need for efficient methods for calculating credit risk measures becomes even more pressing. The aim of this paper is to analyze whether a Fourier based ap-

[1] Department of Banking, University of Cologne, Albertus-Magnus-Platz, 50923 Cologne, Germany, email: grundke@wiso.uni-koeln.de.

proach can be such an efficient tool in the context of a credit portfolio model with integrated market risk. This technique has already been successfully applied to pure market risk portfolio models, but there are only a few papers which are concerned with the application of this method to credit portfolio models, especially with integrated market risk. Exceptions are Merino and Nyfeler (2002), Duffie and Pan (2001) and Reiß (2003). Merino and Nyfeler apply the technique to a standard default mode portfolio model. Duffie and Pan as well as Reiss work within an intensity-based framework with integrated market risk factors, whereas in this paper an extended CreditMetrics model is employed. Furthermore, Duffie and Pan need, among others, the assumption that a delta-gamma approximation of the market risk component of the credit portfolio value is sufficiently accurate which seems problematic for longer time horizons such as one year usually employed for credit risk management.

This paper is structured as follows: In section II a short overview of the computational approach is given. Then, in section III, the CreditMetrics model is extended by correlated interest rate and credit spread risk and the Fourier based approach is explained. The results of a numerical example are presented in section IV. Section V contains a short discussion of possible extensions of the analysis presented here. Finally, in section VI the main results are summarized.

II. The General Computation Approach

The characteristic function of a continuous random variable X with density function $f(x)$ is a complex-valued function defined as:

$$\varphi_X(s) := E\left[e^{isX}\right] = \int_{-\infty}^{\infty} e^{isx} f(x)dx \quad (s \in \mathbb{R},\ i = \sqrt{-1}). \tag{1}$$

In a non-probabilistic context the characteristic function $\varphi_X(s)$ is called the Fourier transform of the (density) function $f(x)$. Two fundamental properties of characteristic functions are used in the following: First, the characteristic function of a sum of independent random variables equals the product of the characteristic functions of the individual random variables. Second, the characteristic function of a random variable uniquely determines its probability distribution, which can be recovered from the characteristic function for example by the following inversion formula (see Davies 1973):

$$P(X < x) = \frac{1}{2} - \frac{1}{\pi}\int_{0}^{\infty} \mathrm{Im}\left(\frac{e^{-isx}\varphi_X(s)}{s}\right) ds\,. \tag{2}$$

With $\mathrm{Im}(\cdot)$ the imaginary part of a complex number is denoted and usually the integral appearing in this inversion formula has to be calculated numerically.

III. An Integrated Market and Credit Portfolio Model

In this section the usual CreditMetrics framework is extended by correlated interest rate and credit spread risk and applied to a portfolio of defaultable zero coupon bonds.

It is assumed that the return R_n on firm n's assets can be described by a normally distributed random variable which is – without loss of generality – standardized:

$$R_n = \sqrt{\rho_V - \rho_{r,V}^2}\, Z + \rho_{r,V} X_r + \sqrt{1-\rho_V}\, \varepsilon_n \; (\rho_{r,V}^2 \le \rho_V, n \in \{1,\ldots,N\}) \tag{3}$$

where Z, X_r, ε_1, …, ε_N are mutually independent standard normally distributed stochastic variables. The variables Z and X_r represent systematic credit risk, by which all firms are affected, whereas the ε_n's stand for idiosyncratic credit risk.

The risk-free short rate is modeled for simplicity as a mean-reverting Ornstein-Uhlenbeck process introduced already by Vasicek 1977, which implies the following representation of the short rate at the risk horizon H :

$$r(H) = \theta + (r(0) - \theta)\exp(-\kappa H) + \sqrt{\frac{\sigma_r^2}{2\kappa}\left(1 - \exp(-2\kappa H)\right)}X_r \tag{4}$$

where $\kappa, \theta, \sigma_r \in \mathbb{R}_+$ and $X_r \sim N(0,1)$ enters the definition (3) of the firms' asset returns.

It is assumed that there are K possible ratings, where K denotes the default state and 1 is the best rating. The simulation of the ratings 1,…, K of the obligors at the risk horizon proceeds exactly as described in the technical document of CreditMetrics (see Gupton et al. 1997).

The price of a zero coupon bond at the risk horizon whose issuer has not defaulted until this time is calculated by discounting the future cash flow with the risk-adjusted spot yield appropriate for the issuer's rating at the risk horizon. This yield is composed of the stochastic risk-free spot yield (evolving according to the Vasicek model) and the stochastic credit spread of the respective rating class. The credit spreads are assumed to be multivariate normally distributed. Furthermore, it is assumed that the random variable X_r, which drives the term structure of risk-free interest rates, and the systematic credit risk factor Z respectively are both correlated with the credit spreads. For the sake of simplicity, these correlation parameters are set equal to constants regardless of the rating grade or the remaining time to maturity. Besides, it is assumed that the idiosyncratic credit risk factors are independent of the credit spreads.

If an issuer of a zero coupon bond has already defaulted until the risk horizon, the value of the bond is set equal to a constant fraction δ of the value of a risk-free but otherwise identical zero coupon bond. The assumption of a constant recovery rate is again for simplicity. The Fourier based approach would also work with an independent beta-distributed recovery rate (as in CreditMetrics) or with systematic recovery rate risk.

Next, it is sketched how the Fourier based approach can be applied to the above model in order to derive the probability distribution of the credit portfolio value. Essential is the computation of the characteristic function of the credit portfolio value at the risk horizon which has to be inserted into the inversion formula (2). For applying the first property of characteristic functions, we observe that the future credit portfolio value is just the sum of the individual bond values and that these, conditional on the realisation of the systematic credit risk factors Z and X_r and the credit spreads, are independent. Thus, first, the conditional characteristic function of a single bond value at the risk horizon is determined. Then, the conditional characteristic function of the whole credit portfolio value is calculated as the product of these individual conditional characteristic functions. Next, the unconditional characteristic function of the credit portfolio value has to be computed by averaging the conditional counterparts. Unfortunately, the unconditional characteristic function can not be calculated in closed-form, but has to be simulated for each grid point of the numerical integration rule which is used for computing the integral in the inversion formula (2). This is by far the most time-consuming part of the calculations. That's why it is crucial to employ a numerical integration rule which only needs a moderate number of grid points for yielding accurate results.

IV. Numerical Example

In order to get an idea about the accuracy and the speed of the Fourier based method, a simulation study is carried out. For this, the 0.1%-, 1%- and 5%-percentiles of the distribution of an homogeneous portfolio of defaultable zero coupon bonds are computed with two methods. On the one hand, the Fourier based approach combined with the Gauss-Legendre integration rule is implemented.[2] For calculating the unconditional characteristic function fifty thousand draws of Quasi Random Numbers (QRN), a 9 dimensional Halton sequence, are used. QRN seem to be especially appropriate for this problem because calculating the unconditional characteristic function means calculating expected values and for this purpose QRN are well suited and their ability to improve the rate of convergence is well documented in the literature. In contrast, the benefit of QRN when estimating tails of probability distributions is reported to be limited due to the high dimensionality of the problem. With this implementation of the Fourier based approach, three types of error are introduced: First, the simulation error of the unconditional characteristic function, and, second and third, the usual errors resulting from numerical integration, which means the error caused by cutting the

---

[2] For the Gaussian integration $n = 96$ grid points are applied on each of the intervals [0,1], (1,3], (3,10] and (10,50].

integration interval at some maximum value and the error caused by the finite number of grid points used by numerical integration rule. On the other hand, in order to control for these three types of error, a full Monte Carlo simulation of the future credit portfolio distribution with a very high number of simulation runs, one million, is carried out. The resulting percentiles are assumed to be the correct ones.

For both methods the (mean) percentile values are close together indicating that the discretization and the truncation error of the Fourier based method is, at least for the considered portfolio composition, not too large. The standard error resulting from an application of the Monte Carlo simulation is smaller than that one resulting from an application of the inversion formula (2), but the Fourier based method combined with Gaussian integration is in this example seven times faster than the Monte Carlo simulation. In order to evaluate the performance of both methods more properly, the run time is plotted against the standard error of the 0.1%- and 1%-percentile estimators. For further reducing the run time of the Fourier based approach a different number of simulation runs is used for different integration intervals in (2): On intervals where the integrand is large more draws of QRN are employed and vice versa. Nevertheless, as the (run time, standard error)-curves show, the Fourier based approach is in general not superior to the crude Monte Carlo simulation when computing risk measures (without figures). This is especially true when Value-at-Risk values with high confidence levels are needed and/or the asset return correlation is large.

V. Further Analysis

Of course, the results of these numerical experiments are rather indicative and do not allow a concluding evaluation of the potential of the Fourier based approach when applied to integrated market and credit portfolio models.

Extensions of the analysis presented here should for example check the robustness of the method to inhomogeneities in the portfolio composition, to the instrument type or the number of systematic risk factors. The use of alternative numerical integration rules, such as inverse Fast Fourier Transform, should be tried, too.

Finally, in order to ensure a 'fair' comparison, the performance of the Fourier based method should be compared with that one of Monte Carlo simulations enhanced by suitable variance reduction techniques. Here, Importance Sampling seems to be especially promising (see Glasserman and Li 2005, Kalkbrener et al. 2004, Merino and Nyfeler 2004, Grundke 2005b).

VI. Conclusions

In this paper it is analyzed whether a Fourier based approach can be an efficient tool for calculating risk measures in the context of a credit portfolio model with integrated market risk factors. Fur this purpose, this technique is applied to a version of the well-known credit portfolio model CreditMetrics extended by correlated interest rate and credit spread risk. Unfortunately, the characteristic function of the credit portfolio value at the risk horizon can not be calculated in closed-form. As a consequence, in the considered numerical examples the performance of the Fourier based approach is not superior to that one of a full Monte Carlo simulation. However, the trade-off between accuracy and speed of the Fourier based approach for real-world credit portfolios, compared to other computational approaches, will have to be explored by future research in more detail.

Bibliography

Barnhill Jr., T. M., and W.F. Maxwell (2002). Modeling Correlated Market and Credit Risk in Fixed Income Portfolios. Journal of Banking & Finance, vol. 26, pp. 347-374.

Davies, R.B. (1973). Numerical inversion of a characteristic function. Biometrika, vol. 60, no. 2, pp. 415-417.

Duffie, D., and J. Pan (2001). Analytical Value-At-Risk with Jumps and Credit Risk. Finance and Stochastics, vol. 5, no. 2, pp. 155-180.

Glasserman, P. and J. Li (2005). Importance Sampling for Portfolio Credit Risk. Working Paper, Columbia Business School.

Grundke, P. (2004). Integrating Interest Rate Risk in Credit Portfolio Models. Journal of Risk Finance, vol. 5, no. 2, pp. 6-15.

Grundke, Peter (2005a). Risk Measurement with Integrated Market and Credit Portfolio Models. Journal of Risk, vol. 7, no. 3, pp. 63-94.

Grundke, Peter (2005b). Importance Sampling for Integrated Market and Credit Portfolio Models. Working Paper, University of Cologne.

Gupton, G.M., C.C. Finger and M. Bhatia (1997). CreditMetrics™ – Technical Document. New York.

Kalkbrener, M., H. Lotter and L. Overbeck (2004). Sensible and efficient capital allocation for credit portfolios. Risk, January 2004, pp. 19-24.

Kiesel, R., W. Perraudin and A. Taylor (2003). The structure of credit risk: spread volatility and ratings transitions. Journal of Risk, vol. 6, no. 1, pp. 1-27.

Kijima, M., and Y. Muromachi (2000). Evaluation of Credit Risk of a Portfolio with Stochastic Interest Rate and Default Processes. Journal of Risk, vol. 3, no. 1, pp. 5-36.

Merino, S., and M. Nyfeler (2002). Calculating portfolio loss. Risk, August 2002, pp. 82-86.

Merino, S. and M. Nyfeler (2004). Applying importance sampling for estimating coherent credit risk contributions. Quantitative Finance, vol. 4, pp. 199-207.

Reiß, O. (2003). Mathematical Methods for the Efficient Assessment of Market and Credit Risk. PhD thesis, Department of Mathematics, University of Kaiserslautern.

Vasicek, O.A. (1977). An Equilibrium Characterization of the Term Structure. Journal of Financial Economics, vol. 5, no. 2, pp. 177-188.

Dynamic Replication of Non-Maturing Assets and Liabilities

Michael Schürle

Institute for Operations Research and Computational Finance,
University of St. Gallen, Bodanstr. 6, CH-9000 St. Gallen, Switzerland
michael.schuerle@unisg.ch

1 Introduction

Non-maturing assets and liabilities (NoMALs) are positions in a bank's balance with no contractual maturity such as savings or sight deposits. Clients have the option to add or withdraw investments or credits while the bank may adjust the customer rate anytime. It is often observed that the volume of NoMALs fluctuates significantly as clients react to changes in the customer rate or to the relative attractiveness of alternative investment opportunities.

Although there is no explicit maturity, banks must assign a fix maturity profile to a NoMAL: *(a)* It defines the transfer price at which the margin is split between a retail unit and the treasury. *(b)* The treasury manages the interest rate risk based on such a transformation of uncertain cash flows into (apparently) certain ones. To this end, most banks construct a replicating portfolio of fixed-income instruments where maturing tranches are always renewed at constant weights. The latter are determined from historical data by minimizing the tracking error between the cash flows of the portfolio (coupon payments) and those of the NoMAL (given by client rate and volume changes) during the sample period. Then, the transfer price is equivalent to the average margin between the yield on the portfolio and the client rate.

In the practical implementation, one can often observe that the portfolio composition and the corresponding margin are sensitive to the sample period. This induces the considerable model risk of an improper transformation of the variable position with direct implications on the "correct" transfer price and hedge-ability of the position. One possible result may also be that the replicating portfolio with the least volatile margin provides an income that does not cover the costs of holding the account. Then, it cannot be seen as the strategy with the lowest risk if it leads to a sure loss.

As an alternative to fitting the portfolio composition to a single *historic* scenario, this paper proposes a multistage stochastic programming model where the optimal allocation of new instruments is derived from some thousand scenarios of *future* interest rates, client rates and volumes. Instead of constant portfolio weights, the amounts invested or financed in each maturity are frequently adjusted taking future transactions and their impact on today's decision explicitly into account. Furthermore, risk is defined as the downside deviation of not meeting a specified minimum target for the margin.

2 Formulation as Multistage Stochastic Program

2.1 Notation

For simplicity, the following description is restricted to a model for deposits. An equivalent formulation for active products can easily be derived when investing is replaced by borrowing and vice versa. Let D be the longest maturity used for the construction of the replicating portfolio. $\mathcal{D} = \{1, \ldots, D\}$ denotes the set of dates where fixed-income securities held in the portfolio mature. The maturities of traded standard instruments that can be used for investment are given by the set $\mathcal{D}^+ \subseteq \mathcal{D}$. Furthermore, instruments in the set $\mathcal{D}^- \subseteq \mathcal{D}$ may be squared prior to maturity (modelled as borrowing funds of the corresponding maturities).

It is assumed that the joint evolution of random data (market rates, client rate and volume of the NoMAL) is driven by a stochastic process $\omega := (\omega_t;\, t = 1, \ldots, T)$ in discrete time defined on a probability space $(\Omega, \mathcal{F}, P)$ which satisfies the usual assumptions (with $\Omega := \Omega_1 \times \ldots \times \Omega_T$). The random vector $\boldsymbol{\omega}_t := (\boldsymbol{\eta}_t, \boldsymbol{\xi}_t) \in \Omega_t^\eta \times \Omega_t^\xi =: \Omega_t \subseteq \mathbb{R}^{K+L}$ can be decomposed into two components: $\boldsymbol{\eta}_t \in \Omega_t^\eta \subseteq \mathbb{R}^K$ is equivalent to the state variables of a K-factor term structure model and controls market rates, client rate and volume. The latter is also affected by the process $\boldsymbol{\xi}_t \in \Omega_t^\xi \subseteq \mathbb{R}^L$ that may represent additional economic factors with impact on the savings volume or a residual variable for non-systematic variations. The relevant stochastic coefficients in the optimization model at stage t depend on the histories of observations $\boldsymbol{\eta}^t := (\boldsymbol{\eta}_1, \ldots, \boldsymbol{\eta}_t)$ and $\boldsymbol{\omega}^t := (\boldsymbol{\omega}_1, \ldots, \boldsymbol{\omega}_t)$:

$r_t^{d,+}(\boldsymbol{\eta}^t)$	bid rate per period for investing in maturity $d \in \mathcal{D}^+$,
$r_t^{d,-}(\boldsymbol{\eta}^t)$	ask rate per period for financing in maturity $d \in \mathcal{D}^-$,
$c_t(\boldsymbol{\eta}^t)$	client rate paid per period for holding the deposit account,
$v_t(\boldsymbol{\omega}^t)$	volume of the non-maturing account position.

In the sequel, the dependency of the coefficients on $\boldsymbol{\omega}^t$ or $\boldsymbol{\eta}^t$ will not be stressed in the notation for simplicity. Interest rates, client rate and volume for $t = 0$ are deterministic and can be obtained from current market observations.

2.2 Optimization Model

At each stage $t = 0, \ldots, T$, where T denotes the planning horizon, decisions are made on the amount $x_t^{d,+}$ invested in maturity $d \in \mathcal{D}^+$ and the amount $x_t^{d,-}$ financed in maturity $d \in \mathcal{D}^-$. The totally invested volume of the replicating portfolio, i.e., the sum of investments minus borrowings over all stages up to t plus instruments held in the initial portfolio that have not yet matured, has to match the stochastic volume of the position at all points in time:

$$\sum_{\tau=0}^{t} \sum_{\substack{d \in \mathcal{D}^+ \\ d > \tau}} x_{t-\tau}^{d,+} - \sum_{\tau=0}^{t} \sum_{\substack{d \in \mathcal{D}^- \\ d > \tau}} x_{t-\tau}^{d,-} + \sum_{d=t+2}^{D} x_{-1}^{d} = v_t \ , \tag{1}$$

where x_{-1}^d denotes a position in the initial portfolio maturing at time d. Negative holdings are not allowed, i.e., the amount squared in a certain maturity must not exceed the investments made earlier with the same maturity date:

$$x_t^{d,-} \leq \sum_{\substack{\tau=1 \\ d+\tau \in \mathcal{D}^+}}^{t} x_{i,t-\tau}^{d+\tau,+} - \sum_{\substack{\tau=1 \\ d+\tau \in \mathcal{D}^-}}^{t} x_{i,t-\tau}^{d+\tau,-} + x_{-1}^{t+1+d} \qquad \forall d \in \mathcal{D}^- \ . \tag{2}$$

Transactions in t result in a surplus x_t^S that is defined as the difference between the income from the positions held in the replicating portfolio which have not matured minus the costs of holding the account:

$$x_t^S = \sum_{\tau=0}^{\tau^+} \sum_{\substack{d \in \mathcal{D}^+ \\ d > \tau}} r_{t-\tau}^{d,+} \cdot x_{t-\tau}^{d,+} - \sum_{\tau=0}^{\tau^-} \sum_{\substack{d \in \mathcal{D}^- \\ d > \tau}} r_{t-\tau}^{d,-} \cdot x_{t-\tau}^{d,-} + cf_{-1}^{t+2} - (c_t + \alpha_0) \cdot v_t \ , \tag{3}$$

where $\tau^+ := \min\{t, \max\{\mathcal{D}^+\} - 1\}$, $\tau^- := \min\{t, \max\{\mathcal{D}^-\} - 1\}$ and cf_{-1}^t is the sum of all coupon payments from positions in the initial portfolio with maturity t. The costs of holding the position consist not only of payments made to clients but also of non-interest expenses of serving the deposit. Hence, the forth term on the right-hand-side of (3) contains the constant α_0 to specify a target for the margin that must be achieved in addition to the client rate. The objective of the stochastic program is to minimize the *expected* downside deviation of not meeting the specified target over all stages:

$$\begin{aligned}
&\min \int_\Omega \sum_{t=0}^{T} x_t^M \, dP(\boldsymbol{\omega}) \\
&\left.\begin{aligned}
\text{s.t. } & \text{equations (1)–(3)} \\
& x_t^M \geq -x_t^S \\
& 0 \leq x_t^{d,+} \leq \ell^{d,+} \quad \mathcal{F}_t\text{-measurable} \quad \forall d \in \mathcal{D}^+ \\
& 0 \leq x_t^{d,-} \leq \ell^{d,-} \quad \mathcal{F}_t\text{-measurable} \quad \forall d \in \mathcal{D}^- \\
& x_t^S \in \mathbb{R};\ x_t^M \geq 0 \ \ \mathcal{F}_t\text{-measurable}
\end{aligned}\right\} \quad t = 0, \ldots, T;\ \text{a.s.}
\end{aligned} \tag{4}$$

Decision and state variables herein for $t > 0$ are stochastic since they depend on observations of the random data process $\boldsymbol{\omega}^t$ up to time t. Therefore, they are adapted to the filtration $\mathcal{F}_t$ that specifies the information structure, i.e., they are taken only with respect to the information available at this time *(nonanticipativity)*. All constraints must hold *almost surely (a.s.)*, i.e., for all $\omega \in \Omega$ except for sets with zero probability. A common way to make the stochastic program (4) computationally tractable is the generation of a *scenario tree* as an approximation of the vector stochastic process ω in $(\Omega, \mathcal{F}, P)$. The resulting deterministic problem is a large-scale linear program that can be solved with standard algorithms like Cplex (see [2] for a recent introduction to stochastic programming methods).

3 Scenario Generation

The model for the stochastic evolution of risk factors consists of three components: Its core is a *term structure model* with $K = 2$ factors for the level and steepness of the yield curve that fluctuate around long term means $\theta_1 := \theta$ and $\theta_2 := 0$. Their dynamics are described by the stochastic differential equations

$$d\eta_{it} = \kappa_i(\theta_i - \eta_{it})\, dt + \sigma_i\, dz_{it}\ , \qquad i = 1, 2, \tag{5}$$

where dz_1 and dz_2 are the increments of two uncorrelated Wiener processes. κ_i controls the speed at which the i-th factor reverts to its mean and σ_i controls the instantaneous volatility. It is known that under specification (5) both factors are normally distributed. This is an extension of the well-known Vasicek model [5]. Explicit formulae exist to derive the yield curve as a function of the two factors. Compared to alternative term structure models in an investigation for the Swiss market [1], this turned out to be the most suitable one for scenario generation. Parameter estimates are derived from historic interest rates using the maximum likelihood method described in [3].

The second component models the *client rate* as a (deterministic) function of the level factor η_1 to reflect specific characteristics of the relevant NoMAL and the dependency on interest rates. In case of Swiss savings accounts, banks adjust the customer rate only at discrete increments, typically a multiple of 25 basis points (bp). Furthermore, the adjustment is asymmetric when market rates rise or fall, and there is a (political) cap where higher rates are no longer passed to depositors. Let $\delta_0 < \ldots < \delta_n$ be the possible increments (including the value 0) and $\gamma_0 < \gamma_1 < \ldots < \gamma_n < \gamma_{n+1}$ some threshold values ($\gamma_0 := -\infty$, $\gamma_{n+1} := \infty$). Then, the client rate changes by $\Delta c_t = \delta_i$ if the latent control variable $c_t^* = \beta_0\, c_{t-1} + \beta_1\, \eta_{1,t} + \ldots + \beta_{m+1}\, \eta_{1,t-m}$ is realized between the threshold values γ_i and γ_{i+1}. The coefficients of the control variable process and the threshold values are estimated jointly from historical data using a maximum likelihood procedure.

Finally, relative changes in the *volume* v_t are modelled by

$$\ln v_t = \ln v_{t-1} + e_0 + e_1 t + e_2\eta_{1t} + e_3\eta_{2t} + \xi_t\ . \tag{6}$$

The constant e_0 and the time component $e_1 t$ reflect that the volume exhibits a positive trend (nominal balances can be expected to increase in the long run due to inflation). The factors η_1 and η_2 of the term structure model are included as explanatory variables because market rates influence the account when clients transfer volume from or to other interest rate sensitive investments like savings certificates. An additional stochastic factor ξ ($L = 1$) which is uncorrelated with the market rate model factors takes into account that the latter do not fully explain the observed evolution of the balance. Equation (6) can easily be estimated by ordinary least squares regression, and the volatility σ_ξ of the residuum factor ξ is immediately derived from the standard error.

For the generation of a scenario tree, the multivariate normal distribution of the random vector $\omega_t := (\eta_{1t}, \eta_{2t}, \xi_t)$ at $t = 1, \ldots, T$ is approximated by a *multinomial* distribution. This provides finite sets of samples and corresponding probabilities at each stage that preserve the expectations and covariance matrix implied by the term structure and volume model (5)–(6) after a transformation (details are described in [4]).

4 Results

The model was tested with client rate and volume data of a real Swiss savings deposit position for the period January 1989 to December 2001. Market rates during this time showed a phase of inverse term structures at high level at the beginning, followed by an abrupt change to a period of low interest rates and normal yield curves in the second half. Interbank market instruments with maturities from 1 to 10 years were used for the construction of the replicating portfolio. Positions were squared only if the amount of maturing tranches was not sufficient to compensate a drop in volume. The margin target for the downside minimization was set to $\alpha_0 = 200$ bp. Figure 1 shows that the stochastic programming model was able to meet this value almost anytime during the sample period while the margin of a static replicating portfolio that was calculated as benchmark collapsed after the drop in market rates.

model	avg. margin	std. dev.	avg. maturity
dynamic replication	2.23 %	0.32 %	2.37 yrs.
static benchmark	1.93 %	0.49 %	1.81 yrs.

According to the table, the average margin was increased by 30 bp while simultaneously the volatility was reduced. Therefore, the dynamic strategy provides the more efficient replication. The corresponding portfolio has also a higher duration, indicating that the "true" maturity of the NoMAL position is approx. 0.5 years longer than implied by the static replication. Note that an extension of the duration by half a year in a portfolio with constant weights would yield at most a gain of 10 bp at larger volatility. This allows the conclusion that the higher margin achieved here can mainly be attributed to the added value of dynamic management.

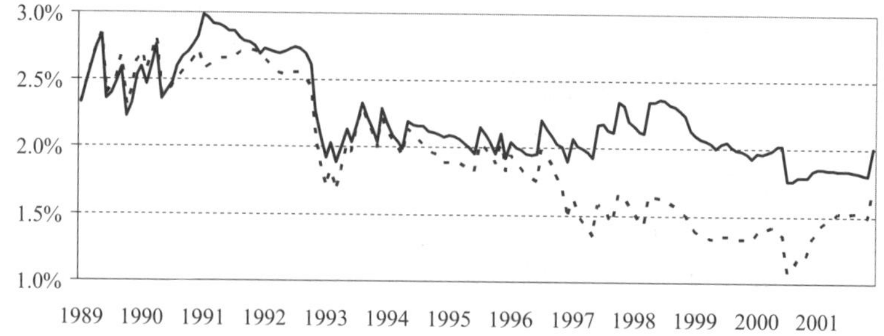

Fig. 1. Margin evolution: Dynamic (solid) vs. static replication (dashed)

5 Conclusions and Outlook

Dynamic strategies have turned out to be more efficient for the replication of variable banking products than the classical static approach. A further analysis reveals that the portfolio compositions are by far less sensitive to changes in the input parameters due to different sample periods. This observation is typical for stochastic programming models [3]. From a practical point of view, no other data are required for the calibration of the risk factor models than for the determination of the (constant) weights in static replication.

The description of the stochastic programming model here was restricted to a basic version (4). The implementation of the complete model contains also alternative objective functions (e.g., minimization of volatility instead of downside risk) and additional constraints for the portfolio structure, admissible transactions and risk. Current research is directed towards the assessment of "end effects", i.e., the truncation of the planning horizon at some finite number of stages T although it is infinite for NoMAL management by definition. Approximations of the true infinite horizon problem are being developed that allow a quantification of the impact of events after time T on earlier decisions. First results imply that these techniques may have some potential for an additional improvement of the model performance and robustness.

References

1. Frauendorfer K, Schürle M (2000) Term structure models in multistage stochastic programming. Annals of Operations Research 100:189–209
2. Kall P, Mayer J (2005) Stochastic Linear Programming. Springer, New York
3. Schürle M (1998) Zinsmodelle in der stochastischen Optimierung. Haupt, Berne
4. Siede H (2000) Multi-Period Portfolio Optimization. PhD thesis, University of St. Gallen
5. Vasicek O (1977) An equilibrium characterization of the term structure. Journal of Financial Economics 5:177–188

Portfolio Optimization Under Partial Information and Convex Constraints in a Hidden Markov Model

Jörn Sass

RICAM, Austrian Academy of Sciences, Altenberger Str. 69, A-4040 Linz
joern.sass@oeaw.ac.at

Summary. In a continuous-time hidden Markov model (HMM) for stock returns we consider an investor who wishes to maximize the expected utility of terminal wealth. As a means to deal with the resulting highly risky strategies we impose convex constraints on the trading strategies covering e.g. short selling restrictions. Based on HMM filtering methods we show how to reformulate this model with partial information as a model with full information. Then results on portfolio optimization under constraints are used to give a verification result. By its application an optimal trading strategy can be computed. Numerical results are provided.

1 Introduction

We consider a multi-stock market model where prices satisfy a stochastic differential equation with instantaneous rates of return modeled as an unobservable continuous time, finite state Markov chain. We assume that only the stock prices are observable, hence we have a model with partial information. In [4] it is shown that on real market data utility maximization strategies based on this simple Markov chain model for the drift process can outperform strategies based on the assumption of a constant drift parameter (Merton strategy). But these strategies are very risky since they lead to extreme long and short positions. This can result in a poor performance if we trade only daily, even bankruptcy might occur. In [5] better results are obtained using a suitable non-constant volatility model. The better performance is due to the fact that the positions are less extreme than for constant volatility. A similar effect can be expected from introducing convex constraints in the model with constant volatility. These cover e.g. short selling restrictions and allow to impose bounds on long positions. At least for one stock their implementation is much simpler.

Heuristically, for one stock we might cut off the optimal strategy given in terms of the risky fraction (the fraction of wealth invested in the stock). We compute in Table 1 for initial capital 1 the averages of 500 applications of different strategies over one-year (20 DJII stocks for 25 years): 'stock' is the pure stock investment, 'Merton' is the optimal strategy assuming constant drift, 'HMM' is the optimal strategy assuming a two-state continuous-time Markov chain for the drift, 'HMM(0,1)' is the strategy where we cut off the risky fraction at 0 and 1, meaning neither short selling nor borrowing are allowed, and 'HMM(-1,4)' less restrictive with cut off at -1 and 4. The HMM- and the Merton strategy lead 12 times and 8 times, respectively, to bankruptcy, hence for logarithmic utility we have to assign $-\infty$ to the average utility. No bankruptcy occurs for the constrained strategies HMM(0,1) and HMM(-1,4) and we see that they outperform the pure stock investment.

Table 1. Terminal wealth and logarithmic utility for different trading strategies: Averages over 500 observations of real market prices (stock prices over one year)

	stock	Merton	HMM	HMM(0,1)	HMM(-1,4)
average terminal wealth X_T:	1.153	1.306	1.520	1.172	1.357
av. utility = av. $\log(X_T)$:	0.116	$-\infty$	$-\infty$	0.141	0.157

Motivated by this example we shall provide in this short paper some technical background how to use in the partial information context results known from models under full information, see [1]. After introducing the model, we transform it in Sect. 2 to a model with full information by filtering techniques, replacing the drift by its filter and the Brownian motion by the innovations process which is an observable Brownian motion under the original measure. This is in line with the separation principle of Genotte [3] which says that we can do the filtering first and the optimization afterwards. In Sect. 3 we then formulate the optimization problem and formulate a verification result based on [1]. In Remark 1 we give a hint how optimal strategies for general utility functions can be derived. But the very technical analysis for this general case, involving the use of Malliavin calculus, is deferred to a future publication. Here we will compute strategies explicitly for logarithmic utility which corresponds to maximizing the expected rate of return. For one stock we will see that the strategies of the form we used in Table 1 above are indeed optimal under the corresponding constraints. We provide some simulation results.

2 Filtering in an HMM for the Stock Returns

Let $(\Omega, \mathcal{A}, P)$ be a complete probability space, $T > 0$ the terminal trading time, and $\mathcal{F} = (\mathcal{F}_t)_{t\in[0,T]}$ a filtration in $\mathcal{A}$ satisfying the usual conditions. We consider a *money market* with interest rates equal 0 (to simplify notation) and n stocks whose *prices* $S = (S_t)_{t\in[0,T]}$, $S_t = (S_t^1, \dots, S_t^n)^\top$ evolve according to

$$\mathrm{d}S_t = \mathrm{Diag}(S_t)(\mu_t\,\mathrm{d}t + \sigma\,\mathrm{d}W_t)\,, \quad S_0 \in I\!R^n\,,$$

where $W = (W_t)_{t\in[0,T]}$ is an n-dimensional Brownian motion w.r.t. $\mathcal{F}$, and σ is the non-singular $(n \times n)$-volatility-matrix. The *return process* $R = (R_t)_{t\in[0,T]}$ is defined by $\mathrm{d}R_t = (\mathrm{Diag}(S_t))^{-1}\,\mathrm{d}S_t$. We assume that $\mu = (\mu_t)_{t\in[0,T]}$, the drift process of the return, is given by $\mu_t = B\,Y_t$, where $Y = (Y_t)_{t\in[0,T]}$ is a stationary, irreducible, *continuous time Markov chain* independent of W with state space $\{e_1,\dots,e_d\}$, the standard unit vectors in $I\!R^d$. The columns of the state matrix $B \in I\!R^{n\times d}$ contain the d possible states of μ_t. Further Y is characterized by its rate matrix $Q \in I\!R^{d\times d}$, where $\lambda_k = -Q_{kk} = \sum_{l=1,l\neq k}^{d} Q_{kl}$ is the rate of leaving e_k and Q_{kl}/λ_k is the probability that the chain jumps to e_l when leaving e_k.

We consider the case of *partial information* meaning that an investor can only observe the prices. Neither Y nor W are observable. Only the events of $\mathcal{F}^S$, the augmented filtration generated by S, can be observed and hence all investment decisions have to be adapted to $\mathcal{F}^S$.

Since $\sigma^{-1}\mu_t = \sigma^{-1}B\,Y_t$, $t \in [0,T]$, is uniformly bounded, the density process $(Z_t)_{t\in[0,T]}$ defined by $\mathrm{d}Z_t = -Z_t(\sigma^{-1}B\,Y_t)^\top\,\mathrm{d}W_t$, $Z_0 = 1$, is a martingale. By $\mathrm{d}\tilde{P} = Z_T\mathrm{d}P$ we define the *reference measure*. $\tilde{E}$ will denote expectation with respect to $\tilde{P}$. Girsanov's Theorem guarantees that $\mathrm{d}\tilde{W}_t = \mathrm{d}W_t + \sigma^{-1}BY_t\,\mathrm{d}t$ defines a $\tilde{P}$-Brownian motion. The definition of R yields

$$R_t = \int_0^t B\,Y_s\,\mathrm{d}s + \sigma\,W_t = \sigma\,\tilde{W}_t\,, \quad t \in [0,T]\,. \tag{1}$$

Note that $\mathcal{F}^S = \mathcal{F}^R = \mathcal{F}^{\tilde{W}}$. By (1) we are in the classical situation of *HMM filtering* with signal Y and observation R, where we want to determine the filter $\eta_t = E[Y_t\,|\,\mathcal{F}_t^S]$ for Y_t. By Theorem 4 in Elliott (1993), Bayes' Law, and using $\mathbf{1}_d^\top Y_t = 1$ we get using the unnormalized filter $\mathcal{E}_t = \tilde{E}[Z_T^{-1}Y_t\,|\,\mathcal{F}_t^S]$ and the conditional density $\zeta_t = E[Z_t\,|\,\mathcal{F}_t^S]$

Theorem 1. *$\eta_t = \zeta_t\,\mathcal{E}_t$, $\zeta_t^{-1} = \mathbf{1}_d^\top\mathcal{E}_t$, and*

$$\mathcal{E}_t = E[Y_0] + \int_0^t Q^\top\mathcal{E}_s\,\mathrm{d}s + \int_0^t \mathrm{Diag}(\mathcal{E}_s)B^\top(\sigma\sigma^\top)^{-1}\mathrm{d}R_s\,, \quad t \in [0,T]\,.$$

Furthermore, $\zeta_t^{-1} = \tilde{E}[Z_t^{-1}\mid\mathcal{F}_t^S]$ *and* $\zeta_t^{-1} = 1 + \int_0^t (B\mathcal{E}_s)^\top(\sigma\sigma^\top)^{-1}\mathrm{d}R_s\,.$

We changed for the filtering to the reference measure $\tilde{P}$ under which the return process becomes a Brownian motion and using Theorem 1 we can compute the filter η. Now we can go back under our original probability measure P and express the return process in terms of the filter η and the *innovation process* $V = (V_t)_{t\in[0,T]}$, $\mathrm{d}V_t = \mathrm{d}\tilde{W}_t - \sigma^{-1}B\eta_t\,\mathrm{d}t$, which is an $\mathcal{F}^S$-Brownian motion under P. From (1) and the definition of V we get

$$\mathrm{d}R_t = B\,\eta_t\,\mathrm{d}t + \sigma\,\mathrm{d}V_t\,, \quad t \in [0,T]\,. \tag{2}$$

Thus we are now with respect to $\mathcal{F}^S$ in a market model with full information. In this formulation the *market price of risk* is $\theta_t = \sigma^{-1} B \eta_t$, $t \in [0,T]$. Since the filtrations generated by S and R coincide, we have not lost any information. The optimal solution we obtain in the next section for this model is also optimal for our original problem.

3 Optimal Trading under Convex Constraints

A *trading strategy* $\pi = (\pi_t)_{t\in[0,T]}$ is an n-dimensional $\mathcal{F}^S$-progressively measurable process which satisfies $\int_0^T \|\pi_t\|^2 \mathrm{d}t < \infty$. For initial capital $x > 0$ the corresponding *wealth process* $X^\pi = (X_t^\pi)_{t\in[0,T]}$ is defined by $\mathrm{d}X_t^\pi = \pi_t^\top \mathrm{d}R_t$, $X_0 = x$. π_t is the wealth invested in the stock at time t, $X_t^\pi - \mathbf{1}_n^\top \pi_t$ is invested in the money market. Further we shall denote the fraction of wealth invested in the stocks by $f_t^\pi = \pi_t / X_t^\pi$, $t \in [0,T]$.

The *constraints* we would like to impose on the strategy are given by a closed, convex set $K \subseteq \mathbb{R}^n$ which contains 0. We shall assume that the support function δ of $-K$,

$$\delta(y) = \sup_{x\in K}(-x^\top y)\,, \quad y \in \mathbb{R}^n,$$

is continuous on its effective domain $\tilde{K} = \{y \in \mathbb{R}^n : \delta(y) < \infty\}$. This is true for the following important examples.

Example 1. Constraints on short and long positions for each stock can be imposed by

$$K_1 = \{x \in \mathbb{R}^n : l_i \le x_i \le u_i, i = 1,\dots,n\}\,, \quad \text{where} \quad l_i \le 0 \le u_i$$

with $\delta_1(y) = \sum_{i=1}^n (u_i y_i^- - l_i y_i^+)$, $\tilde{K}_1 = \mathbb{R}^n$. Or we can put a bound on the fraction u_0 of the wealth we borrow from the bank by

$$K_2 = \{x \in \mathbb{R}^n : 1 - \mathbf{1}_n^\top x \ge u_0 \text{ and } x_i \ge 0, i = 1,\dots,n\}, \quad \text{where} \quad u_0 \le 1$$

with $\delta_2(y) = \max_{i=1}^n(-(1-u_0)y_i)$ if $y_i < 0$ for some i, $\delta_2(y) = 0$ otherwise.

A *utility function* $U : (0,\infty) \to \mathbb{R}$ is strictly increasing, strictly concave, continuously differentiable, and satisfies $\lim_{x\to\infty} U'(x) = 0$, $U'(0+) = \infty$. Further, $I : (0,\infty) \to (0,\infty)$ denotes the inverse function of U'.

For initial capital $x > 0$, $\mathcal{A}_K(x)$ is the class of K-admissible trading strategies π which satisfy $P(X_t^\pi \ge 0 \text{ for all } t \in [0,T]) = 1$, $f_t^\pi \in K$ for $t \in [0,T]$, $X_0^\pi = x$, $E[U^-(X_T^\pi)] < \infty$.

Optimization Problem For given $x > 0$ and utility function U determine

$$J(x) = \sup\{\, E[U(X_T^\pi)] \,:\, \pi \in \mathcal{A}_K(x)\,\}$$

and find a K-admissible $\hat{\pi}$ which satisfies $E[U(X_T^{\hat{\pi}})] = J(x_0)$. For the optimal strategy $\hat{\pi}$ we denote $\hat{X} = X^{\hat{\pi}}$, $\hat{f} = f^{\hat{\pi}}$. We shall always assume $J(x) < \infty$.

While the market with respect to $\mathcal{F}_T^S$-measurable claims is complete in the unrestricted case, the constraints result in an incomplete market. So we cannot work with a unique martingale measure. To describe the possible measures we introduce the set $\mathcal{H}$ of $\mathcal{F}^S$-progessively measurable $\tilde{K}$-valued processes satisfying $E[\int_0^T (\|\nu_t\|^2 + \delta(\nu_t))\,\mathrm{d}t] < \infty$, corresponding to the shadow prices due to incompleteness. With $\mathcal{H}_b$ we denote the subset of uniformly bounded processes. For $\nu \in \mathcal{H}$ we introduce the modified market price of risk $\theta_t^\nu = \theta_t^\nu + \sigma^{-1}\nu_t$ and the density process $\zeta^\nu = (\zeta_t^\nu)_{t\in[0,T]}$ by $\mathrm{d}\zeta_t^\nu = -\zeta_t^\nu(\theta_s^\nu)^\top \mathrm{d}V_s$, $\zeta_0^\nu = 1$. If $\nu \in \mathcal{H}_b$, then $\mathrm{d}P^\nu = \zeta_T^\nu \mathrm{d}P$ defines a probability measure and $\mathrm{d}W_t^\nu = \mathrm{d}V_t + \theta_s^\nu\,\mathrm{d}s$ a Brownian motion under P^ν.

Proposition 1. *For given $x > 0$ a trading strategy $\pi \in \mathcal{A}_K(x)$ is optimal, if for some $y > 0$, $\nu \in \mathcal{H}$*

$$X_T^\pi = I(y\,\gamma_T^\nu\zeta_T^\nu)\,, \quad \delta(\nu_t) = -(f_t^\pi)^\top\nu_t\,,\, t\in[0,T]\,, \quad E^\nu[\gamma_T^\nu X_T^\pi] = x\,, \tag{3}$$

where $\gamma_t^\nu = \exp(-\int_0^t \delta(\nu_s)\,\mathrm{d}s)$ is the modified discount factor.

For a proof we refer to [1, Lemma 11.6]. There also conditions are given which guarantee the existence of a solution. For our purpose, the computation of an optimal strategy, the above proposition is sufficient.

Remark 1. If $y > 0$ and $\nu \in \mathcal{H}_b$ exist which satisfy the last two equations in Proposition 1 and $\gamma_T^\nu I(y\gamma_T^\nu\zeta_T^\nu)$ is sufficiently smooth for the application of the Malliavin derivative with respect to P^ν, then similar as in [4],

$$f_t^{\hat{\pi}} = (\gamma_t^\nu)^{-1}(\sigma^\top)^{-1}E^\nu[D_t(\gamma_T^\nu I(y\gamma_T^\nu\zeta_T^\nu)\,|\,\mathcal{F}_t^S]\,.$$

Note that ν depends on $f^{\hat{\pi}}$, hence this representation is not explicit, making the constrained case much more complex than the unconstrained case.

We shall now provide some explicit results for logarithmic utility $U = \log$ for which we know from [4] that the optimal strategy without constraints is of the form $f_t^o = (\sigma\sigma^\top)^{-1}B\eta_t$. With constraints we can use Proposition 1: For $U(x) = \log(x)$ we have $U'(x) = 1/x$, hence $I(y) = 1/y$. If an optimal $\nu \in \mathcal{H}_b$ exists, the first and the last condition in (3) imply $\hat{X}_T = x(\gamma_T^\nu\zeta_T^\nu)^{-1}$, hence

$$\gamma_T^\nu \hat{X}_T = x(\zeta_T^\nu)^{-1} = x + \int_0^T x(\zeta_t^\nu)^{-1}(\theta_t^\nu)^\top \mathrm{d}W_t^\nu\,. \tag{4}$$

On the other hand, using the second condition in (3)

$$\mathrm{d}(\gamma_t^\nu \hat{X}_t) = \gamma_t^\nu \hat{X}_t\left((\hat{f}_t{}^\top B\eta_t - \delta(\nu_t))\mathrm{d}t + \hat{f}_t{}^\top\sigma\mathrm{d}V_t\right) = \gamma_t^\nu\hat{X}_t\hat{f}_t{}^\top\sigma\,\mathrm{d}W_t^\nu\,,$$

showing that $\gamma^\nu\hat{X}$ is a martingale under P^ν, hence $\gamma_t^\nu\hat{X}_t = x(\zeta_t^\nu)^{-1}$ and a comparison with (4) yields $\hat{f}_t = (\sigma^\top)^{-1}\theta_t^\nu = (\sigma\sigma^\top)^{-1}(B\eta_t + \nu_t)$. Comparing with f_t^o, we thus have to solve (observing $\hat{f}_t \in K$, $\nu_t \in \tilde{K}$)

$$\hat{f}_t = f_t^o + (\sigma\sigma^\top)^{-1}\nu_t\,, \quad \hat{f}_t{}^\top\nu_t = -\delta(\nu_t)\,, \quad t\in[0,T]\,. \tag{5}$$

Example 2. For $n = 1$, (5) yields for the constraints given by K_1 of Example 1

$$\hat{f}_t = u_1 \text{ if } f_t^o > u_1 , \quad \hat{f}_t = f_t^o \text{ if } l_1 \leq f_t^o \leq u_1 , \quad \hat{f}_t = l_1 \text{ if } f_t^o < l_1 .$$

Then ν is given by (5), in particular, ν is bounded, since f^o is bounded.

So the strategies HMM(0,1) and HMM(-1,4) defined in the introduction are optimal for the corresponding constraints. In Table 2 we compare these strategies with the unconstrained strategy HMM for different trading periods on 500 simulated prices (each over one year). We use $x = 1$, $d = 2$, $\sigma = 0.4$, $B_{11} = 2.5$, $B_{12} = -1.5$, $\lambda_1 = 60$, $\lambda_2 = 40$. This choice corresponds to an average drift 0.1. Table 2 shows the same effect we observed for the real market data. The unconstrained strategy is too risky for infrequent trading. Already for daily trading bankruptcy occurred 2 times. In parentheses we provide the average utilities when we assign a utility of -1 to bankruptcy. Even then the constrained strategies give better results for less frequent trading. The pure stock investment gives for this sample an average utility of 0.027.

Table 2. Utility for different trading strategies: Averages over 500 simulated prices

strategy	10/day	5/day	4/day	2/day	daily	every 2 days
HMM	1.101	1.055	0.925	0.717	$-\infty$ (0.13)	$-\infty$ (-0.67)
HMM(0,1)	0.261	0.256	0.246	0.230	0.192	0.165
HMM(-1,4)	0.899	0.888	0.787	0.691	0.460	0.308

For more stocks the optimal strategy can still be obtained from (5), but positions in other stocks may be affected when one position has to be adjusted.

References

1. Cvitanić J (1997) Optimal trading under constraints. In: Financial Mathematics (Bressanone, 1996). Lecture Notes in Mathematics 1656: 123–190
2. Elliott RJ (1993) New finite-dimensional filters and smoothers for noisily observed Markov chains. IEEE Transactions on Information Theory 39: 265–271
3. Genotte G (1986) Optimal portfolio choice under incomplete information. The Journal of Finance 41: 733–749
4. Sass J, Haussmann UG (2004) Optimizing the terminal wealth under partial information: The drift process as a continuous time Markov chain. Finance and Stochastics 8: 553–577
5. Sass J, Haussmann UG (2004) Portfolio optimization under partial information: Stochastic volatility in a hidden Markov model. In: Ahr D, Fahrion R, Oswald M, Reinelt G (eds) Operations Research Proc. 2003, Springer, Berlin: 387–394

Robuste Portfoliooptimierung: Eine kritische Bestandsaufnahme und ein Vergleich alternativer Verfahren

Ulf Brinkmann

Universität Bremen, Fachbereich 7, Wirtschaftswissenschaft, Lehrstuhl für Finanzwirtschaft, Hochschulring 4, 28359 Bremen ulfb@uni-bremen.de

1 Einleitung

Im Mittelpunkt des Portfoliomanagementprozesses steht die Portfoliorealisierung, deren Schwerpunkt die systematische Aufteilung des Anlagebetrages auf Anlageobjekte, die sog. Asset Allocation, ist (vgl. [7], S. 25). Eine mögliche Methode zur Bewältigung dieser Problemstellung ist der von Markowitz entwickelte, sog. Mean-Variance-Ansatz (kurz M-V-Ansatz), der in Kap. 2 präsentiert wird. Zudem werden vier alternative Ansätze der Portfoliobildung, der Ansatz nach Kataoka, ein robuster Optimierungsansatz, die Optimierung mit der Kosemivarianz sowie ein "mittelwertbasierter" Ansatz (in Anlehnung an [5]), vorgestellt. Der M-V-Ansatz weist beim praktischen Einsatz Schwächen auf, die in Kap. 3 geschildert werden. Anhand einer Fallstudie werden die vorgestellten Optimierungsansätze bezüglich ihrer Robustheit überprüft. Das Kap. 4 schließt die Ausführungen mit einem Fazit ab.

2 Portfoliobildung

2.1 Mean-Variance-Framework

Für die Bestimmung eines anlegerindividuell optimalen Portfolios nach Markowitz muss die (Risiko-)Nutzenfunktion des Investors bekannt sein (vgl. [4]). In Wissenschaft und Praxis wird wegen der sich bei der Bestimmung der Nutzenfunktion ergebenden Probleme oft von der in 1 dargestellten vereinfachten Zielfunktion ausgegangen (vgl. [7], S. 382 ff. oder [6], S. 148 f.).

$$ZF = \mathbf{w}^T\mathbf{r} - \lambda \cdot \mathbf{w}^T\mathbf{V}\mathbf{w} \rightarrow \; max! \tag{1}$$

mit $\mathbf{V}$: Varianz-Kovarianzmatrix der Assetrenditen
$\mathbf{w}$: Vektor der Anteilsgewichte im Portfolio

$\mathbf{r}$	:	Vektor der erwarteten Assetrenditen
λ	:	Risikoaversionsparameter

Die Optimierung wird unter Beachtungung der Budgetbeschränkung in 2 und einem Leerverkaufsverbot (vgl. 3) durchgeführt. Beide Nebenbedingungen gelten für alle Optimierungsansätze und werden nicht mehr explizit aufgeführt.

$$\sum_{i=1}^{n} w_i = 1 \tag{2}$$

$$w_i \geq 0; \forall i = 1, ...n \tag{3}$$

mit

w_i	:	Anteilsgewicht des Assets i im Portfolio
n	:	Anzahl der Assets

2.2 Optimierungsansatz nach Kataoka

Dem M-V-Ansatz liegt ein zweiseitiges Risikoverständnis zugrunde. Es werden Situationen als "risikobehaftet" empfundenen, bei denen die eintretende Rendite unterhalb sowie oberhalb des Erwartungswertes liegt. Als "Risiko" kann alternativ nur die erstgenannte Situation, die Ausfallsituation, verstanden werden. Bei der Portfoliooptimierung mit der Ausfallwahrscheinlichkeit kommen die als "safety-first" bezeichneten Ansätze zum Einsatz, zu denen der von [3] entwickelte Ansatz zählt. Er ermittelt die maximale Portfoliorendite, bei der die Wahrscheinlichkeit einer Unterschreitung des zu maximierenden Renditeanspruchsniveaus τ^* höchstens einen Wert von α beträgt (vgl. 4 und 5). Das Risikoniveau ist individuell festzulegen (hier wird $\alpha = 0,05$ gewählt).

$$\tau^* \rightarrow \text{max!} \tag{4}$$

$$P(r_P \leq \tau^*) \leq \alpha \tag{5}$$

2.3 Optimierung mit symmetrischer Kosemivarianz

Ein weiteres ausfallorientiertes Risikomaß ist die sog. Kosemivarianz, von der hier die symmetrische Kosemivarianz (kurz Kosemiv.) betrachtet wird. Die Optimierung wird auf Basis der Zielfunktion 1, aber unter Verwendung einer symmetrischen Kosemivarianzmatrix durchgeführt (vgl. [7], S. 426f.).

$$ZF = \mathbf{w}^T\mathbf{r} - \lambda \cdot \mathbf{w}^T\mathbf{sVw} \rightarrow \ max! \tag{6}$$

mit

$\mathbf{sV}$	:	symmetrische Kosemivarianzmatrix der Assetrenditen

2.4 Robuster Portfoliooptimierungsansatz

Idee der robusten Optimierung ist, die Optimierungsprogramme so zu formulieren, dass sie für alle möglichen Realisierungen der unsicheren Inputparameter "gute" Lösungen liefern. Zur Vorgehensweise bei der Bestimmung der Menge aller möglichen Renditevektoren sowie Varianz-Kovarianzmatrizen vgl. [8], S. 168 f. Ein (lösungs-)robustes Optimierungsmodell für die Bestimmung des anlegerindividuell optimalen Portfolios wird in [8] formuliert. Es handelt sich um ein zirkuläres Optimierungsproblem (vgl. 7). Die ursprüngliche Zielfunktion 1 wird in Abhängigkeit der Rendite und Varianz minimiert und gleichzeitig in Bezug auf die Anteilsgewichte maximiert.

$$max_{\mathbf{w}} \left\{ min_{\mathbf{r}\in U_{\mu};\mathbf{V}\in U_{\mathbf{V}}} \mathbf{w}^T\mathbf{r} - \lambda \cdot \mathbf{w}^T\mathbf{V}\mathbf{w} \right\} \tag{7}$$

mit $U_{\mathbf{V}}$: Menge aller Varianz-Kovarianzmatrizen
$U_{\boldsymbol{\mu}}$: Menge aller Renditevektoren

Unter Annahme eines Leerverkaufsverbotes kann der Optimierungsansatz zu 8 vereinfacht werden (vgl. hierzu [8], S. 161 ff.).

$$ZF = \mathbf{w}^T\mathbf{r}^L - \lambda \cdot \mathbf{w}^T\mathbf{V}^U\mathbf{w} \rightarrow max \tag{8}$$

mit $\mathbf{r}^L$: minimales Element der Menge $U_{\boldsymbol{\mu}}$
$\mathbf{V}^U$: maximales Element der Menge $U_{\mathbf{V}}$

Unter diesen Annahmen ist eine Bestimmung des minimalen Elementes der Menge $U_{\boldsymbol{\mu}}$ möglich. Zudem ist es notwendig, dass es sich bei der Matrix $\mathbf{V}^U$ um eine positiv semidefinite Matrix handelt (vgl. [8], S. 164).

2.5 Mittelwertsansatz

Bei diesem Ansatz werden die Anteilsgewichte der Portfolios als Mittelwerte der Anteilsgesichte von 1.000 Portfolios ermittelt, welche unter Verwendung der Zielfunktion 1 auf Basis eines Bootstrapverfahrens aus dem jeweiligen zugrundeliegenden Datensample generiert wurden (in Anlehnung an [5]). Zur Vorgehensweise des Bootstrapverfahrens vgl. bspw. [8], S. 168 f.

3 Probleme des Mean-Variance-Framework

Die praktische Anwendung des M-V-Ansatzes birgt einige Probleme, die auch in der Literatur kritisch diskutiert werden. So sind die resultierenden Portfoliostrukturen des M-V-Ansatzes oft extrem, instabil, ökonomisch unplausibel und für den Anwender nicht intuitiv nachvollziehbar (vgl. z.B. [1], S. 28, [2],

S. 204. oder [5], S. xiv.) Ein Grund hierfür ist die hohe Sensitivität der durch die Optimierung ermittelten Portfoliogewichte in Bezug auf Veränderungen der Inputparameter. Hinzu kommen mögliche Schätzfehler bei den prognostizierten Inputparametern. Eine Systematisierung der Problembereiche wird bei [2], S. 206 ff. geleistet. Bei prognostizierten Inputparametern stellt sich dem Anwender auch die Frage, wie gut sich das konstruierte Portfolio bei von diesen Prognosen abweichenden Inputparametern, beispielsweise einem "worst-case" (wc) Szenario, entwickelt.

Zur Illustration der Problembereiche wird ein Beispiel der Portfolioallokation herangezogen. Das Anlageuniversum besteht aus zehn fiktiven Assets, deren Renditen multivariat normalverteilt sind (erwartete Rendite von 5% p.M.; zukünftige Standardabweichung von 10% p.M.; Korrelation der Assets untereinander ist 0,2; $\lambda = 3$). Für das optimale Portfolio ergibt sich bei Kenntnis der wahren Verteilungsparameter ein gleichgewichtetes Portfolio. In der praktischen Anwendung sind diese aber aus Beobachtungen zu schätzen. Für jedes der zehn Assets wurden pro Asset 60 Zufallszahlen aus der angegebenen Verteilung gezogen, für die jeweils die "empirischen" Mittelwerte und "empirischen" Standardabweichungen berechnet wurden. So wurden insgeamt 10.000 Datensamples konstruiert, für die für jeden Optimierungsansatz jeweils das optimale Portfolio bestimmt wurde. Für die Anwendung des robusten Optimierungsansatzes wird die Menge der möglichen Varianz-Kovarianz-Matrizen, sowie die Menge der möglichen Renditevektoren mit einem Bootstrapverfahren aus dem jeweiligen Basisdatensample gebildet. Zur Vorgehensweise vgl. bspw. [8], S. 168 f. Kennzahlen für die resultierenden durchschnittlichen Anteilsgewichte der Assets zeigt die Tab. 1. Für die einzelnen Ansätze sind die über alle zehn Assets durchschnittlichen, maximal resultierenden Anteilsgewichte (Max) und die Standardabweichungen der resultierenden Portfoliorenditen (Std), als Maß für die Stabilität der Portfoliostrukturen aufgeführt.

	Markowitz	Kataoka	Kosemiv.	Robust	Mittel.
⌀ Max	0,89683	0,54455	1	0,83640	1
⌀ Std	0,14213	0,09523	0,25557	0,12546	0,09096

Tabelle 1. Kennzahlen Anteilsgeichte

Die durchschnittlichen Assetgewichte über die 10.000 Samples liegen bei allen Ansätzen nahe an den 10% für das wahre Portfolio. Der Optimierungsansatz unter Verwendung der Kosemiv. und der Mittelwertansatz (bedingt durch das Bootsrapverfahren) sind mehr als der M-V-Ansatz durch extreme Anteilsgewichte geprägt (Vgl. durchschnittliche Maximalwerte der Anteilsgewichte. Das Minimum der Anteilsgewichte beträgt bei allen Assets und Ansätzen Null.). Der robuste Optimierungsansatz weist kleinere Extremwerte für die Anteilsgewichte aus. Die Werte für den Ansatz nach Kataoka sind deutlich geringer. Der Mittelwertansatz und der Ansatz nach Kataoka weisen gemessen an der durchschnittlichen Standardabweichung der Portfolioanteilsgewichte die stabilsten Strukturen auf. Der robuste Optimierungsanstz zeigt (im Vergleich zu diesen beiden Ansätzen) nur wenig stabilere Strukturen als der M-V-Ansatz. Der kosemivarianzbasierte Ansatz hat die instabilsten Strukturen. Die Tab. 2 enthält Kennzahlen für die mit den verschiedenen Ansätzen generierten wahren Portfoliorenditen und die "worst-case" Portfoliorenditen und die Werte für den t-Test mit der Nullhypothese, dass die Mittelwerte der Portfoliorendite im "worst-case" aus der gleichen Verteilung stammen (der "worst-case" wird hier durch die bei der robusten Optimierung verwendeten Inputparameter repräsentiert) sowie die f-Werte für die Testhypothese, dass bei Verwendung der wahren Inputparameter die Standardabweichungen der Portfoliorenditen der alternativen Ansätze denen des M-V-Ansatzes entsprechen (die Nullhypothese kann in allen Fällen für ein 5%-Niveau nicht verworfen werden). Im worst-case ist in allen Fällen die Nullhypothese auf einem 1%-Niveau abzulehen. Es ist davon auszugehen, dass die alternativen Ansätze zum M-V-Ansatz verschiedene Renditeverteilungen generieren.

	⊘ Rendite	Min	Max	Std	f-Werte	⊘ P.-Std	dSR
Markowitz	0,05	0,05	0,05	0,00000		0,06626	0,75465
Kataoka	0,05	0,05	0,05	0,00000	1,00002	0,05933	0,84274
Kosemiv.	0,05	0,05	0,05	0,00000	1,00004	0,08893	0,56226
Robust	0,05	0,05	0,05	0,00000	1,00000	0,06356	0,78667
Mittel.	0,05	0,05	0,05	0,00000	1,00003	0,05876	0,85091
worst-case					t-Werte		
Markowitz	0,04651	0,01327	0,08800	0,00843		0,06667	0,69767
Kataoka	0,04170	0,00924	0,07170	0,00766	-187,53112	0,05759	0,72400
Kosemiv.	0,05120	0,01431	0,09042	0,00916	195,84637	0,09506	0,53868
Robust	0,04567	0,01321	0,08460	0,00825	-81,74203	0,06058	0,75385
Mittel.	0,04295	0,01091	0,07868	0,00794	-252,70071	0,05848	0,73437

Tabelle 2. Kennzahlen der Portfoliorenditen

Die Portfoliorenditen bei Verwendung der wahren Inputparameter sind durch den Aufbau der Simulation für alle Ansätze gleich. Die durchschnittliche Portfoliostandardabweichung des kosemivarianzbasierten Ansatzes ist höher als beim M-V-Ansatz. Die anderen drei Ansätze weisen eine geringere durchschnittliche Portfoliostandardabweichung aus. Gemessen an der durchschnittlichen Sharpe-Ratio (dSR) weist die Optimierung unter Anwendung der Kosemiv. den geringsten Wert auf (unter Annahme eines Zinssatzes von Null für die risikofreie Anlage). Der Ansatz nach Kataoka und der Mittelwertansatz erzielen höhere dSR als der M-V-Ansatz. Die robuste Optimierung liegt nur knapp oberhalb der dSR des M-V-Ansatzes. Im "worst-case" weisen die robuste Optimierung, aber auch der Ansatz nach Kataoka sowie der Mittelwertansatz bessere Werte für die dSR (gegenüber dem M-V-Ansatz und der Optimierung mit der Kosemiv.) auf, wobei die robuste Optimierung den höchsten Wert für die dSR erreicht. Bei der Betrachtung der dSR hat die Optimierung mit der Kosemiv. deutlich den geringsten Wert (sogar unterhalb des M-V-Ansatzes). In den Simulationsergebnissen (insbesondere Tab.

2) erscheint der M-V-Ansatz vergleichsweise gut, es ist jedoch zu beachten, dass bedingt durch den Aufbau der Simulation nur geringe Unterschiede bei Portfoliorendite und -standardabweichung resultieren, jedoch die Generierung extremer Anteilsgewichte gut gelingt. Die Betrachtung der dSR deutet auf die Vorteilhaftigkeit alternativer Ansätze im worst-case-Szenario hin (die Ergebnisse sind jedoch von der Definition des worst-case abhängig). Dies bedarf weiterer Untersuchungen, insbesondere mit variierten Basisdaten, unterschiedlichen worst-case Definitionen und Risikoaversionsparametern.

4 Fazit

Durch den robusten Optimierungsansatzes, den Ansatz nach Kataoka sowie den Mittelwertansatz ist es unter den gesetzten Annahmen möglich, in dem Sinne robustere Portfoliostrukturen zu erzeugen, dass die generierten Portfolios in einem "worst-case-Szenario" noch gute dSRs hervorbringen und die Portfoliostrukturen insgesamt weniger schwanken als bei den beiden anderen Ansätzen. Die robusten Portfolios sind bei Annahme der wahren Inputparameter vergleichsweise nur geringfügig ineffizient (gemessen an der dSR). Die M-V-optimalen Portfolios sind im worst-case in einem höheren Maße ineffizient (vgl. [8], S. 3.). Der Einsatz des Optimierungsansatzes nach Kataoka führt im "worst-case" zwar nicht zu gleich guten Ergebnissen wie die robuste Optimierung hat jedoch bei den wahren Inputparametern höhere Ergebnisse (in Bezug auf die dSR). Wichtig für die Wahl eines Optimierungsansatzes ist eine genaue und zweckmäßige Definition des Begriffes "robust" ("worst-case" Betrachtung und/oder stabile Portfoliostrukturen).

Literaturverzeichnis

1. Black F, Litterman R (1992) Global Portfolio Optimization Financial Analysts Journal 48:28–43
2. Drobetz W (2003) Einsatz des Black-Littermann-Verfahrens in der Asset Allocation. In: Dichtl H, Kleeberg JM, Schlenger C (Herausgeber) Handbuch Asset Allocation: Innovative Konzepte zur systematischen Portfolioplanung. Uhlenbruch Verlag, Bad Soden/Ts.
3. Kataoka S (1963) A Stochastic Programming Model Econometrica 31:181–196
4. Markowitz HM (1952) Portfolio Selection The Journal of Finance VII:77–91
5. Michaud RO (1998) Effizient asset management: a practical guide to stock portfolio optimization and asset allocation. Harvard Business School Press, Boston
6. Rehkugler H, Schindel V (1990) Entscheidungstheorie. Verlag V. Florenz GmbH, München
7. Schmidt-von Rhein A (1996) Die Moderne Portfoliotheorie im praktischen Wertpapiermanagement. Uhlenbruch Verlag, Bad Soden/Ts.
8. Tütüncü RH, Koenig M (2004) Robust Asset Allocation Annals of Operations Research 132:157–187

Effizienzanalyse deutscher Banken mit Data Envelopment Analysis und Stabilitätsanalysen

Armin Varmaz[1]

Armin Varmaz, Lehrstuhl für Finanzwirtschaft, Universität Bremen, Hochschulring 4, 28359 Bremen varmaz@uni-bremen.de

1 Einleitung

Die Data Envelopment Analysis (DEA) ist eine Methode zur Effizienzmessung von unabhängigen Entscheidungseinheiten (EE), die im angelsächsischen Raum eine hohe Verbreitung in Wissenschaft und Praxis erreicht hat. Die hohe Akzeptanz von DEA beruht nicht nur auf der Möglichkeit der Effizienzmessung, sondern der gleichzeitigen Möglichkeit von Zielvorgaben durch die Identifizierung von "benchmarking targets" bzw. "best practice"-EE auf der Effizienzgrenze. Da DEA auf ein Lineares Programm (LP) mit einer deterministischen Struktur reduziert wird, können die Ergebnisse sensitiv auf Permutationen der ursprünglichen Daten reagieren. Insbesondere zufällige Ereignisse oder einfache Messfehler könnten Ergebnisse in der Weise verfälschen, dass nicht nur die Effizienzmessung ungenau wird, sondern auch die "best practice"-EE ineffizient werden und somit keine erstrebenswerte Merkmale mehr aufweisen. In diesem Beitrag werden nach der Vorstellung eines DEA-Basismodells im Kap. 2 neue Möglichkeiten von Sensitivitätsanalysen im Rahmen von DEA (Kap. 3) vorgestellt und im Rahmen einer empirischen Fallstudie zur Effizienzanalyse deutscher Kreditgenossenschaften (Kap. 4) umgesetzt.

2 Data Envelopment Analysis

Die Effizienzberechnung mit DEA beruht auf dem ökonomischen Prinzip. Danach soll mit einem gegebenen Mitteleinsatz der größtmögliche Zielerreichungsgrad erreicht (Maximalprinzip) oder ein gegebener Zielerreichungsgrad mit geringstmöglichem Mitteleinsatz realisiert werden (Minimalprinzip). Im Folgenden wird die Operationalisierung des Minimalprinzips betrachtet.

Angenommen seien n EE. Jede EE_j $(j = 1, \ldots, n)$ stellt s verschiedene Produkte (Output) y_{rj} $(r = 1, \ldots, s)$ mit Hilfe von m verschiedenen Einsatzfaktoren (Input) x_{ij} $(i = 1, \ldots, m)$ her. Zur Berechnung der inputorientierten

(= kostenminimierenden) technischen Effizienz einer beliebigen EE_0 wird das DEA–Optimierungsproblem in 1 gelöst (vgl. [1]),

$$\begin{aligned} &\min_{\theta_0,\lambda_j} \quad \theta_0 && (1)\\ s.t. \quad &\sum_{j=1}^{n} \lambda_j x_{ij} + s_i^- = \theta_0 x_{i0} \qquad i = 1,\ldots,m \\ &\sum_{j=1}^{n} \lambda_j y_{rj} - s_r^+ = y_{r0} \qquad r = 1,\ldots,s \\ &\theta_0, \lambda_j, s_i^-, s_r^+ \geq 0 \end{aligned}$$

wobei x_{i0} und y_{r0} der i-te Input bzw. r-te Output der gerade betrachteten EE_0 sind. Die Variablen s_i^- sowie s_r^+ repräsentieren Schlupfvariablen, die jeweils eine Verschwendung bzw. Nichterreichung von Input- bzw. Outputmengen anzeigen. θ_0 gibt die technische Effizienz von EE_0 unter der Annahme der konstanten Skalenerträge wieder, wobei $\theta_0 \in [0,1]$ gilt. Die Abbildung 1a veranschaulicht die Effizienzmessung mit DEA. In der Abbildung produzieren alle EE eine Einheit Output mit Hilfe von zwei unterschiedlichen Inputmengen.

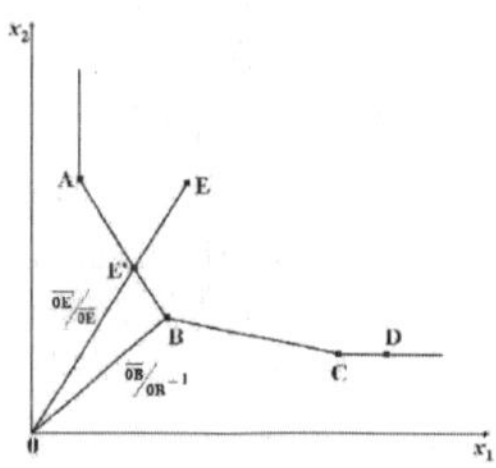

Abb. 1a. Grafische Veranschaulichung von DEA

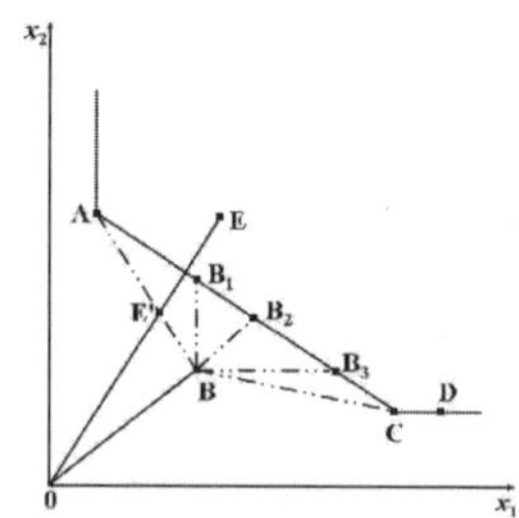

Abb. 1b. Grundidee einer DEA-Stabilitätsanalyse

Die EE A, B und C bilden die Effizienzgrenze und sind effizient ($\theta = 1$), da sie von keiner anderen Input–Output–Alternative dominiert werden. Die EE D liegt zwar auf der Effizienzgrenze mit $\theta = 1$, aber es ist offensichtlich, dass D von C dominiert wird, da C weniger von Input x_1 einsetzt. Daher sichert im DEA-Kontext ein Effizienzwert von $\theta = 1$ mindestens die sog. "schwache" Effizienz. Um das in der mikroökonomischen Produktionstheorie wichtige Konzept der Pareto-Koopmans-Effizienz zu berücksichtigen, müssen die Input- und Outputslacks berücksichtigt werden, die eine Verschwendung in verschiedenen Input- und Outputbereichen nach einer proportionalen Senkung auf die Effizienzgrenze signalisieren. Eine Pareto-Koopmans-effiziente EE wird durch das LP 1 ermittelt, nur wenn $\theta = 1$, $s_i^- = 0$ und $s_r^+ = 0$ gilt.

Die EE E wird dominiert und ist ineffizient ($\theta < 1$). θ gibt das Niveau an, auf welches EE E sein Input bei effizienter Produktion proportional senken könnte. Grafisch wird die Effizienz von E durch das Verhältnis der Streckenzüge $\overline{0E'}$ und $\overline{0E}$ definiert. Die imaginäre EE E' ist eine Linearkombination der "best practice"–EE A und B, die in LP 1 den Gewichtevektor λ_j bilden. Da DEA eine Extremwertanalyse darstellt, die alleine auf den zugrundeliegenden Daten ohne eine Möglichkeit der Angabe von Konfidenzintervalen oder Inferenzstatistiken beruht, können Änderungen der ursprünglichen Daten durch Messfehler oder zufällige Ereignisse die Effizienzmessung beeinflussen. Um mögliche Einflüsse von Störfaktoren zu isolieren, wird im folgenden Abschnitt eine Stabilitätsanalyse von DEA vorgestellt.

Eine Erweiterung des DEA-Modells in 1 auf variable Skalenerträge oder auf Messung von allokativer Effizienz ist einfach möglich (vgl. [1]). Da sich die grundlegenden Gedanken einer Stabilitätsanalyse jedoch nicht ändern, konzentriert sich dieser Beitrag auf die Betrachtung des LP 1.

3 Stabilität

Im Rahmen einer Stabilitätsanalyse von DEA wird die Stabilität der errechneten Effizienzgrenze auf Permutationen von Input- und/oder Outputdaten überprüft. Danach sollte eine effiziente EE nach einer (kleinen) Störung der Daten weiterhin auf der Effizienzgrenze bleiben (vgl. [4]). Eine Stabilitätsanalyse effizienter EE ist besonders wichtig, da diese die Effizienzgrenze konstruieren, an der alle anderen EE gemessen werden. Eine mögliche Stabilitätsanalyse wird durch Nutzung metrischer Konzepte eröffnet. Die Grundidee hinter diesem Konzept liegt in der Nutzung der Norm eines Vektors (= "Distanz"), um "Stabilitätsradien" zu bestimmen. Innerhalb dieser Radien wird eine Variation von Daten nicht zu einer Herabstufung einer effizienten zu einer ineffizienten EE führen (und umgekehrt). Der besondere Beitrag in dieser Arbeit liegt in der Betrachtung von simultaner Variation aller Inputdaten. Dabei werden nicht nur eine Verschlechterung der Inputdaten von effizienten EE, sondern auch eine gleichzeitige Verbesserung der Inputdaten anderer EE berücksichtigt. Da eine Verringerung von Input den Status einer effizienten EE_0 nicht ändern kann, wird die Stabilität effizienter EE_0 bei steigendem Input und gleichzeitig sinkendem Input anderer EE ("worst case"-Szenario) betrachtet.

Sei $\mathbf{I}$ die Untermenge von Input, die im Rahmen von der Stabilitätsstudie geändert werden sollen. Die Änderung der Daten von EE_0 und EE_j $(j \neq 0)$ kann geschrieben werden als:

$$\text{für } \mathrm{EE}_0 \qquad \begin{cases} \widehat{x}_{i0} = \delta_i x_{i0} = x_{i0} + (\delta_i - 1)x_{i0} & \delta_i \geq 1, i \in \mathbf{I} \\ \widehat{x}_{i0} = x_{i0} & i \notin \mathbf{I}. \end{cases} \tag{2}$$

und

$$\text{für } \mathrm{EE}_{j,j\neq 0} \begin{cases} \widehat{x}_{ij} = x_{ij}/\widetilde{\delta}_i = x_{ij} - \frac{\widetilde{\delta}_i - 1}{\widetilde{\delta}_i} x_{ij} & \widetilde{\delta}_i \geq 1, i \in \mathbf{I} \\ \widehat{x}_{ij} = x_{ij} & i \notin \mathbf{I}. \end{cases} \tag{3}$$

Nach [3] wird die Stabilität einer EE_0 durch LP 4 geprüft.

$$\begin{aligned} \min_{\theta_0,\lambda_j} \quad & \theta_0^I && && (4)\\ s.t. \quad & \sum_{j=1,j\neq 0}^{n} \lambda_j x_{ij} + s_i^- = \theta_0^I x_{i0} && i \in \mathbf{I} \\ & \sum_{j=1,j\neq 0}^{n} \lambda_j x_{ij} + s_i^- = x_{i0} && i \notin \mathbf{I} \\ & \sum_{j=1,j\neq 0}^{n} \lambda_j y_{rj} - s_r^+ = y_{r0} && r = 1,\ldots,s \\ & \theta_0^I, \lambda_j (j \neq 0), s_i^-, s_r^+ \geq 0 \end{aligned}$$

Wenn im LP 4 alle Inputdaten in die Menge **I** aufgenommen werden, spricht man von einem Supereffizienz-DEA-Modell. Im Modell 4 ist zu beachten, dass die gerade betrachtete EE_0 bei der Konstruktion der Effizienzgrenze nicht berücksichtigt wird ($j \neq 0$). Gleichzeitig kann der Wert von θ größer Eins werden. Die Wirkung des Modells 4 wird in der Abbildung 1b für die EE B veranschaulicht. Da die EE B nicht in die Konstruktion der Effizienzgrenze einbezogen wird, wird die "neue" Effizienzgrenze durch EE A und C definiert. Durch LP 4 wird das Niveau θ_0^I angezeigt, auf welches die EE B seine Inputmengen erhöhen könnte, ohne ineffizient zu werden. Wird eine Gerade vom Ursprung durch den Punkt B zum Punkt B_2 auf der neuen Effizienzgrenze gezogen, wird das Niveau angezeigt, auf welches alle Inputmengen von B proportional erhöht werden könnten. Ist man nur an Auswirkungen einer Erhöhung von Input x_1 oder x_2 interessiert, wird EE B auf die Punkte B_3 bzw. B_1 projiziert. Das Dreieck zwischen den Punkten B, B_1 und B_3 bildet die sog. Stabilitätsregion.

Mit der Lösung des Modells 4 sind zwei Betrachtungsweisen möglich. Zum einen kann die Stabilitätsregion für das Modell 1 gebildet werden. Innerhalb dieser Region bleibt eine EE effizient nach einem Anstieg von Inputmengen. Aus Sicht der Modellbildung ist diese Information wertvoll, da sie die Stabilität des Modells anzeigen. Darüber hinaus erhält das verantwortliche Management einer EE die Möglichkeit "was, wenn"-Szenarioanalyen durchzuführen, um z.B. mögliche Auswirkung neuer Strategien zu testen.

Mit Modell 4 kann gleichzeitig der Fall berücksichtigt werden, bei dem simultan eine effiziente EE_0 Inputmengen erhöht und $EE_{j,j\neq 0}$ Inputmengen senken. In [4] wurde gezeigt, dass die optimale Lösung von θ_0^I mit der quadratischen Funktion $\theta_0^I = \delta_i \widetilde{\delta}_i$ in Datenänderung für EE_0 und Datenänderung für $EE_{j,j\neq 0}$ zerlegt werden kann. In Gleichungen 2 und 3 repräsentieren $g_0 = \delta - 1$ den möglichen Anstieg der Inputmengen für x_{i0} und $g = (\widetilde{\delta}_i - 1)/\widetilde{\delta}_i$ die mögliche Senkung der Inputmengen für $x_{ij,j\neq 0}$. In [4] wurde gezeigt, dass für $g_0 = \sqrt{\theta_0^I} - 1$ und $g = (\sqrt{\theta_0^I} - 1)/\sqrt{\theta_0^I}$ auch gilt.

4 Empirische Fallstudie

Das Modell zur Stabilitätsanalyse wird in einer empirischen Fallstudie zur Effizienzbeurteilung von 477 Genossenschaftsbanken für das Jahr 2004 herangezogen. Die Festlegung von Input- und Outputdaten zur Effizienzanalyse von Banken ist Teil einer andauernden Diskussion in der bankwirtschaftlichen Literatur (vgl. [2]). Hier werden folgende Inputdaten angenommen: Sachkapital, Anzahl der Mitarbeiter, Einlagen, Provisionsaufwendungen sowie Kreditrisikoprovisionen. Im Gegensatz zu bisherigen Effizienzstudien deutscher Banken wird hier das Risiko aus dem Kreditengagement berücksichtigt. Das Kreditrisiko kann als ein unerwünschter Output der Bankenproduktion angesehen werden. Die Einbeziehung von unerwünschten Output (bzw. Input) in DEA ist methodisch problematisch, da annahmegemäß der Input zu minimieren und der Output zu maximieren wäre. Eine Maximierung von Kreditrisiko ist jedoch unerwünscht. In der Literatur werden verschiedene Ansätze vorgeschlagen, von denen der gängigste Vorschlag angewendet wird, der eine Einbeziehung des Kreditrisikos als zu minimierenden Input vorsieht. Als Output der Bankenproduktion gelten fest und nicht fest verzinsliche Anlagen, Kredite an Kunden, Kredite an Kreditinstitute sowie Provisionserträge.

Die Ergebnisse einer inputorientierten Effizienzuntersuchung sind in der Tabelle 1 zusammengefasst. Der erste Teil von Tabelle 1 gibt die Ergebnisse des Modells 1 mit und ohne Einbeziehung des Kreditrisikos wieder. Die Berücksichtigung von Kreditrisiko hat zu höherem durchschnittlichem Effizienzwert geführt. Dieses Ergebnis ist plausibel, da die Banken, die sehr viele Kredite vergeben, einen höheren Output als risikoaverse Banken produzieren. Somit erscheinen risikoaverse Banken als weniger effizient. Gleichzeitig müssen risikoaverse Banken weniger Risikovorsorgemittel bereithalten, was sich dann in höhren Effizienzwerten manifestiert. Insgesamt sind mit Berücksichtigung von Risiko 70 (14,7% aller) Banken effizient. Im Durchschnitt könnten die Kreditgenossenschaften bei effizienter Produktion die Inputmengen auf ein Niveau von 84,41% senken, um den gleichen Output herzustellen. Im unteren Teil der Tabelle 1 sind die Ergebnisse einer Stabilitätsanalyse aus Darstellungsgründen nur für eine effiziente Bank präsentiert. Für die Untersuchung wurde eine Bank herangezogen, die in dem gewöhnlichen DEA-Modell von 402 ineffizienten Banken als "benchmarking target" identifiziert wurde. Insofern handelt es sich um eine für die Effizienzanalyse sehr wichtige Bank. In der Stabilitätsanalyse wurden sechs Fälle von Datenänderungen unterstellt. Im ersten Fall wurde eine Erhöhung der Inputs der effizienten Bank bei gleichzeitigem Absenken der Inputs anderer Banken betrachtet. In anderen fünf Fällen wurden die Effekte separater Änderungen einzelner Inputmengen analysiert.

In der Zeile θ_0^I sind die optimalen Lösungen des Modells 4 zusammengefasst, die eine maximale Steigerung der Inputmengen repräsentieren, wenn man nur Datenänderungen bei der gerade betrachteten EE_0 zulässt. Insgesamt könnte somit diese Bank den Input proportional um 7,54% erhöhen und immer noch effizient operieren. In den nächsten Zeilen wird simultan der In-

	Effizienzanalyse 2004				
	∅	StdAbw	Maximum	Minimum	Anzahl
ohne Risiko	0,8289	0,1049	1,0000	0,4115	52
mit Risiko	0,8441	0,1050	1,0000	0,4893	70

	Stabilitätsanalyse					
	Alle	**Sachkapital**	**Mitarbeiter**	**Einlagen**	**ProvAufwand**	**Risikovorsorge**
θ_0^{I}	1,0754	2,8564	1,0955	Infeasible	1,5663	1,1266
EE$_0$	0,03701	0,6901	0,0467	∞	0,2515	0,0614
EE$_{j,j\neq 0}$	0,03569	0,4083	0,0446	∞	0,2010	0,0578

Tabelle 1. Ergebnisse von Effizienz- und Stabilitätsanalyse für 2004

put der effizienten Bank erhöht (EE$_0$), während die Inputs anderer Banken gesenkt werden (EE$_{j,j\neq 0}$). In diesem Fall bleibt die Bank effizient bei einem proportionalen Anstieg eigenen Inputs um 3,7% und bei einer gleichzeitigen Senkung des Inputs anderen Banken um 3,57%. Somit erscheint die Klassifikation dieser Bank als recht stabil, da kleine Änderungen der Daten nicht unmittelbar zur Ineffizienz der Bank führen. Für das Bankenmanagement sind darüber hinaus die möglichen Änderungen einzelner Inputmengen alleine von besonderem Interesse. So reagiert diese Bank auf eine Erhöhung der Anzahl der Mitarbeiter besonders sensibel. Auf der anderen Seite ist das LP 4 für den Input "Einlagen" nicht lösbar. Das deutet die Möglichkeit einer unendlichen Erhöhung dieses Inputs bei dieser effizienten Bank an. Dieses Ergebnis könnte allerdings auch eine mögliche Fehlspezifikation des Modells anzeigen, da Einlagen ebenfalls als Output der Bankenproduktion angesehen werden könnten.

5 Zusammenfassung

In dieser Arbeit wurde auf die Problematik der Stabilität der Ergebnisse einer DEA-Effizienzmessung eingegangen. Der besondere Beitrag der Arbeit liegt in der Vorstellung eines DEA-Modells zur Stabilitätsanalyse, mit dem simultan die Änderungen aller Daten für jede EE betrachtet werden kann. Daneben kann dieses Modell in der betrieblichen Praxis zur Simulation verschiedener Szenarien und deren Auswirkung auf die Effizienz benutzt werden. Die Anwendung des Modells wurde in einer Fallstudie demonstriert.

Literaturverzeichnis

1. Cooper W, Seiford L (2000) Data Envelopment Analysis. Kluwer, Boston
2. Poddig Th, Varmaz A (2004) Effizienzprobleme bei Banken: Fusionen und Betriebswachstum als tragfähige Mittel? ZBB, 16:236–247
3. Seiford L, Zhu J (1998) Sensitivity analysis of DEA models for simultaneous changes in all the data. Journal of Operational Research Society 49:1060–1071
4. Zhu J (2001) Super-efficiency and DEA sensitivity analysis. European Journal of Operational Research 12:443–455

Part VII

Artificial Intelligence and Fuzzy Logic

Duality in Fuzzy Multiple Objective Linear Programming

Jaroslav Ramík

Department of Mathematical Methods in Economics
Faculty of Economics, VSB - Technical University Ostrava,
Czech Republic, jaroslav.ramik@vsb.cz

Abstract

A class of fuzzy multiple objective linear programming (FMOLP) problems with fuzzy coefficients based on fuzzy relations is introduced, the concepts of feasible and (alpha ,beta)-maximal and minimal solutions are defined. The class of crisp (classical) MOLP problems can be embedded into the class of FMOLP ones. Moreover, for FMOLP problems a new concept of duality is introduced and the weak and strong duality theorems are derived.

1 Introduction

The problem of duality has been investigated since the early stage of fuzzy linear programming (FLP), see [1], [4], [10]. In this paper we first introduce a broad class of fuzzy multiple objective linear programming problems (FMOLP problems) and define the concepts of -feasible and (,)-maximal and minimal solutions of FMOLP problems. The class of classical MOLP problems can be embedded into the class of FMOLP ones, moreover, for FMOLP problems we define the concept of duality and prove the weak and strong duality theorems - generalizations of the classical ones. The results are compared to the existing literature, see [7], [8], [9].

2 Preliminaries

By $F(X)$ we denote the set of all *fuzzy subsets of X, X* is a subset of $\mathbf{R}^n$. Every fuzzy subset A of X is uniquely determined by the membership function $\mu_A : X \to [0;1]$, and $[0;1] \subset \mathbf{R}$ is a unit interval and $\mathbf{R}$ is the Euclidean space of real numbers. We say that the fuzzy subset A is *crisp* if μ_A is a characteristic function of A, i.e. $\mu_A : X \to \{0;1\}$. It is clear that the set of all subsets of X, $P(X)$, can be isomorphically embedded into $F(X)$.

Let

$$[A] \ = \{x \in X | \mu_A(x) \geq \ \} \text{ for } \ \in (0;1];$$

$$[A]_0 = cl\{x \in X | \mu_A(x) > 0\},$$

where clB means a usual closure of B, $B \subset X$. For $\ \in [0;1]$, $[A]$ are called *-cut,*. $[A]_0$ is called a *support of A*.

A fuzzy set A with the membership function μ_A is called the *fuzzy quantity* if there exist $a; b; c; d \in \mathbf{R}$; $-\infty < a \leq b \leq c \leq d < +\infty$, such that

$\mu_A(t) = 0$ if $t < a$ or $t > d$;

$\mu_A(t)$ is strictly increasing if $a < t < b$;

$\mu_A(t) = 1$ if $b \leq t \leq c$;

$\mu_A(t)$ is strictly decreasing if $c < t < d$.

The set of all fuzzy quantities is denoted by $F_0(\mathbf{R})$, or shortly F_0. This set contains well known classes of fuzzy numbers: crisp (real) numbers, crisp intervals, triangular fuzzy numbers, trapezoidal and bell-shaped fuzzy numbers etc.

A fuzzy subset $\tilde{P}$ of $F(X) \times F(X)$ is called a *fuzzy relation on X*, i.e. $\tilde{P} \in F(F(X) \times F(X))$.

A fuzzy relation $\tilde{Q}$ on X is called a *fuzzy extension of relation P*, if for each $x, y \in X$, it holds

$$\mu_{\tilde{Q}}(x,y) = \mu_P(x,y) \ . \tag{1}$$

Let A, B be fuzzy sets with the membership functions $\mu_A : \mathbf{R} \to [0;1]$, $\mu_B : \mathbf{R} \to [0;1]$, respectively. We shall consider

$$\mu_{Pos}(A,B) = \sup\{\min(\mu_A(x), \mu_B(y)) | \ x \leq y, \ x,y \in \mathbf{R}\}, \tag{2}$$

$$\mu_{Nec}(A,B) = \inf\{\max(1 - \mu_A(x), \ 1 - \mu_B(y))\} | \ x > y, \ x,y \in \mathbf{R}\}. \tag{3}$$

Here (2) is called the *possibility relation*, (3) is called the *necessity relation*. Possibility and necessity relations (2) and (3) have been originally introduced as *possibility and necessity indices* in [2], where also mathematical analysis and interpretation has been discussed. We write alternatively

$$\mu_{Pos}(A,B) = (A \leq^{Pos} B), \ \mu_{Nec}(A,B) = (A <^{Nec} B), \tag{4}$$

where μ_{Pos} and μ_{Nec} are the membership functions of the fuzzy relations on $\mathbf{R}$. By $A \geq^{Pos} B$ or $A >^{Nec} B$ we mean $B \leq^{Pos} A$ or $B <^{Nec} A$, respectively. It can be easily verified that the possibility and necessity relations are fuzzy extensions of the classical binary relation $\leq$, see [8].

3 Multiple Objective Linear Programming Problem with Fuzzy Coefficients

In this section we introduce a fuzzy multiple objective linear programming problem (FMOLP problem) where coefficients are fuzzy quantities.

Let $K = \{1,2,...,k\}$, $M = \{1,2,...,m\}$, $N = \{1,2,...,n\}$, k, m, n be positive integers. The *multiple objective linear programming problem* (MOLP problem) is a problem

maximize $\quad z_q = c_{q1}x_1 + ... + c_{qn}x_n$, $q \in K$,

subject to

$$a_{i1}x_1 + ... + a_{in}x_n \quad b_i,\ i \in M,$$
$$x_j \geq 0,\ j \in N. \tag{5}$$

Here, we assume that the weights of the criteria are known, i.e. we have positive numbers $v_q > 0$ for all $q \in K$, such that $\sum_{q \in K} v_q = 1$. In practical situations the weights are interpreted as the relative importances of the objectives (or, criteria) and can be identified e.g. by using the well known Saaty's pairwise comparison method. Setting

$$c_j = \sum_{q \in K} v_q c_{qj},\ j \in N, \tag{6}$$

we obtain an associated LP problem

maximize $\quad z = c_1x_1 + ... + c_nx_n$,

subject to

$$a_{i1}x_1 + ... + a_{in}x_n \quad b_i,\ i \in M,$$
$$x_j \geq 0,\ j \in N. \tag{7}$$

The following result is well known, see e.g. [3].

Theorem 1 *If $x = (x_1,...,x_n)$ is an optimal solution of LP problem (7), then it is a Pareto-optimal solution of MOLP problem (5).*
Moreover, if $x = (x_1,...,x_n)$ is a Pareto-optimal solution of MOLP problem (5), then there exist positive numbers $v_q > 0$, for all $q \in K$ with $\sum_{q \in K} v_q = 1$ such that x is an optimal solution of LP problem (7) with (6).

Knowing relative importances, i.e. the weights v_q for all $q \in K$, of the objectives in some MOLP problem one can solve this problem as an associated LP problem. Here, we utilize this result in building duality theory for fuzzy MOLP problems.

Applying the Extension principle we can easily obtain the following property: If $\tilde{c}_{qj}, \tilde{a}_{ij} \in F_0(R), x_j \geq 0$, $q \in K$, $i \in M$, $j \in N$, then the fuzzy sets $\tilde{c}_{q1}x_1 \tilde{+} ... \tilde{+} \tilde{c}_{qn}x_n$, $\tilde{a}_{i1}x_1 \tilde{+} ... \tilde{+} \tilde{a}_{in}x_n$ are again fuzzy quantities.

Let $\tilde{d}$ be a fuzzy quantity, i.e. $\tilde{d} \in F_0(\mathbf{R})$, $\in [0;1]$. We use the following notation: $\tilde{d}^L(\beta) = \inf\{t \mid t \in [\tilde{d}]_\beta\}$, $\tilde{d}^R(\beta) = \sup\{t \mid t \in [\tilde{d}]_\beta\}$. □

Let $\tilde{P}$ be a fuzzy relation - fuzzy extension of the usual binary relation on **R**. The *fuzzy multiple objective linear programming problem* (FMOLP problem) is denoted as

"maximize" $\tilde{z}_q = \tilde{c}_{q1}x_1 \tilde{+} \ldots \tilde{+} \tilde{c}_{qn}x_n$, $q \in K$,

"subject to"

$$(\tilde{a}_{i1}x_1 \tilde{+} \ldots \tilde{+} \tilde{a}_{in}x_n)\tilde{P}\tilde{b}_i \text{ , } i \in M,$$

$$x_j \geq 0, \ j \in N. \tag{8}$$

In (8) the value $\tilde{a}_{i1}x_1 \tilde{+} \ldots \tilde{+} \tilde{a}_{in}x_n \in F_0(\mathbf{R})$ is compared with a fuzzy quantity $\tilde{b}_i \in F_0(\mathbf{R})$ by some fuzzy relation $\tilde{P}$. The maximization of the objective functions denoted by "maximize" $\tilde{z} = \tilde{c}_{q1}x_1 \tilde{+} \ldots \tilde{+} \tilde{c}_{qn}x_n$ (in quotation marks) will be investigated later on. Setting

$$\tilde{c}_j = \sum_{q \in K} v_q \tilde{c}_{qj} \text{ , } j \in N, \tag{9}$$

we obtain an associated FLP problem

"maximize" $\tilde{z} = \tilde{c}_1 x_1 \tilde{+} \ldots \tilde{+} \tilde{c}_n x_n$

"subject to"

$$(\tilde{a}_{i1}x_1 \tilde{+} \ldots \tilde{+} \tilde{a}_{in}x_n)\tilde{P}\tilde{b}_i \text{ , } i \in M,$$

$$x_j \geq 0, \ j \in N. \tag{10}$$

Now, we shall deal with the constraints of FMOLP problem (8), or, (10), see also [6], [5], or [9].

4 Feasible Region, -Feasible Solution

A fuzzy set $\tilde{X}$, whose membership function $\mu_{\tilde{X}}$ is defined for all $x \in \mathbf{R}$ by

$$\mu_{\tilde{X}}(x) = \begin{cases} \min\{\mu_{\tilde{P}}(\tilde{a}_{11}x_1 + \ldots + \tilde{a}_{1n}x_n, \tilde{b}_1), \ldots, \mu_{\tilde{P}}(\tilde{a}_{m1}x_1 + \ldots + \tilde{a}_{mn}x_n, \tilde{b}_m)\} \\ \qquad\qquad\qquad \text{if } x_j \geq 0 \text{ for all } j \in N, \\ 0 \text{ otherwise,} \end{cases} \tag{11}$$

is called the *fuzzy set of feasible region* of the FMOLP problem (8).

For $\beta \in (0;1]$, a vector $x \in [\tilde{X}]_\beta$ is called the *β-feasible solution* of the FMOLP problem (8), or, FLP (10). Notice that the feasible region $\tilde{X}$ of FMOLP problem (10) is a fuzzy set. On the other hand, β-feasible solution is a vector belonging to the β-cut of the feasible region $\tilde{X}$. It is not difficult to show, that if all coefficients $\tilde{a}_{ij}$ and $\tilde{b}_i$ are crisp fuzzy quantities, i.e. they are isomorfic to the corresponding real numbers, then the fuzzy feasible region is isomorfic to the set of all feasible solutions of the corresponding classical LP problem, see [6], or [7].

5 Maximizing the Objective Function

Now we look for the "best" fuzzy quantities $\tilde{z}_q$ with respect to the given fuzzy constraints, or, in other words, with respect to the fuzzy set of feasible region of (8). Knowing the weights v_q, $q \in K$, of the objectives we shall deal with the associated problem (10), particularly, with the single objective function $\tilde{z} = \tilde{c}_1 x_1 \tilde{+} \ldots \tilde{+} \tilde{c}_n x_n$, where $j \in N$. We define special relations.

Let $\tilde{\preceq}$ be a fuzzy relation on **R**, let $\tilde{a}, \tilde{b}$ be fuzzy sets of **R** and let $\in (0;1]$.

We say that $\tilde{a}$ is *-less than $\tilde{b}$ with respect to* $\tilde{\preceq}$ and write

$$\tilde{a} \tilde{\preceq}_\alpha \tilde{b} \tag{12}$$

if

$$\mu_{\tilde{\preceq}}(\tilde{a}, \tilde{b}) \geq \alpha \text{ and } \mu_{\tilde{\preceq}}(\tilde{b}, \tilde{a}) < \alpha. \tag{13}$$

We call $\tilde{\preceq}_\alpha$ the *-relation on* R *with respect to* $\tilde{\preceq}$.

Notice that $\tilde{\preceq}_\alpha$ is a binary relation on the set of fuzzy sets $F(\mathbf{R})$ being constructed from a fuzzy relation $\tilde{\preceq}$ on the level of $\in (0;1]$. If $\tilde{a}$ and $\tilde{b}$ are crisp numbers corresponding to real numbers a and b, respectively, and $\tilde{\preceq}$ is a fuzzy extension of relation · , then $\tilde{a} \tilde{\preceq}_\alpha \tilde{b}$ if and only if $a \leq b$.

Now, modifying the well known concept of efficient solution in multi-criteria optimization we define maximization (or equivalently minimization) of the objective function of FLP problem (10). We shall consider a fuzzy relation $\tilde{\preceq}$ on **R** being a fuzzy extension of the usual binary relation $\leq$ on **R**, see also [9].

Here, we allow for independent, i.e. different satisfaction levels: $\neq \beta$, where is considered for the objective functions and β for the constraints.

Let $\tilde{c}_j$, $\tilde{a}_{ij}$ and $\tilde{b}_i$, $i \in M$, $j \in N$, be fuzzy quantities on **R**. Let $\tilde{\preceq}$ be a fuzzy relation on **R**, being also a fuzzy extension of the usual binary relation $\leq$ on **R**. A β-feasible solution of (10) $x \in [\tilde{X}]_\beta$ is called the *(,β)-maximal solution of* (10) *with respect to* $\tilde{\preceq}$ if there is no $x' \in [\tilde{X}]_\beta$ such that $\tilde{c}^T x \tilde{\preceq}_\alpha \tilde{c}^T x'$.

Clearly, if all coefficients of FLP problem (10) are crisp fuzzy quantities, then (,β)-maximal solution of this problem is isomorphic to the classical Pareto-optimal solution of the corresponding LP problem (5).

Analogically, we define *(,β)-minimal solution of* (10) *with respect to* $\tilde{\preceq}$. Clearly, if all coefficients of FLP problem (10) are crisp fuzzy quantities, then (,β)-maximal (minimal) solution of this problem is isomorphic to the classical Pareto-optimal solution of the corresponding LP problem (5).

6 Dual Problem and Duality Theorems

In this section we shall investigate the well known concept of duality in LP for FMOLP problems based on possibility and necessity fuzzy relations $\leq^{Pos}$ and $<^{Nec}$. Here, we present some extension of the weak and strong duality theorems which extend the known results for LP problems. Consider the following FLP problem

(P)

"maximize" $\tilde{z} = \tilde{c}_1 x_1 \tilde{+} \ldots \tilde{+} \tilde{c}_n x_n$

"subject to"

$$(\tilde{a}_{i1} x_1 \tilde{+} \ldots \tilde{+} \tilde{a}_{in} x_n) \tilde{P} \tilde{b}_i, \; i \in M,$$

$$x_j \geq 0, \; j \in N. \tag{14}$$

an associated problem to FMOLP problem (8), where $\tilde{c}_j$ is defined by (9).

FLP problem (14) will be called the *primal FMOLP problem (P).* The feasible region of (P) and (,β)-maximal solution has been defined before. The *dual FMOLP problem (D)* can be formulated as follows

(D)

"minimize" $\tilde{w} = \tilde{b}_1 y_1 + \ldots + \tilde{b}_m y_m$

"subject to" $\tilde{c}_j \tilde{Q} (\tilde{a}_{1j} y_1 + \ldots + \tilde{a}_{mj} y_m$, $j \in N$,

$$y_i \geq 0, \; i \in M, \tag{15}$$

Here, we consider either $\tilde{P} = \leq^{Pos}, \tilde{Q} = <^{Nec}$ or $\tilde{P} = <^{Nec}, \tilde{Q} = \leq^{Pos}$. In problem (P), "maximization" is considered with respect to fuzzy relation $\tilde{P}$. On the other hand, "minimization" in problem (D) is considered with respect to fuzzy relation $\tilde{Q}$, which can be formulated analogically.

In the following duality theorems we present two versions: (i) for fuzzy relation $\leq^{Pos}$ in problem (P) and fuzzy relation $<^{Nec}$ in problem (D), and (ii), for fuzzy relation $<^{Nec}$ in problem (P) and fuzzy relation $\leq^{Pos}$ in problem (D). In order to prove duality results we assume $= \beta$. Otherwise, the duality theorems in our formulation do not hold, for more details see [9]. Moreover, we assume that each objective function is associated with a weight $v_q > 0$, $q \in K$, such that $\sum_{q \in K} v_q = 1$, where v_q may be interpreted as a relative importance of the q-th objective function. The corresponding proofs can be found in [9].

Theorem 2 *Weak Duality Theorem. Let* $\tilde{c}_{qj}$, $\tilde{a}_{ij}$ *and* $\tilde{b}_i$ *be fuzzy quantities for all* $q \in K$, $i \in M$, $j \in N$, $\in (0;1)$.

(i) Let $\tilde{X}$ *be a feasible region of FMOLP problem (14)* $\tilde{P} = \leq^{Pos}$, $\tilde{Y}$ *be a feasible region of FMOLP problem (15) with* $\tilde{Q} = <^{Nec}$. *If a vector* $x = (x_1, \ldots, x_n)$ *be-*

longs to $[\tilde{X}]_{\alpha}$, $y = (y_1,...,y_m)$ *belongs* $[\tilde{Y}]_{1-\alpha}$, *and* $q \in K$, *then*

$$\sum_{j\in N}\tilde{c}_{qj}^{R}(\alpha)x_j \leq \sum_{i\in M}\tilde{b}_i^{R}(\alpha)y_i \tag{16}$$

(ii) Let $\tilde{X}$ *be a feasible region of FMOLP problem (14)* $\tilde{P} =<^{Nec}$, $\tilde{Y}$ *be a feasible region of FMOLP problem (15) with* $\tilde{Q} =\leq^{Pos}$. *If a vector* $x = (x_1,...,x_n)$ *belongs to* $[\tilde{X}]_{1-\alpha}$, $y = (y_1,...,y_m)$ *belongs* $[\tilde{Y}]_{\alpha}$, *and* $q \in K$, *then*

$$\sum_{j\in N}\tilde{c}_{qj}^{L}(\alpha)x_j \leq \sum_{i\in M}\tilde{b}_i^{L}(\alpha)y_i. \tag{17}$$

Theorem 3 *Strong Duality Theorem. Let* $\tilde{c}_{qj}$, $\tilde{a}_{ij}$ *and* $\tilde{b}_i$ *be fuzzy quantities for all* $q \in K$, $i \in M$, $j \in N$, *let* $v_q > 0$, *for all* $q \in K$ *with* $\sum_{q\in K} v_q = 1$.

(i) Let $\tilde{X}$ *be a feasible region of FMOLP problem (14) with* $\tilde{P} =\leq^{Pos}$, $\tilde{Y}$ *be a feasible region of FMOLP problem (15) with* $\tilde{Q} =<^{Nec}$. *If for some* $\in (0;1)$ $[\tilde{X}]_{\alpha}$ *and* $[\tilde{Y}]_{1-\alpha}$ *are nonempty, then there exists* $x = (x_1,...,x_n) \in [\tilde{X}]_{\alpha}$ *- an (,)-maximal solution of (P) with respect to* $\leq^{Pos}$, *and* $y = (y_1,...,y_m) \in [\tilde{Y}]_{1-\alpha}$, *(1- ,1-)-minimal solution of (D) with respect to* $<^{Nec}$, *such that*

$$\sum_{q\in K}\sum_{j\in N}v_q\tilde{c}_{qj}^{R}(\alpha)x_j = \sum_{i\in M}\tilde{b}_i^{R}(\alpha)y_i. \tag{18}$$

(ii) Let $\tilde{X}$ *be a feasible region of FMOLP problem (14)* $\tilde{P} =<^{Nec}$, $\tilde{Y}$ *be a feasible region of FMOLP problem (15) with If for some* $\in (0;1)$ $[\tilde{X}]_{1-\alpha}$ *and* $[\tilde{Y}]_{\alpha}$ *are nonempty, then there exists* $x = (x_1,...,x_n) \in [\tilde{X}]_{1-\alpha}$ *- an (1- ,1-)-maximal solution of (P) with respect to* $<^{Nec}$ *and* $y = (y_1,...,y_m) \in [\tilde{Y}]_{\alpha}$, *(,)-minimal solution of (D) with respect to* $\leq^{Pos}$, *such that*

$$\sum_{q\in K}\sum_{j\in N}v_q\tilde{c}_{qj}^{L}(\alpha)x_j = \sum_{i\in M}\tilde{b}_i^{L}(\alpha)y_i. \tag{19}$$

Remarks.

1. In the crisp and single-objective case, the above stated theorems are standard LP (Weak, Strong) Duality Theorems.

2. Usually, $\geq 0{,}5$. Then $[\tilde{X}]_{\alpha} \subset [\tilde{X}]_{1-\alpha}$, $[\tilde{Y}]_{\alpha} \subset [\tilde{Y}]_{1-\alpha}$, hence we can assume $x \in [\tilde{X}]_{\alpha}$ and $y \in [\tilde{Y}]_{\alpha}$. Evidently, the statement of the theorem remains unchanged.

3. Theorem 3 provides only the existence of the (,)-maximal solution (or (1- ,1-)-maximal solution) of FMOLP problem (P), and (1- ,1-)-minimal solution ((,)-minimal solution) of FMOLP problem (D) such that (18) or (19) holds.

However, the proof of the theorem gives also the method for finding the solutions by solving (MO)LP problems (P) and (D), see [9].

7 Conclusion

It is possible to investigate duality in FLP problems even in more general settings. There exist several ways of generalization. For instance, it is possible to extend the duality results to some other classes of fuzzy relations, or, to find some necessary conditions that fuzzy relations for comparing fuzzy numbers should satisfy in order to provide a duality result, or, eventually a duality gap. Moreover, in [7], the concept of dual couples of t-norms and t-conorms has been formulated and dual fuzzy relations have been defined. The role of dual relations in the couple of dual FLP problems should be also clarified and a more general duality theory could be derived. The other way of generalization is based on introducing interactive fuzzy coefficients, or oblique fuzzy vectors, see e.g. [7].

References

[1] G.B. Dantzig, Linear Programming and Extensions, Princeton University Press, Princeton, N.J., 1963.
[2] D. Dubois and H. Prade, Ranking fuzzy numbers in the setting of possibility theory. Inform. Sci. 30, (1983), 183 - 224.
[3] M. Ehrgott, Multicriteria optimization. Springer, Lecture Notes in Economics and Math. Systems 491, Berlin-Heidelberg 2000.
[4] H. Hamacher, H. Leberling and H.-J. Zimmermann, Sensitivity analysis in fuzzy linear programming. Fuzzy Sets and Systems 1 (1978), 269-281.
[5] M. Inuiguchi, H. Ichihashi and Y. Kume, Some properties of extended fuzzy preference relations using modalities. Inform. Sci. 61, (1992) 187 - 209.
[6] M. Inuiguchi, J. Ramik, T. Tanino and M. Vlach, Satis¯cing solutions and duality in interval and fuzzy linear programming. Fuzzy Sets and Systems 135 (2003), 151-177.
[7] J. Ramik and M. Vlach, Generalized Concavity in Fuzzy Optimization and Decision Analysis. Kluwer Acad. Publ., Dordrecht - Boston - London, 2002.
[8] J. Ramik, Duality in Fuzzy Linear Programming: Some New Concepts and Results. Fuzzy Optimization and Decision Making, Vol.4, (2005), 25-39.
[9] J. Ramik, Duality in Fuzzy Linear Programming with Possibility and Necessity Relations. Fuzzy Sets and Systems, to appear.
[10] W. Rodder and H.-J. Zimmermann, Duality in fuzzy linear programming. Extremal Methods and System Analysis, A.V. Fiacco and K.O. Kortanek, Eds., Berlin - New York, 1980, 415-429.

Variable Subset Selection for Credit Scoring with Support Vector Machines

Ralf Stecking and Klaus B. Schebesch

Department of Economics, University of Bremen, D-28359 Bremen, Germany
stecking@uni-bremen.de, kbsbase@gmx.de

Summary. Support Vector Machines (SVM) are very successful kernel based classification methods with a broad range of applications including credit scoring and rating. SVM can use data sets with many variables even when the number of cases is small. However, we are often constrained to reduce the input space owing to changing data availability, cost and speed of computation. We first evaluate variable subsets in the context of credit scoring. Then we apply previous results of using SVM with different kernel functions to a specific subset of credit client variables. Finally, rating of the credit data pool is presented.

1 Introduction

Classification models related to credit scoring may use up to a hundred potential input features. However, some of these input variables may or may not be present in actually recorded data sets, which can also vary widely in the number of clients or records. We propose to use Support Vector Machines (SVM) in evaluating subsets of inputs to credit scoring classification models in order to use them for different reduced data sets. For the description of SVM models and their extensive use on credit scoring data with full input features we refer to our past work [3] [6] [7]. We first evaluate the performance of classification models on subsets of inputs and show how to proceed from random input choices to more informed input subsets. We then turn to a much bigger data set which was recorded on a given subset of all input variables and compare the results using SVM with different kernels and by using additional similarity and stability measures.

2 Variable subset selection

The basic data set for our past credit scoring models is an equally distributed sample of 658 clients for a building and loan credit with a total number of

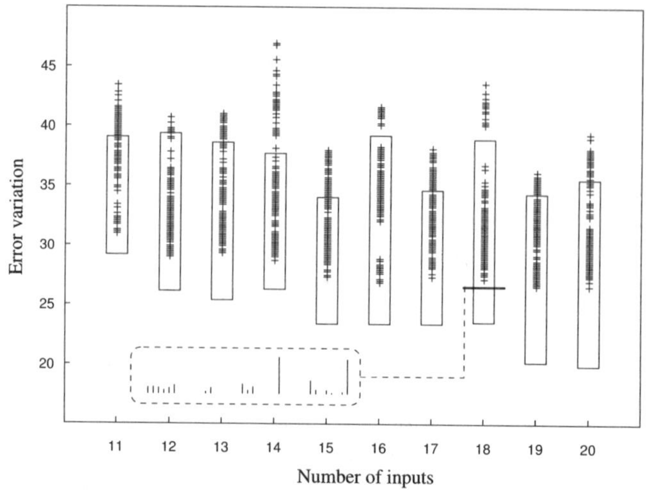

Fig. 1. Error variation over 2000 SVM runs with 100 randomly sampled input combinations using 11 to 20 inputs. Vertical bars stand for the range of training errors and stars show the leave-one-out errors.

40 input variables. This data set contains 49.1% defaulting and 50.9% non defaulting credit clients (the two classes). In order to forecast the defaulting behavior of new credit clients, a variety of SVM models were constructed, in part also for addressing special circumstances like asymmetric costs of misclassification and unbalanced class sizes in the credit client population [3] [6] [7].

Next we describe a procedure to find a subset out of these input variables, which improve over the best of SVM models on randomly chosen subsets of inputs. Automatically evaluating such models must ensure the validity of the results. Therfore, on every trained model a leave-one-out validation procedure is performed, the result of which (expected percentage of misclassification) is a good estimator of the true out of sample performance. However, this substantially increases the computational cost per model, which calls for the use of time-out restrictions. When varying the SVM kernel parameters, the radial basis function (RBF) kernel has the least variation in SVM optimization runtime [7]. Hence, we use RBF-models and if a time-out threshold is exceeded in training, the model is excluded from evaluation. First, ten groups of inputs with lengths of 11 to 20 (out of 40) are formed. Every group contains ten input combinations of the given length and on each input combination 20 differently parameterized SVM models are trained and evaluated. Figure 1 shows the variation of leave-one-out and training errors on the 2000 SVM runs. As expected, the best errors decrease in the number of inputs, and more moderately so for the leave-one-out error. During the runs, each single input is evaluated. If the input is used in a model, one minus the leave-one-out error of the model is added to a contribution vector. Finally, these sums are divided by the number of times the single input was effectively used in all runs.

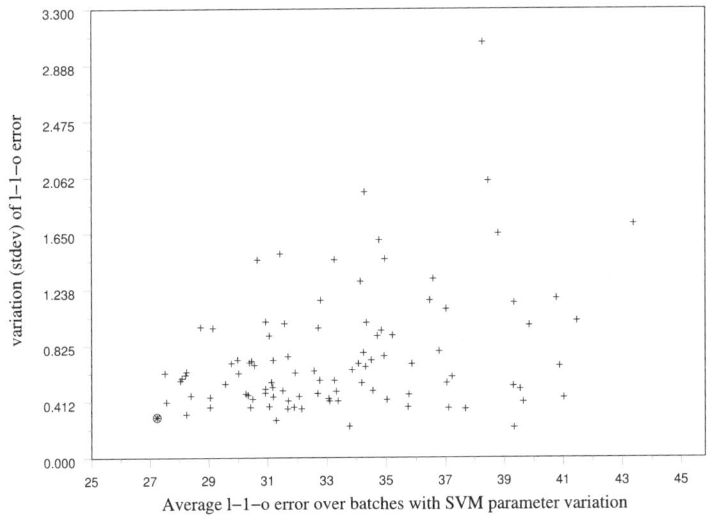

Fig. 2. Average error against error variation over the 100 run batches (over all inputs lengths). The new input proposal leads to the leftmost non-dominated point.

Finally, the contribution of inputs which exceed the average contribution of all inputs are proposed for a new input combination. The new input combination (lower left inlet of fig.1) uses 18 inputs. When evaluated, the best model of this input combination has lower leave-one-out error than the best model from the group of models with 18 inputs. When plotting the average leave-one-out errors against their variation over all runs within the 100 different input combinations, the new input combination is "efficient", by not being dominated by any model in both measures (figure 2).

3 Fixed variable subset evaluation for Credit Scoring

In the sequel a variable subset of the original credit scoring data set is used, consisting of 15 out of 40 variables. These (and only these) 15 variables are available for the whole credit portfolio of about 140 thousand clients. We are looking for the credit scoring function that gives good predictive results on the full data set as well as on the subset. Subsequently, the credit scoring function, which selects the credit client, can be used to establish a rating class system for the whole credit client portfolio. Subset and full data set model performance is compared using (*i*) Kernel Alignment, (*ii*) Vector Norm, (*iii*) number of support vectors and (*iv*) leave one out error. In a second step the subset classification models are used to compute a real valued output for each of the 140 thousand clients of the credit portfolio. ROC (Receiver Operating Characteristics) curves then are drawn and evaluated to compare the classification performance of the different models in a cut-off independent way.

Table 1. Evaluation and comparison of six SVM with different kernel functions trained on the full data set with 40 input variables and on the subset selection with 15 input variables.

SVM-Kernel	No. of Inputs	No. of SVs	No. of UBSVs	Kernel Alignment	Vector Norm	Leave one out Error
Linear	40	357	41		4.0962	27.20 %
	15	455	50	0.6657	4.0985	36.63 %
Sigmoid	40	561	17		10.5478	27.05 %
	15	637	4	0.3513	9.3847	36.63 %
Polynomial	40	455	63		0.5868	26.29 %
$(d = 2)$	15	567	19	0.9934	0.5221	30.24 %
Polynomial	40	427	216		0.3909	26.44 %
$(d = 3)$	15	540	49	0.9933	0.3241	31.76 %
RBF	40	431	179		26.6666	25.08 %
	15	484	60	0.9270	15.7441	33.59 %
Coulomb	40	554	186		16.4808	24.92 %
	15	531	83	0.9676	11.3886	33.28 %

SVM with six different kernel functions are used for classifying good and bad credit clients. Detailed informations about kernels, hyperparameters and tuning can be found in [7]. In table 1 for each kernel the number of support vectors (SVs), the number of unbounded support vectors (UBSVs), the vector norm and the leave one out error is shown for both the full data set with 40 variables and the subset with 15 variables in the input space. Subset selection methods are often based on comparing the vector norm $\|w\|^2$ between full set and subset [1]. Small differences usually are interpreted as minimal information loss when replacing full set with subset for model building. Here, this is true for linear, sigmoid and the polynomial kernels. Kernel alignment between K_1 (kernel of the full data set) and K_2 (kernel of the subset) is computed as $A(K_1, K_2) = \frac{\langle K_1, K_2 \rangle}{\sqrt{\langle K_1, K_1 \rangle \langle K_2, K_2 \rangle}}$ with $\langle K_1, K_2 \rangle = \mathrm{tr}(K_1^T K_2)$ as *Frobenius inner product*. Kernel alignment $A(K_1, K_2)$ in general can be interpreted as a Pearson correlation coefficient between two random variables $K_1(u,v)$ and $K_2(u,v)$ [5]. High alignment can be found for the polynomial kernels, the RBF kernel and the Coulomb kernel. The leave one out error as an unbiased estimator of the true generalization error finally shows a high increase of around 8 to 9 percentage points for linear, sigmoid, RBf and Coulomb kernel SVMs, whereas there only is a moderate increase of about 4 to 5 percentage points

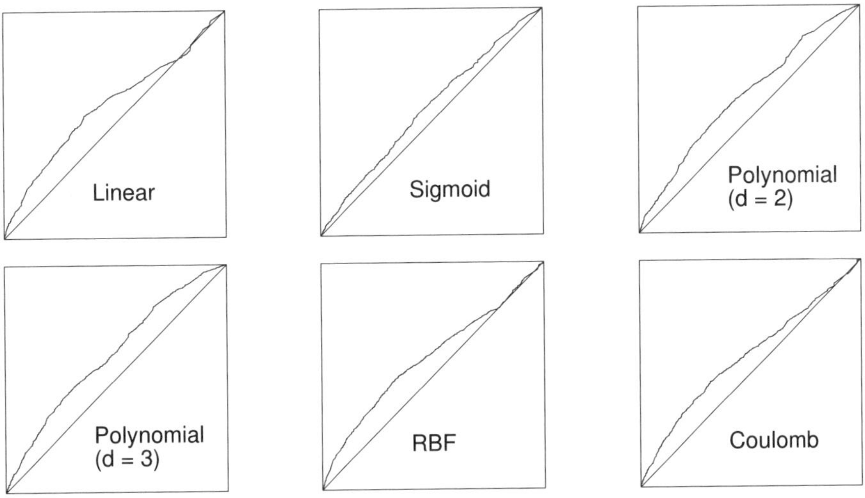

Fig. 3. ROC curves for different kernel functions. The ordinate of the ROC curve denotes the hit rate (= the proportion of rejected bad compared to all bad credit clients). The abscissa denotes the false alarm rate (= the proportion of rejected good compared to all good credit clients). Test variable is "Basel II default".

for the two polynomial kernel SVMs when changing from the full data set to the subset.

Table 2. ROC areas of SVM and traditional models for different default definitions.

	Cancellation			**Basel II default**		
	ROC	95% bound		ROC	95% bound	
Kernel	area	Lower	Upper	area	Lower	Upper
Linear	0.583	0.573	0.594	0.582	0.575	0.590
Sigmoid	0.559	0.550	0.569	0.556	0.549	0.563
Polynomial (d=2)	0.603	0.594	0.612	0.599	0.592	0.606
Polynomial (d=3)	0.613	0.604	0.622	0.609	0.602	0.616
RBF	0.576	0.566	0.587	0.579	0.571	0.587
Coulomb	0.570	0.559	0.580	0.573	0.565	0.581
Log. Reg.	0.582	0.572	0.592	0.583	0.575	0.591
LDA	0.584	0.574	0.594	0.585	0.577	0.592

In a second step the six SVM models trained on the variable subset can be used to classify the credit portfolio data, consisting of 139951 credit clients and the 15 variables of the selected subset. ROC curves are used for evaluation. There are two alternative default definitions available: (*i*) the loan was canceled in 3692 cases and (*ii*) there is at least a 90 days delay-in-payment in 6393 cases, which is the tighter Basel II default definition. ROC curves for the Basel II default definition can be seen in figure 3. The area between the ROC curve and the diagonal is a measure for the overall ability of the models to separate "good" and "bad" credit clients regardless of the cut off value. Table 2 also shows a remarkably bigger ROC area for the two SVMs with polynomial kernel, when compared to others. ROC areas for the traditional benchmark models Logistic Regression and Linear Discriminant Analysis are also reported. They are similar to the results of the linear SVM.

4 Conclusions and outlook

Credit scoring often has to deal with changing data availability. In this work it was shown how to compare and evaluate free-choice and fixed subset selection for different SVM models. It was found that small vector norm differences lead to misleading results, when there is weak kernel alignment, and that models with similar kernels *and* similar weight vectors yield more stable out of sample classification results and are better suited to construct a more reliable rating class system for the credit client portfolio.

References

1. RAKOTOMAMONJI, A. (2003): Variable Selection Using SVM-based Criteria. *Journal of Machine Learning Research, 3, 1357-1370.*
2. SCHEBESCH, K.B. and STECKING, R. (2005): Support Vector Machines for Credit Scoring: Extension to Non Standard Cases. In: Baier, D. and Wernecke, K.-D. (Eds.): *Innovations in Classification, Data Science and Information Systems.* Springer, Berlin, 498-505.
3. SCHEBESCH, K.B. and STECKING, R. (2005): Extracting Rules from Support Vector Machines. In: Fleuren, H., den Hertog, D. and Kort, P. (Eds.): *Operations Research Proceedings 2004.* Springer, Berlin 408-415.
4. SCHÖLKOPF, B. and SMOLA, A. (2002): *Learning with Kernels.* The MIT Press, Cambridge.
5. SHAWE-TAYLOR, J. and CRISTIANINI, N. (2004): *Kernel Methods for Pattern Analysis.* Cambridge University Press, Cambridge.
6. STECKING, R. and SCHEBESCH, K.B. (2003): Support Vector Machines for Credit Scoring: Comparing to and Combining with some Traditional Classification Methods. In: Schader, M., Gaul, W. and Vichi, M. (Eds.): *Between Data Science and Applied Data Analysis.* Springer, Berlin, 604-612
7. STECKING, R. and SCHEBESCH, K.B. (2005): Comparing and Selecting SVM-Kernels for Credit Scoring, submitted to Proceedings of the 29th Annual Conference of the GfKl 2005.

Genetically Constructed Kernels for Support Vector Machines

Stefan Lessmann[a], Robert Stahlbock[a], Sven Crone[b]

[a] University of Hamburg, Inst. of Information Systems, Von-Melle-Park 5, 20146 Hamburg, Germany
[b] Lancaster University Management School, Dept. of Management Science, Lancaster, LA1 4YX, United Kingdom

Abstract

Data mining for customer relationship management involves the task of binary classification, e.g. to distinguish between customers who are likely to respond to direct mail and those who are not. The support vector machine (SVM) is a powerful learning technique for this kind of problem. To obtain good classification results the selection of an appropriate kernel function is crucial for SVM. Recently, the evolutionary construction of kernels by means of meta-heuristics has been proposed to automate model selection. In this paper we consider genetic algorithms (GA) to generate SVM kernels in a data driven manner and investigate the potential of such hybrid algorithms with regard to classification accuracy, generalisation ability of the resulting classifier and computational efficiency. We contribute to the literature by: (1) extending current approaches for evolutionary constructed kernels; (2) investigating their adequacy in a real world business scenario; (3) considering runtime issues together with measures of classification effectiveness in a mutual framework.

1 Introduction

The support of managerial decision making in marketing applications is a common task for corporate data mining with classification playing a key role in this context [2]. The SVM [9] is a reliable classifier that has been successfully applied to marketing related decision problems, e.g. [1; 10]. Like other learning algorithms such as neural networks, the SVM algorithm offers some degrees of freedoms that have to be determined within the data mining process. The selection of suitable parameters is crucial for effective classification. Therefore, we propose a data driven heuristic to determine the SVM parameters without manual intervention.

The remainder of this paper is organised as follows: Following a brief introduction to SVM theory we present our combination of GA and SVM (GA-SVM) in Section 3. The potential of GA-SVM is evaluated in a real world scenario of direct marketing in Section 4. Conclusions are given in Section 5.

2 Support Vector Machines

The SVM is a supervised learning machine to solve linear and non-linear classification problems. Given a training set $S = \{\mathbf{x}_i; y_i\}_{i=1}^{m}$ where $\mathbf{x}_i$ is a n-dimensional real vector and $y_i \in \{-1, +1\}$ its corresponding class label, the task of classification is to learn a mapping $\mathbf{x}_i \mapsto y_i$ from *S*, that allows the classification of new examples with unknown class membership.

The SVM is a linear classifier of the form

$$y(\mathbf{x}) = sgn(\mathbf{w} \cdot \mathbf{x} + b), \tag{1}$$

which strives to maximise the margin of separation between the two classes [9]. The parameters $\mathbf{w}$ and b realising such a maximal margin hyperplane can be found by solving a quadratic optimisation problem with inequality constraints; e.g. [3].

In order to derive more general, non-linear decision surfaces SVMs implement the idea to map the input data into a high-dimensional feature space via an a priori chosen non-linear mapping function. Due to the fact, that the SVM optimisation problem contains the input patterns only as dot products, such a mapping can be accomplished implicitly by introducing a kernel function [3; 9]

$$K(\mathbf{x}_i, \mathbf{x}_j) = \Phi(\mathbf{x}_i) \cdot \Phi(\mathbf{x}_j). \tag{2}$$

Beside the selection of an appropriate kernel and its corresponding kernel parameters, see Section 3, the SVM classifier offers one additional regularisation parameter *C* which controls the trade off between maximising the margin of separation and classifying the training set without error.

3 Genetic algorithms for SVM model selection

The classification performance of SVM depends heavily on the choice of a suitable kernel function and an adequate setting of the regularisation parameter *C*.

Consequently, we develop a data driven approach to determine the kernel K and its corresponding kernel parameters together with C by means of GA. Using the five basic kernels of Table 1, we construct a combined kernel function as

$$K_{poly}^{1} \otimes K_{rad}^{\alpha} \otimes K_{sig}^{\beta} \otimes K_{imq}^{\gamma} \otimes K_{anova}^{1} , \tag{3}$$

with $\otimes \in \{+;\cdot\}$, where we exploit the fact that if K_1 and K_2 are kernels, $K_1 + K_2$ and $K_1 \cdot K_2$ are valid kernels as well [3].

Table 1. Basic SVM kernel functions

Polynomial kernel	$K_{poly}\left(\boldsymbol{x}_i,\boldsymbol{x}_j\right)=\left(a(\boldsymbol{x}_i\cdot\boldsymbol{x}_j)+b\right)^c$
Radial kernel	$K_{rad}\left(\boldsymbol{x}_i,\boldsymbol{x}_j\right)=\exp\left(-a\left\|\boldsymbol{x}_i-\boldsymbol{x}_j\right\|^2\right)$
Sigmoidal kernel	$K_{sig}\left(\boldsymbol{x}_i,\boldsymbol{x}_j\right)=\tanh\left(a(\boldsymbol{x}_i\cdot\boldsymbol{x}_j)+b\right)$
Inverse multi-quadratic kernel	$K_{imq}\left(\boldsymbol{x}_i,\boldsymbol{x}_j\right)=1/\sqrt{\left\|\boldsymbol{x}_i-\boldsymbol{x}_j\right\|^2+b^2}$
Anova kernel	$K_{anova}\left(\boldsymbol{x}_i,\boldsymbol{x}_j\right)=\left(\sum_j \exp\left(-a\left(\boldsymbol{x}_i-\boldsymbol{x}_j\right)\right)^2\right)^c$

To encode (3) into a structure suitable for GA based optimisation we use five integer genes for the kernel exponents in (3), four binary genes for the kernel combination operator $\otimes$ and sixteen real-valued genes for the specific kernel parameters (three per kernel) as well as the regularisation parameter C. The complete structure is given in Fig. 1. This coding is inspired by [7] and extends their approach to five kernels and the inclusion of C into the GA based optimisation.

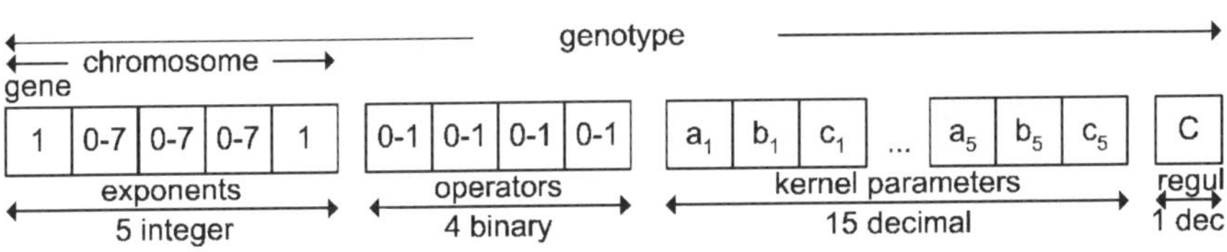

Fig. 1. Structure of the genotype for SVM kernel construction

The GA is implemented in accordance with [8] and utilises a uniform crossover for the five kernel exponent genes. That is, all genes between two random points within this string are interchanged between two genotypes representing parents for the resulting two new genotypes. The mutation operator is implemented as a simple bit swap for the four kernel combination genes and a random increment or decrement for all integer and real value genes. Crossover and mutation probabilities have been determined through pre-tests to 0.7 and 0.3 respectively.

4 Empirical evaluation

4.1 Experimental setup

The simulation experiment aims at comparing genetically constructed SVM with conventional ones to assess capabilities of GA to support SVM model selection.

We consider the case of repeat purchase modelling in a direct marketing setting, see e.g. [1; 10], using real world data from a German publishing house. The data set consists of 300,000 customer records that have been selected for a past mailing campaign to cross-sell an additional magazine subscription to customers that have subscribed to at least one periodical. Each customer is described by a 28-dimensional vector of 9 numerical and 19 categorical attributes describing transactional and demographic customer properties. The number of subscriptions sold in this campaign is given with 4,019, leading to a response rate of 1.35% which is deemed to be representative for the application domain. An additional target variable indicates the class membership of each customer (class 1 for subscribers and class -1 for non subscribers) facilitating the application of supervised learning algorithms to model a relationship between customer attributes and likelihood of responding to direct mail.

Classifiers are evaluated applying a hold-out method of three disjoint datasets to control over-fitting and for out-of-sample evaluation. While training data is used for learning, i.e. determining the decision variables $\boldsymbol{w}$ and b, see (1), a validation set is used to steer the GA. That is, a classifier's performance on the validation set represents its fitness and is used to select items for the mating pool within the GA [4]. The trained and selected classifiers are finally tested on an unknown hold-out set to evaluate their generalisation ability on unknown data.

In order to assure computational feasibility and with regard to the vast imbalance between class 1 and class -1 membership within our data set, we apply an undersampling approach [11] to obtain a training and validation data set of 4,144 and 2,070 records respectively with equal class distributions. The test set consists of 65,000 records containing 912 class 1 customers, reflecting the original unequal distribution of the target variable.

4.2 Experimental results

In order to deliver good results GA usually require a large population size that ensures sufficient variability within the elements in the gene pool [8]. For GA-SVM we select a population size of 50 and monitor the progress in classification quality for 15 generations. Thus, 750 individual SVMs with genetic kernel are constructed on the training set, assessed on the validation set and finally evaluated on the test set. Since the skewed class distribution of the target variable prohibits the application of standard performance metrics of classification accuracy [11], we used the G-metric instead [6]. Striving to maximise the class individual accuracies while keeping them balanced the G-metric is calculated as the geometric mean between class individual accuracies. Consequently, higher values indicate improved predictive accuracy.

Results at the generation level are given in Table 2 where each value is calculated on the basis of the 50 individual GA-SVM classifiers within a generation.

Table 2. Results of GA-SVM at the generation level over 15 generations

	Mean runtime per SVM [min]		SVM performance by means of G-metric on training set		validation set		test set	
Generation	mean	std.dev.	mean	std.dev.	mean	std.dev.	mean	std.dev.
0	91.3	53.4	0.596	0.306	0.544	0.277	0.444	0.225
1	71.2	37.0	0.731	0.158	0.661	0.145	0.534	0.111
2	78.1	38.8	0.687	0.236	0.633	0.215	0.496	0.168
3	77.8	27.9	0.754	0.158	0.685	0.142	0.528	0.110
4	79.6	31.1	0.736	0.192	0.668	0.172	0.516	0.132
5	76.0	27.8	0.759	0.158	0.684	0.142	0.527	0.110
6	68.8	16.6	0.786	0.025	0.713	0.019	0.549	0.013
7	77.3	31.9	0.785	0.030	0.714	0.015	0.547	0.012
8	67.8	22.8	0.775	0.114	0.703	0.102	0.537	0.078
9	65.1	21.7	0.768	0.115	0.696	0.105	0.539	0.079
10	67.8	25.0	0.784	0.034	0.711	0.027	0.552	0.012
11	64.2	11.2	0.795	0.008	0.721	0.012	0.551	0.009
12	62.2	12.5	0.796	0.008	0.720	0.015	0.552	0.009
13	59.6	12.5	0.791	0.014	0.716	0.019	0.553	0.010
14	59.4	12.6	0.789	0.014	0.720	0.015	0.553	0.008

Our results show a generally increasing average performance from generation to generation over all data sets. However, vast improvements are obtained only when moving from generation 0 to 1, indicating that a saturation level is reached early in the evolutionary process. In fact, while a oneway analysis of variance confirmed a highly significantly difference in mean performance over all data sets at the 0.001 level, a Tukey post hoc test revealed that only the generations 0 and 2 differ from the remaining ones significantly at the 0.01 level.

The decrease in standard deviation is more explicit and illustrates a higher similarity within the gene pool. Interestingly, the average runtimes decrease tremendously, meaning that the high quality kernels of later generations are also computationally more efficient. The best kernel was found in generation 14 with a test set G-value of 0.585 incorporating all base kernels but the anova kernel.

To compare our approach with standard SVM we calculate solutions for the radial and polynomial SVM classifier, conducting an extensive grid search [5] in the range $\log(C) = \{-4; 4\}$ and $\log(a) = \{-4; 4\}$ with a step size of one for the radial kernel and $\log(C) = \{-2; 3\}$, $\log(a) = \{-2; -1\}$, $b = \{0; 1\}$, $c = \{2; 7\}$ for the polynomial kernel to obtain and average G-value of $G_{radial} = (0.70; 0.58; 0.53)$ and $G_{polynomial} = (0.71; 0.65; 0.54)$ on training, validation and test sets. As expected, the higher flexibility of the combined kernel in GA-SVM allows a purer separation of the training set. Regarding generalisation, GA-SVM consistently outperforms classical SVM in later generations, providing superior results on the validation set from generation 3 and on the test set from generation 10 onwards.

5 Conclusions

We investigated the potential of SVMs with GA-optimised kernel functions in a real world scenario of corporate decision making in marketing. Solving more than 750 evolutionary constructed SVMs, the GA proved to be a promising tool for kernel construction, enhancing the predictive power of the resulting classifier. However, the vastly increased computational cost might be the main obstacle for practical applications. Most radial SVMs needed less than a minute to construct a solution and the runtime of polynomial SVMs ranged from 12 to 60 minutes. In contrast, we observed average GA-SVM runtimes of 60 to 90 minutes.

Since the task of model selection shifts from setting SVM parameters to determining the parameters of the utilised search heuristic, the proposed GA is a promising candidate for SVM tuning, offering only four degrees of freedom on its own (crossover and mutation probabilities, population size, termination criterion e.g. number of generations).

Further research involves the application of GA-SVM to other data sets as well as a detailed analysis and comparison of the constructed kernels per generation.

References

[1] Baesens B, Viaene S, Van den Poel D, Vanthienen J, Dedene G (2002) Bayesian neural network learning for repeat purchase modelling in direct marketing. European Journal of Operational Research 138(1):191-211

[2] Berry MJA, Linoff G (2004) Data mining techniques: for marketing, sales and customer relationship management, 2. edn. Wiley, New York

[3] Cristianini N, Shawe-Taylor J (2000) An introduction to support vector machines and other kernel-based learning methods. Cambridge University Press, Cambridge

[4] Goldberg DE (1989) Genetic Algorithms in Search, Optimization and Machine Learning. Addison-Wesley, Reading

[5] Keerthi SS, Lin C-J (2003) Asymptotic Behaviours of Support Vector Machines with Gaussian Kernel. Neural Computation 15(7):1667-1689

[6] Kubat M, Holte RC, Matwin S (1998) Machine Learning for the Detection of Oil Spills in Satellite Radar Images. Machine Learning 30(2-3):195-215

[7] Nguyen H-N, Ohn S-Y, Choi W-J (2004) Combined Kernel Function for Support Vector Machine and Learning Method Based on Evolutionary Algorithm. In: Pal NR, Kasabov N, Mudi RK (eds) Proc. of the 11th Intern. Conf. on Neural Information Processing, Calcutta, India, pp 1273-1278

[8] Stahlbock R (2002) Evolutionäre Entwicklung künstlicher neuronaler Netze zur Lösung betriebswirtschaftlicher Klassifikationsprobleme. WiKu, Berlin

[9] Vapnik VN (1995) The Nature of Statistical Learning Theory. Springer, New York

[10] Viaene S, Baesens B, Van Gestel T, Suykens JAK, Van den Poel D, Vanthienen J, De Moor B, Dedene G (2001) Knowledge discovery in a direct marketing case using least squares support vector machines. International Journal of Intelligent Systems 16(9):1023-1036

[11] Weiss GM (2004) Mining with rarity: a unifying framework. ACM SIGKDD Explorations Newsletter 6(1):7-19

Optimierung von Warteschlangensystemen durch Approximation mit Neuronalen Netzen

Frank Köller, Michael H. Breitner

Institut für Wirtschaftsinformatik, Universität Hannover, Königsworther Platz 1, 30167 Hannover, {koeller;breitner}@iwi.uni-hannover.de

Kurzfassung

Verschiedene Inbound Call Center Probleme haben gezeigt, dass künstliche neuronale Netze sehr gut in der Lage sind, Kennzahlen für Warteschlangenprobleme zu approximieren. Dabei wurden bisher Vergleiche mit analytischen Lösungen z. B. für die mittlere Wartezeit angestellt. Dieser Aufsatz beschreibt den nächsten Schritt: Künstliche neuronale Netze können ebenfalls für grundlegenden Warteschlangenproblemen eingesetzt werden, für die keine exakten Lösungen berechnet werden können. In der Praxis werden alle Warteschlangenprobleme entweder mit komplexen, diskreten Simulationen gelöst, oder die Probleme werden soweit vereinfacht, dass sie analytisch lösbar werden. Im Gegensatz dazu wird hier die Problemstruktur für das Training der neuronalen Netze nicht vereinfacht und es werden auch nur wenige Simulationspunkte im Vergleich zu Standardsimulationen generiert. Das unvermeidliche Rauschen in den Simulationsdaten wird durch die kontinuierliche, approximierte Lösung geglättet, d. h. die Kennzahlen sind wesentlich schneller und auch genauer verfügbar. So kann z. B. die Anzahl der Call Center Agenten in Echtzeit optimiert werden.

1 Einleitung

Der überwiegende Anteil, etwa drei Viertel des Gesamtbudgets, in einem Call Center sind personalbezogene Ausgaben (vgl Henn et al. 1998 und Call Center-Benchmark Kooperation 2004). In der Praxis erfolgt gegenwärtig die Personalbedarfsermittlung und -einsatzplanung in der Regel in den folgenden drei Schritten (vgl. Helber und Stolletz 2004):

1. Prognose des Anrufaufkommens je Periode (häufig 30- oder 60-Minutenintervalle).
2. Ermittlung der erforderlichen Zahl von Agenten je Periode für einen vorgegebenen Servicegrad hinsichtlich der Wartezeit (meist mit dem M/M/c-Modell).
3. Zeitliche Einplanung der Mitarbeiter über die Perioden (oder zeitliche Einplanung „anonymer“ Schichten mit anschließender Zuordnung der Mitarbeiter zu den Schichten).

An die Personaleinsatzplanung im Schritt 3 schließt sich noch eine Echtzeit-Steuerung an, in der in Abhängigkeit des aktuellen Systemzustandes z. B. die Pausen der Agenten, Besprechungen oder Trainingsmaßnahmen zeitlich festgelegt werden.
Im ersten Schritt, der Prognose, ist ein Anrufaufkommen vorherzusagen, das zwar innerhalb eines Tages oder einer Woche hochgradig variabel ist, dabei aber häufig wiederkehrende Muster aufweist (vgl. Abbildung 1). Für das Training der neuronalen Netze mit dem Neurosimulator FAUN[1] können nun diese Muster als reale Datengrundlage dienen, um eine effizientere Personaleinsatzplanung zu ermöglichen.

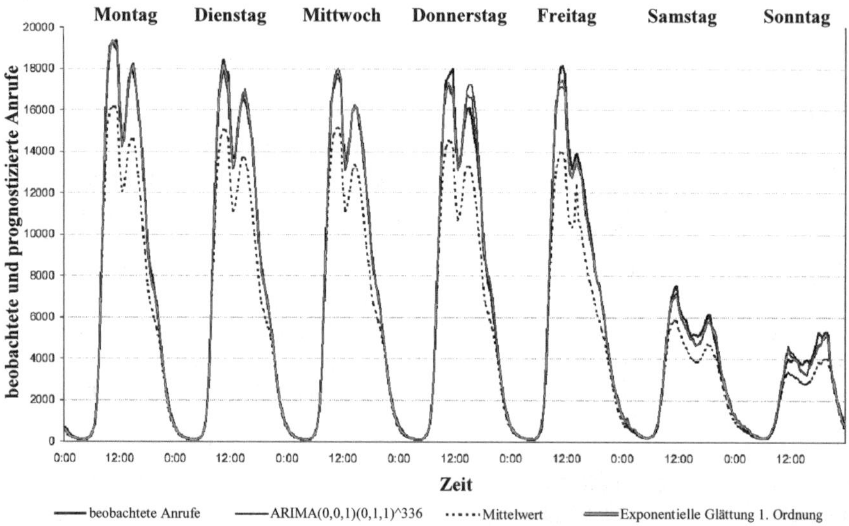

Abb. 1. Anrufaufkommen und Prognosen in Halbstundenintervallen in den Call Centern des Auskunftsdienstes der Deutschen Telegate AG vom 2. – 8.11.1998. Deutlich sind die Auswirkung der Mittagspausen und des Wochenendes zu erkennen (Helber und Stolletz 2004).

2 Approximation von Kennzahlen für Warteschlangen

Anhand von Simulationen für Inbound Call Center kann gezeigt werden, dass künstliche neuronale Netze Kennzahlen von Warteschlangenproblemen, bei denen

[1] „**F**ast **A**pproximation with **U**niversal **N**eural Networks". Neurosimulator bezieht sich nicht auf die Simulation von Warteschlangen, sondern auf die komfortable, GUI-unterstützte Simulation gehirnanaloger Vorgänge, die als überwachtes Training bzw. Lernen von künstlichen neuronalen Netzen bekannt ist (Breitner 2003).

analytische Lösungen existieren, sehr gut approximieren können[2]. In einem weiteren Schritt wird dann untersucht, ob künstliche neuronale Netze auch auf allgemeine Warteschlangenprobleme angewendet werden können, für die keine exakten, expliziten Lösungen für die Warteschlangenkennzahlen existieren[3], indem reale Daten als Grundlage für das Training der neuronalen Netze dienen.

2.1 Approximation der Kennzahlen durch FAUN mittels diskreter Simulationsdaten

Um die Qualität der durch den Neurosimulator FAUN[4] approximierten Lösungen für die Warteschlangenkennzahlen bestimmen zu können, werden zunächst Warteschlangenprobleme betrachtet, bei denen eine analytische Lösung bekannt ist. Deshalb wird im ersten Schritt dem Training der neuronalen Netze eine Simulation anhand des M/M/*c*-Modells vorangestellt, die aber im Vergleich zu einer „flächendeckenden" Auswertung nur aus wenigen Simulationspunkten bestehen muss[5].

Wesentliche Vorteile der Approximation gegenüber der einfachen diskreten Simulation von Kennzahlen sind,

- dass eine kontinuierliche Funktion zur Kostenminimierung generiert wird, und
- dass die approximierte Funktion eine bessere Annäherung an die analytische Lösung aufweist als die Simulationsdaten.

Letzteres ist dadurch begründet, dass die Simulationsdaten immer ein Rauschen aufweisen und die approximierte Funktion in diesen Daten liegt (vgl. Abb. 2). Da auch stärkere Schwankungen der verwendeten Musterdatensätze durch das neuronale Netz ausgeglichen werden, ist die Simulation, die der Approximation durch den Neurosimulator FAUN vorangestellt ist, zeitlich wesentlich weniger aufwändig, als wenn die gewünschte Kennzahl allein durch Simulation bestimmt werden soll. Wichtig ist jedoch, dass die zugrundeliegende Simulation annähernd den stationären Zustand erreicht und somit der analytischen Lösung hinreichend nahe ist. Da der weitere Arbeitsschritt durch die Approximation mit FAUN nur wenige Sekunden dauert, entsteht hierdurch kein wesentlicher Nachteil.

Neuronale Netze mit mehreren inneren Neuronen neigen dazu zwischen den Daten zu oszillieren, um diese auswendig zu lernen. Diese Oszillation ist aber in vielen Praxisanwendungen, so auch hier, nicht erwünscht und damit liefern neuronale

---

[2] Für das M/M/1-Modell teilweise untersucht in Barthel (2003), einer Diplomarbeit betreut durch die Autoren und für das M/M/*c*-Modell in Köller und Breitner (2005).

[3] Meist können obere und untere Schranken bestimmt werden, die die Bandbreiten für Warteschlangenkennzahlen begrenzen. Somit ist überprüfbar, ob die approximierten Kennzahlen innerhalb dieser Bandbreiten liegen.

[4] Siehe auch http://www.iwi.uni-hannover.de/faun.html.

[5] Vertiefende Beispiele und Erläuterungen zu den Simulationen von Warteschlangenproblemen sind in Zimmermann (1997), Domschke (2002) und Siegert (1991) zu finden.

Netze mit nur wenigen inneren Neuronen trotz eines höheren Trainingsfehlers bessere Ergebnisse. Daher ist eine graphische Analyse empfehlenswert bzw. muss in einer mathematischen Analyse der Krümmungstensor möglichst klein sein (vgl. Abb. 2 und Breitner 2003).

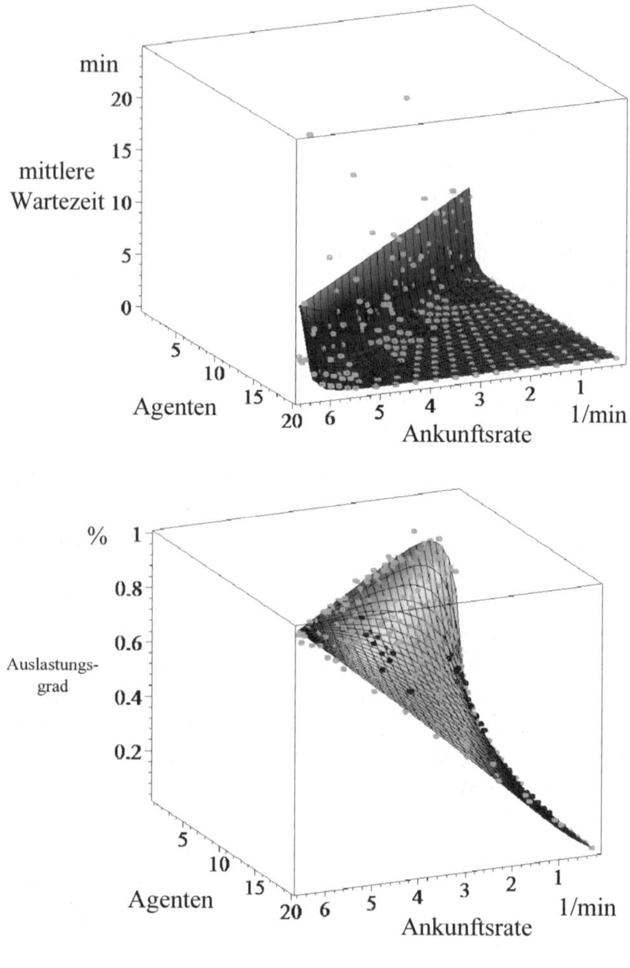

Abb. 2. Approximierte durchschnittliche Wartezeit der Kunden in der Warteschleife (oben) und der dazugehörige approximierte Auslastungsgrad (unten) mit den jeweiligen Simulationspunkten.

2.2 Approximation der Kennzahlen durch FAUN mittels realer Daten

Für ein Inbound Call Center sind einschränkende Annahmen bei dem M/M/c-Wartesystem, dass die Zwischenankunftszeiten ebenso wie die Bearbeitungszeiten unabhängig exponentialverteilt, alle Anrufer geduldig und der Warteraum unendlich groß seien[6]. Auf viele Inbound Call Center treffen diese Annahmen des Erlang-C-Modells nicht zu. Meist steht nur eine begrenzte Zahl an Wartepositionen zur Verfügung, das heißt, wenn dieser Warteraum voll ist, erhält der Anrufer ein Besetztzeichen. Zudem weisen Call Center mehrere Klassen von Anrufern und Agenten auf oder die Anrufer sind ungeduldig und legen vorzeitig auf. Sind die Zwischenankunftszeiten und die Bearbeitungszeiten nicht exponentialverteilt, so ist es in der Regel nicht möglich, eine analytische Lösung zu finden. Abhilfe können hier neuronale Netze schaffen, indem reale Daten als Basis für das Training dienen. Dabei sind grundsätzlich zwei Wege möglich.

Zum einen können aus den realen Daten die Verteilungsfunktionen für den Ankunftsprozess und für die verschiedenen Bedienprozesse in Abhängigkeit von den entsprechenden Geschäftsprozessen bestimmt werden. Mit den so gewonnen, praxisnahen Verteilungen können dann realistischere Trainingsdaten simuliert werden.

Zum anderen können reale Daten direkt als Eingabegrößen für das Training der neuronalen Netze dienen, indem z. B. der Wochentag und die Urzeit mitberücksichtigt werden. Dies hat den Vorteil, dass zeitabhängige Zustände des Systems bzw. des Call Centers, wie z. B. ein erhöhtes Anruferaufkommen zu immer den gleichen Zeiten und Tagen, mittrainiert werden kann (vgl. Abb. 1).

3 Fazit

Der Neurosimulator FAUN bietet eine Möglichkeit, für alle Warteschlangensysteme eine approximierte, explizite Lösung für deren Kennzahlen zu generieren. Dieser Aufsatz zeigt anhand des Standardmodells M/M/*c* und der Benutzung realer Daten, wie dies möglich wird. Die aus dem Vergleich mit dem M/M/c-Modell gewonnenen Erkenntnisse können auf Modelle ohne analytische Lösung übertragen werden, die bisher nur mit Simulationen gelöst werden können.

Vorteile der Approximation von Warteschlangenkennzahlen bei schwierigen Warteschlangenproblemen gegenüber der Analyse durch diskrete Simulationen bestehen darin,

- dass eine analytische Funktion zur Personaleinsatzplanung und Kostenminimierung generiert wird, die extrem schnell auswertbar ist, und
- dass das unvermeidliche Rauschen in den Simulationsdaten geglättet wird, d. h. die Kennzahlen genauer verfügbar sind bzw. deutlich weniger Simulationen nötig sind.

Es brauchen nicht besonders viele Punkte zum Training der neuronalen Netze simuliert werden im Vergleich zu Standardsimulationen. Daraus ergibt sich ein erheblicher Zeitvorteil, da der zusätzliche Schritt des FAUN-Trainings in der Regel nur wenige Sekunden bis wenige Minuten dauert.

Ein weiterer Vorteil ist, dass bei der Simulation der Muster für das Training mit FAUN unterschiedlichste Verteilungen für die Ankunfts- und Bedienrate, so wie

[6] Ausführliche Darstellungen zur Warteschlangentheorie findet man z.B. in Schassberger (1973), Bolch (1989), Meyer und Hansen (1996, S. 210 ff.) oder Hillier und Lieberman (1997, S. 502 ff.)

sie in der Praxis tatsächlich vorkommen, eingesetzt werden können. Beispielsweise kann so anhand des realen Anrufer-aufkommens in einem Call Center die tatsächliche Verteilung über einen längeren Zeitraum bestimmt und für die Simulation verwendet werden. Analog kann mit dem Bedienprozess verfahren werden. Aus realen Daten können dann Simulationspunkte für das Training künstlicher neuronaler Netze generiert werden, um so die Abläufe in einem Call Center durch realistischere Bestimmung der Warteschlangenkennzahlen wesentlich praxisnäher abzubilden. Es muss also nicht das M/M/*c*-Modell mit all seinen Einschränkungen als Grundlage für die Mustergenerierung dienen. Dieses aus der Praxis gewonnene Datenmaterial kann durchaus verrauscht sein, da neuronale Netze mit wenigen inneren Neuronen sich „in die Daten legen" und so oft ein gleichmäßiges, oft weißes Rauschen ausgleichen.

Literatur

Barthel A (2003) Effiziente Approximation von Kenngrößen für Warteschlangen mit dem Neurosimulator FAUN 1.0. Diplomarbeit am Institut für Wirtschaftswissenschaft der Universität Hannover, Königsworther Platz 1, D-30167 Hannover

Bolch G (1989) Leistungsbewertung von Rechensystemen mittels analytischer Warteschlangenmodelle. Teubner, Stuttgart

Breitner MH (2003) Nichtlineare, multivariate Approximation mit Perzeptrons und anderen Funktionen auf verschiedenen Hochleistungsrechnern. Akademische Verlagsgesellschaft Aka GmbH, Berlin

Call Center-Benchmark Kooperation (2004) Kooperationsprojekt: Purdue University, Universität Hamburg, Initiator der profiTel MANAGEMENT CONSULTING. http://www.callcenter-benchmark.de/index3.html. Letzter Abruf: 10.10.2004

Domschke W, Drexl A (2002) Einführung in Operations Research, 5. Aufl. Springer, Berlin

Helber S, Stolletz R (2004) Call Center Management in der Praxis: Strukturen und Prozesse betriebswirtschaftlich optimieren. Springer, Berlin

Henn H, Kruse JP, Strawe OV (1998) Handbuch Call Center Management (2. Aufl.). Telepublic Verlag, Hannover

Hillier FS, Lieberman GJ (1997) Operations Research, 5. Aufl. Oldenbourg, München Wien

Köller F, Breitner MH (2005) Optimierung von Warteschlangensystemen in Call Centern auf Basis von Kennzahlenapproximation. In Günther HO, Mattfeld DC, Suhl L (2005) Entscheidungsunterstützende Systeme in Supply Chain Management und Logistik (S.459 - 482). Physica, Heidelberg

Meyer M, Hansen K (1996) Planungsverfahren des Operations Research (4. Aufl.). Vahlen, München

Schassberger R (1973) Warteschlangen. Springer, Berlin

Siegert HJ (1991) Simulation zeitdiskreter Systeme. Oldenbourg, München Wien

Zimmermann W (1997) Operations Research: quantitative Methoden zur Entscheidungsvorbereitung (8. Aufl.). Oldenbourg, München Wien

Aktienkursprognose anhand von Jahresabschlussdaten mittels Künstlicher Neuronaler Netze und ökonometrischer Verfahren

Thorsten Poddig und Oxana Enns

1 Einleitung

Für die Prognose zukünftiger Aktienkursentwicklung eines Unternehmens muss unter anderem eine Betrachtung der aktuellen Situation des Unternehmens erfolgen. Der Jahresabschluss ist oft die einzige öffentlich verfügbare und somit die wichtigste Informationsquelle, aus der ein Anleger sein Wissen über die Vermögens-, Finanz- und Ertragslage eines Unternehmens schöpfen kann (vgl. [3] S.34).

Zahlreiche Studien befassen sich mit der Frage der Brauchbarkeit von Jahresabschlussdaten für die Prognose von Aktienkursen (vgl. [2], [8]). Während in den USA bereits seit den siebziger Jahren der Querschnittszusammenhang zwischen Aktienrenditen und verschiedenen potentiellen Einflussgrößen untersucht wurde (vgl. [8] S.27), ist diese Frage in Deutschland erst seit einigen Jahren Gegenstand der Diskussion.

In dieser Arbeit werden mit Hilfe Neuronaler Netze die Zusammenhänge zwischen den Jahresabschlusskennzahlen und der Aktienkursrendite untersucht. Als Benchmark werden lineare Regressionsmodelle einbezogen. Wenn ein Zusammenhang zwischen den Jahresabschusskennzahlen und der Aktienkursrendite festgestellt werden kann, wird es möglich sein, aufgrund der Güte der Modelle die Linearität oder Nichtlinearität des existierenden Zusammenhangs zu erkennen. Es bleibt zu überprüfen, welche Kennzahlen einen Zusammenhang mit der Aktienkursrendite in verschiedenen Jahren zeigen. Wenn sich herausstellt, dass es in den verschiedenen Jahren die gleichen Kennzahlen sind, wird es möglich sein, über die Stabilität der Modelle und die Möglichkeit der Prognose zu sprechen. Abschließend wird untersucht, ob die Methode der Neuronalen Netze zu besseren Ergebnissen als andere statistische Verfahren führen kann. Dieser Beitrag stellt eine Ergänzungsstudie einer umfangreichen Untersuchung von *Petersmeier* (vgl. [4]

S.363ff) dar, in der die oben genannte Fragenstellung mit Hilfe der nichtparametrischen Regression untersucht wurde.

2 Der Datensatz

Der folgenden empirischen Untersuchung liegen Jahresabschlüsse aus der Hoppenstedt-Bilanzdatenbank für insgesamt 279 deutsche Unternehmen aus verschiedenen Branchen zugrunde (vgl. [7] S.5). Der Untersuchungszeitraum erstreckt sich von 1990 bis 2001 (der Untersuchungszeitraum wurde so festgelegt, um eine Vergleichbarkeit mit den Ergebnissen der Studie von Petersmeier zu ermöglichen). Wegen der bilanziellen Besonderheiten bei Jahresabschlüssen von Banken und Versicherungen wurden deren Daten nicht berücksichtigt (vgl. [1] S.43ff). Weiterhin beschränkt sich die Auswahl auf börsennotierte Unternehmen. Es wurden die Jahresabschlüsse mit dem gleichem Bilanzstichtag, dem 31.12, in die Stichprobe einbezogen.

Neben originären Jahresabschlüssen beinhaltet die Hoppenstedt-Bilanzdatenbank auch aufbereitete Daten, wie z.B. Jahresabschlusskennzahlen. In der Untersuchung wurden Kennzahlen als erklärende Variablen verwendet. Einige Kennzahlen der Hoppenstedt-Bilanzdatenbank sind nicht korrekt berechnet, weswegen diese Werte durch eigene Berechnungen ersetzt wurden. Insgesamt wurden 43 Kennzahlen einbezogen, die auch in der Studie von *Petersmeier* berücksichtigt wurden (vgl. [4] S.436).

Die Aktienkurse wurden Datastream entnommen. Die als zu erklärende Variable betrachtete Aktienkursrendite wurde als diskrete Rendite modelliert.

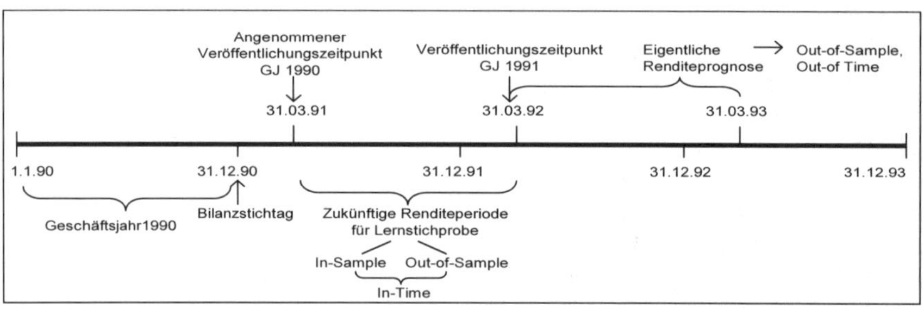

Abbildung 1: Schematische Darstellung der Datenmengen für die Jahre 90-93

Abbildung 1 stellt schematisch die zeitliche Struktur der Untersuchungsdaten am Beispiel der Jahre 1990-1993 dar. Zwischen dem Bilanzstichtag und dem Veröffentlichungszeitpunkt wurde ein dreimonatiges Veröffentlichungslag berücksichtigt. D.h., die Jahresabschlusskennzahlen, die am 31.03.91 bekannt geworden sind, wurden als erklärende Variablen definiert. Die Rendite wurde entsprechend für den Zeitraum zwischen dem 31.03.91 und dem 31.03.92 berechnet und als zu erklärende Variable in die Untersuchung einbezogen. Die aus diesen Daten erstellte Lernstichprobe wurde in zwei Mengen aufgeteilt: Schätzmenge (In-Sample, In-Time) und Testmenge (Out-of-Sample, In-Time). Nachdem die Parameter des funktionalen Zusammenhangs zwischen den Kennzahlen und der Aktienkursrendite geschätzt wurden, wurden die Jahresabschlusskennzahlen des Geschäftsjahres 1991, die am 31.03.92 veröffentlicht wurden, in das Modell eingesetzt und somit die eigentliche Renditeprognose für das Jahr 1992/1993 erstellt (Out-of-Sample, Out-of-Time).

3 Modelle

Für die vorliegende Problemstellung wurde als Netzwerkmodell eine Feedforward-Architektur des Multilayer-Perceptrons ausgewählt, die mit Hilfe eines Backpropagation-Algorithmus trainiert wurde (vgl. [5] S.288ff). Das Modell besteht aus einer Input-, einer Hidden- und einer Outputschicht. Die Dimension der Inputschicht ist durch die Anzahl der erklärenden Variablen bestimmt. Die Hiddenschicht wurde mit 5 Neuronen ausgestattet. Die Outputschicht besteht aus einem Neuron.

Die Schätzmenge wurde ferner in Trainings- und Validierungsmenge unterteilt. Die Validierungsmenge wird nicht zum Training herangezogen, sondern aufgrund dieser Daten wird die Entwicklung des Schätzfehlers überwacht. Zur Evaluierung der Modelle wurden zwei Typen von Generalisierungsmengen benutzt (vgl. [5] S.435): Out-of-Sample, In-Time sowie Out-of-Sample, Out-of-Time (siehe oben).

Während des Trainings wurde Overlearning dadurch diagnostiziert, dass der Fehler auf Grundlage der Validierungsmenge anstieg, während er auf Grundlage der Trainingsmenge bereits gesunken war. Zur Optimierung des Netzes und um die Komplexität des Neuronalen Netzes zu reduzieren, sowie Overlearning einzugrenzen, wurde versucht, die Anzahl der Verbindungen zwischen den Neuronen zu reduzieren (Weight-Pruning). Dabei wurde die Methode des Statistischen Prunens angewendet (vgl. [10] S.70).

Um zu beurteilen, ob sich der Aufwand zur Entwicklung der Modelle mit Neuronalen Netzen überhaupt lohnt und um zu prüfen, ob zwischen den Kennzahlen und Renditen eher ein linearer oder nichtlinearer Zusammenhang vorliegt, wurden auch lineare Regressionsmodelle gebildet. Zur Überprüfung der Güte der Prognosemodelle wurde als Benchmark eine naive Prognose in Form der Same-Change-Variante angewendet (vgl. [5] S.456). Als geschätzte Werte der Renditen wurden dabei einfach die Renditen des vergangenen Jahres angenommen.

Zur Beurteilung der Güte der Modelle wurden die Schätzungen mit den tatsächlich eingetretenen Realisationen verglichen. Zur Evaluierung der Modelle wurden die folgenden vier Gütemaße benutzt: Trefferquote (TQ), Mean Squared Error (MSE), Rangkorrelationskoeffizient (rr) und Separationsmaß (SEP) (zur Beschreibung der Gütemaße siehe [6] S.439, [9] S.159f).

Auf Basis der Prognose aller Modelle wurden Aktienportfolios gebildet. Für diesen Zweck wurden aufgrund der Prognoseergebnisse die 25% der besten und die 25% der schlechtesten Aktien ausgewählt. Weiterhin wurden aus den 25% der besten Aktien nur diejenigen ausgesucht, die positiv prognostiziert wurden. Auf gleiche Weise wurden aus den 25% der schlechtesten Aktien diejenigen Aktien mit negativ prognostizierten Renditen selektiert. Aus diesen Aktien wurde pro Jahr und Prognosemodell jeweils ein gleichgewichtetes Portfolio gebildet, bei dem eine Long-Position für die besten Aktien und eine Short-Position für die schlechtesten Aktien eingenommen wurden. Zum Performancevergleich wurden die Portfoliorenditen berechnet, die sich bei einer wirklichen Realisation dieser Portfolios ergeben hätten.

4 Ergebnisse

Tabelle 1 enthält die Ergebnisse hinsichtlich der Gütemaße (Out-of-Sample, In-Time) der Untersuchung des Zusammenhangs zwischen Jahresabschlusskennzahlen und Aktienkursrendite. Es lässt sich allgemein sagen, dass die Ergebnisse sehr enttäuschend sind. Die Trefferquoten schwanken zwischen 29,2% und 63,2%, die Rangkorrelationskoeffizienten zwischen -0,303 und 0,271. Das Separationsmaß SEP deutet fast in allen Jahren auf keine signifikante Trennfähigkeit hin. Nur mit Modellen der Neuronale Netze der Jahre 1993 und 2000 lassen sich gute Ergebnisse erzielen. Nach der Betrachtung der einzelnen Modelle wurde festgestellt, dass in verschiedenen Jahren jeweils unterschiedliche Kennzahlen selektiert wurden.

Modell	Jahr	TQ	MSE	SEP	rr	Jahr	TQ	MSE	SEP	rr
Lin.Regr	1991	0,459	0,145	0,795	0,063	1996	0,436	0,183	0,384	0,178
NNetz		0,625	1,941	1,465	0,251		0,568	1,342	0,544	0,182
Lin.Regr	1992	0,447	0,902	-0,424	0,041	1997	0,462	0,188	1,367	0,078
NNetz		0,514	1,289	0,272	0,159		0,543	2,529	-0,512	0,128
Lin.Regr	1993	0,436	1,893	-1,049	-0,109	1998	0,392	0,077	-0,489	0,080
NNetz		0,632	1,572	1,972	0,265		0,550	0,678	1,122	0,112
Lin.Regr	1994	0,462	0,309	0,671	0,086	1999	0,390	0,137	1,255	0,266
NNetz		0,476	0,958	0,585	0,162		0,474	0,969	0,643	0,032
Lin.Regr	1995	0,590	0,185	0,491	0,048	2000	0,292	1,984	-1,367	-0,303
NNetz		0,581	0,966	1,343	0,125		0,615	0,854	1,639	0,527

Tabelle 1. Gütermaße basierend auf der Testmenge (Out-of-Sample, In-Time)

In der Tabelle 2 sind die Prognoseergebnisse der Modelle für die Renditen aus Platzgründen nur für Jahre 1997/1998 bis 2000/2001 dargestellt (in der Tabelle als St1 bezeichnet), für welche die Jahresabschlusskennzahlen der Geschäftsjahre 1991 bis 1999 als erklärende Variablen bei der Prognose eingesetzt wurden (Out-of-Sample, Out-of-Time). Dabei wurden zum Vergleich auch die Ergebnisse der Studie *Petersmeier* (vgl. [4] S.412-413) angegeben (in der Tabelle als St2 bezeichnet). Die Ergebnisse sind auch wie bei der ersten Untersuchung als enttäuschend anzusehen, obwohl diejenigen der Modelle von 1992/1993 und 1995/1996 bessere Werte besitzen. Die Rangkorrelationsmaße liegen fast bei allen Modellen in diesen Jahren über 0,1. Die Trefferquoten liegen um 0,5; das Modell Neuronaler Netze im Jahr 1995/1996 hat jedoch 60% erreicht. Alle Modelle in diesen Jahren führen zu Portfolios, die eine positive Rendite aufweisen, und die Modelle Neuronaler Netze überboten die Rendite der Benchmarkmodelle.

Jr	Modell	Ren.		TQ		MSE		SEP		rr	
		St1	St2	St1	St2	St1	St2	St1	St2	St1	St2
97/98	Benchmark	12,4	14,2	0,58	0,57	0,27	7,07	2,85	0,09	0,22	0,19
	Lin.Regression	9,65	14,2	0,54	0,56	0,24	7,05	0,59	-0,9	0,13	0,02
	NNetz/Nichtp.R.	7,61	21,8	0,46	0,53	2,33	7,04	4,36	-0,5	-0,1	0,09
98/99	Benchmark	3,42	2,62	0,47	0,57	0,70	7.03	0,83	0,88	0,04	0,04
	Lin.Regression	-7,7	-0,1	0,50	0,52	0,48	1,62	2,96	-0,2	-0,1	0,01
	NNetz/Nichtp.R.	-8,5	4,32	0,53	0,60	2,24	0,54	2,96	0,20	0,05	0,05
99/00	Benchmark	-1,6	8,68	0,54	0,53	0,67	0,74	-0,4	1,62	-0,1	0,13
	Lin.Regression	-0,2	5,60	0,56	0,51	0,67	0,58	2,98	0,64	-0,1	-0,1
	NNetz/Nichtp.R.	-4,1	-7,3	0,60	0,52	2,85	0,67	2,98	0,09	-0,1	-0,1
00/01	Benchmark	-0,5	-1,9	0,49	0,47	0,41	0,87	-0,1	-0,5	0,05	-0,1
	Lin.Regression	4,9	-8,1	0,57	0,40	0,14	0,59	6,23	-2,6	0,25	-0,1
	NNetz/Nichtp.R.	-1,7	-7,1	0,48	0,41	1,62	0,64	6,23	-2,1	0,01	-0,1

Tabelle 2. Gütemaße der Prognosemodelle (Out-of-Sample, Out-of-Time)

Zusammenfassend lässt sich festhalten, dass das Benchmarkmodell am häufigsten zu den höheren Renditen geführt hat. Das lineare Regressionsmodell erweist sich nur im Jahr 2000/2001 als geeignetes Prognosemodell. Hinsichtlich der Rendite zeigt sich das Modell Neuronaler Netze in den Jahren 1992/1993, 1994/1995 und 1995/1996 als prognosegeeignet. Über die Vorteilhaftigkeit der Modelle lassen sich also keine generellen Aussagen formulieren. Die Modelle der Neuronalen Netzen oder der Nichtparametrischen Regression liefern keine systematisch besseren Ergebnisse, wodurch keines der beiden Verfahren als dominierend bezeichnet werden kann. Bezüglich der Trefferquote unterscheiden sich die Ergebnisse höchstens um 0,076. Lediglich in Jahren 1994/1995 und 1998/1999 erzielen die Modelle vollkommen unterschiedliche Renditewerte.

5 Zusammenfassung

In diesem Beitrag wurde untersucht, ob Jahresabschlusskennzahlen zur Erklärung und Prognose von Aktienkursrenditen geeignet sind. Es wurde, wie auch in der Studie von *Petersmeier* (vgl. [4] S.363ff), festgestellt, dass mit den genannten

Kennzahlen und den ausgewählten Modellen keine leistungsstarke Prognose erzielt werden kann. Die Gründe dafür können in dem geringen Umfang des Datenmaterials liegen. Besonders für Neuronale Netze ist die maximale Anzahl mit 200 Jahresabschlüssen sehr klein.

Außerdem ist es fraglich, ob die Jahresabschlussinformationen durch die untersuchten Jahresabschlusskennzahlen in vollem Maß erfasst wurden. Es bleibt zu überprüfen, ob sich die Prognose bei der Einbeziehung weiterer Einflussgrößen in die Untersuchung verbessern wird.

Die hier betrachteten Modelle sind statisch aufgebaut, d.h. es werden die Einflüsse der Jahresabschlusskennzahlen auf die Aktienkursrendite nur aus den Daten eines Jahres berücksichtigt. Deswegen wäre es sinnvoll ein Modell zu entwickeln, das die in den Jahresabschlusskennzahlen versteckten Informationen über mehrere Jahre, also dynamisch, betrachtet.

Literaturverzeichnis

1. Baetge J, Kirsch HJ, Thiele S (2001) Bilanzen, 5. Auflage, Düsseldorf
2. Booth GG, Loistl O (1998) "Aktienkursprognosen auf der Basis von Jahresabschlussdaten", Handbuch Portfoliomanagement (Hrsg. J.M. Kleeberg und H. Rehkugler), Bad Soden/Ts., S. 297-313
3. Coenenberg AG (2000) Jahresabschluss und Jahresabschlussanalyse, 17. Auflage, Landsberg/Lech
4. Petersmeier K (2003) Kerndichte- und Kernregressionsschätzungen im Asset Management, Bad Soden/Ts
5. Poddig T (1999) Handbuch Kursprognose: quantitative Methoden im Asset Management, Bad Soden/Ts
6. Poddig T, Dichtl H, Petersmeier K (2003) Statistik, Ökonometrie, Optimierung-Methoden und praktische Anwendungen in Finanzanalyse und Portfoliomanagement, 3. erweiterte Auflage, Bad Soden/Ts
7. Verlag Hoppenstedt (1997) Hoppenstendt CD-ROM Bilanzdatenbank (Handbuch), Darmstadt
8. Wallmeier M (2000) „Determinanten erwarteter Renditen am deutschen Aktienmarkt- Eine empirische Untersuchung anhand ausgewählter Kennzahlen", Zeitschrift für Betriebswirtschaftliche Forschung 52,2:27-57
9. Wolberg JR (2000) Expert Trading System, New York
10. Zimmermann HG (1994) „Neuronale Netze als Entscheidungskalkül", Neuronale Netze in der Ökonometrie: Grundlagen und finanzwirtschaftliche Anwendungen (Hrsg. H. Rehkugler und H. G. Zimmermann), München, S. 1-88

Part VIII

Discrete and Combinatorial Optimization

On the Computational Performance of a Semidefinite Programming Approach to Single Row Layout Problems *

Miguel F. Anjos[1] and Anthony Vannelli[2]

[1] Department of Management Sciences, University of Waterloo, Waterloo, Ontario, Canada N2L 3G1 anjos@stanfordalumni.org
[2] Department of Electrical & Computer Engineering, University of Waterloo, Waterloo, Ontario, Canada N2L 3G1 vannelli@cheetah.vlsi.uwaterloo.ca

1 Introduction

The single-row layout problem (SRLP) is concerned with the arrangement of a given number of rectangular facilities next to each other along a line so as to minimize the total weighted sum of the center-to-center distances between all pairs of facilities. This problem is a special case of the unequal-area facility layout problem, and is also known in the literature as the one-dimensional space allocation problem, see e.g. [12]. An instance of the SRLP consists of n one-dimensional facilities, denoted $1, \ldots, n$, with given positive lengths $\ell_1, \ldots, \ell_n$, and pairwise weights c_{ij}. The objective is to arrange the facilities so as to minimize the total weighted sum of the center-to-center distances between all pairs of facilities. If all the facilities have the same length, the SRLP becomes a special case of the quadratic assignment problem, see e.g. [11]. Several applications of the SRLP have been identified in the literature. One such application arises in the area of flexible manufacturing systems, where machines within manufacturing cells are often placed along a straight path travelled by an automated guided vehicle, see e.g. [10]. Furthermore, the SRLP is closely related to the linear ordering problem, which also has a number of practical applications, see e.g. [5, 6, 13].

The SRLP was first studied by Simmons [14] who proposed a branch-and-bound algorithm. Subsequently, Picard and Queyranne [12] developed a dynamic programming algorithm, and mixed integer linear programming models have also been proposed, most recently in [1]. While these algorithms

* Research partially supported by grant NAL/00636/G from the Nuffield Foundation (UK), grants 312125, 314668 and 15296 from the Natural Sciences and Engineering Research Council of Canada, and a Bell University Laboratories Research Grant.

are guaranteed to find the global optimal solution, they have very high computational time and memory requirements, and are unlikely to be effective for problems with more than about 20 facilities. Several heuristic algorithms for the SRLP have also been proposed. We refer the reader to the recent paper of Solimanpur et al. [15], and the references therein. However, these heuristic algorithms do not provide a guarantee of global optimality, or even some measure of the possible distance from optimality. Progress in obtaining such a measure was recently reported in [2], where non-trivial global lower bounds were obtained using a semidefinite programming relaxation.

Semidefinite programming (SDP) refers to the class of optimization problems where a linear function of a symmetric matrix variable X is optimized subject to linear constraints on the elements of X and the additional constraint that X must be positive semidefinite. This includes linear programming problems as a special case, namely when the matrix variable is diagonal. A variety of algorithms for solving SDP problems, including polynomial-time interior-point algorithms, have been implemented and benchmarked, and several excellent solvers for SDP are now available. We refer the reader to the SDP webpage [7] as well as the books [4, 16] for a thorough coverage of the theory and algorithms in this area, as well as of several application areas where SDP researchers have made significant contributions.

The application of SDP to the SRLP was initiated in the aforementioned paper [2] which proposed an SDP relaxation as well as a heuristic which extracts a feasible solution to the SRLP from the optimal matrix solution to the SDP relaxation. Therefore, this SDP-based approach yields both a feasible solution to the given SRLP instance as well as a guarantee of how far it is from global optimality. The majority of the results reported in [2] were for fairly large instances, and were obtained using the spectral bundle solver SB [8, 9] which is able to handle very large SDPs, but with the drawback that it must run for several hours. The results reported in [2] showed that the SDP-based approach yields layouts that are consistently a few percentage points from global optimality for randomly generated instances with up to 80 facilities.

In this paper, we further study the computational performance of the SDP relaxation by considering smaller problems for which the SDP relaxation can be solved using an interior-point solver. This allows us to test a branch-and-bound (B&B) algorithm for the SRLP that solves the SDP relaxation from [2] at each node. Our results show that it is possible to compute solutions that are provably very close to global optimality (typically less than 1% gap) for randomly generated instances of the SRLP with up to 40 facilities with a reasonable amount of computational effort. More interestingly, the results also show that branching does not provide a significant improvement in the results obtained at the root node of the B&B tree, suggesting that although the SDP approach efficiently computes layouts that are provably very close to optimality, it would require a significant increase in computational effort to attain provable global optimality for this problem.

2 The Semidefinite Programming Relaxation

Let $\pi = (\pi_1, \ldots, \pi_n)$ denote a permutation of the indices $[n] := \{1, 2, \ldots, n\}$ of the facilities, so that the leftmost facility is π_1, the facility to the right of it is π_2, and so on, with π_n being the last facility in the arrangement. Given a permutation π and two distinct facilities i and j, the center-to-center distance between i and j with respect to this permutation is $\frac{1}{2}\ell_i + D_\pi(i,j) + \frac{1}{2}\ell_j$, where $D_\pi(i,j)$ denotes the sum of the lengths of the facilities between i and j in the ordering defined by π. The problem is therefore to

$$\min_{\pi \in \Pi} \sum_{i<j} c_{ij} \left[\frac{1}{2}\ell_i + D_\pi(i,j) + \frac{1}{2}\ell_j \right]$$

where Π denotes the set of all permutations of $[n]$.

Simmons [14] observed that if we rewrite the objective function as

$$\min_{\pi \in \Pi} \sum_{i<j} c_{ij} D_\pi(i,j) + \sum_{i<j} \frac{1}{2} c_{ij}(\ell_i + \ell_j)$$

where the second summation is a constant independent of π, then it is clear that the crux of the problem is to minimize $\sum_{i<j} c_{ij} D_\pi(i,j)$ over all permutations π. Furthermore, it is clear that $D_\pi(i,j) = D_{\pi'}(i,j)$, where π' denotes the permutation symmetric to π, defined by $\pi'_i = \pi_{n+1-i}, i = 1, \ldots, n$. This shows that we can exchange the left and right ends of the layout and obtain the same objective value. Hence, it possible to simplify the problem by considering only the permutations for which, say, facility 1 is on the left half of the arrangement. This type of symmetry-breaking strategy is important for reducing the computational requirements of most algorithms, including those based on linear programming or dynamic programming. One noteworthy aspect of the SDP-based approach is that it implicitly accounts for these symmetries, and thus does not require the use of additional explicit symmetry-breaking constraints.

The SDP relaxation for the SLRP proposed in [2] is obtained as follows. Define a binary ± 1 variable for each pair i, j of facilities with $i < j$ such that

$$R_{ij} := \begin{cases} 1, & \text{if facility } i \text{ is to the right of facility } j \\ -1, & \text{if facility } i \text{ is to the left of facility } j \end{cases}$$

In this definition, the order of the subscripts matters, and $R_{ij} = -R_{ji}$. To accurately formulate the SRLP, it is further required that the R_{ij} variables represent a valid arrangement of the n facilities. Therefore we require that if $R_{ij} = R_{jk}$ then $R_{ik} = R_{ij}$, a necessary transitivity condition which can be formulated as a set of quadratic constraints:

$$R_{ij}R_{jk} - R_{ij}R_{ik} - R_{ik}R_{jk} = -1 \text{ for all triples } i < j < k. \tag{1}$$

This leads to the following formulation of the SLRP:

$$\min K - \sum_{i<j} \frac{c_{ij}}{2} \left[\sum_{k<i} \ell_k R_{ki} R_{kj} - \sum_{i<k<j} \ell_k R_{ik} R_{kj} + \sum_{k>j} \ell_k R_{ik} R_{jk} \right]$$

s.t.

$$R_{ij} R_{jk} - R_{ij} R_{ik} - R_{ik} R_{jk} = -1 \text{ for all triples } i < j < k$$
$$R_{ij}^2 = 1 \text{ for all } i < j$$

where $K := \left(\sum_{i<j} \frac{c_{ij}}{2} \right) \left(\sum_{k=1}^{n} \ell_k \right)$. Note that if every R_{ij} variable is replaced by its negative, then there is no change whatsoever to the formulation. This is how our formulation, and the subsequent SDP relaxation, implicitly take into account the natural symmetry of the SRLP.

To formulate the SLRP in the space of real symmetric matrices, let P denote the set of all pairs (i, j) such that $i < j$. Fixing an ordering of the elements of P, we can define the vector

$$v := (R_{p_1}, \ldots, R_{p_{\binom{n}{2}}})^T,$$

where each p_k denotes a distinct element of P. Using v, we construct the rank-one matrix $X := vv^T$ whose rows and columns are indexed by P according to the ordering fixed above. By construction, $X_{p_i, p_j} = R_{p_i} R_{p_j}$ for all $p_i, p_j \in P$, and therefore we can formulate the SRLP as:

$$\min K - \sum_{i<j} \frac{c_{ij}}{2} \left[\sum_{k<i} \ell_k X_{ki,kj} - \sum_{i<k<j} \ell_k X_{ik,kj} + \sum_{k>j} \ell_k X_{ik,jk} \right]$$

s.t.

$$\begin{aligned} & X_{ij,jk} - X_{ij,ik} - X_{ik,jk} = -1 \text{ for all triples } i < j < k \\ & \operatorname{diag}(X) = e \\ & \operatorname{rank}(X) = 1 \\ & X \succeq 0 \end{aligned} \tag{2}$$

where $\operatorname{diag}(X)$ represents a vector containing the diagonal elements of X, e denotes the vector of all ones, and $X \succeq 0$ denotes that X is symmetric positive semidefinite. Removing the rank constraint yields the SDP relaxation. Note that in general the SDP problem only provides a lower bound on the optimal value of the SRLP, and not a feasible solution, unless the optimal matrix X^* happens to have rank equal to one. A rounding scheme specific to this problem was proposed in [2], and is discussed in the next Section.

3 Algorithm Description and Computational Results

The algorithm we report on here is a standard B&B algorithm that solves the SDP relaxation at every node. After solving each SDP, we apply the following extension of the rounding procedure from [2]. If X^* is the optimal solution to the SDP relaxation, then each row of X^* corresponds to a specific pair of (i_1, j_1) of facilities. Therefore, for each row, if we set $R_{i_1 j_1} = +1$, then we can scan the other entries of the row and assign the value $X_{i_1 j_1, i_2 j_2}$ to the variable R_{i_2, j_2}, for every pair $(i_2, j_2) \neq (i_1, j_1)$. Using these values, we compute

$$\omega_k = \frac{1}{2}\left(n + 1 + \sum_{j \neq k} R_{kj}\right)$$

for $k = 1, \ldots, n$, and hence obtain a permutation of $[n]$ by sorting these (in decreasing or increasing order, whichever satisfies $R_{i_1 j_1} = +1$). This approach improves on [2] by considering every row, rather than only the first row, thus obtaining a greater number of candidate layouts[3]. The best layout found so far is updated after each use of this heuristic.

After solving the relaxation and applying the rounding procedure at a node of the B&B tree, if the optimal solution X^* to the SDP problem is not rank-one, and the remaining subtree cannot be pruned, then the algorithm branches as follows. First, with $p_1 = (1, 2)$ being the pair corresponding to the first entry of v in the construction of the SDP, we set $R_{12} = +1$, and scan the first row of X^*, assigning value $X_{12,ij}$ to the variable R_{ij}, for every pair $(i, j) \neq (1, 2)$. Second, among all the variables R_{ij} not yet fixed and with absolute value less than 0.5 according to the assignment we just made, we choose the variable for which c_{ij} is largest. Any remaining ties are broken arbitrarily. This strategy is motivated by the fact that for the rank-one matrices, all the entries equal ± 1; hence entries with small magnitude are less desirable.

All the computational results were obtained on a 2.0GHz Dual Opteron with 16Gb of RAM, and the SDP problems were solved using the interior-point solver CSDP [3]. We ran the algorithm with a time limit of 3600 seconds, and with the possibility of exceeding this limit slightly if there was an SDP being solved when the time limit was reached, in which case the solver was allowed to run to completion. We solved 20 randomly generated instances of the SRLP for each of $n = 25, 30, 35, 40$, and the averages over the 20 instances for each value of n are reported in Table 1.

Our results suggest that using the algorithm described, the SDP-based approach is able to compute solutions that are provably very close to global optimality, typically with a gap of less than 1%, for randomly generated instances of the SRLP with up to 40 facilities. The results also show that branching does not provide a significant improvement in the results obtained at the root node of the B&B tree, suggesting that although the SDP approach efficiently computes layouts that are provably very close to optimality, it would require a significant increase in computational effort to attain provable global optimality for this problem. This last observation clearly illustrates the hardness of the SRLP.

[3] We thank Robert J. Vanderbei for suggesting this extension of the heuristic.

Table 1. Summary of computational results

Size of instances (n)	Time to solve root SDP (sec)	Gap to best layout found at root	Levels below the root completed within 1 hour	Total time to fully solve the completed levels	Gap to best layout found after total time
25	60.4	0.66%	3	2761.9	0.44%
30	260.3	1.11%	2	2498.7	0.97%
35	808.0	0.89%	1	3309.4	0.81%
40	2900.9	0.85%	0	2900.9	0.85%

References

1. A.R.S. Amaral. On the exact solution of a facility layout problem. *Eur. J. Oper. Res.*, to appear.
2. M.F. Anjos, A. Kennings, and A. Vannelli. A semidefinite optimization approach for the single-row layout problem with unequal dimensions. *Discr. Opt.*, 2(2):113–122, 2005.
3. B. Borchers. CSDP, a C library for semidefinite programming. *Optim. Methods Softw.*, 11/12(1-4):613–623, 1999.
4. E. de Klerk. *Aspects of Semidefinite Programming*, volume 65 of *Applied Optimization*. Kluwer Academic Publishers, Dordrecht, 2002.
5. M. Grötschel, M. Jünger, and G. Reinelt. A cutting plane algorithm for the linear ordering problem. *Oper. Res.*, 32(6):1195–1220, 1984.
6. M. Grötschel, M. Jünger, and G. Reinelt. Facets of the linear ordering polytope. *Math. Program.*, 33(1):43–60, 1985.
7. C. Helmberg. `http://www-user.tu-chemnitz.de/~helmberg/semidef.html`.
8. C. Helmberg and K.C. Kiwiel. A spectral bundle method with bounds. *Math. Program.*, 93(2, Ser. A):173–194, 2002.
9. C. Helmberg and F. Rendl. A spectral bundle method for semidefinite programming. *SIAM J. Optim.*, 10(3):673–696 (electronic), 2000.
10. S.S. Heragu and A. Kusiak. Machine layout problem in flexible manufacturing systems. *Oper. Res.*, 36(2):258–268, 1988.
11. W. Liu and A. Vannelli. Generating lower bounds for the linear arrangement problem. *Discrete Appl. Math.*, 59(2):137–151, 1995.
12. J.-C. Picard and M. Queyranne. On the one-dimensional space allocation problem. *Oper. Res.*, 29(2):371–391, 1981.
13. G. Reinelt. *The linear ordering problem: algorithms and applications*, volume 8 of *Research and Exposition in Mathematics*. Heldermann Verlag, Berlin, 1985.
14. D.M. Simmons. One-dimensional space allocation: An ordering algorithm. *Oper. Res.*, 17:812–826, 1969.
15. M. Solimanpur, P. Vrat, and R. Shankar. An ant algorithm for the single row layout problem in flexible manufacturing systems. *Comput. Oper. Res.*, 32(3):583–598, 2005.
16. H. Wolkowicz, R. Saigal, and L. Vandenberghe, editors. *Handbook of Semidefinite Programming*. Kluwer Academic Publishers, Boston, MA, 2000.

On Some Probability Inequalities for Some Discrete Optimization Problems *

Edward Kh. Gimadi

Sobolev Institute of Mathematics, prospekt Akademika Koptyuga 4, 630090 Novosibirsk, Russia

Summary. Probabilistic analysis of algorithms for discrete optimization problems is the subject of many recent investigations (Karp, Frieze, Radhavan at al). Probability inequalities of well known researchers (Chernov, Hoefding at al) are widely exploited in this works. The paper aims at demonstrating some results of probabilistic analysis of algorithms for a number of well known combinatorial problems that use another efficient (productive) probability inequalities (of Bernstein, Borovkov, Petrov, etc.). The results deal with proofs of asymptotical optimality of polynomial approximation algorithms for solving the traveling salesman problem, the bin packing problem, the multi-index assignment problem, the uncapacitated facility location problem, the problem of finding the regular connected subgraph of maximum weight

1 Introduction

Most of discrete optimization problems are NP–hard. Because it is easy to understand numerous attempts to construct fast approximation algorithms solving the problem and to obtain performance estimates of these algorithms. We single out complexity (running-time), relative error and failure probability as the main performance characteristics of an algorithm. Estimates for two latter characteristics are obtained under the assumption that a probabilistic distribution from distribution class is given in the set of all individual problems (i. e. inputs). Thus these characteristics and their estimates depend especially on a distribution class. Apparently the first serious probabilistic analysis was performed by A. Borovkov [1] for solving two discrete optimization problems well known as Traveling Salesman Problem (TSP) and Assignment Problem (AP).

* This research was supported by the Russian Foundation for Basic Research (grant 05-01-00395), program of supporting of leading science schools of Russia (project "Nauchnaya Shkola - 313.2003.1"), and INTAS (grant 04-77-7173)

An approximation algorithm $\mathcal{A}$ has *bounds of relative error* ε_n and *failure probability* δ_n, if: $Pr\left\{\left|\frac{F_{\mathcal{A}}-F^*}{F^*}\right| \geq \varepsilon_n\right\} \leq \delta_n$. An algorithm $\mathcal{A}$ is called *asymptotically optimal*, if there exist sequences $\{\varepsilon_n\} \to 0,\ \ \{\delta_n\} \to$ as $n \to \infty$.

We investigate an asymptotically optimal approach for the solving of well known discrete optimization problems: the traveling salesman problem, the bin packing problem, the multi-index assignment problem, the uncapacitated facility location problems, the problem of finding the regular connected subgraph of maximum weight.

2 Probabilistic analysis of approximation algorithms for solving TSP

The classical Traveling Salesman Problem (TSP) [17] is to find the minimum length route through n cities. A number of results concerning the algorithm "Nearest City" (NC) is obtained under the assumption that elements of the $n \times n$ distance matrix (c_{ij}) are identically distributed independent random variables.

In [15] it is shown that in the case of the discrete distribution function $p_k = Pr\{c_{ij} = k\},\ k = 1,\ldots,r_n,$ the algorithm NC is asymptotically optimal if $\sum_{k=1}^{r_n}(\sum_{i=1}^{k} p_i)^{-1} = o(n)$. In case of continuous random variables $c_{ij} \in [a_n, b_n],\ \ a_n > 0$ the algorithm is asymptotically optimal if $b_n/a_n = o\Big(n/\max\{n\gamma_n, J_n\}\Big)$, and $J_n = \int_{\gamma_n}^{1} \frac{dx}{P(x)} \to \infty$, where $P(x) = Pr\{(c_{ij} - a_n)/(b_n - a_n) < x\}$, γ_n is a root of the equation $P(x) = 1/n,\ 0 \leq x \leq 1$ [11]. Therefore it is clear that for uniform distribution the algorithm NC is asymptotically optimal if $b_n/a_n = o(n/\log n)$.

In [9] it is shown that for the Max TSP with the continuous distribution function $P(\xi)$ the "Farthest City" (FC) algorithm is an asymptotically optimal if $\int_0^{1-1/n} \frac{dx}{1-P(x)} = o(n)$. In the case $P(x) \leq x$ the algorithm FC has performance bounds $\varepsilon_n = \frac{2(\ln n+1)}{n},\ \ \delta_n = \frac{1}{\ln n+1}$, i.e. in this case FC is asymptotically optimal without any additional conditions.

These results were obtained using the classical Chebyshev's probabilistic inequality. Better approximation guarantees can be get by techniques initiated by Chernov [4] and generalized by Hoeffding [14]. But the best bounds are investigated using like-wise Bernstein probability inequalities [16].

3 Using improved probabilistic inequalities for solving TSP in the case of uniform distribution

The result of the section relies on Petrov's theorem [16]:

Theorem 1. *Consider independent random variables $X_1, \ldots, X_n$ and let $S = \sum_{k=1}^{n} X_k$. Let there be positive constants $g_1, \ldots, g_n$ and T such that for all $0 \le t \le T$ $\mathbf{E}e^{tX_k} \le \exp\left\{\frac{g_k t^2}{2}\right\}(k = 1, \ldots, n)$. Let $G = \sum_{k=1}^{n} g_k$. Then $\mathbf{Pr}\{S > x\} \le \exp\left\{-\frac{x^2}{2G}\right\}$, if $0 \le x \le GT$, and $\mathbf{Pr}\{S > x\} \le \exp\left\{-\frac{Tx}{2}\right\}$, if $x \ge GT$.*

We will also need the result of the following lemma:

Lemma 1. *Let ξ_k be a minimum between k random variables with identical distribution function $\mathbf{Pr}\{\xi < x\} = x,\ 0 \le x \le 1$. Then for all $y \ge \left(\frac{11\pi^2}{5} - \frac{37}{12}\right)$ and a sum $S = \sum_{k=1}^{n} \widetilde{\xi}_k$ of variables $\widetilde{\xi}_k = \xi_k - \mathbf{E}\xi_k,\ k = 1, \ldots, n$ the following inequality holds: $\mathbf{Pr}\{S \ge y\} \le \exp\{-\frac{3y}{2}\}$.*

We need some auxiliary facts which are valid for the uniform distribution function of ξ_k : $F_{\xi_k}(x) = 1 - (1-x)^k$.

Obviously ξ_k be the first order statistic in the set of k independent random variables uniformly distributed in $[0,1]$. Note that $\mathbf{E}\xi_k = 1/(k+1)$.

Lemma 2. *In the case $k \ge 1$ we have* $\mathbf{E}e^{t\xi_{(k+1)}} = \frac{k+1}{t}(\mathbf{E}e^{t\xi_k} - 1)$.

Proof. By definition $\mathbf{E}e^{t\xi_{k+1}} = \int_0^1 e^{tx} dF_{\xi_{k+1}}(x) = \int_0^1 e^{tx}(k+1)(1-x)^k dx = \frac{k+1}{t}(k\int_0^1 e^{tx}(1-x)^{k-1}dx - 1) = \frac{k+1}{t}(\mathbf{E}e^{t\xi_k} - 1)$.

Lemma 3. *For all $k \ge 1$ we have* $\mathbf{E}e^{t\xi_k} = \sum_{i=0}^{\infty} \frac{t^i}{\prod_{m=1}^{i}(k+1+m)}$.

Proof. The proof is by induction on k. In the case $k = 1$ we have

$$\mathbf{E}e^{t\xi_k} = \int_0^1 e^{tx}dx = \frac{e^t - 1}{t} = \sum_{i=0}^{\infty} t^i \frac{1}{(i+1)!}. \tag{1}$$

By the induction hypothesis the claim is proved for $k' = 1, \ldots, k$. For $k' = k+1$ using Lemma 3 we have $\mathbf{E}e^{t\xi_{k+1}} = \frac{k+1}{t}\left(\sum_{i=0}^{\infty} \frac{t^i}{\prod_{m=1}^{i}(k+m)} - 1\right) = \frac{k+1}{t}\sum_{i=1}^{\infty} \frac{t^i}{\prod_{m=1}^{i}(k+m)} = \sum_{i=0}^{\infty} \frac{t^i}{\prod_{m=1}^{i}(k+1+m)}$. Lemma 3 is proved.

Lemma 4. *Let $\widetilde{\xi}_k = \xi_k - \mathbf{E}\xi_k$, $1 \le k < n$, be centered random uniform variables. Then for all t, $0 \le t \le 3$, the following inequalities $\mathbf{E}e^{t\widetilde{\xi}_k} \le \exp\left\{\frac{1}{2}g_k t^2\right\}$ hold, where $g_k = 1/12, 11/36$, and $\frac{11/5}{(k+1)^2}$, if $k = 1$, $k = 2$, and $3, \ldots, n-1$ respectively.*

Proof. CASE $k = 1$. Using (1) we have $\mathbf{E}e^{t(\xi_k - \mathbf{E}\xi_k)} = e^{-\frac{t}{2}}\mathbf{E}e^{t\xi_k}$

$$= e^{-\frac{t}{2}}\frac{e^t - 1}{t} = \frac{e^{\frac{t}{2}} - e^{-\frac{t}{2}}}{t} = \sum_{i=0}^{\infty} \frac{1}{(2i+1)!}\left(\frac{t}{2}\right)^{2i} \le \sum_{i=0}^{\infty} \frac{1}{i!}\left(\frac{t^2}{24}\right)^i \le \exp\left\{\frac{t^2}{24}\right\}.$$

CASE $k = 2$. We have $\mathbf{E}e^{t(\xi_k - \mathbf{E}\xi_k)} = e^{-\frac{t}{3}}\mathbf{E}e^{t\xi_k}$. Using (3) we get $\mathbf{E}e^{t\xi_2} = 2(e^t - 1 - t)/t^2$. So we obtain $\mathbf{E}e^{t(\xi_k - \mathbf{E}\xi_k)} = 2e^{-\frac{t}{3}}\frac{e^t-1-t}{t^2}$

$$= 2e^{-\frac{t}{3}}\sum_{i=2}^{\infty}\frac{t^{i-2}}{i!} = e^{-\frac{t}{3}}\left(1+\frac{t}{3}+\frac{t^2}{12}\sum_{i=0}^{\infty}\left(\frac{t}{5}\right)^i\right) = e^{-\frac{t}{3}}\left(1+\frac{t}{3}+\frac{t^2}{12(1-t/5)}\right)$$

(and taking into account that $0 \le t \le 3$)

$$\le e^{-\frac{t}{3}}\left(1+\frac{t}{3}+\frac{t^2}{18}\right)\left(1+\frac{11t^2}{72}\right) \le e^{-\frac{t}{3}}e^{\frac{t}{3}}\left(1+\frac{11t^2}{72}\right) \le \exp\left\{\frac{1}{2}\cdot\frac{11}{36}\cdot t^2\right\}.$$

CASE $k \ge 3$. Putting $\alpha = \frac{t}{k+1}$ for $0 \le t \le 3$ we have

$$Ee^{tX_{(1)}(k)} \le 1+\alpha+\alpha^2\frac{(k+1)}{(k+2)}\sum_{i=0}^{\infty}\left(\frac{t}{k+3}\right)^i = 1+\alpha+\alpha^2\frac{(k+1)}{(k+2)(1-t/(k+3))}$$

$$\le 1+\alpha+\frac{8}{5}\alpha^2 \le \exp\left\{\alpha+\frac{11\alpha^2}{10}\right\} = e^{\frac{t}{k+1}}\exp\left\{\frac{1}{2}\cdot\frac{11/5}{(k+1)^2}\,t^2\right\}.$$

Therefore $\mathbf{E}e^{t(\xi_k - \mathbf{E}\xi_k} \le \exp\left\{\frac{1}{2}\cdot\frac{11/5}{(k+1)^2}\,t^2\right\}$. Lemma 4 is proved.

Let us denote the sum of all g_k and sum of all ξ_k, $1 \le k < n$, by G and S respectively. Then

$$G = \frac{1}{12}+\frac{11}{36}+\frac{11}{5}\sum_{k=3}^{n-1}\frac{1}{(k+1)^2} \le \frac{1}{12}+\frac{11}{36}+\frac{11}{5}\left(\frac{\pi^2}{3}-1-\frac{1}{4}-\frac{1}{9}\right) < \frac{1}{3}\left(\frac{11\pi^2}{5}-\frac{37}{12}\right).$$

For these g_k, G, and $T = 3$ random variables $\widetilde{\xi_k}$ satisfy Petrov's Theorem. So $\mathbf{Pr}\{S \ge y\} \le \exp\left\{-\frac{3y}{2}\right\}$. The proof of Lemma 1 is complete.

Now we can obtain performance bounds for Algorithm NC. Indeed using Lemma 1 we have
$Pr\left\{\left|\frac{F_{\mathcal{A}}-F^*}{F^*}\right| \ge \varepsilon_n\right\} \le Pr\left\{F_{\mathcal{A}} \ge 1+\varepsilon_n)na_n\right\} \le Pr\left\{(n-1)a_n + (b_n - a_n)S + b_n \ge (1+\varepsilon_n)na_n\right\} \le Pr\left\{S \ge \frac{n\varepsilon_n}{b_n/a_n - 1} - 1\right\} \le \exp\left\{-\frac{3}{2}\left(\frac{n\varepsilon_n}{b_n/a_n-1}-1\right)\right\}$.

Now if we put $\varepsilon_n = (b_n/a_n - 1)(\ln n + 1)/n$, we have the following bounds $\varepsilon_n = O\left(\frac{b_n/a_n}{n/\ln n}\right)$, $\delta_n = n^{-3/2}$. Therefore we can conclude that Algorithm NC is asymptotically optimal if $b_n/a_n = o\left(n/\ln n\right)$.

4 Another results

A patching algorithm for solving TSP In [12] was investigated a randomized version of a known patching algorithm for solving TSP on a class of probability distributions in the set of all $n \times n$ -dimensional matrices, relative to which columns ξ_k of the random matrix $C = (\xi_1, \ldots, \xi_n)$ form a sequence of n–dimensional random variables with the symmetric joint distribution function. So distances considered are not independent. Using Petrov's theorem we obtain performance bounds $\varepsilon_n = O\Big(\frac{b_n/a_n}{n/\ln n}\Big)$, $\delta_n = \left(\frac{e}{n}\right)^{0.38}$, and the condition of the asymptotic optimality: $b_n/a_n = o\Big(n/\ln n\Big)$.

Bin packing problem One-dimensional bin packing problem [5] is to pack items of a list $\{w_i \mid i = 1, \ldots, n\}$ into the minimal number of bins of the identical capacity C so that the sum of item weights in each bin is at most of C. We suppose that $w_i \in \{1, 2, \ldots, C\}$. In [7],[13] using Borovkov inequality for sums of binary independent variables [2] were obtained the following performance bounds for the problem $\varepsilon_n = O\left(\sqrt{\frac{C/\rho}{n/\ln n}}\right)$, $\delta_n = o\left(\frac{1}{n \ln n}\right)$, where $\rho = 1$ if function p_r is C–asymmetric and $\rho = \sum_{r=1}^{C} r p_r / C$ in case C–regular function p_r. So the algorithms in [7],[13] are asymptotically optimal if $C/\rho = o\Big(n/\ln n\Big)$. Note that particular case of C–regular function is decreasing function.

Facility Location Problem (FLP) One of a possible formulation of the Ucapacitaited FLP [6] can be written in the the following form:

$$\sum_{i=1}^{m} g_i x_i + \sum_{j=1}^{n} \min_{i|x_i=1} c_{i,j} \rightarrow \min_{(x_i)},$$

where (c_{ij}) is a distance matrix, (g_i) is m–vector, $x_i \in \{0, 1\}$ $i = 1, \ldots, m$.

In [7]–[8] an approximation algorithm was constructed for the uniform distribution when $g_{min}(n)/(b_n - a_n) \geq \psi_n (\log n)^2/n$, $m = o(n^\lambda)$, where λ is a constant, $\lambda \geq 1$. In this case using Petrov's Theorem we prove that the solution gives the relative value of objective function $F_{\mathcal{A}}/F^*$ which is restricted from above by r_n such that $\lim_{n\to\infty} r_n = 1.5 \sqrt{g_{max}(n)/g_{min}(n)}$ with the probability tending to 1 as $n \to \infty$.

The problem of finding the d-regular connected subgraph of maximum weight The problem is generalization of TSP, when $d = 2$. In the case of random instances it is shown that the approximation algorithm in [3]is asymptotically optimal if weights of edges in graph are independent random variables with identical distribution function likewise $P(x) \leq x$, $0 \leq x \leq 1$. Probabilistic analysis gives the following estimates: $\varepsilon_n \leq \frac{3\ln(m+1)+1}{m}$, $\delta_n = \frac{1}{m+1}$, where $m = n/(d-1)$. So the condition of the asymptotic optimality is $d = o(n)$.

Multi-index Assignment Problem (MAP) The MAP is NP–hard for the number of indexes at least of three in axial and planar cases both.

In the axial three-index assignment problem n elements in the matrix must be selected such that in every "cross-section" exactly one element is chosen. The planar three-index assignment problem deals with selection of n^2 elements in a cubic matrix $(c_{ijk})_{n\times n\times n}$. One chooses exactly one element in each line.

(*A line* is the set of n elements with fixed pair of two indexes). The goal is to minimize the sum of the chosen elements.

For the axial MAP was obtained conditions of an asymptotic optimality likewise conditions for TSP. For the planar tree-index Assignment Problem analogous conditions of asymptotic optimality were obtained when the number of layers of the matrix (c_{ijk}) is at most of $O(\ln n)$ [10].

References

1. Borovkov A.A. On probabilistic formulation of two economic problems. Doklady AN SSSR, 1962, 146, N 5 (in Russian).
2. Borovkov A.A. Probability theory. Moscow: Nauka, 1976 (in Russian).
3. Baburin A.E., Gimadi E.Kh. Approximation algorithm for finding a maximum-weight spanning connected subgraph with given vertex degrees. Oper. Res. Proc. 2004, Springer-Verlag, 2005, 343–351.
4. Chernov H. A measure of asymptotic efficiency for tests of a hypothesis based on the sum of observations. Annals of Mathem. Statistics, 23, 1952, 493–509.
5. Coffman E.G., Garey M.R., Johnson D.S. Approximation algorithms for bin–packing — An updated survey. Algorithm Design for Computer System Design. CISM and lectures, 1984, *284,* 49–106.
6. Discrete Location Theory (1990). Ed. P.B.Mirchandani, R.L.Francis. Wiley-Interscience Publication. Wiley and Sons Inc.
7. Gimadi E.Kh. On some mathematical models and methods of planning large–scale projects. Trudy Instituta Matematiki. Sibirskoe Otdelenie, 1988, *10,* 89–115 (in Russian).
8. Gimadi E.Kh. Aprior performance estimates of the approximation solving for the Problem of Standardization. Upravlyaemye sistemy, Novosibirsk, 1987, *27,* 25–29 (in Russian).
9. Gimadi E.Kh. A maximum TSP: conditions of asymptotically optimality for "Visit farthest city" algorithm. Upravlyaemye sistemy. Novosibirsk, 1989, *29,* 11–15 (in Russian).
10. Gimadi E. Kh., Korkishko N.M. On some modifications of three index planar assignment problem. Proc. of the 2-nd Intern. Workshop "Discrete Optimization Methods in Production and Logistics" DOM'2004, Omsk, 2004, 161–165.
11. Gimadi E.Kh., Perepelitsa V.A. An asymptotical approach to solving the traveling salesman problem. Upravljaemye sistemy, Novosibirsk, 1974, *12,* 35–45 (in Russian).
12. Gimadi E.Kh., Glebov N.I., Serdjukov A.I. An Algorithm for finding approximation solution of the traveling salesman problem and its probabilistic analysis. Discrete Analysis and Operation Research (Series: Mathematics and its Applications), Kluwer Academic Publishers, Dordrecht, 1995, V.355, 35–43.
13. Gimadi E.Kh., Zaljubovsky V.V. Asymptotically optimal approach to solving one-dimensional bin packing problem. Upravljaemye sistemy, Novosibirsk, 1984, *25,* 48–57 (in Russian).
14. Hoeffding W. Probability inequalities for sums of bounded random variables. Amer. Statist., 1963, *58,* 39–52.

15. Perepelitsa V.A., Gimadi E.Kh. On the problem of finding the minimal Hamiltonian circuit in a graph with weighted arcs. Diskretnyi Analiz, Novosibirsk, 1969, *15,* 57–65 (in Russian).
16. Petrov V.V.: Limit Theorems of Probability Theory. Oxford Univ. Press, 1995.
17. The Traveling Salesman Problem and its variations. G.Gutin and P.Punnen (Eds.) Kluwer Acad. Publ., 2002, 830 p.

Two-Dimensional Cutting Stock Problem Under Low Demand: a Study Case

Kelly Cristina Poldi, Marcos Nereu Arenales[1], Andrea Carla G. Vianna[2]

[1] USP - University of Sao Paulo, Av. do Trabalhador Sancarlense, 400. 13560-970 - Sao Carlos - SP - Brazil kelly@icmc.usp.br, arenales@icmc.usp.br

[2] UNESP - State University of Sao Paulo, Av. Eng. Luiz Edmundo Carrijo Coube, s/n - 17033-360 - Bauru - SP - Brasil vianna@fc.unesp.br

1 Introduction

A cutting stock problem basically consists of cutting large pieces available in stock to produce smaller pieces (called *items*) in order to meet a given demand. The cutting is planned to minimize waste of material (other objectives can arise). This kind of problem arises in several industries such as paper, aluminum, steel, glass, furniture and so son. The problem can be one-dimensional, e.g., the cutting of rolls; two-dimensional, e.g. the cutting of plates, etc.

There are lots of applications of cutting problems and several approaches to theirs solution which are surveyed in the literature, such as Golden 1976; Dowsland and Dowsland 1992; Sweeney and Paternoster 1992; Dyckhoff and Finke 1992; Dyckhoff et al. 1997.

Cutting stock problems under low demand generally occur in small-scale industries. Such industries have customized order books with only few standards products. Few papers in the literature concern low demand problems, although small-scale industries need good and quick solutions. Greedy solutions are useful but an expert worker might find solutions as good as or better solutions than the solutions obtained by greedy heuristics. Riehme et al. 1996 studied two-stage guillotine two-dimensional cutting problems with big demand variability (i.e., items with low demand and items with high demand). However, the important limitation due to low demand is just partially considered during the construction of cutting patterns.

This study proposes a new heuristic to solve the integer two-dimensional cutting stock problem. A study-case on real-world instances provided by a Brazilian small-scale metallic frameworks industry is carried out.

2 Two-Dimensional Cutting Stock Problem

Assume we have K types of plates of given dimensions (L_k, W_k), where L_k is the length and W_k is the width of plate type k ($k = 1, \ldots, K$), each type is available in quantity e_k, $k = 1, \ldots, K$. Consider, also, we have sets of clients' requirements, with known demand $d_i, i = 1, \ldots, m$, of items of dimensions (ℓ_i, w_i), where ℓ_i and w_i are respectively the length and width of item $i, i = 1, \ldots, m$ (Fig. 1(**a**)). The problem consists of producing the ordered items by cutting the stock plates in order to meet the demand and minimize the material waste. In the literature, this problem is known as a two-dimensional cutting stock problem. The layout in which the items are arranged in a plate is called a cutting pattern (Fig. 1(**b**)) which is used to cut many plates.

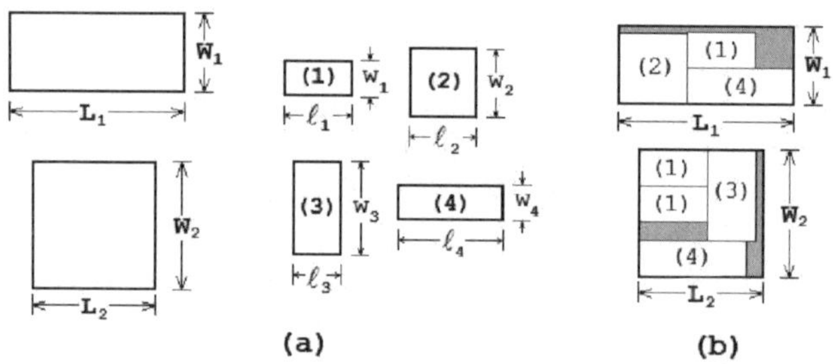

Fig. 1. (**a**)Cutting problem with K types of plates ($K = 2$) and m types of items ($m = 4$), (**b**) examples of two-dimensional cutting pattern.

A cutting stock problem is called *unconstrained* when there is no limitation on the number of each type of item in a cutting pattern. Otherwise, the problem is called *constrained.* Although this is just a light detail in the formulation, the limitation on the number of types of items causes considerable difficulties in the problem's resolution. Note that in this work, which deals with low demand, we have a constrained cutting stock problem.

When a rectangular plate is cut into two others rectangles, i.e., the plate is cut vertically or horizontally, the cutting pattern is called a *guillotine* cutting pattern. For instance, the cutting pattern shown in Fig.1(**b**) is a guillotine cutting pattern.

In order to formulate the mathematical model for the two-dimensional cutting stock problem we consider the following data and parameters.

Data:

K: number of types of stock plates;
L_k: length of plate $k, k = 1, \ldots, K$;
W_k: width of plate $k, k = 1, \ldots, K$;
e_k: stock availability of plate $k, k = 1, \ldots, K$;
m: number of types of items;
ℓ_i: length of item $i, i = 1, \ldots, m$;

w_i: width of item $i, i = 1, \ldots, m$;
d_i: demand of item $i, i = 1, \ldots, m$.
Parameters:
N_k: number of cutting patterns to plate type k, for all k;
a_{ijk}: number of item i in the jth cutting pattern for plate k, for all i, j and k;
c_{jk}: waste of cutting plate k according to j^{th} cutting pattern, for all j and k, which is given by $c_{jk} = L_k W_k - \sum_{i=1}^{m} \ell_i w_i a_{ijk}$.

The problem's decision variable is x_{jk}, which represents the number of plates type k cut according to the cutting pattern j. The mathematical model is as follows.

$$
\begin{array}{ll}
minimize & f(\mathbf{x}) \displaystyle\sum_{k=1}^{K} \sum_{j=1}^{N_k} c_{jk} x_{jk} \\
subject\ to: & \left\{ \begin{array}{l} \displaystyle\sum_{k=1}^{K} \sum_{j=1}^{N_k} \mathbf{a}_{jk} x_{jk} = \mathbf{d} \\ \displaystyle\sum_{j=1}^{N_k} x_{jk} \leq e_k, k = 1, \ldots, K \\ x_{jk} \geq 0, integer, j = 1, \ldots, N_k, k = 1, \ldots, K. \end{array} \right.
\end{array}
\tag{1}
$$

The objective function in (1) is to minimize the total waste. The first set of constraints refer to demand, they assure that the total number of produced items exactly meets demand. The m-vector $\mathbf{a}_{jk} = (a_{1jk}, a_{2jk}, \ldots, a_{mjk})^t$ corresponds to the cutting pattern j to a plate k.

For one-dimensional cutting stock problems, a cutting pattern can be modeled by a knapsack problem. For the two-dimensional problem, we used an AND/OR-graph approach to build a cutting pattern; this approach is commented in section 3.2. The second set of constraints (in model (1)) are due to the stock plates availability. They assure that the used amount of each type of plate does not exceed its availability $e_k, k = 1, \ldots, K$.

3 Solution Methods

3.1 Simplex Method with Column Generation

Integrality constraints on x_{jk} turn the problem (1) difficult to be computationally solved, even for medium-size instances. Another difficulty is the huge number of possible cutting patterns.

Practical approaches to solve the problem (1) consist of relaxing the integrality constraints and solve the relaxed problem (linear programming problem) by the column generation technique, proposed by Gilmore and Gomory 1963, so it is not necessary to generate a priori all the columns (cutting patterns). The optimal solution for the relaxed problem (1) is, in general, non-integer, and then rounding heuristics have been developed to determine a good integer solution. In each simplex iteration, k cutting patterns are generated (one for each type of stock plate available), however, only the one with the minimum relative cost will be used. Let (π, ν) be the simplex multiplier associated with the current basis. The relative cost of variable x_{jk} is given by:

$$L_k W_k - \sum_{i=1}^{m} \ell_i w_i a_{ijk} - \sum_{i=1}^{m} \pi_i a_{ijk} - \nu_k = L_k W_k - \sum_{i=1}^{m} (\ell_i w_i + \pi_i) a_{ijk} - \nu_k \quad (2)$$

To determine the minimum relative cost we must solve K subproblems:

$$\begin{aligned} &maximize \ \sum_{i=1}^{m} (\ell_i w_i + \pi_i) a_i \\ &subject\ to: \begin{cases} (a_1, \ldots, a_m)^t corresponds\ to\ a\ pattern\ for\ plate\ k \\ 0 \leq a_i \leq d_i,\ integer,\ j = 1, \ldots, m \end{cases} \end{aligned} \quad (3)$$

3.2 AND/OR-Graph Approach

To solve the constrained two-dimensional cutting stock problem (3) an AND/OR-graph approach was used (Morabito and Arenales 1996). A AND/OR-graph can be defined to represent all possible cutting patterns. To solve the problem it has been used an implicit enumeration method which search strategy combines backtracking and hill-climbing.

3.3 Constructive Greedy Heuristic

A simple way of determining an integer solution to the cutting stock problem is a constructive greedy heuristic, which is also called exhaustion repetition heuristic (Hinxman 1980). It is a classic procedure, widely used in literature and in practice: Hinxman 1980; Wscher and Gau 1996; Poldi and Arenales 2005; and others. A greedy heuristic consists of generating a good two-dimensional cutting pattern (by the AND/OR-graph approach) and utilizing this cutting pattern as many times as possible (until demand is met or there is no stock plate available). Repeat the steps until all demand is met.

3.4 A Residual Heuristic

Poldi and Arenales 2005 developed a heuristic approach to round the continuous solution of the one-dimensional cutting stock problem (solved by the simplex method with column generation) to an integer solution. In this work this approach is extended to solve two-dimensional cutting stock problems. Briefly, the heuristic developed by Poldi and Arenales 2005, in each iteration, (a) solves the continuous relaxation of the cutting stock problem; (b) sorts the frequencies (x_{jk}) of the cutting patterns in a special way; (c) tries first to round up the cutting pattern's frequency (one by one), if this produces oversupply, the frequency is reduced; (d) update data. Details in Poldi and Arenales 2005.

4 Computational Results

The 20 instances used in the computational experiments were kindly provided by a Brazilian small-scale metallic frameworks industry. The instances consists of a one-week production, the number of types of items (m) are between 2 and 16, the number of types of stock plates are between 1 and 3, and the stock availability is pretty low.

Table 1. Percentage of material waste

Instance	C G	Developed	Greedy	Industry
1	6.96	14.71	14.71	14.71
2	24.75	28.93	28.93	28.93
3	4.78	4.78	4.78	4.78
4	7.88	**8.19**	11.35	**8.19**
5	5.37	**6.04**	**6.04**	10.52
6	1.69	4.21	4.21	4.21
7	2.80	**3.02**	4.06	5.09
8	3.40	**9.29**	**9.29**	15.77
9	5.51	**7.87**	7.92	**7.87**
10	5.44	6.90	6.90	6.90
11	2.57	**3.89**	**3.89**	10.84
12	2.70	8.21	8.21	8.21
13	2.84	6.38	6.38	6.38
14	2.81	**7.44**	**7.44**	13.61
15	5.82	**10.01**	**10.01**	14.74
16	5.26	**7.44**	**7.44**	15.49
17	2.35	**6.16**	**6.16**	12.86
18	4.70	**4.70**	**4.70**	10.30
19	2.18	**3.69**	**3.69**	8.51
20	3.42	**8.02**	31.02	**8.02**

Table 1 shows the results of the 20 real-world instances from the metallic frameworks industry. The second column in Table 1 shows the material waste percentage in the optimal solution given by the simplex method with column generation (CG); although this is a non-integer solution (infeasible in practice), it represents a lower bound for the integer solution. Third (Developed Heuristic) and fourth (Greedy Heuristic) columns show the waste percentage in the solution given by the methods studied, which are the approach from Poldi and Arenales and the constructive greedy heuristic, respectively. In the fifth column is shown the waste percentage of the handmade solutions given by the industry expert worker, which were used in practice.

Also in Table 1, it is highlighted in bold "the best solutions" that could be found. Note that for all the 20 instances, the heuristics studied obtained the same or better results than the expert worker in the industry. Our approach obtained better solutions than the expert in 10 of 20 instances. The average gain of the proposed heuristic over the expert's solution is 5.60%.

5 Conclusions and Future Research

This work revised a heuristic to determine an integer solution to one-dimensional cutting stock problems and extend it to two-dimensional problems. Real-world instances from a Brazilian metallic frameworks industry were solved. The proposed heuristic was able to improve several instances' solution. For future research we intend to analyze more instances from the industry and also test some modifications on the heuristic proposed which were successful when applied to solve one-dimensional problems. Also, we intend to compare the proposed approach with the one by Riehme et al. 1996.

Acknowledgements: This work was sponsored by FAPESP and CNPq.

References

1. Dowsland K, Dowsland W (1992) Packing problems. European Journal of Operational Research 56:2–14
2. Dyckhoff H, Fink U (1992) Cutting and packing in production and distribution: topology and bibliography, Springer-Verlag Co, Heidelberg
3. Dyckhoff H, Scheithauer G, Terno J (1977) Cutting and Packing. In: Amico M, Maffioli F, Martello S (eds) Bibliografies in combinatorial optimization. Wiley, New York
4. Gilmory PC, Gomory RE (1963) A linear programming approach to the cutting stock problem - Part II. Operations Research 11: 863-888
5. Golden B (1976) Approaches to the cutting stock problem. AIIE Transactions 8:265–274
6. Hinxman A (1980) The trim loss and assortment problems: a survey. European Journal of Operational Research 5:8–18
7. Morabito R, Arenales MN (1996) Staged and constrained two-dimensional guillotine cutting problems: an AND/OR-graph approach. European Journal of Operational Research 94:548–560
8. Poldi KC, Arenales MN (2005) Dealing with small Demand in integer cutting stock problems with limites different stock lengths. Thecnical Report 85, University of Sao Paulo, Sao Paulo
9. Riehme J, Scheithauer G, Terno J (1996) The solution of two-stage guillotine cutting stock problems having extremelly varying order demands. European Journal of Operational Research 91:543–552
10. Sweeney P, Paternoster E (1992) Cutting and packing problems: a categorized application-oriented research bibliograph. Journal of the European Operational Research Society 43:691–706
11. Waescher G, Gau T (1996) Heuristics for the integer one-dimensional cutting stock problem: a computational study. OR Spektrum 18:131–144

Length-Bounded and Dynamic k-Splittable Flows*

Maren Martens and Martin Skutella

Universität Dortmund, Fachbereich Mathematik, Lehrstuhl V,
Vogelpothsweg 87, 44227 Dortmund, Germany,
{maren.martens,martin.skutella}@math.uni-dortmund.de,
http://www.mathematik.uni-dortmund.de/lsv

Summary. Classical network flow problems do not impose restrictions on the choice of paths on which flow is sent. Only the arc capacities of the network have to be obeyed. This scenario is not always realistic. In fact, there are many problems for which, e.g., the number of paths being used to route a commodity or the length of such paths has to be small. These restrictions are considered in the length-bounded k-splittable s-t-flow problem: The problem is a variant of the well known classical s-t-flow problem with the additional requirement that the number of paths that may be used to route the flow and the maximum length of those paths are bounded. Our main result is that we can efficiently compute a length-bounded s-t-flow which sends one fourth of the maximum flow value while exceeding the length bound by a factor of at most 2. We also show that this result leads to approximation algorithms for dynamic k-splittable s-t-flows.

1 Introduction

Problem Definition and Motivation

We consider generalizations of the classical maximum s-t-flow problem where flow must be sent through a given network (digraph) $G = (V, E)$ with arc capacities $c : E \to \mathbb{R}^+$ from a source $s \in V$ to a sink $t \in V$.

k-Splittable Flows. The NP-hard *maximum k-splittable s-t-flow problem* introduced by Baier, Köhler, and Skutella [3] asks for a maximum s-t-flow which can be decomposed into flow on at most k paths. Here k is either a fixed constant or part of the input[2]. A feasible solution to this problem is called *k-splittable s-t-flow*; it is specified by a collection $\mathcal{P} = (P_1, \dots, P_k)$ of k paths from s to t with corresponding nonnegative flow values $f_1, \dots, f_k$ such that arc capacities are obeyed: $\sum_{i:\, e \in P_i} f_i \leq c(e)$

* This work was partially supported by DFG Focus Program 1126, "Algorithmic Aspects of Large and Complex Networks", grant no. SK 58/4-1 and SK 58/5-3.

[2] Since every s-t-flow can be decomposed into flow on at most $|E|$ paths and cycles, we always assume that $k \leq |E|$.

for all $e \in E$. The value of this flow is $\sum_{i=1}^{k} f_i$. Notice that some of the values f_i can be zero such that less than k paths are actually used to send flow.

A *uniform k-splittable s-t-flow* is a k-splittable s-t-flow where every path carries the same amount of flow f, i.e., $f = f_1 = \cdots = f_k$. The s-t-paths in collection $\mathcal{P}$ are not necessarily distinct, that is, $\mathcal{P}$ may contain several copies of the same s-t-path such that the total amount of flow being sent along this path is a multiple of the common value f.

The notion of k-splittable flows is motivated by transportation problems where divisible goods have to be shipped through a network using a bounded number of containers and each container must be routed along some path through the network. In the more general context of multicommodity flows, k-splittable flows generalize the notion of unsplittable flows which were introduced by Kleinberg [6]. A natural restriction in the area of transportation is to bound the length of paths that might be used to ship some commodity from its source to its destination. We therefore consider a generalization of k-splittable s-t-flows by imposing bounds on the lengths of the paths in $\mathcal{P}$.

Length-Bounded Flows. In addition to the setting described above we assume that there are also *arc lengths* $\ell : E \to \mathbb{R}^+$. Then, an s-t-flow specified by a collection of s-t-paths $\mathcal{P} = (P_1, \ldots, P_k)$ and corresponding flow values $f_1, \ldots, f_k$ is called *L-length-bounded* for some $L \in \mathbb{R}^+$ if $\sum_{e \in P_i} \ell(e) \leq L$ for $i = 1, \ldots, k$, that is, no path in $\mathcal{P}$ is longer than L. Baier [2] gives an extensive survey of what is known for length-bounded flows. We consider the *maximum length-bounded k-splittable s-t-flow problem*: Given k and L, find a maximum k-splittable s-t-flow among the ones which are L-length-bounded. This constitutes a natural combination and generalization of k-splittable and length-bounded s-t-flows.

Dynamic Flows. A crucial characteristic of network flows occurring in real-world applications is flow variation over time and the fact that flow does not travel instantaneously through a network but requires a certain amount of time (*transit time*) to travel through each arc. Both characteristics are captured by *dynamic flows* which specify a flow rate for each arc and each point in time. The *quickest s-t-flow problem* is to send a given amount of flow from s to t such that the last unit of flow arrives at the sink t as early as possible, i.e., within minimum time T. We consider the *quickest k-splittable s-t-flow problem* where, as in the static setting described above, the number of s-t-paths used to send flow is bounded by k. This dynamic flow problem is NP-hard since already its 'static' counterpart is NP-hard [3].

Related Results from the Literature

As mentioned above k-splittable flows are introduced in [3]. Among other results it is shown there that a maximum uniform k-splittable s-t-flow can be computed in polynomial time by a variant of the classical augmenting path algorithm. In contrast, it is NP-hard to find a maximum k-splittable s-t-flow. The value of a maximum k-splittable s-t-flow is at most twice as large as the value of a maximum uniform k-splittable s-t-flow. That is, computing a maximum uniform k-splittable s-t-flow yields a $1/2$-approximation algorithm for the maximum k-splittable s-t-flow problem. Other results on k-splittable flows have been found, e.g., by Bagchi [1]. Ford and Fulkerson [5] introduce dynamic s-t-flows. It follows from their work that the quickest s-t-flow problem can be solved in polynomial time. Fleischer and Skutella [4] show that certain NP-hard generalizations of the quickest s-t-flow problem (with multiple commodities or costs) can efficiently be approximated with constant performance guarantees via static length-bounded flow computations.

Contribution of this Paper

In Section 2 we present the following bicriteria approximation result for computing maximum length-bounded k-splittable s-t-flows.

Theorem 1. *There is a polynomial-time algorithm that computes a $2L$-length-bounded k-splittable s-t-flow whose flow value is at least one fourth of the value of a maximum L-length-bounded k-splittable s-t-flow.*

In Section 3 we apply a variant of this result in order to obtain the following approximation for the quickest k-splittable s-t-flow problem.

Theorem 2. *There is a $(3+2\sqrt{2})$-approximation algorithm for the quickest k-splittable s-t-flow problem.*

Due to space limitations, we only give an intuitive outline of the proof of Theorem 2.We conclude by presenting an interesting open problem.

2 Length-Bounded k-Splittable Flows

In this section we derive a simple combinatorial algorithm with the property stated in Theorem 1. Throughout this section, lengths of arcs are also interpreted as cost coefficients. We first show that a k-splittable s-t-flow obeying the given length bound L only on average can be found in polynomial time.

Lemma 1. *For given k and L, a maximum uniform k-splittable s-t-flow with average path length at most L can be computed in polynomial time.*

The proof of Lemma 1 is similar to the proof of [3, Theorem 6]. It is based on the insight that a uniform k-splittable s-t-flow with flow value kf is an f-integral s-t-flow (a flow is called *f-integral* for some $f \in \mathbb{R}^+$ if the flow value on each arc is an integral multiple of f). Moreover, any f-integral s-t-flow of value kf induces a uniform k-splittable s-t-flow of the same value by constructing an f-integral decomposition into paths and cycles and ignoring the cycles.

Proof (of Lemma 1). Consider a maximum uniform k-splittable s-t-flow with average path length at most L. There exists at least one arc $e \in E$ with a tight capacity constraint since otherwise a better solution can be obtained by increasing the common flow value f on all k paths. Hence, f is equal to the capacity $c(e)$ of arc e divided by the number of paths using the arc. Thus, $f = c(e)/i$ for some arc $e \in E$ and some $i \in \{1, \dots, k\}$. Based on this insight we formulate an algorithm: For all $e \in E$ and $i \in \{1, \dots, k\}$, compute a $c(e)/i$-integral min-cost s-t-flow of value $F_{e,i} := kc(e)/i$ or find out that no such flow exists. Among all computed flows whose total cost is at most $F_{e,i}L$ output one with largest flow value. If no such flow exists then output the zero flow. The running time of this algorithm is dominated by $k|E|$ min-cost s-t-flow computations. □

As discussed above, the flow described in Lemma 1 is an f-integral s-t-flow of value kf and cost at most kfL for some $f \in \mathbb{R}^+$. It can be turned into a $2L$-length-bounded k-splittable s-t-flow while decreasing the flow value only by a factor $1/2$.

Lemma 2. *Given an f-integral s-t-flow of value kf and cost at most kfL, a $2L$-length-bounded uniform k-splittable s-t-flow of value $kf/2$ can be found in polynomial time.*

Proof. The algorithm works as follows. First, the given flow is made acyclic by repeatedly canceling flow on cycles. Notice that this step does not increase cost since all cost coefficients (arc lengths) are nonnegative. Next, we cancel $f/2$ units of flow along the currently longest flow-carrying s-t-path and repeat this step k times. The resulting s-t-flow is $f/2$-integral and has flow value $kf/2$. Moreover, the length of any flow-carrying s-t-path is at most $2L$. Otherwise, all paths on which flow was canceled have length strictly larger than $2L$. Since $kf/2$ flow units were deleted from these paths, the cost of the initial flow must have been strictly larger than kfL—a contradiction. □

Notice that the computed s-t-flow is not only $2L$-length-bounded but has the stronger property that the length of *any* flow-carrying path is at most $2L$. This means that *any* path decomposition of this flow has the nice property of being $2L$-length-bounded. It is easy to come up with examples showing that this does not hold for arbitrary length-bounded s-t-flows.

We can now state the bicriteria approximation algorithm mentioned in Theorem 1: In the first step, a maximum uniform k-splittable s-t-flow with average path length at most L is computed (see Lemma 1). The second step turns this flow into a $2L$-length-bounded uniform k-splittable s-t-flow (Lemma 2). It remains to show the performance guarantee $1/4$ for the value of the computed flow. This follows from Lemma 2 and the following result.

Lemma 3. *The value of a maximum L-length-bounded k-splittable s-t-flow is at most twice as large as the value of a maximum L-length-bounded uniform k-splittable s-t-flow.*

The proof of this result is identical to the proof of [3, Theorem 12] and therefore omitted. Since the problem solved in Lemma 1 is a relaxation of the maximum L-length-bounded uniform k-splittable s-t-flow problem, the value of the flow computed in the first step of our algorithm is at least half as large as the optimum. Since the second step decreases the flow value by another factor $1/2$, this concludes the proof of Theorem 1.

Using the same technique as in the proof of Lemma 2 one can show for any ϵ with $0 < \epsilon < 1$ that given an f-integral s-t-flow of value kf and cost at most kfL, a $(1/\epsilon)L$-length-bounded uniform k-splittable s-t-flow of value $(1-\epsilon)kf$ can be found in polynomial time. This result can be used in order to show that there is a polynomial-time algorithm that computes a $(1/\epsilon)L$-length-bounded k-splittable s-t-flow whose flow value is at least $(1-\epsilon)/2$ times the value of a maximum L-length-bounded k-splittable s-t-flow.

A Note on the Complexity of Length-Bounded k-Splittable Flows

To emphasize that the maximum length-bounded k-splittable s-t-flow problem is indeed harder than the usual maximum length-bounded s-t-flow problem, we want to point out that it is possible to find a maximum length-bounded s-t-flow in polynomial time, if all arc lengths are equal to 1 (see, e.g., [2]). It is also shown in [2] that it is NP-complete to decide whether a digraph has a given number of length-bounded arc-disjoint s-t-paths with respect to unit arc lengths. This implies the following remark.

Proposition 1. *Even in a network with unit arc lengths it is NP-complete to decide whether there exists a length-bounded k-splittable s-t-flow of given value.*

Proof. It is easy to see that the problem is in NP. We reduce the NP-complete length-bounded arc-disjoint s-t-paths problem to it. One can decide whether a digraph has a given number M of length-bounded arc-disjoint s-t-paths with respect to unit arc lengths by checking if there exists a length-bounded M-splittable s-t-flow of value M in the network based upon this digraph with unit capacities. □

3 Dynamic k-Splittable Flows

The approximation algorithm in Theorem 1 can be used to construct an approximation algorithm for the quickest k-splittable s-t-flow problem. An instance of this problem consists of the same input as an instance of the maximum k-splittable s-t-flow problem. In addition, we are given transit times $\tau : E \to \mathbb{R}^+$ on the arcs and a prescribed demand value D. The task is to send D units of flow from the source s to the sink t on at most k paths within minimal time horizon T. For an exact definition of dynamic s-t-flows we refer to [5, 4]. We sometimes use the notion 'static flow' in order to emphasize that some flow is not dynamic. It follows from the work of Fleischer and Skutella [4] that a dynamic (k-splittable) s-t-flow of value D with time horizon T yields a static T-length-bounded (k-splittable) s-t-flow of value D/T (here we interpret transit times of arcs also as lengths). This static flow can be obtained by essentially averaging the dynamic flow over time. In particular, if the dynamic flow sends flow along at most k paths, then the same holds for the resulting static flow. On the other hand, a static T-length-bounded (k-splittable) s-t-flow of value d can be transformed into a dynamic (k-splittable) s-t-flow of value D

with time horizon $T + D/d$. The underlying transformation sends flow according to the given static flow pattern into the network for D/d time units. Then one has to wait for another T time units until the last unit of flow (traveling on a path of length, i.e., transit time, at most T) has arrived at the sink. Notice that the resulting dynamic flow uses exactly the same s-t-paths as the underlying static flow. For further details we refer to [4].

We can now prove Theorem 2. The time horizon of an optimum solution to the quickest k-splittable s-t-flow problem is denoted by T^*. Thus, there exists a static T^*-length-bounded k-splittable s-t-flow of value D/T^*. If T^* was known, one could compute a $2T^*$-length-bounded k-splittable s-t-flow of value at least $D/(4T^*)$; see Theorem 1. By slightly modifying the algorithm presented in Section 2 we can find $T \leq T^*$ and a $2T$-length-bounded k-splittable s-t-flow of value at least $D/(4T)$ in polynomial time. (We omit further details due to space limitations.) Applying the result of [4] thus yields a dynamic k-splittable s-t-flow of value D with time horizon $2T + 4T \leq 6T^*$.

Analogously we can use the general bicriteria approximation for length-bounded k-splittable s-t-flows in order to obtain a $(1 + \epsilon)/(\epsilon - \epsilon^2)$-approximation algorithm for the dynamic k-splittable flow problem for every ϵ with $0 < \epsilon < 1$. Optimizing over ϵ we obtain a minimum for $\epsilon = \sqrt{2} - 1$ which yields a $(3+2\sqrt{2})$-approximation with $3 + 2\sqrt{2} \approx 5.828$. This concludes the proof of Theorem 2.

Concluding Remark

We conclude by presenting a challenging open problem. Given a network with capacities and lengths on the arcs, a single source node s, and k sink nodes $t_1, \ldots, t_k$ with demand values $d_1, \ldots, d_k$. It is NP-hard to find an unsplittable flow that sends d_i units of flow from s to t_i along a single path of length at most L for $i = 1, \ldots, k$. It is an open problem to find a bicriteria approximation algorithm which sends a constant fraction of each demand d_i along a single s-t_i-path of length $O(L)$.

References

1. A. Bagchi. *Efficient Strategies for Topics in Internet Algorithmics*. PhD thesis, The Johns Hopkins University, October 2002.
2. G. Baier. *Flows with Path Restrictions*. PhD thesis, TU Berlin, 2003.
3. G. Baier, E. Köhler, and M. Skutella. On the k-splittable flow problem. *Algorithmica*, 42:231–248, 2005.
4. L. Fleischer and M. Skutella. The quickest multicommodity flow problem. In *Proceedings of the 9th Conference on Integer Programming and Combinatorial Optimization*, pages 36–53, 2002.
5. L. R. Ford and D. R. Fulkerson. Constructing maximal dynamic flows from static flows. *Operations Research, 6*, pages 419–433, 1958.
6. J. M. Kleinberg. *Approximation Algorithms for Disjoint Path Problems*. PhD thesis, Massachusetts Institute of Technology, May 1996.

Locating and Sizing Bank-Branches by Opening, Closing or Maintaining Facilities

Marta S. Rodrigues Monteiro[1,2] and Dalila B. M. M. Fontes[2]

[1] DMCT - Universidade do Minho
Campus de Azurém, 4800 Guimarães, Portugal 020414011@fep.up.pt
[2] Faculdade de Economia - LIACC fontes@fep.up.pt

Summary. The bank-branch restructuring problem seeks to locate bank-branches by maintaining, closing, or opening branches, to provide the service required by clients, at minimum total cost. This nonlinear problem, due to the existence of economies of scale, is formulated as a mixed binary, integer linear model. The model obtained can be solved by a ready-available software. However, due to the problem combinatorial nature, only small size instances can be solved. Thus, we also propose a local search heuristic that iteratively improves the solution obtained for a related linear problem by applying drop and swap operations. The computational experiments performed show the effectiveness and efficiency of the proposed heuristic.

Keywords: Bank-branch, Location, Concave Optimization, Heuristics

1 Introduction

Although bank-branch restructures have long been present in the financial world, they have not been the subject of much academic study, particularly from the operational research point-of-view [6]. A similar problem is addressed by chance-constrained goal-programming in [1] where three levels of bank services are considered: ATM, branches, and main branches. In [2] the bank-branch location problem is addressed by a two stage procedure where the number of branches needed to provide the minimum coverage is found by solving a classical covering problem; and then, their exact location is determined by solving a maximal coverage location problem. A budget constrained facility relocation problem is studied in [3] where both opening and closing facilities is considered. Three heuristics were developed: greedy-interchange, tabu search, and lagrangean relaxation.

In this paper, a new heuristic based on local search is presented to solve the bank-branch restructuring problem. This heuristic is divided in two stages: (i) obtaining an initial solution by solving a related linear problem [5]; (ii) improving that solution by applying drop and swap operations. The rest of the paper is organized as follows: in section 2 we describe the bank-branch location and sizing problem considered in this work and give the mathematical model. In section 3 we explain the methodology used. Computational experiments are provided in section 4 and finally, in section 5 some conclusions are drawn.

2 Problem Definition and Mathematical Formulation

The bank-branch restructuring problem seeks to locate branches, such that client needs for banking services are satisfied at a minimum cost. This can be achieved by opening new branches, and closing or resizing existing ones. Client needs need not to be satisfied by a single branch. Costs are incurred by opening, closing, and operating branches, and by providing clients with the required service. For each client we consider an ideal coverage that must be satisfied, and a minimum coverage that may or may not be satisfied. A penalty cost is incurred whenever the coverage provided is below the ideal coverage. This cost is proportional to the difference between these values. Employees are also taken into account in our problem both in terms of costs (hiring and firing costs) and in terms of needs (branches require a pre-specified number of employees to be able to operate). As far as the authors are aware of this aspect has always been neglected in the literature. We consider that banks operate in different areas, named counties, and that each of these counties is divided into smaller regions, called parishes. We assume that all clients of a parish are located at its geographical centre. The same applies to branches thus, there can only exist a single branch per parish. Different branch sizes with different service capacity are considered.

Let C be the set of counties and D the set of parishes, where $D = \cup_j D_j$ with $j \in C$. Let also K be the set of branch sizes. Since we may take decisions on whether to open new branches and whether to close existing branches we have defined the following decision variables.

- $y_{ij}^k = \begin{cases} 1, \text{ if a branch of size } k \text{ is closed in parish } i \text{ of county } j, \\ \quad \text{where } j \in C, i \in CB_j, k \in K, \\ 0, \text{ otherwise.} \end{cases}$
- $z_{ij}^k = \begin{cases} 1, \text{ if a branch of size } k \text{ is opened in parish } i \text{ of county } j, \\ \quad \text{where } j \in C, i \in D_j \backslash NCB_j, k \in K, \\ 0, \text{ otherwise.} \end{cases}$
- $x_{ij}^k = \begin{cases} 1, \text{ if a branch of size } k \text{ is operating in parish } i \text{ of county } j, \\ \quad \text{where } j \in C, i \in D_j, k \in K, \\ 0, \text{ otherwise.} \end{cases}$
- $he_j \geq 0$, number of employees hired in county j
- $fe_j \geq 0$, number of employees fired in county j
- q_{ij}^{lm}, number of service units provided by branch in parish i of county j to client in parish l of county m.

$$\min \sum_{j\in C}\sum_{i\in D_j}\sum_{k\in K} f_{ij}^k(x) + \sum_{j\in C}\sum_{i\in CB_j}\sum_{k\in K} g_{ij}^k(y) +$$
$$\sum_{j\in C}\sum_{i\in D_j\setminus NCB_j}\sum_{k\in K} h_{ij}^k(z) + \sum_{j\in C} T_j \times he_j + \sum_{j\in C} CMP_j \times fe_j +$$
$$\sum_{m\in C}\sum_{l\in D_m} P_{lm} \times (\overline{W}_{lm} - \sum_{j\in C}\sum_{i\in D_j} q_{ij}^{lm}) + \sum_{j\in C}\sum_{i\in D_j}\sum_{m\in C}\sum_{l\in D_m} q_{ij}^{lm} \times v_{ij}^{lm}\,. \tag{1}$$

subject to:

$$x_{ij}^{k_i} = 1, \quad \forall j \in C, \forall i \in NCB_j, k_i = k(i,j), \tag{2}$$
$$x_{ij}^{k\neq k_i} = 0, \quad \forall j \in C, \forall i \in NCB_j, \forall k \neq k_i \in K, \forall k_i = k(i,j), \tag{3}$$
$$x_{ij}^{k_i} = 1 - y_{ij}^{k_i}, \quad \forall j \in C, \forall i \in CB_j, k_i = k(i,j), \tag{4}$$
$$x_{ij}^{k\neq k_i} = z_{ij}^{k\neq k_i}, \quad \forall j \in C, \forall i \in CB_j, \forall k \neq k_i \in K, \forall k_i = k(i,j), \tag{5}$$
$$\sum_{k\in K} z_{ij}^k \leq 1, \quad \forall j \in C, \forall i \in D_j \setminus B_j, \tag{6}$$
$$x_{ij}^k = z_{ij}^k, \quad \forall j \in C, \forall i \in D_j \setminus B_j, \forall k \in K, \tag{7}$$
$$\underline{W}_{lm} \leq \sum_{j\in C}\sum_{i\in D_j} q_{ij}^{lm} \leq \overline{W}_{lm}, \quad \forall m \in C, \forall l \in D_m, \tag{8}$$
$$q_{ij}^{lm} \leq a_{ij}^{lm} \times \sum_{k\in K} k \times x_{ij}^k, \forall j, m \in C, \forall i \in D_j, \forall l \in D_m, \tag{9}$$
$$\sum_{m\in C}\sum_{l\in D_m} q_{ij}^{lm} \leq \alpha \times a_{ij}^{lm} \sum_{k\in K} k \times x_{ij}^k, \forall j \in C, \forall i \in D_j, \tag{10}$$
$$\sum_{j\in C}\sum_{i\in D_j}\sum_{k\in K} \epsilon_{ij}^k(x) \times x_{ij}^k = E + \sum_{j\in C} he_j - \sum_{j\in C} fe_j, \tag{11}$$
$$\sum_{i\in CB_j}\sum_{k\in K} \epsilon_{ij}^k(y) \times y_{ij}^k - fe_j \geq 0, \quad \forall j \in C, \tag{12}$$
$$\sum_{i\in D_j\setminus B_j}\sum_{k\in K} \epsilon_{ij}^k(z) \times z_{ij}^k - he_j \geq 0, \quad \forall j \in C, \tag{13}$$
$$he_j, fe_j, q_{ij}^{lm} \geq 0, \text{ integer, and } \quad x_{ij}^k, y_{ij}^k, z_{ij}^k \in \{0,1\}\,. \tag{14}$$

The objective function (1), minimizes the total cost, which is made up four components: branch costs (operating, closing, and opening costs); employee costs (hiring and firing costs); penalty costs; and service costs. The objective function is concave as it is given by the sum of linear and concave components (operating costs and service costs). The functions f_{ij}^k, g_{ij}^k, and h_{ij}^k, are non linearly dependent on several factors, see [6] for more details. Constraints (2) and (3) are related to the existing branches that are not allowed to be closed, while constraints (4) and (5) are related to the existing branches for which a closing decision is possible. Constraints (6) and (7) guarantee that at most one branch is opened at each new potential location and that it is operated. Constraint (8) guarantees that the service provided to each client is within the limits required, while constraints (9) and (10) are boundaries for the service provided by each branch to a single client and to all clients allocated to it, respectively. Constraints regarding the number of employees needed, fired, and hired are given by (11) to (13).

3 Solution Methodology

The above model has been set-up in a format such that CPLEX could be used to solve it. However, given that CPLEX works with matrices derived from the mathematical model that has $3 \times n_p \times n_c \times k$ binary variables and $2 \times n_c + n_c^2 \times n_p^2$ integer variables, the memory requirements are large and grow rapidly with problem size. Therefore, many of our problem instances could not be solved by CPLEX. In order to solve larger instances, which realistically banks are faced with, we have developed the following local search heuristic.

3.1 Initial Solution

In order to find an initial feasible solution we have solved a related linear programming problem that covers all demand locations at a minimum service cost. The objective function for this problem is,

$$\min \sum_{j \in C} \sum_{i \in D_j} \sum_{m \in C} \sum_{l \in D_m} \phi_{ij}^{lm} \tag{15}$$

As before the service provided to each client must satisfy lower and upper limits, as in (8). The overall coverage capacity for each branch must be at most α times the maximum branch capacity if the branch is to be opened; or α times the existing branch capacity, otherwise. Similar constraints are imposed, but now to the covers that can be provided to a single client.

We successively solve this LP model with updated cost function ϕ_{ij}^{lm}. At each iteration the cost function is updated by using the information of the solution to the previous iteration. This approach is based on [5].

$$(\phi_{ij}^{lm})^T = \sum_{m \in C} \sum_{l \in D_m} ((v_{ij}^{lm})^T - P_{lm}) \times q_{ij}^{lm}.$$

Initially, we only consider the linear cost, i.e. $(v_{ij}^{lm})^0 = v_{ij}^{lm}$. The cost function is updated as follows:

$$(\phi_{ij}^{lm})^{T+1} = \begin{cases} (\overline{v}_{ij}^{lm})^T + \dfrac{h_{ij} + f_{ij}}{(q_{ij}^{lm})^T \times \varphi_{ij}^T}, & \text{if } (q_{ij}^{lm})^T > 0 \\ (\overline{v}_{ij}^{lm})^R, & \text{otherwise.} \end{cases}$$

where φ_{ij}^T is the number of demand locations serviced by branch in parish i of county j, at iteration T and R is the index of the last iteration where $(q_{ij}^{lm})^T > 0$. The update procedure stops whenever either the solution of two consecutive iterations is the same, or the maximum number of iterations is reached. The initial solution is provided by the best solution obtained at the end of the procedure.

3.2 Improving the Initial Solution

The initial feasible solution is improved further by consecutively applying the following steps.

- Step 1 Attempt to drop branches that serve only one client, (a) as long as the minimum coverage is provided; (b) by distributing the service units provided to their clients by other branches still having available capacity.
- Step 2 Try to eliminate branches which are not using all service capacity.
- Step 3 Attempt to downsize branches, (a) as long as the minimum coverage is provided; (b) by distributing some service to other branches with available capacity.
- Step 4 Try to swap branches of different locations.

We do not consider adding branches since typically the initial solution completely satisfies the ideal coverage. To compute the cost variation for each of the above steps all components of the original cost function must be included. Furthermore, the cost function to be used is the original.

4 Computational Experiments

The proposed local search heuristic has been implemented in Visual C++ 6.0. Computational experiments were carried out on a 1.8-GHz Pentium4 with 256 MB of RAM. The MIP model given in Sect. 2 has been implemented in CPLEX. The optimality gap is given by $Error = \frac{(x - \overline{x})}{\overline{x}} \times 100$. In Table 1 we report on the variation of the number of employees E; the percentage ratio Q between covers provided and ideal coverage; the number of operating branches B; and the computational time required to solve the problem, in CPU seconds, both for CPLEX and Heuristic. Overall 180 problems have been solved. For each entry of the table we report the number of parishes M, and the average number of counties N for the 30 problems we have generated. In average, the heuristic is quicker to solve a problem and, although the solution is usually more expensive it provides better service than the CPLEX solution, since more coverage is provided and more branches exist. As it can be seen in Fig. 1 the variation on the number of counties does not seem to affect the error, while the error gap increases with the number of parishes.

Table 1. Average quality of the solutions

m	n	E	Q%	B	Time	E	Q%	B	Time	Error%
			CPLEX				Heuristic			
15	13	-52	99	6	1	-45	98	7	1	6.24
25	16	-38	99	9	7	-30	99	10	3	5.99
35	23	-73	98	13	38	-63	100	14	3	3.81
45	37	-150	99	16	397	-138	100	18	5	3.42
55	36	-104	99	20	688	-92	100	22	8	4.16
65	16	55	97	25	1966	75	99	24	11	6.18
Average		**-60**	**99**	**15**	**516**	**-49**	**99**	**16**	**5**	**5**

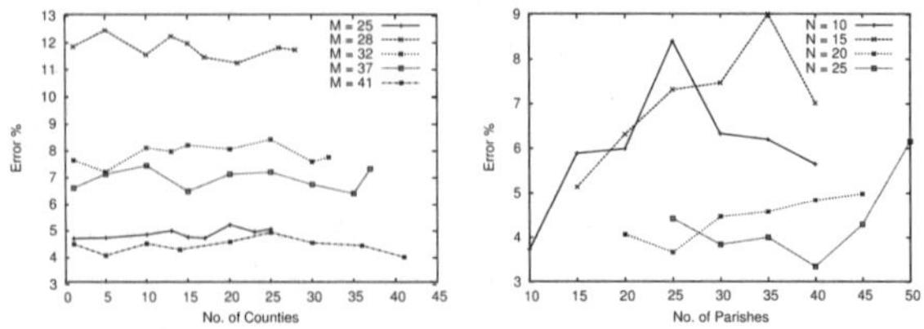

Fig. 1. Average error for varying number of (**a**) Counties (**b**) Parishes

5 Conclusion

We have developed a local search heuristic to solve the bank-branch location and sizing problem with concave cost functions. The heuristic is based on the solution to a related linear integer programming problem, iteratively improved by applying drop and swap operations. The computational experiments indicate that our heuristic is faster and that the number of counties does not affect the solution. The number of parishes affects the optimality gap due to the combinatorial nature of the problem.

References

1. Min H, Melachrinoudis E (2001) The three-hierarchical location-allocation of banking facilities with risk and uncertainty. Int Trans Oper Res 8:381-401
2. Miliotis P, Dimopoulou M, Giannikos I (2002) A hierarchical location model for locating bank branches in a competitive environment. Int Trans Oper Res 9:549-565
3. Wang Q, Batta R, Bhadury J, Rump CM (2003) Budget constrained location problem with opening and closing of facilities. Comput Oper Res 30:2047-2069
4. Kim D, Pardalos PM (1999) A Solution Approach to the Fixed Charge Network Flow Problem Using a Dynamic Slope Scaling Procedure. Oper Res Lett 24:195-203
5. Kim D, Pardalos PM (2000) Dynamic slope scaling and trust interval techniques for solving concave piecewise linear network flow problems. Networks 35(3):216-222
6. Monteiro MSR (2005) Bank-branch location and sizing under economies of scale. Master Thesis, Faculdade de Economia do Porto, Portugal

Simulated Annealing Based Algorithm for the 2D Bin Packing Problem with Impurities

B. Beisiegel[1], J. Kallrath[2], Y. Kochetov[3], A. Rudnev[4]

[1]B2 Software-Technik GmbH
45472 Mülheim an der Ruhr Germany

[2]Am Mahlstein 8
67273 Weisenheim am Berg Germany

[3,4]Sobolev Institute of Mathematics
630090 Novosibirsk Russia

1 Introduction

In the classical 2D rectangular bin packing problem [2] we are given a set of two dimensional rectangular items and an unlimited number of identical large rectangular bins. We need to place the items into a minimal number of bins. The orientation of the items is parallel to the bounds of the bins. Overlaps of items are not allowed.

In this paper we consider a more complicated real-world problem originating in the steel industry. The bins are inhomogeneous sheets with impurities. We assume that each impure area is rectangle. For each bin we are given a set of impurities, size, and location of each impurity into the bin. As a consequence now the bins are not identical anymore and the number of bins is finite. Moreover, we introduce the linear order on the set of bins. First of all, we have to use the first bin. If we need additional bins we use the second bin and so on. The items have the attribute whether they can be located in the area with impurities. The goal is to find solutions with a minimal number of bins.

For solving this NP-hard problem we have developed a tailored Simulated Annealing algorithm (SA). Feasible solutions are presented by the directed root tree encoding scheme. It has linear decoding time if the maximal number of impurities per bin is a constant. The initial solution is built by a greedy algorithm. It is a polynomial time heuristic which allows us to start SA with low temperature. The SA algorithm packs the items separately in each bin. It uses two types of neighborhoods. The first one changes the structure of the directed root tree. The second one swaps two items in the vertices of the tree. Computational experiments show that the algorithm produces feasible solutions with small deviations from the lower bound within a few minutes.

2 Representation of solutions

There are many encoding schemes for the 2D Rectangular Packing problems [3]. In this paper we use the oriented tree representation.

2.1 Encoding

Let us consider a feasible solution and show how to generate an oriented tree for each bin.

Definition 1. A feasible solution is called compacted if there is no item that can be shift left or bottom from its original position with other items fixed.

In *Figure* 1 we can see compacted and non-compacted feasible solutions. Further we consider the compacted solutions only.

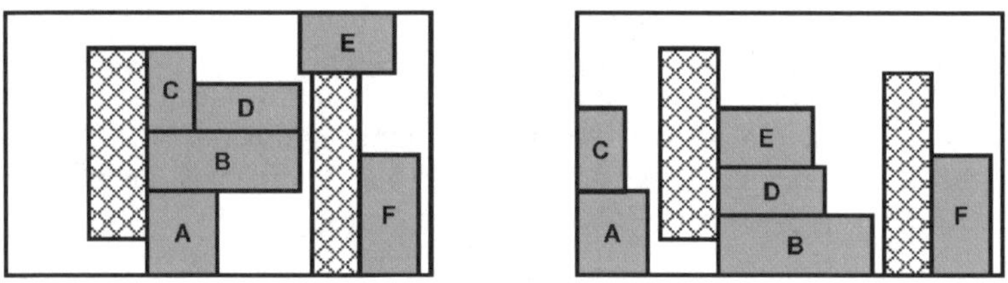

Fig. 1. An example of non-compacted (left) and compacted (right) placements for a bin

Definition 2. The item B is in the horizontal relation to the item A if

1. B is to the right of A.
2. The projections of A and B on the vertical axis are overlapped.
3. A and B are either adjunct or are divided by impurities only.

The oriented tree is built as follows. The set of nodes is the set of items in the bin with an additional node representing the root of the tree. The root corresponds to a dummy item placed on the left bound of the bin. The height of this item is the height of the bin. Node *A* is the parent of node *B* if item *B* is in horizontal relation to item *A*. If *B* is in horizontal relation with several items then the lowest one is the parent for *B*. In *Figure* 2 an example of oriented tree is presented.

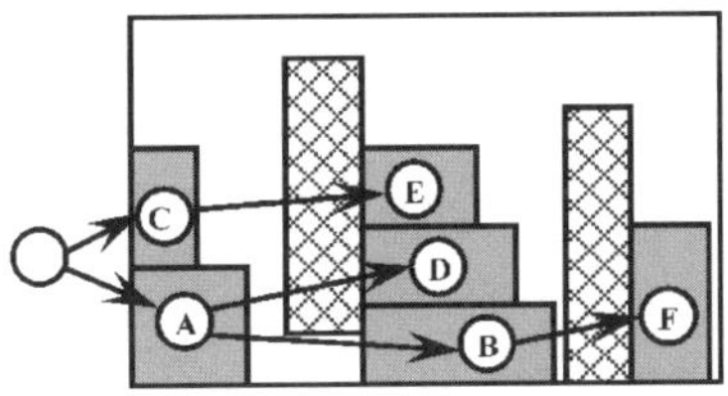

Fig. 2. The oriented root tree for the packing

2.2 Decoding

For a given oriented tree we generate packing by the following way. The root dummy rectangle is placed on the left side of the bin. According to the depth-first rule for the tree, we place items one by one in such a way that the left side of each item and the right side of its parent are on the same vertical line. The *y*-coordinate is defined by the previous packed items. Roughly speaking, we "drop" the current item on the right of its parent. If the item overlaps an impurity and cannot use it we consider two new positions for the item: above and right of the impurity. In the first case the item is shifted upwards and put on the impurity. In the second case the item is shifted to the right and put after the impurity. In the last case the new vertical position is defined by the previous packed items again. If the new position is overlapped with other impurities we put the item on the impurities. So we have two positions for the item. The lowest one is selected for packing. In *Figure* 3 we illustrate the idea of this algorithm. The time complexity of the algorithm is linear if the number of impurities per bin is a constant.

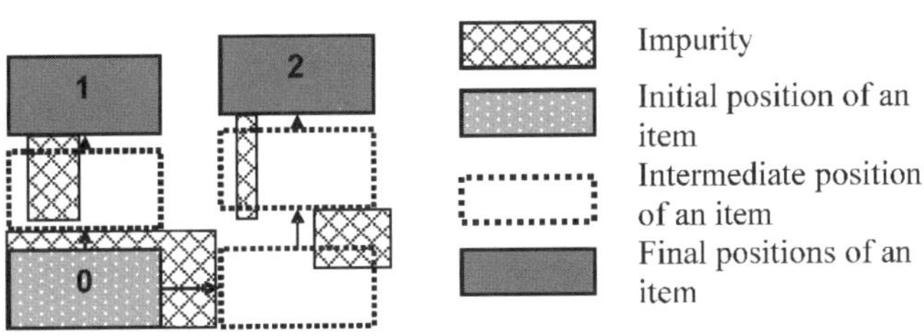

Fig. 3. Packing of the item

3 Neighborhood

Let us consider an oriented root tree. Each node of the tree corresponds to an item. The set of neighbors for the tree consists of two parts. In the first part we include all trees, which can be obtained by moving a leaf to another position. It gives us $O(n^2)$ neighbors, n is number of nodes. The second part of this set contains $(n-1)(n-2)/2$ neighbors. They are obtained by swapping items for the non root nodes. This neighborhood has the following property. We can reach an arbitrary oriented root tree from an arbitrary starting tree in a finite number of steps by moving from a tree to a neighboring one.

4 Simulated annealing

We apply the Simulated Annealing algorithm [4] for each bin in parallel. The algorithm is shown in *Figure* 4. The initial solution, i.e. initial set of trees, can be selected at random but we develop a polynomial time heuristic to get near optimal solutions. It allows us to start SA with low temperature. The objective function $F(T)$ which we wish to minimize is the used part of the bin with penalties. We apply the penalties for infeasible solutions when some items go out from the bin. For fixed temperature $t > 0$ we apply the random search during prescribed number of iterations for all bins (*Step* 3.1) and decrease the temperature (*Step* 3.2, $r < 1$). If the current solution is feasible we apply the Unload algorithm to change the set of items into the bins. This algorithm tries to unload the last bins by moving or swapping "large" items.

1. Find initial tree for every bin
2. Set initial temperature $t > 0$
3. Repeat until stopping criteria is true
    - 3.1. Repeat loop given number of times for every bin
        - 3.1.1. Chose random tree T' from neighborhood $N(T)$
        - 3.1.2. Set $\Delta = F(T') - F(T)$
        - 3.1.3. If $\Delta \le 0$, then $T = T'$
        - 3.1.4. If $\Delta > 0$, then $T = T'$ with probability $e^{-\Delta/t}$
    - 3.2. Set $t = rt$
    - 3.3. Execute *Unload* algorithm
4. Return set of trees

Fig. 4. Simulated annealing

5 The initial solution

Let us remove all impurities of the bins and compute the lower bound N_{LB} for the minimal number of bins [1]. Put the items in decreasing order of their areas and declare all bins are closed. Open the first N_{LB} bins. We place the items into the bins one by one: the first item into the first bin, the second item into the second bin and so on. At the (N_{LB} + 1) step we put the current item into the N_{LB} bin, the next one we put into the (N_{LB} – 1) bin and so on. We try to distribute "large" items through the N_{LB} bins. If we spend all items then we have an optimal solution. Otherwise we repeat this approach for the unpacked items. In *Figure* 5 we show the idea of this algorithm. It is polynomial time heuristic where we generate oriented root trees for all bins step by step.

1. Sort items list L by non-increasing area
2. Close all bins to use
3. Find the lower bound N_{LB} of optimal bins number for list L
4. Open N_{LB} empty bins to use
5. Place items from L into the open bins in prescribed order
6. Remove placed items from list L
7. If L is empty then stop
8. If set of closed bins is not empty, then go to *step* 3, else place remained items in additional infinitely large bin

Fig. 5. Heuristic for initial solution

6 Unload algorithm

This algorithm is used in our SA after decreasing the temperature in *Step* 3.3. It tries to select small items for the last bins and next to unload them. The algorithm uses two operations:

1. Swap operation. It swaps "small" items in the current bin by "large" items in the last bins without violation of feasibility.

2. Move operation. It moves the items from the current bin into the previous bins.

The algorithm consists of two stages. At the first stage we use swap operation for every bin starting from the end of the bin list. At the second stage we use move operation starting from the beginning of the list. The time complexity of the algorithm is $O(n^3)$.

7 Experimental results

The developed algorithm is coded in DELPHI environment and tested on real world and random generated instances. Our real world instances have small dimension, $n \leq 30$. The algorithm finds optimal solutions for all of them. To study the algorithm for large dimension we generate identical bins 200×300 and cut them to get the set of items. So, we have optimal solution if the impurities are absent, otherwise we have a lower bound only. The impurities are generated at random for every bin. The total number of impurities is $n/5$. The sizes of impurities are generated at random as items. Number of items, which can use the impurities, is selected as $n/5$. Computational results for three classes of instances are presented in *Table* 1. In class *Dl* every bin contains *l* items in average. For $n = 100$ the lower bound N_{LB} is quite far from the optimal solution and relative errors are large. For large scale instances the heuristic solutions is not too far from the lower bound and the relative errors are small.

Table 1. Deviation from the lower bound, %

	$n = 100$	$n = 150$	$n = 200$	$n = 300$	$n = 500$	$n = 700$	$n = 1000$
*D*3	**24**	**24**	**25**	**12**	**8.4**	**6.8**	**4**
*D*5	**35**	**30**	**27**	**13**	**8**	**6.4**	**6**
*D*10	**30**	**26**	**30**	**13**	**10**	**5.7**	**8**

References

[1] Boschetti M. A., Mingozzi A. (2003) The two-dimensional finite bin packing problem. Part I: New lower bounds for the oriented case. 4OR, vol 1, 1, pp 27–42.

[2] Dyckhoff H. (1990) A typology of cutting and packing problems. European J. Oper. Res. vol 44, 2, pp 145–159.

[3] Guo P.–N., Cheng C.–K., Yoshimura T. (1999) An O-tree representation of non-slicing floorplan and its applications. Proc. DAC, pp. 268–273.

[4] Johnson D. S., Aragon C. R., McGeoch L. A., Schevon C. (1989) Optimization by simulated annealing: An experimental evaluation, part I (graph partitioning). Operations Research, vol 37, 6, pp 865–892.

LP-based Genetic Algorithm for the Minimum Graph Bisection Problem

Michael Armbruster[1], Marzena Fügenschuh[2], Christoph Helmberg[1], Nikolay Jetchev[2], and Alexander Martin[2]

[1] Chemnitz University of Technology
Department of Mathematics, D-09107 Chemnitz, Germany
michael.armbruster@mathematik.tu-chemnitz.de

[2] Darmstadt University of Technology
Department of Mathematics, D-64289 Darmstadt, Germany
mfuegenschuh@mathematik.tu-darmstadt.de

Summary. We investigate the minimum graph bisection problem concerning partitioning the nodes of a graph into two subsets such that the total weight of each set is within some lower and upper limits. The objective is to minimize the total cost of edges between both subsets of the partition. This problem has a variety of applications, for instance in the design of electronic circuits and in parallel computing. We present an integer linear programming formulation for this problem. We develop a primal heuristic based on a genetic algorithm, incorporate it in a branch-and-cut framework and present some computational results.

1 The Minimum Graph Bisection Problem

We consider a weighted graph $G = (V, E)$ with edge weights $w_e \in \mathbb{R}_+$, $e \in E$, and node weights $f_i \in \mathbb{Z}_+$, $i \in V$. A pair (V_1, V_2) satisfying $V_1 \cup V_2 = V$ and $V_1 \cap V_2 = \emptyset$ is called *bipartition*, if $V_1 \neq \emptyset$ and $V_2 \neq \emptyset$. V_1 and V_2 are called *clusters*. Given a real number $\tau \in [0, 1]$ we define bounds l_τ and u_τ such that $u_\tau - l_\tau = \tau \sum_{i \in V} f_i$ and $l_\tau + u_\tau = \sum_{i \in V} f_i$. A bipartition (V_1, V_2) with $l_\tau \leq \sum_{i \in V_k} f_i \leq u_\tau$, $k = 1, 2$ is called *bisection*. $\Delta(V_1, V_2)$ denotes the set of edges incident to nodes in different clusters and is called *bisection cut*. The minimum graph bisection problem is to find a bisection (V_1, V_2) that minimizes $\sum_{e \in \Delta(V_1, V_2)} w_e$. This problem is known to be NP-hard [6]. An integer programming formulation of the minimum graph bisection problem can be stated as follows. We introduce binary variables y_{ij} for all $ij \in E$, such that $y_{ij} = 1$ if nodes i and j are in different clusters and $y_{ij} = 0$ otherwise. Consider a cycle $C = (E_C, V_C)$ in G and $F \subset E_C$ such that $|F|$ is an odd number, then

$$\sum_{e \in F} y_e - \sum_{e \in E_C \setminus F} y_e \leq |F| - 1$$

is a valid inequality for

$$P = \text{conv}\{y \in \{0,1\}^{|E|} \mid y \text{ is an incidence vector of a bisection cut in } G\},$$

see [2]. We select a node $s \in V$ and extend E so that s is adjacent to all other nodes in V, setting the weights w_{is} of new edges to zero. For all nodes $i \in V$, which are in the same cluster as s, holds

$$l_\tau \leq f_s + \sum_{i \in V \setminus \{s\}} f_i(1 - y_{is}) \leq u_\tau. \tag{1}$$

We obtain the following model of the minimum graph bisection problem.

$$\min \sum_{e \in E} w_e y_e$$

s.t.

$$l_\tau \leq \sum_{i \in V \setminus \{s\}} f_i y_{is} \leq u_\tau \tag{2}$$

$$\sum_{e \in F} y_e - \sum_{e \in E_C \setminus F} y_e \leq |F| - 1, \quad \forall C \subset G,\ F \subset E_C, |F| \text{ is odd} \tag{3}$$

$$y_e \in \{0,1\}, \quad \forall e \in E.$$

Note that (2) is a reformulation of (1). The constraint (2) assures that the total weight of nodes in each cluster stays within the given lower and upper bounds. The constraints (3) can be considered as the feasibility check for bisection cuts: each cycle in G must contain an even number of edges from the cut. It was also shown in [2] that inequalities (3) can be separated in polynomial time.

Graph bisection and its generalizations, e.g. when V is partitioned into more than two subsets [4], have considerable practical significance, especially in the areas of VLSI design and parallel computing. Exact methods such as branch-and-cut, however, are too slow to solve these problems for instances of practical interest. On the other hand, genetic algorithms are known to find good solutions to graph partitioning problems, see e.g. [3, 9, 10]. This motivated us to incorporate a primal heuristic based on a genetic algorithm in a branch-and-cut framework.

2 LP-based Genetic Algorithm

Genetic algorithms solve optimization problems in an analogous manner to the evolution process of nature [7]. A solution of a given problem is coded into a string of *genes* termed *individual*. New solutions are generated by operations called *crossover* and *mutation*. In a crossover two individuals called *parents* are drawn from the current population and parts of their genes are exchanged

Table 1. LP-based genetic algorithm.

input	y' current fractional LP solution, p number of individuals in population, k factor of population growth, g number of generations, φ number of fitness loops, M mutation type, m mutation rate
	(1) based on y' create initial population with p individuals (2) *while* sufficient improvement on fitness and number of loops less than g *do* (2a) perform crossover and mutation M till population grows to kp individuals (2b) evaluate fitness of each individual (2c) select p best individuals *done*
output	individual with the best fitness value

resulting in new, so called, *child* solutions. A mutation is an adequate transformation of a single individual. Individuals delivering the best objective value are selected to create the next generation.

In our heuristic, outlined in Table 1, an individual is a vector $v \in \{0,1\}^n$, $n = |V|$. If $v_i = v_j$, nodes i and j are in one cluster. To create the initial population of solutions we use a method based on the idea of the heuristic *Edge* introduced in [5]. Suppose that y' is the current fractional LP solution. Given a small value ε, if $y'_{ij} < \varepsilon$ nodes i and j will be most likely in the same cluster. We compute a minimum spanning forest on $\bar{E} := \{e \in E : y'_e < \varepsilon\}$. The initial two clusters are the two components of $(V_{\bar{E}}, \bar{E})$ with the greatest weighted sum of nodes. We complete the clusters using a bin packing algorithm and obtain individuals for the initial population. To create a new generation we apply the following four mutations for a given individual $v \in \{0,1\}^n$.

1 We select randomly a subset of v coordinates and set their values to the opposite, e.g. if the selected v_i is equal 0, we set it to 1.

2 For $ij \in E$ we consider the value $p_{ij} := 1 - y'_{ij}$ as the probability that nodes i and j are in the same cluster. If p_{ij} exceeds a randomly selected number from $[0,1]$ and $v_i \neq v_j$, we set either $v_i := v_j$ or $v_j := v_i$ at random.

3 We consider the cut Δ corresponding to the bipartition represented by the vector v. We select randomly an edge ij from Δ. It holds $v_i \neq v_j$. If the total weight of nodes in the cluster containing i is less that l_τ, we set $v_j := v_i$, otherwise $v_j := v_i$.

4 We assign a randomly selected v_i to such a value that in the new solution node i belongs to the cluster with the majority of nodes adjacent to i.

Note that the last three mutations are problem specific. The last operation is some sort of neighborhood search. The percentage of exchanged coordinates in a mutation is controlled by the parameter called *mutation rate*.

For the crossover operation we apply a one-point crossover: we select at random two parents from the present population and a crossover point from the numbers $\{1, \ldots, n-1\}$. A new solution is produced by combining the pieces of the parents. For instance, suppose parents $(u_1, u_2, u_3, u_4, u_5)$, $(v_1, v_2, v_3, v_4, v_5)$ and crossover point 2 are selected. The child solutions are $(u_1, u_2, v_3, v_4, v_5)$ and $(v_1, v_2, u_3, u_4, u_5)$.

To each individual we assign a fitness value. If it corresponds to a feasible bisection, i.e., the total node weight in both clusters stays within the limits l_τ and u_τ, we take the inverted objective function value in the incidence vector of the corresponding bisection cut. Otherwise we take a negated feasibility violation, i.e., the weight of the cluster which is greater than u_τ. The p fittest individuals are selected from the expanded population and the next generation is created. The fitness value of the best individual is stored. It defines the fitness of the generation. We consider two stopping criteria of the genetic algorithm. One is the fitness loop number φ, it defines the limit of loops we perform without increase in the generation fitness. The second limit is the maximal number of loops we perform in one heuristic's round. The output is the fittest individual. If it corresponds to a feasible bisection cut its fitness value gives an upper bound for the objective function value.

3 Computational results

In our empirical investigations we generally set $\tau = 0.05$. We consider four graph instances from the sample presented in [8]. As a branch-and-cut framework we use SCIP [1]. The computations are executed on a 1GHz Pentium III processor with 1 GB main storage. In the presented test cases we set $g = 300$, $\varphi = 60$ and $m = 1\%$, since these values brought the most success during our entire tests. If optimality is not achieved, we terminate SCIP after 3600 CPU seconds.

On one side we investigate the performance of our heuristic without SCIP's standard separators and primal heuristics. We vary the parameters p - population size, k - population growth and M - mutation type as listed in Table 2. We obtain that small instances can be most efficiently solved applying just the standard mutation type (M=1). In these cases the solution time is proportional to the parameters p and k. On the contrary, the bigger instances perform better with problem related mutation types (M= 2,3,4) and the biggest choice of solutions, i.e., the population growth. Due to this observation we apply all mutation types at random but uniformly distributed ($M = 5$). This delivers the best upper bounds.

To see the impact of our heuristic on the whole solution process we run SCIP with the standard settings on separators and primal heuristics, with and without the genetic algorithm (left and right hand side in Table 3, respectively). For each instance we set those values to parameters p, k and M, which appear to be the most efficient in the previous tests, i.e., either they yield solutions in the shortest period of time or the best upper bounds. Generally we obtain the best primal solutions of the entire test, in less average time of one heuristic's round (T_h/r), as well as the best lower bounds.

Table 2. Performance of the LP-based genetic algorithm.

		p=100	k=2			p=200	k=2			p=150	k=5		
Instance.n.m	M	r	T_h	b_U	b_L	r	T_h	b_U	b_L	r	T_h	b_U	b_L
taq.170.573	1	9	17	55	55	6	37	55	55	5	36	55	55
	2	8	41	55	55	3	68	55	55	7	184	55	55
	3	7	30	55	55	8	133	55	55	6	117	55	55
	4	7	19	55	55	5	58	55	55	4	43	55	55
	5	8	36	55	55	8	139	55	55	6	107	55	55
taq.228.903	1	2	5	63	63	2	20	63	63	2	24	63	63
	2	5	46	63	63	2	70	63	63	2	98	63	63
	3	2	14	63	63	2	52	63	63	2	73	63	63
	4	2	9	63	63	2	33	63	63	2	49	63	63
	5	2	14	63	63	2	56	63	63	2	79	63	63
diw.681.3752	1	39	857	4537	169	23	1055	4385	159	22	1338	3278	158
	2	20	970	4078	162	18	2722	3589	146	11	2419	2628	133
	3	24	935	3545	166	18	2060	3821	161	18	2626	2697	149
	4	29	852	4251	165	20	1495	4066	151	20	1949	2970	149
	5	26	1124	3764	163	20	2411	3486	155	16	2621	2741	143
taq.1021.6356	1	16	1058	8463	196	17	1812	8245	247	15	2016	7116	232
	2	14	1574	7945	211	9	2724	7761	219	7	2947	6630	137
	3	14	1306	7237	220	11	2438	7342	188	10	2672	6570	230
	4	15	1225	7841	234	13	2119	7898	229	12	2397	6669	226
	5	16	1789	7466	227	11	2650	7125	217	8	2661	6464	144

n number of nodes, m number of edges, r number of heuristic rounds, T_h total heuristic's running CPU time in seconds, b_U upper bound, b_L lower bound

4 Conclusion

In practice, graph partitioning problems are tackled heuristically. Exact solutions for large instances using state-of-the-art solvers are still unattainable in a reasonable amount of time. In the presented paper we combine both approaches to solve the minimum graph bisection problem. We develop a genetic algorithm as a primal heuristic routine in a branch-and-cut framework. It appears that the dual information in form of a fractional LP solution can be a significant help for the algorithm to deliver good feasible solutions.

Table 3. Performance of the LP-based genetic algorithm.

Instance.$n.m$	M	p	k	r	T_h	b_U	b_L	$T_{b\&b}$	b_U	b_L
taq.170.573	1	50	2	5	5	55	55	47	55	55
taq.228.903	1	50	2	3	6	63	63	94	63	63
diw.681.3752	5	150	5	10	772	2641	219	3600	*inf*	191
taq.1021.6356	5	150	5	4	570	6309	258	3600	*inf*	178

$T_{b\&b}$ total SCIP's running CPU time in seconds

Aknowledgments. This work is supported by German Research Foundation (DFG).

References

1. T. Achterberg. SCIP - a framework to integrate constraint and mixed integer programming. *ZIB-Report*, 2004.
2. F. Barahona and A. R. Mahjoub. On the cut polytope. *Math. Programming*, 36(2):157–173, 1986.
3. T. N. Bui and B. R. Moon. Genetic algorithm and graph partitioning. *IEEE Trans. Comput.*, 45(7):841–855, 1996.
4. C. E. Ferreira, A. Martin, C. C. de Souza, R. Weismantel, and L. A. Wolsey. Formulations and valid inequalities for the node capacitated graph partitioning problem. *Math. Programming*, 74:247–267, 1996.
5. C. E. Ferreira, A. Martin, C. C. de Souza, R. Weismantel, and L. A. Wolsey. The node capacitated graph partitioning problem: A computational study. *Math. Programmming*, 81(2):229–256, 1998.
6. M. R. Garey and D. S. Johnson. *Computers and Intractability*. W.H. Freeman and Company, 1979.
7. H. H. Hoos and T. Stützle. *Stochastic Local Search: Foundations and Applications*. Morgan Kaufmann, San Francisco (CA), 2004.
8. M. Jünger, A. Martin, G. Reinelt, and R. Weismantel. Quadratic 0/1 optimization and a decomposition approach for the placement of electronic circuits. *Math. Programmming B*, 63(3):257–279, 1994.
9. K. Kohmoto, K. Katayaman, and H. Narihisa. Performance of a genetic algorithm for the graph partitioning problem. *Math. Comput. Modelling*, 38(11-13):1325–1333, 2003.
10. H. Maini, K. Mehrotra, C. Mohan, and S. Ranka. Genetic algorithms for graph partitioning and incremental graph partitioning. In *Supercomputing '94: Proceedings of the 1994 ACM/IEEE conference on Supercomputing*, pages 449–457, New York, NY, USA, 1994. ACM Press.

Scheduling Departures at Airports – a MILP Approach

Florian Büchting, Petra Huhn

Institute for Mathematics, Clausthal University of Technology,
Erzstr. 1, 38678 Clausthal-Zellerfeld, Germany.

1 Introduction

In the last years the amount of air traffic considerably increased, so that nowadays many airports are running at their highest possible capacity and building new runways and gates is not always an option. Therefore the optimal scheduling of departures and arrivals at airports is one option to handle the increasing traffic. Besides airports[1] general transportation system engineering for all kinds of traffic obtains more and more attention from scientists.

In this paper we look at the aspect of scheduling the departure of planes into a stream of landings. Therefore, landings and departures will be handled differently as landings usually have priority while departures easily can be delayed. Nevertheless, also landings can be rescheduled and delayed by holding pattern or accelerated. The number of gates and runways is considered to be fixed. Moreover, a flight plan for departures and landings is given and contains basic information on the number and expected time windows for departing and landing aircrafts.

The main task is to assign starting times to the aircrafts which are ready for takeoff. But this problem is more than a simple assignment problem, where departing aircrafts are assigned to given time windows, as with real data many of these problems are infeasible or many departures are scheduled after the last landing to get a feasible solution. This is caused by the large number of constraints that need to be taken into account. Different criteria for optimization like waiting times for starting and/or landing aircrafts, deviations from the planned schedule, the time of the last runway operation (landing or start) or unoccupied times of runways and weighted combinations are considered. We will consider a single runway.[2] The runway capacity is mainly determined by the time an airplane takes to land resp. to start and occupies the runway. In addition minimum distances need to be kept between successive airplanes for

[1] cf. [ABKS04, BTYZ03, AC02] and references in there.

[2] The formulation can be generalized for multiple runways.

security reasons as a leading plane causes turbulences, called wakes, behind itself and a following plane can crash if it comes too close. Usually large planes cause major wakes whereas small planes are more sensitive to those turbulences. Because of that the ICAO[3] distinguishs the classes *light*, *medium*, and *heavy* according to the maximal takeoff weight of a plane. Similar to other air traffic simulation models (cf. [BTYZ03, ABKS04]) we assume that the time a landing resp. departing aircraft blocks the runway is determined by its wake category.[4] Another difficulty arises if for the departures only a single queue is considered so that the aircrafts have to take off in a fixed order. Again, for security reasons aircrafts are not allowed to overtake other planes on the taxiways or runways. Waiting at the gate and blocking it is not desirable. So, if a heavy aircraft waits at the head of that single queue, many time windows suitable for small aircrafts pass unused and the heavy aircraft blocks the entire queue. For that reason we use the idea of waiting queues from [Win99] to allow overtaking.[5] In our air traffic simulation model the time windows available for takeoffs are basically defined by the time period between two successive landings. To make full use of large time windows we try to schedule more than one departing aircraft to those slots by splitting large time slots into smaller ones and inserting artificial landings which can freely be shifted from the start to the end of that time window. We also introduce dummy departures to be assigned to time windows being too small for the departure of any aircraft if there is no other possibility. But as we are allowed to reschedule landings we try to shift time windows and landings in such a way that at least some of those small time windows are enlarged and can be used for a departure.

2 A mixed integer linear programming formulation

We give a detailed mathematical description of the problem as a mixed integer linear programming problem and start with the introduction of the parameters of the model.

$N, L =$ number of starting resp. landing aircrafts ($N, L \in \mathbb{N}$);
$R =$ number of queues, where aircrafts wait for the takeoff clearance ($R \in \mathbb{N}$);
$M =$ length of a queue ($M \in \mathbb{N}$);
$\mathcal{A} = \{1, \ldots, N\} \cup \{N+1, \ldots, RM\}$ set of starting aircrafts (real and dummy ones);
$\mathcal{L} = \{l_1, \ldots, l_L\}$ set of real landings $\subseteq \mathcal{D} = \{1, \ldots, RM\}$ set of time slots;
$\mathcal{P} = \{1, ..., RM\} = \{\mathcal{P}_1 \cup \ldots \cup \mathcal{P}_R\}$ set of waiting positions, where $\mathcal{P}_r$ denotes denotes the rth queue with $(r-1)M + 1$ as the head and rM as tail position;

[3] International Civil Aviation Organisation

[4] The minimum distances for airborne aircrafts are usually given in nautic miles. To get starting times and time windows we use a transformation method proposed in Appendix A of [FAA-GUT] to get time periods instead of distances.

[5] In [Win99] a tram dispatch problem is discussed: at the end of a day trams arrive in a certain sequence at a depot with a given number of parallel queues and these trams have to leave the next morning in a specific order to serve different lines.

The idea is to assign each (real and dummy) aircraft to a position in one of those queues first, and then any aircraft being at the head of a queue can get the clearance and start, while all aircrafts in a specific queue have to keep the ordering of that queue during the following assignment of starting times.

$E_j =$ earliest takeoff time of aircraft j $\forall j \in \mathcal{A}$, $(E_j \in \mathbb{R}_{\geq 0})$;
$s_j =$ time period that starting plane j blocks the runway $\forall j \in \{1, \ldots, N\} \subset \mathcal{A}$ (s_j depends on the wake category, $s_j \in \mathbb{R}_{\geq 0}$ and $s_j = 0$ $\forall j \geq N+1$);
$g_j =$ length of the jth time slot (for departure) $\forall j \in \mathcal{D}$ $(g_j \in \mathbb{R}_{\geq 0})$;
$T_j =$ starting time of the jth time slot (for departure) $\forall j \in \mathcal{D}$ $(T_j \in \mathbb{R}_{\geq 0})$; to guarantee feasibility the starting times must satisfy $T_j + g_j \leq T_{j+1}$ $\forall j$;
$\Delta_j^+, \Delta_j^- =$ maximal time period a landing can be delayed resp. brought forward $\forall j \in \mathcal{D}$, $(\Delta_i^-, \Delta_i^+ \in \mathbb{R}_{\geq 0})$;
$\mathcal{G}_{ij}^{SS}$, $\mathcal{G}_l^{LL}$, $\mathcal{G}_{li}^{LS}$, $\mathcal{G}_{il}^{SL}$ = minimal distances for two aircrafts, where S resp. L denotes a starting resp. landing aircraft, $\forall i, j = 1, \ldots, N, l = 1, \ldots L$,[6] $(\mathcal{G}_{pq}^{XY} \in \mathbb{R}_{\geq 0})$;
$\delta_{\min} := -\max_{j \in \mathcal{D}}\{\Delta_i^-\}$ and $\delta_{\max} := \max\{\max_{j \in \mathcal{D}}\{\Delta_i^+\}, \Delta_{WC}\}$ with
$\Delta_{WC} := N(\max_i s_i + \max_{ijl}\{\mathcal{G}_{ij}^{SS}, \mathcal{G}_l^{LL}, \mathcal{G}_{li}^{LS}, \mathcal{G}_{il}^{SL}\} + \max_i \Delta_i^+)$
$\mathcal{M}$ large constant ("big M", $\mathcal{M} \approx 2\max\{\delta_{\min}, \delta_{\max}, RM+1\}$).

The decision variables to assign aircrafts to queue positions and to encode the assignment of aircrafts to time windows and the ordering are as follows

$x_{iq} = \{1,$ if the aircraft $i \in \mathcal{A}$ departs from position $q \in \mathcal{P}$, and 0, otherwise;
$y_{jq} = \{1,$ if time slot $j \in \mathcal{D}$ is served from position $q \in \mathcal{P}$, and 0, otherwise
$\alpha_{ij} = \{1,$ if aircraft $i \in \mathcal{A}$ starts in time slot $j \in \mathcal{D}$, and 0, otherwise;
$\beta_{ij} = \{1,$ if aircraft $i \in \mathcal{A}$ starts before aircraft $j \in \mathcal{A}$, and 0, otherwise;

For the shifting of time windows that can effect all following slots we use

$\delta_i =$ amount of time landing i is shifted;
$\widetilde{\delta_p^i} = \{\delta_{p-1}$ if aircraft $i \in \mathcal{A}$ starts in slot $p \in \mathcal{D}$ $(\alpha_{ip} = 1)$, and 0 otherwise.

We look at different objectives, i.e. the waiting time for starting aircrafts $\min \sum_{i=1}^N \left(\sum_{j=1}^{RM} \alpha_{ij} T_j - E_i\right)$, the shifting of landing aircrafts $\min \sum_{j \in \mathcal{L}} \delta_j$ and the maximal waiting time $\min \max_{i=1}^N \left(\sum_{j=1}^{RM} \alpha_{ij} T_j - E_i\right)$.

Now, we state the constraints and start with the assignment constraints.

$$\sum_{i=1}^{RM} x_{iq} = 1, \sum_{j=1}^{RM} y_{jq} = 1 \;\forall q=1,\ldots,RM, \qquad \sum_{q=1}^{RM} x_{iq} = 1, \sum_{p=1}^{RM} y_{ip} = 1 \;\forall i=1,\ldots,RM.$$

The following constraints preserve the ordering in each queue for real aircrafts, whereas dummy aircrafts are allowed to overtake other aircrafts when they are entering a queue. The term $\sum_{i=1}^N i x_{iq}$ gives the index of aircraft i.

$$\sum_{i=1}^{N} i x_{iq} - \sum_{i=1}^{N} i x_{i(q+n)} \leq \left(1 - \sum_{i=1}^{N} x_{i(q+n)}\right)\mathcal{M}$$
$$\forall q=(r-1)M+1,\ldots,rM-n, \forall n=1,\ldots,M-1, \forall r=1,\ldots,R,$$
$$\sum_{j=1}^{RM} j y_{jq} - \sum_{j=1}^{RM} j y_{jq+1} \leq 0 \qquad \forall q=(r-1)M+1,\ldots,rM-1, \forall r=1,\ldots,R.$$

[6] Although the preprocessing during the air traffic simulation guarantees that two landing aircrafts have minimum distance $\mathcal{G}_{ij}^{LL}$, potential shifts can cause problems.

An aircraft can only be routed to a time window of appropriate length g_j, which can be effected by shifts δ_j and δ_{j-1}. Minimum distances and earliest takeoff times have to be guaranteed. These constraints look complicated but the main difficulty is to distinguish between real and artificial aircrafts.

$$\alpha_{ij} \cdot s_i \leq g_j + \delta_j - \delta_{j-1} \qquad \forall i{=}1,\ldots,N,\ j{=}1,\ldots,RM{-}1,$$

$$\alpha_{i(l_k+1)} \cdot (s_i + \mathcal{G}_{ki}^{LS}) \leq g_{l_k+1} + \delta_{l_k+1} - \delta_{l_k} \qquad \forall i{=}1,\ldots,N,\ k{=}1,\ldots,L,\ l_k{+}1{\neq}RM,$$

$$\sum_{j=1}^{RM} \alpha_{ij} T_j \geq E_i \qquad \forall i{=}1,\ldots,N,$$

$$g_{l_{i+1}} + \delta_{l_{i+1}} - \delta_{l_i} \geq \mathcal{G}_i^{LL} + T_{l_i+1} - T_{l_{i+1}} \qquad \forall i{=}1,\ldots,L{-}1,$$

$$(T_p + \delta_{p-1}) - (T_{l_k+1} + \delta_{l_k}) \geq \alpha_{ip} \mathcal{G}_{ki}^{LS} \qquad \forall i{=}1,\ldots,N,\ p{=}l_k{+}2,\ldots,l_{k+1},\ k{=}1,\ldots L,$$

$$T_{l_k} + \delta_{l_k} + g_{l_k} - T_{l_{k-1}+1} - \alpha_{i(l_{k-1}+1)} s_i - \delta_{l_{k-1}} - \alpha_{i(l_{k-1}+1)} \mathcal{G}_{(k-1)i}^{LS}$$
$$\geq \alpha_{i(l_{k-1}+1)} \mathcal{G}_{ik}^{SL} \quad \forall i{=}1,\ldots,N,\ k{=}1,\ldots L,$$

$$T_{l_k} + \delta_{l_k} + g_{l_k} - (T_p + \alpha_{ip} s_i + \delta_{p-1}) \geq \alpha_{ip} \mathcal{G}_{ik}^{SL} \quad \forall i{=}1,\ldots,N,\ p{=}l_{k-1}{+}2,\ldots,l_k,\ k{=}1,\ldots L,$$

$$\sum_{p=1}^{RM} \Big((\alpha_{jp} T_p + \widetilde{\delta_p^j}) - (\alpha_{ip} T_p + \widetilde{\delta_p^i}) \Big) + \sum_{k=1}^{L} \Big(\alpha_{j\, l_k+1} \mathcal{G}_{kj}^{LS} - \alpha_{i\, l_k+1} \mathcal{G}_{ki}^{LS} \Big)$$
$$\geq \beta_{ij} (\mathcal{G}_{ij}^{SS} + s_i) - \beta_{ji} \mathcal{M} \qquad \forall i,j{=}1,\ldots,N, i{\neq}j.$$

The remaining constraints define the relation between binary variables and bounds on the amounts of shifting

$$\alpha_{ij} \geq x_{iq} + y_{jq} - 1 \quad \forall i{=}1,\ldots,N,\ j,q{=}1,\ldots,RM \quad \text{and} \quad \sum_{j=1}^{RM} \alpha_{ij} = 1 \quad \forall i{=}1,\ldots,N,$$

$$\beta_{ij} + \beta_{ji} = 1 \quad \text{and} \quad \beta_{ij} \geq \frac{1}{RM} \cdot \Big(\sum_{p=1}^{RM} p \cdot (\alpha_{jp} - \alpha_{ip}) \Big) \quad \forall i,j{\in}\{1,\ldots,N\}, i{\neq}j,$$

$$\delta_i - \delta_{i+1} \leq g_{i+1} \quad \forall i{=}0,\ldots,RM{-}2 \quad \text{and} \quad -\Delta_i^- \leq \delta_i \leq \Delta_i^+ \quad \forall i{=}1,\ldots RM{-}1,$$

$$-\alpha_{ip} \cdot \delta_{\min} \leq \widetilde{\delta_p^i} \leq \alpha_{ip} \cdot \delta_{\max} \qquad \forall i{=}1,\ldots,N,\ p{=}1,\ldots,RM,$$

$$\delta_{p-1} - (1 - \alpha_{ip}) \mathcal{M} \leq \widetilde{\delta_p^i} \leq \delta_{p-1} + (1 - \alpha_{ip}) \mathcal{M} \qquad \forall i{=}1,\ldots,N,\ p{=}1,\ldots,RM,$$

and finally $x_{iq}, y_{jq}, \alpha_{ij}, \beta_{ij} \in \{0,1\}\ \forall i,j,q$ and $\delta_j, \widetilde{\delta_p^i} \in \mathbb{R}\ \forall i,j,p$. The starting times for takeoffs are not explicitly formulated but can be calculated from the starting times T_j of the time windows and shifts d_j. The number of variables resp. constraints is bounded by $O(R^2M^2 + N^2 + RMN)$ resp. by $O(RM^2N + L)$, but the constraint matrix is not dense.

3 Computational Results

To generate a typical scenario flight timetables of Munich Airport and Frankfurt am Main International Airport have been analyzed. We found relative frequencies of 0.2%/96(95)%/3.8(4.8)% for the categories light/medium/heavy of starting (landing) aircrafts in Munich and rel. frequencies of 2.1%/71%/26.9% in Frankfurt.[7] A common approach is to assume a Poisson process with param-

[7] Munich Airport serves mostly European destinations, whereas at Frankfurt International Airport has a large number of intercontinental flights (heavy) and the two percent of business jets (light).

eter α_A for the arrivals and an independent Poisson process with parameter α_D for departures. Then the waiting time for the next aircraft has an exponential distribution with an expected waiting time of $1/\alpha_A$ resp. $1/\alpha_D$.[8] The frequency of the wake categories for the landing and starting aircrafts are generated independently from the Poisson processes according to the data of Frankfurt am Main International Airport. These wake categories give the minimum distances $\mathcal{G}_{pq}^{XY}$. If the data of an instance resp. scenario are generated according to these realistic settings it may be infeasible because the minimum distances are not considered while generating the instances. This is checked and handled by a preprocessing procedure. For example the resulting times of the arrival process can be shifted or the reschedule of landings is allowed up to a sufficiently large amount of time. In addition, we do not consider grounding orders at night as hard constraints so that all landings and departures for light/medium/heavy aircrafts can be operated. The objective is to minimize the waiting time of starting aircrafts.[9]

The computational results[10] for the complete model as introduced in Section 2 are not really encouraging. In Fig. 1 the results for test sets with $L = 5$ landings, $N = 5$ departures and $R = 3$ ($R = 4$) queues are given. The column *time (std)* refers to CPU times where we used the standard parameter settings for CPLEX while the column *best time* refers to CPU times with different options for branching strategies and priority rules. All problems were solved to optimality.[11] But thinking of about 1000 resp. 1500 flight operations (starts and landings) each day at Munich resp. Frankfurt Airport or 50-100 flight operations each hour we are far away from solving problems of practical size – even for online-optimization problems embedded in a decision support system with a short time horizon of about 15-30 minutes resp. 10-50 aircrafts.

test set	rows	colums	nonzeros	RM	time (std)	best time
fs4.kl1	3044	987	14242	21	3,80 sec	3,78 sec
fs4.kl2	3039	987	14227	21	1:44 h	12,46 sec
fs4.kl3	3064	987	14302	21	1:29 min	14,17 sec
fs4.kl4	2337	738	10484	18	3,91 sec	3,89 sec
fs4.kl5	3852	1272	17662	24	1:13 min	31,74 sec

Fig. 1: Computational results for the complete model

So the idea is to solve the model approximately by polynomial time algorithms (heuristics) or to reduce the complexity of the model. We developed different heuristics for this model [12] and a simple idea to reduce complexity.

[8] Typical parameters for the mentioned airports are $1/\alpha_A \approx 1/\alpha_D \approx 180$[sec.]. But there are usually dependencies as aircrafts landing at an airport are cleaned and depart later that day.

[9] Objectives as the minimization of the deviations seem to be very difficult.

[10] We have used CPLEX 9 on a Linux PC (Pentium IV, 2,8 GHz, 512 MB).

[11] Solving a problem not to optimality but up to predefined tolerance avoids extraordinary long computation times (i.e. the cpu time for test set fs4.kl2 goes down to 1:04 min), but does not really help much otherwise.

[12] Due to space limitations we will not discuss this approach in this paper.

The assignment of an aircraft to a queue position and then to a time window can be simplified by matching the aircrafts and time windows directly and we allow overtaking in the reduced model which still considers the shifting of time windows. The hope is that the time windows and earliest takeoff times will exclude too many overtaking maneuvers. Afterwards the aircrafts can hopefully be lined up in different queues to implement the optimal schedule – although finding the minimum number of queues necessary for all overtaking maneuvers is another difficult combinatorial optimization problem. Figure 2 shows computational results for small problems[13] and some real size problems[14] with $30 = L + N$ runway operations.

test set	rows	col.	non zeros	RM	time (std)	**test set**	rows	col.	non zeros	RM	time (std)
fs4li.k1	766	120	4943	24	0,08 sec	fs4li.m1	2946	450	32714	45	7:34 min
fs4li.k2	981	150	6529	30	0,31 sec	fs4li.m2	3507	540	39971	48	3,92 sec
fs4li.k3	963	155	6115	30	3,14 sec	fs4li.m3	2381	360	26015	36	12,52 sec
fs4li.k4	599	90	3826	18	0,12 sec	fs4li.m4	2811	420	31211	24	13,46 sec
fs4li.k5	1506	225	9814	45	17,63 sec	fs4li.m5	6767	1050	76777	105	16:31 h

Fig.2: Computational results for the reduced model

These results are quite motivating. So, integrating heuristics and problem specific branching rules could be a good approach to solve practical problems.

References

[ABKS04] D. Abramson, J.E. Beasley, M. Krishnamoorthy, and Y.M. Sharaiha, *Scheduling aircraft landings - the static case*, Transportation Science 34, pp.180-197, 2000.

[ABKS04] D. Abramson, J.E. Beasley, M. Krishnamoorthy, and Y.M. Sharaiha, *Displacement problem and dynamically scheduling aircraft landings*, Journal of the Operational Research Society 55, pp.54-64, 2004.

[AC02] I. Anagnostiakis, J.-P. Clarke, *Runway Operations Planning: A Two-Stage Heuristic Algorithm*, Technical Report, American Institute of Aeronatics and Astronautics, MIT, 2002.

[BTYZ03] A.M. Bayen, C.J. Tomlin, Y. Ye, and J. Zhang, *MILP Formulation and Polynomial Time Algorithm for an Aircraft Scheduling Problem*, Proceedings of The 42nd IEEE Conference on Decision and Control (CDC), USA, 2003.

[DLR-G92] Deutsches Zentrum für Luft- und Raumfahrttechnik (DLR), Institut für Flugführung, Braunschweig, *Kapazitätsstudie G9.2*, 2001.

[FAA-GUT] C. Hüttenmoser, F. Knabe, H. Offerman, J. Reichmuth, C. Oliva, *Qualitätsbeurteilung des FAA-Gutachtens "An Investigation of the Present And Potential Future Capacity of Frankfurt am Main International Airport"*, 1999.

[Win99] Th. Winter, TU Braunschweig, *Online and Real-Time Dispatching Problems*, Dissertation, 1999.

[13] Test set fs4li.kx with $N = L = 5$, $R = 3$ resp. $R = 4$

[14] Test set fs4li.mx with $N = L = 15$, $R = 4$ resp. $R = 5$

Optimization of Sheet Metal Products

Herbert Birkhofer[1], Armin Fügenschuh[2], Ute Günther[2], Daniel Junglas[2], Alexander Martin[2], Thorsten Sauer[1], Stefan Ulbrich[2], Martin Wäldele[1], and Stephan Walter[1]

[1] Fachgebiet *Produktentwicklung und Maschinenelemente Darmstadt*, Fachbereich Maschinenbau, Technische Universität Darmstadt, Magdalenenstraße 4, D-64289 Darmstadt

[2] Arbeitsgruppe *Diskrete und Kontinuierliche Optimierung*, Fachbereich Mathematik, Technische Universität Darmstadt, Schlossgartenstraße 7, D-64289 Darmstadt

Summary. Linear flow splitting enables the forming of branched sheet metal products in integral style. To optimize those products design parameters have to be based on market requirements. We show that methods that are also used in Operations Research can, in principle, be applied to solve these optimization problems. For this, engineers provide constructive parameters that describe the demands of customers in a mathematical way. Based on these descriptions, we develop a two-stage model. First, a topology and shape optimization of branched sheet metal products is carried out, where the best-possible product is automatically designed by solving some OR models. Then, in stage two, we deal with the problem of how to incorporate manufacturing constraints for sheet metal products. The solution to this model corresponds to a construction plan. The entire approach is demonstrated in the design and construction of a cable conduit.

1 Introduction

Sheet metal is one of the most commonly used semi-finished products in metalworking. Countless everyday products are constructed from it. Its main characteristic is the ability to be formed and shaped up to high deformation degrees. In many cases branches like stringers, which are often used in aviation industry, give sheet metal the needed rigidity. Nowadays stringers or branches are welded on the material. This differential construction has many disadvantages, such as shape distortion or worsened heat transfer. These properties are far better in monolithic systems. A newer massive forming process, "linear flow splitting" [3], provides the opportunity to form branched profiles in an integral style out of sheet metal. The new roll forming process uses obtuse angled splitting rolls and supporting rolls to increase the surface of the band edge, which form the workpiece in discrete work steps up to a profile with the final geometry (Figure 1). The produced geometries are characterized by varying stiffness, surface hardness, surface roughness and heat transfer. Every additional branch leads to a new geometry and a whole set of new properties of the produced part. Handling the amount of possible product variants requires a methodical procedure.

Fig. 1. Process Principle and produced simple geometry

Methodical product design is the step-by-step synthesis from vague customer wishes and requirements to the final product shape of a technical product. During the task-clarifying phase of the design process a requirements-list is derived, which describes the properties the customer is looking for in a product (e. g. "low bending") [13]. These outer-properties cannot be established in a direct way by the engineer. The engineer has to choose parameters which are related to the outer-properties and which can be established in a direct way (e. g. material and geometry parameters) [10]. Thus, product design can be seen as the optimization of design parameters to fulfill defined outer-properties [8]. To this end, the designer has to know which properties are design parameters (resp. inner-properties) and how the definition of these properties interacts with the outer-properties (Figure 2).

In the following sections of this paper we will focus on how methods that are also known in Operations Research can be used to provide the optimization of sheet metal products with branches of higher order. For simplicity's sake, we start with two-dimensional profiles. As a guiding example we pose the following construction task: Construct a cable conduit with three separate channels, so that its maximal bending is minimal. The conduit is defined by the given sizes of the three separate channels, the size of the design envelope, the length and the type of sheet metal.

Thus we have the stiffness defined by the bending w and the momentum $I_{\bar{y}}$ as outer-properties. The outer-properties are coupled to inner-properties (the material and the topology) by the following equations [4]:

$$w = \frac{q_0 \cdot l^4}{8 \cdot E \cdot I_{\bar{y}}}, \tag{1}$$

and

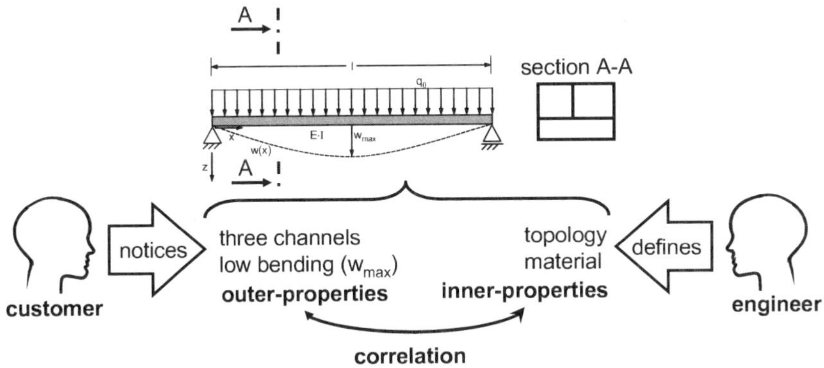

Fig. 2. The correlation between inner and outer properties

$$I_y = \iint_G (y - s_{\bar{y}})^2 \, dxdy. \tag{2}$$

Here, G is the topology of profile, q_0 is the uniform load, l is the length, $s_{\bar{y}}$ is the y-coordinate of the profile's center of mass, and E is a constant depending on the type of metal.

2 Models for Topology and Shape Optimization

The interrelation of design parameters and requirements on branched sheet metal products can in principle be expressed in terms of functional equations. Each feasible parameter configuration can be rated by a suitable objective function. These functional relations are generally non-linear equations, such as heat transfer or bending, which makes the corresponding models very difficult to solve.

For a fast computation of feasible or optimal solutions we develop a two-stage decomposition of the optimization problem. In stage one, we solve a coarse mixed-integer programming (MIP) model with linearized functional relations to find the overall topology of the product. In stage two, a detailed non-linear continuous shape optimization model is formulated and solved by non-linear optimization methods to obtain a more detailed product geometry.

2.1 Topology Optimization Models

To get an overall idea what the sheet metal product should look like, we introduce a rectangular pixel grid as a discretization of the design envelope. This pixel grid corresponds to a graph (V, E), where the vertex set V represents the pixels and the edge set E the horizontal or vertical neighbors of the pixels. Each pixel can either be assigned with metal, or it belongs to the inside (i.e., some chamber or channel) or the outside of the product. For each pixel $v \in V$ we introduce a binary (decision) variable $\delta_v \in \{0, 1\}$ with $\delta_v = 1$ if and only if pixel v is filled with metal.

Using these variables, the continuous and non-linear integral in (2) can be re-formulated as a linear equation as follows:

$$\iint_G (y - s_{\bar{y}})^2 dxdy \approx \sum_{v=(x,y) \in V} \delta_v \cdot (y - s_{\bar{y}})^2 \cdot \Delta x \Delta y = I_{\bar{y}}, \tag{3}$$

where Δx and Δy are the horizontal and vertical sizes of the pixels in the grid, respectively. With this, equation (1) can also be linearized:

$$8 \cdot E \cdot w \cdot I_y \leq q_0 \cdot l^4. \tag{4}$$

In addition, the model contains constraints that force the channels to be non-intersecting and closed (surrounded by metal). Moreover, each channel must have the exact given size. The center of mass of the whole profile must lie close to the specified center coordinate $s_{\bar{y}}$. All these constraints can be formulated as linear equalities or inequalities. Hence, we end up with a MIP model to describe all possible topologies of the sheet metal products that fulfill the required specifications. The objective is to find a feasible solution with minimal bending, that is, the objective is $\min w$. Note that this is equivalent to the (linear) objective $\max I_{\bar{y}}$.

Using a state-of-the-art solver for MIPs (Cplex9 [9]) and a fast computer (2.6GHz Pentium-IV), we are able to optimally solve the example from above on a pixel grid of size 8×10 within one hour. Some feasible solutions and the optimal solution are shown in Figure 3. For finer pixel grids, the solution time increases drastically. In the future, new strong valid or possibly facet-defining cutting planes should be incorporated into the branch-and-cut solution process to speed-up the computations.

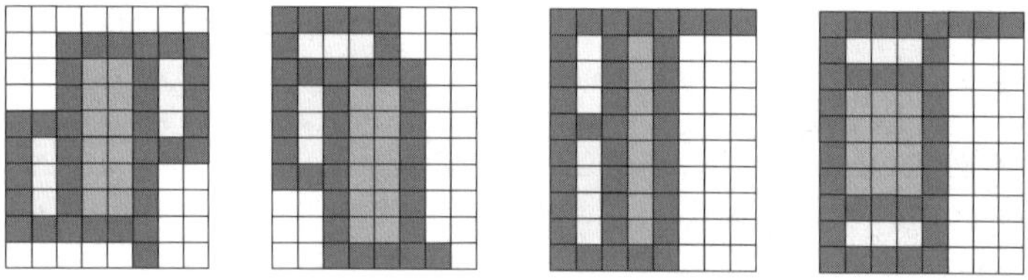

Fig. 3. Feasible and optimal solution of the topology optimization model (bendings in mm)

2.2 Geometry Optimization Models

A further improvement of the design requires more refined methods in comparison to discrete optimization. At this stage, continuous optimization comes into play. As input data this method requires the coarse topology that was found after solving the MIP to optimality. Within this framework, a fine-tuning of the geometry is possible, where more detailed models can additionally be taken into account. For example, the somehow simple stiffness from above is replaced by elasticity equations. The topology of the profile is given by a set of branching nodes and sheet metal in between them. Thus each sheet metal i of the profile between two nodes can be described as a rectangle $q_i := (x_i, y_i, l_i, d_i)$ having a certain length l_i, a width d_i, and a coordinate for the center (x_i, y_i). We parameterize the geometry of the profile by a vector $q = (q_1, \ldots, q_N)$. Hence the topological structure of the profile can be encoded in a linear equality system $A \cdot q = b$. Bounds of the width, length, and positions of the sheet metal pieces are described by a linear inequality system $B \cdot q \leq d$. So far, the refined model is a linear program. The advantage is that we can take constraints such as (2) directly into the model, without a discretization such as in (3). This yields a non-linear optimization problem that can be solved using recent SQP- [1, 15], interior point [16, 17] or semi-smooth Newton methods [6, 14]. Its solution corresponds to the detailed and refined geometry of the product's profile, see Figure 4. A further advantage of this

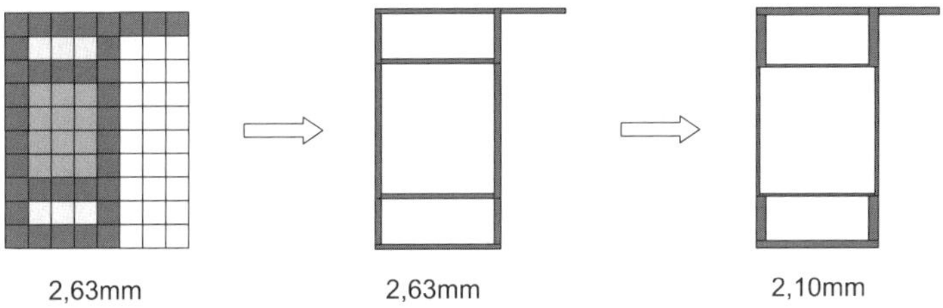

Fig. 4. Optimal solution of the refined model (bendings in mm)

method is that more detailed models for the description of physical effects concerning sheet metal products can additionally be taken into account, e.g., more detailed elasticity models in three dimensions or physical heat transfer models (for heat transfer between different channels).

3 Models for Manufacturing Constraints

Given a profile with optimal product geometry, we still have to determine how it can be constructed. Every branch in the profile can be obtained by either splitting up the piece of sheet metal or by gluing two ends together. Hence there are many ways to construct one and the same component. In order to decide where to glue and where to cut, we introduce a graph for every profile. To obtain this graph we proceed in two steps:

First, we introduce an edge in the graph for each segment in the profile. If two segments of the profile are connected, the corresponding edges have a common end vertex. All those edges will be called *non-pseudo edges.* In the second step, we replace every node of the graph by the complete graph K_n, where n is the degree of the node. The edges introduced in the second step will be called *pseudo edges.* Figure 5 illustrates this procedure.

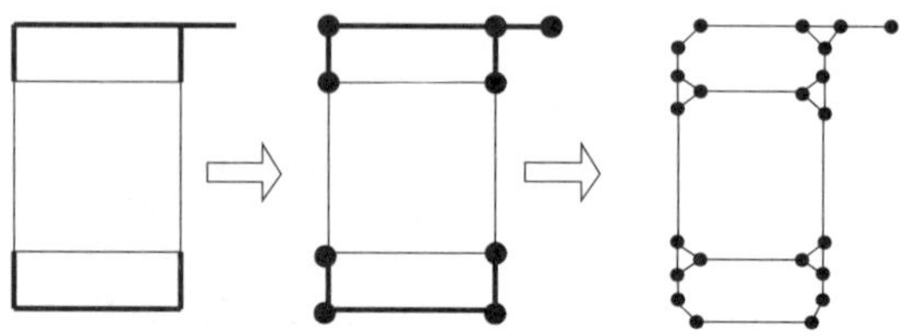

Fig. 5. Creating the graph

If we now compute a spanning tree T that contains every non-pseudo edge, we can interpret the tree as a possible unrolling in the following way: For every node of degree $d_T(n) > 2$ the sheet metal is split $d_T(n) - 2$ times in the corresponding place in the profile. Whenever a node is incident to pseudo edges in the graph but not in the tree, we only glue at the corresponding places.

Now every valid unrolling corresponds to exactly one spanning tree in this graph. On the other hand not every spanning tree corresponds to a valid unrolling as there are several manufacturing restrictions which have to be adhered to.

First of all the length of the piece of sheet metal that can be processed is limited. In terms of the graph model, this corresponds to the MDST problem, i.e. to the problem of finding a tree with limited diameter. In [7], Hassin and Tamir establish that the MDST Problem can be reduced to the so-called *Absolute 1-Center Problem* which has been proven to be solvable in polynomial time (see [5]). Furthermore the number of flanges that can be generated at a certain point on the sheet metal is limited as well. This can be modelled via a degree constraint for the tree. If k is the maximal number of flanges that can be generated then we have to find a spanning tree containing with all nodes having a degree of less or equal than $k - 1$. This is known as the d-MST problem which belongs to the group of $\mathcal{NP}$-complete problems (see [2]). Further restrictions, such as the fact that flanges cannot be arbitrarily resplit,often impose additional constraints to the spanning tree, making the overall problem of finding a valid unrolling $\mathcal{NP}$-hard.

Whereas most of these problems have been studied separately, there are no algorithms known to handle them all at once. In order to deal with all the manufacturing restrictions at once, we formulate the graph theoretical problem of finding a spanning tree with additional constraints from above as a Mixed Integer Problem. If the size of the profile becomes too large to solve the resulting MIP using a MIP solver, a possible way of approaching a solution is the Lagrangian relaxation (LR) heuristic (see [11] e.g.). Here the constraints which make the problem of finding a valid unrolling difficult are neglected and instead added to the objective function with certain correctional parameters. Combining the two manufacturing restrictions mentioned above, the degree constraint can be relaxed so that it only remains to repeatedly solve the MDST-problem. In case the optimal product geometry developed in Section 2 is not producible due to manufacturing restrictions we have to restart the optimization process and try to find a valid unrolling for the second best product geometry. If there is no valid unrolling for the second best, we try to find one for the third best, etc. This means, in case of our cable conduit, the profile with the lowest bending might be not producible, whereas the second best profile is producible. A possible unrolling of it can be found in Figure 6. Using the LR heuristic, it is also imaginable to relax a number of manufacturing restrictions that we are not yet aware of.

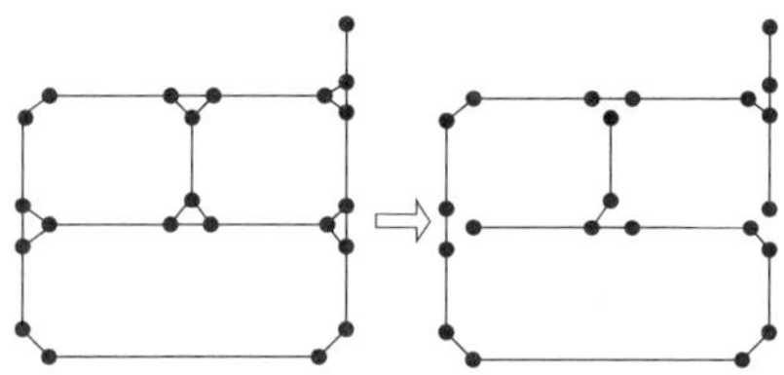

Fig. 6. From the graph towards a producible unrolling

The best producible unrolling of the two-dimensional profile computed in the previous step is then transformed into a three-dimensional construction plan of the new sheet metal product. Here, further design elements can be incorporated by the engineer, such as mounts or clearances, see Figure 7.

4 Further Work

Besides the optimization of sheet metal products according to market-driven inner- and outer-properties, technological findings from the production process and the evaluation, such as structural and materials testing also have a great influence on sheet metal products and their properties. In conventional product development these technological findings from the downstream

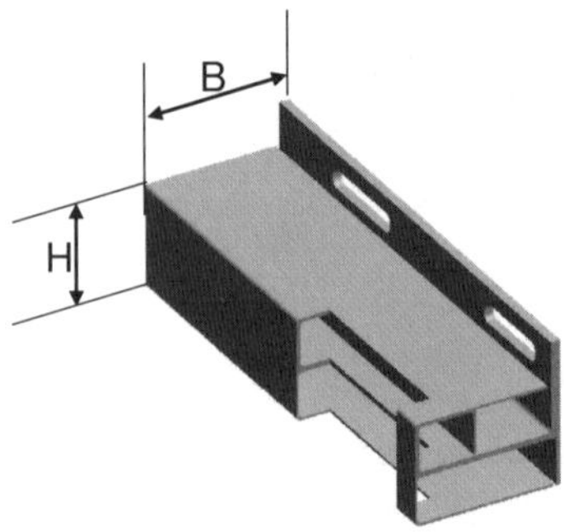

Fig. 7. Construction plan of the new product

production line are mostly regarded as time-consuming and iterative improvements. Minimizing the number of iterations and the required expenditure, as well as supporting the mathematical optimization of the product development process, are intended areas of future work. One of the primary tasks will be to detect the influence of technological findings from the downstream production line (branches of production subsequent to product development). Regarding the variation of the representation of the information (protocols, diagrams, series of measurements etc.), these findings will be standardized. They will be edited and provided toward the mathematical optimization of the development process as design parameters, just like the design parameters interacting with outer-properties. In this way, the transformation of market requirements into design parameters described above will be supplemented by the transformation and feedback of technological findings into design parameters. We will use results from earlier stages and expand upon and complement them. The earlier model, which describes the transformation from customer requirements to design parameters, will be the basis for further steps. This model will be extended by the transformation description of technological findings to design parameters and their back flow into the production process.

A further advantage is the profit of these findings for a technology-driven development of new, yet not customer- or market-demanded products. This so-called technologypush approach [12] shall especially be used to achieve unique selling points and competitive advantages for the manufacturers. The second focus of the work will be to use technological findings to develop a methodology in order to systematically derive potentials for the fulfillment of customer and market-expectations as well as to discover new fields of application (Figure 8).

We thank the German Research Association (DFG) for partially funding this work (Research Grant SFB 666).

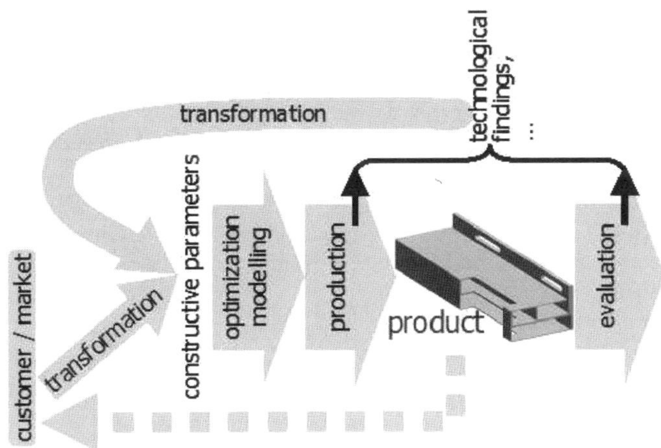

Fig. 8. Transformation into design parameters

References

1. R. Fletcher, N.I.M. Gould, S. Leyffer, Ph.L. Toint, A. Wächter. Global convergence of a trust-region SQP-filter algorithm for general nonlinear programming. *SIAM J. Optim.* 13: 635–659, 2002.
2. M.R. Garey, D.S. Johnson. Computers and Intractability: A Guide to the Theory of NP-completeness. W.H. Freeman, San Francisco, 1979.
3. P. Groche, G. v. Breitenbach, M. Jöckel, A. Zettler. New tooling concepts for the future roll forming applications. ICIT Conference. Bled, Yugoslvia, 2003.
4. D. Gross, W. Hauger, W.Schnell. Technische Mechanik, Bd. 2: Elastostatik. Springer, Berlin, 2005.
5. O.S. Hakimi. Optimal Locations of Switching Centers and Medians of a Graph. *Operations Research* 12:450–459, 1964.
6. M. Hintermüller, K. Ito, K. Kunisch. The primal-dual active set strategy as a semismooth Newton method. *SIAM J. Optim.* 13: 865–888, 2002.
7. J.-M. Ho, D.T. Lee, C.-H. Chang, C.K. Wong. Minimum diameter spanning trees and related problems. *SIAM J. Computing* 20: 987 - 997, 1991.
8. V. Hubka, E. W. Eder. Theorie technischer Systeme - Grundlagen einer wissenschaftlichen Konstruktionslehre. Springer-Verlag, Hamburg, 1984.
9. ILOG CPLEX Division. Suite 279, 930 Tahoe Blvd., Bldg 802, Incline Village, NV 89451, USA. Information available via WWW at URL http://www.cplex.com.
10. T. Sauer, M. Wäldele, H. Birkhofer. Providing Examples for Students and Designers. *Proceedings of the NordDesign 2004 Conference*, 340–349. Tampere, Finnland.
11. A. Schrijver. Theory of Linear and Integer Programming. Wiley, Chichester, 1986.
12. G. Specht, C. Beckmann. F&E Management, Schäffer Poeschel Verlag. Stuttgart, 1996.
13. N.P. Suh. Axiomatic Design - Advances and Applications. Oxford University Press, New York, 2001.

14. M. Ulbrich. Semismooth Newton methods for operator equations in function spaces. *SIAM J. Optim.* 13: 805–842, 2003.
15. S. Ulbrich. On the Superlinear Local Convergence of a Filter-SQP Method. *Mathematical Programming* 100:217–245, 2004.
16. S. Ulbrich, M. Ulbrich, L.N. Vincente. A Globally Convergent Primal-Dual Interior Point Filter Method for Nonconvex Nonlinear Programming. *Mathematical Programming* 100:379–410, 2004.
17. A. Wächter, L.T. Biegler. On the Implementation of an Interior-Point Filter Line-Search Algorithm for Large-Scale Nonlinear Programming. *To appear in Mathematical Programming*, 2005.

Modellierung von Entscheidungsproblemen in der Lehre - Ein Erfahrungsbericht

von

Prof. Karel Vejsada
Hochschule Pforzheim

Die OR-Ausbildung in wirtschaftswissenschaftlichen Studienrichtungen beschränkt sich oft auf Methoden und Algorithmen. Die erworbenen Kenntnisse der Absolventen stellen jedoch keine optimalen Voraussetzungen für die Anwendung im Betrieb dar. Dieser Beitrag beschreibt eine erfolgreiche Lehrveranstaltung, in der die Bildung linearer Modelle unterschiedlicher Komplexität und EDV-Berechnung / Auswertung im Vordergrund standen.

Klassische Studieninhalte

Der angehende Betriebswirt erwartet von der Ausbildung an einer Fachhochschule vor allem den Bezug zur Praxis. Diese Erwartung wird nur teilweise erfüllt. Quantitative Methoden zählen zu den abstrakten, unliebsamen Fächern, weil die Studierenden nicht immer zum deduktiven Denken geführt werden. Eine abstrakt erscheinende Formel als das Modell eines betriebswirtschaftlichen Zusammenhangs zu sehen, das überfordert die meisten.

Mathematik endet nicht selten mit einer Klausur, in der zuvor eingeübte Ableitungen oder Integrale zu berechnen sind. Mathematische Zusammenhänge und Konklusionen darf man meistens nicht erwarten, geschweige denn zu prüfen.

Die OR-Ausbildung in wirtschaftswissenschaftlichen Studienrichtungen leidet unter diesen Defiziten und beschränkt sich daher oft auf Methoden und Algorithmen. Es werden manuelle Berechnungen des Simplextableaus u. ä. verlangt. Kreatives OR, bei dem das Modell einer Entscheidungssituation entwickelt wird, ist manchmal in der Vorlesung, relativ selten jedoch in der Prüfung anzutreffen. Die algorithmische Ausrichtung scheint auch der Grund zu sein, warum in den meisten Betrieben OR-Anwendungen recht wenig fruchten: Es fehlen praxistaugliche Spezialisten.

Fraglich bleibt der Anwendungswert einer praktizierten Ausbildung. Diese Frage stellt sich vor allem deshalb, weil OR-Programme für die unterschiedlichsten Anwendungsgebiete für wenig Geld oder gar als Share- /Freeware erhältlich sind.

Die Studienordnung der Hochschule Pforzheim sieht je vier Semesterwochenstunden (SWS) in Mathematik und Statistik vor. Es schließt sich das Fach Operations Research mit vier SWS an. Einige der Studierenden wählen einen Wahlpflichtfach-Block mit drei Fächern zu je 2 SWS, in dem überwiegend quantitative Methoden, darunter auch das „modellorientierte" OR behandelt werden.

Der Verfasser versucht seit Jahren, das im vierten Semester angesiedelte Fach OR auf die Belange der Anwendung auszurichten. In letzter Zeit wurde an der Pforzheimer Hochschule das Software-Bündel „Lingo – Lindo – What'sBest" als Fakultätslizenz implementiert. Das war der entscheidende Schritt zum Durchbruch. Die Handhabung der Software ist leicht erlernbar, der Zeitaufwand ist gering. So können sich die Studierenden auf das Modelldesign konzentrieren.

Der Modellansatz

Es zeigte sich als sinnvoll, anfangs die Unterschiede zwischen Variablen, Konstanten und Parametern zu verdeutlichen. Schulabgänger kennen diese Unterschiede kaum. In einem nächsten Schritt wurden sehr einfache Muster von Beziehungen zwischen den Variablen formuliert. So wird beispielsweise gezeigt, wie aus mehreren „Zutaten" Finalprodukte als Mischungen entstehen.

Nachdem die drei Grundtypen der linearen Restriktionen behandelt und eine einfache Zielfunktion formuliert wurden, folgen Modellberechnungen mit Auswertungen. Die in zahlreichen Lehrbüchern verwendeten symbolischen Variablenbezeichnungen, meistens $x_1, \dots, x_n$, werden durch lange, „sprechende" Benennungen ersetzt. In dieser Phase müssen die Studierenden eine weitere Hürde nehmen – die eineindeutige Zuordnung einer Bezeichnung zu einer Modellgröße. Es scheint auf den ersten Blick banal. Zu oft sah sich der Verfasser mit regelrecht chaotischen Bezeichnungen konfrontiert, die den Durchblick in der Aufgabe verschleierten und letztendlich auf völlig falsche Modelle hinaus führten.

Die Benennung von Restriktionen ist für die Studierenden obligatorisch. Durch diese einfachen, aber konsequent angewendeten Regeln werden die Kursteilnehmer schnell in die Lage versetzt, die Ergebnisse eines Modells zu verstehen. Somit kann man sich auf die Verifizierung des Modelldesigns anhand der Analyse der EDV-Ausgabe konzentrieren. Die Unsicherheiten und die Fehleranfälligkeit der Auswertung wegen unzureichender Modelldokumentation werden eingedämmt.

Das führt dazu, dann man sich bereits in einer frühen Phase mit Schlupfvariablen befassen und dabei erkennen kann, welche Engpässe im Modell vorhanden sind. In den früheren Jahren musste der Verfasser immer wieder die Erfahrung machen, dass die Studierenden Schlupfvariablen erkennen, nicht jedoch ihre Bedeutung. Die Vernetzung des oft auswendig gelernten Wissens war noch nicht vorhanden, die Begriffe standen mit kaum greifbarem Inhalt isoliert im Raum.

Unmittelbar nach den ersten Auswertungen wird mit der Sensitivitätsanalyse fortgefahren. Somit wird das Wissen vertieft und gefestigt. Die Bedeutung der Schlupfvariablen wird durch einfache Veränderungen bis hin zur Gültigkeitsgrenze des jeweiligen RHS-Koeffizienten deutlich, ebenfalls der Zusammenhang zwischen Engpass bzw. Untergrenze und Schlupfvariablen.

Ferner wird in dieser Phase die Bedeutung der Opportunitätskosten behandelt, ohne dabei auf die duale Simplexmethode einzugehen. Die Beobachtung der Veränderungen der Schattenpreise infolge der Veränderungen der RHS-Koeffizienten liefert eine gute Basis für deren Verständnis.

Die Sensitivitätsanalyse hat in der Ausbildung noch einen positiven Nebeneffekt: Die aus der Sicht der Lernenden etwas problematische Interpretation der Schattenpreise von Gleichungen wird damit erleichtert. Die Auswirkung einer Datenänderung ist unmittelbar sichtbar.

Folgen von Modell- und Lernmustern

Der nächste Schritt ist die Verfeinerung der eingangs erwähnten Modelle der Mischungsprobleme. Sobald die Handhabung der Software und die Auswertung der Ergebnisse geläufig sind, werden die Aufgaben komplizierter. Zugleich bekommen die Studierenden zur Auflage, die Lösungen selbständig zu erarbeiten und sie der Gruppe im Intranet zur Diskussion zu stellen.

Zu diesen Aufgaben gehören etwa das Mischen von Zutaten nach dem bereits bekannten Muster mit Zusatzbedingungen, etwa einem eingetretenen Schwund (Abfall, Ausschuss, sonstige Verluste).

Nach und nach wird die Aufgabenstellung komplexer, das ursprüngliche Muster eines Aufgabentyps bleibt jedoch erhalten. Die Teilnehmer lernen durch eigene Formulierungen und vor allem durch EDV-Auswertungen, die einzelnen Modelltypen zu klassifizieren.

Die Vorlesungszeit wird nun genutzt, um Diskussionen im Plenum zu ermöglichen. Diese Phase ist besonders effizient. Eine Aufgabe kann zahlreiche, vom Rechenergebnis her äquivalente Lösungsvarianten haben. Die Aussagekraft des Modells kann dabei je nach Verdichtungsgrad sehr unterschiedlich sein. Es wird anhand konkreter Modellvarianten gezeigt, dass der Umfang eines Modells und dessen Verständlichkeit im direkten Verhältnis zueinander stehen.

Ferner werden einige „Feinheiten“ der Modellformulierung gezeigt, etwa die Einbindung von produktfixen Kosten mit Hilfe der gemischt-ganzzahligen LP.

Ein weiteres Muster wird aus dem Herstellungsprozess abgeleitet. Die Produktion wird zunächst als eigenständige Modellkomponente am Beispiel der Verwendung von Maschinenzeit erklärt. Die Zuteilung von Ressourcen kann direkt (Zeitbedarf je ME des Produkts) oder indirekt (z. B. Rüstzeiten) erfolgen. Ferner werden Kosten in die Zielfunktion aufgenommen.

Anschließend wird die Problematik von alternativen Herstellungsprozessen angegangen: „Make or Buy“ wird zur exakt kalkulierbaren Entscheidung. Auf der gleichen Ebene spielt sich das Problem von Ersatz- bzw. Ausweichmaschinen innerhalb des eigenen Unternehmens ab.

Diese Ausbildungsphase ist durch eine erhöhte Motivation gekennzeichnet. Diskussionen werden fachkompetent geführt. Der Lehrende greift immer weniger ein und lässt Diskussionen von studentischen Moderatoren in Wechsel leiten.

Es folgt eine Erweiterung der Produktionsmodelle um Personalpools. Das Personal als Ressource ist i. d. R. vielseitig einsetzbar. Ferner kann Mehrmaschinenbedienung eingebaut werden. Unterschiedliche Kostenfaktoren, etwa infolge einer Ausgleichs- oder Gefahrenzulage, können berücksichtigt und von den Studierenden verstanden werden.

Dynamische Modelle

Des Weiteren wurden zeitabhängige Prozesse behandelt. Das einfachste Modell stellt eine primitive Lagerbewirtschaftung dar, die durch die Gleichung

$$lagerbestd\_neu = lagerbestd\_alt + zugang - abgang$$

charakterisiert ist. Ein Mehrperioden-Modell entsteht simpel durch einen Index. Hierbei wird die Eigenschaft des Programms Lingo ausgenutzt: Es reicht, wenn eine Restriktion formuliert und mit einem Index versehen wird. Das Programm generiert die gewünschte Anzahl von Restriktionen. Dies ist eine wichtige Komponente. Die Studierenden konzentrieren sich auf den Sachverhalt und werden nicht durch eine oft sehr umfangreiche Eingabe abgelenkt. Die Codierung

```
SETS:
WOCHE /0..12/: LAGERBESTD, ZUGANG, ABGANG;
ENDSETS

@FOR (WOCHE(I) | I#GT#0: [Lager_aktuell]
LAGERBESTD(I) = LAGERBESTD(I-1) + ZUGANG(I) - ABGANG(I)
);
```

bewirkt, dass für einen 3-monatigen Planungszeitraum 12 Restriktionen generiert und mit dem generischen Namen „Lager_aktuell (i)" versehen werden.

Nach dieser leicht verständlichen Einführung wird das Fortschreibungsmuster des Lagerbestandes erweitert. Die ursprüngliche Funktion des Lagers als Puffer für schwankende Entnahmen wird mit dem Kundenbedarf und dem Produktionsniveau kombiniert und unter diversen Aspekten als Varianten durchgespielt. Die Grundgleichung lautet:

$$produktionsmenge + abgang - zugang = bedarf - fehlmenge$$

Die Zu- und Abgänge vom Lager werden mittels einer zweiten Gleichung mit dem Lagerbestand gekoppelt. Eventuelle Lagerkapazitäten (Gewicht, Volumen) oder Lagerschwund können leicht eingebaut werden. In der Zielfunktion werden bestandsabhängige Kosten der Lagerhaltung und im Falle eines entfernten Lagers die Transportkosten berücksichtigt.

Zum Schluss wird auf die in dieser Phase als „Blackbox" dargestellte Produktionsmenge fokussiert. Dabei werden die im vorherigen Abschnitt beschriebenen Modelle ausgenützt und in Mehrperioden-Modelle umgewandelt.

Aus Zeitgründen mussten die dynamischen Modelle auf kurze Produkt-Durchlaufzeiten beschränkt werden, da Modelle langer Durchlaufzeiten ungleich komplexer sind. Die Periodenübergänge wurden als Lageranfangs- und Endbestände (Sicherheitsbestände) formuliert. Auf Überlappungen von Auftragsbearbeitung und mittelfristige, periodenübergreifende Ressourcenauslastung musste verzichtet werden.

Die Leistungskontrolle

Der Leistungsnachweis wurde als Projektarbeit deklariert. Die Teilnehmer arbeiteten im Alleingang oder sie bildeten 2-er Teams. Die Prüfungsaufgaben wurden zugeteilt, die Lösungen mussten termingerecht ins Intranet eingestellt werden. Am Ende fanden kurze Referate mit Diskussionen statt, in denen die Ausarbeitungen präsentiert wurden. Eine Modelllösung galt als vollständig, wenn die Ergebnisse des Programms einschließlich der Sensitivitätsanalyse ausgewertet und interpretiert wurden.

Fazit

Die resultierenden Modelle hatten zwischen ca. 20 und 200 Variablen und 10 bis 100 Restriktionen. Dadurch, dass ein Standardmuster immer wieder ausgebaut und ergänzt wurde und mehrere solche Muster (etwa Stücklisten, Produktionsprozesse, Lager, Transport) miteinander kombiniert wurden, empfanden die Studierenden ein stufenweise aufgebautes, relativ komplexes Modell als aufwändig, aber nicht darüber hinaus als schwierig.

Die Umfrage nach der Prüfung ergab, dass diese Methode als „außergewöhnlich arbeitsintensiv“ eingestuft wurde. Der Dozent war angenehm überrascht, als die Teilnehmer einhellig die bisher eher als „respektvoll“ angesehene OR-Veranstaltung sehr positiv und deren Nutzwert als hoch beurteilten: „Es war anstrengend, hat aber viel Spaß gemacht“ war die Resonanz der ganzen Gruppe.

Quellenverzeichnis:

a) Studentische Arbeiten (Hochschule Pforzheim, Sommersemester 2005)

Produktion mit Schwankungen (zwei Varianten; erarbeitet von B. Giebels und J. Schmidtobreick sowie von C. Gundelach und T. Stegner): Ein Mehrperioden-Modell mit Produktionsschwankungen und produktionsmengenabhängigen, intervalllinearen Umstellungskosten, mit Lagerbewirtschaftung, variablen Lagerhaltungs- und Transportkosten.

Optimale Ressourcenausnutzung und Variantenfertigung (erarbeitet von Th. Eberhardt und M. Engel): Bestehende Vorräte an Edelmetallen sollen aufgebraucht werden. Die Produkte (Schmuckserien) bestehen teilweise aus einem Stück, etwa Ring, teilweise sind es materialbedingte Varianten von Kollektionen, etwa Collier + Armband + Ohrringe.

Eindimensionaler Zuschnitt (erarbeitet von W. Moor): Das Branch-and-Bound Verfahren bestimmt die relevanten Zuschnittmuster, die als Varianten in das ganzzahlige Modell der Abfallminimierung einfließen.

Personalplanung (erarbeitet von M. Biesinger und S. Walzhauer): Ein vorhandener Personalpool mit teilweise spezialisierten, teilweise universell einsetzbaren Mitarbeitern soll in mehrstufiger Produktion (Kuppelproduktion mit Fertigungsvarianten) kostenminimal eingesetzt und den Arbeitsgängen zugewiesen werden.

Jahresplan I (erarbeitet von J. Kroschel und F. Walter): Ein 12-Perioden-Modell mit saisonal schwankendem Bedarf, beschränkten Produktionskapazitäten, Losgrößen-Umstellungskosten, Mehrarbeit und Lagerhaltung mit dem Ziel der Kostenminimierung.

Jahresplan II (erarbeitet von T. Stegner): Wie oben, zusätzlich ist das Finalprodukt nur befristet lagerungsfähig. Die besonders pfiffig gelöste Aufgabe besaß über 200 Variablen.

b) Sonstige Quellen

LINDO Systems Inc.: LINGO the modeling language and optimizer (LINDO Systems, Chicago, 2004)

Vejsada, Karel: Operations Research, Script und Blattsammlung zur Vorlesung (Hochschule Pforzheim, 1988 – 2001; vergriffen)

Anm.: Die studentischen Modelle und deren Lösungen mit dem Programm Lingo können auf der Homepage des Verfassers unter http://or.vejsada.com eingesehen werden.

A Column Generation Approach to Airline Crew Scheduling*

Ralf Borndörfer[1], Uwe Schelten[2], Thomas Schlechte[1], and Steffen Weider[1]

[1] Konrad-Zuse-Zentrum für Informationstechnik Berlin, Takustr. 7, 14195 Berlin, Germany; Email: {borndoerfer, schlechte, weider}@zib.de
[2] Lufthansa Systems Berlin, Fritschestraße. 27-28, 10585 Berlin, Germany; Email: uwe.schelten@lhsystems.com

Summary. The airline crew scheduling problem deals with the construction of crew rotations in order to cover the flights of a given schedule at minimum cost. The problem involves complex rules for the legality and costs of individual pairings and base constraints for the availability of crews at home bases. A typical instance considers a planning horizon of one month and several thousand flights. We propose a column generation approach for solving airline crew scheduling problems that is based on a set partitioning model. We discuss algorithmic aspects such as the use of bundle techniques for the fast, approximate solution of linear programs, a pairing generator that combines Lagrangean shortest path and callback techniques, and a novel "rapid branching" IP heuristic. Computational results for a number of industrial instances are reported. Our approach has been implemented within the commercial crew scheduling system NetLine/Crew of Lufthansa Systems Berlin GmbH.

1 The Airline Crew Scheduling Problem

The Airline Crew Scheduling Problem (CSP) plays a prominent role in the operations research literature not only because of its economic significance, but also because of its influence on the development of important mathematical techniques, among them branch-and-cut [7], branch-and-price [1], shortest path algorithms [4], stabilization [5], aggregation [10], and heuristics [11], see also the book of Yu [12] for a general overview.

The CSP can be described in terms of a *pairing digraph* $N = (V, A)$. Its nodes V are called *tasks*. They can be subdivided into *legs* $L \subseteq V$ that model flights and have to be assigned to crews, *supplementary tasks* that model crew activities such as check-in and check-out, deadheading (flying as a passenger), and ground transports, and *artificial tasks*, among them two tasks s and t that model the beginning and the end of a pairing. The arcs A are called *links*.

* Supported by Lufthansa Systems Berlin.

They connect tasks that can be performed consecutively by a single crew. A digraph as just described is known as a leg-on-node network; we assume that it is acyclic.

Associated with N is a set R of *pairing resources* (flight duty time, landings, etc.), a set K of *pairing types*, and a set B of *base resources* (no of crews, production days, etc.). The links are labeled with costs $c \in \mathbb{Q}^A$, and pairing and base resource consumptions $w \in \mathbb{Q}^{A\times R}$ and $d \in \mathbb{Q}^{A\times B}$, respectively (node costs and resource consumptions can be adding to adjacent arcs). There are pairing resource limits $u_k \in \mathbb{Q}^R$ for each pairing type $k \in K$, and base resource limits $\ell \in \mathbb{Q}^B$.

A path p in N has cost $c_p := \sum_{a\in p} c_a$ and consumes pairing and base resources $w_p := \sum_{a\in p} w_a$ and $d_p := \sum_{a\in p} d_a$, respectively. A path p is a *pairing* of pairing type k if $w_p \leq u_k$. We assume w.l.o.g. (by introducing pairing type resources) that each pairing is of exactly one type. A *cover* is a set of pairings that contains each leg exactly once; a cover C is a *schedule* if $d(C) := \sum_{p\in C} d_p \leq \ell$, its cost is $c(C) := \sum_{p\in C} c_p$. The CSP is to find a schedule of minimum cost.

Denoting by $\mathcal{P}$ the set of all pairings and introducing decision variables x_p for each pairing, and slack variables s_b and costs c_b for each base resource, the CSP can be stated as

$$\text{(CSP)} \qquad \min \sum_{p\in\mathcal{P}} c_p x_p + \sum_{b\in B} c_b s_b$$

$$\sum_{p\ni v} x_p = 1 \qquad \forall\, v \in L \tag{1a}$$

$$\sum_{p\in\mathcal{P}} d_{bp} x_p - s_b \leq \ell_b \qquad \forall\, b \in B \tag{1b}$$

$$0 \leq x_p \leq 1 \qquad \forall\, p \in \mathcal{P} \tag{1c}$$

$$s_b \geq 0 \qquad \forall\, b \in B \tag{1d}$$

$$x_p \in \{0,1\} \qquad \forall\, p \in \mathcal{P}. \tag{1e}$$

Here, the *partitioning constraints* (1a) guarantee that every leg is covered exactly once; to ensure feasibility, we assume that there is a "slack" pairing type with single-leg pairings of high cost. Let $\mathcal{S} \subseteq \mathcal{P}$ be the set of these pairings, one for each leg. Similarly, the slack variables ensure feasibility of the *base constraints* (1b), which control base resource consumption; strict compliance can be forced by choosing sufficiently high costs c_b. Using the leg-pairing incidence matrix $A := (a_{vp})$, i.e., $a_{vp} = 1$ if $v \in p$ and 0 otherwise, and collecting costs and base resource consumptions in vectors $c_{\mathcal{P}} := (c_p)$, $c_B := (c_b)$, $c := (c_{\mathcal{P}}, c_B)$, and a matrix $D = (d_{bp})$, (CSP) reads

$$\text{(CSP)} \qquad \min\ c^{\mathsf{T}}(x,s) \quad Ax = \mathbb{1},\ Dx - s \leq \ell,\ 0 \leq x \leq \mathbb{1},\ s \geq 0,\ x \in \{0,1\}^{\mathcal{P}}.$$

2 A Column Generation Algorithm

We use a column generation algorithm to solve (CSP). Denote by $\mathcal{S} \subseteq \mathcal{P}' \subseteq \mathcal{P}$ some subset of pairings, by $A' := A_{\cdot\mathcal{P}'}$ the submatrix of A restricted to the pairings in $\mathcal{P}'$, and similarly c', x', and B', by $A_p := A_{\cdot\{p\}}$ and $D_p := D_{\cdot\{p\}}$, and by

$$\begin{array}{ll}\text{(MLP)} & \min\ c^{\mathsf{T}}(x,s),\quad Ax = \mathbb{1},\quad Dx - s \leq d,\quad 0 \leq x \leq \mathbb{1}, s \geq 0 \\ \text{(RMLP)} & \min\ c'^{\mathsf{T}}(x',s),\ A'x' = \mathbb{1},\ D'x' - s \leq d,\ 0 \leq x' \leq \mathbb{1}, s \geq 0\end{array}$$

the LP-relaxation associated with (CSP), the *master LP*, and the *restricted master LP*, respectively. Denoting for a given dual solution (π, μ) to (RMLP) (where π is associated with the partitioning and μ with the base constraints) by $\overline{c}_p := c_p - \pi^{\mathsf{T}} A_p + \mu^{\mathsf{T}} D_p$ the *reduced cost* of pairing p and by

$$\text{(PRICE)}\quad \min\nolimits_{p\in\mathcal{P}}\ \overline{c}_p$$

the *pricing problem* associated with (RMLP), our method can be outlined as follows. It tries to solve the master LP in a first phase. In a second phase, a plunging heuristic is started that fixes and generates pairings to (hopefully) produce a feasible solution. Such a method is known as a branch-and-generate algorithm [9]. We will sketch in this section three important components.

2.1 LP Solution

The proximal bundle method [8, 6] is a fast subgradient-type method for convex programming that can be used to solve Lagrangean relaxations of linear programs. It computes Lagrangean multipliers of the relaxed constraints, an approximate primal solution, and a bound on the optimum objective value of the original LP. We consider the Lagrange function arising from (CSP) by relaxing the partitioning and base constraints

$$L(\pi,\mu) := \pi^{\mathsf{T}} 1 - \mu^{\mathsf{T}} \ell + \min_{0\leq x\leq 1} (c_{\mathcal{P}}{}^{\mathsf{T}} - \pi^{\mathsf{T}} A + \mu^{\mathsf{T}} D)x + \min_{s\geq 0} (c_B - \mu)^{\mathsf{T}} s$$

Applying the bundle method to compute $\max_{\pi \text{ free}, \mu\geq 0} L(\pi,\mu)$, the main work turns out to be the computation of the expressions $\pi^{\mathsf{T}} A$ and $\mu^{\mathsf{T}} D$ in the function $L(\pi,\mu)$. We use an active set method to speed up this step.

It restricts the evaluation of L to a subset $I \subset \mathcal{P}$ of pairings. This set I, the *active set*, gives rise to a function L_I by replacing A, D, and $c_{\mathcal{P}}$ by submatrices $A_{\cdot I}$, $D_{\cdot I}$, and c_I. We have $L_I(\pi,\mu) \geq L(\pi,\mu)$ for all π and for all $\mu \geq 0$, and it is easy to see that $L_I(\pi,\mu) = L(\pi,\mu)$ holds if $\overline{c}_p = c_p - \pi^{\mathsf{T}} A_p + \mu^{\mathsf{T}} D_p \geq 0$ for all pairings p. We use this observation to restrict I to pairings p with reduced cost $\overline{c}_p \leq \epsilon$ for some threshold $\epsilon > 0$. We update the active set I only if the so-called *stability center* of the bundle method changes. This can lead to situations where the active set does not contain all columns with $\overline{c}_p < 0$, which can result in a model of L_I that overestimates the real value of L at some points. If we notice that, we repair the model of L_I.

2.2 Column Generation

As all pairings end in the non-leg task t, we can define the *reduced cost of an arc* $ij \in A$ as $\bar{c}_{ij} := c_{ij} - \sum_{v=i} \pi_v + \sum_{b \in B} \mu_b d_{ij,b}$ and the pricing problem to construct a pairing of type k of negative reduced cost becomes a constrained shortest path problem in the acyclic digraph N:

$$\text{(PRICE)} \qquad \min \sum_{a \in A} \bar{c}_a x_a$$

$$\sum_{a \in \delta^{\text{out}}(v)} x_a - \sum_{a \in \delta^{\text{in}}(v)} x_a = \delta_{st}(v) \qquad \forall\, v \in V \tag{2a}$$

$$\sum_{a \in A} w_{ar} x_a \le u_{kr} \qquad \forall\, r \in R \tag{2b}$$

$$0 \le x_a \le 1 \qquad \forall\, a \in A \tag{2c}$$

$$x_a \in \{0,1\} \qquad \forall\, a \in A. \tag{2d}$$

Here, $\delta_{st}(v) = 1$ if $v = s$, $\delta_{st}(v) = -1$ if $v = t$ and $\delta_{st}(v) = 0$ else. We solve this problem using a branch-and-bound algorithm similar to [2], using lower bounds derived from a Lagrangean relaxation of the resource constraints (2b), see [3] for more details. Using configurable classes of linear resource constraints and multilabel methods, we can handle most pairing construction rules directly. Some rules, however, are so complex, that these techniques would become unwieldy or require too much customization. For such cases, we use a callback mechanism, that is, we ignore the rule in our pricing model, construct a pairing, and send it to a general *rule verification oracle* that either accepts or rejects the pairing.

2.3 IP Heuristic

The idea of our "rapid branching" heuristic is to produce a solution quickly by iteratively solving the (RMLP) using the bundle method and fixing large numbers of pairing variables to one. The fixed variables are selected according to their x-value, i.e., the closer x_p is to 1, the higher the probability that x_p is fixed to 1. In order to have a large number of variables with values close to 1 available, the heuristic perturbs in each iteration the objective function according to the formula $c_p := c_p(1 - \alpha x_p^2)$ (α is a control parameter), which favors such pairings; we call this method "perturbation branching", see also [11] for a similar idea. Between fixes, pairings are generated to complement the fixes; backtracks, i.e., unfixings of pairings, are also performed sometimes. The frequency and intensity of column generation and backtracks is controlled by *targets*, i.e., estimates on the increase of the objective function under fixings, see also [9]. If the objective develops as expected, no pairings are generated; if it increases more than expected, we try to correct the problem by generating new pairings, if this does not work, we backtrack, and if we are out of time, we output the best solution found.

Table 1. Test Scenarios.

Name	Scenario 1	Scenario 2
#Days	14	31
#Home Bases	3	2
#Pairing Types	4	2
#Legs	4104	2154
#Tasks	32832	14373
#Links	1438659	168352

3 Computational Results

We now present computational results on industrial data provided by Lufthansa Systems Berlin, see Table 1. Scenario 1 is a 14-days problem with linear pairing rules; the main objective was to minimize the number of production days, secondary objectives were to minimize flight transports and rest periods. We consider two variants, an unconstrained scenario and one with 56 base constraints on the distribution of crews at home bases and of pairing types for each day. Scenario 2 is a 31-day instance with more complex rules, some of which had to be handled with callbacks. All computations were made single threaded on a Dell Precision 650 PC with 2GB of main memory and a dual Intel Xeon 3.2 GHz CPU running SUSE Linux 9.3.

Comparing the bundle method to an exact LP solver within a column generation algorithm is not straightforward. Bundle solves an individual LP clearly faster, but it produces approximate solutions, which may lead to more pairing generation iterations. Table 2 reports the results of running our op-

Table 2. Solving the Reduced Master LP.

Scenario 1	Unconstrained with Bundle	Unconstrained with CPLEX	Base Constraints with Bundle	Base Constraints with CPLEX
LP-Value	833	836	859	860
IP-Value	835	836	862	864
#Production days	1055	1057	1089	1091

Table 3. Constructing Pairings with Lagrangean Pricing and Callbacks.

Scenario 2	with Deadheads	without Deadheads
LP-Value	2977	3757
IP-Value	2981	3762
#Production days	1668	1660
Time	8h	38 min.
Callbacks failed/overall	174130/270378	0/58648
in %	64.40	0.00

timizer on scenario 1, solving RMLPs with our bundle code and with the barrier implementation of CPLEX 9.0; the overall time limit was 2 days. It can be seen that there is no loss of solution quality using the bundle method, and that the rapid branching heuristic constructs a solution with an objective value that almost equals that of the master LP.

Table 3 gives some details on pairing construction for scenario 2. This scenario involves a non-linear rule on the reduction of resources in the presence of deadheads, which was modeled using callbacks. It can be seen that the percentage of rejected pairings is relatively high, because the pairing generator tends to produce infeasible pairings repeatedly. The scenario can, however, be handled successfully.

It is not only possible to model this particular rule in a linear way. We are currently working on more general classes of linear and multilabel rules in order to cover all important rules directly in the pairing generator. We are confident that we can improve the performance of our optimizer significantly in this way in the future.

References

1. C. Barnhart, E. L. Johnson, G. L. Nemhauser, M. W. P. Savelsbergh, and P. H. Vance, *Branch-and-price: Column generation for solving huge integer programs*, Operations Res., 46 (1998), pp. 316–329.
2. J. E. Beasley and N. Christofides, *An algorithm for the resource constrained shortest path problem*, Networks, 19 (1989), pp. 379–394.
3. R. Borndörfer, M. Grötschel, and A. Löbel, *Duty scheduling in public transit*, in Mathematics – Key Technology for the Future, W. Jäger and H.-J. Krebs, eds., Springer, 2003, pp. 653–674.
4. M. Desrochers, *A new algorithm for the shortest path problem with resource constraints*, Tech. Rep. 421A, Centre de Recherche sur les Transports, Univ. Montréal, 1986.
5. O. du Merle, D. Villeneuve, J. Desrosiers, and P. Hansen, *Stabilized column generation*, Discrete Math., 194 (1999), pp. 229–237.
6. C. Helmberg and K. C. Kiwiel, *A spectral bundle method with bounds*, Math. Prog., 2 (2002), pp. 173–194.
7. K. Hoffman and M. W. Padberg, *Solving airline crew scheduling problems by branch-and-cut*, Management Sci., 39 (1993), pp. 657–682.
8. K. C. Kiwiel, *Proximal bundle methods*, Math. Prog., 46 (1990), pp. 105–122.
9. R. Marsten, *Airline crew scheduling*. Talk at the 15th Int. Symp. Math. Prog., 1994.
10. D. Villeneuve, J. Desrosiers, M. E. Lübbecke, and F. Soumis, *On compact formulations for integer programs solved by column generation*, Tech. Rep. No. 2003/25, TU Berlin, 2003.
11. D. Wedelin, *An algorithm for a large scale 0-1 integer programming with application to airline crew scheduling*, Ann. Oper. Res., 57 (1995), pp. 283–301.
12. G. Yu, *Operations Research in the Airline Industry*, Kluwer, 1997.

A Flexible Model and Efficient Solution Strategies for Discrete Location Problems

Alfredo Marín[1], Stefan Nickel[2], Justo Puerto[3] and Sebastian Velten[4]

[1] Department of Statistics and Operations Research, University of Murcia, 30100 Murcia, Spain, amarin@um.es
[2] Department of Operations Research and Logistics, Saarland University, 66041 Saarbrücken, Germany, s.nickel@orl.uni-saarland.de
[3] Department of Statistics and Operations Research, University of Seville, 41012 Seville, Spain, puerto@us.es
[4] Department of Operations Research and Logistics, Saarland University, 66041 Saarbrücken, Germany, s.velten@orl.uni-saarland.de

Abstract. We propose a new formulation for the Discrete Ordered Median Problem (DOMP) which has first been introduced in [5]. For this new formulation we present several variable fixing strategies and one family of valid inequalities which are used in a specialized branch & cut procedure. Extensive computational results show that using this method, problems can be solved, which are more than twice as large as those that can be solved by existing solution approaches.

1 Introduction and Problem Definition

Let V be a set of M discrete locations, which represent clients as well as potential facility locations. Moreover, let $C = (c_{ij})$ $(i, j = 1, \ldots, M)$ be a non-negativ $M \times M$ cost matrix, whereas c_{ij} denotes the cost of satisfying the total demand of client i from a facility at location j. In addition, we assume that $c_{ii} = 0$ $(\forall\ i = 1, \ldots, M)$ ($\rightarrow$ free self–service (FSS)) and we have no capacities.

Given N as the number of facilities to locate and $\Lambda = (\lambda_1, \ldots, \lambda_M)$ with $\lambda_i \geq 0\ \forall\ 1, \ldots, M$, the Discrete Ordered Median Problem (DOMP) is defined as follows:

$$\min_{\substack{X \subset N \\ |X|=N}} f_\Lambda(X) = \sum_{i=1}^{M} c_{\sigma(i)}(X) \cdot \lambda_i. \tag{1}$$

Here, $\sigma(\cdot)$ is a permutation of $\{1, \ldots, M\}$ such that (with $c_i(X) = \min_{j \in X}\{c_{ij}\}$)

$$c_{\sigma(1)}(X) \leq \cdots \leq c_{\sigma(M)}(X),$$

and we set

$$c_{\leq}(X) = (c_{\sigma(1)}(X), \ldots, c_{\sigma(M)}(X)),$$

or for short

$$c_{\leq}(X) = (c_{(1)}(X), \ldots, c_{(M)}(X)).$$

By using appropriate values for Λ, nearly all classical discrete facility location problems, like center, median and cent-dian can be modeled. In addition, a wide range of new location objectives can be derived (see [2] and [6]).

Since the (DOMP) contains the discrete N–median problem, which is $\mathcal{NP}$–hard (see [4]), as a special instance, it is $\mathcal{NP}$–hard, too. Moreover, the above formulation is non–linear. Different formulations have been proposed which are quadratic (see [5]) or even linear (see [1], [5] and [6]). However, none of these approaches leads to satisfactory results concerning the solution times of larger problem instances. Therefore, we provide a different formulation which is based on the idea of [3] for modeling the p–center problem.

2 A New Formulation

Let G be the number of different non–zero elements of C. Then we can order these values in non–decreasing sequence:

$$c_{(0)} := 0 < c_{(1)} < c_{(2)} < \cdots < c_{(G)} := \max_{1 \leq i,j \leq M} \{c_{ij}\}.$$

Given a feasible solution (i.e. $X \subset V$, $|X| = N$), we can use this ordering to define the following variables ($i = 1, \ldots, M-N$ and $k = 1, \ldots, G$):

$$x_{jk} := \begin{cases} 1 \text{ if the } j\text{–th smallest allocation cost (w.r.t. } X) \text{ is at least } c_{(k)} \\ 0 \text{ otherwise} \end{cases}$$

With respect to this definition the j–th smallest allocation cost element (w.r.t. X) is equal to $c_{(k)}$ if and only if $x_{jk} = 1$ and $x_{jk+1} = 0$. Therefore, we can reformulate the objective function of the (DOMP) as follows:

$$f_\Lambda(X) = \sum_{j=1}^{M-N} \sum_{k=1}^{G} \lambda_{N+j} \cdot (c_{(k)} - c_{(k-1)}) \cdot x_{jk}. \tag{2}$$

Next, we are interested in variables reflecting the location decisions of X. Let, for each row i ($i = 1, \ldots, M$) of C, G_i be the number of different non-zero elements in this row. Thus, we obtain, as for the whole matrix C, the ordering

$$c^i_{(0)} := 0 < c^i_{(1)} < \cdots < c^i_{(G_i)} := \max_{j=1,\ldots,M} \{c_{ij}\},$$

and we can define the following variables ($i = 1, \ldots, M$ and $k = 1, \ldots, G_i$):

$$z_{ik} := \begin{cases} 1 \text{ if the allocation cost for location } i \text{ is at least } c^i_{(k)} \\ 0 \text{ otherwise} \end{cases}$$

Note that this definition implies that a facility is open at location i if and only if $z_{i1} = 0$.

Using these variables the (DOMP) can be reformulated, as a binary linear program, in the following way:

$$\text{Min} \quad \sum_{j=1}^{M-N} \sum_{k=1}^{G} \lambda_{N+j} \cdot (c_{(k)} - c_{(k-1)}) \cdot x_{jk} \tag{3}$$

$$\text{s.t.} \quad \sum_{i=1}^{M} z_{i1} = M - N \tag{4}$$

$$z_{ik} \geq 1 - \sum_{\substack{j=1,\dots,M \\ c_{ij} < c^i_{(k)}}} (1 - z_{j1}) \qquad i = 1, \dots, M;\ k = 1, \dots, G_i \tag{5}$$

$$\sum_{j=1}^{M-N} x_{jk} = \sum_{\substack{i=1,\dots,M \\ l^i_k \leq G_i}} z_{il^i_k} \qquad k = 1, \dots, G \tag{6}$$

$$x_{jk} \geq x_{j-1k} \qquad j = 2, \dots, M;\ k = 1, \dots, G \tag{7}$$

$$x_{jk} \in \{0, 1\} \qquad j = 1, \dots, M;\ k = 1, \dots, G \tag{8}$$

$$z_{ik} \in \{0, 1\} \qquad i = 1, \dots, M;\ k = 1, \dots, G_i \tag{9}$$

In the above formulation, Constraint (4) ensures that exactly N facilities are opened. Moreover, Constraints (5) assure that $z_{ik} = 0$ if an open facility can be reached from location i within a cost radius of $c^i_{(k)}$ and, that $z_{ik} = 1$ otherwise. Constraints (6) ensure that the values of the x– and z–variables are consistent. Therefore, we define

$$l^i_k := \begin{cases} \min\{s : c^i_{(s)} \geq c_{(k)}\} \text{ if } c_{(k)} \leq c^i_{(G_i)} \\ \quad G_i + 1 \qquad\qquad \text{otherwise} \end{cases}, \tag{10}$$

and the number of locations allocated with cost at least $c_{(k)}$ represented via the x–variables has to be equal to the same number represented via the z–variables. At last, Constraints (7) make sure that if the $(j-1)$–th smallest allocation cost is at least $c_{(k)}$, then the j–th smallest allocation cost has the be at least $c_{(k)}$, too.

Since the proposed formulation contains $O(M^2)$ binary variables and $O(M^2)$ constraints, fast solution times for larger problem instances, using standard software–tools, seem to be very unlikely. Therefore, we propose a specialized branch & cut procedure and several variable fixing strategies, exploiting the special structure of the presented model.

3 Variable Fixing and Valid Inequalities

First of all observe that the following proposition holds because of (3) and Constraints (5) and (6):

Proposition 1. *If* $z_{i1} \in \{0,1\}$, *then* $z_{ik} \in \{0,1\}$ *for* $i = 1,\ldots,M$ *and* $k = 2,\ldots,G_i$.

Therefore, the binary restrictions on z_{ik} ($i = 1,\ldots,M$; $k = 2,\ldots,G_i$) can be replaced by $0 \leq z_{ik} \leq 1$.

Furthermore, some of the x– and z–variables can be fixed to 0 or 1 in a preprocessing step. For some of these variable fixing strategies an upper bound on the optimal objective value is needed, which can be obtained, for example, by the Variable Neighborhood Search presented in [2] and [6].

Fixing x–variables to 1

Obviously $x_{j1} = 1$ ($j = 1,\ldots,M-N$), because if a location has to be allocated, the cost is at least $c_{(1)}$. Now assume that $x_{jk} = 0$ (with $1 \leq j \leq M-N$, $2 \leq k \leq G$ fixed) and let $L = \{i = 1,\ldots,M : l_k^i = 1\}$. Then, for a feasible solution it holds:

$$M - N - j \geq \sum_{j'=1,\ldots,M-N} x_{j'k} = \sum_{\substack{i=1,\ldots,M \\ l_k^i \leq G_i}} z_{il_k^i} \geq |L| - N \tag{11}$$

Hence, if $M - j < |L|$, x_{jk} has to be equal to 1. Otherwise, the solution cannot be feasible.

Fixing x–variables to 0

Assume $x_{jk} = 1$. Then it follows from Constraints (7) that $x_{lk} = 1$ for all $l = j+1,\ldots,M-N$. Thus, and since $x_{j1} = 1$ for all $j = 1,\ldots,M-N$, the following expression provides a lower bound on the optimal objective value for the case $x_{jk} = 1$:

$$c_{(1)} \cdot \left(\sum_{l=N+1}^{j-1} \lambda_l \right) + c_{(k)} \cdot \left(\sum_{l=j}^{M-N} \lambda_l \right). \tag{12}$$

This lower bound can be compared to the upper bound mentioned above, and in case it is higher, x_{jk} has to be equal to 0.

Fixing z–variables to 0

Assume $z_{ik} = 1$ and let W be the set of locations outside cost–radius $c_{(k)}^i$ around i. These are the only locations where a facility can be open if $z_{ik} = 1$. If, on the one hand, the cardinality of this set is less than N, it is easy to see that z_{ik} has to be equal to 0. On the other hand, if the cardinality of W is greater than or equal to N, we define

$$\tilde{c}_l(W) = \begin{cases} \min\limits_{j \in W \setminus \{l\}} \{c_{lj}\} & \text{if } l \in W \\ \min\limits_{j \in W} \{c_{lj}\} & \text{otherwise} \end{cases}, \tag{13}$$

change the N largest values of $c(W)$ to 0 and sort this vector in non–decreasing sequence ($\rightarrow c_{\leq}(W)$). Then $\langle \Lambda, c_{\leq}(X) \rangle$ provides a lower bound on the optimal objective value for the case $z_{ik} = 1$. As above, this lower bound can be compared to a given upper bound and in case it is higher z_{ik} has to be equal to zero.

Valid Inequalities

In designing the specialized branch & cut procedure for the new formulation of the (DOMP) the following family of valid inequalities proved to be quite useful:

$$\sum_{j=1}^{|A|} x_{(M-N+1-j)\,k} \geq \sum_{\substack{i \in A \\ l_k^i \leq G_i}} z_{i\,l_k^i} \quad \forall\, A \subset V,\ |A| \leq M-N;\ k = 2, \ldots, G. \qquad (14)$$

Note that these valid inequalities can directly be derived from Constraints (6) and (7) and the fact that all variables are binary.

Proposition 2. *If a feasible solution satisfies $z_{ik} \in \{0,1\}$ ($i = 1, \ldots, M$, $k = 1, \ldots, G_i$), then the Valid Inequalities (14) enforce $x_{jk} \in \{0,1\}$ ($j = 1, \ldots, M-N$, $k = 1, \ldots, G$).*

Branch & Cut

After applying the variable fixing strategies presented above, the new formulation for the (DOMP) can be solved by a branch & cut procedure using the Valid Inequalities (14). Thereby, the steps for solving a subproblem of the branching–tree are:

1. Solve the LP–relaxation.
2. Add for each k ($2 \leq k \leq G$) **one** violated cut of type (14) (if there is any). If any cuts have been added, goto step 1.
3. If $z_{i1} \in \{0,1\}$ for all $i = 1, \ldots, M$, the subproblem can be pruned by optimality. Otherwise, check if it can be pruned by bound. If not, choose a non–integral z_{i1} variable and build two new subproblems.

Observe that, due to Proposition 1 and 2, branching is only necessary for M binary variables (z_{i1} ($i = 1, \ldots, M$)).

4 Computational Results

The branch & cut procedure described in the previous section has been implemented using *Visual C++* 7.0. *ILOG Concert Technology* 2.1 and *ILOG CPLEX* 9.1 has been used for the implementation and solution of linear programs. All computational studies have been performed on a PC with a *Pentium IV* processor with 2.4 GHz and 512 MB of RAM.

Problem instances have been tested for eight different Λ–vectors, whereas Median (row 1 of Table 1), Center (row 2 of Table 1), k–Centra (row 3 of Table 1) and k_1+k_2–Trimmed Mean (row 4 of Table 1) are well known special cases. Moreover, $M = 20 - 70$ and $N = 3 - 8$. Five problem instances have been randomly generated for each combination of M and N and each choice of Λ. Some of these computational results are given in Table 1.

Table 1. Sample of Average Solution Times (in s) (Preprocessing/**Total**).

Λ	$M = 30,\ N = 5$	$M = 50,\ N = 5$	$M = 70,\ N = 5$
$(1,\dots,1)$	1.02/**1.29**	4.88/**28.22**	10.11/**130.73**
$(0,\dots,0,1)$	1.11/**1.55**	4.46/**12.66**	8.45/**202.82**
$(0,\dots,0,1,\dots,1)$	1.13/**13.86**	4.63/**1462.70***	–/–**
$(0,\dots,0,1,\dots,1,0,\dots,0)$	1.12/**1.91**	4.36/**43.69**	9.46/**153.70**
$(0,1,0,1,\dots,0,1,0,1)$	0.99/**2.79**	4.84/**51.34**	9.39/**227.83**
$(1,0,1,0,\dots,1,0,1,0)$	1.15/**2.54**	4.14/**57.73**	8.99/**203.78**
$(\dots,0,1,1,0,1,1)$	1.19/**2.47**	4.38/**54.35**	9.91/**241.09**
$(\dots,0,0,1,0,0,1)$	1.12/**5.64**	4.84/**83.26**	10.24/**443.48**

* Only 3 of 5 problems could be solved within the time limit of 3600s
** No problem could be solved within the time limit of 3600s

It can be observed that, for nearly all choices of Λ, problem instances with up to $M = 70$ can be solved in reasonable time. These results are quite promising, since this M–value is more than twice as large as the one of the problems which can be solved by existing solution approaches (see [1] and [6]).

References

1. N. Boland, P. Dominguez-Marin, S. Nickel, and J. Puerto, *Exact procedures for solving the discrete ordered median problem*, To appear in Computers and Operations Research (2005).
2. P. Dominguez-Marin, *The discrete ordered median problem: Models and solution methods*, Ph.D. thesis, University of Kaiserslautern, 2003.
3. S. Elloumi, M. Labbe, and Y. Pochet, *A new formulation and resolution method for the p-center problem*, INFORMS Journal on Computing **16** (2004), 84–94.
4. O. Kariv and S.L. Hakimi, *An algorithmic approach to network location problems. II: The p-medians*, SIAM Journal on Applied Mathematics **37** (1979), 539–560.
5. S. Nickel, *Discrete ordered weber problems*, Operations Research Proceedings 2000, Springer Verlag, 2001, pp. 71–76.
6. S. Nickel and J. Puerto, *Facility location - a unified approach*, Springer Verlag, 2005.

Finding Feasible Solutions to Hard Mixed-integer Programming Problems Using Hybrid Heuristics

Philipp M. Christophel[1], Leena Suhl[1], and Uwe H. Suhl[2]

[1] DS&OR Lab, University of Paderborn, Warburgerstr. 100, 33098 Paderborn, [pmc | suhl]@dsor.de

[2] Institut für Produktion, Wirtschaftsinformatik und OR, Freie Universität Berlin, Garystr. 21,14195 Berlin (Dahlem), suhl@wiwiss.fu-berlin.de

Abstract: In current mixed-integer programming (MIP) solvers heuristics are used to find feasible solutions before the branch-and-bound or branch-and-cut algorithm is applied to the problem. Knowing a feasible solution can improve the solutions found or the time to solve the problem very much. This paper discusses hybrid heuristics for this purpose. Hybrid in this context means that these heuristics use the branch-and-bound algorithm to search a smaller subproblem. Several possible hybrid heuristics are presented and computational results are given.

1 Introduction

Nowadays, a wide variety of optimization problems is modeled as linear mixed-integer programming (MIP) problems. Many of these problems can be solved to optimality using specialized algorithms or MIP solvers. However, there are still many problems which can not be solved in acceptable time.

One way to improve the solution behavior of MIP solvers is to use primal heuristics for finding feasible solutions before starting the solution process or to use improvement heuristics to search the vicinity of a known feasible solution for a better one. This usually results in a good primal bound that speeds up the algorithm or at least results in a better solution when the MIP solver run is aborted, e.g. when a time limit is reached.

In literature a number of publications on primal heuristics and improvement heuristics are known. Among the most important are [1], [2], [3], [4] and [5]. In this paper, we discuss a class of primal heuristics that are similar to the improvement heuristics RINS [5] and Local Branching [4]. Further details and more computational tests about these heuristics are given in [6].

2 Hybrid Heuristics

Hybrid heuristics combine exact methods for solving MIP problems with heuristic methods. In the context of this paper, this means that a branch-and-bound (or branch-and-cut) algorithm is used to conduct a local search [7] on a certain part of the solution space. This part of the solution space is defined by restricting some or all variables of the problem to create a subproblem that potentially holds a good solution.

The hybrid heuristics for MIP problems published so far ([4] and [5]) are improvement heuristics, i.e. they try to find better solutions based on a known feasible solution. In this paper, hybrid heuristics that do not need a feasible solution are discussed. Such a hybrid heuristic has two parts. First, some of the integer variables are fixed based on the linear programming (LP) relaxation solution to define a *relaxation-based search space*. This search space then is explored by a MIP solver.

2.1 Relaxation-based Search Spaces

A relaxation-based search space is defined by fixing some integer variables to values obtained from an LP relaxation solution. Some variables already take integer values in the LP relaxation, others may have to be rounded. The result is a subproblem of the original MIP problem that concentrates on a certain part of the search tree.

Two factors have to be considered when a relaxation-based search space is defined. The first is how many variables are fixed. For each variable fixed the remaining subproblem becomes smaller and therefore potentially easier to solve. But smaller MIP problems are not necessarily easier to solve and in a smaller search space the chance to find a *good* solution decreases.

The second factor is which variables to fix. Several approaches for this question can be thought of. One is to fix those variables that already take integer values in the relaxation solution. It also makes sense to take a look at the reduced-cost values of the non-basic variables before fixing them. A high reduced-cost value indicates that the variable is likely to take this value in a good solution. Another approach is to fix variables that already are close to integer values.

The order in which the variables are fixed also influences the resulting search space if indirect fixings are taken into account. Indirect fixings occur if it is checked whether fixing a variable leads to more fixings based on the implications found in the preprocessing phase. This is done in the MOPS® (Mathematical OPtimization System) MIP solver (see [8] and [9]), which is used for the results in this paper.

2.2 Examples for Relaxation-based Search Spaces

In this section, three different relaxation-based search spaces are introduced. The first two of these are very simple, but section 3 shows that even these very simple approaches can achieve good results. The third relaxation-based search space is currently used in the primal heuristic of the MOPS MIP solver.

The first search space presented here is defined by fixing all binary variables that take integer values in the LP relaxation. A heuristic based on this search space can be seen as applying the RINS [5] heuristic to the root node under the assumption that a feasible solution exists where all variables that take integer values in the LP relaxation solution also take integer values. Further on, this search space and the corresponding hybrid heuristic are called RSS01.

The drawback of this first search space is that it might be too small for certain problems—especially if a problem has many non-basic variables. In a MIP problem with binary variables, fixing a variable to one usually has a larger impact on the solution of the problem than fixing a variable to zero. Fixing a variable to one also usually results in many indirect fixings of variables to zero. Therefore, the second search space suggested here is defined by fixing all binary variables that take the value one in the LP relaxation solution. The result typically is a larger search space than RSS01. It is further on called RSS1.

The third search space is the one used by the heuristic in the current versions of the MOPS MIP solver (see [10]). Here, the first step is to fix the general integer variables, if the problem contains any, to their nearest integer value. The next step is to fix those binary variables with the largest reduced-cost values. To do this, the non-basic binary variables that have positive reduced-cost values are sorted. The largest seventy percent of these variables are then fixed to their value in the LP relaxation. Then all basic binary variables that are close to one or zero are fixed. This means, a variable j with a relaxation solution value x_j is fixed to one if $x_j \geq 1 - \texttt{tolqi}$ or to zero if $x_j \leq \texttt{tolqi}$. The default value for `tolqi` in MOPS is 0.05. Further on this search space and the resulting heuristic are called MOPSHEU.

The search spaces presented so far focus on binary variables, because binary variables in a MIP problem usually have a strong influence on the solution. For problems that do not include any binary variables RSS1 and RSS01 do not have any effect.

2.3 Exploration of Relaxation-based Search Spaces

The second part of the hybrid heuristics presented in this paper is about searching the search space defined in the first part. This is done by starting to solve the resulting subproblem with a MIP solver.

For a MIP solver, many parameters can be configured, and each parameter can have a strong impact on the results. One important parameter is the stopping criterion. Even if the search space is chosen very carefully, it may happen that a subproblem is too hard to be solved to optimality and for a heuristic this is also not needed. Therefore, it is reasonable to use a time limit, a node limit or to stop after the first solution has been found. The last case may be chosen if an improvement heuristic is used. Time limits have the drawback of not taking into account that some problems have harder to solve LP relaxations and therefore all LP based methods need more time. Node limits serve better because they take this fact into account.

Another important parameter that can be set for MIP solvers is the node selection strategy. The MOPS heuristic uses a LIFO branch-and-bound and a node limit of 100 to find solutions very quickly, because it is intended to be used on easy and hard problems. The LIFO branch-and-bound is a depth-first strategy. For hard problems and larger node limits we suppose that using the default branch-and-bound (or branch-and-cut) of a MIP solver is advisable.

3 Computational Results

This section lists results for the three hybrid heuristics mentioned in the last section (see table 1). The heuristics were run as primal heuristics after preprocessing as part of the MOPS MIP solver. The test problems are those problems from MIPLIB 2003 [11] that are stated to take more than one hour to solve with a commercial solver or are unsolved. Some problems had to be omitted because MOPS was not able to complete preprocessing or solve the LP relaxation in acceptable time. The machine used was a 1.8 GHz AMD Athlon™ processor, 512 MB RAM and a Windows® 2000 operating system.

The MOPSHEU uses the LIFO branch-and-bound (depth-first) with a node limit of 100 to search the defined search space. The other two heuristics were configured with a node limit of 1000 and the default branch-and-bound (best projection node selection).

Table 1. Computational results for three different hybrid heuristics.

		MOPSHEU			RSS01			RSS1		
	best Obj.	Obj.	Ratio	Time/sec	Obj.	Ratio	Time/sec	Obj.	Ratio	Time/sec
a1c1s1	1,16E+04*			0,031	1,54E+04	1,333	26,141	1,48E+04	1,277	25,359
arki001	7,58E+06*			0,016			42,375			26,094
dano3mip	694,00*	747,62	1,077	110,875	694,00	1,000	1275,109	724,79	1,044	1493,844
danoint	65,67	66,50	1,013	1,734	66,27	1,009	18,344	66,27	1,009	18,312
glass4	1,46E+09*			0,125			0,172			1,141
harp2	7,39E+07			0,656			47,781			14,906
liu	1448,00*	4720,00	3,260	3,266	4540,00	3,135	7,562	1846,00	1,275	8,641
markshare1	1,00	314,00	314,000	0,016	206,00	206,000	0,000	134,00	134,000	0,016
markshare2	1,00	206,00	206,000	0,031	226,00	226,000	0,016	116,00	116,000	0,359
mkc	-563,85	-359,28	1,569	44,297	-367,82	1,533	59,047	-407,92	1,382	85,594
net12	214,00*			805,547			34,469			707,438
noswot	-41,00			0,016	-37,00	1,108	1,703	-40,00	1,025	1,875
nsrand-ipx	5,14E+04*	5,47E+04	1,065	97,641	5,78E+04	1,125	531,312	5,89E+04	1,146	238,875
opt1217	-16,00*	-8,00	2,000	0,094	-16,00	1,000	0,031	-16,00	1,000	1,625
roll3000	1,29E+04*			0,031			82,344	1,42E+04	1,105	46,953
seymour	423,00	444,00	1,050	18,859	431,00	1,019	325,656	436,00	1,031	448,375
sp97ar	6,68E+08*	7,14E+08	1,070	256,516	7,00E+08	1,048	2062,656	7,10E+08	1,063	777,250
timtab1	7,65E+05			0,016			5,156	1,23E+06	1,613	3,750
timtab2	60,00*			0,016			7,984			7,469
tr12-30	1,31E+05			0,141			4,109			3,969

In the best objective column, a * indicates that this is not a proven optimal objective value.

The results in table 1 show that for many problems the presented heuristics deliver good results. But it also can be seen that they may take a long time and still do not find a feasible solution. Surprisingly, the RSS1 heuristic performs best on the given problems compared with the other two heuristics.

The MOPS heuristic has the advantage that it altogether takes less time than the other two heuristics. For more computational results see [6].

To show that using a very simple hybrid heuristic can improve the solution behavior of a MIP solver, figure 1 shows the performance profiles for the MOPS MIP solver after one hour with the RSS1 heuristic and without any heuristic. Performance profiles are a method for benchmarking optimization algorithms described in [12]. The same machine and test problems as before are used. The ratio r for the profile does not indicate time but solution quality.

$$r_{p,s} = \frac{\text{objective value of configuration } s}{\text{best objective value of the two compared configurations}}$$

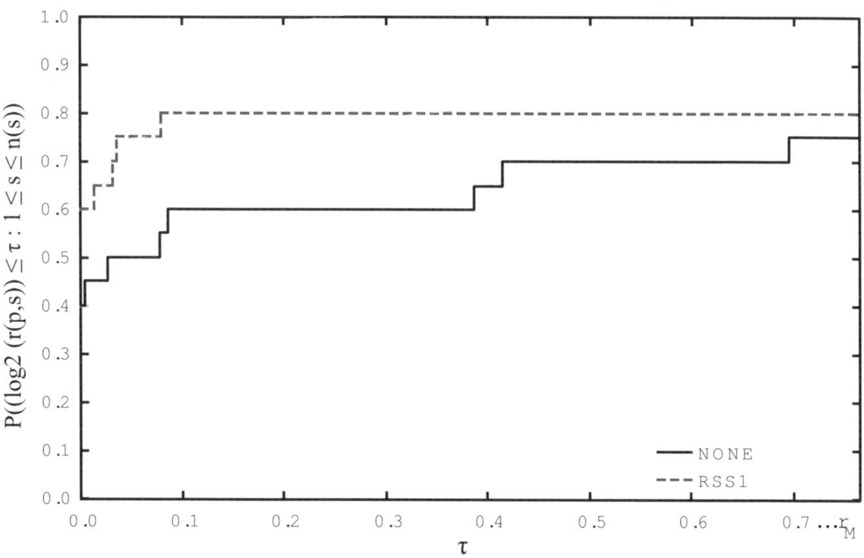

Fig. 1. Performance profile for MOPS with no heuristic and with the RSS1 heuristic.

The performance profiles in figure 1 show that MOPS with the RSS1 heuristic delivers the best objective value after one hour for 60% of the problems and for 80% of the problems a solution is found within an hour. Therefore it performs better than MOPS without any heuristic because more and better solutions are found.

4 Conclusions

This paper shows that the ideas from RINS [5] and Local Branching [4] can be used to search for feasible solutions. The resulting hybrid heuristics have the advantage that they are easy to implement and can be used for many different types of problems.

Future research has to show if relaxation-based search spaces can be defined that result in outstanding results or if all search spaces have advantages and disadvantages. Another point of interest is to find out which configuration for the MIP solver should be used when exploring a relaxation-based search space. It would also be advisable to integrate techniques from local search like intensification and diversification (see [7]).

Some of these points can be covered by searching more than one search space in a heuristic. In [6], an example for such a multiple search space hybrid heuristic is given.

References

[1] BALAS, Egon ; MARTIN, Clarence H.: Pivot and Complement – A Heuristic for 0–1 Programming. In: *Management Science* 26(1) (1980), S. 86–96

[2] LØKKETANGEN, Arne ; GLOVER, Fred: Solving Zero-one Mixed Integer Programming Problems Using Tabu Search. In: *European Journal of Operations Research* 106(2–3) (1998), S. 624–658

[3] BALAS, Egon ; CERIA, Sebastián ; DAWANDE, Milind ; MARGOT, Francois ; PATAKI, Gábor: OCTANE: A New Heuristic For Pure 0–1 Programs. In: *Operations Research* 49(2) (2001), S. 207–225

[4] FISCHETTI, Matteo ; LODI, Andrea: Local Branching. In: *Mathematical Programming* 98(1–3) (2003), S. 23–47

[5] DANNA, Emilie ; ROTHBERG, Edward ; LE PAPE, Claude: Exploring Relaxation Induced Neighborhoods to Improve MIP Solutions. In: *Mathematical Programming* 101(1) (2004), S. 71–90

[6] CHRISTOPEL, Philipp M.: *An Improved Heuristic for the MOPS Mixed-integer Programming Solver*, University of Paderborn - DS&OR Lab, Diplomarbeit, 2005

[7] YAGIURA, M. ; IBARAKI, T.: Local Search. In: PARDALOS, P. M. (Hrsg.) ; RESENDE, M. G. C. (Hrsg.): *Handbook of Applied Optimization.* Oxford University Press, Oxford, 2002

[8] SUHL, Uwe H.: MOPS - Mathematical OPtimization System. In: *European Journal of Operations Research* 72 (1994), S. 312–322

[9] SUHL, Uwe H. ; WAUE, Veronika: Fortschritte bei der Lösung gemischt-ganzzahliger Optimierungsmodelle. In: SUHL, Leena (Hrsg.) ; VOSS, Stefan (Hrsg.): *Quantitative Methoden in ERP und SCM.* 2004 (DSOR Beiträge zur Wirtschaftsinformatik, Band 2)

[10] SUHL, Uwe H.: Solving Large-scale Mixed-Integer Programs with Fixed Charge Variables. In: *Mathematical Programming* 32 (1985), S. 165–182

[11] ACHTERBERG, Tobias ; MARTIN, Alexander ; KOCH, Thorsten: *The Mixed Integer Problem Library: MIPLIB 2003.* Version: 2003. `http://miplib.zib.de`. – Online–Ressource, Abruf: 2005-07-28

[12] DOLAN, Elizabeth D. ; MORÉ, Jorge J.: Benchmarking Optimization Software with Performance Profiles. In: *Mathematical Programming* 91 (2002), S. 201–213

Optimisation of the Variant Combination of Control Units Considering the Order History

Bernd Hardung[1] and Thomas Kollert[2]

[1] AUDI AG, I/EE-81
August-Horch-Str.
85045 Ingolstadt, Germany
bernd.hardung@audi.de

[2] Darmstadt University of Technology
FG Operations Research
Hochschulstr. 1
64289 Darmstadt, Germany
kollert@bwl.tu-darmstadt.de

Summary. In modern cars, an increasing number of functions are integrated in single control units. Some of the functions can be ordered as optional equipment by the customer. To reduce costs, variants of the control units are created differing in hardware and software. Variants are created by not populating sections of a circuit board. Each additional variant of a control unit causes expenses for logistics and development. Today the process for the determination of the variants is not automated. Non-technical dependencies like equipment packages or common ordered equipment combinations can only partially be taken into account. In this article we formulate this problem and show how existing information on the manufacturer's side can serve as a base for an optimisation of the variants. We also show how the problem can be transformed into a warehouse location problem, so that it can be solved in short time. Finally, the results of an application example are presented.

1 Introduction

The functionality of modern cars is steadily increasing. Most of these functions are designed to increase comfort and safety. In order to cope with the trend towards more functions and the cost pressure, an increasing number of functions are integrated in single control units.

Functions that can be ordered as optional equipment by the customer are often performed by electronic control units, which have to be put in every car. In an effort to reduce costs, variants of these control units are created, which differ in hardware and software. The variants are designed, for example, by not populating sections of a circuit board. Each additional variant of a control unit demands high efforts in logistics and development. For this reason, it is mostly not possible to install tailor-made control unit variants for all possible orders.

Today the determination of the variants is based on equipment order rates and technical inter-dependencies. This process is complex and not automated. Non-technical dependencies like equipment packages or common ordered equipment combinations can only partially be taken into account.

This article describes how the cost-optimal variants can be determined. In Sect. 2, the problem will be described more in detail. Afterwards (Sect. 3), it will be explained how data, required for a variant combination optimisation, can be gathered. The mathematical problem description will be presented in Sect. 4, before the transformation of the variant combination problem into a warehouse location problem is shown (Sect. 5). Finally the cost reduction potential of the variant optimisation will be presented (Sect. 6).

2 Problem Description

This section gives an overview of the variant combination problem (VCP). After the definition of special terms, it will be described how the number of variants can be reduced by deactivating not ordered functions in control units. Furthermore, it will be explained how the average costs for a manufacturer can be calculated.

2.1 Terms and Definitions

The electronic system in a vehicle has to implement several *functions* or *features* f. During the development of a car, for each *electronic control unit (ECU)*, the features F_{ECU}, which have to be supported, are defined. The customer in turn decides which combination of features and properties his car should cover – the configuration of his vehicle. This influences the features that every ECU has to implement. The according set of features $F_l \subseteq F_{ECU}$ for each single car and per ECU is called *feature combination.* In this article, only optional features, which do not belong to the standard equipment, are considered.

An ECU can also be seen as composed of technical components. Each *component* c_k of an ECU supports one or more features and is therefore required by these features. On the other hand, a feature sometimes requires more than one component. As for the features, in the following, only optional components are considered, since the optimisation result is not influenced by components that all vehicles contain. For convenience, the word *optional* is left out, and it is just spoken of components and features.

Based on the relations between features and components of an ECU, it is possible to determine all components that are required to support the ordered feature combination F_l of an ECU. This combination of components is called *(minimal) component combination* $C_{F_l}^{\min}$, because no component can be removed without losing the ability to support all ordered features.

Variants v of ECUs can also be considered as combinations of components. They are built in cars to satisfy the demand for features. The variant being installed in an ordered car has to be a superset of the component combination. This means that variants sometimes support more components than required. The components in that ECU can be deactivated without being visible to the customer. A set of variants used for all orders of a vehicle type is called variant combination V_q. A variant combination is valid if for each component combination, at least one variant can be found being a superset.

2.2 Quality of a Combination of Variants

The sum of the average hardware costs and the handling costs for the variants reflect the quality of a variant combination. Both cost types depend on the number of variants. The more variants are managed, the lower the material costs $\tilde{c}$, but the higher the overall handling costs $\tilde{f}^c = \tilde{f} \cdot m$ (see Fig. 1).[3] The optimal number of variants m^* can be found where the sum $\tilde{F}$ of both costs has its minimum.

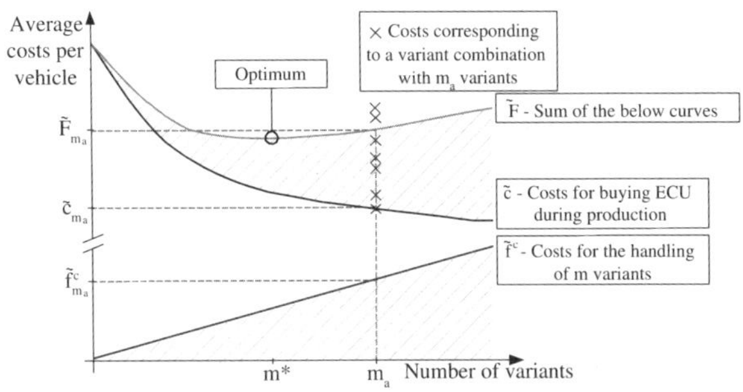

Fig. 1. Average costs dependent on the number of variants

The determination of the hardware costs depends on the chosen variant combination V_q. For each valid variant combination V_q, the corresponding cost $\tilde{c}^{V_q} = \sum_{\forall v_i \in V_q} p_i \cdot r_i$ can be calculated using the variant price p_i and the according installation rate r_i of all variants. The installation rates r_i depend on the customer orders. For each order the cheapest variant of V_q that supports all required functions is built in.

Usually, there is more than one valid variant combination that contains a defined number of variants. In Fig. 1 "x" indicates the costs $\tilde{c}^{V_q}$ for dif-

[3] For simplification, it is assumed that each additional variant causes the same order-independent costs $\tilde{f} = \tilde{f}^c/m$, where m represents the number of variants in the variant combination.

ferent variant combinations with m_a elements. The minimum of the costs corresponding to these variant combinations with m_a elements belongs to the optimal variant combination V^{m_a} with exactly m_a elements. Connecting the cost points of all optimal variant combinations V^m $(m = 1, 2, \dots)$, the curve $\tilde{c}$ in Fig. 1 can be constructed.

3 Data Determination

This section explains how all information for the determination of the optimal variant combination can be gathered:

- *The predicted "order numbers" n_j of all component combinations $C_j^{\min}$*
  For the calculation of the order numbers n_j, the feature combinations F_l of the future car orders must be known. To create an order forecast, past customer orders can be used. Some additional information is necessary in order to consider new features and customer behaviour. Using the relations between features and components (see Sect. 2.1), each feature combination F_l can be transformed into a corresponding component combination $C_j^{\min}$. For all $C_j^{\min}$ the according order number n_j can be determined as well by aggregation.
- *The prices p_i of the variants v_i*
  We assume that the price p_i of a variant v_i is composed of a base price p^{basis} for the control unit with the standard components and the respective prices p^{c_k} of the components c_k that are implemented additionally. It is sufficient, to add up only the prices p^{c_k} of the implemented components $c_k \in v_i$ because the constant base price does not influence the optimisation $(p_i = \sum_{c_k \in v_i} p^{c_k})$.
- *The cost relations $\tilde{c}_{ij}$ between each component combination $C_j^{\min}$ and each possible variant v_i*
  For a specific component combination $C_j^{\min}$, only those variants containing at least all components of $C_j^{\min}$ can be installed. For all these variants v_i, the product of their price p_i and the order number n_j of the component combination $C_j^{\min}$ equals the costs $\tilde{c}_{ij}$ that arise if in all cars that require the component combination $C_j^{\min}$, variant v_i is installed. For all other couples ij, the corresponding cost $\tilde{c}_{ij}$ can be set to a high value M. The high value ensures that the optimisation algorithm does not assign a "not allowed" variant v_i to the component combination $C_j^{\min}$. Thus, the cost $\tilde{c}_{ij}$ can be explained as follows: $\tilde{c}_{ij} = \begin{cases} n_j \cdot p_i & \text{if } C_j^{\min} \subseteq v_i, \\ M & \text{else.} \end{cases}$
- *The costs $\tilde{f}$ for handling an additional variant*
  The costs $\tilde{f}$ for the development, administration, and handling of an additional variant can be estimated using past values. This can be difficult, since various divisions of a car manufacturer get in contact with ECUs, and so all these divisions have to evaluate the costs caused by the variants.

4 Mathematical Problem Description

Besides the identifiers for the data values (see above), the variables y_i and x_{ij} are used to describe the problem. The meaning of these variables in a valid solution of the VCP is:

- $y_i = 1$, if variant i belongs to the variant combination, $y_i = 0$ otherwise,
- x_{ij}, percentage of the orders n_j of C_j^{Min} that are satisfied by the installation of v_i. Should be 1 or 0 after the optimisation.

With these variables and identifiers, the VCP can be formulated as follows:

$$\text{Min. } \mathrm{F}(\mathbf{x},\mathbf{y}) = \sum_i \sum_j \tilde{c}_{ij} \cdot x_{ij} + \sum_i \tilde{f} \cdot y_i \tag{1}$$

with the constraints

$$x_{ij} \leq y_i \qquad \forall i \text{ and } \forall j \tag{2}$$

$$\sum_i x_{ij} = 1 \qquad \forall j \tag{3}$$

$$y_i \in \{0,1\} \qquad \forall i \tag{4}$$

$$x_{ij} \geq 0 \qquad \forall i \text{ and } \forall j \tag{5}$$

The constraints (2) ensure that orders of component combinations are only satisfied by variants that are elements of the variant combination. For each component combination the corresponding constraint (3) guarantees, that all orders of this component combination are fulfilled. The ranges of the variables are restricted by the constraints (4) and (5).

5 VCP as a Warehouse Location Problem

Similar to our VCP is the (uncapacitated, simple) warehouse location problem (WLP)[4]. In a WLP, customers with demands n_j for one homogeneous product are given. They are supplied from some warehouses, which are not established yet. There exist some places where warehouses could be established. Opening a warehouse at place i causes costs equal to $\tilde{f}$. The transportation of one unit of the product from the possible place i of a warehouse to customer j costs p_{ij} and the transportation of the whole demand n_j costs $\tilde{c}_{ij} = p_{ij} \cdot n_j$. The question is at what places warehouses should be opened and from which opened warehouse the customers should be delivered, so that the overall costs are minimal. The VCP can be seen as a WLP. Therefore, the following transformations have to be done:

[4] This problem is also called the uncapacitated facility location problem or simple plant location problem (SPLP).

Table 1. Transformation of a VCP into a WLP

	VCP	**WLP**
j	component combination j	customer j
i	possible variant i	possible warehouse location i
y_i	equals one, if variant i belongs to the variant combination, else zero	equals one, if at location i a warehouse has to established, else zero
x_{ij}	$x_{ij} \cdot 100$ percent of the orders of component combination j is satisfied by the installation of variant i	$x_{ij} \cdot 100$ percent of the demand of the customer j is satisfied by the warehouse at the location i
$\tilde{c}_{ij}$	costs if variant i is built in all cars that require comp. combination j	costs if warehouse of location i covers the whole demand of customer j
$\tilde{f}$	costs for developing and handling the additional variant i	costs for building and administration of the additional warehouse i

Thus, both problems can be described mathematically in the same way ((1)-(5) see also [1, p. 52] and all solution methods that are available for the WLP can be used to solve the VCP. Besides heuristic methods, especially Branch & Bound procedures are available to solve the WLP [1, p. 78]. An efficient and exact one that was developed by Erlenkotter is presented in ([2, p. 215ff.]).

6 Conclusions

The determination of the cost-optimal variants of a control unit can be simplified and improved by a computer-aided optimisation. One of the first applications of this procedure was the variant optimisation of a control unit with 10 (optional) components. In comparison to a previously realised manual variant determination, using computer-aided optimisation can save about one euro per car if the future orders exactly match the forecast. We have also tested the results using varying forecasts. It can be shown that the results are relatively stable and represent a big improvement over the manual determination of the variants.

References

1. Domschke, W., Drexl, A.: Logistik - Standorte. 4th edn. Volume 3. Oldenbourg, München - Wien (1996)
2. Körkel, M.: Effiziente Verfahren zur Lösung unkapazitierter Standort-Probleme. VWF, Berlin (1999)

Solving a Dynamic Real-Life Vehicle Routing Problem

Asvin Goel and Volker Gruhn

Chair of Applied Telematics and e-Business, Computer Science Faculty, University of Leipzig, Klostergasse 3, 04109 Leipzig, Germany
{goel,gruhn}@ebus.informatik.uni-leipzig.de

Summary. Real-life vehicle routing problems encounter a number of complexities that are not considered by the classical models found in the vehicle routing literature. In this paper we consider a dynamic real-life vehicle routing problem which is a combined load acceptance and generalised vehicle routing problem incorporating a diversity of practical complexities. Among those are time window restrictions, a heterogeneous vehicle fleet with different travel times, travel costs and capacity, multi-dimensional capacity constraints, order/vehicle compatibility constraints, orders with multiple pickup, delivery and service locations, different start and end locations for vehicles, route restrictions associated to orders and vehicles, and drivers' working hours. We propose iterative improvement approaches based on Large Neighborhood Search. Our algorithms are characterised by very fast response times and thus, can be used within dynamic routing systems where input data can change at any time.

1 Introduction

In this paper we present algorithms for solving a dynamic real-life problem. The problem incorporates various practical complexities among which some have received only little attention in the vehicle routing literature. The problem is dynamic and information can change during the transportation process. We propose iterative improvement approaches based on Large Neighborhood Search. The algorithms we present are characterised by two features: they are capable of handling the practical complexities and they have very fast response times and thus, are suitable for dynamic optimisation.

2 Problem Formulation

This work is motivated by a practical problem arising in air-cargo transport. Most of the air-cargo within Europe is transported by so-called *road feeder services (RFS)*, that is the transport is done on roads, see [8]. In the problem considered not all transportation requests are known before load acceptance and planning starts. Instead, transportation requests may become known at any time. In contrast to many other commonly known routing problems not all transportation requests have to be assigned to a vehicle, instead a so-called *make-or-buy* decision is necessary to determine whether a transportation request should be assigned to some vehicle (make) or not (buy).

A transportation request is specified by a nonempty set of locations which have to be visited in a particular sequence by the same vehicle, the time windows in which these locations have to be visited, and the revenue gained when the transportation request is served. Furthermore, some characteristics can be specified which constrain the possibility of assigning the transportation requests to certain vehicles due to compatibility constraints and capacity constraints. At each of the locations some shipment(s) with several describing attributes can be loaded or unloaded.

A fleet of heterogeneous vehicles is available to serve the transportation requests. The vehicles can have different capacities, as well as different travel times and travel costs between locations. The vehicles can transport shipments which require some of the capacity the vehicle supplies. Instead of assuming that each vehicle becomes available at a central depot, each vehicle is given a start location where it becomes available at a specific time and with a specific load. Furthermore, the vehicles do not have to return to a central depot and for each vehicle a final location is specified, which has to be reached within a specific time and with a specific load. Each vehicle may have to visit some locations in a particular sequence between leaving its start and reaching its final location. All locations have to be visited within a specific time window. If the vehicle reaches one of these locations before the begin of the time window, it has to wait. At each of these locations some shipment(s) may have to be loaded or unloaded. Drivers' working hours are regulated by EU Council Regulation No 3820/85. After a certain amount of driving an obligatory daily rest period is necessary before the driver(s) may continue driving. The maximal time allowed driving between two consecutive daily rest periods depends on whether a vehicle is manned by one or two drivers.

A tour of a vehicle is a journey in accordance with EU social legislation starting at the vehicles start location and ending at its final location, passing all other locations the vehicle has to visit in the correct sequence, and passing all locations belonging to each transportation request assigned to the vehicle in the correct respective sequence. A tour is *feasible* if and only if for all orders assigned to the tour compatibility constraints hold and at each point in the tour time window and capacity restrictions hold. The objective is to find distinct feasible tours maximising the profit, which is determined by the accumulated revenue of all served transportation requests, reduced by the accumulated costs for operating these tours.

3 Related work

The dynamic real-life problem discussed in this paper is a generalisation of the vehicle routing problem (VRP) and the pickup and delivery problem (PDP), see [3], and [12] and secondary literature given there. Some of the generalisations have been discussed by [7], however, no model in literature considers all aspects of the RFS problem. Several extensions of the VRP have been widely studied in previous works, as the VRP with time windows, see [2], and the capacitated VRP, see [10]. In many cases it is assumed that load is accepted before planning begins and tours are generated assuming that all accepted transportation requests must be served. Work regarding load acceptance issues for the travelling salesman problem (TSP) has been surveyed by [5], but only few attempts have been made to tackle extensions of this problem. Although some work addresses certain aspects of the complexities resulting from restrictions to drivers' working hours, see [15], the only work known to the authors explicitly regarding drivers' working hours is given by [17]. A comprehensive discussion of dynamic vehicle routing can be found in [13] and [14]. Dynamic real-life problems often require rich models, in most of the literature on dynamic routing problems however, some simplifying assumptions are made. The dynamic full-truckload pickup and delivery problem for example has been studied by [6] and [18]. The only work known to the authors regarding rich VRP with multiple pickup and delivery locations in a dynamic context is presented by [15].

4 Large Neighborhood Search

Large Neighborhood Search (LNS) has been introduced for the VRP with time windows by [16] and can be interpreted as a special case of Iterated Local Search, described in [11]. The LNS method starts with an initial solution s. In each iteration k transportation requests are removed from their tours in the current solution s. A new solution s^* is then generated by inserting unscheduled transportation requests. The new solution is accepted as the next current solution if the objective value is improved. The number of removals can be adjusted before the next iteration. If no termination criterion is fulfilled, the algorithm continues with the next iteration.

LNS algorithm

0. s := initialsolution()
1. s' := removeorders(s, k)
2. s^* := insertorders(s')
3. if s^* is better than s set $s := s^*$
4. adjust parameters
5. goto 1 or stop

[9] have shown that LNS is well suited for the VRP with several additional constraints. We will show that LNS can also be used to solve the dynamic real-life problem considered in this paper. To ensure very fast response times we propose fast insertion methods.

The first method is a sequential insertion method. In the sequential insertion method unscheduled transportation requests are randomly chosen and all feasible and *efficient* insertion possibilities are determined. We assume an infinite incremental cost if no feasible insertion is possible and say that an insertion possibility is *efficient* if the incremental cost is smaller than the revenue of the order. If an efficient insertion possibility is found the transportation request is inserted to a tour with high efficiency.

The second insertion method is based on the auction method for the vehicle routing problem with time windows by [1]. This method is illustrated in figure 1. In the first phase all unscheduled orders request and receive from each vehicle an insertion possibility and the efficiency of insertion. In the second phase each unscheduled order, which did receive an efficient insertion possibility, chooses a vehicle with low incremental costs and sends a proposal for insertion to this vehicle. In phase three each vehicle which received a proposal chooses an order for insertion to the tour. The method stops if no order can be efficiently inserted and continues otherwise.

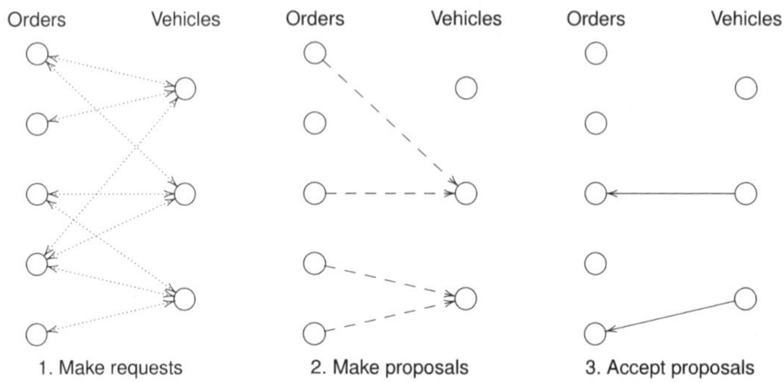

Fig. 1. Illustration of the auction method

For step 1 of the LNS algorithm [16] propose to only unschedule transportation requests which are *related* to each other. For the VRP a relatedness measure based on geographical closeness of customer locations has been applied to increase the opportunity for the reinsertion to achieve some improvement in the schedule. A concept similar to geographical closeness in the VRP however, does not exist for vehicle routing problems with pickups and deliveries. Thus, geographical closeness cannot be used for the problem considered in this work and transportation requests are unscheduled randomly in step 1 of the LNS algorithm.

5 Computational experiments

Computational experiments were performed on test cases derived from the real-life problem. We generated test problem with $|\mathcal{V}|$ vehicles and $|\mathcal{O}_0|$ orders which are known at the beginning of the planning horizon. In every hour of the planning horizon of one week $|\mathcal{O}_t|$ new orders become known. We generated a heterogeneous vehicle fleet and transportation requests with different requirements to the vehicles and pickup and delivery locations distributed as indicated in figure 2. The length of all time windows at the locations is denoted by τ.

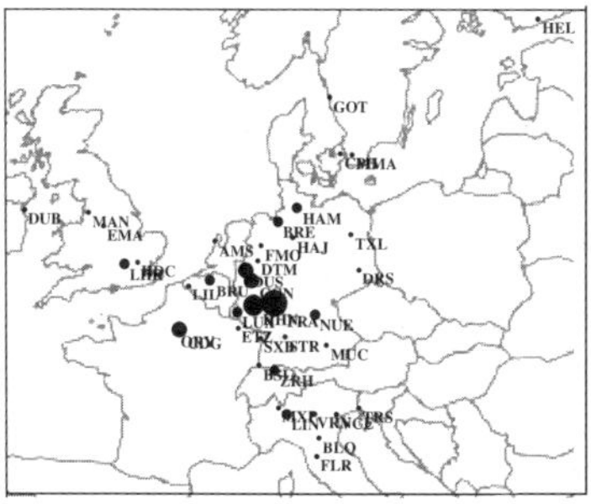

Problem				Solver		
$\|\mathcal{V}\|$	$\|\mathcal{O}_0\|$	$\|\mathcal{O}_t\|$	τ	LNS-A	LNS-AS	LNS-S
50	150	5	2h	53.54	57.93	54.67
100	300	10	2h	41.42	39.52	34.03
250	750	25	2h	24.30	21.42	17.19
500	1500	50	2h	13.19	9.49	8.77
50	150	5	12h	58.57	49.92	54.49
100	300	10	12h	39.00	35.53	32.37
250	750	25	12h	29.50	24.73	25.38
500	1500	50	12h	13.75	13.06	9.38

Fig. 2. Problem characteristics and results

At each timestep all transportation requests which were unscheduled were permanently rejected before new transportation requests were added to the problem. New transportation requests were inserted to the tours by the auction method afterwards. The solution obtained hereby was used as a reference solution. To test our algorithms we only allowed 30 seconds of computation time per timestep on a personal computer with Intel Pentium 4 processor with 3.00 GHz and linux operating system. The average time per iteration of our LNS algorithms was below one second for all test problems except for those with 500 vehicles where the average time per iteration was below 1.75 seconds. In figure 2 we show the percentage of improvement over the reference solutions. The LNS method using the auction method for reinsertion is denoted by LNS-A, the LNS method using the sequential method for reinsertion is denoted by LNS-S. The LNS method denoted by LNS-AS randomly switches between the sequential and the auction method. We can see that in most cases LNS-A outperforms the other algorithms. As diversification is high, the sequential insertion methods LNS-S and LNS-AS can also in certain cases produce very good results. No significant changes in the performance of our algorithms can be identified between the test cases with very short time windows and longer time windows.

6 Conclusions

In this paper we considered a dynamic real-life problem which is a combined load acceptance and vehicle routing problem. The problem incorporates some practical complexities which received only little attention in the vehicle routing literature. We presented algorithms based on Large Neighborhood Search which are capable of handling these complexities. Our computational experiments have shown that the algorithms perform well for problems with hundreds of vehicles and several hundreds of transportation requests and response times were often less than a second. The combination of fast response times and the capability of handling the practical complexities allows the use of our algorithms in dynamic routing systems.

References

1. Antes J and Derigs U (1995). A new parallel tour construction algorithm for the vehicle routing problem with time windows. Department of Information Systems and Operations Research, University of Cologne, Cologne, Germany.
2. Cordeau J-F, Desaulniers G, Desrosiers J, Solomon MM, and Soumis F (2002). VRP with time windows. In P. Toth and D. Vigo, editors, The Vehicle Routing Problem, pages 157-193. SIAM Monographs on Discrete Mathematics and Applications, Philadelphia.
3. Cordeau J-F, Gendreau M, Hertz A, Laporte G, and Sormany J-S (2004). New heuristics for the vehicle routing problem. Les cahiers du GERAD G-2004-33, Université de Montreal HEC, Montréal, Canada.
4. Council of the European Communities (1985). Council Regulation (EEC) No 3820/85 of 20 December 1985 on the harmonization of certain social legislation relating to road transport.
5. Feillet D, Dejax P, and Gendreau M (2005). Traveling Salesman Problems with Profits. Transportation Science, 39(2):188-205.
6. Fleischmann B, Gnutzmann S, and Sandvoß E (2004). Dynamic vehicle routing based on on-line traffic information. Transportation Science, 38(4):420-433.
7. Hasle G (2003). Heuristics for rich VRP models. Presented at the Seminar at GERAD, 30.10.2003, Montréal, Canada.
8. Heckmann M (2002). DV-gestütztes Geschäftsprozeßmanagement in der Luftfrachtlogistik. Dissertation, Shaker Verlag Aachen.
9. Kilby P, Prosser P, and Shaw P (2000). A comparison of traditional and constraint-based heuristic methods on vehicle routing problems with side constraints. Constraints, 5:389-414.
10. Laporte G and Semet F (2002). G. Classical heuristics for the capacitated VRP. In Toth P and Vigo D, editors, The Vehicle Routing Problem, pages 109-128. SIAM Monographs on Discrete Mathematics and Applications, Philadelphia.
11. Lourenço H, Martin O, and Stützle T (2002). Iterated Local Search. In Glover F and Kochenberger G, (eds), Handbook of Metaheuristics. Kluwer, pp 321-353.
12. Mitrović-Minić S (1998). Pickup and delivery problem with time windows: A survey. Technical report TR 1998-12, School of Computing Science, Simon Fraser University, Burnaby, BC, Canada.
13. Psaraftis HN (1988). Dynamic vehicle routing problems. In B.L. Golden and A.A. Assad, editors, Vehicle routing: Methods and studies, pages 233-248. North-Holland Amsterdam.
14. Psaraftis HN (1995). Dynamic vehicle routing: Status and prospects. Annals of Operations Research, 61:143-164.
15. Savelsbergh MWP and Sol M (1998). DRIVE: dynamic routing of independent vehicles. Operations Research, 46:474-490.
16. Shaw P (1997). A new local search algorithm providing high quality solutions to vehicle routing problems. Technical report, APES group, Department of Computer Sciences, University of Strathclyde, Glasgow, Scottland.
17. Xu H, Chen Z-L, Rajagopal S, and Arunapuram S (2003). Solving a practical pickup and delivery problem. Transportation Science, 37(3):347-364.
18. Yang J, Jaillet P, and Mahmassani H (2004). Real-time multi-vehicle truckload pickup-and-delivery problems. Transportation Science, 38(2):135-148.

Heuristic Enhancements to the k-best Method for Solving Biobjective Combinatorial Optimisation Problems

Sarah Steiner and Tomasz Radzik

Dept. of Computer Science, King's College London, Strand, London, WC2R 2LS
{sarah|radzik}@dcs.kcl.ac.uk

1 Introduction and Preliminaries

Combinatorial optimisation problems with multiple objectives are natural extensions, practical as well as theoretical, of single objective problems. However, generalising a single objective model to include multiple objectives often dramatically increases the computational complexity of the problem; a polynomial-time single objective problem often turns into an NP-hard problem when we add more objectives [3].

A Biobjective Combinatorial Optimisation (BOCO) problem can be described as:

$$\text{`}\min_{S \in X}\text{'}(f^1(S), f^2(S)),$$

where X is the feasible set, S a feasible solution and $f^i(S)$ an objective function ($i \in \{1, 2\}$). The 'min' is in quotes as there are different ways of considering the optimal solution. In this paper we consider the problem of efficiency or Pareto Optimality.

An efficient solution is one which no other solution is better on all objectives:

$S \in X$ is *efficient* if:
$\nexists\, S' \in X : f^i(S') \leq f^i(S)\ \forall\, i \in \{1, 2\}$ and $f^i(S') < f^i(S)$ for at least one i.

We also use the term *dominating* to refer to a solution S' that prevents solution S from being efficient. It should be noted that a dominating solution need not necessarily be efficient, and that the set of efficient solutions is also the set of non-dominated solutions.

The set of all efficient solutions can be divided into two types: *supported/extreme* and *non-supported/non-extreme*. The extreme solutions are those that can be found by using weighted sums of the objectives; the non-extreme solutions are all those that cannot. When considering just two objectives we can assert that the extreme efficient solutions are found on the convex hull of the feasible space while the non-extreme efficient solutions are found in the triangles formed by considering the areas that adjacent extreme solutions dominate (see Fig. 1). To confirm that a non-extreme solution is efficient, we first have to know its relationship with the other solutions found in the same triangle (see a_2 and d_1 in Fig. 1). Assessing these relationships makes finding the non-extreme efficient solutions considerably harder than finding the extreme ones.

Biobjective problems are commonly solved using two phase methods, where the first phase finds the extreme solutions and the second phase finds the non-extreme solutions. In the past the non-extreme efficient solutions have been neglected in favour of finding the more conspicuous extreme solutions. However it is well known that in general there are a large number of non-extreme efficient solutions [4] and that is why our effort is directed towards improving a second phase method.

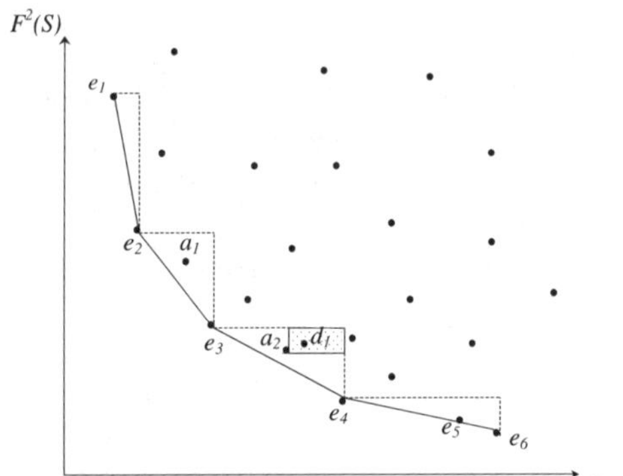

Fig. 1. The feasible space: $e_1, \ldots, e_6$ are extreme efficient solutions; a_1, a_2 are potential non-extreme efficient solutions; the shaded region is the area within the triangle that a_2 dominates; d_1 is a dominated solution inside a triangle.

2 Algorithm

The k-best method is a second phase method which uses a k-best algorithm to find the non-extreme solutions of biobjective problems. It was proposed in [1] within the context of the biobjective shortest path problem and later implemented and tested for the same problem in [2]. It was discussed with reference to the BMST problem in [7] and later experimentally tested and compared to a branch and bound method in [9].

2.1 The k-best Method

The k-best method computes all the non-extreme efficient solutions given a lexicographically ordered list of extreme efficient solutions as an input. The

k-best method finds solutions in such a way that we can declare that any solutions found must be efficient and so explicit dominating checks are not needed.

The k-best method systematically searches the triangles formed by pairs of consecutive extreme efficient solutions $e_i = (x', y')$ and $e_{i+1} = (x'', y'')$. The biobjective edge costs are transformed by (1):

$$f(e) = f^1(e)(y' - y'') + f^2(e)(x'' - x'), \tag{1}$$

and a k-best algorithm is used to compute solutions in increasing order of these transformed costs. This corresponds with searching away from the line joining the two extreme efficient solutions i.e. the hypotenuse of the triangle. The viable region is initially the whole interior of the triangle determined by points e_i, e_{i+1} and $r = (x'', y')$. During the computation the shape of the viable region will change as new solutions are found and areas of the triangle discarded from consideration (see shaded area in Fig. 1). As the search proceeds systematically from the hypotenuse, each time a solution is found in the current viable region, it must be efficient: none of the previously found solutions dominate it, and none of the solutions found subsequently could dominate it.

The k-best algorithm is applied without a fixed value for k, and is run for as long as there is still a chance to catch another solution (to the transformed single objective problem) within the current viable region. Further details and an algorithm to find all the extreme efficient solutions can be found in [9].

2.2 Heuristic Enhancements

In this section we present heuristic enhancements to this k-best method. The performance of the method described in the previous section is affected by the fact that a solution found during the computation (whether efficient or not) may actually be recomputed several times. Our heuristics aim to reduce these recomputations.

Consider the application of the k-best algorithm to the current triangle T. When a solution is discovered which is not in the current viable region $R \subseteq T$, it may be an efficient solution belonging to some subsequent triangle T'. If we have not previously found a solution in T', then this new solution must be efficient due to the systematic manner in which the solutions are found using the k-best algorithm. If we continue in this way, recording efficient solutions found in T' and maintaining the viable region $R' \subseteq T'$, and manage to pass over the whole region R', then we find all solutions for the triangle T' while the algorithm is considering the triangle T. If this happens, then there is no need to consider triangle T' itself. We propose that by maintaining other viable regions, in addition to the viable region of the current triangle, we should be able to speed up the method as we may bypass some triangles and reduce repetitions of solutions.

We propose heuristics to incorporate this idea. The original algorithm, PHASE2-KB1, considers each triangle in the order given by the lexicographic ordering of the extreme efficient solutions. The first heuristic, PHASE2-KB2, considers one additional triangle T' adjacent to the triangle we are considering T. The second, PHASE2-KB3, considers a list of all triangles $T^1, \ldots, T^j$ that could potentially be completed by considering T (in natural order). The third and fourth, PHASE2-KB4 and PHASE2-KB5, process the triangles in the order determined by their sizes. For the current triangle T, PHASE2-KB4 considers lists of triangles to both the right and left of T, while PHASE2-KB5 considers only one triangle to the right and one triangle to the left of T.

3 Computational Experiments

We have tested these heuristics on two different types of BOCO Problems. The solutions to these problems each form a different structure and so are well placed to test the general efficacy of the heuristics. In this paper we have considered only integer costs.

We have considered the Biobjective Minimum Spanning Tree (BMST) problem and the Biobjective Minimum s-t Cut (BMCUT) problem. These problems are formulated by having two costs associated with each edge instead of one.

We implemented the k-best MST algorithm of Gabow [5] as a subroutine for solving the BMST problem, and the k-best minimum cut algorithm of Hamacher and Queyranne [6] for the BMCUT problem. They both work by partitioning the feasible solutions once an initial optimal solution has been found.

We implemented all algorithms described in this paper in C++ using the C++ Library of Efficient Data Types and Algorithms (LEDA) [8]. For the BMST problem we have used three types of graph generators to compare the performance of our implementations: planar, grid and random simple; while for the BMCUT problem we have used a LEDA generator designed for network flow problems. We have chosen costs randomly and assigned costs to edges in both an independent and correlated way. Heuristic KB5 was not tested for the BMST problem.

The same pattern of results was observed across all the graph generators for the BMST problem. For all class of graph and type of costs at least one heuristic improved the times produced by the original method. KB2 and KB4 were the best performing methods overall, with KB4 resulting in the largest improvements (up to 25% faster than KB1) but KB2 giving the most consistent improvements across all classes of graphs (average improvement 15% faster than KB1).

Figure 2 shows the results for the random generator (density 0.25) with strongly negatively correlated costs and the results for the grid generator with independent costs. The first graph shows the three heuristics KB2, KB3 and

KB4 all performing better than KB1, the original k-best method, which is the slowest. The second graph shows KB2 and KB4 again performing better than KB1.

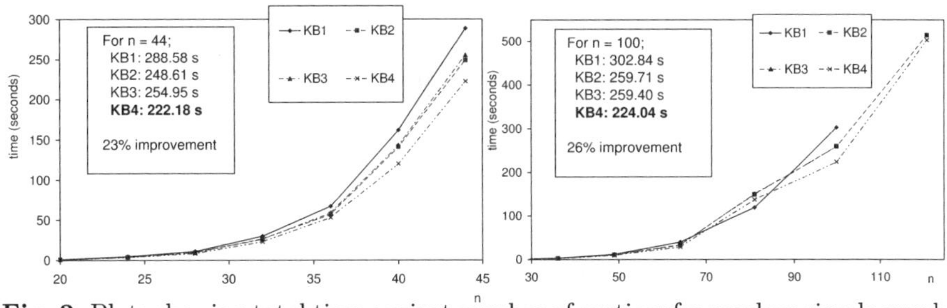

Fig. 2. Plots showing total time against number of vertices for random simple graph with negatively correlated costs, and grid graphs with independent costs (BMST problem).

For the BMCUT problem the results show a different pattern. Problem instances with strongly positively correlated costs show the KB2 heuristic outperforming the original algorithm and the other heuristics (being on average 15% faster than KB1). The other heuristics do not improve on the original KB1 algorithm for any other instance.

Figure 3 shows the results for instances with strongly positively correlated costs. KB2 consistently improves on the original method while KB4 shows a similar performance to KB1. Table 1 shows the percentage improvement of the best performing heuristic for denser problem instances.

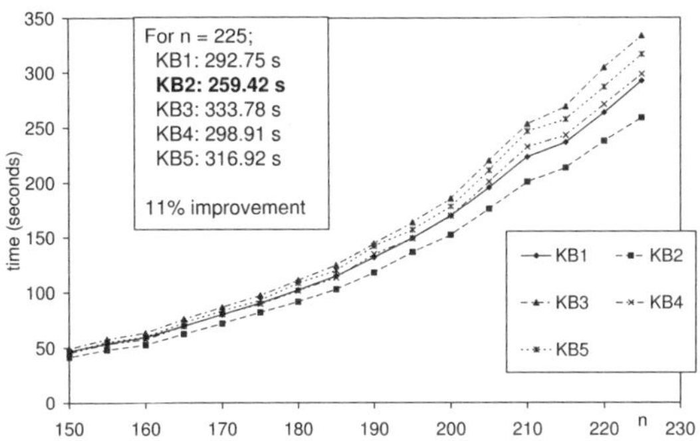

Fig. 3. Plot showing total time against number of vertices for the general generator with strongly positively correlated costs (BMCUT problem).

4 Conclusions

Our computational study shows that the heuristics we propose work well for all problem instances of the BMST problem and for positively correlated

Table 1. Table showing percentage improvement of the best performing heuristic as compared to the original algorithm, KB1, for a set of BMCUT problems with strongly positively correlated costs.

n	KB1	KB2	KB3	KB4	KB5	% improvement
60	0.4	0.36	0.38	0.36	**0.34**	15%
80	1.11	0.99	1.03	0.99	**0.94**	15%
100	3.7	**3.28**	3.72	3.38	3.37	11%
120	7.05	5.97	6.37	**5.74**	5.81	19%
140	15.94	**13.98**	15.4	14.78	15.13	12%
160	33.3	**30.22**	34.97	31.64	33.02	9%
180	73.45	**68.13**	83.81	72.28	77.39	7%

instances of the BMCUT problem. Our results have shown a general improvement of 20% for the BMST problem and 15% for the BMCUT problem.

Any k-best algorithm can be used in the PHASE2-KB algorithm therefore any problem that has a k-best algorithm can be solved using this method. The heuristic enhancements are also general and can therefore be used for any problem but as we have shown here the levels of improvement may vary between problems.

References

1. J C N Clímaco and E Q V Martins. A bicriterion shortest path algorithm. *European Journal of Operational Research*, 11:399–404, 1982.
2. J M Countinho-Rodrigues, J C N Clímaco, and J R Current. An interactive bi-objective shortest path approach: searching for unsupported nondominated solutions. *Computers & Operations Research*, 26:789–798, 1999.
3. M Ehrgott. Approximation algorithms for combinatorial multicriteria optimization problems. *International Transactions in Operational Research*, 7:5–31, 2000.
4. M Ehrgott and X Gandibleux. A survey and annotated bibliography of multiobjective combinatorial optimization. *OR Spektrum*, 22:425–460, 2000.
5. H N Gabow. Two algorithms for generating weighted spanning trees in order. *Siam Journal on Computing*, 6(1):139–150, 1977.
6. H W Hamacher and M Queyranne. k best solutions to combinatorial optimization problems. *Annals of Operations Research*, 4:123–143, 1985.
7. H W Hamacher and G Ruhe. On spanning tree problems with multiple objectives. *Annals of Operations Research*, 52:209–230, 1994.
8. K Melhorn and S Näher. *LEDA: A platform for combinatorial and geometric computing*. Cambridge University Press, 1999.
9. S Steiner and T Radzik. Computing all efficient solutions of the biobjective minimum spanning tree problem. *To appear: Computers & Operations Research*.

Part IX

Routing and Networks

Sollen Anschlussverbindungen bei Verspätungen unterbrochen werden? - Ein Ansatz zur Formulierung der Fragestellung in der Theorie des Option Pricing

Ina Bauerdorf

Universität Göttingen, Platz der Göttinger Sieben 3, 37073 Göttingen
bauerdorf.ina@web.de

Über die Einhaltung vorgegebener Anschlussverbindungen können sich Verspätungen von Zubringerzügen auf die wartenden Anschlusszüge vererben. Auch wenn die Fortpflanzung solcher Verspätungen in stabilen Fahrplänen durch Pufferzeiten sukzessive abgefangen wird, stellt sich bis zum Ausgleich der Verspätungen bei entsprechender Rechtslage die Frage, ob bzw. in welchem Anschlusspunkt es unter Berücksichtigung der zu erwartenden Schadenersatzverpflichtungen im Falle einer Unterbrechung der Anschlussverbindung und der zu erwartenden kumulierten Entschädigungen bei sich fortpflanzenden Verspätungen geraten sein könnte, die Verbindung zu unterbrechen. Zur Abbildung dieser Fragestellung wird ein Modell auf der Grundlage der diskreten Martingaltheorie und der Optionspreistheorie entwickelt und anhand eines Beispiels verdeutlicht.

1 Einleitung

Die Formulierung des Entscheidungsproblems hinsichtlich der Unterbrechung von Anschlussverbindungen bei Verspätungen als Stoppproblem erscheint nicht nur aus theoretischer, sondern auch aus anwendungsorientierter Sicht interessant, denn im Zuge der intensiven Untersuchung der Theorie des optimalen Stoppens im Rahmen der Anwendung in der Finanzmathematik bzw. der Optionspreistheorie steht ein breites Instrumentarium zur Implementierung und (numerischen) Behandlung solcher Modelle zur Verfügung. Zur Anwendung des Konzepts auf die vorliegende Fragestellung werden weiterhin Resultate auf der Basis der im Kontext diskreter Ereignissysteme nicht selten verwendeten Max-Plus Algebra herangezogen ([1],[3]). Auch diesbezüglich stehen etwa mit den Arbeiten der INRIA MaxPlus Working Group anwendungsreife Umsetzungen zur Verfügung [4].

Im folgenden Abschnitt 2 wird zunächst die Snell-Envelope als Basismodell allgemein vorgestellt. Daran anschließend findet eine Übertragung dieses Konzepts auf die vorliegende Problemstellung statt. Der Beitrag schließt mit einer kurzen Bewertung des vorgestellten Modells.

2 Die Snell-Envelope

Die im Folgenden dargestellte Formulierung eines Stoppproblems bildet die Grundlage für ein diskretes Bewertungsmodell amerikanischer Optionen [5]. Um den Rahmen des vorliegenden Beitrages nicht zu sprengen, sei hinsichtlich einer detaillierten Darstellung der stochastischen Grundlagen und der Grundlagen der Stopptheorie auf die Literatur verwiesen ([6],[5]). Die Darstellung in diesem Abschnitt orientiert sich an [5].

Kernstück des Modells ist die Konstruktion eines den zu stoppenden stochastischen Prozess von oben einhüllenden Supermartingals. Im Folgenden bezeichne $K = \{1, ..., N\}$ eine diskrete Indexmenge, $(\mathcal{F}_k)_{k \in K}$ eine Filtration und $(X_k)_{k \in K}$ einen $\mathcal{F}_k$-adaptierten stochastischen Prozess mit $X_k \geq 0$ fast sicher. Aus diesem kann eine $\mathcal{F}_k$-adaptierte Sequenz $(Z_k)_{k \in K}$ gemäß der folgenden Rückwärtsinduktion gewonnen werden:

$$\begin{aligned} Z_N &= X_N \\ Z_{N-1} &= \max(X_{N-1}, E(Z_N \mid \mathcal{F}_{N-1})) \\ &\vdots \end{aligned} \tag{1}$$

Hierin wird durch $E(Z_{N-(s+1)} \mid \mathcal{F}_{N-s})$ der bedingte Erwartungswert von $Z_{N-(s+1)}$ bezeichnet. Der so konstruierte stochastische Prozess $(Z_k)_{k \in K}$ heißt Snell-Envelope. Sie hat folgende wichtige Eigenschaft: Bezeichne $\mathcal{T}$ die Menge aller Stoppregeln. Für den gestoppten Prozess X^{τ} und die Stoppzeit $\tau^* = \min(t \geq 0 \mid X_t = Z_t)$ gilt die Beziehung:

$$E(X^{\tau^*}) = \sup_{\tau \in \mathcal{T}} E(X^{\tau}). \tag{2}$$

Der Zeitpunkt der ersten Berührung des stochastischen Prozesses X_k mit seiner Snell-Envelope Z_k markiert also den optimalen Stoppzeitpunkt in dem Sinne, dass der Erwartungswert bzgl. X_k maximal ist.

3 Die Entscheidung über die Unterbrechung von Anschlussverbindungen als Stoppproblem

Tritt in einem Verkehrsnetz mit einem stabilen Fahrplan eine einmalige exogene Störung auf, so können die bis zum Ausgleich der Verspätungen anfallenden erwarteten kumulierten Schadenersatzzahlungen vermindert werden, indem die Anschlussverbindungen unterbrochen werden, allerdings um den Preis der dann anfallenden Schadenersatzverpflichtungen. Grundgedanke der folgenden Darstellung dieser Fragestellung als Stoppproblem ist, die Fortpflanzung der Verspätungen dann zu unterbrechen, wenn die Kostenersparnis, gemessen als die Differenz zwischen den erwarteten kumulierten Schadenersatzzahlungen bei Verspätungen und den erwarteten Schadenersatzverpflichtungen infolge der Unterbrechung der Anschlussverbindungen, am größten ist.

Zur Verdeutlichung der vorzustellenden Konstruktion soll die Darstellung durch das folgende Beispiel begleitet werden: Zwei Linien 1 und 2 verbinden die Orte A und B, eine Linie 3 startet und endet in Ort A. Die ersten Abfahrtszeiten $d_i(0)$, Fahrzeiten und Takt (T) der Linien können der Tabelle 1 entnommen werden. Alle Linien seien fahrplanmäßig als Anschlussverbindungen vorgesehen.

Tabelle 1. Die Daten des Fahrplans

Linie	von	nach	$d_i(0)$	Fahrzeit	T
1	B	A	00	25 min.	30 min.
2	A	B	01	22 min.	30 min.
3	A	A	02	24 min.	30 min.

Grundlegend ist zunächst die Darstellung der Entwicklung der exogen auftretenden Verspätung in einem Verkehrsnetz. Der Literatur folgend soll dies auf der Basis der Max-Plus Algebra geschehen ([1],[3]). Wird durch $x_i(k)$ bzw. $d_i(k)$ die tatsächliche bzw. die fahrplanmäßige k-te Abfahrtszeit des i-ten Zuges bezeichnet, ergibt sich die Verspätung bei der k-ten Abfahrt durch $x_i(k) - d_i(k)$. Die Fortpflanzung einer einmalig auftretenden exogenen Verspätung kann durch eine rekursive Gleichung abgebildet werden [2]:

$$x_i(k) - d_i(k) = \max\left[\max_{1\leq j\leq m}\left(r_{ij} + x_j(k-1) - d_j(k-1)\right), 0\right], \tag{3}$$

mit

$$r_{ij} = d_j(k-1) - d_i(k) + a_{ij} \quad \text{und} \quad d_i(k) = d_i(0) + k \cdot T, \tag{4}$$

wobei a_{ij} die planmäßige Fahrzeit des Zuges j von seinem Startpunkt bis zu der Station angibt, an der der Anschlusszug i wartet. Ist der Zug i kein Anschlusszug, hat a_{ij} den Wert $\epsilon = -\infty$. Für das Beispiel lassen sich hierdurch die folgenden Werte $R = (r_{ij})$ berechnen:

$$R = (r_{ij}) = \begin{pmatrix} \epsilon & -7 \ (= d_2(0) - d_1(0) - 30 + 22) & \epsilon \\ -6 & \epsilon & -5 \\ -7 & \epsilon & -6 \end{pmatrix}. \tag{5}$$

Verspätet sich nun die Linie 2 bei der p-ten Ankunft aufgrund einer exogenen Störung um 20 Minuten, pflanzt sich dies unter Einhaltung aller Anschlussvorschriften in der folgenden Weise fort:

$$\begin{pmatrix} 0 \\ 20 \\ 0 \end{pmatrix}, \begin{pmatrix} 13 \\ 0 \\ 0 \end{pmatrix}, \begin{pmatrix} 0 \\ 7 \\ 6 \end{pmatrix}, \begin{pmatrix} 0 \\ 1 \\ 0 \end{pmatrix}, \begin{pmatrix} 0 \\ 0 \\ 0 \end{pmatrix}. \tag{6}$$

Bei der k=p+4 Ankunft ist die Verspätung somit vollständig ausgeglichen.

Die Schadenersatzverpflichtungen hängen nun neben den gesetzlichen Vorschriften von der zu erwartenden Entwicklung der Anzahl der Kunden ab, die bei jeder Ankunft k aussteigen bzw. umsteigen wollen. Um diese Daten in dem Modell zu erfassen, sei ω_{ik}^U die Zufallsvariable, die die Anzahl der Kunden angibt, die bisher mit der Linie i reisen und bei der k-ten Ankunft in eine andere Linie umsteigen wollen, und ω_{ik}^V die Zufallsvariable, die die Anzahl der Kunden angibt, die bisher mit der Linie i reisten und bei der k-ten Ankunft an ihrem Ziel angelangt sind.

Hinsichtlich der Schadenersatzzahlung soll beispielsweise festgelegt sein, dass ein Kunde, der seinen Anschlusszug nicht erreicht, eine Entschädigung von 30 Euro erhält und einem Kunden, der sich in der Ankunft verspätet, 2 Euro pro Minute Verspätung gezahlt werden. Damit ergeben sich die Kosten einer Unterbrechung der Anschlüsse bei der k-ten Ankunft aus dem mit 30 multiplizierten Skalarprodukt des Vektors der Verspätungen mit dem Zufallsvektor der umsteigenden Kunden

$$U_k = 30 \cdot \left(\left(\delta\big(x_i(k) - d_i(k)\big) \right)_{i=1,\dots,n} \right)^T \omega_k^U \text{ für } k = p, ..., N-2, \tag{7}$$

denn genau die Kunden in den verspäteten Linien versäumen bei einer Unterbrechung ihren Anschluss. In der Formel gilt $\delta(x) = 1$ für $x > 0$ und $\delta(x) = 0$ sonst.

Die Kosten der Verspätung ergeben sich als erwartete kumulierte Kosten durch

$$V_k = 2 \cdot E \left(\sum_{l=p+1}^{N-1} (x(l) - d(l))^T (\omega_l^V) \mid \mathcal{F}_k \right) \text{ für } k = p, ..., N-2. \tag{8}$$

Die in p fälligen Schadenersatzzahlungen bleiben hierbei unberücksichtigt, denn sie können nicht mehr durch eine Unterbrechung beinflusst werden. In $k = N$ ist die Verspätung ausgeglichen. Es treten daher weder Verspätungs- noch Unterbrechungskosten auf und die Zufallsvariablen haben daher den Wert Null. Theoretisch wäre es möglich, noch in $k = N - 1$ die Verbindungen zu unterbrechen. Unter dem Kostenaspekt macht dies jedoch keinen Sinn, da keine Verspätungskosten mehr vermieden und daher (unnötige) Unterbrechungskosten anfallen würden. Um diese Überlegung in das Modell einzubinden, wird den Zufallsvariablen V_{N-1} und U_{N-1} jeweils der Wert Null zugewiesen.

In dem Beispiel sollen hinsichtlich der Entwicklung der Zahlen der aussteigenden bzw. umsteigenden Fahrgäste die in Tabelle 2 wiedergegebenen Werte angenommen werden. Jeder Zustand soll zwei Folgezustände haben, wobei der Übergang jeweils mit einer Wahrscheinlichkeit von 0, 5 erfolgt. Für jeden Zustand müssen natürlich nur die relevanten Werte prognostiziert werden, d.h. die Fahrgastzahlen hinsichtlich der verspäteten Linien. Die Werte aller übrigen Variablen ($\bullet$) sind beliebig und könnten gleich Null gewählt werden.

Tabelle 2. Die Entwicklung der relevanten Zahlen der Fahrgäste im Planungszeitraum

$(\omega_{i,0}^V/\omega_{i,0}^U)_{i=1,2,3}$	$(\omega_{i,1}^V/\omega_{i,1}^U)_{i=1,2,3}$	$(\omega_{i,2}^V/\omega_{i,2}^U)_{i=1,2,3}$	$(\omega_{i,3}^V/\omega_{i,3}^U)_{i=1,2,3}$
		$(\bullet/\bullet, 50/74, 40/40)$	$(\bullet/\bullet, 100/\bullet, \bullet/\bullet)$
			$(\bullet/\bullet, 50/\bullet, \bullet/\bullet)$
	$(80/110, \bullet/\bullet, \bullet/\bullet)$	$(\bullet/\bullet, 50/70, 20/30)$	$(\bullet/\bullet, 80/\bullet, \bullet/\bullet)$
			$(\bullet/\bullet, 30/\bullet, \bullet/\bullet)$
$(\bullet/\bullet, \bullet/80, \bullet/\bullet)$		$(\bullet/\bullet, 40/35, 10/10)$	$(\bullet/\bullet, 60/\bullet, \bullet/\bullet)$
	$(30/40, \bullet/\bullet, \bullet/\bullet)$		$(\bullet/\bullet, 20/\bullet, \bullet/\bullet)$
		$(\bullet/\bullet, 20/30, 20/10)$	$(\bullet/\bullet, 50/\bullet, \bullet/\bullet)$
			$(\bullet/\bullet, 10/\bullet, \bullet/\bullet)$

Entsprechend der Berechnungsvorschriften ergeben sich für das Beispiel die in Tabelle 3 wiedergegebenen Werte für die Zufallsvariablen U_k und V_k für $k = p, p+1, p+2$, wobei auf die Angabe der Werte für $k = p+3, p+4$ verzichtet wurde, da sie nach obigen Überlegungen per Definition den Wert Null haben. Nach diesen Vorbereitungen kann die Zufallsvariable X_k nun definiert werden als $X_k = (V_k - U_k)^+ = \max(V_k - U_k, 0)$ und die Snell-Envelope ergibt sich nach der in (1) genannten Vorschrift. Auch die Werte für X_k und Z_k sind für $k = p, p+1, p+2$ in der Tabelle 3 abgebildet, wobei die übrigen Werte wiederum gleich Null sind.

Tabelle 3. Die Werte der Zufallsvariablen V_t und U_t

V_p	V_{p+1}	V_{p+2}	U_p	U_{p+1}	U_{p+2}	X_p	X_{p+1}	X_{p+2}	Z_p	Z_{p+1}	Z_{p+2}
		3410			3420			0			0
	3270			3300			0			65	
		3130			3000			130			130
2360			2400			0			157, 50		
		1540			1350			190			190
	1450			1200			250			250	
		1360			1200			160			160

Aus der Konstruktion von X_k folgt hinsichtlich der Umsetzung der Stoppstrategie unmittelbar, dass die Anschlussverbindungen bei der k-ten Abfahrt nur dann unterbrochen werden sollten, wenn $X_k = Z_k > 0$ im ersten Berührungspunkt k gilt.

4 Kritische Würdigung

Der vorgestellte Ansatz orientiert sich an den Erwartungswerten der Kosten und setzt daher Risikoneutralität bei den Entscheidungsträgern voraus. Weiterhin geht er von stabilen Fahrplänen aus, d.h. exogen auftretende Verspätungen werden im Zeitablauf sukzessive ausgeglichen. Hinsichtlich der Entstehung der Verspätungen wird von einer einmal auftretenden exogenen Störung ausgegangen.

Nach Abschnitt 2 muss die Konstruktion von X_k lediglich zu einer fast sicher positiven Zufallsvariablen führen, sodass sich ein breites Spektrum sowohl hinsichtlich der möglichen Verteilungen der Fahrgastzahlen als auch hinsichtlich der möglichen Ausgestaltung der Vorschriften für die Schadenersatzverpflichtungen eröffnet. Das grundsätzliche Problem der Prognose der Fahrgastzahlen kann jedoch nicht vereinfacht werden.

Ebenfalls hinsichtlich der Konstruktion von X_k ist zu bemerken, dass es sich in der Max-Plus Algebra bei der Berechnung der Entwicklung der Verspätungen gemäß (3) - (4) lediglich um wiederholte Matrixmultiplikationen handelt, die beispielsweise mit der Software der INRIA einfach implementiert werden können [4].

Literaturverzeichnis

1. Bacelli F, Cohen G, Olsder GJ, Quadrat JP (2001) Synchronization and Linearity - An Algebra for Discrete Event Systems. Springer, Berlin Heidelberg New York
2. Bauerdorf I (2005) Die Auswirkungen von Verspätungen in Verkehrsnetzen - Ansätze einer Analyse auf der Grundlage der Max-Plus Algebra. In: Günther HO, Mattfeld DC, Suhl L (Hrsg) Supply Chain Management und Logistik - Optimierung, Simulation, Decision Support. Springer, Berlin Heidelberg New York
3. Braker JG (1993) Algorithms and Applications in Timed Discete Event Systems. Dissertation, Faculty of Technical Mathematics and Informatics, University Delft, Delft
4. INRIA Max Plus Working Group (2003) MAXPLUS ForScilab2.7. http://scilabsoft.inria.fr/contribution/displayContribution.php?fileID=174. Letzter Abruf: 10.10.2004
5. Elliott RJ, Kopp PE (1999) Mathematics of Financial Markets. Springer, Berlin Heidelberg New York
6. Williams D (1989) Probability with Martingales. Cambridge University Press, Cambridge

Some Remarks on the GIST Approach for the Vehicle Routing Problem with Pickup and Delivery and Time Windows (VRPPDTW)

Ulrich Derigs and Thomas Döhmer

Department of Information Systems and Operations Research (WINFORS),
University of Cologne, Pohligstr. 1, 50969 Cologne, Germany
derigs@informatik.uni-koeln.de, thomas.doehmer@uni-koeln.de

1 Introduction

During the last years there has been extensive research on extensions of the classical Vehicle Routing Problem (VRP) with respect to additional constraints which occur in real-world applications. The family of those "new" and difficult VRP-variants is often referred to as Rich Vehicle Routing Problems (RVRP). In [1] we have presented a rather specific, non-literature standard Pickup and Delivery Vehicle Routing Problem with Time Windows (VRP-PDTW) together with an effective heuristic and system called ROUTER. This work was motivated by a system development project. In this work we focus on solving the standard VRPPDTW which is one of the RVRP-challenges discussed in literature.

The algorithms for VRPPDTW, and for vehicle routing problems in general, which have been published in the scientific literature were developed with a focus on accuracy and speed. Yet, especially in a decision support system (DSS) development project simplicity and flexibility are of dominant importance because of two reasons: In the initial requirement phase the problem owner will in general not be able to explicit all aspects of the problem, i.e. all constraints and objectives, and thus system development can only lead to a useful system if the development process is based on a sequence of models with associated system prototypes. Such a process of "rapid prototyping" requires flexible modeling tools and heuristics.

In [3] the general GIST-framework (GIST = Greedy Indirect Search Technique) for solving rich/hardly constrained combinatorial optimization problems has been described, which copes with these requirements of DSS development projects. This approach has been implemented successfully in a DSS for course scheduling [2] and also ROUTER can be viewed as an implementation of this concept for a specific RVRP.

The GIST-framework is based on the concept of indirect (evolutionary/local) search (see Gottlieb [5]). That is, we work in a search space of genotypes obtained by encoding the phenotypes and we manipulate encoded solutions by a metaheuristic control. Conceptually, we distinguish between a (problem independent) metaheuristic local search engine (MLS-engine) which manipulates abstract encoding schemes like permutations, bit-strings etc., and a (problem specific) decoder-function which encodes the domain specific knowledge to identify feasible solutions and which constructs feasible solutions using a simple and fast greedy insertion rule.

In this work we present first results on applying this framework or modifying ROUTER to the standard VRPPDTW. By solving a standard literature problem, like the VRPPDTW, we want to show that this approach can also compete with other heuristics with respect to accuracy and speed.

2 Problem Formulation

In a VRP a set R of transportation requests must be satisfied by a set of vehicles, each with a limited capacity Q. A solution, also called a schedule consists of a partition of the requests into clusters/tours which are assigned to the vehicles and an ordering/routing of the requests within a cluster. In the VRPPDTW each transportation request $r \in R$ combines a pickup location r^+ and a delivery location r^- and there is a load q_r which has to be carried from r^+ to r^-. With every location i there is associated a service time s_i for loading/unloading and a time window $[e_i, l_i]$, which specifies the time in which this service has to start. When arriving to early at a location i, i.e. before e_i, the vehicle is allowed to wait until e_i to start the service. Now a feasible schedule has to obey the following properties:

- Each request $r \in R$ is assigned to exactly one vehicle/tour.
- For each request $r \in R$ location r^+ is visited before r^-.
- Each location i is visited not later than l_i.
- For each route the load of the vehicle must not exceed Q at any time.

Each route has to start and to end at a specific location, the depot, and there is a distance $d_{i,j}$ between any two locations i and j. Now, the optimization problem is to construct a feasible solution where the number of tours is minimal and the total length of the routes is minimal. Obviously these objectives may conflict. In most practical problems the number of vehicles available is pre-specified and there may be a fixed cost for using (additional) vehicles. For most literature problems minimizing the number of tours/vehicles is the primary objective. Note that a partial schedule is a feasible solution for a subset of requests.

3 GIST-Implementation for VRPPDTW

3.1 Encoding

Let $R = \{1, \ldots, N\}$ be the set of transportation request. Then we encode a solution/schedule by a permutation p of the numbers $1, \ldots, N$. This permutation determines the sequence in which the requests are scheduled.

3.2 The Greedy-Decoder

The decoder gets a permutation $p \in P$ as input and generates a feasible schedule $X(p)$ using a deterministic sequential construction procedure which starting from the empty schedule inserts the requests one by one. The core of the decoder function $(X, C) \leftarrow \text{DECODER}(p, S)$ is a scheduling rule $S(r, X)$ which given a partial schedule X and an unscheduled requests r as input specifies a tour $t(r)$ and positions for the corresponding pickup location $\text{pos}(r^+)$ and delivery location $\text{pos}(r^-)$ within the associated routing as well as the increase of the objective function value Δ. It should be mentioned here that the scheduling rule may extend the given partial schedule by introducing a new tour with r being the only request. In this case Δ also includes the cost for using a (previously unused) vehicle.

The decoder is generic in the sense that the scheduling rule S, i.e. the rule that decides at which positions the locations of a given request are inserted into a partial schedule is interchangeable and therefore it is a problem specific parameter of the procedure. In the ROUTER-implementation for VRPPDTW we have implemented a specific scheduling rule *cheapest insertion* $CI(r, X)$, where for the given request r those positions for inserting r^+ and r^- are determined which lead to a minimal increase in transportation cost/travel distance.

3.3 The MLS-Engine

The basic concept of local search is the definition of a so-called *neighborhood topology* on the search space X. Now a local search heuristic H for solving a problem $\min\{f(x) | x \in X\}$ is controlled by three functions:

- SELECTNEIGHBOR(x, x', N, H) where given a solution $x \in X$ an element $x' \in N(x)$ is selected according to the heuristic H's selection strategy.
- EVALUATENEIGHBOR(x, x', H) where the cost of solution x' is calculated.
- ACCEPTNEIGHBOR(x, x', H) where based on $\Delta = f(x) - f(x')$ and the heuristic H's acceptance criteria it is decided whether solution x is replaced by solution x' or not.

Also, depending on the implementation of these functions, a STOPCRITERION has to be specified to guarantee finiteness of the procedure.

Since in our indirect search approach the search space X equals the set P of permutations, we have to define for every $p \in P$ a set $N(p) \subseteq P$ of *neighbors*. In our implementation we have used the well known *2-opt-neighborhood*, i.e. we obtain a neighbor q of p by selecting two elements/requests i and j and replacing $q(i)$ by $p(j)$ and $q(j)$ by $p(i)$. As selection strategy we implemented a simple *random selection* of the two requests. The acceptance strategy and stop-criterion which we have implemented in ROUTER is a variant of the well known *threshold accepting strategy* (cf. Dueck and Scheuer [4]). The ROUTER-local search algorithm is depicted in algorithm 1.

Algorithm 1 ROUTER-local search procedure

input: initial permutation p, threshold $0 \leq T \leq 1$, maxIter > 0 for maximal number of iterations without improvement
output: feasible schedule X^* and schedule cost C^*

```
procedure ROUTER(p, T, maxIter)
    (X*, C*) ← DECODER(p, CI)
    threshold ← T · C*
    counter ← 0
    C ← C*
    while counter < maxIter do                          ▷ STOPCRITERION
        counter ← counter + 1
        choose q ∈ N(p) randomly                        ▷ SELECTNEIGHBOUR
        (X, C') ← DECODER(q, CI)
        if C' < C + threshold then                      ▷ ACCEPTNEIGHBOUR
            C ← C'
            p ← q
            if C < C* then                              ▷ Improvement
                (X*, C*) ← (X, C')
                threshold ← T · C*
                counter ← 0
            end if
        end if
    end while
    return (X*, C*)
end procedure
```

Every improvement heuristic has to be initiated from a given initial solution/permutation and, usually, it will be applied several times with different initial solutions. We have implemented a specific strategy. ROUTER-local search is applied sequentially to different initial permutations p as long as `maxLS` consecutive local search phases did not result in an improvement. Here `maxLS` is a parameter of ROUTER. Starting with all requests unscheduled an initial permutation p is constructed as follows: Until all requests are scheduled we determine for all unscheduled requests $r \in R$ the cheapest positions for insertion. Then we select r_0 randomly among the cheapest k unscheduled requests and we insert r_0^+ and r_0^- at their cheapest insertion positions, with $k \in \mathbb{N}$ chosen appropriately. Then the resulting sequence of insertions defines the initial permutation p.

4 Computational results

In the following we present computational results comparing the ROUTER-implementation with the results presented by Li and Lim [6] and Pankratz [7] on the set of VRPPDTW-instances which have been compiled and published by Li and Lim [6]. While for ROUTER and the implementation of Li and Lim the primary objective is to minimize the total number of vehicles (NV), the primary objective for the implementation of Pankratz is to minimize total travel cost (TC).

class	Li and Lim			Pankratz				
	NV	TC	CT	TC^{best}	NV^{best}	TC^{avg}	NV^{avg}	CT^{avg}
LC1	89	7488.77	2030	7445.42	90	7445.49	90.00	735.93
LC2	24	4713.83	1570	4718.87	24	4749.14	24.00	1209.80
LR1	143	14666.41	4453	14642.51	144	14744.52	147.06	1246.86
LR2	30	10712.59	20640	10662.29	31	10892.29	33.13	3582.73
LRC1	93	11100.76	2090	11088.34	93	11132.89	94.51	765.83
LRC2	26	9502.55	12289	9081.05	28	9226.79	29.60	1816.51
sum	405	58184.91	43072	57638.48	410	58191.22	418.30	9357.66

Table 1. Reference results

ROUTER as well as the Li and Lim procedure is implemented in C++ while the Pankratz-procedure is implemented in JAVA. Our experiments were performed on a 1-GHz P3 machine while the times reported by Pankratz were obtained on a 2-GHz P4 machine and Li and Lim state that they were using a 686 PC.

We used the the same experimental setting as Pankratz [7], i.e. 30 runs of ROUTER were performed on each benchmark instance. The control parameters for ROUTER have been set as follows: $T = 0.08$, $\texttt{maxIter} = 100 \cdot \lceil \log_2 N \rceil$ and $\texttt{maxLS} = 2$.

In table 1 and table 2 we display the aggregated values for NV and TC for the instances within the six different subclasses of the benchmark-set. For Pankratz and ROUTER we display the average and best results obtained over the sample of 30 runs per instance. We also report results on the computation time (CT reap. CT^{avg}), yet, these numbers are hard to interpret because of the different programming languages and computers employed.

class	ROUTER				
	TC^{best}	NV^{best}	TC^{avg}	NV^{avg}	CT^{avg}
LC1	7486.83	89	7500.76	89.30	196.23
LC2	4718.87	24	4753.06	24.00	161.09
LR1	14636.03	143	14770.29	146.33	411.49
LR2	10658.34	31	10939.69	32.53	1478.36
LRC1	11094.46	92	11126.05	92.63	224.54
LRC2	9064.98	26	9271.41	27.63	742.14
sum	57659.51	405	58361.26	412.43	3213.85

Table 2. ROUTER results

Table 2 shows that ROUTER is competitive with special purpose developments for the VRPPDTW with respect to both objectives, NV and TC, and with respect to computation time CT.

Due to the the choice of the objective in the specific implementation of ROUTER we were not able to achieve lower TC^{best} values, yet, when spending more computing time by increasing `maxLS` to 3 we were able to improve the TC^{avg} value to $58090,85$ (with $NV^{avg} = 410,37$ and $CT^{avg} = 4616,34$).

References

1. Derigs U, Döhmer T (2004) ROUTER: A fast and flexible local search algorithm for a class of rich vehicle routing problems. Operations Research Proceedings 2004:144–149
2. Derigs U, Jenal O (2005) A GA-based decision support system for professional course scheduling at Ford Service Organisation. OR Spectrum 27(1):147–162
3. Derigs U, Döhmer T, Jenal O (2005) Indirect Search With Greedy Decoding Technique (GIST) – An Approach for Solving Rich Combinatorial Optimization Problems. MIC 2005
4. Dueck G, Scheuer T (1990), Threshold Accepting: A General Purpose Optimization Algorithm Appearing Superior to Simulated Annealing. Journal of Computational Physics 90:161–175
5. Gottlieb J (2000) Evolutionary Algorithms for Constrained Optimization Problems. Shaker Verlag, Aachen.
6. Li H, Lim A (2003) A Metaheuristic for the Pickup and Delivery Problem with Time Windows. International Journal on Artificial Intelligence Tools 12(2):173–186
7. Pankratz G (2005) A Grouping Genetic Algorithm for the Pickup and Delivery Problem with Time Windows. OR Spectrum 27:21–41

Analyse der Beschleunigung des A*-Verfahrens durch verbesserte Schätzer für die Restdistanz

Felix Hahne

Universität Hildesheim, Institut für Betriebswirtschaft

Wegsuche in digitalen Karten mit dem A*-Verfahren

Das A*-Verfahren, dessen erste formale Beschreibungen sich etwa in (Hart et al. 1968, 1972; Nilsson 1971) finden, zählt zu den am häufigsten für die Suche von kürzesten Wegen zwischen einem Start- und einem Zielpunkt in Graphen (*OLSP-Problem*, siehe etwa (Djidjev 1997, S. 152)) eingesetzten Verfahren.

In Verbindung mit digitalen Karten ergeben sich zahlreiche praktische Anwendungen wie beispielsweise Längenmessungen in Geoinformations- oder Routenberechnungen in Fahrzeugnavigationssystemen.

Die vorliegende Untersuchung geht von einer digitalen Karte aus, die sich als Digraph interpretieren lässt und bei der zusätzlich die geographischen Koordinaten der Knoten bekannt sind.

Zur Rolle des Restdistanzschätzers beim A*-Verfahren

Der Aufwand, den das A*-Verfahren betreiben muss, hängt wesentlich von der Qualität einer Schätzung für die Restdistanz zum Ziel ab.

Unter allen demnächst zu untersuchenden Kanten[1] (*Kandidatenliste, open list*), die um den Rand des bisher untersuchten Gebiets (*Suchbaum, closed list*) liegen, wird in der nächsten Iteration diejenige Kante e betrachtet, die den Ausdruck

$$f(e) = g(e) + h(e), \tag{1}$$

mit $g(e)$ Länge des kürzesten bisher bekannten Wegs vom Start zu e und $h(e)$ als Schätzung für die Restdistanz von e zum Ziel, minimiert.

Falls $h(e)$ die echte Restdistanz $ED(e)$ nie überschätzt, d.h. stets

$$h(e) \leq ED(e) \tag{2}$$

gilt, findet das A*-Verfahren die optimale Lösung.

Im Falle einer möglichen Überschätzung der echten Restdistanz durch $h(e)$ mutiert das Verfahren zu einer Heuristik. Mit dem Grad der Überschätzung steigt der „Greedy-Charakter" des Verfahrens: Tendenziell wird der Suchbaum des Verfahrens ausgedünnt, d.h. der Aufwand verringert sich, und die Länge der gelieferten Wege steigt, d.h. die Lösungsqualität sinkt (vgl. (Hahne 2000, S. 71ff)).

Die Kenntnis der echten Restdistanz für jede Kante des Graphen ist gleichzusetzen mit der Kenntnis des kürzesten Weges: In diesem Fall wählt das Verfahren

---

[1] Es wird im Folgenden von einer kantenorientierten Sichtweise, d.h. einer Iteration über Kanten, ausgegangen. Die getroffenen Aussagen lassen sich aber auch auf eine knotenorientierte Vorgehensweise übertragen; vgl. (Hahne 2000, S. 26f.).

in jeder Iteration rekursiv unter den Nachfolgerkanten diejenige aus, welche die Restdistanz minimiert. Damit besteht bei Abbruch des Verfahrens der Suchbaum nur aus dem optimalen Weg und die open list aus den nicht benutzten Abbiegemöglichkeiten von diesem Weg.

Abbildung 1 zeigt das Resultat einer Wegsuche von Startpunkt *s* zu Zielpunkt *t*, wobei der optimale Weg schwarz und die Kanten der open list grau markiert sind.

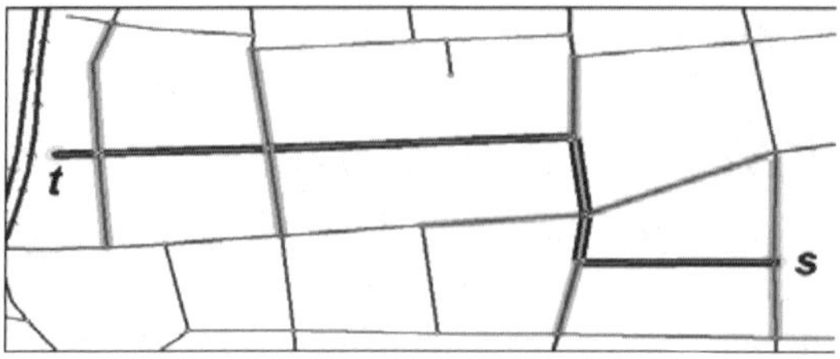

Abb. 1 Minimaler Suchbaum bei Kenntnis der echten Restdistanz

Als Standardschätzer für die Restdistanz wird in digitalen Straßenkarten die Restluftdistanz (*RLD*) verwendet und das sich damit ergebende Verfahren als *Standard-A*-Verfahren* bezeichnet. (Hart et al. 1968) und (Gelperin 1977) zeigen, dass das A*-Verfahren diese Information am besten unter allen Verfahren nutzt.

Die Restluftdistanz ist aus der Geometrie der Karte leicht zu berechnen und erfüllt die Eigenschaft (2). Als praktisch relevantes Intervall für die exakte Suche nach kürzesten Wegen mit dem A*-Verfahren lässt sich damit für den Schätzer der Restdistanz angeben:

$$h(e) \in [RLD(e), ED(e)] \tag{3}$$

Ziel dieser Arbeit ist eine quantitative Untersuchung, welchen Einfluss eine unterschiedliche Wahl von $h(e)$ aus diesem Intervall auf die Größe des Suchbaums hat; mithin eine Abschätzung, ob sich zusätzlicher Aufwand zur Ermittlung eines gegenüber der Restluftdistanz verbesserten Schätzers lohnt, da ein kleinerer Suchbaum einen geringeren Speicherplatzbedarf und eine geringere Laufzeit indiziert.

Ansätze zur Verbesserung des Restdistanzschätzers

Praktische Erfahrungen, z. B. in (Schäfer 1998), zeigen, dass eine „on-the-fly"-Berechnung eines verbesserten Schätzers während des Verfahrenslaufs zumindest in Bezug auf die Laufzeit keinen Vorteil bringt: Der Vorteil durch den verkleinerten Suchbaums im A*–Basisverfahren wird durch den Aufwand zur Ermittlung des Schätzers mehr als aufgezehrt.

Daher konzentrieren sich die Lösungsansätze häufig auf eine Vorverarbeitung (*Preprocessing*) der Karte, die sie mit zusätzlichen Informationen anreichern, um eine im Vergleich zur Berechnung von $RLD(e)$ nicht zu aufwändige Berechnung eines Schätzers zu ermöglichen.

In (Dubois und Semet 1995) werden Trennlinien zur Modellierung von Hindernissen wie Bergen, Flüssen oder Fußgängerzonen, die zu großen Fehlabschätzungen der Restdistanz durch $RLD(e)$ führen können, eingeführt.

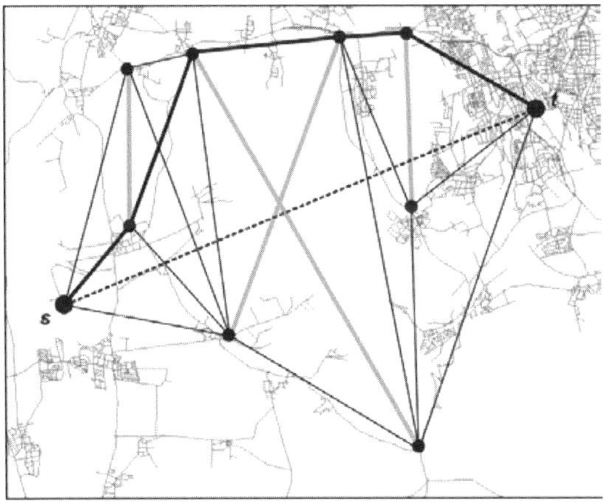

Abb. 2 Verbesserte Restdistanzschätzung durch Trennlinien, aus (Hahne 2000, S. 227)

In Abbildung 2 sind einige Trennlinien (graue dicke Linien) eingefügt. Die Abschätzung der Distanz von *s* nach *t* erfolgt durch die Berechnung des kürzesten Weges in einem Hilfsgraphen, dessen Knotenmenge neben *s* und *t* alle Endpunkte von Trennlinien umfasst (dicke schwarze Punkte) und dessen Kanten alle direkten Verbindungen zwischen den Knoten sind, die keine Trennlinie schneiden (schwarze Linien). Der kürzeste Weg im Hilfsgraphen (dicke schwarze Linie) stellt eine deutlich bessere Abschätzung der Restdistanz dar als die Luftlinie (gestrichelte schwarze Linie).

Weitere Ansätze ergeben sich aus der Ermittlung von besonderen Kanten (z. B. Brücken oder Passstrassen), die auf kürzesten Wege zwischen zwei Regionen immer benutzt werden sowie der Verwendung hierarchischer Karten, die auf aggregierter Ebene das schnelle Abschätzen von Distanzen erlauben.

Quantitative Messung der Beschleunigung von A*

Aufbau und Vorgehensweise der Benchmarks

Es werden je zwei Benchmarks auf zwei digitalen Karten des Stadtgebiets und des Landkreises Hildesheim durchgeführt. Die Karten umfassen ca. 9000 bzw. 23 000 Kanten; Details siehe (Hahne 2000, S. 21ff).

Auf den Karten werden Testpunkte, zwischen denen jeweils paarweise kürzeste Wege berechnet werden, annähernd homogen verteilt. Der Testpunktsatz *Hi20* umfasst 20 Testpunkte (= 380 Verbindungen) auf der Stadtgebietskarte; der Testpunktsatz *Lk13* verwendet 13 Testpunkte in der Landkreiskarte (= 156 Verbindungen).

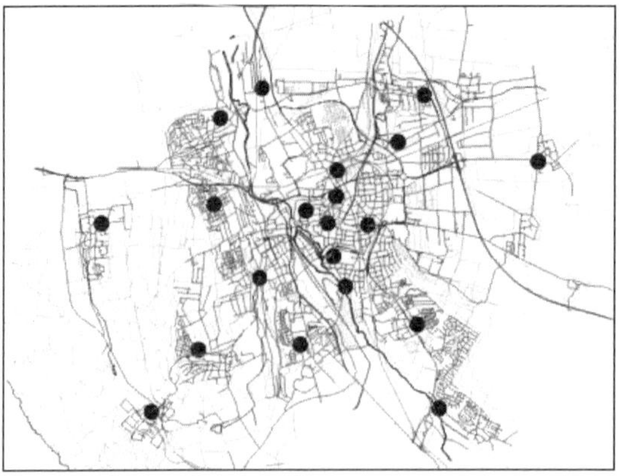

Abb. 3 Lage der Testpunkte von Testpunktsatz Hi20

Zunächst wurde jeweils der vollständige Suchbaum inklusive der open list, der sich beim A*-Verfahren unter Verwendung von $RLD(e)$ als Schätzer ergibt, für alle Verbindungen in einer Datenbank abgespeichert. Für jede Kante aus dieser Menge wurde, erneut unter Verwendung des A*-Verfahrens, die echte Restdistanz ermittelt und ebenso in der Datenbank abgelegt.

In Benchmark 1 wird als Schätzer verwendet:

$$h_1(e, \gamma) = (1 - \gamma)\, RLD(e) + \gamma\, ED(e), \text{ mit } \gamma \in [0, 1] \tag{4}$$

Benchmark 2 trägt der Tatsache Rechnung, dass die Qualität der Schätzer, z.B. die der im vorigen Abschnitt vorgestellten Ansätze, in der Regel kein fester Anteil der echten Restdistanz ist, sondern je nach Datenlage schwanken kann. Dies wird modelliert durch eine Gleichverteilung auf dem Teilintervall $[\gamma_1, \gamma_2]$:

$$h_2(e, \gamma_1, \gamma_2) = h_1(e, \gamma), \text{ mit } 0 \leq \gamma_1 \leq \gamma_2 \leq 1 \text{ und } \gamma \in \mathbf{U}[\gamma_1, \gamma_2] \tag{5}$$

Die Messgrößen, die über alle Verbindungen gemittelt werden, sind

- **#Iter** (Anzahl Iterationen): Diese Größe ist in Umgebungen von Interesse, in denen die Rechengeschwindigkeit die Laufzeit bestimmt. Ein praktisches Beispiel ist ein mobiles Navigationssystem auf einem Pocket-PC mit Kartendaten im Hauptspeicher bzw. auf einer Speicherkarte.
- **#Kanten** (Anzahl der *verschiedenen* Kanten in der open und der closed list). Auf Systemen mit beschränktem Hauptspeicher, die Kartendaten erst bei Bedarf von einem langsamen Massenspeicher nachladen, ist die Anzahl der einzulesenden Kanten für Laufzeit und Speicherplatzbedarf entscheidend. Zu dieser Art von Systemen zählen Navigationssysteme mit optischen Massenspeichern oder Handy-basierte Navigationssysteme.

Ergebnisse der Testläufe

Dargestellt werden aus Platzgründen nur die Ergebnisse der Benchmarks auf dem Testpunktsatz Hi20; die des Testpunktsatz Lk13 liefern sehr ähnliche Aussagen.

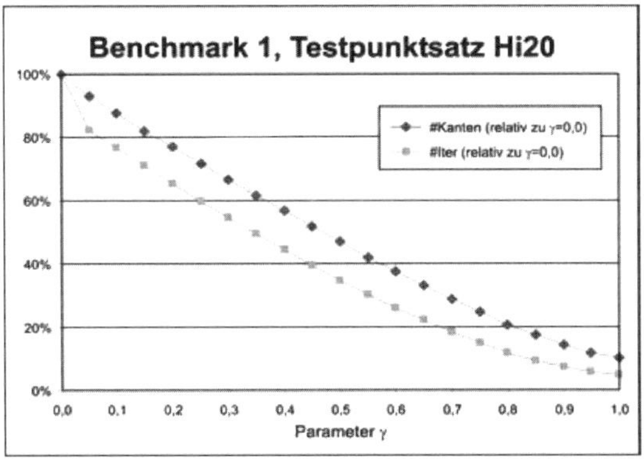

Abb. 4 Ergebnisse von Benchmark 1 auf Testpunktsatz Hi20

Im Vergleich zum Standard-A*-Verfahren ($\gamma = 0{,}0$) sinkt bei vollständiger Kenntnis der echten Restdistanz ($\gamma = 1{,}0$) die Anzahl der Iterationen auf 4,7 % und die der Anzahl Kanten auf 9,9 %. Diese Werte sind abhängig von der Anzahl der Kanten auf dem optimalen Weg sowie der Anzahl der Abbiegemöglichkeiten.

Der Verlauf zwischen diesen Endpunkten ist unterlinear[2]. Bei Verwendung eines Schätzers, der das arithmetische Mittel aus *RLD*(*e*) und *ED*(*e*) liefert ($\gamma = 0{,}5$), würde die Kantenzahl auf 46,9 % und die der Iterationen auf 34,7 % sinken.

Bei Benchmark 2 werden die fünf Schätzer $h_2(e, 0.00, 0.25)$, $h_2(e, 0.25, 0.50)$, $h_2(e, 0.50, 0.75)$, $h_2(e, 0.75, 1.00)$ und $h_2(e, 0.00, 1.00)$ verwendet.

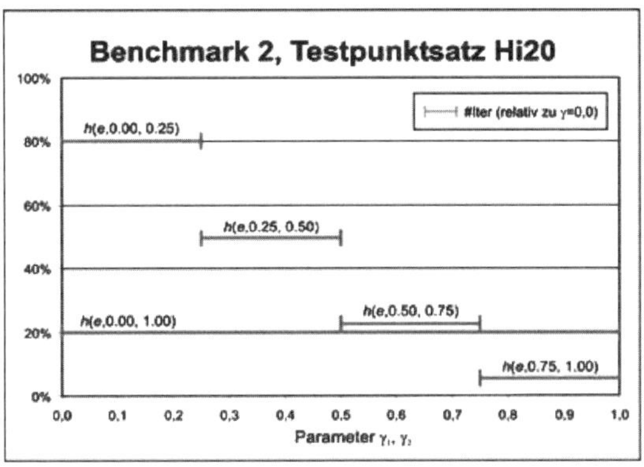

Abb. 5 Ergebnisse von Benchmark 2 auf Testpunktsatz Hi20

Die Ergebnisse von Benchmark 2 – dargestellt wird aus Gründen der Übersichtlichkeit nur die Anzahl der Iterationen – zeigen, dass sich die Schwankungen in der Qualität der ersten drei Schätzer im Mittel weitestgehend ausgleichen. Das Verfahren benötigt beispielsweise bei einem auf dem Intervall [0.50, 0.75] gleich-

---

[2] Genau: unterhalb einer gedachten geraden Verbindungslinie.

verteilten Parameter γ noch 23,2 % der Iterationen des Standard-A*-Verfahrens. Dieser Wert ließe sich auch als arithmetisches Mittel der Ergebnisse von $\gamma = 0{,}50$, $\gamma = 0{,}55$, ..., bis $\gamma = 0{,}75$ annähern (24,3 %).

Erstaunlich gut ist sind die Ergebnisse der letzten beiden Schätzer. Bei einer Gleichverteilung auf dem Intervall [0.00, 1.00] werden nur 19,8 % der Iterationen des Standard-A*-Verfahrens benötigt. Der Grund kann erneut in der in (Hahne 2000, S. 73ff) beobachteten Eigenschaft des A*-Verfahrens liegen, dass schlechte Abschätzungen nur selten große Fehler nach sich ziehen, sondern oft lokal beschränkt bleiben, eine Suche auf einer Kante in die falsche Richtung durch eine nachfolgende gute Abschätzung deshalb lokal wieder repariert werden kann.

Fazit und Ausblick

Die Kenntnis eines besseren Schätzers für das A*-Verfahren bei OLSP-Problemen auf digitalen Straßenkarten ist für Umgebungen mit beschränkter Rechenleistung bzw. geringer Zugriffsgeschwindigkeit auf die Kartendaten attraktiv, da hier große Einsparungen (bis zu Faktor 20) möglich sind.

Ein praktischer Nutzen ist nur von solchen Schätzern zu erwarten, die durch Preprocessing verbessert wurden, da die im Standard–A*-Verfahren verwendete Abschätzung durch die Luftdistanz in digitalen Karten nahezu „kostenlos" (sehr geringer Rechenaufwand, kein zusätzlicher Speicherplatz) zur Verfügung steht. Neben der schnellen Berechenbarkeit dürfen die Informationen aus dem Preprocessing selber keinen zu großen Speicherplatzbedarf haben.

Einen Anreiz für die Entwicklung solcher Schätzer liefern die empirischen Ergebnisse der Benchmarks: ein Schätzer, der eine um p % (exakt oder bei Gleichverteilung im Mittel) genauere Abschätzung der Restdistanz liefert, bewirkt beim A*-Verfahren für einen großen Wertebereich von p bei der Anzahl der Iterationen und der Anzahl der verwendeten Kanten eine Reduktion um mehr als p %.

Literatur

Djidjev HN (1997) *Efficient Algorithms for Shortest Path Queries in Planar Digraphs. Lecture Notes in Computer Science* 1973:51–165

Dubois N, Semet F (1995) *Estimation and determination of shortest path length in a road network with obstacles. European Journal of Operational Research* 83:105–116

Gelperin D (1977) *On the Optimality of A**. Artificial Intelligence 8:69-76.

Hahne F (2000) *Kürzeste und schnellste Wege in digitalen Straßenkarten.* Dissertation, Universität Hildesheim

Hart PE, Nilson, NJ, Raphael, B (1968) *A formal basis for the heuristic determination of minimal cost paths.* IEEE transactions on SSC 4:100–107.

Hart PE, Nilson NJ, Raphael, B (1972) *Correction to: „A formal basis for the heuristic determination of minimal cost paths".* Sigart newsletters 37:28–29

Nilsson N (1971) *Problem-Solving Methods in Artificial Intelligence.* McGraw-Hill, New York

Schäfer A (1998) *Kürzeste Wege in digitalen Straßenkarten unter Berücksichtigung von Gedächtnissen*, Diplomarbeit, Universität Hildesheim

Modelling Transport Networks by Means of Autonomous Units*

Karsten Hölscher, Peter Knirsch, and Hans-Jörg Kreowski

University of Bremen, Department of Mathematics and Computer Science
(hoelscher,knirsch,kreo)@informatik.uni-bremen.de

Summary. The concept of autonomous units to model distributed logistic processes and their interactions in a transport network is introduced. Autonomous units provide a general approach with rigorous semantics that allow the visual modelling of logistic processes in the transport domain in a systematic and structured way. Differing from existing models it especially incorporates the specification of autonomous or self-controlled behaviour of the participating actors. It means that the respective actions are not always predefined but allow for autonomous choice. By example in this paper a negotiation based approach is introduced. Due to this approach being formal and well-defined it supports testing and verification of required properties of the modelled systems at the level of specification.

1 Introduction

Transport logistics deals with the problems of how to transport load from one place to another while minding a set of constraints. Time frames for the delivery have to be kept in mind. The fleet size is restricted and so are the drivers capacities. In general not only the feasibility of the transport is of importance but also economic constraints. It is a well-known result from graph theory and complexity theory that those scheduling problems are hard to solve. The travelling sales person problem [LK73], for instance, is NP-complete although it seems to be quite simple compared to realistic scenarios. Having small scheduling problems and an idealised environment, exact solution can be computed in time. But if schedules become large, the runtime of such exact algorithms increases dramatically and make them practically not applicable. As a result, it is most likely that there is no efficient, exact algorithm computing such tours

* Research partially supported by the EC Research Training Network SegraVis (Syntactic and Semantic Integration of Visual Modeling Techniques) and by the German Research Foundation (DFG) as part of the Collaborative Research Centre 637 *Autonomous Cooperating Logistic Processes — A Paradigm Shift and its Limitations.*

in a reasonable time. But concerning the transport logistics it is the everyday business of a carrier to schedule trucks that pickup and deliver loads, and return to a depot afterwards. Heuristics that compute good solutions instead of optimal are the way out of this dilema. In [SS95] an introduction to the pickup and delivery problem can be found.

Today the structural and dynamic complexity of transport networks is increasing. The demands for transports are hardly predictable. If demand changes occur many plans are invalidated and the scheduling has to start again. Central planning is a bottleneck in the decentralised global world. A new challenge is to pass autonomy to the actors that have capabilities to adapt to changes at runtime. This is the scope of the Collaborative Research Centre CRC 637 *Autonomous Cooperating Logistic Processes – A Paradigm Shift and its Limitations.*

In this work a methodology is sketched to formally model transport networks by means of autonomous units that allow for autonomous adaptations. Although formal modelling seems to be extra work load in business it has many advantages. Using formal models one can specify all processes that are valid in a certain transport network where processes are regarded as sequences of operations. From all valid processes the best can be chosen. A model additionally facilitates the understanding of the processes especially if it has a visual representation. A model allows fast adaptations and algorithms can easily be derived thereof.

The concept of autonomous units generalises graph transformation units as studied in Kreowski, Kuske, and Schürr [KK99, KKS97] to structure large rule-based systems. It is a rule-based instantiation of the idea of agents and agent systems (see, e.g., [WJ94]) as first introduced in Knirsch and Kreowski in [KK00]. Graph transformation (see, e.g., [Roz97]) is a well-suited rule based mechanism to change graphs in a well-defined way.

An autonomous unit may represent any active component of a logistic system. In the particular context of transport logistics, it represents a vehicle, load, or even an RFID tag. Autonomous units have access to a common environment, in which they may cooperate or compete. Depending on the application domain such an environment can consist of all relevant places, e.g. cities, ports, stations, airports, etc., and relations between them, e.g. roads, railways, waterways, and communication channels. Additionally, in the pickup and delivery scenario the loads and available vehicles are part of the environment.

The autonomous units define the operational capabilities of the components. They run in a potentially non-deterministic way. In general they have a choice when performing the next action. Each unit controls itself autonomously to cut down this non-determinism. It is not controlled from outside. How the choice of the next action is done depends on the type of autonomy specified in the autonomous unit.

2 Autonomous Units

An *autonomous unit* is defined as $unit = (g, U, P, c)$ where g is a *goal* (formulated in a proper logic or language), U is a set of identifiers naming *used autonomous units* (that are imported in order to use their capabilities), P is a set of rules specifying the operational capabilities, and c is some control condition, restricting the possible orders of actions.

The goal g describes what the unit is trying to achieve or what is meant to become true. An autonomous unit acts in a specific environment, which it may change by choosing a suitable rule from P and applying it. Via these changes a unit may act directly towards the given goal or it may communicate with other units (on which actions it might depend). It may also use other units and their functionalities from U. The unit is considered autonomous in the sense that the next action is selected non-deterministically and not controlled from outside the unit. The simplest form would be to let the unit randomly decide on the next action. A more sophisticated form of autonomy can be achieved by using the control condition. The control condition may be very restricting and thus eliminating the non-determinism completely. It may as well be less restrictive to leave some room for non-deterministic choice. A typical kind of control conditions forces the order of rule applications or of calls of helping units.

3 Basic Modelling of Transport Networks

In the context of this work the environment of autonomous units is a transport network, consisting of places like depots or airports connected by different relations like roads or railways. Such a network of places and relations is naturally visualised as graphs with nodes representing places and edges representing relations. For this reason the rules of autonomous units are here graph transformation rules. For the pickup and delivery scenario more rules are specified in [KKK02].

Loads have to be picked up at certain places and delivered to other ones. In this very basic network example, the only mode of transport are trucks. Autonomous units are assigned to each of the trucks and loads, respectively, modelling their capabilities and autonomous behaviour.

In a first step a truck unit plans its tour for the day. How this is done is not nearer specified here. Graph transformation rules select non-deterministically (and currently regardless of a concrete goal) place nodes as part of the tour and mark them by inserting special tour nodes. This is done by all truck units that are used in the model. Figure 1 shows an excerpt of a transport network with a tour of one truck and one load. The tour of the truck (represented by the truck-shaped node) is planned to start in `Dortmund` and lead to `Hamburg` via `Bremen`. The tour is visualized by square nodes that are connected (by dotted lines) to the place nodes (depicted as circles) and the truck. The direction of the edges connecting the places with the tour node determines the direction of this tour section. Analogously the direction of the edges connecting the places with the load (represented by the rectangular node) determine the pickup and the delivery place. In the example the load has to be picked up in Hanover and delivered to Hamburg. The solid, undirected edges represent the roads between the places.

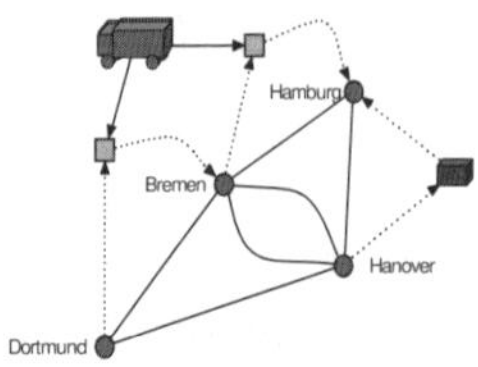

Fig. 1. Transport Network Graph

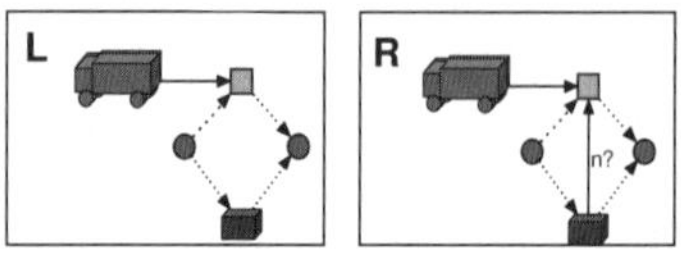

Fig. 2. Rule for a package offer

If a situation is found, where source and target of a tour section coincide with source and target places of a load, a negotiation for transport may commence. It is initiated by a package unit which contains a rule as shown in Figure 2. A graph transformation rule can be applied if the situation specified in its left-hand side L can be found in the environment. The counterparts of items that are only present in L but not in the right-hand side R are deleted from the environment. Items that are only present in R are added to the environment. Thus the application of the rule in Figure 2 yields a new edge connecting the load with the tour node of a truck. The edge is labelled with n?, meaning that the load offers the truck a price n for being transported.

The truck unit in turn has a rule that reacts to this new situation. It checks for an incoming n? edge, and may either accept or reject the offer. This depends on the internal structure of the truck unit. It could e.g. be possible that two packages make a price offer for transport for that tour section, so that the truck may choose the higher offer. In our first basic approach the truck unit decides non-deterministically and regardless of its goal whether to accept or reject an offer. The corresponding rule is depicted in Figure 3. It shows two right-hand sides R1 and R2 for the left-hand side L. This is a shorthand notation for two rules with the same left-hand side. In both cases L specifies a situation where a truck received an offer from a package for a tour section. In R1 the offer is accepted by flipping the edge and relabelling it with n!. In R2 the offer is rejected by deleting the offer edge. The truck units may also reschedule their tours. This involves deleting those tour nodes, that are not necessary to transport all packages with accepted offers. More formally, the control condition of a truck unit is specified in the following way: *plan_tour***;(accept_offer|reject_offer|reschedule_tour)!*

This means that the unit applies the rule *plan_tour* arbitrarily often followed by a choice of three rules. Here offers may be accepted or rejected as explained above or the tour be rescheduled. This choice is iterated as long as

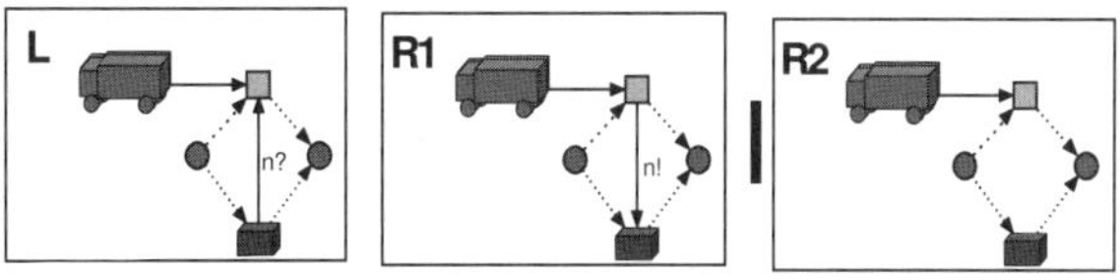

Fig. 3. Rule for accepting or rejecting an offer

possible, i.e. until no offers and also no empty tour sections remain in the environment. The control condition of a package unit is specified as: *make_offer!*

This means that the unit applies the rule *make_offer* as long as possible, i.e. until the corresponding load is scheduled to be transported from the pickup place to the delivery point.

4 Semantics of Autonomous Units

The semantics of the example transport network is given as environment transformations which are composed of rule applications and actions of used units obeyeing the control condition. This operational semantics provides a description of a simulation of the modelled processes and their cooperation.

Let *ENV* be the set of environments, and *unit* be one autonomous unit of a given system *Net* of autonomous units. Furthermore let $CHANGE(unit) \subseteq ENV \times ENV$ be a binary relation of environments describing the changes in the environment that can occur in addition to the changes *unit* can perform while acting autonomously. Then a **computation** of *unit* is defined as a sequence of environments $E_1, \ldots, E_k$, such that (E_i, E_{i+1}) for $i = 1, \ldots, k-1$ is obtained by applying a rule of *unit* or by calling a used unit according to the control condition. In order to account for changes not performed by *unit*, it may also be from *CHANGE(unit)*. Such a computation yields the input/output pair (E_1, E_k). The set of all these pairs is called the **semantic relation** *SEM(unit)* of one *unit*. Analogously a **computation** of *Net* is a sequence $(E_1, \ldots, E_k)$ if it is a computation of every unit of *Net*. This sequence describes the interaction of all the units with each other during a single run of the system, yielding the input/output pair (E_1, E_k) of a system run. The set *SEM(Net)* of all these pairs is called the **semantic relation** of *Net*.

This semantics definition induces a proof schema that allows to verify properties of the semantic relations by induction on the length of computations. This may be used to prove that the goals of a unit are reached (or not reached).

5 Conclusion

In this work we have presented the basic ideas and features of autonomous units and explained by example their application for modelling autonomous processes in a basic transport network. The model we presented so far is simple, since e.g. the trucks plan random tours and randomly accept or reject offers. Additional information like the distance of the regarded tour section or the size and weight of the package, the loading time, priorities, intelligent choice using reasoning and the like may get involved in the decisions. They may also depend on the goal that the truck unit tries to achieve. Detailed case studies are needed in the near future to illustrate and investigate these concepts. Furthermore the idea of autonomous units as a method for modelling should be compared to other modelling approaches that are currently employed in logistic scenarios, like e.g. Petri Nets, UML, or business process models. Currently the operational character of the autonomous units is purely sequential. It should be investigated how parallelism and concurrency can be incorporated into this approach to better reflect the real world, where all the units act simultaneously.

References

[KK99] H.-J. Kreowski and S. Kuske. Graph transformation units and modules. In H. Ehrig, G. Engels, H.-J. Kreowski, and G. Rozenberg, editors, *Handbook of Graph Grammars and Computing by Graph Transformation, Vol. 2: Applications, Languages and Tools*, pages 607–638. World Scientific, Singapore, 1999.

[KK00] P. Knirsch and H.-J. Kreowski. A note on modeling agent systems by graph transformation. In M. Nagl, A. Schürr, and M. Münch, editors, *AGTIVE'99 International Workshop on Applications of Graph Transformation with Industrial Relevance*, volume 1779 of *Lecture Notes in Computer Science*, pages 79–86, Berlin, 2000. Springer-Verlag.

[KKK02] R. Klempien-Hinrichs, P. Knirsch, and S. Kuske. Modeling the pickup-and-delivery problem with structured graph transformation. In H.-J. Kreowski and P. Knirsch, editors, *Proc. Applied Graph Transformation (AGT'02)*, 2002. 119–130.

[KKS97] H.-J. Kreowski, S. Kuske, and A. Schürr. Nested graph transformation units. *International Journal on Software Engineering and Knowledge Engineering*, 7(4):479–502, 1997.

[LK73] S. Lin and B. W. Kernighan. An effective heuristic algorithm for the traveling salesman problem. *Operations Research*, 21(9):498–516, 1973.

[Roz97] G. Rozenberg, editor. *Handbook on Graph Grammars and Computing by Graph Transformation. Vol. 1: Foundations.* World Scientific, Singapore, 1997.

[SS95] M. W. P. Savelsbergh and M. Sol. The general pickup and delivery problem. *Transportation Science*, 29(1):17–29, 1995.

[WJ94] M. Wooldridge and N. R. Jennings. Agent theories, architectures, and languages: A survey. In M. J. Wooldridge and N. R. Jennings, editors, *Intelligent Agents: ECAI-94 Workshop on Agent Theories, Architectures, and Languages*, volume 890 of *Lecture Notes in Artificial Intelligence*, pages 1–39. Springer, Berlin, 1994.

Routing in Line Planning for Public Transport*

Marc E. Pfetsch and Ralf Borndörfer

Konrad-Zuse-Zentrum für Informationstechnik Berlin, Takustr. 7, 14195 Berlin, Germany; Email: {borndoerfer, pfetsch}@zib.de

Summary. The line planning problem is one of the fundamental problems in strategic planning of public and rail transport. It consists in finding lines and corresponding frequencies in a network such that a given demand can be satisfied. There are two objectives. Passengers want to minimize travel times, the transport company wishes to minimize operating costs. We investigate three variants of a multi-commodity flow model for line planning that differ with respect to passenger routings. The first model allows arbitrary routings, the second only unsplittable routings, and the third only shortest path routings with respect to the network. We compare these models theoretically and computationally on data for the city of Potsdam.

1 Introduction

Integer programming methods have become a successful tool for line planning in the last decade. Fixing passenger routes according to a so-called system split and choosing lines from a precomputed pool, Bussieck et al. [3] maximized direct travelers, and Claessens et al. [4] minimized costs; the latter approach was improved by Goossens et al. [6]. Recently, the system-split assumptions were relaxed by Goossens et al. [5] and by Schöbel and Scholl [7, 8], who minimize the number of transfers or transfer times.

In [1, 2] we introduced a basic IP model for the line planning problem in which both lines and passenger routes are generated dynamically, imposing a length restriction on lines. The results can contain multiple passenger paths between the same endpoints and detours, which passengers would not take in practice. The aim of this paper is to study more realistic variants of our model with passenger routing restrictions, namely, unsplittable routings and routings on a shortest path w.r.t. the network. Length restricted routes would lead to additional interesting variants. We have also investigated these, but do not discuss them here due to lack of space.

* Supported by the DFG Research Center MATHEON "Mathematics for key technologies" in Berlin.

2 Line Planning Models

The line planning problem involves a number k of *transportation modes* (bus, tram, subway, etc.), an undirected multigraph $G = (V, E) = (V, E_1 \dot\cup \ldots \dot\cup E_k)$, which we call *transportation network*, *terminal sets* $\mathcal{T}_1, \ldots, \mathcal{T}_k \subseteq V$, operating costs $\boldsymbol{c}^1 \in \mathbb{Q}_+^{E_1}, \ldots, \boldsymbol{c}^k \in \mathbb{Q}_+^{E_k}$, vehicle capacities $\kappa_1, \ldots, \kappa_k \in \mathbb{Q}_+$, and a (not necessarily symmetric) *origin-destination matrix* (OD-matrix) $(d_{st}) \in \mathbb{Q}_+^{V \times V}$ of travel demands, i.e., d_{st} is the number of passengers that want to travel from node s to t. Let $D := \{(s,t) \in V \times V \,:\, d_{st} > 0\}$ be the set of all *OD-pairs*.

A *line* of mode i is a path in the mode graph $G_i := (V, E_i)$ connecting two (different) terminals of $\mathcal{T}_i$. Note that paths are always *simple*, i.e., node repetitions are not allowed. We denote by $\mathcal{L}$ the set of all lines, by $\mathcal{L}_e := \bigcup\{\ell \in \mathcal{L} \,:\, e \in \ell\}$ the set of lines that use edge $e \in E$, by $c_\ell := \sum_{e \in \ell} c_e^i$ the operating cost of line ℓ of mode i, and by $\kappa_\ell := \kappa_i$ its vehicle capacity.

We derive from G a directed *passenger route graph* (V, A) by replacing each edge $e \in E$ with two antiparallel arcs $a(e)$ and $\overline{a}(e)$; conversely, let $e(a) \in E$ be the undirected edge corresponding to $a \in A$. For an OD-pair $(s,t) \in D$, an *(s,t)-passenger path* is a directed path in (V, A) from s to t. Denote by $\mathcal{P}_{st}$ the set of all (s,t)-passenger paths and by $\mathcal{P} := \bigcup\{p \in \mathcal{P}_{st} \,:\, (s,t) \in D\}$ the set of all passengers paths. We are given *travel times* $\tau_a \in \mathbb{Q}_+$ for every arc $a \in A$. The *travel time* of a passenger path p is defined as $\tau_p := \sum_{a \in p} \tau_a$.

Let $\mathcal{P}' \subseteq \mathcal{P}$ and $\mathcal{P}'_{st} := \mathcal{P}' \cap \mathcal{P}_{st}$ be subsets of passenger paths that model routing restrictions. Introducing variables $y_p \in \mathbb{R}_+$ for the fraction of the demand d_{st} traveling from s to t on path p and $f_\ell \in \mathbb{R}_+$ for the frequency of line $\ell \in \mathcal{L}$, and a parameter $0 \leq \lambda \leq 1$ that weights line operating costs and travel times, we can state the following general line planning model:

$$
\text{(LPP)} \qquad \min \lambda\, \boldsymbol{\gamma}^{\mathrm{T}} \boldsymbol{f} + (1-\lambda)\, \boldsymbol{\tau}^{\mathrm{T}} \boldsymbol{y}
$$

$$
\begin{aligned}
\boldsymbol{y}(\mathcal{P}'_{st}) &= 1 && \forall\, (s,t) \in D && \text{(1a)}\\
\sum_{(s,t)\in D} d_{st} \sum_{p: a \in p \in \mathcal{P}'_{st}} y_p &\leq \sum_{\ell: e(a) \in \ell} \kappa_\ell f_\ell && \forall\, a \in A && \text{(1b)}\\
f_\ell &\geq 0 && \forall\, \ell \in \mathcal{L} && \text{(1c)}\\
0 \leq y_p &\leq 1 && \forall\, p \in \mathcal{P}'. && \text{(1d)}
\end{aligned}
$$

Here, we write $\boldsymbol{y}(\mathcal{P}'_{st}) := \sum_{p \in \mathcal{P}'_{st}} y_p$ and similarly for other vectors and sets. Constraints (1a) force that demand d_{st} is routed from s to t. The *capacity constraints* (1b) ensure that all passengers can be transported.

We now derive three variants of the model (LPP). The *multi-path routing* (MPR) model is obtained from (LPP) by setting $\mathcal{P}' := \mathcal{P}$, i.e., by allowing arbitrary passenger routings. The *unsplittable path routing* (UPR) model is derived from (LPP) by setting $\mathcal{P}' := \mathcal{P}$ and by forcing $y_p \in \mathbb{Z}$ for all $p \in \mathcal{P}$, which ensures passenger paths to be unsplittable. The *network path routing* (NPR) model is obtained from (LPP) by letting $\mathcal{P}'$ only contain shortest paths from s to t with respect to the travel times in the transportation network G (independent of the lines) for every $(s,t) \in D$. We assume w.l.o.g. that shortest paths are unique and, therefore, that passengers are routed on a unique shortest path for every OD-pair; note that such a routing is automatically unsplittable.

3 Theoretical Comparison

We study in this section the influence of routing restrictions on the optima and the complexity of the line planning problem. Denote by $\mathrm{opt}(X;I)$ the optimal solution of problem $X \in \{\mathrm{MPR}, \mathrm{UPR}, \mathrm{NPR}\}$ for an instance I, by $\mathrm{opt}_{\mathrm{LP}}(X;I)$ the optimal solution of the corresponding LP relaxation, and by

$$\mathrm{gap}(X,Y) := \sup_I \frac{\mathrm{opt}(X;I)}{\mathrm{opt}(Y;I)}$$

the *gap* between (the optimum of) problem X and Y, where the supremum is taken over all instances I. From the definitions of the problems, by observing that (MPR) and (NPR) are LPs, and that (MPR) is the LP relaxation of (UPR), we obtain immediately for any instance I:

$$\begin{aligned} &\mathrm{opt}(\mathrm{MPR};I) \le \mathrm{opt}(\mathrm{UPR};I) \le \mathrm{opt}(\mathrm{NPR};I) \\ &\mathrm{opt}_{\mathrm{LP}}(\mathrm{MPR};I) = \mathrm{opt}(\mathrm{MPR};I) = \mathrm{opt}_{\mathrm{LP}}(\mathrm{UPR};I) \\ &\mathrm{opt}_{\mathrm{LP}}(\mathrm{NPR};I) = \mathrm{opt}(\mathrm{NPR};I)\,. \end{aligned} \tag{2}$$

We have shown in [2] that (MPR), even though it is an LP, is $\mathcal{NP}$-hard. Our complexity proof uses only unsplittable shortest paths. This implies:

Proposition 1. (MPR), (UPR), *and* (NPR) *are* $\mathcal{NP}$*-hard.*

We can strengthen this result for (UPR) as follows:

Proposition 2. (UPR) *is* $\mathcal{NP}$*-hard, even when the lines are fixed a priori.*

We skip the proof, which works by reduction of the disjoint paths problem, and return to the relations (2). We show that there exist instances for which the inequalities are strict and that, in fact, the gap can be arbitrary large.

Theorem 1. $\mathrm{gap}(\mathrm{UPR}, \mathrm{MPR}) = \infty$.

Proof. Consider the digraph D on the left of Figure 1. It has $2k+2$ nodes. The graph G underlying D gives rise to a line planning problem as follows. We consider a mode i for each edge $\{s,i\}$ and a mode i' for each $\{i',t\}$, $i = 1,\ldots,k$, with edge sets $E_i := \{\{s,i\}\}$ and $E_{i'} := \{\{i',t\}\}$, and terminal sets $\mathcal{T}_i := \{s,i\}$ and $\mathcal{T}_{i'} := \{i',t\}$, i.e., there exists exactly one line path for each such mode and its terminals are the endpoints of the corresponding edge. The costs and

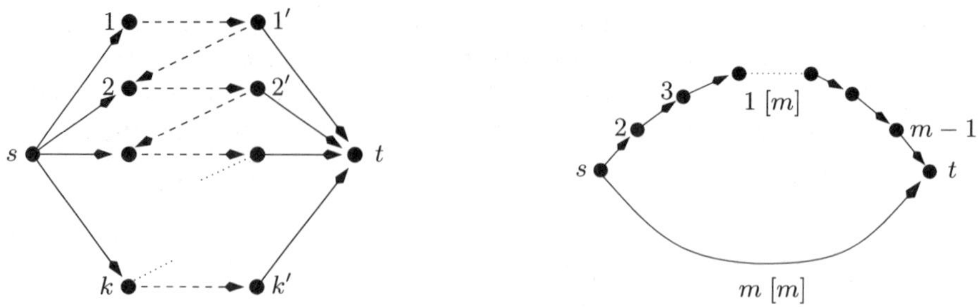

Fig. 1. Constructions for Theorems 1 and 2, respectively.

capacities of these lines ℓ are $c_\ell = 0$ and $\kappa_\ell = 1$, respectively. There is an additional line mode for the zig-zag structure in the middle; its edge set are the zig-zag-edges (the edges shown dashed) and its terminal set is $\{1, k'\}$. This modes also supports exactly one line path on the zig-zag structure. The cost of this line is $c_{\ell'} = 1$ and its capacity is $\kappa_{\ell'} = 1$. The travel times are all zero. We finally set the demand between s and t to k and to zero otherwise.

Consider (MPR) associated with this instance. Its optimal solution is to set all line frequencies to 1 and to route a passenger flow of 1 on the k paths of length three between s and t. We therefore have opt(MPR; I) = 1, since only the zig-zag line contributes to the objective. Now consider the corresponding (UPR). Any solution must choose a single path between s and t and has to route a flow of k on it. Such a path must use the zig-zag line, whose frequency must therefore be k, i.e., opt(UPR; I) = k. This implies

$$\frac{\text{opt(UPR;}\,I)}{\text{opt(MPR;}\,I)} = \frac{k}{1} = k \to \infty,$$

which concludes the proof. □

Note that this shows that the integrality gap for (UPR) is arbitrarily large.

Theorem 2. gap(NPR, UPR) = ∞.

Proof. Consider the digraph D on the right of Figure 1 and number its nodes from left to right as $s = 1, 2, \ldots, m-1, m = t$. It gives rise to an instance I of the line planning problem as follows. D has $m - 1$ arcs at the top and one arc at the bottom. We associate with each arc a mode as in the proof of Theorem 1 that supports exactly one line on this arc. The cost of each such line ℓ is $c_\ell = m$ and its capacity is $\kappa_\ell = 1$. We want to route a demand of 1 from s to t and set the travel times to 1 on the top arcs and to m on the bottom arc.

The optimal (unsplittable) solution of the associated (UPR) sets the frequency of the line on the bottom arc to 1, incurring a line cost of m, and routes all demand on this line with a travel time of m. Hence, opt(UPR; I) = $2m$. The optimal solution of the associated (NPR) routes all demand on the shortest path with respect to G, which is the path through the upper arcs. Since we

Table 1. Comparison of all results for $\lambda = 0.9979$. The CPU time is measured in seconds on a 3.4 Ghz Pentium 4.

	pass. time	line cost	objective	lines/pass.	CPU
Reference solution:	104,977,699.00	479,839.22	699,284.72	61/4854	49.8
MPR:	108,763,392.00	225,062.62	452,993.11	63/4890	191.2
NPR:	92,124,536.00	886,760.85	1,078,360.18	95/4685	89.9
shortest path UPR:	95,270,123.00	652,363.55	851,060.85	67/4685	222.8
thickest path UPR:	108,729,269.00	236,046.53	463,882.29	69/4685	233.6

need $m-1$ lines with frequencies 1 and costs m, the line costs are $m \cdot (m-1)$. The travel time is $m-1$. Hence, $\text{opt}(\text{NPR}; I) = m^2 - 1$. It follows that

$$\frac{\text{opt}(\text{NPR}; I)}{\text{opt}(\text{UPR}; I)} = \frac{m^2 - 1}{2m} = \tfrac{1}{2}(m - \tfrac{1}{m}) \to \infty,$$

which concludes the proof. □

4 Computational Results

We now provide an empirical comparison of the different routing variants of our model on data for the city of Potsdam. The data represents the network of 1998. It has 27 bus lines and 4 tram lines. Including line variants, regional trains, and city railroad, the total number of lines is 80. The preprocessed network has 410 nodes, 106 of which are OD-nodes, and 891 edges. The OD-matrix has 4685 nonzeros and the total demand is 42,796, for a time horizon of 3 hours. No data was available for line costs; we decided on operating costs $c_e^i := 100$ for each edge e and mode i.

We have described in [2] a column generation algorithm for the line planning problem, solving shortest path problems to price the passenger variables $\boldsymbol{y}$ and longest path problems to price the line variables $\boldsymbol{f}$. In our approach, we restrict the set $\mathcal{L}$ of line paths. Namely, we compute the minimal number $k(a, b)$ of edges needed to connect a and b in $G_i = (V, E_i)$ and allow only lines with $k \leq \max\{1.2 \cdot k(a, b), 55\}$ edges. The idea is to produce only lines that do not deviate too much from a shortest path. With this restriction, line pricing can be performed quite fast by enumeration. Passenger paths are priced out by using Dijkstra's algorithm. The master LPs are solved with the barrier algorithm and, towards the end, with the primal simplex algorithm of CPLEX 9.1. Our algorithm can be applied directly to (MPR). Table 1 reports our computational results as well a reference solution, i.e., an optimal solution to (MPR), where the lines were fixed to be the lines of the 1998 Potsdam system (only 61 were active, i.e., had a positive frequency).

To compute solutions for (NPR) we modified our code by fixing the passenger paths to shortest connections between the OD-nodes. The restrictions on the line construction, however, can cause that there are arcs which cannot be covered by lines. We ignore such arcs for the computation of the shortest paths. From the results in Table 1, we see that indeed 4685 passenger paths are needed which equals the number of OD-pairs.

Computing optimal solutions to (UPR) is not only hard from a theoretical, but also from a practical viewpoint. Indeed, the model uses $4,174,335$ binary variables y_a^{st} for our data. This makes a direct integer programming approach impractical. Note that a Lagrangean relaxation of the capacity constraints (1b) will not help, as this does not improve over the LP relaxation solution (MPR). We therefore implemented two heuristics for (UPR).

The first heuristic computes a solution to (MPR) and determines the shortest paths with respect to the computed line system. It then deletes all other passenger paths and re-solves the LP, thereby allowing the pricing of new lines. Table 1 shows that the gap to the MPR solution is 46.8%. The second heuristic chooses for each OD-pair among the paths used by the MPR solution the thickest, i.e., the one carrying the highest number of passengers. It then prices out lines as above. The gap to the MPR solution is only 2.3%. The quality of these heuristics clearly depends on the weighting λ. If λ is very small, the MPR solution will use very short passenger paths and the solutions of both heuristics will be close to that of (MPR), i.e., the gap is small.

Analyzing Table 1, we see that (MPR) improves upon the reference solution (but increases the total travel time). The NPR solution has the shortest total travel time, but the highest objective. The shortest path UPR solution also has very low total travel time and high line costs, while the thickest path UPR solution is very close to the lower bound solution of (MPR).

References

1. R. Borndörfer, M. Grötschel, and M. E. Pfetsch, *Models for line planning in public transport*, Report 04–10, ZIB, 2004.
2. ——, *A path-based model for line planning in public transport*, Report 05–18, ZIB, 2005.
3. M. R. Bussieck, P. Kreuzer, and U. T. Zimmermann, *Optimal lines for railway systems*, Eur. J. Oper. Res., 96 (1997), pp. 54–63.
4. M. T. Claessens, N. M. van Dijk, and P. J. Zwaneveld, *Cost optimal allocation of rail passanger lines*, Eur. J. Oper. Res., 110 (1998), pp. 474–489.
5. J.-W. H. M. Goossens, S. van Hoesel, and L. G. Kroon, *On solving multi-type line planning problems*, METEOR Research Memorandum RM/02/009, University of Maastricht, 2002.
6. ——, *A branch-and-cut approach for solving railway line-planning problems*, Transportation Sci., 38 (2004), pp. 379–393.
7. A. Schöbel and S. Scholl, *Line planning with minimal travelling time.* Preprint, 2005.
8. S. Scholl, *Customer-Oriented Line Planning*, PhD thesis, Universität Göttingen, 2005.

Tourenplanung mittelständischer Speditionsunternehmen in Stückgutkooperationen

Julia Rieck und Jürgen Zimmermann

Technische Universität Clausthal, Abteilung für Unternehmensforschung, Julius-Albert-Str. 2, D–38678 Clausthal-Zellerfeld, `julia.rieck@tu-clausthal.de`

Zusammenfassung. Um gestiegenen Kundenansprüchen gerecht zu werden, arbeiten mittelständische Speditionsunternehmen zunehmend in Stückgutkooperationen zusammen. Für die einzelnen Kooperationspartner ergeben sich dabei eine Reihe neuer Anforderungen bei der Erstellung eines Tourenplans. Neben der Berücksichtigung heterogener Fahrzeuge und Kundenzeitfenster sowie simultaner Auslieferung und Einsammlung sind z.B. ein mehrfacher Fahrzeugeinsatz vorzusehen und Belegungszeiten der Rampen im Depot zu berücksichtigen. Das betrachtete Tourenplanungsproblem wird als gemischt-ganzzahliges lineares Programm formuliert. Da eine Lösung mit Cplex 9.0 nur für kleine Instanzen in akzeptabler Zeit möglich ist, werden heuristische Lösungsverfahren und erste Performance-Ergebnisse vorgestellt.

1 Einführung in die Problemstellung

Die Globalisierung der Märkte und die Internationalisierung der Wertschöpfungsprozesse haben dazu geführt, dass der Transport von Waren zu einer zentralen Aufgabe für Unternehmen geworden ist. Aus dieser Entwicklung resultieren vielschichtige Anforderungen an Logistikdienstleister, z.B. hinsichtlich der Flächendeckung und der Laufzeiten. Insbesondere kleine und mittelständische Speditionsunternehmen haben unter der angespannten Marktsituation zu leiden und schließen sich daher in *Stückgutkooperationen* zusammen. Transportnetze von Stückgutkooperationen bestehen aus den Depots der Kooperationspartner, die ähnlich einer Hub-and-Spoke Struktur miteinander verknüpft sind. Aufgabe des einzelnen Kooperationspartners ist das Ausliefern und Einsammeln von Stückgütern in der Umgebung seines Depots. In den Depots werden die Waren sortiert und gebündelt, um daraufhin im Fernverkehr weiter transportiert zu werden. Eine kostengünstige Prozessabwicklung der operativen Tourenplanung vor Ort ist somit für die einzelnen Speditionsunternehmen von großer Bedeutung.

Wir betrachten im Folgenden eine Erweiterung des Standardproblems der Tourenplanung (Vehicle Routing Problem – VRP). Beim VRP werden ausgehend von einem Depot Auslieferungs- bzw. Einsammlungstouren zu Kunden durchgeführt. Jeder Kunde ist dabei in der Planungsperiode nur einmal anzufahren und die Kapazität der Fahrzeuge sowie eine maximale Tourdauer dürfen zu keiner Zeit überschritten werden. Ziel ist es, einen Tourenplan zu erstellen, bei dem die gesamte zurückgelegte Strecke minimal ist. Um das VRP den Anforderungen mittelständischer Speditionsunternehmen anzupassen, sind folgende praxisrelevante Aspekte zu berücksichtigen:

- Kundenzeitfenster und Depotöffnungszeiten,
- eine simultane Auslieferung und Einsammlung von Stückgütern bei den Kunden,
- heterogene Fahrzeuge, die mehrfach eingesetzt werden und
- Belegungszeiten der Rampen im Depot.

Bei der erweiterten Problemstellung wird davon ausgegangen, dass jeder Kunde nur einmal besucht und eine simultane Auslieferung und Einsammlung durchgeführt wird. Da die Distanzen zwischen den Kunden sowie den Kunden und dem Depot i.d.R. gering sind, ist es zweckmäßig, Fahrzeuge mehrfach einzusetzen. Die Belegungszeiten der Rampen werden berücksichtigt, um Wartezeiten im Depot zu vermeiden und so einen reibungslosen Ablauf an der Schnittstelle zwischen Tourenplanung und Umschlagbetrieb zu gewährleisten.

In den letzten vierzig Jahren wurden eine Vielzahl von Artikeln im Bereich der Tourenplanung veröffentlicht. Als Einstieg in den Themenkomplex verweisen wir bspw. auf [2]. Das hier betrachtete Tourenplanungsproblem stellt eine Kombination aus dem VRP mit Zeitfenstern (VRPTW, vgl. z.B. [4]), dem VRP mit simultaner Auslieferung und Einsammlung (VRPSDP, vgl. z.B. [1]) und dem VRP mit mehrfachem Fahrzeugeinsatz (VRPMU, vgl. z.B. [6]) dar. Zusätzlich zu diesen Erweiterungen des VRP werden Belegungszeiten der Rampen im Depot berücksichtigt.

2 Modellformulierung

Sei $G = (V, A)$ ein gerichteter Graph mit Knotenmenge V und Pfeilmenge A. Die Knotenmenge $V = K \cup R \cup S \cup E$ besteht aus Kundenknoten K, Rampenknoten R sowie Knoten S (start) und E (end), die die Abfahrten der einzelnen Fahrzeuge aus dem Depot bzw. die Ankünfte der Fahrzeuge im Depot beschreiben. Wir nehmen an, dass jedes vorhandene Fahrzeug zweimal in der Planungsperiode eingesetzt werden kann. Somit beinhaltet $S = S^1 \cup S^2$ zwei Abfahrtsknoten pro Fahrzeug, die Menge S^1 besteht aus den ersten und die Menge S^2 aus den zweiten Abfahrtsknoten der Fahrzeuge. Menge E setzt sich analog zusammen. Es gilt $|S| = |E|$. Jeder Kunde $i \in K$ besitzt einen Auslieferungsbedarf $d_i \geq 0$ (delivery) und einen Einsammlungsbedarf $p_i \geq 0$ (pick-up). Für die Knoten $j \in S \cup E$ gilt $d_j := p_j := 0$. Kundenbedarfe werden in abstrahierten Transporteinheiten angegeben, die sich aus Größe und Gewicht des Stückgutes ermitteln lassen. Weiterhin ist für jeden Knoten $i \in V$

eine Menge $\{a_i, b_i, s_i\}$ mit $a_i \leq b_i$ gegeben. Die Anfangszeit der Bedienung in Knoten i muss innerhalb des Zeitfensters $[a_i,b_i]$ liegen. Wir setzen $a_i := \rho$ und $b_i := \xi$ für alle $i \in S \cup E \cup R$, wobei ρ der Öffnungszeit und ξ der Schlusszeit des Depots entspricht. Die Servicezeit $s_i \geq 0$ erhalten wir aus der zu verladenden Gütermenge in Knoten i, multipliziert mit einem Ladefaktor l, der die Verladezeit pro Transporteinheit angibt. Jedem Pfeil $\langle i, j \rangle, i,j \in K \cup S \cup E$ sind eine Entfernung d_{ij} (distance) und eine Fahrzeit t_{ij} (travel time) zugeordnet. Für alle $i,j \in K, i \neq j$, sowie für $i \in S, j \in K$ und $i \in K, j \in E$ gilt $d_{ij} > 0$, $t_{ij} > 0$, andernfalls setzen wir $d_{ij} := t_{ij} := 0$. Jedem Knoten $i \in S$ ordnen wir die Kapazität cap_i des entsprechenden Fahrzeugs zu. Diese Kapazität ist für alle Knoten zu berücksichtigen, die von i aus angefahren werden. Die Kapazitäten der Fahrzeuge sowie eine maximal vorgegebene Fahr- und Servicezeit T_{max} dürfen zu keiner Zeit überschreiten werden. Für jeden Knoten $i \in K \cup S \cup E$ führen wir eine Hilfsvariable h_i ein und setzen $h_i := 1$ für alle $i \in S \cup E$ sowie $h_i := 0$ sonst. Mit den Entscheidungsvariablen

$t_j \geq 0$	Anfangszeit der Bedienung in Knoten $j \in K \cup S \cup E \cup R$
$f_j \geq 0$	benötigte Fahr- und Servicezeit bis zu Knoten $j \in K \cup S \cup E$
$\pi_j \geq 0$	Anzahl der besuchten Depots vor Besuch des Knoten $j \in K \cup S \cup E$
$l_j \geq 0$	Ladungsmenge nach Besuch des Kunden $j \in K$ bzw. Ladungsmenge im Depot $j \in E$
$ld_j \geq 0$	Ladungsmenge, die ab Knoten $j \in K \cup S$ noch auszuliefern ist, inkl. der Menge die bei j ausgeliefert wird
X_{ij}	Binärvariable: 1, wenn von Knoten $i \in K \cup S$ zu Knoten $j \in K \cup E$ gefahren wird; 0 sonst
Y_{ij}	Binärvariable: 1, wenn die Be- bzw. Entladung in Knoten $i \in S \cup E$ vor der Be- bzw. Entladung in Knoten $j \in S \cup E$ durchgeführt wird, wenn an Rampe $i \in R$ die Be- bzw. Entladung von $j \in S \cup E$ durchgeführt wird, wenn nach $i \in S \cup E$ alle weiteren Be- und Entladungen an Rampe $j \in R$ stattfinden sowie wenn an Rampe $i \in R$ keine Verladung stattfindet und mit $j \in R$ fortgefahren wird; 0 sonst (vgl. Abb. 1)

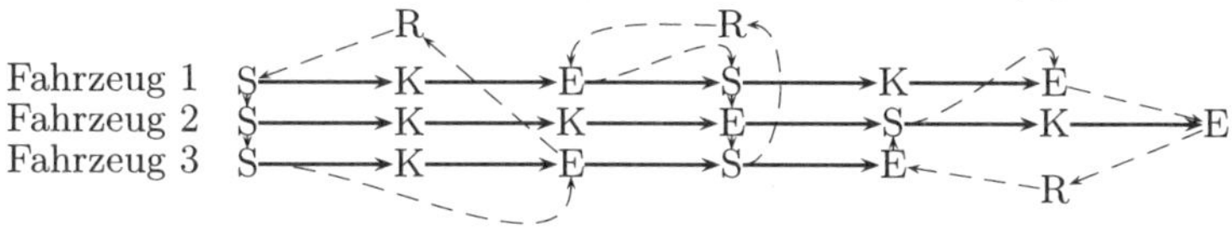

Abb. 1. Ein möglicher Tourenplan

lässt sich das Tourenplanungsproblem wie folgt als gemischt-ganzzahliges lineares Optimierungsproblem formulieren.

Minimiere
$$\sum_{i \in K \cup S} \sum_{j \in K \cup E} d_{ij} X_{ij} \tag{1}$$

unter den Nebenbedingungen:

$$\sum_{j \in K \cup E^\alpha} X_{ij} = 1 \qquad i \in S^\alpha, \alpha = 1,2 \tag{2}$$

$$\sum_{i \in K \cup S^\alpha} X_{ij} = 1 \qquad j \in E^\alpha, \alpha = 1,2 \tag{3}$$

$$\sum_{\substack{i \in K \cup E \\ i \neq j}} X_{ji} = 1; \quad \sum_{\substack{i \in K \cup S \\ i \neq j}} X_{ij} = 1 \qquad j \in K \tag{4}$$

$$\sum_{i \in E^1} X_{ij} = 1 \qquad j \in S^2 \tag{5}$$

$$\sum_{j \in S^2} X_{ij} = 1 \qquad i \in E^1 \tag{6}$$

$$\pi_j \geq \pi_i + h_i - 3(1 - X_{ij}) \qquad i \in K \cup S, j \in K \cup E; i \in E^1, j \in S^2 \tag{7}$$

$$\pi_j \leq 3; \; f_j + s_j \leq T_{max} \qquad j \in E^2 \tag{8a,b}$$

$$f_j \geq f_i + s_i + t_{ij} - T_{max}(1 - X_{ij}) \quad i \in K \cup S, j \in K \cup E; i \in E^1, j \in S^2 \tag{9}$$

$$t_j \geq t_i + s_i + t_{ij} - (\xi - \rho)(1 - X_{ij}) \quad i \in K \cup S, j \in K \cup E; i \in E^1, j \in S^2 \tag{10}$$

$$a_j \leq t_j \leq b_j \qquad j \in K \tag{11}$$

$$s_i = l \cdot (d_i + p_i) \qquad i \in K \tag{12}$$

$$s_i = l \cdot ld_i \qquad i \in S \tag{13}$$

$$s_i = l \cdot l_i \qquad i \in E \tag{14}$$

$$ld_i \geq ld_j + d_i - c_{max}(1 - X_{ij}) \qquad i \in K \cup S, j \in K, i \neq j \tag{15}$$

$$l_j \geq ld_j - d_j + p_j - c_{max}(1 - X_{ij}) \qquad i \in S, j \in K \tag{16}$$

$$l_j \geq l_i - d_j + p_j - c_{max}(1 - X_{ij}) \qquad i \in K, j \in K \cup E, i \neq j \tag{17}$$

$$cap_i \geq cap_j - c_{max}(1 - X_{ij}) \qquad i \in K \cup S, j \in K \cup E; i \in E^1, j \in S^2 \tag{18}$$

$$ld_i \leq cap_i; \; d_i \leq ld_i \qquad i \in K \cup S \tag{19}$$

$$l_i \leq cap_i; \; p_i \leq l_i \qquad i \in K \cup E \tag{20}$$

$$\sum_{\substack{i \in S \cup E \cup R \\ i \neq j}} Y_{ji} = 1; \quad \sum_{\substack{i \in S \cup E \cup R \\ i \neq j}} Y_{ij} = 1 \qquad j \in S \cup E \cup R \tag{21}$$

$$t_j \geq t_i + s_i - (\xi - \rho)(1 - Y_{ij}) \qquad i \in S \cup E \cup R, j \in S \cup E, i \neq j \tag{22}$$

$$t_i = \rho; \; s_i = 0 \qquad i \in R \tag{23}$$

$$\rho \leq t_i; \; t_i + s_i \leq \xi \qquad i \in S \cup E \tag{24}$$

$$X_{ij} \in \{0, 1\} \qquad i \in K \cup S \cup E, j \in K \cup S \cup E \tag{25}$$

$$Y_{ij} \in \{0,1\} \qquad i \in S \cup E \cup R, j \in S \cup E \cup R \tag{26}$$

Zielfunktion (1) minimiert die insgesamt zurückgelegte Strecke. Nebenbedingungen (2) – (8a) gewährleisten, dass jeder Kunde nur einmal bedient und jedes Fahrzeug genau zweimal eingesetzt wird. Bedingungen (8b) – (11) sorgen für die Einhaltung der Kundenzeitfenster und der maximalen Fahr- und Servicezeit T_{max}. Durch die Bedingungen (12) – (14) werden die Servicezeiten in den Knoten festgelegt. Mit den Bedingungen (15) – (17) bestimmen wir die Ladungsmenge der Fahrzeuge nach Besuch des ersten und aller weiteren Kunden. Nebenbedingungen (18) – (20) sichern, dass die Kapazitäten der Fahrzeuge nicht überschritten werden. Bedingungen (21) – (24) gewährleisten, dass den Abfahrts- bzw. Ankunftsknoten Rampen zugeordnet werden, an denen die Be- und Entladung stattfinden kann.

3 Heuristische Lösungsverfahren

Für das Tourenplanungsproblem aus Abschnitt 2 wurden zwei heuristische Lösungsverfahren (SA, RSA) implementiert, die auf Grundlage des Savings-

Verfahrens (vgl. [3]) Touren bilden. Zu Beginn der Verfahren beinhaltet der jeweilige Tourenplan alle Pendeltouren. Da i.d.R. nicht gewährleistet ist, dass die gegebene Fahrzeugflotte alle Pendeltouren durchführen kann, fügen wir in der Initialisierung fiktive Fahrzeuge mit benötigter Kapazität cap_i, $i \in S$ ein. In den Iterationen der Verfahren werden je zwei Touren t_1 und t_2 zu einer Kombinationstour t^* zusammengefasst, indem ein Randkunde i von t_1 mit einem Randkunden j von t_2 verbunden wird. Die Algorithmen bilden die Touren t^* so, dass eine möglichst große Streckenersparnis (Saving) realisiert wird. Algorithmus RSA verwendet randomiserte Savings, so dass mit diesem Verfahren mehrere unterschiedliche Lösungen erzeugt werden können. Zur Überprüfung der Kundenzeitfenster und Depotöffnungszeiten definieren wir wie in [5] $a(t)$ als früheste und $b(t)$ als späteste Ankunft beim ersten Kunden der Tour t. Mit Hilfe von $a(t_i), b(t_i)$, der gesamten Fahr- und Servicezeit $f(t_i)$ sowie der Wartezeit $w(t_i)$ der Touren t_i, $i = 1{,}2$ bestimmen wir das Zeitfenster $[a(t^*), b(t^*)]$ für die Ankunft beim ersten Kunden der Kombinationstour t^*. Die Tour t^* ist zulässig, wenn $a(t_1) \leq b(t_2) - t_{ij} - w(t_1) - f(t_1)$, ansonsten ist t^* zu verwerfen. Danach berechnen wir den frühesten und spätesten Startzeitpunkt (ES_{t^*} – earliest start time, LS_{t^*} – latest start time) der Tour t^*. Sei $j \in S$ der erste Abfahrtsknoten, i_{t^*} der erste Kunde und s_j die Beladezeit der Tour t^*, dann gilt z.B. $ES_{t^*} = \max\{0, a(t^*) - t_{ji_{t^*}} - s_j\}$. Ist $LS_{t^*} < ES_{t^*}$, so ist t^* zu verwerfen. Wurde die Tour t^* angenommen, versuchen wir, die Touren des entstandenen Tourenplans auf die Fahrzeuge und Rampen zu verteilen. Dabei prüfen wir für jede Tour, ob sie von einem realen Fahrzeug übernommen werden kann. Ein Fahrzeug v kann eine Tour t durchführen, wenn es entweder noch nicht in Anspruch genommen wurde oder wenn die Touren, die v bereits zugeordnet sind, die Durchführung der Tour t zulassen. Falls ein Fahrzeug v Tour t übernehmen kann, ist zu prüfen, ob für die Be- und Entladung jeweils eine Rampe zur Verfügung steht. Kann kein reales Fahrzeug gefunden werden, dann versuchen wir t einem fiktiven Fahrzeug zuzuordnen. Wenn für alle Touren eine Zuordnung gefunden wurde, werden t_1 und t_2 aus dem Tourenplan entfernt, t^* aufgenommen und nicht genutzte fiktive Fahrzeuge gestrichen. Das Verfahren terminiert, wenn keine Streckenersparnis mehr realisiert werden kann. Beinhaltet der jeweilige Tourenplan nur reale Fahrzeuge, so liegt eine zulässige Lösung vor.

4 Erste Performance-Ergebnisse

Um die Güte der Heuristiken zu testen, haben wir Testsets mit jeweils 60 Instanzen generiert, bei denen die Anzahl der Kunden, der Fahrzeuge und der Rampen variiert wurde. Die einzelnen Instanzen sind so geartet, dass die Pendeltouren zulässig sind. Cplex 9.0 konnte nur Instanzen mit fünf und zehn Kunden innerhalb einer Zeitspanne von 1.800 sec. lösen. Algorithmus SA fand für 43 Instanzen mit fünf und für 16 Instanzen mit zehn Kunden eine optimale Lösung. Die Abweichung vom Optimum der übrigen 17 bzw. 44

Instanzen betrug 7,70% bzw. 6,57%. Algorithmus RSA generierte in jeweils 100 Läufen für 56 Instanzen mit fünf und für 45 Instanzen mit zehn Kunden eine optimale Lösung. Die Abweichung vom Optimum der übrigen 4 bzw. 15 Instanzen betrug 1,98% bzw. 3,21%. In Tabelle 2 sind die kumulierten Strecken der einzelnen Testsets angegeben. Im Durchschnitt ist der Algorithmus RSA 3,11% besser als SA.

Tabelle 2. Kumulierte Strecken der Testsets

	5 Kunden	10 Kunden	15 Kunden	20 Kunden	30 Kunden	50 Kunden
SA	16.915	28.308	37.089	45.838	65.573	96.486 *
RSA	16.072	27.253	35.021	44.351	64.866	96.419 (99.963)

(* bei 2 Instanzen wurde keine zulässige Lösung gefunden)

5 Zusammenfassung und Ausblick

In diesem Aufsatz wurde eine Erweiterung des VRP betrachtet, die für die operative Tourenplanung mittelständischer Speditionsunternehmen in Stückgutkooperationen relevant ist. Bei der Modellierung wurde ein verallgemeinertes, mit zusätzlichen Restriktionen behaftetes Zuordnungsproblem mit einem weiteren Zuordnungsproblem verknüpft. Mit diesem Modell konnten nur Instanzen mit fünf bzw. zehn Kunden in akzeptabler Zeit gelöst werden. Probleminstanzen mit 50 Kunden konnte Algorithmus SA in weniger als einer Sekunde lösen, Algorithmus RSA benötigte für 100 Läufe ca. 90 Sekunden. Insgesamt können wir festhalten, dass Algorithmus RSA eine gute Entscheidungsunterstützung bei der Tourenplanung mittelständischer Speditionsunternehmen bietet.

Literaturverzeichnis

1. Angelelli, E., Mansini, R. (2002) The Vehicle Routing Problem with Time Windows and Simultaneous Pick–up and Delivery. In: Klose, A., Speranza, M. G., Van Wassenhove, L. N. (Hrsg.) Quantitative Approaches to Distribution Logistics and Supply Chain Management. Springer, Berlin Heidelberg New York
2. Bodin, L., Golden, B, Assad, A., Ball, M. (1983) Routing and Scheduling of Vehicles and Crews: The State of the Art. Computers and Operations Research 10: 79 – 115
3. Clarke, G., Wright, J. W. (1964) Scheduling of Vehicles from a Central Depot to a Number of Delivery Points. Operations Research 12: 568 – 581
4. Desrochers, M., Desrosiers, J., Solomon, M. (1992) A new Optimization Algorithm for the Vehicle Routing Problem with Time Windows. Operations Research 40: 342 – 354
5. Diruf, G. (1980) Kundenzeitschranken in der computergestützten Tourenplanung. Zeitschrift für Operations Research 24: B207 – B220
6. Hajri–Gabouj, S., Darmoul, S. (2003) A Hybrid Evolutionary Appraoch for a Vehicle Routing Problem with Double Time Windows for the Depot and Multiple Use of Vehicles. Studies in Informatics and Control 12: 253 – 268

Closed Networks of Generalized S-queues with Unreliable Servers

Kersten Tippner

Hamburg University, Department of Mathematics, Bundesstrasse 55, 20146 Hamburg, Germany. kersten_eckert@gmx.net

S-queues are models for discrete-time queueing networks in which customers can both arrive and be served in batches. The model is extended by introducing unreliable nodes as well as strategies for customers to get round inactive nodes. It is shown that the equilibrium distribution for these networks has product form; moreover, some availability and performance measures are computed.

1 Introduction

We consider a discrete-time closed queueing network with N nodes and K customers. The nodes are unreliable, i. e. they can break down in the course of a time slot and then have to be repaired. Thus we want the description of the system state to contain some information on which nodes are working and which nodes are currently being repaired. Moreover, we examine the joint queue-length process for the N nodes of the network. The states of the network are thus given in the form

$$x = (\mathfrak{n}, \bar{I}) = (((\mathfrak{n}_1, \ldots, \mathfrak{n}_N), (\bar{I})) : \mathfrak{n}_i \in \mathbb{N}, i = 1, \ldots, N; \\ \sum_i \mathfrak{n}_i = K; \bar{I} \subseteq \bar{N} := \{1, \ldots, N\}),$$

$\mathfrak{n}_i$ being the number of customers at node i, $\bar{I}$ being the set of nodes currently under repair. Changes in the state of the system occur due to
a) breakdowns of active nodes and/or repairs of inactive nodes, and
b) departures of customers from nodes and their arrival to other nodes.
Breakdowns and repairs are assumed to occur independently from the queue-lengths at the various nodes of the network. If, at the beginning of a time slot, the nodes in $\bar{I} \subseteq \bar{N}$ are inactive, the probability that, by the end of the same time slot, the set of inactive nodes is $\bar{J} \subseteq \bar{N}$, is given by $\gamma(\bar{I}, \bar{J})$, γ being a

reversible transition matrix on $Mat(\mathcal{P}(\bar{N}), \mathcal{P}(\bar{N}))$.
Customers whose service times end leave their respective nodes at the end of the respective time slot and are then immediately routed to the next node. We denote the joint queue-length process by $X = ((X_t^1, \ldots, X_t^N) : t \in \mathbb{Z})$; $D = ((D_t^1, \ldots, D_t^N) : t \in \mathbb{Z})$ and $A = ((A_t^1, \ldots, A_t^N) : t \in \mathbb{Z})$ are the sequences of departure and arrival vectors, respectively. By $\mathcal{A}$, we denote the set of all possible departure and arrival vectors, which can be summarized as "transfer vectors" .
The joint queue-length process is defined by

$$X_{t+1} = X_t - D_t + A_t;$$

its state space is given by $S(K, N) := \{(\mathfrak{n}_1, \ldots, \mathfrak{n}_N) : \sum_{i=1}^N \mathfrak{n}_i = K\}$.

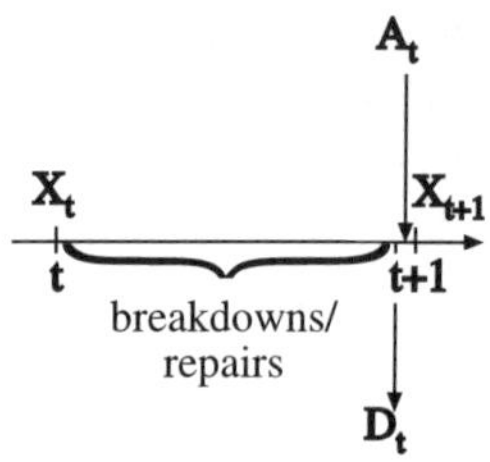

Fig. 1. Inner structure of a time slot

If $\bar{I} = \emptyset$; i. e. if all nodes are active, customers are served at all nodes. The nodes are assumed to be working independently from each other and to be generalized S-queues, that is,
a) customers both arrive at the nodes and depart from them in batches, and
b) D_t^i, $i \in \bar{N}$, does not depend on the past of the system but via X_t^i.
Moreover, we assume that the departure probability $q_i(\mathfrak{n}_i, \mathfrak{a}_i) := P(D_t^i = \mathfrak{a}_i \mid X_t^i = \mathfrak{n}_i)$ is given by

$$q_i(\mathfrak{n}_i, \mathfrak{a}_i) = \frac{\Psi_i(\mathfrak{n}_i - \mathfrak{a}_i)}{(\mathfrak{a}_i)! \Phi_i(\mathfrak{n}_i)}. \tag{1}$$

$$\Longrightarrow q(\mathfrak{n}, \mathfrak{a}) := P(D_t = \mathfrak{a} \mid X_t = \mathfrak{n}) = \prod_{i=1}^N q(\mathfrak{n}_i, \mathfrak{a}_i) = \prod_{i=1}^N \frac{\Psi_i(\mathfrak{n}_i - \mathfrak{a}_i)}{(\mathfrak{a}_i)! \Phi_i(\mathfrak{n}_i)}.$$

Example 1. Infinite server network
All customers are served at the same time as though each of them were the only customer in the network. The service of a customer at node i terminates in the current time slot with probability p_i; with probability $(1 - p_i)$, service continues for at least one more time slot.

$$\Longrightarrow q_i(\mathfrak{n}_i, \mathfrak{a}_i) = \binom{\mathfrak{n}_i}{\mathfrak{a}_i} p_i^{\mathfrak{a}_i}(1-p_i)^{\mathfrak{n}_i - \mathfrak{a}_i}.$$

Setting

$$\Phi_i(\mathfrak{n}_i) := [(\mathfrak{n}_i)! p_i^{\mathfrak{n}_i}]^{-1} \qquad \text{and} \qquad \Psi_i(\mathfrak{n}_i) := \frac{1}{(\mathfrak{n}_i)!}(\frac{1-p_i}{p_i})^{\mathfrak{n}_i},$$

it is obvious that $q_i(\mathfrak{n}_i, \mathfrak{a}_i)$ fulfills (1).

A customer having left node i, $i \in \bar{N}$ is routed to node j, $j \in \bar{N}$ with jump probability $r(i,j)$, independent of the routing of all other customers. The routing probability $\mathfrak{r}(\mathfrak{a}, \mathfrak{a}') := P(A_t = \mathfrak{a}' \mid D_t = \mathfrak{a})$ is thus given by

$$\mathfrak{r}(\mathfrak{a}, \mathfrak{a}') = \sum_{\tilde{\mathfrak{a}}} \prod_{i=1}^{N} \mathfrak{a}_i \prod_{j=1}^{N} \frac{r(i,j)^{\mathfrak{a}_{i,j}}}{(\mathfrak{a}_{i,j})!}$$

where the summation is over all $\tilde{\mathfrak{a}} := ((\mathfrak{a}_{1,1}, \ldots, \mathfrak{a}_{1,N}), (\mathfrak{a}_{2,1}, \ldots, \mathfrak{a}_{2,N}), \ldots, (\mathfrak{a}_{N,1}, \ldots, \mathfrak{a}_{N,N}))$ such that $\sum_{j=1}^{N} \mathfrak{a}_{i,j} = \mathfrak{a}_i$ and $\sum_{i=1}^{N} \mathfrak{a}_{i,j} = \mathfrak{a}'_i$.
If, however, $\bar{I} \neq \emptyset$, that is, if at least one node is inactive, either all nodes immediately stop serving customers ("stalling", see section 3), or just the nodes in $\bar{I}$ interrupt services and reject the arrival of new customers so that one of the re-routing strategies described in section 2 has to be applied; the active nodes continue their services. In either case, though, all customers being served by a node that breaks down have to stay there until the node is repaired and their service time is terminated.

2 Re-routing strategies

Given that the nodes in $\bar{I} \subseteq \bar{N}$ are inactive, $\mathcal{A}_k := \{\mathfrak{a} \in \mathcal{A} : \mathfrak{a}$ consists of k customers$\}$ can be divided into two disjunct subsets, the permitted transfer vectors $(=: \mathcal{A}_k^{\oplus} = \{\mathfrak{a} \in \mathcal{A}_k : \mathfrak{a}_i = 0 \forall i \in \bar{I}\})$ and the prohibited ones $(=: \mathcal{A}_k^{\ominus} = \{\mathfrak{a} \in \mathcal{A}_k : \exists i \in \bar{I}$ such that $\mathfrak{a}_i \neq 0\})$. Departure vectors are (by construction) always permitted; an arrival vector, though, may be either permitted or prohibited.
The re-routing strategies described below are then applied to prohibited arrival vectors.

2.1 Re-routing at customer level

As customers are routed independently, we can determine for each of them whether they are making a permitted or a prohibited jump, i. e. whether they are jumping to an active or an inactive node. Customers making prohibited jumps are then re-routed according to RS-RD ("Repeated Service – Random Destination") or skipping rules.

RS-RD

Each customer making a prohibited jump is sent back to the node they have just left; there they served once more. After having terminated the additional service time, they jump to a node in the network according to the jump matrix r. The jump matrix $r_{\bar{I}}$ for the system with inactive nodes is thus given by

$$\tilde{r}_{\bar{I}}(i,j) = \begin{cases} r(i,j), & i,j \in \bar{N} \setminus \bar{I}, i \neq j \\ r(i,i) + \sum_{k \in \bar{I}} r(i,k), & j = i. \end{cases}$$

Skipping

A customer performing a prohibited jump just makes a virtual jump to the node of their choice; having arrived there, they immediately jump to the next node according to the jump matrix r as though they had just left the respective inactive node. A customer has to jump until they reach an active node.
In that case, the jump matrix for the system with inactive nodes is given by

$$\tilde{r}_{\bar{I}}(i,j) = r(i,j) + \sum_{k \in \bar{I}} r(i,k) \tilde{r}_{\bar{I}}(k,j) i,j \in \bar{N} \setminus \bar{I}$$

and similarly for $i \in \bar{I}, j \in \bar{N} \setminus \bar{I}$.

2.2 Re-routing at transfer-vector level

Instead of re-routing individual customers, though, one can also determine whether an arrival vector is permitted or prohibited and then, if the vector is prohibited, re-transform it according to some specified global RS-RD or global skipping rules.

Global RS-RD

If a departure vector is transformed into a prohibited arrival vector, the transformation is invalidated. That is, all customers, independently of whether they would have made a permitted or a prohibited jump, are sent back to the nodes they have just left; there they are served once again. The routing matrix for the system with inactive nodes $\tilde{\mathfrak{r}}_{\bar{I}}$ is then given by

$$\tilde{\mathfrak{r}}_{\bar{I}}(\mathfrak{a}, \mathfrak{a}') = \begin{cases} \mathfrak{r}(\mathfrak{a}, \mathfrak{a}'), & \mathfrak{a}, \mathfrak{a}' \in \mathcal{A}_k^{\oplus}, \mathfrak{a} \neq \mathfrak{a}' \\ \mathfrak{r}(\mathfrak{a}, \mathfrak{a}) + \sum_{\mathfrak{a}'' \in \mathcal{A}_k^{\ominus}} r(\mathfrak{a}, \mathfrak{a}''), & \mathfrak{a}' = \mathfrak{a} \end{cases}$$

Global Skipping

If a departure vector $\mathfrak{a}$ is transformed into a prohibited arrival vector $\mathfrak{a}'$, the vector $\mathfrak{a}'$ has to be re-transformed into a vector $\mathfrak{a}''$ etc. until a permitted transfer vector $\hat{\mathfrak{a}}$ is reached. In that case, the routing matrix $\tilde{\mathfrak{r}}_{\bar{I}}$ for the system with inactive nodes is given by

$$\tilde{\mathfrak{r}}_{\bar{I}}(\mathfrak{a}, \mathfrak{a}') = r(\mathfrak{a}, \mathfrak{a}') + \sum_{\mathfrak{a}'' \in \mathcal{A}^{\ominus}} \mathfrak{r}(\mathfrak{a}, \mathfrak{a}'')\tilde{\mathfrak{r}}_{\bar{I}}(\mathfrak{a}'', \mathfrak{a}'), \quad \mathfrak{a}, \mathfrak{a}' \in \mathcal{A}^{\oplus}$$

and similarly for $\mathfrak{a} \in \mathcal{A}^{\ominus}, \mathfrak{a}' \in \mathcal{A}^{\oplus}$.

3 Stalling

Stalling, as opposed to the more theoretical re-routing strategies describes above, is a technique already used in practice, for instance in quality control. It implies that, as long as there are nodes under repair in the network, not a single customer is served. All nodes immediately interrupt service. The inactive nodes are repaired; it is, however, possible that further breakdowns occur during a such "idle" period. Services are only continued when all nodes are in active status again.

4 The equilibrium distribution in networks of generalized S-queues with unreliable servers

Theorem 1. *In closed networks of generalized S-queues with departure probabilities in the form* (2) *and with unreliable servers, the equilibrium distribution for the network process is independent of the strategy applied and is*

$$\pi((\mathfrak{n}_1, \ldots, \mathfrak{n}_N), \bar{I}) = \tilde{K}_{K,N}^{-1}\bar{\pi}(\bar{I}) \prod_{i=1}^{N} y(i)^{\mathfrak{n}_i} \Phi_i(\mathfrak{n}_i), \tag{2}$$

$\tilde{K}_{K,N}^{-1}$ *being the norming constant,* $y(\bullet)$ *solving the balance equations*

$$y(i) = \sum_{i=0}^{N} y(i) r(i, j)$$

and $\bar{\pi}(\bullet)$ *being the probability solution of the balance equations*

$$\bar{\pi}(\bar{I}) = \sum_{i=1}^{N} \bar{\pi}(\bar{J})\gamma(\bar{J}, \bar{I}).$$

Proof. (2) fulfills the global balance equations of the system.
(The full proof for this theorem as well as the proofs for the theorems in the following section can be obtained from the author via E-mail.)

5 Availability and performance measures

Knowing the equilibrium distribution π for the network process, we can compute different performance and availability measures:

Theorem 2. *1. The stationary joint point availability of subnetwork $\bar{K} \subseteq \bar{N}$ at time $t \geq 0$ is*

$$Av(\bar{K})(t) = 1 - \sum_{\bar{I} \supseteq \bar{K}} \gamma(\bar{I})$$

2. The mean queue lengths $E(X_i)$, $i \in \{1, \ldots, N\}$, are the same as in the according network without breakdowns and repairs.

Theorem 3. *In a network with unreliable servers working with a re-routing strategy, the throughput at node j, $TH(j)$, is*

$$TH(j) = T\hat{H}(j) \cdot Av(\{j\})(t),$$

$T\hat{H}(j)$ being the throughput at node j in the according network without breakdowns and repairs.
If, however, stalling is applied, the throughput is

$$TH(j) = T\hat{H}(j) \cdot \bar{\pi}(\emptyset).$$

References

1. Daduna, Hans (2001) Queuing Networks with discrete time scale: explicit expressions for the steady state behaviour of discrete time stochastic networks. Springer, Berlin Heidelberg New York
2. Henderson, W. und Taylor, P.G.(1990) Product form in networks of queues with batch arrivals and batch services. Queueing Systems 6 : 71-88
3. Henderson, W. und Taylor, P.G.(1991) Some new results on queueing networks with batch movement. Journal of Applied Probability 28 : 409-421
4. Kelly, F.P.(1979) Reversibility and stochastic networks. Wiley, New York
5. Miyazawa, Masakiyo(1994) On the characterization of departure rules for discrete-time queueing networks with batch movements and its applications, in: Miyazawa, M. und Takagi, H.: Advances in Discrete Time Queues. Vol. 18: Queueing Systems, S. 149-166.
6. Osāwa, Hideo (1994) Quasi-reversibility of a discrete-time queue and related models, in: Miyazawa, M. und Takagi, H.: Advances in Discrete Time Queues. Vol. 18: Queueing Systems, S.133-148 .
7. Sauer, Cornelia und Daduna, Hans (2003) Availability Formulas and Performance Measures for Separable Degradable Networks. Economic Quality Control 18 : 165 - 194
8. Walrand, Jean (1983) A Discrete-Time Queueing Network. Journal of Applied Probability 20 : 903-909

Part X

OR Applications in Health and Life Sciences

A Set Packing Approach for Scheduling Elective Surgical Procedures

R. Velásquez[1,2] and M.T. Melo[1,3]

[1] Fraunhofer Institute for Industrial Mathematics (ITWM), D 67663 Kaiserslautern, Germany
[2] velasque@itwm.fhg.de
[3] melo@itwm.fhg.de

Summary. The efficient scheduling of surgical procedures to operating rooms in a hospital is a complex problem due to limited resources (e.g. medical staff, equipment) and conflicting objectives (e.g. reduce running costs and increase staff and patient satisfaction). A novel approach for scheduling elective surgeries over a short-term horizon is proposed which takes explicit consideration of these aspects. The problem is formulated as a set packing problem and solved optimally through column generation and constraint branching. Good results were obtained for instances from the literature.

1 Introduction

The operating theatre (OT) suite is one of the major service areas in a hospital. Due to highly specialized staff and large capital investments on equipment it is also a high cost service facility. As a result, a substantial degree of efficiency is required in the daily management of an OT suite. This is usually the task of the OT manager who is in charge of regularly creating OT schedules for elective surgeries. As described in [6], an OT schedule is often the outcome of personal expertise or experience as well as the fruit of negotiation with all parties involved such as surgeons, anesthesiologists and nurses. Roughly, the creation of an OT schedule for elective surgeries consists in setting the intervention dates, assigning the surgeries to operating rooms, determining the team members and sequencing all procedures on each particular room and date. Naturally, the availability of the required resources (e.g. various rooms with different opening hours and equipment/devices, heterogeneous work schedules of medical staff and other personnel, etc.) must be taken into account while building a tentative plan. This entails considering several, and often conflicting, objectives. On the one hand, the OT manager would like to keep costs low and resource usage high, while on the other hand a high quality of service provided to patients ought to be accomplished and staff preferences should be satisfied. As highlighted in [3], it is rather difficult to create such an OT schedule. Furthermore, the use of exploratory methods makes it practically impossible to obtain good OT schedules that fulfill all the above goals and that simultaneously take hospital-specific policies into account (e.g. different priorities for scheduling surgeries depending on patient age).

The literature on surgical suite scheduling has considered individual aspects of the problem. For example, the problem of scheduling patients in an ambulatory surgical center such that resources required by the post-anesthesia care unit are minimized was modelled in [4] as a two-stage process shop scheduling problem, and solved with greedy and tabu search heuristics. In [6], a non-linear mixed integer model was proposed for a real-life situation with limited working hours and rooms, equipment conflicts and hospital specific prioritization rules based on age and health condition of patients. This problem was solved using simulated annealing. In contrast, many other studies do not use mathematical approaches for scheduling surgical cases. Instead, simple rules such as "first-come, first-served", "longest-cases-first", "shortest-cases-first" are proposed (see e.g. [2]). Thus, this paper fills the gap resulting from the lack of a far-reaching model for scheduling surgical procedures by using a well-known model as is the set packing formulation. In the next section, a binary mathematical formulation is proposed for the problem. Sect. 3 briefly describes an exact solution method. Sect. 4 reports on the computational experience while Sect. 5 presents some conclusions and directions for future research.

2 Problem Description and Formulation

The problem of scheduling surgical cases is considered on an operational planning level over a time horizon of one or several days (up to one week in advance). It is assumed that the OT manager receives input from all departments regarding surgery requirements and resources' availability. The required notation is introduced as follows:

Index sets S = Set of elective surgeries; R = Set of resources (e.g. rooms, staff, equipment); T = Set of time steps resulting from the discretization of the planning horizon (e.g. an eight hour day is divided into 96 time steps of five minutes each); M_s = Set of possible combinations of resources for surgery $s \in S$ scheduled at a preferred starting time t_m; Δ_m = Set of possible time deviations w.r.t. the preferred starting time t_m of surgery s for each combination $m \in M_s$.

Parameters $c_{s,(m,\delta)}$ = unitary preference for surgery $s \in S$ to use the m-th ($m \in M_s$) resource combination and to be scheduled $\delta \in \Delta_m$ time steps after its preferred starting time t_m; $1^{t,r}_{(s,(m,\delta))}$ = indicator that states if surgery $s \in S$ in its m-th resource combination and time deviation δ w.r.t.

its preferred starting time t_m, uses in time step $t \in T$, resource $r \in R$; $k_{t,r}$ = capacity limit of resource $r \in R$ during time step $t \in T$.

Decision variables $x_{s,(m,\delta)} = 1$ if surgery $s \in S$ is scheduled using resource combination $m \in M_s$ and starts δ time steps after its preferred starting time t_m, and 0 otherwise.

The following set packing formulation is proposed to model the operating theatre scheduling problem (OTSP).

$$\max \quad \sum_{s \in S} \sum_{m \in M_s} \sum_{\delta \in \Delta_m} c_{s,(m,\delta)} \; x_{s,(m,\delta)} \tag{1}$$

$$\text{subject to} \quad \sum_{m \in M_s} \sum_{\delta \in \Delta_m} x_{s,(m,\delta)} \leq 1 \qquad \forall s \in S \tag{2}$$

$$\sum_{s \in S} \sum_{m \in M_s} \sum_{\delta \in \Delta_m} 1_{(s,(m,\delta))^{t,r}} \; x_{s,(m,\delta)} \leq k_{t,r} \qquad \forall t \in T, r \in R \tag{3}$$

$$x_{s,(m,\delta)} \in \{0,1\} \qquad \forall s \in S, m \in M_s \tag{4}$$

The objective function (1) maximizes the satisfaction of preferences for performing the surgeries. The weight parameter $c_{s,(m,\delta)}$ permits a flexible modelling of any type of preference and common practices in the OT suite. The latter include, for example, scheduling children surgeries early in the day and assigning different priorities to surgeries depending on the health condition of adult patients. Inequalities (2) and (3) represent *surgery* and *resource constraints* respectively, with the first group ensuring that only one combination of resources $m \in M_s$ is assigned to each surgery $s \in S$, while the second group models the use of resources $r \in R$ in all time steps $t \in T$, preventing them to be in conflict among surgeries. Furthermore, the resulting coefficient matrix in the OTSP consists of $|S|$ column blocks, one for every elective surgery s. Within each block all possible resource combinations $m \in M_s$ and starting times t for surgery s are listed.

The use of the above formulation presents a number of advantages both from a modelling as well as from a solution viewpoint. Regarding the first aspect, a variety of situations and preferences arising in practice can be easily modelled, like including all kinds of resource types and their corresponding availability (e.g. rooms, staff, equipment, beds in preoperative holding (PHU) and intensive care (ICU) units). In addition, staff preferences such as surgeons desiring to conduct cases in succession and in the same room rather than scattered at different times and rooms during the day, can also be easily modelled. Moreover, resource requirements for given surgeries and the sequence followed by a patient through the OT suite – starting from the PHU until the ICU – are naturally modelled in the OTSP. Observe that resource requirements at a particular time step are represented through non-zero entries in a column of the coefficient matrix along with the corresponding resource constraints. Staff skills, durations of surgeries and individual preferences w.r.t. starting times, equipment, etc. are also taken into consideration. Appropriate settings of the indicator parameter $1_{(s,(m,\delta))}$ in (3) and of the weight parameter $c_{s,(m,\delta)}$ in (1) allow this to be done. In the next section the solution opportunities arising from the set packing formulation are discussed.

3 Column Generation and Constraint Branching

Although a set packing problem (SPP) is in general hard to solve, the particular structure of the OTSP allows to develop special procedures to solve it to optimality even for large instances. Observe that in the LP-relaxation of the OTSP, the surgery constraints (2) form a totally unimodular matrix. However, the desired optimal binary solution of the OTSP is hard to find due to (3). This type of problem was successfully explored in [5] by a method using column generation and constraint branching for solving the airline crew rostering problem. This approach will also be applied to the OTSP.

The column generation scheme consists of a pricing step for each of the possible combinations of resources $m \in M_s$ for surgery $s \in S$. It starts with a reduced form of the LP-relaxation of the OTSP (R-OTSPLP) which includes only one resource combination for each surgery, namely the one with the largest objective function value. Furthermore, it uses a vector π of dual variables obtained after solving the R-OTSPLP, and calculates the reduced costs $\kappa(a_{\cdot m})$ for all columns corresponding to the resource combinations m that were previously not part of the R-OTSPLP. The reduced costs are obtained as in (5), with $a_{\cdot m}$ denoting the column of the coefficient matrix of the OTSP corresponding to resource combination m, and c_m its objective function coefficient.

$$\kappa(a_{\cdot m}) = c_m - \pi^T\, a_{\cdot m} \qquad \forall\, m \in M_s,\;\; s \in S \tag{5}$$

The column associated to the resource combination that yields the largest positive reduced cost is then added to the R-OTSPLP and the procedure is repeated until no resource combination leads to a positive reduced cost. The optimal solution to the R-OTSPLP is also an optimal solution to the LP-relaxation of the OTSP. For further details on column generation, the interested reader is referred to [1].

The branching step uses constraint branching rather than traditional variable branching. Variable branching is highly ineffective for binary optimization problems (BOP) as it leads to a very large and unbalanced tree. The constraint branching proposed in [5] yielded good results in resolving fractional solutions of a set packing problem, and also produced the desired integer optimal solution. It consists of identifying two constraints $\hat{s}$ and $\hat{r}$ such that $0 < \sum_{j\in J(\hat{s},\hat{r})} x^j_{s,(m,\delta)} < 1$ holds. The set $J(\hat{s},\hat{r})$ is defined as the set of columns of the coefficient matrix which have non-zero entries for constraints $\hat{s}$ and $\hat{r}$, i.e. $J(\hat{s},\hat{r}) = \{j | a_{\hat{s},j} = 1 \text{ and } a_{\hat{r},j} = 1\}$. In the OTSP any fractional solution will have at least one such pair of constraints, namely those constraints corresponding to surgery $\hat{s}$ in (2) and $\hat{r}$ in (3), corresponding to the use of resource r during time t, and where a series of resource combinations partially cover $\hat{s}$ and $\hat{r}$. The constraints $\hat{s}$ and $\hat{r}$ are chosen such that the largest fractional value of a surgery requiring a resource at a certain time is obtained. This corresponds to selecting $\sum_{j \in J(\hat{s},\hat{r})} x^j_{s,(m,\delta)}$ as close to one as possible. Branching is then performed by setting $\sum_{j \in J(\hat{s},\hat{r})} x^j_{s,(m,\delta)} = 1$ (1-branch) which states that $\hat{s}$ and $\hat{r}$ are covered by the same resource combinations, and $\sum_{j \in J(\hat{s},\hat{r})} x^j_{s,(m,\delta)} = 0$ (0-branch), which states that $\hat{s}$ and $\hat{r}$ are covered by different resource combinations. A sequence of constraint branches thus leads to a binary solution for all variables. Once again, the interested reader is referred to [5] for a detailed description of the procedure.

4 Computational Experience

The solution approach was implemented in C++ and the LP subproblems were solved with Xpress-MP on a Pentium 4 PC with a 1.7 GHz processor and 512 MB RAM. All data was taken from [6] and stems from a broad study carried out in several Australian hospitals, thus representing real life situations. Table 1 summarizes this information.

Table 1. Index sets based on information in [6]

Set	Data
S	Set of 27 surgeries, detailing age of the patient and surgery priority
R	Set of 14 resources: 4 rooms and 10 mobile pieces of equipment
T	Set of 120 time steps corresponding to a working day from 8 am until 6 pm. Each time step represents five minutes.
M_s	Each of the 27 surgeries is assigned to a room and the required pieces of equipment. $\|M_s\| = 1, \forall s \in S$. Preferred starting time is in time step $t_m = 0$.
Δ_m	Each surgery can start in every time step as long as it is completed within the working day

The resulting OTSP is a BOP with 2885 variables and 1707 constraints. The LP-relaxation of this BOP has 175 fractional variables (29%) and is the root node for the branch and bound tree of the constraint branching method.

The generation of the columns required 48 seconds, while the constraint branching step ran for almost 16 minutes. The branch and bound tree consisted of 1571 nodes and its maximum depth was 35 levels. Following a depth-first 1-branch strategy, the first integer solution produced was also optimal. It should be emphasized that this is not always the case. However, the strategy is expected to provide good feasible solutions early in the branch and bound tree, and thus can be useful as a good lower bound to prune further nodes.

This first integer solution – the optimal one – was obtained 12 seconds after the start of the constraint branching step. The remaining running time was required to complete the branch and bound tree, in order to confirm the optimality of the best solution found.

5 Conclusions and Outlook

The OTSP formulation and solution method proposed in this paper achieved two main goals: (i) to provide a flexible model to handle the planning and scheduling of elective surgeries in hospitals; (ii) to find the optimal schedule within a reasonable time limit. Ongoing research is directed towards further improving the computational time required to solve the OTSP. One way to achieve this goal is to apply preprocessing strategies by exploring the fact that the coefficient matrix has usually a low density and its structure is block-wise. Further research will be directed to analyzing its performance in huge problem instances, especially when completing the branch and bound tree is no longer an option due to the problem size. Finally, the extension of the model to include multiple and conflicting criteria will also be investigated.

References

1. Barnhart C, Johnson EL, Nemhauser GL, Savelsbergh MWP, Vance PH (1998) Branch-and-price: column generation for solving huge integer programs. OR 46(3):316–329
2. Franklin D, Traub RD (2002) How to schedule elective surgical cases into specific operating rooms to maximize the efficiency of use of operating room time. Anesth Analg 94:933–942
3. Hamilton DM, Breslawski S (1994) Operating room scheduling - factors to consider. AORN J. 59(3):665–674
4. Hsu VN, de Mata R, Lee C-Y (2003) Scheduling patients in an ambulatory surgical center. NRL 50:218–238
5. Ryan DM (1992) The solution of massive generalized set partitioning problems in aircrew rostering. JORS 43:459–467
6. Sier D, Tobin P, McGurk C (1997) Scheduling surgical procedures. JORS 48:884–891

Locating Health Facilities in Nouna District, Burkina Faso

Cara Cocking[1], Steffen Flessa[2], and Gerhard Reinelt[1]

[1] Institute of Computer Science, University of Heidelberg
{cara.cocking,gerhard.reinelt}@informatik.uni-heidelberg.de
[2] Department of Health Care Management, University of Greifswald
steffen.flessa@uni-greifswald.de

Summary. The Nouna health district in Burkina Faso, Africa, has a population of approximately 275,000 people living in 290 villages, who are served by 23 health facilities. At present, the time and effort required in travelling to a health facility (especially during the rainy season) is, for many people, a deterrent to seeking proper medical care. As one step toward improving health care for the people in the district, this study focuses on the problem of optimally locating new health facilities. In light of a government goal that every village should have a health facility within 10 kilometers, the basic model we use for this problem is a covering model, which we then further adapt to the specific situation in Nouna. In this paper we will discuss our application of location analysis techniques to this problem, including an analysis of the current situation, the creation of an appropriate model, solution techniques, and final results.

1 Introduction

Location analysis deals with efficiently locating facilities. Using methods of location analysis can have an especially large impact in developing countries where location decisions, commonly being made by local officials and based on political, economic, and cultural factors, can result in a placement of facilities that is far from optimal in terms of geographical accessibility [1]. Location analysis provides a means of evaluating the efficiency of a set of locations as compared to an optimal (based on some criteria) set of locations, as well as of determining which new locations, when additional facilities are to be added to a given fixed set (which could be empty), would effect the best accessbility.

In developing countries, many people still lack basic health care, and a primary contributor to thc lack is inaccessibility, especially in rural areas. Various studies [2] [3] [5] [6] [8] have shown the effectiveness of applying location analysis in these situations.

In the Nouna health district in Burkina Faso, Africa, health care utilization is low [7]. Some villages are located as far away as 45 kilometers (by road) from their assigned health centers and roads are in poor condition. Thus the time and effort required in travelling to a health center can be a deterrent to seeking proper medical care.

In this study we considered the problem of adding new health facilities to the current set in the Nouna health district. We analyzed the locational efficiency of the existing locations and created an IP model for solving the problem of determining where to locate new facilities. In section 2 we discuss the problem setting, in section 3 we discuss our model, in section 4 we discuss results, and in section 5 we conclude.

2 Setting

The setting for the Nouna Location Problem, as we call it, is the Nouna health district in the north-west of Burkina Faso. This district has a population of about 275,000 people living in 290 villages who are served by 23 health facilities. As one step towards improving health care for the people in the district, this study focuses on the problem of optimally locating new health centers in order to improve accessibility.

The following constraints regarding the placement of health centers have been specified by the government and are used as guidelines by the district when determining where a new health center should be built: (1) Every village should be within 10 kilometers of a health facility. (2) A health center should serve a population in the range 10,000 - 20,000. (3) Two health centers should not be closer than 10 kilometers to each other.

These are not, however, strict regulations. The 21 health facilities in the district excluding Nouna and Djibasso (which have larger than usual facilities) each serve between 3,700 and 22,900 people, and four pairs of health centers are closer than 10 kilometers to each other. At present not every village has a health facility within 10 kilometers: this is a goal to aim for as new facilities are built. In addition to the government guidelines, we consider also the goal of minimizing the travel costs of the people from their village to their assigned health facility.

3 Model

The first government goal leads to a maximal covering model with cover distance 10 kilometers. At the same time we would like to minimize the travel cost of the people to their assigned health facility, and this leads to a p-median model. To represent this situation we use a model which maximizes coverage while minimizing the travel cost for uncovered demand.

We use a minimum capacity of 4,000 and a maximum of 20,000 since in reality some health centers serve many fewer people than the target minimum 10,000. We also need a smaller minimum capacity to keep the model from becoming infeasible for larger numbers of facilities. The only exceptions are the larger facilities at Nouna and Djibasso where we use larger capacities.

We did some tests with constraints modeling the third government guideline, but left it out of the model in the end since it wasn't helpful in improving accessibility for the people.

The IP formulation we use (Figure 1) is based on Pirkul and Schilling [4]. The sets I and J represent the clients and facility sites (in both cases, the set of villages), respectively. Variable x_{ij} is 1 if client i is assigned to facility j, and z_j is 1 if a facility is sited at j. The a_i's are the populations and the d_{ij}'s are the distances between villages, with $maxd_{ij}$ being the maximum d_{ij}. We use Euclidean distances since this corresponds to the government's covering goal which drives the model and is common practice for location analysis applications in developing countries where data is hard to come by [1]. The constant S is the cover distance (10 km) and C_{jmax} and C_{jmin} are the capacities of a facility at j.

$$\text{Minimize} \quad \sum_{i\in I}\sum_{j\in J} w_{ij}a_i x_{ij}$$

$$\text{Subject to} \quad \sum_{j\in J} x_{ij} = 1 \qquad \text{for all } i \in I \qquad (1)$$

$$x_{ij} \leq z_j \qquad \text{for all } i \in I, j \in J \qquad (2)$$

$$\sum_{j\in J} z_j = p \qquad (3)$$

$$\sum_{i\in I} a_i x_{ij} \leq C_{jmax} \qquad \text{for all } j \in J \qquad (4)$$

$$\sum_{i\in I} a_i x_{ij} \geq C_{jmin} z_j \qquad \text{for all } j \in J \qquad (5)$$

$$z_j = 1 \qquad \text{for all existing facilities} \qquad (6)$$

$$x_{ij}, z_j \in \{0,1\} \qquad \text{for all } i \in I, j \in J$$

where

$$w_{ij} = \begin{cases} 0 & \text{if } d_{ij} \leq S \\ 1 - \beta\left(\frac{maxd_{ij} - d_{ij}}{maxd_{ij} - S}\right) & \text{if } d_{ij} > S \end{cases}$$

Fig. 1. Our IP formulation.

This formulation is basically a capacitated p-median-like version of a maximal covering problem with a bit of a twist in the objective. The objective minimizes uncovered demand while at the same time (possibly) minimizing the average distance of uncovered demand to its facility, as controlled by the w_{ij} coefficients. For covered demand, w_{ij} is 0, while for uncovered demand, w_{ij}

is either 1 or a number between 0 and 1 based on β and how far the demand is from being covered. We can think of β as representing how much weight we give to the objective of minimizing the average distance of uncovered demand.

The first three sets of constraints are standard p-median constraints which assign each village to exactly one facility (1), assign villages only to open facilities (2), and open p facilities (3). Note that the value of p is the total number of facilities, including both existing facilities and facilities that are to be located. Constraints (4) and (5) enforce the capacities on the facilities. Constraints (6) fix the locations of the facilities that already exist.

4 Results

The IP model described in the previous section was solvable by CPLEX on a 2.8 GHz machine with 2 gigabytes of RAM in anywhere from a few seconds to half an hour, depending on the input parameters. In any case, the model is solvable to optimality in a reasonable amount of time for the Nouna Location Problem, and all results discussed below are optimal solutions.

Before looking at locations for additional facilities, we consider the efficiency of the current set of facilities. We compare the existing facility locations and village assignments to optimal (according to the model described in the previous section) facility locations and village assignments. See Table 1 for the comparison. The huge improvement in the numbers from the first to the third row of the table demonstrates the difference that location analysis can make when taken advantage of in the process of making locational decisions.

Table 1. Comparing the district setup to an optimal setup with $\beta = 0.01$. (All distances are in kilometers.)

	demand covered	avg uncovered dist	avg dist	max dist
district facs, district assign	74%	15.5	7.0	35.2
district facs, optimal assign	79%	14.1	6.6	27.4
optimal facs, optimal assign	97%	13.1	5.7	15.2

Now as we consider adding new facilities to the current set, there are two parameters to the model that we can vary: β and the number of new facilities to add. The cover distance and capacities are fixed at the values previously discussed based on the information we obtained from the district.

The parameter β controls the weight we give to minimizing the distance of uncovered demand. When β is 0, meaning we don't care at all about the distance of uncovered demand, we end up with large distances, as expected. In effect, the uncovered demand is assigned randomly to facilities with no consideration of the distance (only of facility capacities). With even a small value of β, such as 0.01, we get a huge improvement in the distance from uncovered demand to their assigned facilities, as compared with $\beta = 0$, and no decrease in population coverage (at least for $1, 2, ..., 20$ new facilities).

After the huge improvement from $\beta = 0$ to $\beta = 0.01$, increasing β to 1 for a fixed p doesn't produce a comparatively large improvement in the average distance of uncovered demand, nor does it produce much decrease in the covered population. We conclude from this that (for the Nouna Location Problem) as long as β is greater than 0, it doesn't much matter what specific value β takes on.

Next we examine the changes in population coverage and average distance as the number of new facilities to locate increases. Naturally the population covered increases as the number of new facilities increases. What may be less obvious is that the population coverage levels off around 99.4% (regardless of β) starting with 14 new facilities. This is because of the capacities on the facilities which mean we cannot always assign demand to its nearest facility.

The average distances also level off starting at around 14 new facilities; see Figure 2. In this figure we additionally see that the average distance can increase as we add more facilities, even when $\beta = 1$ and we are giving the maximum weight to minimizing the distance of uncovered demand. This may be explained firstly, by the fact that this goal is balanced with that of covering as much demand as possible, and secondly, by the capacity constraints which mean that assignments of villages to facilities have to be shuffled around each time a new facility is added.

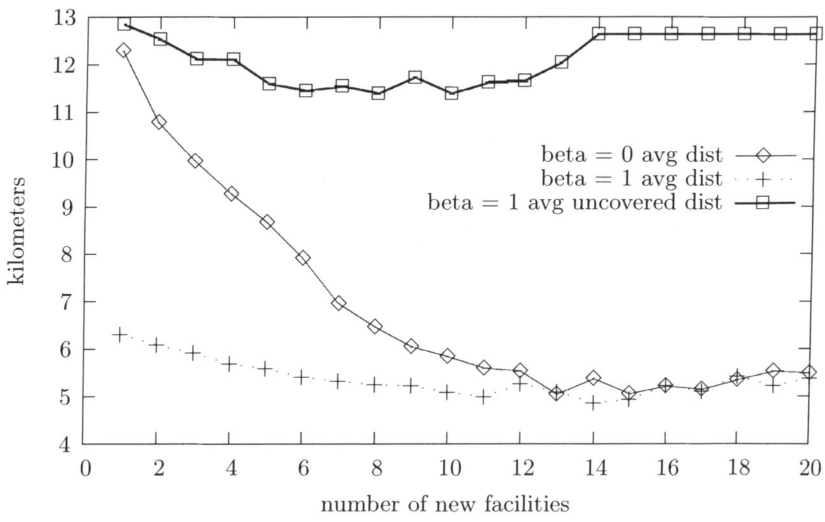

Fig. 2. Average distances from facilities as the number of new facilities increases.

5 Conclusions

Given the results above, we conclude that 14 new facilities will achieve the government's goal of covering everyone (or in reality, as many people as possible given the capacity constraints) within 10 kilometers, while at the same time providing reasonable access for those villages not covered. Even 10 new facilities will cover 98% of the population with a low average distance for uncovered demand. A possible extension of this work would be to consider the costs of the facilities and optimize the number of new facilities to be opened.

Another interesting point is that in this model a village selected to have a facility may not be assigned to its own facility if there is another facility within the cover distance. All covered demand is equally beneficial, whether 0 or 10 kilometers from its assigned facility. Another extension would be to update the model to additionally minimize the distance of covered demand to their facilities; this would be done by manipulating the w constants.

In future work we plan to consider the availability of roads and more precise travel costs which take into account things like the means of transportation, the terrain, and the condition of the roads.

References

1. Kumar N (2004) Changing geographic access to and locational efficiency of health services in two Indian districts between 1981 and 1996. Social Science and Medicine 58:2045–2067
2. Mehretu A, Wittick R, Pigozzi B (1983) Spatial Design for Basic Needs in Eastern Upper Volta. The Journal of Developing Areas 17:383–394
3. Oppong J (1996) Accomodating the rainy season in third world location-allocation applications. Socio-Economic Planning Sciences 30:121-137
4. Pirkul H, Schilling D (1991) The maximal covering location problem with capacities on total workload. Management Science 37:233-248
5. Rahman S, Smith D (2000) Use of location-allocation models in health service development planning in developing nations. European Journal of Operational Research 123:437-452
6. Rushton G (1984) Use of location-allocation models for improving the geographical accessibility of rural services in developing countries. International Regional Science Review 9:217-240
7. Su T, Kouyate B, Flessa S (2005) Household cost-of-illness and its determinants in Nouna health district, Burkina Faso. Not yet published
8. Tien J, El-Tell K (1984) A quasihierarchical location-allocation model for primary health care planning. IEEE Transactions on Systems, Man, and Cybernetics 14:373-380

A Dual Algorithm to Obtain Highly Practical Solutions in Static Multileaf Collimation

Philipp Süss

Fraunhofer Institut für Techno- und Wirtschaftsmathematik, Europaallee 10, 67657 Kaiserslautern suess@itwm.fhg.de

1 IMRT delivery using MLCs

An intensity map $\mathbf{x}$ is a $m \times n$ matrix with nonnegative integral entries corresponding to *beamlets*. We will assume for this work that the MLC has m rows called *channels*, and n columns. One "step" in a sequence can then be coded as a $m \times n$ 0-1 (read "zero one") *shape matrix* with entries corresponding to the entries in the intensity map. A 0 in the shape matrix codes that a leaf from the MLC blocks the region corresponding to the entry, and a 1 marks this region open for this step. The positive entries in a channel must be consecutive, as the opening is coherent. The kth shape in a sequence is denoted by $\mathbf{S}_k$ and the index set of all allowable shape matrices is $\mathcal{K}$. This set is influenced by hardware-specific constraints concerning the separation of leaves. In general, there is a minimum distance that must separate the leaf ends in one channel at all times ($\Delta_{intra} \geq 0$), and the distance from one leaf to an opposing leaf in a neighboring channel must be at least Δ_{inter}. Every shape matrix has a corresponding opening time α_k, called the *monitor unit* of this shape. The general sequencing problem is to express the intensity map as a sum of shapes and monitor units to keep the total treatment time as low as possible to minimize the stress endured by the patient. A major component of the treatment time is the sum of the monitor units, called the *beam-on time* (BOT). Boland et al [1] have formulated the problem to minimize the beam-on time as a network flow problem with side constraints. We will only give the formal definition of a network structure $G_T = (V_T, A_T)$ that is equivalent to the formulations in [1] and the problem statement here. For more details, refer to the works [1] and [2].

The set of nodes and the set of arcs are given by

$$V_T := \left\{(i,l,r)^1, (i,l,r)^2 : 1 \leq i \leq m,\ 0 \leq l < r - \Delta_{intra} \leq n+1\right\} \\ \cup \left\{(i,j) : i = 1,\dots,m,\ i = 0,\dots,n\right\} \cup \left\{D, D'\right\}, \quad (1)$$

$$
\begin{aligned}
A_S := \big\{\big(D,(1,l,r)^1\big) : (1,l,r)^1 \in V_T\big\} \\
\cup \big\{\big((i,l,r)^1,(i,l)\big) : (i,l,r)^1 \in V_T\big\} \\
\cup \big\{\big((i,r-1),(i,l,r)^2\big) : (i,l,r)^2 \in V_T\big\} \\
\cup \Big\{\big((i,l_1,r_1)^2,(i+1,l_2,r_2)^1\big) : 1 \le i \le m,\ (i,l_1,r_1)^2 \in V_T, \\
(i+1,l_2,r_2)^1 \in V_T,\ l_1 < r_2 - \Delta_{inter},\ r_1 > l_2 + \Delta_{inter}\Big\} \\
\cup \big\{\big((m,l,r)^2,D'\big) : (m,l,r) \in V_T\big\} \cup \{(D',D)\}\,. \quad (2)
\end{aligned}
$$

Notice that the hardware-specific separation constraints can be treated with at the time the network is built, and do not have to be considered by the algorithm. This makes the network formulations very flexible models for all static sequencing problems.

The decision variables are the flows f_a on the arcs, and each arc has a cost c_a. To formulate the objective to attain the minimal beam-on time, the cost function is given by $c_{(D',D)} = 1$ and $c_a = 0$ for all other arcs in A_T. In addition, we have a *demand function* $b : V_T \mapsto \mathbb{R}$ for the nodes. The demand for the leaf positions nodes is 0, and the demand of the intensity nodes is given by its level jump $x_{ij} - x_{i\,j-1}$. Given a flow f in G_T, the *excess function* $e_v(f)$ is the net amount that flows into node v after demand. The minimum cost flow problem that solves the min BOT problem is now given by

$$\min_{\mathbf{f}} \quad \mathbf{c}^T\mathbf{f} \qquad (3)$$

$$\sum_{(w,v)\in A_T} f_{(w,v)} - \sum_{(v,w)\in A_T} f_{(v,w)} = b_v \qquad v \in V_T \qquad (4)$$

$$f_{\left((i,l,r)^1,(i,l)\right)} - f_{\left((i,r-1),(i,l,r)^2\right)} = 0 \qquad (i,l,r)^1 \in V_T \qquad (5)$$

$$f_a \ge 0 \qquad a \in A_T \qquad (6)$$

Constraint (4) models the flow conservation constraint and (5) is called the *leaf position matching* constraint [1].

2 Solving the sequencing problem

The leaf positions matching constraints (5) are the culprit to applying a pure network optimization algorithm to the network problem stated above. Notice, however, that it is easy create feasible solutions by fixing the flow on the arcs that are involved in the complicating leaf positions matching constraints. This process effectively decides on the channel openings to realize the intensity map before the shapes are constructed. The only considerations in this preprocessing is that the realized intensity map is indeed the one we want.

By modifying the flow on arcs, the excess function is modified. A necessary condition for the flow to be a feasible solution to the sequencing problem is that the excess at the intensity nodes is 0. After pre-processing, only those leaf positions nodes that are incident to intensity nodes with nonnegative demand may have a nonnegative excess value. To fuse the chosen channel openings to shapes, we now solve the (pure) network optimization problem with the excess function as the demand. The optimal solution to this problem is the sequence with minimal beam-on time given the channel openings that were decided on during pre-processing. Moreover, if the pre-processing preserves integrality of the excess function, there exists an integral solution to the pure network optimization problem [2], and this solution is obtained without an integer formulation of the problem. This results in a lower number of shapes (as was stated in [1]) at the same computing time.

As this method serves to obtain feasible solutions, it will in general not deliver the optimal. However, experience with numerical tests have shown that the solution is optimal in about 97% of all cases. The reason for this high frequency of optimal solutions is that most choices of channel openings that are not too small lead to an optimal beam-on time when combined to shapes. In [2], a strategy based on the information from the dual of the sequencing is developed to price bad choices of channel openings and obtain improved solutions in case of suboptimality. We will now focus on some practical issues that have to be addressed treatment planning.

3 Practical considerations

Implicit stratification

There is intuitive reason to assume that if an intensity map is relatively *smooth*, it can be more easily realized, and the treatment time is consequently shorter. Smoothness can be measured by the level jumps in the map: a map with few and only moderate jumps requires fewer shapes in static sequencing since fewer monitor units have positive values that are larger in magnitude.

Methods to attain intensity maps that are easy to deliver (*stratified maps*) have been proposed and some of them are discussed in [2]. As a measure of quality for a stratified map, the deviation of the dose distributions should be calculated. We assume that the patient's body is discretized into small volume elements called *voxels*, and the dose (in Grey) in each voxel can be calculated using the *dose information matrix* $\mathbf{P}$. Then the vector of dose values $\mathbf{d} \in \mathbb{R}^{|\mathcal{V}|}$, where $\mathcal{V}$ is the index set over all voxels is obtained by $\mathbf{d} = \mathbf{P}\mathbf{x}$. Thus, the entry P_{vb} contains the contribution of the bth beamlet to the vth voxel.

One objective function in [2] to measure the quality of a modified solution $\tilde{\mathbf{x}}$ is to compare the deviation in the realized dose: $\|\mathbf{P}\mathbf{x} - \mathbf{P}\tilde{\mathbf{x}}\|$. We now derive a bound on this error that is based on the deviation of the intensity maps $\mathbf{x}$ and $\tilde{\mathbf{x}}$ only. This allows us to formulate a sequencing strategy that performs the stratification implicitly during sequencing.

The dose error is stated as

$$\|\mathbf{d} - \tilde{\mathbf{d}}\| = \|\mathbf{P}(\mathbf{x} - \tilde{\mathbf{x}})\| \tag{7}$$

$$= \left\| \sum_{b \in \mathcal{B}} \mathbf{P}_b (x_b - \tilde{x}_b) \right\|, \tag{8}$$

where $\mathcal{B}$ is the index set over all beamlets, and $\mathbf{P}_b$ denotes the bth column of $\mathbf{P}$. By the homogeneity of a norm, we obtain the inequalities

$$\leq \sum_{b \in \mathcal{B}} \|\mathbf{P}_b (x_b - \tilde{x}_b)\| \tag{9}$$

$$\leq \sum_{b \in \mathcal{B}} \|\mathbf{P}_b\| \, |x_b - \tilde{x}_b|. \tag{10}$$

If the maximum deviation in the beamlet intensities is known: $|x_b - \tilde{x}_b| \leq \varepsilon \quad \forall b \in \mathcal{B}$, we can bound the error in the realized dose from above.

$$\sum_{b \in \mathcal{B}} \|\mathbf{P}_b\| \, |x_b - \tilde{x}_b| \leq \sum_{b \in \mathcal{B}} \|\mathbf{P}_b\| \, \varepsilon \tag{11}$$

$$= \varepsilon \sum_{b \in \mathcal{B}} \|\mathbf{P}_b\| \tag{12}$$

This approximation is used in [2] to formulate a stratification problem as a location problem on the real line by interpreting the norm of the columns of $\mathbf{P}$ as distance weights. We will use the approximation here to formulate a stopping criterion for the sequencing algorithm.

We will perform the sequencing algorithm on a series of intensity maps that successively approximate the original map with increasing precision. That is, at stage t, we solve the sequencing problem on the intensity map $\mathbf{x}^{(t)}$ that is only slightly more complicated than the last intensity map $\mathbf{x}^{(t-1)}$, i.e. $\|\mathbf{x}^{(t)} - \mathbf{x}^{(t-1)}\| < \rho$, for $\rho > 0$ small. We start the series with $\mathbf{x}^{(0)} = \mathbf{0}$.

Given the solution to the tth problem, we may calculate the *residual map* $\mathbf{r}^{(t)}$ that remains to be delivered:

$$\mathbf{r}^{(t)} = \mathbf{x} - \mathbf{x}^{(t)} = \mathbf{x} - \sum_{k \in \mathcal{K}} \alpha_k^{(t)} \, \mathbf{S}_k \tag{13}$$

Now it is clear that the data for the next problem should be constructed by modifying $\mathbf{x}^{(t)}$ and using the information from $\mathbf{r}^{(t)}$. This is done by identifying the largest level jump in $\mathbf{r}^{(t)}$, and adding this value to the excess function of an appropriate channel opening that removes this level jump. This small change in the excess function can be dealt with reasonably fast by using the previous solution as a dual feasible starting solution. Any dual ascent method (such as the shortest augmenting path method) solves the new problem. The proposed strategy is reminiscent of the idea of *scaling* in network algorithms. Here, the excess function is scaled to obtain closer results to the original problem, while the problems in the sequence do not differ much from each other.

Given that the norm of the residual map $\|\mathbf{r}^{(t)}\|$ can by (12) be used in the approximation of the error in the realized dose, a stopping criterion in terms of this quality is readily available for the sequencing.

Tongue-and-groove

A leaf hinges upon a neighboring leaf by a tongue-and-groove design. Underdosage may occur if two beamlets in neighboring channels are exposed one at a time by different shapes. In such cases, the tongue covers the region between the beamlets, and this region is underdosed.

Most sequencing algorithms consider the tongue-and-groove effects separately from the actual creation of shapes and monitor units. Here we will give a heuristic idea to reduce the tongue-and-groove effects to a large extent. The idea is based on the simple fact that if the channel openings were the same in all channels of a given shape, there would be no tongue-and-groove effects because the tongues would never "stick out". Thus, we modify the cost function to penalize channel openings in one shape that are different from each other. That is, the cost function for the arcs between the leaf positions nodes from different channels is set to the following:

$$c_a = |l_2 - l_1| + |r_2 - r_1|, \quad \forall\, a = \left((i, l_1, r_1)^2, (i+1, l_2, r_2)^1\right) \in A_T. \tag{14}$$

In addition, the arcs connecting closed leaf positions get a positive cost in order to prevent the network algorithm to fuse largely closed shapes in favor of connecting similar openings.

Numerical results have shown that the beam-on time does not increase significantly compared to the original cost structure, but the tongue-and-groove effects as measured by the tongue-and-groove index (TGI) by Que et al [3] are comparably very low (see results below).

4 Conclusions

Formulating the sequencing problem as a network optimization problem has several advantages. The most obvious is the easy inclusion of separation constraints between channels and within a channel. On closer observation, the network structure also creates an extremely flexible solution environment by being able to quickly calculate solutions to slightly modified problems by using a previous solution as a dual feasible starting solution.

In the following are some performance measures for 2 clinical prostate cases. The first case is comprised of 5 beams, and the second case consists of 7 beams. The intensities were rounded to the nearest integer to obtain integral data.

Table 1. Maps 1 to 5 are 11 × 9 intensity maps with up to 20 different levels ranging up to 42, and maps 6 to 13 are 17 × 11 intensity maps with up to 25 different intensity levels up to 105. The first 4 columns show the comparison of the optimal beam-on time $TNMU_{opt}$ to the beam-on time found by the algorithm. NS is the number of shapes in the sequence. The next 3 columns show the performance measures for the solution to the network problem with the cost structure defined in 3. Finally, the last 2 columns depict the improvement over the original cost structure.

Map	TNMU	$TNMU_{opt}$	NS	TGI	$TNMU_{TG}$	NS_{TG}	TGI_{TG}	TGI (%)	NS (no.)
1	35	35	29	1274	35	26	32	98	3
2	56	56	33	747	56	31	0	100	2
3	42	42	32	1810	42	30	137	92	2
4	42	42	36	2263	42	31	46	98	5
5	46	46	27	1608	46	28	16	99	-1
6	47	47	43	6359	47	38	851	87	5
7	148	148	90	20112	148	67	826	96	23
8	54	54	46	6292	54	44	254	96	2
9	53	53	45	4037	53	46	191	95	-1
10	38	38	35	4976	38	33	156	97	2
11	60	60	57	13982	60	49	411	97	8
12	45	45	43	7202	45	37	312	96	6

The network optimization took on average 0.7 seconds per map for the first case, and 4.8 seconds per map for the second case. CPLEX needed about 3 minutes per map for the first case, and about 18 minutes per map for the second case. As can be seen, the reduction in the tongue-and-groove index TGI is almost always in the high 90 percent ranges, showing an excellent improvement in these cases without increasing beam-on time. On top of that, the number of shapes is almost always lower than with the original cost structure.

References

1. N Boland, HW Hamacher and F Lenzen (2004) Minimizing beam-on time in cancer radiation treatment using multileaf collimators. Networks 43(4):226–240
2. P Süss (2005) A dual network simplex algorithm for an extended transshipment problem. Diplomarbeit, Technische Universität Kaiserslautern
3. W Que, J Kung and J Dai (2004) 'Tongue-and-groove' effect in intensity modulated radiotherapy with static multileaf collimator fields. Phys Med Biol 49:399–405

Challenges in the Optimization of Biosystems II: Mathematical Modeling and Stability Analysis of Gene-Expression Patterns in an Extended Space and with Runge-Kutta Discretization

Mesut Taştan[1], Stefan W. Pickl[2], Gerhard Wilhelm Weber[1]

[1] Institute of Applied Mathematics, METU, Ankara, Turkey
[2] Department of Computer Science, Universität der Bundeswehr, München, Germany

1 Introduction

Gene (or *protein*) *expression* is the process by which gene information is converted for producing cell structures and cell functions. There are two main process events: *transcription and translation*. After them, steps like *folding*, *post-translational modification* and *targeting* occur up to the protein product, we leave these details to [12]. Since mRNA is an exact copy of the DNA coding regions, mRNA analysis can be well used to explore the process in coding regions of DNA. More importantly, the measure of gene expression can be determined from the genomic analysis at the mRNA level [15]. Both genomic and environmental factors affect the gene expression levels. For example, the environmental factors including stress, light, temperature and other signals cause some changes in hormones and in enzymatic reactions which influence the gene expression level. That is why mRNA analysis informs us not only about genetic viewpoints of an organism but also about the dynamic changes in environment of that organism. For most genes, protein levels are defined by steady state mRNA levels [16]. Thus, quantitative expressions at mRNA level provide important clues about the underlying dynamics. Peculiar changes in monitoring mRNA levels generally refer to drug treatment, shocks, disease or metabolic states.

1.1 Microarray Technology

Microarray technology is an array-based technology which monitors thousands of different RNA molecules simultaneously revealing their expression patterns and perturbed subsequent cellular pathways. One of the most frequently used microarray applications [2] is to compare gene expression levels of the same cell type like healthy cell and diseased cell under two different conditions. Such application can give vital information on the reasons of diseases.

1.2 Evaluating the Expression Data

Microarray experiments can quickly monitor the expression values for large numbers of genes. The goal researchers have in mind is ultimately to clarify the precise connections of the *genetic network*: mathematically speaking a graph consisting of nodes representing genes and with the edges and their weights representing the influence which the genes mutually exercise. Here, the nodes themselves can also be viewed as a function obtained by combining basic inputs. For each gene it is aimed to find and to predict which and how much other genes influence it. Different mathematical methods have been developed for construction and analyzing such networks. In this study, we refine the model derived from differential equations by adding shift terms and by extending space.

2 Modeling Gene Networks with Ordinary Differential Equations

Since the 1950s, a variety of mathematical identifications have been proposed. In 1952, *Turing* has firstly introduced the idea of a mathematical model for biological systems [12, 56, 62]. In according to this approach, the change of the state of the cell is equal to the sum of all acting forces on that cell. This basic idea is one of the foundations for regulatory systems, but the development of experiments at molecular levels requires extended and computer supported models.

A differential relation between variables of gene networks is generally represented in the form of ordinary differential equations (ODEs)

$$\frac{dE_i}{dt} = f_i(E) \qquad (i = 1, 2, ..., n),$$

where $E = (E_1, ..., E_n)^T$ is the vector of positive concentrations of proteins, mRNAs, or small components, $f_i : \mathbb{R}^n \to \mathbb{R}$ are nonlinear functions and n being the number of genes. First differential equation or dynamic system model consisting of mRNA and protein concentrations was proposed by *Chen, He and Church* [4] in the form of $\dot{E} = ME$, where M is a constant matrix and the vector E comprises the expression level of individual genes. Later on, *De Hoon and Imoto* [10] used this linear model on mRNA data of *Bacillus subtilis* to estimate M with maximum likelihood estimation method. In 2001, *Sakamoto and Iba* [14] proposed more the flexible model

$$\dot{E}_i = f_i(E_1, E_2, ..., E_n),$$

where f_i being functions of $E = (E_1, E_2, ..., E_n)^T$ determined by genetic programming and least squares methods.

The models described above were studied and improved by *Gebert, Lätsch, Pickl, Weber, and Wünschiers* with many ideas. In [7], they regarded the model $\dot{E} = M(E)E$ in which the matrix M, not usually a constant matrix, depends on E. In the same study, to modify the optimization problem the solution space is restricted by assuming number of regulating factors for each gene is bounded.

3 The State of the Art

Let the n-column vector $E = E(t)$ be gene expression patterns at different times t. We denote the given finite set of experimental results as $\bar{E}_0, \bar{E}_1, ..., \bar{E}_{l-1}$, where each $\bar{E}_m \in R^n$ corresponds to the gene profile taken at time $\bar{t}_m$.

Gebert et al. [8] refined the time-continuous model $(\mathcal{CE})$ first formulated by *Chen et al.* by taking into account that interaction between variables is nonlinear but the number of associated regulating influences is bounded. This flexible model was formulated in the multiplicative nonlinear form

$$(\mathcal{CE}) \qquad \dot{E} = M(E)E.$$

Here, we refer to corresponding initial values $E(t_0) = E_0$. Note that, $(\mathcal{CE})$ is homogeneous and autonomous (i.e., the right hand-side depends on the states E but not on time t). This implies that trajectories do not cross themselves. The matrix $M(E)$ is defined component-wise by a family of any class of functions including unknown parameters.

4 Our Generalized Model

The model extended by *Yılmaz et al.* [18] allows the nonlinear interactions and uses affine linear terms as shifts. However, the recursive iteration idea mentioned in [6] is lost by these shift terms, at the first glance. Thus, we again turn to $(\mathcal{CE})$ by making following *affine* addition:

$$(\mathcal{ACE}) \qquad \dot{E} = M(E)E + C(E).$$

Here, we defend that additional column vector $C(E)$ can help us for accounting the environmental changes and capturing the dynamics better. Our approach on $C(E)$ is that it can be written as

$$C(E) = \check{M}(E)\check{E},$$

where

$$\check{M}(E) := diag(\mathcal{C}^T(E)) = \begin{pmatrix} \mathcal{C}_1(E) & & & & 0 \\ & \mathcal{C}_2(E) & & & \\ & & \cdot & & \\ & & & \cdot & \\ & & & & \cdot \\ 0 & & & & \mathcal{C}_n(E) \end{pmatrix} \quad \text{and} \quad \check{E} := \begin{pmatrix} \check{E}_1 \\ \check{E}_2 \\ \cdot \\ \cdot \\ \cdot \\ \check{E}_n \end{pmatrix}.$$

In fact, we shall see by means of the corresponding initial value $\check{E}(t_0) = e$ ($e := (1, 1, ..., 1)^T$) that the time depending variable $\check{E}$ is constant and identically $\check{E} \equiv e$. In this sense, $(\mathcal{ACE})$ is equivalent to

$$\dot{E} = M(E)E + \check{M}(E)\check{E}.$$

Let us define the vector and the matrix

$$\mathbb{E} := \begin{pmatrix} E \\ \check{E} \end{pmatrix} \quad \text{and} \quad \mathbb{M}(\mathbb{E}) := \begin{pmatrix} M(E) & \check{M}(E) \\ 0 & 0 \end{pmatrix}.$$

so that we end up with the following form of the extended initial value problem

$$(\mathcal{ACE})_{ext} \qquad \dot{\mathbb{E}} = \mathbb{M}(\mathbb{E})\mathbb{E} \quad \text{and} \quad \mathbb{E}_0 := \mathbb{E}(t_0) = \begin{pmatrix} E(t_0) \\ \check{E}(t_0) \end{pmatrix} = \begin{pmatrix} E_0 \\ 1 \\ \cdot \\ \cdot \\ \cdot \\ 1 \end{pmatrix}.$$

4.1 Time Discretization

Here, we start with Euler's Method mainly for conceptual reasons.

Euler's Method

Euler's method results from ignoring second and higher order terms in Taylor series to approximate the solution. We apply Euler's method for right hind-of side of $(\mathcal{ACE})_{ext}$ $\mathbb{M}(\mathbb{E})\mathbb{E}$ as follows: for all $k \in \mathbb{N}_0$,

$$\frac{\mathbb{E}_{k+1} - \mathbb{E}_k}{h_k} = M(\mathbb{E}_k)\mathbb{E}_k, \tag{1}$$

$$\Leftrightarrow \qquad \mathbb{E}_{k+1} = (I + h_k M(\mathbb{E}_k))\mathbb{E}_k \tag{2}$$

where h_k is the step-size (i.e, $h_k = t_{k+1} - t_k$). In the extended model, (1.2) means

$$\begin{pmatrix} E_{k+1} \\ \check{E}_{k+1} \end{pmatrix} = (I + h_k \begin{pmatrix} M(E_k) & \check{M}(E_k) \\ 0 & 0 \end{pmatrix}) \begin{pmatrix} E_k \\ \check{E}_k \end{pmatrix}. \quad (3)$$

Let us define

$$\mathbb{M}_k := (I + h_k \begin{pmatrix} M(E_k) & \check{M}(E_k) \\ 0 & 0 \end{pmatrix})$$

so that we obtain difference equation and dynamics

$$(\mathcal{DE}) \qquad \mathbb{E}_{k+1} = \mathbb{M}_k \mathbb{E}_k \qquad (k \in \mathbb{N}_0).$$

Thus, we iteratively approximate the next state from the previous one. Note that, since the experimental results are represented as $\bar{E}_0, \bar{E}_1, ..., \bar{E}_{l-1}$, we denote the approximations by $\widehat{\mathbb{E}}_l, \widehat{\mathbb{E}}_2, ..., \widehat{\mathbb{E}}_{l-1}$. Setting $\widehat{\mathbb{E}}_0 = \bar{E}_0$, the kth approximation is calculated as

$$\widehat{\mathbb{E}}_k = \mathbb{M}_{k-1}(\mathbb{M}_{k-2}...(\mathbb{M}_1(\mathbb{M}_0 \bar{E}_0))) \quad (k \in \mathbb{N}_0).$$

Runga-Kutta Method

While solving ODEs numerically we face with two kinds of errors namely, the rounding error as a result of finite precision of floating-point arithmetic and, secondly, the truncation error associated with the method used. Hence, more symmetric integration methods like *Runge-Kutta method* (*RK*), which takes into account the midpoint of the interval can be applied on the the system $(\mathcal{ACE})_{ext}$. Runge-Kutta methods have the advantage of stability which is closer to the stability of the given time-continuous model. RK methods use only the information at time t_k, which makes them self-starting at beginning of integration, and also makes methods easy to program, which accounts in part for their popularity [9]. Idea of applying RK methods to model of gene expression patterns is first introduced by *Ergenç and Weber* [5]. Here we illustrate applying a different RK Method, the simplest case, which is called *Heun's method* as follows:

$$\mathbb{E}_{k+1} = \mathbb{E}_k + \frac{h_k}{2}(k_1 + k_2), \quad (4)$$

where

$$\begin{aligned} k_1 &= \mathbb{M}(\mathbb{E}_k)\mathbb{E}_k, \quad \text{and} \\ k_2 &= \mathbb{M}(\mathbb{E}_k + h_k k_1)(\mathbb{E}_k + h_k k_1). \end{aligned}$$

More explicitly, instead of (4) we write

$$\mathbb{E}_{k+1} = \mathbb{E}_k + \frac{h_k}{2}\mathbb{M}(\mathbb{E}_k)\mathbb{E}_k + \frac{h_k}{2}\mathbb{M}(\mathbb{E}_k + h_k\mathbb{M}(\mathbb{E}_k)\mathbb{E}_k)(\mathbb{E}_k + h_k\mathbb{M}(\mathbb{E}_k)\mathbb{E}_k),$$

$$\Leftrightarrow \quad \mathbb{E}_{k+1} = [I + \frac{h_k}{2}\mathbb{M}(\mathbb{E}_k) + \frac{h_k}{2}\mathbb{M}(\mathbb{E}_k + h_k\mathbb{M}(\mathbb{E}_k)\mathbb{E}_k)(I + h_k\mathbb{M}(\mathbb{E}_k))]\mathbb{E}_k.$$

Defining

$$\mathbb{M}_k := I + \frac{h_k}{2}\mathbb{M}(\mathbb{E}_k) + \frac{h_k}{2}\mathbb{M}(\ I + \mathbb{E}_k + h_k\mathbb{M}(\mathbb{E}_k)\mathbb{E}_k)(I + h_k\ I + \mathbb{M}(\mathbb{E}_k)),$$

we get the following discrete equation

$$(\mathcal{DE})^2_{ext} \qquad \mathbb{E}_{k+1} = \mathbb{M}_k\mathbb{E}_k.$$

We note that Runge-Kutta discretization of our model equation generates a nonlinear discrete equation for parameters. In our case, the term

$$\mathbb{M}(\mathbb{E}_k + h_k\mathbb{M}(\mathbb{E}_k)\mathbb{E}_k)(I + h_k\mathbb{M}(\mathbb{E}_k))$$

shows the parametrical nonlinearity. If we use implicit Runge-Kutta methods, it may not be possible to get the discrete equation $(\mathcal{DE})^2_{ext}$.
Nevertheless, the stability analysis of $(\mathcal{DE})^2_{ext}$ is still possible in terms of modified *Brayton and Tong* algorithm mentioned in [1, 6] in the extended space.

5 Stability of Matrices in Terms of Polytopes

An important conclusion for stability of a set of matrices are formulated and proved by *Brayton and Tong* [1] as follows

Definition 1. *A set of matrices $\mathcal{M}$ is stable is and only if there exists a bounded neighborhood of the origin $B \subset \mathbb{C}^n$ such that for each $\mathbb{M} \in \mathcal{M}$, $\mathbb{M}B \subseteq B$. Furthermore, B can be chosen to be convex and balanced, i.e, if $z \in B$, then there is also $ze^{i\theta} \in B$, for all $\theta \in B$.*

Lemma 1. *If the set of matrices $\mathcal{M}$ is stable then there exists a norm, $\|.\|$, such that*

$$\|\mathbb{M}\mathbb{E}\| \leq \|\mathbb{E}\| \qquad \forall \mathbb{M} \in \mathcal{M}, \qquad \mathbb{E} \in \mathbb{C}^n.$$

In the pioneering work [6], this stability condition was verified the finite approximating sets $\mathcal{M}$ with the help of *Brayton and Tong* algorithm and implemented by *Pickl* [6]. Very analogously, we apply the same algorithm to $(\mathcal{DE})^2_{ext}$ by introducing extended vectors $\mathbb{E}_k = (E_k^T,\ 1, ...,\ 1)^T$ which force us to consider the bigger sets $\mathbb{B}_k = (B_k^T,\ 1, ...,\ 1)^T$, where we understand "$T$" and Cartesian product element-wise.

6 Conclusion Sensitivity Analysis

There are several methods to apply sensitivity analysis. For example in [17], the sensitivity analysis is applied with differential algebraic equation methods. Another method was developed by *Gebert et al.* [6] in which based on *Brayton*'s constructive method mentioned above, automatically generated Lypunov functions are used to estimate the behavior of a nonlinear system. First numerical results are presented.

References

1. Brayton, R.K., and Tong, C.H., Stability of dynamical systems: A constructive approach, IEEE Transactions on Circuits and Systems 26, 4 (1979) 224-234.
2. Causton, C.H., Quackenbush J., and Brazma, A., *A Beginner's Guide Microarray Gene Expression Data Analysis*, Blackwell Publishing (2003).
3. Carbayo, M.S, Bornman, W., and Cardo C.C., DNA Microchips: technical and practical considerations, Current Organic Chemistry 4, 9 (2000) 945-971.
4. Chen, T., He, H.L., and Church, G.M., Modeling gene expression with differential equations, Proc. Pacific Symposium on Biocomputing (1999) 29-40.
5. Ergenç, T., and Weber, G.-W., Modeling and prediction of gene-expression patterns reconsidered with Runge-Kutta discretization, special issue at the occasion of seventith birthday of Prof. Dr. Karl Roesner, TU Darmstadt, of Journals of Computational Technologies 9, 6 (2004) 40-48.
6. Gebert, J., Lätsch, M., Pickl, S. W., Weber, G.-W., and Wünschiers, R., An algorithm to analyze stability of gene-expression pattern, Discrete Appl. Math., accepted for publication (2005).
7. Gebert, J., Lätsch, M., Pickl, S.W., Weber, G.-W., and Wünschiers, R., Genetic networks and anticipation of gene expression patterns, *Computing Anticipatory Systems: CASYS'03 - Sixth International Conference*, AIP Conference Proceedings 718 (2004) 474-485.
8. Gebert, J., Radde, N., and Weber, G.W., Modeling gene regulatory networks with piecewise linear differential equations, Preprint 37, Middle East Technical University, Institute of Applied Mathematics (2005).
9. Heath, M., *Scientific Computing: An Introductory Survey*, McGraw-Hill (2002).
10. Hoon, M.D., Imoto, S., Kobayashi, K., Ogasawara, N., and Miyano, S. Inferring gene regulatory networks from time-ordered gene expression data of Bacillus subtilis using differential equations, Proc. Pacific Symposium on Biocomputing (2003) 17-28.
11. Jong, H.D., Modeling and simulation of genetic regulatory systems: a literature review, J. Comput. Biol. 9 (2002) 103-129.
12. Klug, W.S., and Cummings, M.R., *Concepts of Genetics,* Prentice Hall (2003).
13. Pickl, S. W., *Der τ-value als Kontrollparameter - Modellierung und Analyse eines Joint - Implementation Programmes mithilfe der kooperativen dynamischen Spieltheorie und der diskreten Optimierung*, Shaker-Verlag (1999).
14. Sakamoto, E., and Iba, H., Inferring a system of differential equations for a gene regulatory network by using genetic programming, Proc. Congress on Evolutionary Computation (2001) 720-726.

15. Schena, M., *DNA Microarrays*, Oxford University Press (2000).
16. Yamamoto, K.R., Steroid receptor regulated transcription of specific genes and gene networks, Ann. Rev. Genetics. 19 (1985) 209-252.
17. Weber, G.-W., Özoğur, S., and Karasözen, B., Challanges in the optimization of biosystems I: parameter estimation of enzymatic reactions with genetic algorithm, Preprint 39, Middle East Technical University, Institute of Applied Mathematics (2005).
18. Yılmaz, F.B., *A Mathemeatical Modeling and Approximation of Gene Expression Patterns by Linear and Quadratic Regulatory Relations and Analysis of Gene Networks*, Institute of Applied Mathematics, METU, MSc Thesis (2004).

Part XI

Continuous Optimization

Wavelet Schemes for Linear–Quadratic Elliptic Control Problems

Angela Kunoth

Institut für Angewandte Mathematik und Institut für Numerische Simulation, Universität Bonn, Wegelerstr. 6, 53115 Bonn, Germany. kunoth@iam.uni-bonn.de

Summary. This paper discusses the efficient numerical solution of a class of continuous optimization problems which are characterized by minimizing a tracking-type quadratic control functional subject to constraints in form of a linear elliptic partial differential equation (PDE) together with appropriate boundary conditions. For such problems, discretizations in terms of (domain-adapted biorthogonal spline-)wavelets offer several favorable properties over conventional approaches: from the evaluation of non-integer Sobolev norms in the objective functional over well-conditioned coupled linear systems of equations up to adaptive solution schemes with provable convergence and optimal convergence rates.

1 Introduction

In continuous optimization, PDE-constrained control problems provide an enormous challenge from a numerics point of view. Specifically, for constraints in form of a linear second order elliptic boundary value problem, standard discretizations on uniform grids based on finite differences or finite elements yield extremely large, ill-conditioned coupled systems of linear equations. The size of these systems mandatorily requires to employ iterative solvers whose speed is determined by the (spectral) condition number of these systems. Unfortunately, these discretizations on a grid with spacing h yield condition numbers which grow proportional to h^{-2} and therefore become even larger for decreasing h and correspondingly larger problem sizes. In order to overcome the conditioning issue for a single elliptic PDE, there are essentially three approaches, all of them based on an underlying multilevel structure, which yield *mesh-independent* condition numbers of the system matrices, namely, multigrid schemes, BPX-type preconditioners and discretizations based on properly scaled wavelets, see [4, 6] for references. These *preconditioners* are asymptotically optimal in the sense that the linear system can be solved in an amount of arithmetic operations that is *linear* in the number of unknowns. For continuous optimization problems constrained by linear elliptic PDEs, the optimality conditions lead to a coupled *system* of PDEs involving additional unknowns for which it is even more tempting to exploit multiscale preconditioners.

In addition to employing such asymptotically optimal solvers, one can try to avoid high cost from the outset by *locally adapting* degrees of freedom subject to singularities in the data, the coefficients or the domain while still guaranteeing an overall desired accuracy. Adaptive schemes based on finite elements have a long tradition for a single elliptic PDE, see, e.g., [1, 8], which have been extended to stationary PDE-constrained control problems in [2]. An adaptive numerical solution scheme based on *wavelets* was shown to converge for a single elliptic boundary value problem in [5]. Most importantly, however, is the fact that such schemes provide *optimal complexity*, meaning that the solution can be computed in a total number of arithmetic operations which is comparable to the wavelet-best N-term approximation of the solution. For the continuous optimization problems considered in this paper and some more classes, this has been derived in [7].

Along these lines, we study the efficient numerical solution for a continuous optimization problem with a quadratic tracking-type objective functional which is constrained by a linear elliptic PDE. In the next section, we introduce the class of PDE-constrained control problems, collect the main ingredients about wavelets needed here and represent the continuous optimization problem in wavelet coordinates. Section 3 is devoted to the numerical solution, discussing wavelet preconditioners on uniform grids as well as adaptive schemes based on wavelets. Some final remarks conclude this note.

2 Control Problems in Wavelet Coordinates

2.1 A Class of PDE-Constrained Control Problems

As a prototype example, consider the following problem. Given some prescribed state y_*, a regularization parameter $\omega > 0$ and a right hand side f, find the solution pair (y, u) which minimizes the tracking-type quadratic functional

$$J(y,u) = \tfrac{1}{2}\|y - y_*\|_Z^2 + \tfrac{\omega}{2}\|u\|_U^2 \tag{1}$$

subject to constraints in form of the linear second order elliptic boundary value problem

$$-\Delta y + y = f + u \qquad \text{in } \Omega, \qquad y = 0 \qquad \text{on } \partial\Omega. \tag{2}$$

Thus, the *state* y of the system, living on a bounded domain $\Omega \subset \mathbb{R}^n$ with piecewise smooth boundary $\partial\Omega$, is to be matched to a given state y_* by enforcing a (distributed) *control* u under minimal cost, measured in terms of the norm $\|\cdot\|_U$. Z and U are appropriate Banach spaces on Ω (or parts thereof) with corresponding norms which model smoothness requirements and whose choice will be discussed below.

While classical finite difference approaches discretize (2) directly, Galerkin schemes start out with a weak formulation of (2) which is in operator form

$$Ay = f + u \qquad \text{in } H'. \tag{3}$$

Here $A : H \to H'$ represents the linear elliptic PDE operator which is defined on the Sobolev space with first order weak derivatives bounded in $L_2(\Omega)$ and zero boundary conditions, $H := H_0^1(\Omega)$ with dual H', see, e.g., [10]. For more general problems than (2), it will be assumed that A is self-adjoint and a bounded linear bijection, i.e.,

$$c_A \|v\|_H \;\le\; \|Av\|_{H'} \;\le\; C_A \|v\|_H \qquad \text{for any } v \in H \tag{4}$$

and for some constants $0 < c_A \le C_A < \infty$. This implies that given any right hand side $f + u \in H'$, there exists a unique $y \in H$ which solves (3).

In the following, we will be concerned with the efficient numerical solution of minimizing (1) subject to (3) in terms of wavelet bases.

2.2 Wavelets

A *wavelet basis* for a Hilbert space H is a collection of functions $\Psi_H := \{\psi_{H,\lambda} : \lambda \in I\!I\} \subset H$ indexed from an infinite set $I\!I$. Each λ comprises different information $\lambda = (j, \mathbf{k}, \mathbf{e})$: the *resolution level* $j =: |\lambda|$, a spatial location $\mathbf{k} \in \mathbb{Z}^n$ and, in more than one space dimensions, information on the *type* of wavelet built by tensor products of univariate functions. We assume that Ψ_H has three crucial properties:

Riesz basis property (R). Every $v \in H$ has a unique expansion $v = \mathbf{v}^T \Psi_H := \sum_{\lambda \in I\!I} v_\lambda \psi_{H,\lambda}$ and its expansion coefficients $\mathbf{v} := (v_\lambda)_{\lambda \in I\!I}$ satisfy for $0 < c_H \le C_H < \infty$ a *norm equivalence*

$$c_H \|\mathbf{v}\|_{\ell_2} \;\le\; \|\mathbf{v}^T \Psi_H\|_H \;\le\; C_H \|\mathbf{v}\|_{\ell_2}, \quad \mathbf{v} \in \ell_2 := \ell_2(I\!I). \tag{5}$$

Locality (L). The $\psi_{H,\lambda}$ have compact support: $\operatorname{diam}(\operatorname{supp} \psi_{H,\lambda}) \sim 2^{-|\lambda|}$.

Cancellation property (CP). This property enables *sparse representations* of differential (and more generally, integral) operators in wavelet bases which is essential for an efficient solution scheme [6, 7].

The elements of Ψ_H are not required to fulfill any orthogonality conditions. There are concrete constructions of wavelets satisfying these properties for Sobolev spaces $H^s(\Omega)$ for a wide range of s (including negative and noninteger smoothness parameters) on bounded domains which are linear combinations of B-splines, see [10] for more details and references. These are typically obtained from an *anchor basis* $\Psi = \{\psi_\lambda : \lambda \in I\!I\}$ which is a Riesz basis for $L_2(\Omega)$ which means that Ψ is scaled such that $\|\psi_\lambda\|_{L_2(\Omega)} \sim 1$. Rescaled versions of Ψ defined by

$$\Psi_s := \{2^{-s|\lambda|} \psi_\lambda : \lambda \in I\!I\} =: \mathbf{D}^{-s} \Psi \tag{6}$$

then form Riesz bases for (closed subspaces of) Sobolev spaces $H^s(\Omega)$ for $0 \le s < \gamma$ (e.g., $\gamma = 3/2$ in the piecewise linear case).

2.3 Wavelet Representations

The mapping property (4) together with the Riesz basis property (R) yield for the representation of A in a (properly scaled) wavelet basis a *well-conditioned* operator $\mathbf{A}$ with spectral condition number *independent* of the grid resolution.

Theorem 1. *The (infinite) matrix* $\mathbf{A}$ *is a boundedly invertible mapping from* ℓ_2 *into itself, and there exists finite positive constants* $c_{\mathbf{A}} \leq C_{\mathbf{A}}$ *such that* $c_{\mathbf{A}}\|\mathbf{v}\|_{\ell_2} \leq \|\mathbf{A}\mathbf{v}\|_{\ell_2} \leq C_{\mathbf{A}}\|\mathbf{v}\|_{\ell_2}$ *for* $\mathbf{v} \in \ell_2$. *The representation of* A *with respect to any finite subset* $\Psi_\Lambda \subset \Psi$ *is a symmetric positive definite matrix with* $\kappa_2(\mathbf{A}_\Lambda) \leq C_{\mathbf{A}} c_{\mathbf{A}}^{-1}$ *uniformly in* Λ.

In view of (5) and (6), consider the following control problem in (infinite) wavelet coordinates: for given data $\mathbf{y}_*, \mathbf{f}$ and $\omega > 0$, minimize

$$\mathbf{J}(\mathbf{y}, \mathbf{u}) := \tfrac{1}{2} \|\mathbf{D}^{-s}(\mathbf{y} - \mathbf{y}_*)\|_{\ell_2}^2 + \tfrac{\omega}{2} \|\mathbf{u}\|_{\ell_2}^2 \tag{7}$$

over $(\mathbf{y}, \mathbf{u})$ subject to (3) in wavelet representation,

$$\mathbf{A}\mathbf{y} = \mathbf{f} + \mathbf{D}^{-t}\mathbf{u}, \tag{8}$$

see [3, 4] for more sophisticated variants involving exact Riesz operators. The scaling in terms of the diagonal matrices $\mathbf{D}^s, \mathbf{D}^t$ corresponds to choosing the norms in (1) as $Z = H^{1-s}(\Omega)$ and $U = (H^{1-t}(\Omega))'$. As long as $s \in [0, 1]$ and $t \geq 0$, (7) together with (8) leads to a well-posed variational problem. The choice $s = t = 1$ yields the classically discussed cases of functionals involving L_2 norms. Here more flexibility is allowed in modeling the control problem which computationally only amounts to the multiplication by diagonal matrices. Applying standard techniques from optimization, problem (7) subject to (8) is equivalent to solve the coupled system

$$\mathbf{A}\mathbf{y} = \mathbf{f} + \mathbf{D}^{-t}\mathbf{u}, \qquad \mathbf{A}^T\mathbf{p} = -\mathbf{D}^{-2s}(\mathbf{y} - \mathbf{y}_*), \qquad \omega\mathbf{u} = \mathbf{D}^{-t}\mathbf{p} \tag{9}$$

involving the additional Lagrange multiplier $\mathbf{p}$, the costate variable. In order to derive convergent iterations and deduce complexity estimates, we use the fact that $\mathbf{A}$ is, according to Theorem 1, a boundedly invertible mapping on ℓ_2. Thus, inverting the first two systems in (9) yields the condensed system

$$\mathbf{Q}\mathbf{u} = \mathbf{g}, \qquad \mathbf{Q} := \mathbf{D}^{-t}\mathbf{A}^{-T}\mathbf{D}^{-2s}\mathbf{A}^{-1}\mathbf{D}^{-t} + \omega\mathbf{I}, \quad \mathbf{g} := \mathbf{D}^{-t}\mathbf{A}^{-T}\mathbf{D}^{-2s}(\mathbf{y} - \mathbf{y}_*) \tag{10}$$

for which the following can be shown [4, 7].

Theorem 2. *The (infinite) matrix* $\mathbf{Q}$ *is a boundedly invertible mapping from* ℓ_2 *into itself, and there exists finite positive constants* $c_{\mathbf{Q}} \leq C_{\mathbf{Q}}$ *such that*

$$c_{\mathbf{Q}}\|\mathbf{v}\|_{\ell_2} \leq \|\mathbf{Q}\mathbf{v}\|_{\ell_2} \leq C_{\mathbf{Q}}\|\mathbf{v}\|_{\ell_2}, \qquad \mathbf{v} \in \ell_2. \tag{11}$$

Since $\mathbf{Q}$ is symmetric positive definite (but yet infinite), finite versions of (10) can be solved by gradient or conjugate gradient iterative schemes, and the convergence speed of any such iteration does *not* depend on the discretization. Of course, in order to make such iterative schemes practically feasible, the explicit inversion of $\mathbf{A}$ in the definition of $\mathbf{Q}$ has to be avoided and replaced by an iterative solver in turn. This is where (9) comes into play. In particular, the third equation in (9) has an interpretation as residual for (10), i.e., $\mathbf{Qu} - \mathbf{g} = \omega \mathbf{u} - \mathbf{D}^{-t}\mathbf{p}$. We have investigated two scenarios.

3 Iterative Solution

3.1 Finite Systems on Uniform Grids

Since all system matrices in (10) or, equivalently, (9) are well-conditioned, finite versions of these corresponding to *uniform discretizations* together with a *nested iteration strategy* can be solved up to discretization error on a finest grid by an asymptotically optimal iterative method which is *linear* in the amount of unknowns on the finest grid and therefore best possible. This result has been proved in [4, 9]. Several numerical examples in up to three spatial dimensions [3, 4] confirm that the size of the constants in (11) is moderate, requiring few constant iterations on each discretization level. The application of $\mathbf{A}$ is in these cases realized by using the Fast Wavelet Transform which is in view of the locality (L) of optimal complexity.

Corresponding extensions to control problems constrained by an elliptic PDE with *Dirichlet boundary control* is investigated numerically in [11].

3.2 Adaptive Schemes

Starting from the condensed system (10) in infinite wavelet coordinates, one can step by step break down the necessary ingredients which are required for a *convergent adaptive solution algorithm*, following the paradigm in [5]. Here the difficulty lies in designing an application routine for $\mathbf{Q}$ which involves two inversions of $\mathbf{A}$ such that the overall scheme is still of *optimal complexity*. As shown in [7] for distributed as well as for Neumann boundary control problems, such a convergent adaptive scheme can indeed be derived. Here fundamental and far-reaching techniques from nonlinear approximation theory come into play. A specific feature of the scheme is that each of the variables $\mathbf{y}, \mathbf{p}$ and $\mathbf{u}$ is approximated *separately*. This is conceptually different from the adaptive schemes bases on finite elements derived in [2] where one underlying grid is used for all variables and where so far neither convergence proofs nor rates are available.

The results can be further extended to *Dirichlet boundary control problems* where the elliptic PDE constraint is formulated as a saddle point problem, introducing yet another interior approximate iteration [9], see also [10].

4 Conclusion and Outlook

Wavelet discretizations for linear-quadratic elliptic control problems with flexible norms in the control functional yield uniformly well-conditioned system matrices and, thus, asymptotically optimal iterative solution schemes. For problems with solutions or controls with singularities, adaptive wavelet schemes can be derived for different types of control. The adaptive wavelet strategy can be further extended to *goal–oriented* functionals where a (local) *functional* of the solution of a linear elliptic PDE is computed up to arbitrary accuracy at possibly minimal cost. This is currently under investigation.

Acknowledgments

I want to thank Roland Pabel for his assistance during the preparation of this manuscript. This work was supported in part by the Deutsche Forschungsgemeinschaft (SFB 611).

References

1. Becker R, Rannacher R (2001) An optimal error control approach to a–posteriori error estimation. Acta Numerica:1–102
2. Becker R, Kapp H, Rannacher R (2000) Adaptive finite element methods for optimal control of partial differential equations: Basic concept. SIAM J. Contr. Optim. 39:113-132
3. Burstedde C (2005) Fast Optimised Wavelet Methods for Control Problems Constrained by Elliptic PDEs. PhD Dissertation. Universität Bonn.
4. Burstedde C, Kunoth A (2005) Fast iterative solution of elliptic control problems in wavelet discretizations. SFB 611 Preprint #127, Universität Bonn; revised June 2005; to appear in: J. Comp. Appl. Maths.
5. Cohen A, Dahmen W, DeVore R (2001) Adaptive wavelet methods for elliptic operator equations — Convergence rates. Math. Comp. 70:27–75
6. Dahmen W (1997) Wavelet and multiscale methods for operator equations. Acta Numerica:55–228
7. Dahmen W, Kunoth A (2005) Adaptive wavelet methods for linear–quadratic elliptic control problems: Convergence rates, SIAM J. Contr. Optim. 43(5):1640–1675
8. Eriksson E, Estep D, Hansbo P, Johnson C (1995) Introduction to adaptive methods for differential equations. Acta Numerica:105–158
9. Kunoth A (2005a) Adaptive wavelet schemes for an elliptic control problem with Dirichlet boundary control. Numer. Algor. 39(1-3):199-220
10. Kunoth A (2005b) Wavelet-based multiresolution methods for stationary PDEs and PDE-constrained control problems. In: "Frontiers in Numerical Analysis", J. Blowey and A. Craig (eds.), Springer, 2005, 1–63.
11. Pabel R (2005) Wavelet Methods for Elliptic PDE Constrained Control Problems with Dirichlet Boundary Control. Diploma Thesis (in English). Universität Bonn.

Part XII

Econometrics, Game Theory and Mathematical Economics

Aggregate Game and International Fishery with Several Countries

Koji Okuguchi
Gifu Shotoku Gakuen University
1-38 Nakauzura, Gifu-shi, Gifu-ken 500-8288, JAPAN
e-mail:okuguchi351013@*ybb.ne.jp*

1. Introduction

Many papers dealing with dynamic analysis of single country or international commercial fishing from a common fishing ground have appeared (see Szidarovszky and Okuguchi1998, 2000, Szidarovszky *et al* 2002, Sandal and Steinshamn 2004, and Szidarovszky *et al* 2004). In these papers each fishing firm or country is assumed to maximize its profits in each harvesting period, taking into account the dynamic equation governing the change of the fish stock over time. In a seminal paper, Levhari and Mirman 1980 have formulated an intertemporal two country model of fishery in which each country is assumed to maximize its total discounted utility from consuming the harvested fish, instead of its total discounted sum of profits, over finite or infinite periods. They have derived, on the basis of dynamic programming approach, a closed- loop solution by considering a simple dynamic equation for the fish stock and a logarithmic utility function. In general, finding a closed- loop solution for dynamic games with two or more players is difficult or impossible. In this paper we will generalize Levhari-Mirmal model to allow for several countries harvesting fish from a common fishing ground. We will, however, analyze only the case of individual country's intertemporal utility maximization as the joint utility maximization by all countries may be analogously analyzed. We will derive closed- loop optimal harvesting rates without much difficulty by taking into account the properties of aggregate games in which equilibrium condition for each country is given by an implicit function of its variable and the sum of all country's variables.This is the equilibrium condition for the so-called aggregate game.Fishing in each fishing period in our model of international fishery is characterized as an aggregate game and its Nash equilibrium is derived using the property of the aggregate game.

2. Model and Solution

We suppose n countries to be harvesting fish from a common fishing ground, harvesting to be costless, and each country to be maximizing its total discounted utility over finite or infinite periods. If β_i and $C_i(t)$ are country i's discount factor and harvesting rate of the fish in period t in the case of m fishing periods, respectively, country i's total discounted utility U_i over periods m is assumed to take a form

$$U_i = \log C_i(1) + \beta_i \log C_i(2) \cdots + \beta_i^{(m-1)} \log C_i(m) \tag{1}$$

where $\beta_i < 1$, $i=1,2,...,n$. Furthermore, if x_t is the fish stock in period t, the fish stock in the absence of fishing is assumed to change according to

$$x_{t+1} = x_t^{\alpha}\ ,\ 0 < \alpha < 1. \tag{2}$$

According to (2) the fish stock converges to 1 in the long run in the absence of fishing. This level of fish stock corresponds to *carrying capacity* of a fishing ground(see Clark 1990).

We will now adopt dynamic programming approach to derive the optimal harvest rate. To do so we have to assume the Nash equilibrium to be attained in each period and each country get the same amount of fish in the last fishing period.

Step 1 : Two fishing periods. In this case country i maximizes its total discounted utility over two periods given by

$$\log C_i^{(1)} + \beta_i \log \frac{(x - \Sigma C_j^{(1)})^{\alpha}}{n}\ ,\quad i=1,2,...,n, \tag{3}$$

where x is the fish stock in period 1 and $C_i^{(1)}$ is the optimal harvest rate in period 1. Maximizing the above expression with respect to $C_i^{(1)}$ and rearranging we get

$$\alpha\beta_i C_i^{(1)} + \Sigma C_j^{(1)} = x\ ,\quad i=1,2,...,n. \tag{4}$$

This gives the equilibrium condition as an implicit function between individual country's variable $C_i^{(1)}$ and the sum of individual country's variable $\Sigma C_j^{(1)}$. Hence the two period game is an aggregate game. Let $C^{(1)} \equiv \Sigma C_j^{(1)}$. We then have from (4)

$$C_i^{(1)} = \frac{x - C^{(1)}}{\alpha\beta_i} \qquad i=1,2,...,n. \tag{5}$$

The summation for all i yields

$$C^{(1)} = \sum_i \frac{x - C^{(1)}}{\alpha\beta_i}.$$

$$C^{(1)} = \frac{\sum_i \frac{x}{\alpha\beta_i}}{1 + \sum_i \frac{1}{\alpha\beta_i}} , \tag{6}$$

$$x - C^{(1)} = \frac{x}{1 + \sum_i \frac{1}{\alpha\beta_i}} , \tag{7}$$

$$C_i^{(1)} = \frac{x}{\alpha\beta_i(1 + \sum_i \frac{1}{\alpha\beta_i})} , \quad \text{i=1,2,...,n.} \tag{8}$$

Substituting (7) and (8) into (3) we have

$$V_i^{(1)}(x) \equiv (1 + \alpha\beta_i)\log x + A_1 \ , \quad i=1,2,...,n, \tag{9}$$

$$A_1 \equiv -\{\log \alpha\beta_i(1 + \Sigma\frac{1}{\alpha\beta_i}) + \alpha\beta_i \log(1 + \Sigma\frac{1}{\alpha\beta_i}) + \beta_i \log_n\}. \tag{10}$$

Step 2 : Three fishing periods. In the case of three fishing periods, the total discounted utility of country *i* is given by

$$\begin{aligned} &\log \boldsymbol{C}_i^{(2)} + \beta_i V_i^{(1)}(\boldsymbol{x} - \sum \boldsymbol{C}_j^{(2)}) = \log \boldsymbol{C}_i^{(2)} + \beta_i\{(1 + \alpha\beta_i) \\ &\times \log(\boldsymbol{x} - \sum \boldsymbol{C}_j^{(2)} + A_1 \ \} \end{aligned} ,$$

$$i=1,2,\ldots,n, \tag{11}$$

where we have taken into account (9) and $C_i^{(2)}$ denotes the first period harvesting rate of country *i*. Maximizing (11) with respect to $C_i^{(2)}$ and defining $C^{(2)} \equiv \sum C_j^{(2)}$.

$$C_i^{(2)} = \frac{x - C^{(2)}}{\beta_i(1 + \alpha\beta_i)} , \quad i=1,2,...,n, \tag{12}$$

Summing (12) for all i and rearranging we have

$$C^{(2)} = \frac{\sum_i \frac{x}{\beta_i(1+\alpha\beta_i)}}{1+\sum_i \frac{1}{\beta_i(1+\alpha\beta_i)}} \tag{13}$$

$$x - C^{(2)} = \frac{x}{1+\sum_i \frac{1}{\beta_i(1+\alpha\beta_i)}} \tag{14}$$

$$C_i^{(2)} = \frac{x}{\beta_i(1+\alpha\beta_i)\{1+\sum_i \frac{1}{\beta_i(1+\alpha\beta_i)}\}}, \quad i=1,2,\ldots,n \tag{15}$$

Substitute (14) and (15) into (11) to yield

$$V_i^{(2)}(x) \equiv (1+\beta_i+\alpha\beta_i^2)\log x + A_2 , \tag{16}$$

Last step: m fishing periods. Arguing similarly as above, we know that country i's total discounted utility over m periods is

$$\begin{aligned} &\log C_i^{(m-1)} + \beta_i\{(1+\beta_i+\beta_i^2+\ldots+\beta_i^{m-3}+\alpha\beta_i^{m-2}) \\ &\times \log(x-\sum C_j^{(m-1)}) + A_{m-2}\} \end{aligned}, \quad i=1,2,\ldots,n. \tag{17}$$

Maximizing this with respect to $C_i^{(m-1)}$ and defining

$C^{(m-1)} = \sum C_j^{(m-1)}$, we finally derive

$$C^{(m-1)} = \frac{\sum_{i=1}^{n} \frac{x}{\beta_i+\beta_i^2+\cdots+\beta_i^{m-2}+\alpha\beta_i^{m-1}}}{1+\sum_{i=1}^{n} \frac{1}{\beta_i+\beta_i^2+\cdots+\beta_i^{m-2}+\alpha\beta_i^{m-1}}}, \tag{18}$$

$$x - C^{(m-1)} = \frac{x}{1 + \sum_{i=1}^{n} \frac{1}{\beta_i + \beta_i^2 + \cdots + \beta_i^{m-2} + \alpha\beta_i^{m-1}}}, \tag{19}$$

$$C_i^{(m-1)} = \frac{x}{(\beta_i + \beta_i^2 + \cdots + \beta_i^{m-2} + \alpha\beta_i^{m-1})} \times \frac{1}{(1 + \sum_{i=1}^{n} \frac{1}{\beta_i + \beta_i^2 + \ldots + \beta_i^{m-2} + \alpha\beta_i^{m-1}})}. \tag{20}$$

In all the steps above we have been able to derive expressions (6)-(8), (13)-(15) and (18)-(20) easily by solving first with respect to $C^{(1)}$, $C^{(2)}$ and $C^{(m-1)}$. Equations leading to $C^{(1)}$, $C^{(2)}$ and $C^{(m-1)}$ show $C_i^{(1)}$, $C_i^{(2)}$ and $C_i^{(m-1)}$ as functions of aggregate variables $C^{(1)}$, $C^{(2)}$ and $C^{(m-1)}$, respectively. In this sense our generalized Levhari-Mirman model of fishery is an example of the so called aggregate games.

3. Conclusion

In this paper we have been able to derive the closed loop solution for international fishery with several countries engaged in fishing to maximize their total discounted utilities over finite fishing periods.Our international fishery model is a generalized version of the well known model of international fish war between two countries. Recognizing that at each stage of maximization,the equilibrium conditions are those of aggregate games,we have derived rather easily the optimal harvesting rates for the first period for any length of harvesting periods,which enables us to determine the optimal harvesting rates for all subsequent periods.

References

Clark C. W. (1990) *Mathematical Bioeconomics*, 2nd edition, John Wiley and Sons, New York and London

Levhari D. and L Mirman (1980) The Great Fish War : An Example using a Dynamic Cournot-Nash Solution, *Bell Journal of Economics* 11, 322-34

Okuguchi K. (2003), Dynamic and Comparative Static Analysis of Imperfectly Competitive International Fishery, *Journal of Economics* 80, 249-265

Sandal L. K. and S. I Steinshamn (2004) Dynamic Cournot-Competitive Harvesting of a Common Pool Resource, *Journal of Economic Dynamics and Control* 28,

1781-1798
Szidarovszky F. and K Okuguchi (1998), An Oligopoly Model of Commercial Fishing, *Seoul Journal of Economics* 11, 321-30
Szidarovszky F. and K Okuguchi (2000), A Dynamic Model of International Fishing, *Seoul Journal of Economics* 13, 471-76
Szidarovszk, F.,K. Okuguchi and M Kopel (2004), International Fishery with Several Countries, Paper presented at NED2004

A Centrist Poverty Index

Gerhard Kockläuner
FB Wirtschaft, FH Kiel, Sokratesplatz 2, 24149 Kiel, Germany
E-Mail: gerhard.kocklaeuner@fh-kiel.de

1 Introduction

Kockläuner (2002) introduces the ethical poverty index

$$C_1^* = \frac{1}{z}(\frac{1}{n}\sum_{i=1}^{n} g_i^{*\alpha})^{1/\alpha}, \alpha > 1. \tag{1}$$

In equation (1) z > 0 denotes a given finite poverty line. For n non-negative incomes y_i censored income gaps are defined as $g_i^* = \max\{z - y_i, 0\}$, i = 1,…,n. The index C_1^* depends on a special measure of inequality, i.e.

$$B_{g^*}^{\alpha} = \frac{1}{\overline{g}^*}(\frac{1}{n}\sum_{i=1}^{n} g_i^{*\alpha})^{1/\alpha} - 1, \alpha > 1, \tag{2}$$

where $\overline{g}^*$ is the mean of all censored income gaps.

Both $B_{g^*}^{\alpha}$ and C_1^* satisfy the axiom of scale invariance. The respective values do not change when z and all y_i are multiplied by the same positive constant. According to Kolm (1976) the measures $B_{g^*}^{\alpha}$ and C_1^* can therefore be called rightist.

Arguing against the property of scale invariance, Kolm (1976) proposes a centrist measure of inequality. With respect to censored income gaps the respective measure reads as

$$A_{g^*}^{\alpha}(\xi) = 1 + \frac{\xi}{\overline{g}^*} - \frac{1}{\overline{g}^*}(\frac{1}{n}\sum_{i=1}^{n} \tilde{g}_i^{*\alpha})^{1/\alpha}, \alpha < 1, \xi \geq 0, \tag{3}$$

where $\tilde{g}_i^* = g_i^* + \xi$, i = 1,…,n. Kolm (1976) characterizes $\overline{g}^* A_{g^*}^{\alpha}(\xi)$ axiomatically.

In the following such a centrist perspective will be transferred to the poverty measurement above. The centrist poverty index resulting will include both a rightist and a leftist view. One assumes the axiom of translation invariance to hold for the latter. Here the amounts of inequality and poverty do not change when the same positive constant is added to z and all y_i.

2 The Poverty Index $C_1^*(\xi)$

Contrary to equation (3) the poverty index C_1^* depends on an α-mean with α > 1. Such a mean can also be introduced with respect to centrist inequality measurement. Being dual to $A_{g^*}^{\alpha}(\xi)$ an extension of equation (2) amounts to

$$B_{g^*}^{\alpha}(\xi) = \frac{1}{\overline{g}^*}\left(\frac{1}{n}\sum_{i=1}^{n}\tilde{g}_i^{*\alpha}\right)^{1/\alpha} - \frac{\xi}{\overline{g}^*} - 1, \alpha > 1, \xi \geq 0. \tag{4}$$

An axiomatical characterization of $\overline{g}^* B_{g^*}^{\alpha}(\xi)$ follows from Kolm (1976) by simply changing minus to plus in his axioms of welfare independence.

The centrist measure of inequality $B_{g^*}^{\alpha}(\xi)$ immediately leads to an extension of equation (1), i.e. the centrist poverty measure

$$C_1^*(\xi) = \frac{1}{z}\left[\left(\frac{1}{n}\sum_{i=1}^{n}\tilde{g}_i^{*\alpha}\right)^{1/\alpha} - \xi\right], \alpha > 1, \xi \geq 0. \tag{5}$$

Obviously ξ = 0 in equation (5), i.e. $C_1^*(0) = C_1^* = C_{1r}^*$ gives the rightist case already discussed. Following Kolm (1976) the leftist case results when both α → ∞ and ξ → ∞ accompanied by α/ξ → c, a finite positive constant. Consequently the measure $B_{g^*}^{\alpha}(\xi)$ approaches

$$I_l = \frac{1}{\overline{g}^* c}\ln\left(\frac{1}{n}\sum_{i=1}^{n} e^{cg_i^*}\right) - 1, c > 0. \tag{6}$$

Equation (6) now gives the inequality measure $\overline{g}^* I_l$ which is translation invariant, i.e. leftist. Interestingly, in the present context translation invariance additionally holds with respect to I_l itself as well as with respect to the measures $A_{g^*}^{\alpha}(\xi)$ and $B_{g^*}^{\alpha}(\xi)$ in equations (3) and (4) respectively. This result is however only obtained because income gaps are used instead of incomes.

The poverty index corresponding to the leftist measure $\overline{g}^* I_l$ is

$$C_{1l}^* = \frac{1}{zc}\ln\left(\frac{1}{n}\sum_{i=1}^{n} e^{cg_i^*}\right), c > 0. \tag{7}$$

Note that $C_1^*(\xi)$ and all its special cases are poverty measures relative to the poverty line z. Translation invariance – in the context of equations (7) and (1) – therefore demands absolute poverty indexes such as zC_{1l}^* or zC_{1r}^*.

The index $C_1^*(\xi)$ satisfies classical poverty axioms such as focus, symmetry, replication invariance, increasing poverty line, non-poverty growth, strong monotonicity, strong transfer, weak transfer sensitivity, increasing poverty aversion and subgroup consistency. Compare Kockläuner (2002) with respect to $C_1^*(0)$ where additionally strong transfer sensitivity holds. Strong transfer sensitivity demands strong continuity as well as weak transfer sensitivity (Zheng 1999). In some instances $\alpha > 2$ is needed. The corresponding results can be easily extended to $\xi > 0$. In case of poverty and of $0 < \xi < \infty$ the value of $C_1^*(\xi)$ increases whenever z and all y_i are multiplied by the same constant $\lambda > 1$. It decreases whenever the same constant $\lambda > 0$ is added to z and all y_i. A similar decrease occurs when ξ increases.

References

Kockläuner G (2002) Revisiting two poverty indexes. Allg Statist Archiv 86: 299-305.

Kolm S (1976) Unequal inequalities I. J Econ Theory 12: 416-442.

Zheng B (1999) On the power of poverty orderings. Soc Choice Welfare 16: 349-371.

Does a Market Sensitive Price Strategy Pay Off in an Oligopoly Market Disturbed by Competitors Without Any Concept?

Vera Hofer[1] and Klaus Ladner[1]

Department of Statistics and Operations Research, Karl-Franzens University Graz, Universitätsstraße 15, 8010 Graz, Austria
vera.hofer@uni-graz.at; klaus.ladner@uni-graz.at

Summary. We investigate the performance of four price strategies in a heterogeneous closed oligopoly with three firms: overhead calculations, target pricing, discount prices and random prices. Using simulation we try to find out, whether market sensitive prices are more successful than others.

1 Introduction

The aim of a firm's price policy is to determine profit maximizing prices for each brand offered. In the course of time a comprehensive literature on the question how to determine prices has emerged. Many details have not been completely clarified till this day. In general internal information as well as information on demand and competitors are essential for determination of prices [8, p. 529].

In operational practice price maximizing rules of the classical price theory cannot be applied, since the model assumptions are mostly not realistic [8, p. 546]. As the reactions of consumers and competitors can hardly be predicted, alternative rules for determining prices are used. These rules are mainly based on three aspects: costs, demand and competition [5, p. 498f], [8, p. 547ff], [2, p. 767].

Based on these three aspects the three price strategies overhead calculations, target pricing and discount prices are designed in the present paper. Furthermore we consider random prices. We investigate which of these price strategies are most successful in various competitive situations on a closed heterogenous oligopoly. In particular we try to find out whether market sensitve price strategies pay off.

2 Description of the Market

The model of the oligopoly market is taken from the management game SINTO-Market which was developed by Becker and Selten [1] in 1967. For our analysis the original model was slightly simplyfied.

Three competitive firms try to maximize their owned capital over 15 periods. Each of the three firms can offer up to 5 brands of a synthetic protein product. For each brand j, three product specific decision parameters must be fixed: fineness r_j, tartness s_j and production y_j. The decision parameters r_j, s_j are discrete variables which must be fixed at one of ten levels numbered from 0 to 9.

Firms cannot offer an unlimited amount of each brand. For each unit of production one unit of capacity k is necessary. Production capacity is determined by the number n of brands offered, and the value of productive assets A such that

$$k = 0.002\,A\,(1 - 0.03\,(n - 1)). \tag{1}$$

Without reinvestment productive assets decrease by 20 % per period. By means of further investment a firm can increase its productive assets and therefore its production capacity. Where productive assets are disposed of, they are sold at 50 % of book value.

All three firms are faced with the same cost structure. Variable costs are 100. The taste characteristics r_j and s_j do not influence costs. Sales costs are 50 for each unit sold. A sales tax of 4 % of sales values adds $0.04\,p_j$ to the unit costs of brand j. Inventory costs are 20 per unit. If a brand is discontinued or if one of the parameters r_j or s_j is changed, the stock becomes useless and must be destroyed. Each firm has general fixed costs of 300,000 and additional fixed costs of 100,000 for each brand offered.

Each of the three firms is a subsidy of a big company. 90 % of the profits are transferred to the parent firm, which also bears 90 % of the losses. The firms have accounts with their parent companies, where financial deficits and surplusses are accumulated. The interest rate is 5 %, regardless of whether the balance is positive or negative. The parent firms do not allow investments which would make productive assets more than three times as big as owned capital.

At the beginning of the game, each of the firms has a capital endowment of 700,000. This includes productive assets of $A = 500,000$ and a positive balance of 200,000 on the account of the parent company. At the end of each period the firms receive a balance sheet and other bookkeeping information. Costs are imputed according to a direct costing scheme. The firms also receive a *market overview* where the variables r_j, s_j, p_j and the sales quantities x_j are listed for all brands on the market in the last period.

The participants have complete information about the cost situation, but only qualitative information about the demand function. Especially they know that total demand increases over time, first slowly, then rapidly until a saturation level is reached.

3 Decisions Concepts

To investigate the performance of various price strategies all market participants considered have the same decision strategies for the number of brands, positioning, quantity and investment. These strategies are sales oriented. For further description confer [4].

In the present paper we compare the following four price strategies: overhead calculations and random prices which are supposed not to be market sensitive, and the two market sensitive strategies target pricing and discount prices.

To get the prices by overhead calculations the variable costs are increased by a factor that takes into account fixed costs and profits. This factor lies between 200% and 400%, linearly depending on the gap between offered and sold quantity. We do not allow an increase of prices by more than 20% per period. Various authors warn against overhead calculations since prices may be too high or low [6, p. 3] [8, p. 547] [2, p. 763].

Target pricing is a demand oriented price determination [6, p. 4] [3, p. 226] which is based on the consumers' willingness to pay [5, p. 498] [7, p. 92]. Our target prices depending on the sales figures in the period before. Prices remain the same if the gap between offert and sold quantity exceeds 15% and lies below 70%. Elsewise prices are linearly changed to a maximum increase of 10% and a maximum decrease of 30%.

A simple and realistic strategy is the determination of discount prices. This strategy is competitors oriented [8, p. 553]. In our case the discount prices are 5% below the smallest price of neighbouring brands. As a lack of market information can lead to unsystematic trial, we also consider random decisions. Our random prices are uniformly distributed between variable costs including proportional fixed costs and twice the full prices.

4 Simulation and Results

The impact of the price strategies described before on the firms' economic success is investigated in homogeneous and heterogeneous constitution of the market participants and in disturbed markets. For this purpose we carry out eight series of simulations with 100 games each. For each of the four price strategies one series each is simulated in homogeneous constitution of the market participants, i. e. all three firms apply the same price strategy. To investigate the heterogenous constitution overhead calculations, target pricing and discount prices are considered. In this series each of the three market participants is assigned a different strategy. Three series are used for simulating the performance of the three price strategies overhead calculations and target pricing in a disturbed market. In each of these three series two firms are assigned the same price strategy. The third firm serves as market disturbance. This firm is designed as a random player, i.e. all decisions occudfdfdfdfgfgh-hdghsgkfgkrring are made randomly.

4.1 Homogeneous Constitution

Figure 1 shows the average prices and the average owned capital in markets where all three firms follow the same price strategy. The average prices on markets where the firms determine the prices according to target pricing mirrors the market growth, underlying the model, with a small time lag. Generally speaking, we observe growing prices for all price strategies except discount prices. These results are expected in a growing market. If all three firms apply a discount price strategy, the prices naturally decline even if demand increases.

The economic success of the price strategies considered is measured by owned capital reached. The market participants applying overhead calculations or target pricing are able to achieve high profits, whereas discount and random prices lead to remarkable losses and to the firm's dissappearance from the market in the long run.

The differences of the owned capital between overhead calculations and target pricing are not significant in the 15th period (p-value of wilcoxon rank sum test = 0.3640). Thus, in the situation of homogeneous constitutions of market participants it is not obvious that market sensitive strategies pay off.

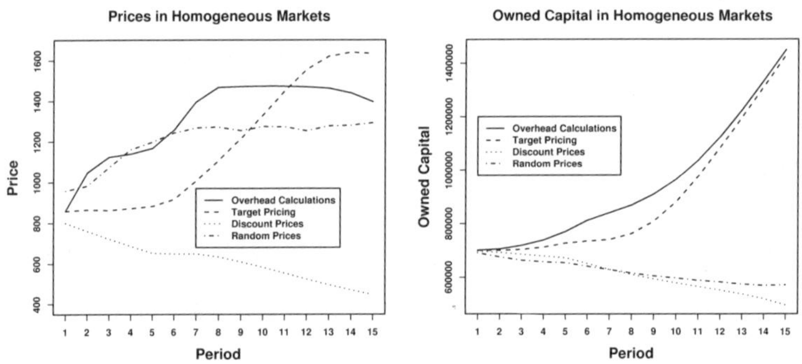

(a) (b)

Fig. 1. Prices (a) and owned capital (b) in homogeneous constitution of market participants

4.2 Heterogeneous Constitution

Figure 2 shows the average prices and the average owned capital in markets with a heterogenoues constitution of the participants. Target pricings leads to a rather smooth development of mean prices, in contrast to overhead calculations.

From the development of owned capital it can be seen that all three market participants can survive in the long run. Market sensitive price strategies turn out to be superior. Despite the fact of higher prices, in the first half of the game target pricing does not lead to lower profits than discount prices. Owned capital for the firm applying target pricing does not significantly differ from owned capital of the firm with discount prices up to period 10 (p-value of wilcoxon rank sum test > 0.3).

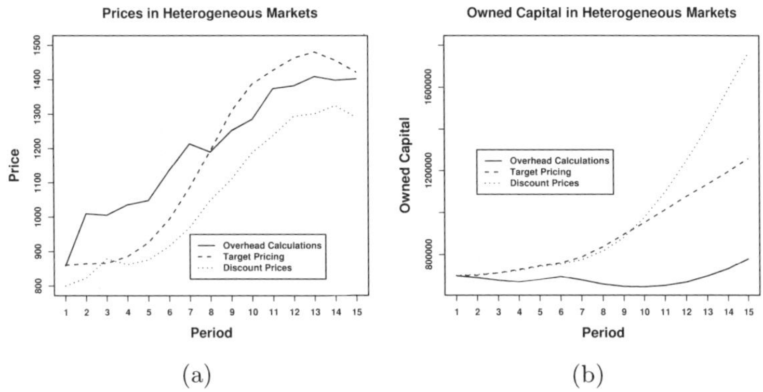

Fig. 2. Prices (a) and owned capital (b) in heterogeneous constitution of market participants

4.3 Disturbed Markets

Figure 3 shows the average prices and the development of the average owned capital in markets where two firms are assigned the same price strategy and the third firm serves as market disturbance. Dispite the disturbances the development of prices of the two firms applying target pricing remains rather smooth und is similar to the homogeneous constitution, in contrast to firms using overhead calculations. Prices derived by overhead calculation react sensitively to disturbances. Target pricing leads to better performance in the second half of the game, which can be seen from the development of owned capital. After the 15th period owned capital is significantly higher for target pricing than for overhead calculations (p-value of wilcoxon rank sum test < 0.0001). Thus, in a disturbed market market sensitive prices pay off in our model.

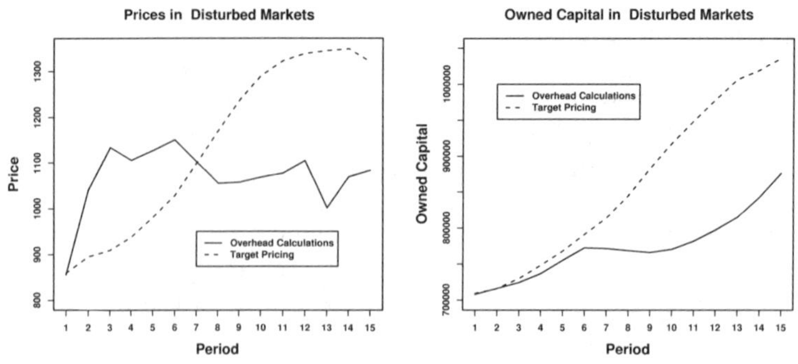

(a) (b)

Fig. 3. Prices (a) and owned capital (b) in a disturbed market (two firms with the same price strategies and one random player)

5 Conclusions

In our model of a closed heterogeneous oligopoly it turns out that the market sensitive price strategies we considered are generally better in heterogeneous constitution and in disturbed markets than non market sensitive ones. The very simple discount price strategy which is based on watching the competitors, can even lead to higher profits than target pricing which is based on watching the consumers. However, this is only true if there are not too many market participants applying a discount price strategy. Overhead calculations prove to be sensitve against market disturbances.

References

1. Becker, O., Selten R., 1970. Experiences with the management game SINTO-market, in: Sauermann H. (Hrsg.), Contribution to experimental economics Vol. 2, Mohr J. C. B. (Paul Siebeck) Tbingen, p. 136–150
2. Corsten H., 2000. Lexikon der Betriebswirtschaftslehre, 4. Auflage, Oldenbourg
3. Diller H., 200. Preispolitik, Kohlhammer
4. Hofer V., Ladner K., 2004. Unterschiedliche Verhaltensweisen am heterogenen Oligopolmarkt bei unbekannter Nachfrage, Working Paper 03/2004, Universität Graz
5. Lechner K., Egger A., Schauer R., 1999. Einführung in die Betriebswirtschaftslehre, 18. Auflage, Linde
6. Nagle T. T,. Holden R. K., 2002. The strategy and tactics of pricing – a guide to profitable decision making, Prentice Hall
7. Simon H., Dahlhoff D., 1998. Target Pricing und Target Costing mit Conjoined Meassurement – Wege zum Preiskonsens zwischen Controlling und Marketing, Controlling, 2, p. 92–96
8. Wöhe G., 2002. Einführung in die Betriebswirtschaftslehre, 21. Auflage, Verlag Vahlen

Order Stable Solutions for Two-sided Matching Problems

Zbigniew Switalski

Department of Operations Research
University of Economics, Al. Niepodleglosci 10
60-967 Poznan, Poland
zbigniew.switalski@ae.poznan.pl

Abstract

We concern the problem of matching the agents from two disjoint sets, such that the agents from the first set have preferences over the agents from the second set and vice versa. Typical problems of this sort are: college admissions problem, matching workers with firms, men with women (in a matrimonial agency) and so on. Classical approach to this problem comes from Gale and Shapley (1962). Recently, a far reaching generalization of Gale-Shapley approach was proposed by Alkan and Gale (2003). In our paper we present another generalization of the Gale-Shapley method, in which we use different concept of stability than the one in Alkan and Gale (the so-called order stability). We show that in some cases the order-stable solutions may be treated as more "fair" than in the Alkan and Gale's approach. We formulate conditions under which the generalized Gale-Shapley algorithm leads to order-stable and optimal solutions and prove that these conditions are independent.

Keywords: Two-sided matching, Stable matching, College admissions, Gale-Shapley algorithm, Choice function.

1 College admissions problem

Typical example of two-sided matching problem is the college admissions problem presented in the classical paper of Gale and Shapley (1962). In our paper we will use, without loss of generality, only the language of college admissions, similarly as in Gale and Shapley (1962). Let $S = \{S_1, S_2, ..., S_n\}$ be a set of students, which apply to colleges from the set $C = \{C_1, C_2, ..., C_m\}$. Every student S_i has preferences in the set of colleges, which are represented by a strict linear order $P(S_i)$ (ordering of all colleges according to $P(S_i)$ will be also called a preference list of S_i). Every college C_j has its preference ordering in the set S denoted by $P(C_j)$ ($P(C_j)$ is a strict linear order). We assume also that every college C_j has a quota number q_j (representing maximal number of students which can be admitted to C_j). We want to assign students to colleges in such a way that the following stability condition is satisfied (we assume that every student is assigned to no more than one college and to every college is assigned at least one student):

(S) There are no students S_i, S_l and colleges C_j, C_p such that: (1) S_i is assigned to C_j, (2) S_l is assigned to C_p, (3) C_p is better than C_j for S_i, (4) S_i is better than S_l for C_p.

Gale and Shapley (1962) constructed an algorithm which always leads to stable solution. They proved also that the obtained solution is "optimal" in the sense that it is better than any other stable solution (i. e. is not worse than any other stable solution for every student and better for at least one student).

2 Choice functions

In some cases colleges (schools, universities) do not use strict linear orders and "hard" quotas to choose candidates. There may be ties between candidates and the quotas may be "soft". For example, if the initial quota is 100, but in the set of applicants there are 99 of them which are ranked best and 3 others which have equal number of scores, then the school will often "expand" the initial quota and admit 102 students instead of 100.

Hence a general model of admission process should include some kind of "choice functions" instead of quotas and preference rankings of candidates. A choice function (admission function) for a college C_j is a mapping

$$A_j : \Pi(S) \to \Pi(S)$$

such that $A_j(U) \subset U$ for every $U \subset S$ ($\Pi(S)$ denotes the set of all subsets of U). We interpret $A_j(U)$ as the set of students admitted by a college C_j, if U is the set of all students which applied to C_j. The function A_j may represent any rule which can be used by C_j in the admission process. If the preferences of C_j are represented by a choice function A_j, then the generalized Gale-Shapley algorithm (GGS algorithm for short) can be described in the following way:

1. Assign every student to college which is in the first place on his/her preference list.
2. For every college find the set $A_j(T_j)$, where T_j is the set of students which are assigned to C_j.
3. For every student in the set $T_j \setminus A_j(T_j)$ delete C_j from his/her preference list. Do this for all $j = 1,2,\ldots,m$.
4. Repeat the procedure with the new preferences.

3 AG-stability

Alkan and Gale (2003) constructed a general matching model for firms and workers and defined stable matchings within their model. It can be proved that the stability condition of Alkan and Gale (AG-stability for short) translated into our model (with general choice functions) is equivalent to the condition:

(AGS): There are no students S_i, S_l and colleges C_j, C_p such that: (1) S_i is assigned to C_j, (2) S_l is assigned to C_p, (3) C_p is better than C_j for S_i, (4) there is a set of students U such that $\mu^{-1}(C_p) \cup \{S_i\} \subset U$ and $S_i \in A_p(U)$.

In (4) $\mu^{-1}(C_p)$ denotes the set of all students assigned to college C_p.

The condition (4) means that if we add S_i (and perhaps some other students) to the set of students assigned to C_p, then S_i will be chosen by C_p.

We will now present an example which shows that in some cases the AG-stability condition may not agree with our intuition.

Example 1. Let $S = \{S_1, S_2, S_3\}$ and $C = \{C_1, C_2\}$. Assume that for each of the colleges the candidate S_1 is the best and candidates S_2 and S_3 are indifferent. Both colleges can admit no more than two candidates and all the candidates have the same preferences, namely C_1 is better than C_2. The admission function for both the colleges is the following: $A(U) = U$ if card $(U) \le 2$, $A\{S_1, S_2, S_3\} = \{S_1\}$. When we use the GGS algorithm in this situation, we obtain the solution:

$$S_1 \text{ in } C_1 \text{ and } \{S_2, S_3\} \text{ in } C_2, \tag{3.1}$$

It is easy to see that this solution is not AG-stable. The only AG-stable solutions here are

$$\{S_1, S_2\} \text{ in } C_1 \text{ and } S_3 \text{ in } C_2, \tag{3.2}$$

$$\{S_1, S_3\} \text{ in } C_1 \text{ and } S_2 \text{ in } C_3, \tag{3.3}$$

Observe that AG-stable solutions (3.2) and (3.3) are in some sense "unfair". For example in (3.2) the student S_3 is not admitted by C_1 although he/she is indifferent to S_2, which is admitted by C_1.

4 Order stability

Let A be an admission function in the set of students S. We introduce now some "preference" relation in S associated with A. Let $S_l, S_i \in S$. We say that S_l *is better than* S_i with respect to A if there is a set $U \subset S$ such that

$$S_l, S_i \in U, \quad S_l \in A(U), \quad S_i \notin A(U). \tag{4.1}$$

In other words, S_l is better than S_i (we write $S_l >_A S_i$) if, in some set U, S_l is admitted in U and S_i is not admitted in U. We will use also the symbol $S_l >_p S_i$ if $A = A_p$ for some college C_p.

Definition 1. We say that assignment of students to colleges is *order stable* if there are no students S_i, S_l and colleges C_j, C_p such that: (1) S_i is assigned to C_j, (2) S_l is assigned to C_p, (3) C_p is better than C_j for S_i, (4) it is not true that $S_l >_p S_i$.

It can be proved that the assignment (3.1) from the preceding section (i. e. $S_1 \to C_1$, $S_2 \to C_2$, $S_3 \to C_2$) is order stable. It can be also shown that the assignments (3.2) and (3.3) are order unstable. The above example shows that the concepts of AG-stability and order stability are essentially different concepts.

Now we can formulate conditions under which the *GGS* algorithm leads to order stable and optimal (optimal is defined here with respect to order stability) solutions.

Let A: $\Pi(S) \to \Pi(S)$ be the admission function for some college. We define "rejection function" as

$$R(U) = S \setminus A(U).$$

Definition 2. A rejection function R is *additive* if for any $U, V \subset S$ we have

$$A(U) \subset V \quad \Rightarrow \quad R(U \cup V) = R(U) \cup R(V).$$

Definition 3. A rejection function is *monotonic* if for any $U,V \subset S$ we have

$$U \subset V \quad \Rightarrow \quad R(U) \subset R(V).$$

Definition 4. An admission function A is *independent* if for any $U,V \subset S$ we have

$$A(U) \cap V = \emptyset \quad \Rightarrow A(U \setminus V) = A(U).$$

Definition 5. An admission function A is *asymmetric* if the relation $>_A$ defined by (4.1) is asymmetric.

It is easy to see that additivity implies monotonicity. We can also prove that monotonicity is equivalent to Alkan and Gale's persistency and independency is equivalent to consistency (see Alkan and Gale (2003)).

Our main results are the following:

Theorem 1. *The conditions of additivity, independency and asymmetry are independent i. e. no one of them is implied by the others.*

Theorem 2. *If for every j the function R_j is additive, then the GGS algorithm leads to an order stable assignment and if all R_j are monotonic and all A_j are independent and asymmetric then the obtained assignment is optimal (i. e. is better than any other order stable assignment).*

References

Alkan A, Gale D (2003) Stable schedule matching under revealed preference. J Econ Theory 112: 289-306

Gale D, Shapley L (1962) College admissions and the stability of marriage. Amer Math Monthly 69: 9-15

Data Mining for Big Data Macroeconomic Forecasting: A Complementary Approach to Factor Models

Bernd Brandl, Christian Keber, Matthias G. Schuster

Faculty of Business, Economics, and Statistics, University of Vienna, Brünner Straße 72, A-1210 Vienna, bernd.brandl@univie.ac.at, christian.keber@univie.ac.at, matthias.schuster@univie.ac.at

1. Introduction

The fact that data mining becomes increasingly interesting for applied macroeconomic problems can be explained by the ever increasing size of databases together with the availability of computing power and algorithms to analyze them. There is no doubt that the last few decades have witnessed the phenomenon that more and more data is available for the economics profession. In economic literature this phenomenon is frequently labeled as the 'big data phenomenon', see, e.g. (Diebold 2003). Most methods in econometrics focusing on data analysis issues have problems in handling large data sets. Even though macroeconomic literature is aware of the big data problem, data mining is a neglected area of research and frequently misunderstood since data mining is often confused with data snooping. For the presented data mining forecasting problem we apply a genetic algorithm (GA). GAs are traditionally categorized as data mining methods and are optimization and search techniques using nature-based concepts such as genetic combination, mutation, and natural selection. In our contribution we show that the use of a GA can produce effective forecast models. To illustrate this effectiveness we focus our results on four different variables to show the flexibility of the GA as a tool to solve some parts of the big data problem. Therefore, section 2 briefly focuses on traditional techniques and frequently applied methods in econometrics for analyzing big data sets. Section 3 shows empirical forecasting results for German macroeconomic variables by applying our GA. Finally, we summarize our results and give conclusions in section 4.

2. Econometric forecasting with big data sets

Most quantitative forecasters have to deal with big data sets and therefore apply factor analysis or build factor models. Since the so-called dynamic approach, based on frequency domain analysis, was proposed by (Forni and Reichlin 1998) there have been many attempts to apply this approach to forecast macroeconomic variables, as evident from the literature. See for example (Stock and Watson 1998) and (Artis et al. 2002). In general, factor models have the advantage that they are able to summarize the information that is available in a big data set by a small number of factors. These factors are weighted linear combinations of all variables in the data set. Factor models are based on the idea that the variance of time series can be described by the sum of two mutually orthogonal components. Popular estimation procedures for the components are, for example, the principal component method, state space models and cointegration frameworks. One disadvantage of the estimated components and factors is that they are artificial: the results are usually very hard to interpret, and can hardly be compared with real time series. However, depending on the quality and length (as well as cross-sections) of data in macroeconomic applications usually 2 to 4 factors are extracted from about 100 to 500 variables. These factors are ones that claim to describe aggregate dynamics. See for example (Forni and Lippi 1997) and (Stock and Watson 2002). As recent literature shows, factor analysis for macroeconomic forecasting applications reveals some more problems. In this respect one problem is that more data results in "unexpected" outcomes. For example, (Boivin and Ng 2003) have shown that factors extracted from as few as 40 series often yield satisfactory or even better results than using more series. This might sound implausible as statistic theory teaches us that more data always improves statistical efficiency. However, such results exemplify the problems of using factor analyses for big data sets as little is known about the impact of size and composition of data on estimating factors. Another problem stressed by (Diebold 2003) is that factor analysis has difficulties in handling dynamically changing databases and non-linearity. Therefore, some authors applied artificial neural networks for forecasting in big data sets. The idea behind factor models and artificial neural networks is the same. The only difference is that artificial neural networks focus on non-linear combinations and, thus, can lead to better forecasts. For attempts in forecasting macroeconomic time series using artificial neural networks, see for example (Heravi et al 2004). However, the problem of estimating non-linear combinations for all variables in the data set is also valid for the artificial neural network approach.

Because of the disadvantage of both the factor and the artificial neural network approach we suggest applying a GA for model selection for big data problems in macroeconomic forecasting. Although we use our GA in combination with linear regression models and, thus, only exploit linear dependencies for forecasting purposes, our approach can easily be adopted to use with any other, i.e., nonlinear, model as well. While we suppose that in big data sets not all the information is necessary to make successful forecasts, no assumptions have to be made about how many time series or independent variables in regression models are necessary to explain future movements. We employ a GA to identify the set of those variables which have - in combination with other selected variables - a high explanatory power. We therefore try to overcome the problem of factor models and artificial neural networks that many (non-)linear combinations have to be estimated. The applied GA is new in economic literature as individuals represent the explanatory variables in an OLS regression and our fitness is defined by a combination of an in- and an out-of-sample measure. The GA is applied to find an optimal forecasting model by minimizing both the one step ahead out-of-sample forecasting error and the in sample goodness of fit measured by the Mean Absolute Error (*MAE*). However, as our focus is on finding well-performing forecast models, we emphasized the out-of-sample measure by a weight of 0.7 (and the in-sample measure by 0.3).

3. Empirical forecasting results

Our goal is to show the effectiveness of applying a GA to find forecast models for macroeconomic variables based on a big data set. We exemplify the success of our approach by forecasting different German variables, such as industrial production (IP), long-term government bond yield (BO), unemployment rate (U%) and inflation (IN); all on a monthly frequency. Our data set consists of 159 variables (including a lag structure of 3 periods) of German key economic indicators including leading indicators, series of main aggregates such as money supply, interest rates, production and more. Because of statistical reasons, series have been subject to transformation such as taking logarithms and/or determining growth rates when necessary. We also paid attention to publication lags so that our results meet real demands. We chose an out-of-sample evaluation period of 24 observations (September 2000 to August 2002) and an in-sample of up to 95 observations (February 1993 to August 2000) since the GA was allowed to adapt the series length automatically within a range of 20 and 95 observations for the estimation of coefficients. This has the advantage that in addition to the optimization of the set of variables, the length of the time series is optimized simultaneously. Table 1 summarizes the out-of-sample forecasting results.

Table 1. Out-of-sample forecasting and in-sample results

Variable	Length	F-Test sig.	R^2	Adj. R^2	DW	MAE
IP	40	0.0009	0.66	0.57	2.2588	0.0011
BO	63	0.0000	0.93	0.92	1.9960	0.0960
U%	34	0.0000	0.93	0.91	1.8389	0.1672
IN	20	0.0003	0.92	0.87	2.0600	0.1085

As can be seen from Table 1, the GA provided very small forecast errors for all variables. As different series are completely different in their predictability and forecastability, results differ for each series. The predictability of all forecast equations (dependent variable at time *t+1* and independent variables at time *t* to *t-3*) is high and expressed by the R^2 and by the adjusted R^2. Furthermore, all models have a high significance and no autocorrelation as can be seen from the *F*-test and the Durbin Watson test (DW), respectively. It is also interesting to see that different variables have different demands for the length of series used (see column *Length*) to estimate coefficients. For BO a number of 63 observations has been identified as optimal (explainable by the long maturity) whereas the IN forecast resulted in 20 observations which corresponds to the lower bound on observations in the GA. More interesting is the high forecastabiltiy expressed by the MAE. Fig. 1 shows the improvements of fitness during the GA optimization process.

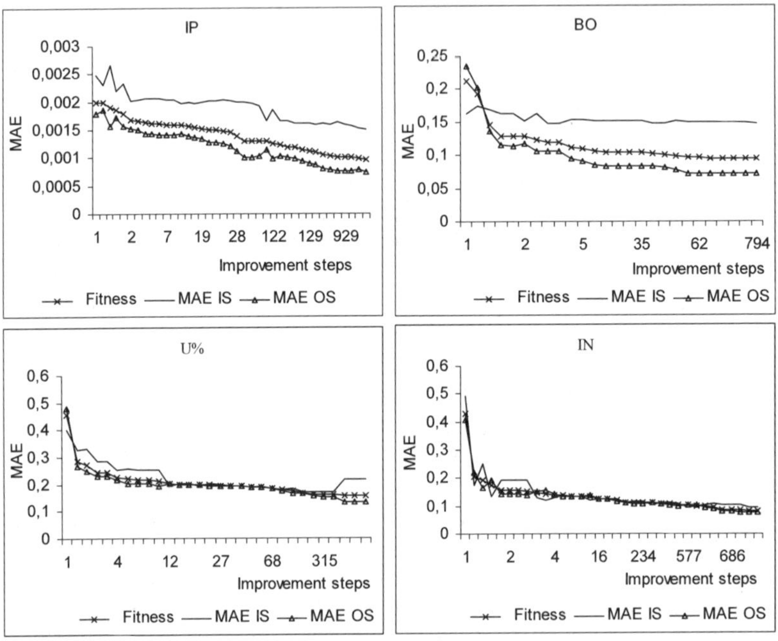

Fig. 1 Improvements of the fitness, in- and out-of-sample MAE during the optimization process

As can be seen from Fig. 1 the fitness and, thus, the quality of the forecasts improved steadily over time. Each graph shows the development of the MAE for the in-sample and for the out-of-sample as well as the weighted combination of both (the fitness values) for one of the four variables. Not surprisingly, the in-sample error is usually higher than the out-of-sample error since we gave more weight to the out-of-sample measure in the fitness function. Nevertheless, both measures are very close to each other indicating well-balanced forecast models. It is also interesting to see that in the first generations substantial improvements are achieved whereas in later generations only slight ones are found. This is highly compatible with GA-theory and indicates that little computational capacity is necessary to gain high forecasting accuracy. As regards the explanatory variables in the optimized forecast models, Table 2 presents a summary.

Table 2. Description and average significance of the forecast models

Dep. Var.	Const.	Independent Variables							
IP_{t+1}	C	XT_t	UE_t	LC_t	CP_t	UE_{t-1}	$I3_{t-2}$	NO_{t-2}	IM_{t-2}
p-value	0.031	0.002	0.028	0.000	0.001	0.117	0.042	0.134	0.030
BO_{t+1}	C	SH_t	BO_{t-1}	VB_{t-1}	CP_{t-1}	BO_{t-2}	CP_{t-2}	VB_{t-3}	HO_{t-3}
p-value	0.071	0.126	0.000	0.467	0.125	0.013	0.348	0.347	0.113
$U\%_{t+1}$	C	$I3_t$	SH_t	$U\%_{t-1}$	CU_{t-1}	IN_{t-1}	IP_{t-3}	CP_{t-3}	$C\$_{t-3}$
p-value	0.009	0.360	0.050	0.000	0.298	0.248	0.572	0.001	0.000
IN_{t+1}	C	IN_t	BK_{t-2}	CP_{t-2}	XU_{t-3}	XT_{t-3}	PO_{t-3}	CP_{t-3}	$C\$_{t-3}$
p-value	0.551	0.000	0.277	0.115	0.432	0.086	0.093	0.234	0.137

As can be seen from Table 2, the GA search process resulted in specifications including 8 independent variables for all four models. For a description of the variables see Table 3 in the appendix. Table 2 shows the independent variables with the corresponding average *p*-values. Average *p*-values are shown as for every step in the forecasting process coefficients of the variables are estimated. It is also interesting to see that the GA made use of a dynamic structure which is expressed by including various lagged variables. However, all forecast models are not artificial in the sense that the specifications are hard to interpret. One strong criticism of factor models and artificial neural networks is the fact that they are hard to interpret and can hardly be compared with real time series. Our results have the advantage of high forecasting performance and, additionally, of being clear in their structure and interpretability.

4. Conclusions

Against the background that methods for efficient use of big data sets become increasingly important in applied macroeconomic forecasting literature we presented a forecast model selection approach based on a GA which tries to overcome problems of alternative quantitative methods, e.g., factor analysis and artificial neural networks. The need for new methods is caused by using big data sets for which the use of GAs (as a typical data mining method) seems to be appropriate. Starting from a big data set with typical macroeconomic variables such as German leading indicators and key indicators our goal was to make forecasts for the German industrial production, a long maturity bond, inflation and unemployment. We employed a GA to optimize forecast models. Our results meet all forecasting requirements and stress the advantages of our approach as opposed to alternative methods.

References

Artis JM, Banerjee A, Marcellino M (2002) Factor forecasts for the UK. CEPR paper 3119

Boivin J, Ng S (2003) Are more data always better for factor analysis?. NBER Working Paper 9829

Diebold FX (2000) 'Big Data' Dynamic Factor Models for Macroeconomic Measurement and Forecasting. Mimeo, University of Pennsylvania, Department of Economics

Forni M, Lippi M (1997) Aggregation and the microfoundation of dynamic macroeconomics, Oxford University Press, Oxford

Forni M, Reichlin L (1998) Let's get real: a factor analytical approach to disaggregate business cycle dynamics. Review of Economic Studies 65: 453-473

Heravi S, Osborn DR, Birchenhall CR (2004) Linear versus neural network forecasts for European industrial production series International J Forecasting 20/3: 435-446

Stock JH, Watson MW (2002) Macroeconomic forecasting using diffusion indexes. J Business and Economic Statistics 20:2: 147-162

Appendix

Table 3. Abbreviations and explanations of variables in the optimized forecast models

Variable	Abbreviation	Transformation
Construction orders received - residential buildings	HO	Δ Ln
Consumer Price Index	CP	Δ Ln
Consumer Price Index in US dollar	C$	Δ Ln
current account balance	CU	Δ
DAX share price index	SH	Δ
FIBOR 3 month	I3	-
German Mark effective exchange rate index	XT	Δ
German Mark exchange rate	XU	Δ
Import price index	IM	Δ Ln
Industrial production including construction	IP	Δ Ln
Inflation rate	IN	Ln
Lending enterprises and individuals - short term	BK	Δ Ln
Long term government bond yield (9-10 years maturity)	BO	-
Manufacturing orders	NO	Δ Ln
PPI - Industrial products	PO	Δ Ln
Registered Unemployed	UE	Δ Ln
Unemployment rate - dependent labor	U%	-
Visible trade balance	VB	Δ
Wages and salaries per unit of output (producing sector)	LC	Δ

Dominance and Equilibria in the Path Player Game

Anita Schöbel and Silvia Schwarze

University of Göttingen schwarze@math.uni-goettingen.de

Summary. This paper investigates the relation between Nash equilibria and non-dominated solutions in a special class of games, namely *path player games*. Nash equilibria are situations in a game where none of the players is able to obtain a better outcome by himself. On the other hand, a situation is *non-dominated* if there does not exist a situation which is really better for one of the players, and at least the same for all others. We provide two classes of path player games in which each non-dominated situation is a Nash equilibrium, and one class in which also the reverse is true.

1 Introduction

Path player games are non-cooperative network games, introduced in [7]. In such a game, each player owns one path in the network. The player receives a benefit depending on the total flow on his path if the sum of all flows meets a given capacity bound. Path player games model situations in which several providers of a commodity are sharing a network. Applications can be found in public transport, telecommunication or information networks.
The path player game is related to *routing games*, which analyze the point of view of the flow units, see e.g. [8, 2, 1]. Other types of games coping with resources in networks are *bandwidth allocation games* and *path auctions* (see e.g. [6, 3] for recent results). Our paper is organized as follows. Section 2 introduces the path player game and Section 3 the concept of dominance of multicriteria optimization. Section 4 provides our results on the relation between non-dominated solutions and Nash equilibria.

2 The path player game

We consider a given network $G = (V, E)$ with vertices $v \in V$ and edges $e \in E$. A path P in G is given by a sequence of edges $e \in E$: $P = (e_1, ..., e_k)$. We denote with $\mathcal{P}$ the set of all paths P in G from the single source s to the single sink t. The paths $P \in \mathcal{P}$ in the network G represent the players[1] of the game. Each player proposes an amount of flow f_P that he wants to be routed along his path. The number of strategies is infinite as a player is allowed to choose any nonnegative real number. Let $f \in \mathbb{R}_+^{|\mathcal{P}|}$ contain the proposed flows f_P for all players $P \in \mathcal{P}$. Under the assumption of sufficient demand, each player implements his proposed flow and receives a benefit which depends on the flow routed along the edges of his path and on the total flow in the network. To describe the benefit, we first determine the flow on an edge $e \in E$ as the sum of the flows on paths that contain e, i.e., $f_e = \sum_{P: e \in P} f_P$. Each edge e is associated with a cost function $c_e(\cdot)$, that depends on the flow $x = f_e$ on e. This cost function represents the income of the edge owners and we assume these functions to be continuous and nonnegative for nonnegative loads, i.e. $c_e(x) \geq 0$ for $x \geq 0$. The cost of a path P (i.e. the income of the path owner) is then given by the sum of costs of the edges belonging to P, i.e., $c_P(f) = \sum_{e \in P} c_e(f_e)$.

To avoid infinite flows in the case of increasing cost functions c_e we bound the sum of all flows by a *flow rate* r, that can be interpreted as a network capacity. We will call a flow f *feasible* if $\sum_{P \in \mathcal{P}} f_P \leq r$. If the flow created by the decisions of the players exceeds the flow rate and hence is not feasible, the benefit of each player will be $-M$, with M being a large number. Summarizing, the benefit function of the path player game is the following.

Definition 1. *The* benefit function *of player $P \in \mathcal{P}$ in a path player game for $f \geq \mathbf{0}_{|\mathcal{P}|}$ is given as:*

$$b_P(f) = \begin{cases} c_P(f) & if \ \sum_{P \in \mathcal{P}} f_P \leq r \\ -M & if \ \sum_{P \in \mathcal{P}} f_P > r \end{cases} .$$

To determine Nash equilibria of this game we investigate the benefit of a player, assuming that all other players have fixed their strategies. We define $f_{-P} \in \mathbb{R}_+^{|\mathcal{P}|-1}$ as the vector containing the proposed flows $f_{P'}$ of all players $P' \in \mathcal{P} \setminus \{P\}$. Then, a flow f^* is a *Nash equilibrium* if for all players $P \in \mathcal{P}$ and for all $f_P \geq 0$ we have that $b_P(f^*_{-P}, f^*_P) \geq b_P(f^*_{-P}, f_P)$.

It will be convenient to use the *one-dimensional benefit* for a player P with respect to the fixed flow f_{-P}, which is given by $\tilde{b}_P(f_P) = b_P(f_{-P}, f_P)$. The *one-dimensional cost function* of a player $P \in \mathcal{P}$ for a fixed flow f_{-P} is given by $\tilde{c}_P(f_P) = c_P(f_{-P}, f_P)$. A player P will only receive c_P as benefit if f is feasible. For a fixed flow f_{-P}, the largest feasible flow that player P can propose is called the *decision limit* d_P of player P. It is given by $d_P = r - \sum_{P' \in \mathcal{P} \setminus P} f_{P'}$.

The one-dimensional benefit function can thus be described as:

$$\tilde{b}_P(f_P) = \begin{cases} \tilde{c}_P(f_P) & if \ \ f_P \leq d_P \\ -M & if \ \ f_P > d_P \end{cases} .$$

[1] In the course of this paper we will denote both, the path and the corresponding player with P, as both notations are handled equivalently.

Given the competitors flow f_{-P}, a player P will choose a best reply f_P contained in his *best reaction set* $f_P^{max} = \{f_P \geq 0 : f_P \text{ maximizes } \tilde{b}_P(f_P)\}$. For continuous cost functions c_e the best reaction sets are nonempty, see [7]. In the same paper it is shown that for path player games feasible Nash equilibria exist in pure strategies, and that a flow f^* is a Nash equilibrium if and only if $f_P^* \in f_P^{max}$ for all $P \in \mathcal{P}$. We will also need the following characterization of Nash equilibria in the case of strictly monotone cost functions.

Lemma 1 ([7]). *Consider a path player game with strictly increasing cost functions c_e on all edges $e \in E$. Then a flow f is a feasible Nash equilibrium if and only if $\sum_{P \in \mathcal{P}} f_P = r$.*

3 Dominance in the path player game

Nash equilibria in path player games may be disadvantageous situations for all players, even an infeasible flow (with benefit $-M$ for all players) may be a Nash equilibrium. This fact rises the question of dominance among Nash equilibria and general flows. When do we, considering the interests of all players, prefer one flow more than another? We surely prefer a flow f rather than $\hat{f}$ if the benefit for f is for all players not lower and at least for one player higher than for $\hat{f}$. This observation is summarized in the definition of dominance among flows, where we will use the following convention of comparing vectors: Let $u = (u_1, \ldots, u_k)$ and $v = (v_1, \ldots, v_k)$ be k-dimensional vectors. We write $u \gneqq v$ if $u_i \geq v_i \; \forall \, i = 1, \ldots, k$ and there exists one index j such that $u_j > v_j$. Furthermore, let $b(f) = (b_P(f))_{P \in \mathcal{P}}$ denote the vector of benefits of the players.

Definition 2. *A feasible flow $\hat{f}$ is called* dominated *if there exists a* dominating flow, *i.e. a flow f such that $b(f) \gneqq b(\hat{f})$. Otherwise, $\hat{f}$ is called* non-dominated.

Note that Harsanyi and Selten propose in [5] a stronger definition of dominance which they call *payoff-dominance.* In the path player game, an infeasible flow will never dominate any other flow and on the other hand is dominated by each feasible flow. Hence we will only consider feasible flows. The following example illustrates dominance for Nash equilibria.

Example 1. Consider a network consisting of two edges that link the nodes s and t as illustrated in Figure 1. Let the flow rate $r = 1$ and the cost functions be $c_1(x) = x$ and $c_2(x) = 1$. The flow $f^* = (0.5, 0.5)$ with $b(f^*) = (0.5, 1)$ is a feasible Nash equilibrium as $f_1^{max} = \{0.5\}$ and $f_2^{max} = [0, 0.5]$. This flow is dominated e.g. by the flows $f^{**} = (0.75, 0.25)$ and $f = (0.7, 0.25)$ as $b(f^{**}) = (0.75, 1) \gneqq b(f^*)$ and $b(f) = (0.7, 1) \gneqq b(f^*)$ holds. Note, that f^{**} is a Nash equilibrium itself, while f is not. Furthermore, note that the flow $\bar{f} = (1, 0)$ with $b(\bar{f}) = (1, 1)$ is a Nash equilibrium which is non-dominated.

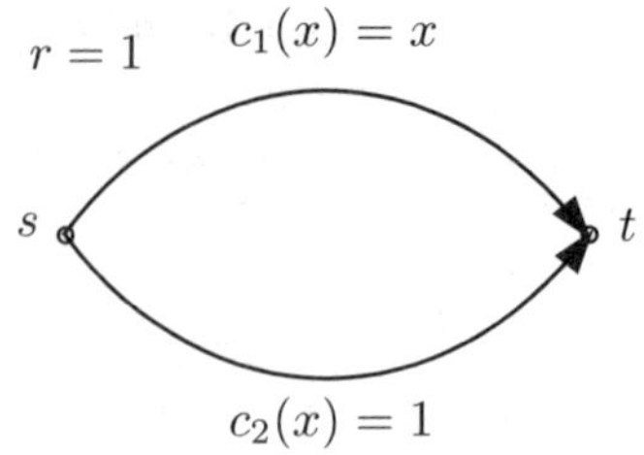

Fig. 1. Game network for Example 1

This example illustrates that in path player games, there may exist dominated Nash equilibria which can be improved to situations that are more preferable for the overall system. However, these situations need not be Nash equilibria and may hence be unstable. A famous example for a situation where the (unique) Nash equilibrium represents a disadvantageous situation for the players is *Prisoner's Dilemma* (see e.g. [4]). However, the only situation, dominating the Nash equilibrium is no Nash equilibrium itself and hence unstable. With these observations the following question appears: What is the relation between the Nash equilibria and the non-dominated flows of a game? To analyze this, let us denote for a game Γ, the set of Nash equilibria with $NE(\Gamma)$ and the set of non-dominated flows with $ND(\Gamma)$.

4 Non-dominated flows and Nash flows

We provide two different classes of path player games satisfying that each non-dominated flow is a Nash equilibrium. In the first case we require strictly increasing cost functions. In the second case we need no restriction on the cost function but a path-disjoint network.

Theorem 1. *Let Γ be a path player game with strictly increasing cost functions c_e for all $e \in E$. Then $ND(\Gamma) \subseteq NE(\Gamma)$.*

Proof. Consider a non-dominated flow f^*. Assume, f^* is no Nash equilibrium, i.e. $\sum_{P \in \mathcal{P}} f_P^* < r$ (see Lemma 1). Take an arbitrary path $\hat{P}$ and set $f_{\hat{P}} = f_{\hat{P}}^* + r - \sum_{P \in \mathcal{P}} f_P^* > f_{\hat{P}}^*$. Due to the strictly increasing cost functions it follows that $b_{\hat{P}}(f) > b_{\hat{P}}(f^*)$ and $b_P(f) \geq b_P(f^*)$ $\forall$ $P \in \mathcal{P} \setminus \{\hat{P}\}$. Thus, $b_P(f) \gneqq b_P(f^*)$ which contradicts $f^* \in ND(\Gamma)$. Hence, $f^* \in NE(\Gamma)$.

The following example illustrates that the reverse of Theorem 1 does not hold in general.

Example 2. Consider the game illustrated in Figure 2. with two paths going from s to t. Path P_1 is sharing all its edges with P_2, while P_2 owns some edges exclusively. Consider a flow f^* being Nash equilibrium, i.e. $\sum_{P \in \mathcal{P}} f_P = r$. Set

$f_{P_1} = f^*_{P_1} - \delta$ and $f_{P_2} = f^*_{P_2} + \delta$. Due to the strictly increasing cost functions, the benefit of P_2 will increase, while the benefit of P_1 is unchanged, i.e. f is dominating f^*: $b(f) \gneqq b(f^*)$.

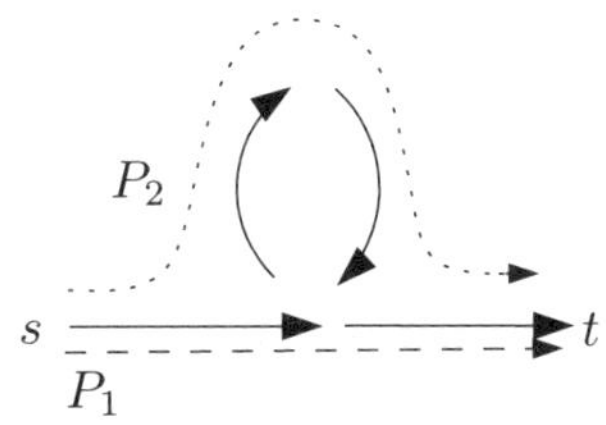

Fig. 2. Game network for Example 2

In the second case we assume a *path-disjoint network*, where each edge belongs to exactly one path. In such networks it holds that the cost of a path only depends on the flow on that path, such that $c_P(f) = c_P(f_P)$.

Lemma 2. *Consider a game on a path-disjoint network with continuous cost functions c_e. For each feasible flow f that is no Nash equilibrium there is a feasible Nash equilibrium f^* dominating f.*

Proof. Consider a feasible flow f that is no Nash equilibrium, i.e. $\exists\ P : f_P \notin f_P^{max}$. We construct the required Nash equilibrium f^* by repeating the following iteration until $f_P \in f_P^{max}\ \forall\ P \in \mathcal{P}$. Note that $f_P^{max} \neq \emptyset$ for continuous cost functions.
Iteration: Choose some $\bar{P} : f_{\bar{P}} \notin f_{\bar{P}}^{max}$. Set f' such that $f'_P = f_P\ \forall\ P \neq \bar{P}$ and $f'_{\bar{P}} \in f_{\bar{P}}^{max}$. Set $f := f'$ and repeat.
Due to the path-disjoint network $c_P(f) = c_P(f_P)$, hence, the one-dimensional cost function $\tilde{c}_P(f)$ will not change in any iteration for any of the non-active players $P \neq \bar{P}$. However, the set of feasible flows $[0, d_p]$ for a player P might change. But since $\bar{P}$ will not choose an infeasible flow, we still have $d_P \geq f_P$ for all P. Hence, $b_P(f') = b_P(f)$ for all $P \neq \bar{P}$.
As $b_{\bar{P}}(f') > b_{\bar{P}}(f)$, it follows that f' dominates f and consequently dominates the flows of all previous iterations.
Using again that the one-dimensional benefit functions for *feasible flows* do not change, it can be shown that $f'_{\bar{P}} < f_{\bar{P}}$ can only occur when $\bar{P}$ is chosen for the first time. Moreover, if there are $|\mathcal{P}|$ subsequent iterations in which no f_P has decreased, the process has found a Nash equilibrium. Together, the above procedure is finite and ends with the required Nash equilibrium f^*.

The following result follows immediately.

Theorem 2. *Let Γ be a path player game on a path-disjoint network with continuous cost functions. Then $ND(\Gamma) \subseteq NE(\Gamma)$.*

Example 1 shows that the reverse of Theorem 2 is not true in general.

We conclude this section by providing a class of path player games in which the set of Nash equilibria equals the set of non-dominated flows. In this nice situation we can be sure, that each equilibrium is "advantageous". On the other hand each non-dominated situation is stable.

Lemma 3. *Let Γ be a path player games on a path-disjoint network with strictly increasing cost functions. Then $ND(\Gamma) = NE(\Gamma)$.*

Proof. $ND(\Gamma) \subseteq NE(\Gamma)$ is true by Theorem 1. For the reverse inclusion, consider a feasible Nash equilibrium f^*. According to Lemma 1 $\sum_{P \in \mathcal{P}} f_P^* = r$. Assume f^* is dominated, i.e. $\exists f : b(f) \gneqq b(f^*)$. Then there exists P such that $b_P(f) > b_P(f^*)$, hence $f_P > f_P^*$ as c_e are strictly increasing and the paths are disjoint. Moreover, there exists $\hat{P} : f_{\hat{P}} < f_{\hat{P}}^*$, otherwise feasibility would be violated. Consequently, $b_{\hat{P}}(f) < b_{\hat{P}}(f^*)$, a contradiction.

5 Conclusion

It is possible to construct path player games Γ in which $ND(\Gamma) \supsetneqq NE(\Gamma)$ holds. Also, path player games do exist in which neither $ND(\Gamma) \subseteq NE(\Gamma)$ nor $ND(\Gamma) \supseteq NE(\Gamma)$ holds. Even situations as in *Prisoner's Dilemma* with $ND(\Gamma) \cap NE(\Gamma) = \emptyset$ may occur. Examples will be presented in [9]. Note that in [7] path player games are defined more general, including a security payment which is given to the players if they route very few flow. In our future work we will extend our results to this case. Moreover, the relation between the sets $ND(\Gamma)$ and $NE(\Gamma)$ is under research for other types of games.

References

1. J.R. Correa, A.S. Schulz, and N.E. Stier-Moses. Selfish routing in capacitated networks. *Mathematics of Operations Research*, 29(4):961–976, 2004.
2. A. Czumaj and B. Voecking. Tight bounds for worst case equilibria. In *Proc. 13th ACM-SIAM Symp. on Discrete Alg.*, pages 413–420. ACM Press, 2002.
3. E. Elkind, A. Sahai, and K. Steiglitz. Frugality in path auctions. In *Proc. 15th ACM-SIAM Symp. on Discrete Alg.*, pages 701–709. ACM Press, 2004.
4. D. Fudenberg and J. Tirole. *Game Theory.* MIT Press, 1991.
5. J.C. Harsanyi and R. Selten. *A general theory of equilibrium selection in games.* MIT Press, 1988.
6. R. Johari and J.N. Tsitsiklis. Efficiency loss in a network resource allocation game. *Mathematics of Operations Research*, 29(3):407–435, 2004.
7. J. Puerto, A. Schöbel, and S. Schwarze. The path player game: Introduction and equilibria. Technical Report 2005-18, NAM, University of Göttingen, 2005.
8. T. Roughgarden and E. Tardos. How bad is selfish routing? *Journal of the ACM*, 49(2):236–259, 2002.
9. S. Schwarze. Phd thesis. To be submitted.

Exact Solution to a Class of Stochastic Resource Extraction Problems

Sum T.S. Cheng[1] and David W.K. Yeung[2]

[1] Department of Finance & Decision Sciences, Hong Kong Baptist University tscheng@phmars.hkbu.edu.hk

[2] Department of Finance & Decision Sciences, Hong Kong Baptist University Centre for Game Theory, St Petersburg State University wkyeung@hkbu.edu.hk

Abstract

In this paper, a class of resource extraction problems involving stochastic dynamics and randomly fluctuating non-autonomous payoffs are developed. An empirically meaningful theory of optimization must therefore incorporate uncertainty in an appropriate manner. The introduction of this stochastic specification lead to a novel approach to solve dynamic problems in terms of properties and solution concepts not explored in the literature before. Exact solution to this stochastically complicated problem is presented. Computer simulations are provided. The analysis could be applied to various practical problems involving complex uncertainties.

1 Introduction

Dynamic analysis of the exploitation of a renewable resource asset was initiated by the work of Plourde (1970) and Clark (1976). Since then time, it has become increasingly important in research into the optimal extraction and pricing of resources. Subsequent work include Clemhout and Wan (1985), Dockner and Kaitala (1989), Plourde and Yeung (1989), Jørgensen and Sorger (1990), Fischer and Mirman (1992), and Kaitala (1993), Dockner et. al. (1989) and Jørgensen and Yeung (1996). A common feature of these works is that the payoffs are assumed to be known with certainty. This paper supplies an analysis of optimal pricing of a renewable resource asset where future payoffs are not known with certainty, and where the evolution of the resource stock over time is stochastic. The introduction of this stochastic specification lead to a novel approach to solve dynamic problems in terms of properties and solution concepts not explored in the literature before. Exact solution to this stochastically complicated problem is presented. The analysis could be applied to various practical problems involving complex uncertainties. Section

2 presents the basic model of a stochastic resource extraction problem. Section 3 characterizes the optimal solution to the problem. Section 4 develops computer algorithm for the derivation of an exact solution. Section 5 provides computer simulated results. Concluding remarks are given in Section 6.

2 The Basic Model

Consider an economy endowed with a single renewable resource, with a resource extractor (firm). Let $u(s)$ denote the rate of resource extraction of the firm at time s. The firm controls its rate of extraction. Let U be the set of admissible extraction rates, and $x(s)$ the size of the resource stock at time s. In particular, we have $U \in R^+$ for $x > 0$, and $U = \{0\}$ for $x = 0$.

The firm's cost of extraction depends on the quantity of resource extracted $u(s)$, the resource stock size $x(s)$, and a parameter c. In particular, this cost can be specified as follows:

$$C = \frac{c}{x(s)^{1/2}} u(s). \tag{1}$$

The market price of the resource at any point of time depends on the total amount extracted and brought to the market. We assume that the price-output relationship at time s is given by the following downward sloping inverse demand curve:

$$P(s) = \theta_{a_0} Q(s)^{-1/2}, \qquad \text{for } s \in [t_0, t_1), \tag{2}$$

$$P(s) = \theta^h Q(s)^{-1/2}, \qquad \text{for } s \in [t_h, t_{h+1}), \tag{3}$$

where $Q(s) = u(s)$ is the amount of resource marketed at time s.

θ^h is a random variable realizable in the time interval $[t_h, t_{h+1})$, and θ^h, for $h = 1, 2, \ldots$, are independent and identical random variables with range $\{\theta_1, \theta_2, \ldots, \theta_m\}$ and corresponding probabilities $\{\lambda_1, \lambda_2, \ldots, \lambda_m\}$. In particular, $\theta_{a_k+1} > \theta_{a_k}$ and $\theta_{a_0} \in \{\theta_1, \theta_2, \ldots, \theta_m\}$ is known by the firm in the time interval $[t_0, t_1)$. Given steady state branching, the lengths of the interval $[t_h, t_{h+1}]$ are the same for $h = 0, 1, 2, \ldots$.

The resource firm is awarded an infinite lease for extraction, beginning at time t_0. The objective of the firm is to maximize expected profits:

$$E_{t_0} \left\{ \int_{t_0}^{t_1} \left[\theta_{a_0} (u(s))^{1/2} - \frac{c}{x(s)^{1/2}} u(s) \right] e^{-r(s-t_0)} ds \right.$$
$$\left. + \sum_{h=1}^{\infty} \int_{t_h}^{t_{h+1}} \left[\theta^h (u(s))^{1/2} - \frac{c}{x(s)^{1/2}} u(s) \right] e^{-r(s-t_0)} ds \right\}, \tag{4}$$

where r is the common discount rate. The resource stock evolves according to:

$$dx(s) = \left[ax(s)^{1/2} - bx(s) - u(s)\right] ds + \sigma x(s)\, dz(s), \qquad x(t_0) = x_0. \quad (5)$$

The above dynamics is also employed in Jørgensen and Yeung (1996).

3 Optimal Solution

Following Theorem 2.1 in Yeung (2001), we obtain

Theorem 1. *A set of controls* $\left\{u_{a_k}^{[k]*}(t) = \varphi_{a_k}^{[k]*}(t,x)\right\}$ *contingent upon the events* θ_{a_k}*, for* $a_k = 1, 2, \ldots, m$*, constitutes an optimal solution for the problem (4)–(5), if there exist suitably smooth function* $W^{(k,a_k)}(t,x) : [0,T] \times R^n \to R$*, for* $a_k = 1, 2, \ldots, m$*, which satisfy the partial equation:*

$$-W_t^{(k,a_k)} - \frac{1}{2}\sigma^2 x^2 W_{xx}^{(k,a_k)} =$$
$$\max_{\phi \in U} \left\{ \left[\theta_{a_k} (\varphi(t,x))^{1/2} - \frac{c}{x^{1/2}} \varphi(t,x)\right] e^{-rt} + W_x^{(k,a_k)} \left[ax^{1/2} - bx - \varphi(t,x)\right] \right\}, \text{ and}$$

$$W^{(k,a_k)}(T,x) = e^{-rT} \sum_{a_{k+1}=1}^{m} \lambda_{a_{k+1}} W^{(k,a_{k+1})}(0,x). \quad (6)$$

The value function $W^{(k,a_k)}(t,x)$ in Theorem 1 yields admits a solution

$$W^{(k,a_k)}(t,x) = e^{-rt}\left[A_{a_k}(t)\, x^{1/2} + B_{a_k}(t)\right], \text{ for } a_k = 1, 2, \ldots, m \quad (7)$$

where $A_{a_k}(t)$ and $B_{a_k}(t)$ satisfy

$$\dot{A}_{a_k}(t) = \left[r + \frac{1}{8}\sigma^2 + \frac{b}{2}\right] A_{a_k}(t) - \frac{1}{4} \frac{(\theta_{a_k})^2}{\left(c + \frac{A_{a_k}(t)}{2}\right)}, \quad (8)$$

$$\dot{B}_{a_k}(t) = rB_{a_k}(t) - \frac{a}{2} A_{a_k}(t), \quad (9)$$

$$A_{a_k}(T) = \sum_{a_{k+1}=1}^{m} A_{a_{k+1}}(0), \quad (10)$$

$$B_{a_k}(T) = \sum_{a_{k+1}=1}^{m} B_{a_{k+1}}(0). \quad (11)$$

The optimal strategy of the firm at any time t in the time interval $[0,T)$, given that $\theta^k = \theta_{a_k}$, can be obtained as:

$$\phi_{a_k}^{[k]*}(t,x) = \frac{(\theta_{a_k})^2 x}{\left[c + A_{a_k}(t)/2\right]^2}, \qquad \text{for } a_k \in \{1, 2, \ldots, m\}. \quad (12)$$

4 Computer Algorithm for the Derivation of Exact Solution

In this section, we developed computer algorithms to solve the stochastic control problem (4)–(5). First the system of ordinary differential equations (8) is solved by numerical method.

The time interval $[t_0, T]$ is divided into M partitions of equal length $\Delta t = \frac{T-t_0}{M}$. In particular, we form the sub-intervals $[t_0, t_1], [t_1, t_2], \ldots, [t_{M-1}, t_M]$.

The general step for Euler Method is

$$y_{a_k}(t_{j-1}) = y_{a_k}(t_j) - \dot{y}_{a_k}(t_j)\,\Delta t \qquad \text{for } j = 1, 2, \ldots, M$$

with $y_{a_k}(t_M) = y_{a_k}(T)$ is given. However, in (10) the initial and final values of $A_{a_k}(t)$ are inter-related. Therefore the problem is neither an initial value problem nor a boundary value problem.

To obtain a solution of $A_{a_k}(t)$ satisfying (10), we adopt an iterative methodology involving

$$\begin{aligned} A_{a_k}^{[i]}(t_{j-1}) &= A_{a_k}^{[i]}(t_j) - \dot{A}_{a_k}^{[i]}(t_j)\,\Delta t \\ &= A_{a_k}^{[i]}(t_j) - \left[\left(r + \frac{1}{8}\sigma^2 + \frac{b}{2}\right) A_{a_k}^{[i]}(t_j) - \frac{1}{4}\frac{(\theta_{a_k})^2}{\left(c + \frac{A_{a_k}^{[i]}(t_j)}{2}\right)}\right]\Delta t, \quad (13) \end{aligned}$$

for $j = 1, 2, \ldots, M$, and

$$A_{a_k}^{[i+1]}(T) = \sum_{a_{k+1}=1}^{m} \lambda_{a_{k+1}} A_{a_{k+1}}^{[i]}(0), \qquad (14)$$

with iteration $i = 1, 2, \ldots, H_A$, and the final iteration H_A will be determined as follows.

First, adopt an initial guess on $A_{a_k}^{[1]}(T)$, corresponding values of $A_{a_k}^{[1]}(0)$ for each $a_k \in \{1, 2, \ldots, m\}$ can be found with the use of (13). Using $A_{a_k}^{[1]}(0)$ for each $a_k \in \{1, 2, \ldots, m\}$ and (14), $A_{a_k}^{[2]}(T)$ can be found.

Using $A_{a_k}^{[2]}(T)$, corresponding values of $A_{a_k}^{[2]}(0)$ for each $a_k \in \{1, 2, \ldots, m\}$ can be found with the use of (13). Using $A_{a_k}^{[2]}(0)$ for each $a_k \in \{1, 2, \ldots, m\}$ and (14), $A_{a_k}^{[3]}(T)$ can be found.

The process is repeated until the following condition is fulfilled:

$$\varepsilon_A^{[H_A-1]} > \varepsilon \text{ and } \varepsilon_A^{[H_A]} < \varepsilon$$

where $\varepsilon_A^{[i+1]} = \left| A_{a_k}^{[i+1]}(T) - A_{a_k}^{[i]}(T) \right|$ and ε is chosen to arbitrarily small. The mechanism works if $\varepsilon_A^{[i]}$ is strictly decreasing as i increases. Then $A_{a_k}^{[H_A]}(t_j)$ is an approximation to the solution of $A_{a_k}(t_j)$.

After solving $A_{a_k}(t_j)$, $B_{a_k}(t_j)$ can be derived in a similar manner as below,

$$B^{[i]}_{a_k}(t_{j-1}) = B^{[i]}_{a_k}(t_j) - \dot{B}^{[i]}_{a_k}(t_j)\,\Delta t$$
$$= B^{[i]}_{a_k}(t_j) - \left[rB^{[i]}_{a_k}(t_j) - \frac{a}{2}A_{a_k}(t_j)\right]\Delta t, \text{ for } j = 1,2,\ldots,M, \quad (15)$$

$$B^{[i+1]}_{a_k}(T) = \sum_{a_{k+1}=1}^{m} \lambda_{a_{k+1}} B^{[i]}_{a_{k+1}}(0), \quad (16)$$

with iteration $i = 1,2,\ldots,H_B$, and the final iteration H_B will be determined as follows. First, adopt an initial guess on $B^{[1]}_{a_k}(T)$, corresponding values of $B^{[1]}_{a_k}(0)$ for each $a_k \in \{1,2,\ldots,m\}$ can be found with the use of (15). Using $B^{[1]}_{a_k}(0)$ for each $a_k \in \{1,2,\ldots,m\}$ and (16), $B^{[2]}_{a_k}(T)$ can be found.

Using $B^{[2]}_{a_k}(T)$, corresponding values of $B^{[2]}_{a_k}(0)$ for each $a_k \in \{1,2,\ldots,m\}$ can be found with the use of (15). Using $B^{[2]}_{a_k}(0)$ for each $a_k \in \{1,2,\ldots,m\}$ and (16), $B^{[3]}_{a_k}(T)$ can be found.

The process is repeated until the following condition is fulfilled:

$$\varepsilon_B^{[H_B-1]} > \varepsilon \text{ and } \varepsilon_B^{[H_B]} < \varepsilon$$

where $\varepsilon_B^{[i+1]} = \left|B^{[i+1]}_{a_k}(T) - B^{[i]}_{a_k}(T)\right|$ and ε is chosen to arbitrarily small. The mechanism works if $\varepsilon_B^{[i]}$ is strictly decreasing as i increases. Then $B^{[H_B]}_{a_k}(t_j)$ is an approximation to the solution of $B_{a_k}(t_j)$.

Substituting $A_{a_k}(t_j)$ into the optimal control (12) yields:

$$\phi^{[k]*}_{a_k}(t_j, x) = \frac{(\theta_{a_k})^2 x}{[c + A_{a_k}(t_j)/2]^2}, \text{ for } a_k \in \{1,2,\ldots,m\}. \quad (17)$$

The resource stock $x(t)$ is then given by

$$x(t_j) = x(t_{j-1}) + \Delta x(t_{j-1})$$
$$= x(t_{j-1}) + \left[ax(t_{j-1})^{1/2} - bx(t_{j-1}) - \phi^{[k]*}_{a_k}(t_{j-1}, x(t_{j-1}))\right]\Delta t$$
$$+\sigma x(t_{j-1})\,\Delta z(t_{j-1})$$
$$= x(t_{j-1}) + \left[ax(t_{j-1})^{1/2} - bx(t_{j-1}) - \frac{(\theta_{a_k})^2}{[c + A_{a_k}(t_{j-1})/2]^2} x_{j-1}\right]\Delta t$$
$$+\sigma x(t_{j-1})\, z\sqrt{\Delta t}, \qquad x(t_0) = x_0, \qquad \text{for } j = 1,2,\ldots,M$$

where $\Delta z = z\sqrt{\Delta t}$ and z follows standard normal distribution (i.e. $z \sim N(0,1)$). Random generator is need for simulating z.

Finally, $\phi^{[k]*}_{a_k}(t_j, x)$ can be obtained as:

$$\phi^{[k]*}_{a_k}(t_j, x_j) = \frac{(\theta_{a_k})^2}{[c + A_{a_k}(t_j)/2]^2} x(t_j), \text{ for } j = 0,1,2,\ldots,M. \quad (18)$$

5 Computer Simulation

This section provides a computer simulation with the following parameters: $m = 10, r = 0.05, \sigma = 0.5, b = 1, a = 1.5, c = 1.2, \theta_1 = 1, \theta_2 = 2, \theta_3 = 2.5, \theta_4 = 3, \theta_5 = 5, \theta_6 = 5.5, \theta_7 = 6.5, \theta_8 = 7, \theta_9 = 7.5, \theta_{10} = 8, \lambda_1 = 0.2, \lambda_2 = 0.05, \lambda_3 = 0.1, \lambda_4 = 0.15, \lambda_5 = 0.1, \lambda_6 = 0.1, \lambda_7 = 0.15, \lambda_8 = 0.05, \lambda_9 = 0.05, \lambda_{10} = 0.05, T = 3, t_0 = 0, M = 3000, \Delta t = 0.001, x_0 = 10$, and the chosen ε is 10^{-7}. The results are depicted graphically in the full paper.

6 Concluding Remarks

In this paper, a class of resource extraction problems involving stochastic dynamics and randomly fluctuating non-autonomous payoffs are developed. A general solution mechanism is characterized and computer algorithms for solving the exact solution are developed. Exact solution to this stochastically complicated problem is presented. A wider spectrum of possibilities of uncertainty – like various types of branching processes – can be adopted.

References

1. Clark CW (1976) Mathematical bioeconomics. John Wiley & Sons, New York
2. Clemhout S, Wan HY. Jr. (1985) Dynamic common-property resources and environmental problems. Journal of Optimization Theory and Applications 46: 471–481
3. Dockner E, Kaitala V (1989) On efficient equilibrium solutions in dynamic games of resource management. Resource and Energy 11: 23—34
4. Dockner EJ, Feichtinger G, Mehlmann A (1989) Noncooperative solutions for a differential game model of fishery. Journal of Economic Dynamics and Control 13: 1–20
5. Fischer R, Mirman L (1992) Strategic dynamic interactions: fish wars. Journal of Economic Dynamic and Control 16: 267-287
6. Jørgensen S, Sorger G (1990) Feedback nash equilibria in a problem of optimal fishery management. Journal of Optimization Theory and Applications: 64, 293–310
7. Jørgensen S, Yeung DWK (1996) stochastic differential game model of a common property fishery. Journal of Optimization Theory and Applications 90: 391–403
8. Kaitala V (1993) Equilibria in a stochastic resource management game under imperfect information. European Journal of Operational Research 71: 439–453
9. Plourde C, Yeung DWK (1989) Harvesting of a transboundary replenishable fish stock: a non cooperative game solution. Marine Resource Economics 6: 57–70
10. Plourde CG (1970) A simple model of replenishable natural resource exploitation. American Economic Review 60: 518–22
11. Yeung DWK (2001) Infinite horizon stochastic differential games with branching payoffs. Journal of Optimization Theory and Applications 111: 445–460

Investment Attraction and Tax Reform: a Stochastic Model

Vadim I. Arkin*, Alexander D. Slastnikov**, Svetlana V. Arkina***

[1] Central Economics and Mathematics Institute, Moscow, Nakhimovskii pr. 47 arkin@cemi.rssi.ru
[2] Central Economics and Mathematics Institute, Moscow, Nakhimovskii pr. 47 slast@cemi.rssi.ru
[3] University Paris I, svetlana.arkina@malix.univ-paris1.fr

Summary. We study a model of the behavior of a potential investor (under risk and uncertainty) who wishes to invest in a project of creating a new enterprise and chooses an investment time (timing problem). This model takes the tax environment exhaustively into account. An optimal rule of investment and its dependence on parameters of tax system are obtained. Investigation is based on solving an optimal stopping problem for two-dimensional geometric Brownian motion. We apply Feinmann-Kac formula and variational inequalities as basic methods for deriving the closed-form formulas for optimal investment time and expected tax revenues from future enterprise into budgets of different levels. Based on those formulas, an analysis of the Russian reform of corporate profit taxation (2002) is undertaken, as well as of the tax cuts in VAT (2004) and Unified Social Tax (UST) Rates (2005).

1 Introduction

The present paper is devoted to the construction of a model of investor behavior in the Russian fiscal environment under risk and uncertainty. It intends to study the influence of the fiscal system on the investment activity in the real sector of the Russian economy. We will take special attention to the recent evolutions in the tax system. Indeed, the Russian corporate tax system has been reformed recently, firstly resulting in the cut of the corporate income tax rate (from 35 to 24%) and the suppression of granted fiscal advantages for newly created enterprises (tax holidays and accelerated depreciation allowances). Regional authorities are still able to control and modulate a share of the tax rate (14.5 out of 24%) for specific categories of taxpayers within strict limits (no more than 4% of cut). Additionally, Russia experienced important cuts

* Central Economics and Mathematics Institute RAS, Moscow; arkin@cemi.rssi.ru
** Central Economics and Mathematics Institute RAS, Moscow; slast@cemi.rssi.ru
*** University Paris I; svetlana.arkina@malix.univ-paris1.fr

in VAT and UST rates. With tax reform still being implemented, it seems relevant to compare the achievements of the former and reformed systems.

For such a comparison we apply in this paper the general investment waiting model. This model describes the behavior of the potential investor who wishes to invest in a project of creating a new enterprise (firm), which produces certain goods, consuming certain resources. This enterprise will work within the framework of the Russian tax system under uncertainty. Investment necessary to the realization of the project (creation and start of a new firm) is supposed to be instantaneous and irreversible. We assume that the opening of the new firm will take place shortly after the investment (investment lag). The starting point for the model presented in this paper is the McDonald-Siegel model [3], which deals with the real options theory.

An important feature of the considered model is the assumption that at any time the investor can either *accept* the project and start with the investment or *delay* the decision until he obtains new information about its environment (prices, demand etc.).

2 Investment waiting model with taxes

Let us suppose that the investment in the project starts at moment τ, the cost of necessary investment (without VAT) is I_τ, and l is the lag during which the investment is consumed and firm is created.

Cash flows structure

At time τ the investor faces additional costs related to the payment of VAT $\gamma_{va} I_\tau$ (where γ_{va} stands for the VAT rate) for the purchase of goods and services necessary to the creation of the new enterprise. At present taxpayers in Russia have the right to claim reimbursement of VAT (as tax deduction) after the end of the capital construction. But in practice this reimbursement is rarely carried out. In the framework of our model this means that at time $\tau + l$ the firm receives from the federal budget (instantaneously) payments $\phi \gamma_{va} I_\tau$ as VAT reimbursement, where ϕ, $0 \le \phi \le 1$ is the "reimbursement coefficient" ($\phi = 1$ corresponds to the full VAT reimbursement, and $\phi = 0$ means no reimbursement).

At time $\tau + l + t$, $t \ge 0$ the before-tax profit of the firm is equal to

$$(1 + \gamma_{va})\pi^\tau_{\tau+l+t} - S^\tau_{\tau+l+t}, \tag{1}$$

where $\pi^\tau_{\tau+l+t}$ is value added (i.e. the difference between income and material costs without VAT), and $S^\tau_{\tau+l+t}$ is payroll cost.

Taxes, which are paid by the firm, consist of value added tax $\gamma_{va}\pi^\tau_{\tau+l+t}$, payroll tax (called in Russia "unified social tax") $\gamma_s S^\tau_{\tau+l+t}$ where γ_s is the relevant tax rate, asset (or property) tax $P^\tau_{\tau+l+t}$ whose base is the residual cost of assets, and corporate profit tax which base is net income (1) minus

VAT, depreciation charges $D^{\tau}_{\tau+l+t}$ and other costs $M^{\tau}_{\tau+l+t}$ (including asset tax and unified social tax), and the rate of this tax is equal to γ_i [4]. We do not take into account other taxes paid by the firm, since they are either minor or applicable to specific kinds of production (excises). Moreover, as far as tax entries into the budget are concerned, we will take into account the personal income tax $\gamma_{pi}S^{\tau}_{\tau+l+t}$ (where γ_{pi} is the relevant tax rate).

Tax payments from the firm and employees are splitted each year between both regional and federal budgetary levels. Hence, federal budget receives VAT, a part of the UST (at the rate γ^f_s, the remaining goes to the non-budgetary funds of social and medical insurances), and the federal part of the corporate income tax (at rate γ^f_i out of γ_i). Regional budget receives enterprise property tax, personal income tax and the regional part of corporate profit tax (at the remaining rate γ^r_i).

Thus, at time $\tau + l + t$ coming from the firm, in the federal budget, we have the following cash flow

$$\gamma_{va}\pi^{\tau}_{\tau+l+t} + \gamma^f_i(\pi^{\tau}_{\tau+l+t} - S^{\tau}_{\tau+l+t} - D^{\tau}_{\tau+l+t} - M^{\tau}_{\tau+l+t}) + \gamma^f_s S^{\tau}_{\tau+l+t}, \qquad (2)$$

and, in the regional budget –

$$\gamma^r_i(\pi^{\tau}_{\tau+l+t} - S^{\tau}_{\tau+l+t} - D^{\tau}_{\tau+l+t} - M^{\tau}_{\tau+l+t}) + P^{\tau}_{\tau+l+t} + \gamma_{pi}S^{\tau}_{\tau+l+t}. \qquad (3)$$

After-tax cash flow of the firm at time $\tau + l + t$ is equal to

$$(1-\gamma_i)(\pi^{\tau}_{\tau+l+t} - S^{\tau}_{\tau+l+t} - M^{\tau}_{\tau+l+t}) + \gamma_i D^{\tau}_{\tau+l+t}. \qquad (4)$$

Depreciation

The base for depreciation charges and connected taxes in our model will be the balance cost of assets which we will relate to the cost of investments I_τ (without VAT), necessary for the start of the project at time τ.

We divide all assets into two aggregated parts: one of them ("active" part) refers to machinery, tools, equipment etc.(its share in the balance costs of all assets will be denoted as ψ, $0 \le \psi \le 1$); and the other ("inactive" part) refers to buildings and structures, whose useful lifetime is long enough.

Depreciation charges at time $\tau + l + t$ for the project started at τ will be $D^{\tau}_{\tau+l+t} = \psi I_\tau a_t + (1-\psi)I_\tau b_t$, $t \ge 0$, where $(a_t,\ t \ge 0)$, $(b_t,\ t \ge 0)$ are the depreciation "densities" of active and inactive parts of assets such that

$$a_t, b_t \ge 0, \quad \int_0^\infty a_t\,dt = \int_0^\infty b_t\,dt = 1.$$

[4] Actually, the corporate income tax equals zero if its tax base (1) is negative. Nevertheless, we shall write the term (1) even if it is negative. This can be viewed as an approximation of the principle of losses carry forward (like deductions from tax base in the future)

Uncertainty, tax holidays, investment timing

Since the economic environment can be subject to the influence of various random factors (uncertainty in market prices, demand, etc.), we will consider that the cost of required investment $(I_t,\ t \geq 0)$ is a random process, and the value added $(\pi^{\tau}_{\tau+l+t},\ t \geq 0)$ is modelled by a family (in $\tau \geq 0$) of random processes, given on some probability space $(\Omega, \mathbb{F}, \mathbf{P})$ with the flow of σ-fields $\mathcal{F} = (\mathcal{F}_t,\ t \geq 0)$. $\mathcal{F}_t$ can be interpreted as the observable information about the system up to time t, and random processes are assumed to be $\mathcal{F}$-adapted.

According to the Russian Tax Code the region can reduce (within certain limits) the corporate profit tax rate in its regional part. Let ν be the length of the time interval (after the firm's creation) during which lower regional profit tax rate $\bar{\gamma}_i^r$ is applicable. We will refer to such a period of time as tax holidays, even if this term is absent in modern tax laws. Concerning the federal part of the profit tax rate, we suppose that during tax holidays it is also subject to lowering and is equal to $\bar{\gamma}_i^f$.

According to the above model and formula (2.4), the expected present value of the firm (discounted to the investment time) can be expressed as

$$V_\tau = \phi\gamma_{va} I_\tau e^{-\rho l} + \mathbf{E}\left(\int_0^\nu [(1-\bar{\gamma}_i)(\pi^\tau_{\tau+l+t} - S^\tau_{\tau+l+t} - M^\tau_{\tau+l+t}) + \bar{\gamma}_i D^\tau_{\tau+l+t}] e^{-\rho(l+t)}\, dt \right.$$

$$\left. + \int_\nu^\infty [(1-\gamma_i)(\pi^\tau_{\tau+l+t} - S^\tau_{\tau+l+t} - M^\tau_{\tau+l+t}) + \gamma_i D^\tau_{\tau+l+t}] e^{-\rho(l+t)}\, dt \,\middle|\, \mathcal{F}_\tau \right), \tag{5}$$

where ρ is the discount rate.

The behavior of the investor is supposed to be rational in the sense that he chooses the time for investment τ (investment rule), in order to maximize his expected net present value (NPV):

$$\mathbf{E}\left[V_\tau - I_\tau(1+\gamma_{va})\right] e^{-\rho\tau} \to \max_\tau, \tag{6}$$

where the maximum is considered over all Markov times τ (with regard to the flow of σ-fields $\mathcal{F}$).

Simultaneously one can calculate (using formulas (2) and (3)), the present tax revenue into federal and regional budgets from the firm after investment.

Main assumptions

The amount of *required investment* I_t is described by geometric Brownian motion $I_t = I_0 + \int_0^t I_s(\alpha_1\, ds + \sigma_1\, dw^1_s), \quad t \geq 0$, where $(w^1_t,\ t \geq 0)$ is a Wiener process, α_1 and σ_1 are real numbers $(\sigma_1 \geq 0)$, and I_0 is a given initial state of the process. The dynamics of *value added* $\pi^\tau_{\tau+l+t},\ t \geq \tau$ is specified by a family of stochastic equations $\pi^{\tau}_{\tau+l+t} = \pi_{\tau} + \int_{\tau}^{t} \pi^{\tau}_{s}(\alpha_2\,ds + \sigma_2\,dw^2_s), \quad t \geq \tau$, where π_{τ} is $\mathcal{F}_{\tau}$-measurable random variable, $(w^2_t,\ t \geq 0)$ is a Wiener process, α_2 and σ_2 are real numbers $(\sigma_2 \geq 0)$. The pair (w^1_t, w^2_t) is two-dimensional Wiener process with correlation r. We assume that at any moment τ, observing the current prices on both input and output production one can calculate $\pi_{\tau} = \pi^{\tau}_{\tau}$, which is the value added at the "initial moment" of creation of firm, and, hence, can evaluate the future profits from the project before the actual creation of the firm. We suppose that the process π_{τ} is a geometric Brownian motion with parameters (α_2, σ_2).

Share of active part of assets ψ is constant over time.

The payroll fund $S^{\tau}_{\tau+l+t}$ is supposed to be proportional to the value added $\pi^{\tau}_{\tau+l+t}$, i.e. $S^{\tau}_{\tau+l+t} = \mu\pi^{\tau}_{\tau+l+t}$, where μ is a given constant ("labor intensity", wage per unit of value added). Such a hypothesis is in accord with the principle of dependence between wages and production activity.

Optimal investment time and tax revenues

The optimal timing problem (6) faced by the investor is an optimal stopping problem for the two-dimensional stochastic process (π_t, I_t) with the reward function defined by formulas (6), (5).

Let β be a positive root of the quadratic equation $\frac{1}{2}\tilde{\sigma}^2\beta(\beta-1) + (\alpha_2 - \alpha_1)\beta - (\rho - \alpha_1) = 0$, where $\tilde{\sigma}^2 = \sigma_1^2 - 2r\sigma_1\sigma_2 + \sigma_2^2$ is "total" volatility of the investment project.

The following theorem specifies completely an optimal rule for investing as well as expected present tax revenues into the federal budget $(\mathcal{T}^f)$ and regional budget $(\mathcal{T}^r)$ under the optimal behavior of the investor.

Theorem. *Let the amount of required investment I_t and value added π_t be described by geometric Brownian motions with parameters (α_1, σ_1) and (α_2, σ_2), respectively. Let us suppose that $\tilde{\sigma}>0$, $\alpha_2 - \frac{1}{2}\sigma_2^2 \geq \alpha_1 - \frac{1}{2}\sigma_1^2$, and $\rho > \max(\alpha_1, \alpha_2)$. Then the optimal investment time for the problem (6) is $\tau^* = \min\{t \geq 0 : \pi_t \geq p^* I_t\}$, where the optimal investment threshold is defined as*

$$p^* = \left\{1+\gamma_{va}(1-\phi e^{-\rho l}) - e^{-\rho l} H_p(\nu, \bar{\gamma}_i, \gamma_i, \gamma_p, \rho, \psi)\right\} \cdot \frac{\tilde{\rho}\exp\{\tilde{\rho} l\}}{[1-(1+\gamma_s)\mu](1-\widehat{\gamma}_i)} \cdot \frac{\beta}{\beta-1},$$

and tax revenues are:

$$\mathcal{T}^f = I_0 \left(\frac{\pi_0}{I_0 p^*}\right)^{\beta} \left[\gamma^f \frac{e^{-\tilde{\rho} l}}{\tilde{\rho}} p^* - e^{-\rho l} H_f(\nu, \bar{\gamma}_i, \gamma_i, \gamma_p, \rho, \psi)\right];$$

$$\mathcal{T}^r = I_0 \left(\frac{\pi_0}{I_0 p^*}\right)^{\beta} \left[\gamma^r \frac{e^{-\tilde{\rho} l}}{\tilde{\rho}} p^* - e^{-\rho l} H_r(\nu, \bar{\gamma}_i, \gamma_i, \gamma_p, \rho, \psi)\right],$$

where $H_p(\dots), H_f(\dots), H_r(\dots)$ are the certain (explicit) functions, $\tilde{\rho}=\rho-\alpha_2$, $\widehat{\gamma}_i=\bar{\gamma}_i+(\gamma_i-\bar{\gamma}_i)e^{-\tilde{\rho}\nu}$.

These formulas can be derived similarly to those in [1], [2] (for simpler model).

Within the framework of proposed model it was consecutively possible to analyze the latest modifications of the tax treatment of enterprises in Russia, including the corporate profit tax reform of 2002 and the cuts in the rates of VAT and the Unified Social Tax put in practice in 2004–2005. The optimal investment threshold which characterizes the moment of the arrival of the investor, expected tax revenues into the federal and the regional (as well as the consolidated) budgets and investor's NPV were all considered as criteria for comparison. It was possible to study the impact of those tax reforms on the Russian investment climate. In particular, we separate investment projects into different groups (depending on the projects' volatility, growth rate, labor intensity and technological performance) which benefited or, on the contrary, suffered from the reforms of each one of the above mentioned taxes.

3 Conclusions

The proposed model allowed us to obtain explicit formulas for expected tax revenues from future enterprise into budgets at federal and regional levels.

Concerning the reform of the corporate profit taxation, it was concluded that the new tax scheme significantly outperforms the former one, resulting, for properly technically rigged projects, in an earlier arrival of investors (increased expected net present value of the project) and improved tax revenues in the federal budget. It was nevertheless shown that differences between those two tax schemes vanish with an increase in volatility.

As far as the cut in the VAT rate is concerned it has been shown that results depend greatly on the enforcement of the VAT reimbursement for costs supported during the creation of the enterprise.

Finally it has been observed that the cut in the rate of UST was mainly profitable in the case of projects with high labor intensity. For this type of projects this cut indeed implies an earlier arrival of the investor and impressive positive effects on budget revenues at both the federal and the regional levels.

Acknowledgement. This work is supported by RFH (grant 04-02-00119).

References

1. *Arkin V.I., Slastnikov A.D., Arkina S.V.* Investment stimulation by a depreciation mechanism. — Working Paper No. 02/05. Moscow: EERC, 2003.
2. *Arkin Vadim, Slastnikov Alexander.* Optimal stopping problem and investment models. — In: Dynamic Stochastic Optimization. Lecture Notes in Economics and Mathematical Systems. 2004, v. 532, p. 83-98.
3. *Dixit A.K., Pindyck R.S.* Investment under Uncertainty. Princeton: Princeton University Press, 1994.

Bayesian Versus Maximum Likelihood Estimation of Term Structure Models Driven by Latent Diffusions

Manfred Frühwirth[1,2], Paul Schneider[2], and Leopold Sögner[3]

[1] Weatherhead Center for International Affairs, Harvard University, 1737 Cambridge Street, Cambridge, MA-02138. mfruehwirth@wcfia.harvard.edu
[2] Department of Corporate Finance, Vienna University of Economics and Business Administration, Nordbergstraße 15, A-1090 Vienna. paul.schneider@wu-wien.ac.at
[3] Department of Management Science, Vienna University of Technology, Theresianumgasse 27, A-1040 Vienna. soegner@imw.tuwien.ac.at

1 Introduction

This article presents an econometric analysis of parameter estimation for continuous-time affine term structure models driven by latent Markovian factors. In this setting either methodology, frequentist or Bayesian, is confronted with two major problems: First, each parameter set implies a time series of latent factors the transition densities of which determine the likelihood of the parameters themselves. Thus, an estimation procedure has to be capable of dealing with data that changes *for each likelihood evaluation.* Second, in contrast to the continuous-time model formulation, data are available only in discrete time and formulae for transition densities are known only for a very small subset of the affine term structure family.

An estimation based on a simple Euler approximation of the stochastic differential equations results in poor performance for any estimation methodology, especially when the step-width is large. Recent literature (both frequentist and Bayesian) has developed alternative tools. Many schemes augment the observations using latent terms. With this augmented set of observations, both maximum likelihood based methods and Bayesian methods deliver improved estimates (see e.g. Elerian, Chibb, and Shephard, 2001; Eraker, 2001; Duffie, Pedersen, and Singleton, 2003). An alternative to the above simulation based likelihood approximations has been recently proposed by Aït-Sahalia (2001, 2002) and Aït-Sahalia and Kimmel (2002), who provide closed-form expansions of the *true* transition densities. The existence of closed-form solutions makes likelihood expansions particularly suitable for numerically intensive work with Markov Chain Monte Carlo methods (MCMC) (see e.g. Robert and Casella, 1999) and for global optimization routines like genetic algorithms for the beginning stage of both MCMC and maximum likelihood estimation (ML).

The goal of this paper is a fair comparison between MCMC and ML estimation. Our investigation is based on a canonical affine two factor short rate model where one factor is a square root process and the other is Gaussian and performed with simulated data.

2 The Model

We work in a frictionless and arbitrage-free market setting in continuous time. The empirical probability measure and an equivalent martingale measure (risk-neutral measure) will be abbreviated by $\mathcal{P}$ and $\mathcal{Q}$, respectively. In this paper, under $\mathcal{P}$, we investigate the canonical representation of the model used in Duffie, Pedersen, and Singleton (2003):

$$\begin{pmatrix} dX^1(t) \\ dX^2(t) \end{pmatrix} = \begin{pmatrix} \kappa_{11} & 0 \\ \kappa_{21} & \kappa_{22} \end{pmatrix} \cdot \begin{pmatrix} \alpha - X^1(t) \\ -X^2(t) \end{pmatrix} + \begin{pmatrix} \sqrt{X^1(t)} & 0 \\ 0 & \sqrt{1+\beta X^2(t)} \end{pmatrix} \cdot \begin{pmatrix} dW^1(t) \\ dW^2(t) \end{pmatrix}$$

with $\kappa_{11}, \alpha, \beta \geq 0$. If $\kappa_{11}\alpha \geq 1/2$, $X^1(t)$ is nonnegative and stationary. The short rate $r(t)$ is affine in the latent state variables, $r(t) = \delta_0 + \delta_1 X^1(t) + \delta_2 X^2(t)$, where $\delta_2 \geq 0$. For affine models the *zero* yield with maturity τ, $y(t,\tau)$, (τ year *spot rate*) is related to the zero-coupon bond price $v(t,\tau)$ by:

$$y(t,\tau) = -\frac{\log v(t,\tau)}{\tau} = -\frac{A(\tau)}{\tau} + \frac{B(\tau)'}{\tau} X(t) \ , \tag{1}$$

where $X(t) = (X^1(t), X^2(t))'$ and $A(\tau)$ and $B(\tau)$ are solutions of the deterministic ordinary differential equations derived in Duffie and Kan (1996). We assume a completely affine market price of risk $\lambda = (\lambda_1, \lambda_2)'$ (see e.g. Dai and Singleton, 2000)).

3 Estimation

We analyze estimation of the continuous-time diffusion stipulated in Section 2 with discretely sampled data. We work with a constant step-width of $\Delta = 1/52$ yielding N observations at $t_1, \ldots, t_N$. The n-th observation of the continuous-time process, $X(t_n)$, is denoted by X_n. To generate our data we start with true parameters θ_0 which come from Bayesian estimation based on USD LIBOR interest rates in the observation period from January 6, 1998 to January 6, 2003 and two zero yields for maturities of 6 months and 5 years from the same sample. Given these yields, we invert equation (1), resulting in initial values X_1. Departing from X_1 we simulate 769 realizations of the latent state variables $X_2, \ldots, X_{770}$ by means of the Euler[4] scheme. By repeatedly applying equation (1) to $X_n, \quad n = 2, \ldots, 770$, we obtain a time series of zero yields for maturities 0.5 and 5 years, i.e. $y_{n,0.5}$ and $y_{n,5}$. We assume that the econometrician is able to observe the 6-month and the 5-year zero yields without any measurement error. We additionally use yields for maturities 0.25, 1, 2, 3 and 4 years in our estimation which can also be obtained from equation (1). However we assume that these additional zero yields are subject to measurement error and cannot be observed precisely (e.g. caused by minor liquidity). Instead of the true yields the econometrician observes only distorted yields $\tilde{y}_{n,\tau}$ which are modeled according to:

$$\tilde{y}_{n,\tau} = y_{n,\tau} + \varepsilon_{n,\tau} \tag{2}$$

where $\varepsilon_{n,\tau} \sim \mathcal{N}(0, \sigma^2(\tau))$, $\sigma^2(\tau) = \exp(a_0 + a_1\tau + a_2\tau^2)$ and $\varepsilon_{n,\tau}$ and X_n are independent. This parametrization of the error variance together with its justification can be found in Brandt and He (2002).

The set of parameters $\theta = \{\kappa_{11}, \kappa_{21}, \kappa_{22}, \alpha, \beta, \lambda_1, \lambda_2, \delta_0, \delta_1, \delta_2, a_0, a_1, a_2\}$. Based on the 0.25, 0.5, 1, 2, 3, 4 and 5 years zero yield time series these parameters are estimated by means of MCMC and ML. For both methodologies the transition densities of the yields are required. For the transition density $\pi(X_n|X_{n-1})$ we work with second-order closed-form approximations as derived in Aït-Sahalia (2001). For any continuously differentiable one-to-one transformation $y_{n,\tau} = G_\tau(X_n)$, the transition density is given by the change of variables formula

$$\pi(y_{n,\tau}|y_{n-1,\tau}) = \pi(X_n|X_{n-1}) \frac{1}{det|JG_\tau(.)|} \tag{3}$$

where $det|JG_\tau(.)|$ is the determinant of the Jacobian of the function $G_\tau(.)$. In our application, G is defined by equation (1). For the distorted zero yields $\tilde{y}_{n,\tau}$, with $\tau = \{0.25, 1, 2, 3, 4\}$, in addition $\pi(\varepsilon_{n,\tau})$ is required, which follows immediately from the distribution of $\varepsilon_{n,\tau}$.

To begin our investigation we start 100 genetic algorithms with suitable penalty functions for the constraints to search the parameter space for reasonably good starting values. From the 100 procedures the best three parameter sets (according to their likelihood score) serve as starting values for both ML and MCMC estimation.

3.1 Maximum Likelihood Estimation

Based on the initial parameter estimates from the genetic algorithm and fixing the vector of state variables, we employ a simplex based solver (this method is used in a similar context in Duffee (2002); Cheridito, Filipović, and Kimmel (2005)) followed by a gradient based solver. The resulting parameter estimates imply a different time series of state variables. In case the new state variables

[4] In the Euler approximation we used a grid of $\Delta/28$.

contain negative realizations of the square root process, the estimation attempt is discarded, otherwise the new (feasible) state variables serve as input for the next iteration. This procedure is repeated until the parameters converge. If the parameters do not converge, the maximization routine terminates when one more step does not result in a sufficient increase in the likelihood.

3.2 Markov Chain Monte Carlo (MCMC) Estimation

To implement MCMC estimation, we take recourse to the Bayes theorem that provides us with the posterior distribution of the parameters $\pi(\theta|Y) \propto \pi(Y|\theta)\pi(\theta)$ where Y is the data observed. The conditional density $\pi(Y|\theta)$ is derived from a product of conditionals provided by equation (3) and $\pi(\varepsilon_{n,\tau})$. As regards the priors $\pi(\theta)$, for each non-negative element of θ (α, β, κ_{11} and δ_2) we use a gamma prior $\mathcal{G}(0.01, 1/1000)$. For parameters with support on the real line we apply a normal prior $\mathcal{N}(0, 10000)$. Moreover, we include in the prior the restriction that $\kappa_{11}\alpha \geq 1/2$ and θ has to result in feasible $X(t)$.

Since all conditional distributions are well defined, Markov chain Monte Carlo methods can be applied (see e.g. Robert and Casella (1999)). For the underlying model, the updating sweep m from $\theta^{[m-1]}$ to $\theta^{[m]}$ is split up into the following steps:

$$\begin{aligned}
&\text{Step 1: } \kappa_{11}^{[m]} \text{ from } \pi(\kappa_{11}|Y, \kappa_{21}^{[m-1]}, \kappa_{22}^{[m-1]}, \alpha^{[m-1]}, \sigma^{[m-1]}, \ldots) \\
&\text{Step 2: } \kappa_{21}^{[m]} \text{ from } \pi(\kappa_{21}|Y, \kappa_{11}^{[m]}, \kappa_{22}^{[m-1]}, \alpha^{[m-1]}, \sigma^{[m-1]}, \ldots), \text{ etc.}
\end{aligned}$$

After each step the latent state variables are updated. This procedure is repeated until the Markov chain has reached or is supposed to be near its invariant distribution. To improve the properties of the sampler some of the above update steps are performed in one block. When updating one parameter block $\theta_+ \in \theta$, the Metropolis-Hastings algorithm is applied to derive samples of the conditionals $\pi(\theta_+|Y, \theta_-)$. E.g. when θ_+ is updated, we propose θ_+^{new} from a proposal density $q(\theta_+^{new}|\theta)$ and accept this proposal with probability $min(1, r_p)$, where

$$r_p = \frac{\pi(Y|\theta_+^{new}, \theta_-)\pi(\theta_+^{new})}{\pi(Y|Y|\theta_+^{old}, \theta_-)\pi(\theta_+^{old})} \frac{q(\theta_+^{old}|\theta_+^{new}, \theta_-)}{q(\theta_+^{new}|\theta_+^{old}, \theta_-)} \,. \tag{4}$$

θ_+^{new} replaces the corresponding $\theta_+^{old} = \theta_+^{[m-1]}$ in case of acceptance, i.e. $\theta_+^{[m]} = \theta_+^{new}$, otherwise $\theta_+^{[m]} = \theta_+^{old}$. For $q(.|.)$ we use random walk proposals.

4 Results and Conclusions

From the simulated time series of zero yields we estimated the 13 parameter, continuous-time two factor term structure model with latent driving state variables outlined before. The estimation methodologies used were MCMC and ML (Method of Moment estimators fail for our model class as reported in Brandt and Chapman (2002)).

Table 1 compares percentage deviations between the estimates and the true parameter values for the three estimations with the alternative starting values (1, 2, 3) for both ML and MCMC estimation. As can be seen, the MCMC estimates exhibit a more consistent behavior, i.e. the Markov chain shows a stable convergence behavior to the limit distribution.

Table 1. Numbers are given in percentage deviations from the true parameters $\theta_0 = \{0.73, 1.4, 0.4, 2, 0.05, -0.28, -2.5, 0.047, 0.002, 0.007, -13, 0.25, -0.25\}$. MCMC estimates are based on 5,000,000 draws of the posterior distribution.

	κ_{11}	κ_{21}	κ_{22}	α	β	λ_1	λ_2	δ_0	δ_1	δ_2	a_0	a_1	a_2
ML 1	-56	-90	56	43	-568	-23	46	35	-493	-1	0	-9	-2
ML 2	97	25	-73	-53	-100	80	-53	33	63	86	-1	-2	0
ML 3	49	15	-38	-71	-100	51	-78	14	1291	112	0	4	1
MCMC 1	33	-6	-13	-10	85	70	-1	34	-26	-8	0	7	1
MCMC 2	22	-3	-15	9	100	39	1	11	-21	-8	0	7	1
MCMC 3	21	-7	-13	14	86	40	2	9	-41	-9	0	7	1

For the MCMC procedure we observe that for κ_{11}, δ_2 and a_2 the deviation from the true parameters is less than two times the estimated standard deviation. For the remaining parameters this criterion is fulfilled for even one times the estimated standard deviation. Those criteria are not even remotely matched by the ML estimates. However, we have to remark that the standard deviations of β and λ_1 are relatively large which can be explained by the effect that the identification of λ is strongly related to β (see Dai and Singleton, 2000), an effect that is more pronounced with the MCMC estimates. For both estimation methodologies we observe a negative correlation between the estimates of the level parameters α (which directly controls the unconditional mean of the square root process) on the one hand and the estimates of δ_0, δ_1 and δ_2 (which indirectly determine the levels of the yields via the ODEs).

The success of MCMC can most probably be attributed to the fact that the latent state variables are updated for every sampled parameter vector. By contrast, simplex based and gradient based solvers used with ML estimation can not handle the discontinuities arising from the ever changing data set of latent state variables. Even solving this problem by applying an iterative ML procedure where the latent state variables are temporarily held fixed (as done by e.g. Duffee (2002) and Yu (2005)) results in parameter estimates farther away from the true parameters than with MCMC estimation and less consistent among different ML runs.

Despite the fact that the MCMC estimates do not exactly match the true parameter values it is noteworthy that the sampled values of the likelihood function match the value that is generated by the true parameters. It is interesting that both estimation methodologies resulted in realizations of the latent state variables where the product of the conditionals $\pi(X_n|X_{n-1})$ was higher than with the true processes and the true parameter values. Thus, given the time series of yields, both technologies identified latent state variables *that were more likely than the true data generating process.*

References

Aït-Sahalia, Y., 2001, "Closed-Form Likelihood Expansions for Multivariate Diffusions," Working paper, Princeton University and NBER.

Aït-Sahalia, Y., 2002, "Maximum-Likelihood Estimation of Discretely-Sampled Diffusions: A Closed-Form Approximation Approach," *Econometrica*, 70, 223–262.

Aït-Sahalia, Y., and R. Kimmel, 2002, "Estimating Affine Multifactor Term Structure Models Using Closed-Form Likelihood Expansions," Working paper, Princeton University and NBER.

Brandt, M. W., and D. A. Chapman, 2002, "Comparing Multifactor Models of the Term Structure," Working paper, Wharton School, McCombs School and NBER.

Brandt, M. W., and P. He, 2002, "Simulated Likelihood Estimation of Affine Term Structure Models from Panel Data," Working paper, University of Pennsylvania.

Cheridito, P., D. Filipović, and R. Kimmel, 2005, "Market Price of Risk Specifications for Affine Models: Theory and Evidence," *Journal of Financial Economics*, forthcoming.

Dai, Q., and K. J. Singleton, 2000, "Specification Analysis of Affine Term Structure Models," *Journal of Finance*, 55, 1943–1978.

Duffee, G. R., 2002, "Term Premia and Interest Rate Forecasts in Affine Models," *Journal of Finance*, 57, 405–443.

Duffie, D., and R. Kan, 1996, "A Yield-Factor Model of Interest Rates," *Mathematical Finance*, 6, 379–406.

Duffie, D., L. H. Pedersen, and K. J. Singleton, 2003, "Modeling Sovereign Yield Spreads: A Case Study of Russian Debt," *Journal of Finance*, 53, 119–159.

Elerian, O., S. Chibb, and N. Shephard, 2001, "Likelihood Inference for Discretely Observed Nonlinear Diffusions," *Econometrica*, 69, 959–993.

Eraker, B., 2001, "MCMC Analysis of Diffusion Models with Application to Finance," *Journal of Business and Economic Statistics*, 19, 177–191.

Robert, C., and G. Casella, 1999, *Monte Carlo Statistical Methods*, Springer, New York.

Yu, J., 2005, "Closed-Form Likelihood Estimation of Jump-Diffusions with an Application to the Realignment Risk of the Chinese Yuan," Working Paper, Columbia University.

Exit in Duopoly Under Uncertainty and Incomplete Information

Makoto Goto[1] and Takahiro Ono[2]

[1] Department of Industrial & Management Systems Engineering, Graduate School of Science and Engineering, Waseda University, 3-4-1 Okubo, Shinjuku-ku, Tokyo, 169-8555, Japan. E-mail: mako.50@ruri.waseda.jp

[2] Department of Industrial & Management Systems Engineering, School of Science and Engineering, Waseda University, 3-4-1 Okubo, Shinjuku-ku, Tokyo, 169-8555, Japan. E-mail: ohno@ohno.mgmt.waseda.ac.jp

1 Introduction

We analyze exit from a duopoly market usign a real options approach and game theory. While entry into markets has received considerable attention in this field, exit from markets has only been dealt with in recent years.

Murto [3], an important literature, analyzes an asymmetric game with complete information. His main result shows that there are two kinds of equilibria, ordinary equilibrium and 'gap equilibrium.' While the weaker firm must exit earlier and the stronger firm can enjoy a monopoly in ordinary equilibrium, when uncertainty is higher gap equilibrium may appear, so that one cannot forecast which firm will exit first. On the contrary, from the aspect of game theory, few analyses of models of incomplete information exist in this field.

Consequently, we have expanded Murto [3] by introducing incomplete information and analyzed exit under uncertainty and incomplete information. Our main result suggests that one cannot forecast exit behavior even under ordinary equilibrium with incomplete information.

2 Complete Information Model [3]

In this section, we illustrate a complete information model, Murto [3].

Let the initial market condition be a duopoly. We assume that the firms are risk neutral, value maximizing and discount with constant factor $\rho > 0$. We denote one firm by i, the other firm by j, with $i, j \in \{1, 2\}$ and $i \neq j$.

Firm i receives revenue flow $X_t D_i$ against cost flow C_i at time $t > 0$. For a monopoly, the profit flow of i is $X_t M_i - C_i$, and $0 < D_i < M_i$. That is, D_i

and M_i are parameters which denote the intensity of duopoly and monopoly, respectively.

X_t is an index which denotes the market general profitability, and follows a geometric Brownian motion

$$\mathrm{d}X_t = \alpha X_t \mathrm{d}t + \sigma X_t \mathrm{d}W_t, \quad X_0 = x > 0, \tag{1}$$

where $\alpha < \rho$ is the mean growth rate of X_t, $\sigma > 0$ is the volatility of X_t, and W_t is the standard Brownian motion.

Firm i is free to exit permanently at any moment by paying a fixed exit cost, U_i. To ensure that it can be optimal for a firm to exit, we assume that $U_i < C_i/\rho$, $\forall i = 1, 2$. Therefore, the decision of i is when to exit.

Definition 1 *The strategy of i is a closed stopping set $S_i \subset \mathbb{R}_+$, which defines actions of i as long as firm j is still in the market: If $X_t \in S_i$, then firm i exits immediately, otherwise it stays in the market. We denote the strategy profile containing the strategies of both firms as $S = \{S_1, S_2\}$.*

The exit time of i is

$$\tau_i = \tau(x, S_i) = \inf\{t > 0 : X_t \in S_i\}, \tag{2}$$

and the game ends at time $\tilde{\tau} = \min\{\tau_1, \tau_2\}$. Given x and S, the value function of i is

$$\begin{aligned} V_i^{\mathrm{D}}(x, S) = \mathbb{E}\Bigg[& \int_0^{\tilde{\tau}} \mathrm{e}^{-\rho t}\Big(X_t D_i - C_i\Big)\mathrm{d}t \\ & + \mathrm{e}^{-\rho\tilde{\tau}}\left(\mathbf{1}_{\{\tau_i \le \tau_j\}} \cdot (-U_i) + \mathbf{1}_{\{\tau_i > \tau_j\}} \cdot V_i^{\mathrm{M}}(X_{\tilde{\tau}})\right)\Bigg], \end{aligned} \tag{3}$$

where $\mathbf{1}_A$ is an indicator function, and the terminal value $V_i^{\mathrm{M}}(\cdot)$ is the monopoly value of i, as derived in the next section. The decision problem of i is to choose its strategy S_i such that equation (3) is maximized.

2.1 Optimistic Situation (Monopoly) and Pessimistic Situation

Since firm j has exited, firm i can solve the maximizing problem independently. Therefore, the monopoly value of i is

$$V_i^{\mathrm{M}}(x) = \begin{cases} -U_i & \text{if } x \le X_i^{\mathrm{M}} \\ v_i x^{\beta_2} + \frac{x M_i}{\rho - \alpha} - \frac{C_i}{\rho} & \text{if } x > X_i^{\mathrm{M}}, \end{cases} \tag{4}$$

where

$$v_i = \frac{1}{1-\beta_2}\left(\frac{C_i}{\rho} - U_i\right)\left(\frac{1}{X_i^{\mathrm{M}}}\right)^{\beta_2}, \tag{5}$$

$$X_i^{\mathrm{M}} = \frac{\beta_2(\rho - \alpha)}{M_i(\beta_2 - 1)}\left(\frac{C_i}{\rho} - U_i\right). \tag{6}$$

In contrast to monopoly, consider the pessimistic situation where firm j will never exit before i exits, i.e. $\tau_i < \tau_j$. Then firm j can not influence the value of i, therefore we have the optimal exit level of i in pessimistic situation

$$X_i^{\mathrm{D}} = \frac{\beta_2(\rho-\alpha)}{D_i(\beta_2-1)}\left(\frac{C_i}{\rho} - U_i\right). \tag{7}$$

2.2 Equilibria

Now, we assume that

1. $X_1^{\mathrm{M}} < X_2^{\mathrm{M}}$, i.e. firm i is relatively 'stronger,' w.l.o.g.,
2. $X_i^{\mathrm{D}} \notin R_i\big((0, X_j^{\mathrm{D}}]\big)$, $\forall i \in \{1,2\}$, $j \neq i$,

where $R_i(S_j)$ is the best response of i to S_j.

Then, two kinds of equilibria are derived under the above assumptions:[3]

Proposition 1 (Ordinary equilibrium).

$$\{S_1 = (0, X_1^{\mathrm{M}}], S_2 = (0, X_2^{\mathrm{D}}]\} \tag{8}$$

is always in equilibrium, then relatively 'weaker' firm 2 must exit first.

Proposition 2 (Gap equilibrium). *Let be $\alpha \geq 0$, and $\sigma \in (\overline{\sigma}, \infty)$ for some $\overline{\sigma}$. If and only if there is $\underline{x} > X_2^{\mathrm{M}}$ such that $R_1(\underline{x})$ exists and $\underline{x} = R_2\big(R_1(\underline{x})\big)$, then*

$$\{S_1 = (0, X_1^{\mathrm{M}}] \cup [\overline{x}, X_1^{\mathrm{D}}], S_2 = (0, \underline{x}]\} \tag{9}$$

is equilibrium, where $\overline{x} = R_1(\underline{x})$, $\underline{x} = R_2(\overline{x})$. Then, one cannot forecast which firm will exit first.

3 Incomplete Information Model

In this section, we introduce incomplete information in the model of Murto [3]. However we assume $\sigma \in (0, \overline{\sigma})$, so that market uncertainty is lower; then gap equilibrium does not arise. Therefore, we restrict the strategy of i S_i to the form of $(0, X_i]$.

3.1 Introduction of Incomplete Information

As do Fudenberg and Tirole [1], we assume that neither firm knows only the cost flow of its rival. Instead, firm i has prior briefs about C_j which are represented by a distribution $F_j(C_j)$ whose support is $[\underline{C}_j, \overline{C}_j]$ for $0 < \underline{C}_j < \overline{C}_j$. Then, we assume that the strategy, i.e. the exit level, of j is a continuous

[3] See Murto [3] for the proofs of proposition 1 and 2.

and increasing function $X_j(C_j)$, because larger cost flow causes earlier exit and vice versa. Moreover, since firm i has all information of j but C_j, it knows X_j given C_j. So, the function $X_j(C_j)$ is common knowledge for firm i.

Therefore, the distribution function $F_j(C_j)$ can be transformed to

$$F_j(C_j) = F_j(C_j(X_j)) = G_j(X_j), \tag{10}$$

where $C_j(X_j)$ is the inverse function of $X_j(C_j)$. Then, we assume that a distribution function $G_j(X_j)$ is continuously differentiable to a density function $g_j(X_j)$. And let $X_j(\underline{C}_j) = \underline{X}_j \leq X_j^{\mathrm{M}}$ and $X_j(\overline{C}_j) = \overline{X}_j \geq X_j^{\mathrm{D}}$, so $G_j(X_j)$ has its support $[\underline{X}_j, \overline{X}_j]$.

Now, the value function should depend on the set of distribution functions $G = \{G_i(\cdot), G_j(\cdot)\}$ in addition to x, S. Since $\tau_i \leq \tau_j$ iff $X_i \geq X_j$ and $\tau_i > \tau_j$ iff $X_i < X_j$, the value function of i is

$$\begin{aligned} V_i(x, S, G) = \Pr(X_j \leq X_i) \times \mathbb{E}\left[\int_0^{\tau_i} \mathrm{e}^{-\rho t}\Big(X_t D_i - C_i\Big)\mathrm{d}t - \mathrm{e}^{-\rho\tau_i} U_i\right] \\ + \int_{X_i}^{\overline{X}_j} \mathbb{E}\left[\int_0^{\tau_j} \mathrm{e}^{-\rho t}\Big(X_t D_i - C_i\Big)\mathrm{d}t + \mathrm{e}^{-\rho\tau_j} V_i^{\mathrm{M}}(X_{\tau_j})\right] g_j(X_j)\mathrm{d}X_j. \end{aligned} \tag{11}$$

Let the expectation of the first and second term in the right part of equation (11) be $A_i(x, X_i)$, $B_i(x, X_j)$ respectively, then we have

$$A_i(x, X_i) = \begin{cases} -U_i & \text{if } x \leq X_i \\ a_i x^{\beta_2} + \frac{x D_i}{\rho - \alpha} - \frac{C_i}{\rho} & \text{if } x > X_i, \end{cases} \tag{12}$$

$$B_i(x, X_j) = \begin{cases} V_i^{\mathrm{M}}(x) & \text{if } x \leq X_j \\ b_i x^{\beta_2} + \frac{x D_i}{\rho - \alpha} - \frac{C_i}{\rho} & \text{if } x > X_j, \end{cases} \tag{13}$$

where

$$a_i = \left(\frac{C_i}{\rho} - U_i - \frac{X_i D_i}{\rho - \alpha}\right)\left(\frac{1}{X_i}\right)^{\beta_2}, \tag{14}$$

$$b_i = \frac{1}{1-\beta_2}\left(\frac{C_i}{\rho} - U_i\right)\left(\frac{1}{X_i^{\mathrm{M}}}\right)^{\beta_2} - \frac{X_j(D_i - M_i)}{\rho - \alpha}\left(\frac{1}{X_j}\right)^{\beta_2}. \tag{15}$$

3.2 Equilibria

We confirm the existence of the best responsef of i to S_j

$$R_i(S_j) = \underset{S_i \subset \mathbb{R}_+}{\arg\sup}\, V_i(x, S, G). \tag{16}$$

The first order condition which maximizes equation (11), $\partial V_i(x, S, G)/\partial X_i = 0$ is

$$g_j(X_i)\Big(A_i(x,X_i) - B_i(x,X_i)\Big) + G_j(X_i)A_i'(x,X_i) = 0, \tag{17}$$

where

$$\begin{aligned} A_i'(x,X_i) &= \frac{\partial A_i(x,X_i)}{\partial X_i} \\ &= \begin{cases} 0 & \text{if } x \leq X_i \\ \left[\frac{(\beta_2-1)D_i}{\rho-\alpha} - \frac{\beta_2}{X_i}\left(\frac{C_i}{\rho} - U_i\right)\right]\left(\frac{x}{X_i}\right)^{\beta_2} & \text{if } x > X_i. \end{cases} \end{aligned} \tag{18}$$

Then, we have the following proposition:

Proposition 3 (The existence of the best response). *There is at least one $X_i = \hat{X}_i$ in $[X_i^{\mathrm{M}}, X_i^{\mathrm{D}}]$, which satisfies equation (17).*

Proof. It is sufficient that we consider $x > X_i$. Let $f_i(X_i) = \partial V_i(x,S,G)/\partial X_i$, then we confirm the sign of $f_i(X_i)$ at both ends of $[X_i^{\mathrm{M}}, X_i^{\mathrm{D}}]$ to establish the existence of $\hat{X}_i$ such that $f_i(\hat{X}_i) = 0$. First,

$$f_i(X_i^{\mathrm{M}}) = G_j(X_i^{\mathrm{M}})(1-\beta_2)\left(\frac{M_i - D_i}{\rho-\alpha}\right)\left(\frac{x}{X_i^{\mathrm{M}}}\right)^{\beta_2} \geq 0, \tag{19}$$

where the equality holds strictly when $G_j(X_i^{\mathrm{M}}) = 0$.

Second, note the limitation of $\beta_2 < 0$, then

$$\begin{aligned} & f_i(X_i^{\mathrm{D}}) \\ &= \frac{g_j(X_i^{\mathrm{D}})}{\frac{D_i}{\rho-\alpha}} \frac{\frac{C_i}{\rho} - U_i}{1-\beta_2}\left(\frac{x}{X_i^{\mathrm{D}}}\right)^{\beta_2}\left[\frac{D_i}{\rho-\alpha}\left(1-\left(\frac{M_i}{D_i}\right)^{\beta_2}\right) + \beta_2\left(\frac{M_i - D_i}{\rho-\alpha}\right)\right] \\ &\leq 0, \end{aligned} \tag{20}$$

where the equality holds strictly when $g_j(X_i^{\mathrm{D}}) = 0$.

Finally, if $G_j(X_i^{\mathrm{M}}) \neq 0$ and $g_j(X_i^{\mathrm{D}}) \neq 0$, then $f_i(X_i^{\mathrm{M}}) > 0$ and $f_i(X_i^{\mathrm{D}}) < 0$, so that the result follows from the continuity of $G_j(X_i)$ and $g_j(X_i)$. □

Let $\hat{X}_i$ which maximizes equation (11) be X_i^*. the best response depends on $G_j(\cdot)$, especially the relation between $[\underline{X}_j, \overline{X}_j]$ and $[X_i^{\mathrm{M}}, X_i^{\mathrm{D}}]$. While there are several relations, we focus on $\underline{X}_j < X_i^{\mathrm{M}} < X_i^{\mathrm{D}} < \overline{X}_j$ here. Then, we have the following proposition about equilibrium:

Proposition 4 (Equilibrium with incomplete information). *When $\underline{X}_j < X_i^{\mathrm{M}} < X_i^{\mathrm{D}} < \overline{X}_j$, $G_j(X_i^{\mathrm{M}}) \in (0,1)$ and $g_j(X_i^{\mathrm{D}}) \neq 0$, so that $X_i^* = X_i^{\mathrm{P}} \in (X_i^{\mathrm{M}}, X_i^{\mathrm{D}})$. Then,*

$$S^{\mathrm{a}} = \left\{S_1 = (0, X_1^{\mathrm{P}}], S_2 = (0, X_2^{\mathrm{P}}]\right\} \tag{21}$$

is ex-ante equilibrium; ex-post equilibrium is

$$S^{\mathrm{p}} = \left\{S_i = (0, X_i^{\mathrm{P}}], S_j = (0, X_j^{\mathrm{M}}]\right\}, \quad X_i^{\mathrm{P}} > X_j^{\mathrm{P}}. \tag{22}$$

Therefore, the relatively stronger firm could exit first (reversal of the order of exit).

Proof. This is the case where the relation with the competitor is not revealed at all under incomplete information. Then $G_j(X_i^{\mathrm{M}}) \in (0,1)$ and $g_j(X_i^{\mathrm{D}}) \neq 0$ from definition of $G_j(\cdot)$, so that $f_i(X_i^{\mathrm{M}}) > 0$, $f_i(X_i^{\mathrm{D}}) < 0$ from the proof of proposition 3. Therefore, there is maximum $X_i^* = X_i^{\mathrm{P}} \in (X_i^{\mathrm{M}}, X_i^{\mathrm{D}})$.

Since equation (11) does not depend on X_j, S^{a} consists of the best responses, so that is equilibrium. While firm 1 is relatively stronger, it is assumed that $X_2^{\mathrm{P}} < X_1^{\mathrm{P}}$ for the anticipation distribution of each firm. Then, firm 2 does not know X_1^{P} until $X_t \leq X_1^{\mathrm{P}}$, at the moment it can observe the exit of firm 1 and enjoy a monopoly. Therefore, it should change strategy X_2^{P} to the optimal monopoly strategy $S_2 = (0, X_2^{\mathrm{M}}]$ after the game ends, and then there is reversal of the order of exit. □

4 Comparison of the Two Models

In the ordinary equilibrium with complete information model, the relation with the competitor is revealed, so that relatively weaker firm must exit first at X_i^{D}. On the other hand, in equilibrium with incomplete information, a firm may exit first at $X_i^{\mathrm{P}} < X_i^{\mathrm{D}}$. Since this does not satisfy the smooth-pasting condition, it is not an optimal decision. Therefore, it shows that incomplete information makes the value of a firm lower.

Moreover, since the relation between X_i^{P} and X_j^{P} is determined by $G_i(\cdot)$ and $G_j(\cdot)$, it could be that $X_i^{\mathrm{P}} > X_j^{\mathrm{P}}$ while $X_i^{\mathrm{M}} < X_j^{\mathrm{M}}$. Therefore incomplete information could cause the order of exit to reverse while market uncertainty is lower.

5 Conclusion

We have shown, by assuming for simplicity that market uncertainty is lower, that in the incomplete information model, there is reversal of exit order in spite of the exclusion of gap equilibrium. Therefore, equilibrium is influenced not only by market uncertainty but also by uncertainty of information about the competitor.

While we assume that the strategy is a continuous and increasing function, there may be a more natural assumption. Also, considering constraint of the distribution function may ensure the uniqueness of the best response.

References

1. Fudenberg D, Tirole J (1986) A theory of exit in duopoly. Econometrica 54:943–960
2. Lambrecht B, Perraudin W (2003) Real options and preemption under incomplete information. Journal of Economic Dynamics and Control 27:619–643
3. Murto P (2004) Exit in duopoly under uncertainty. RAND Journal of Economics 35:111–127

Real Option Approach on Implementation of Wind-diesel Hybrid Generators

Hideki Honda[1] and Makoto Goto[2] and Takahiro Ono[3]

[1] Department of Industrial & Management Systems Engineering, Graduate School of Science and Engineering, Waseda University, 3-4-1 Okubo, Shinjuku-ku, Tokyo, 169-8555, Japan. E-mail: hideki-honda@suou.waseda.jp

[2] Department of Industrial & Management Systems Engineering, Graduate School of Science and Engineering, Waseda University, 3-4-1 Okubo, Shinjuku-ku, Tokyo, 169-8555, Japan. E-mail: mako.50@ruri.waseda.jp

[3] Department of Industrial & Management Systems Engineering, School of Science and Engineering, Waseda University, 3-4-1 Okubo, Shinjuku-ku, Tokyo, 169-8555, Japan. E-mail: ohno@ohno.mgmt.waseda.ac.jp

1 Introduction

At the present time, the smaller, more isolated Japanese islands use diesel generators for power. Because of high diesel fuel costs, this technology may not be as efficient as wind-diesel hybrid generators. Wind power generators have the potential to reduce fuel costs, but it is necessary to combine wind power generators and diesel generators owing to the wind power generators' undependable nature. Hybrid generators have operational flexibility (real option); they can reduce generation when the wind blows. This operational flexibility increases the operational profit of hybrid generators.

However, hybrid generators may or may not be as profitable as diesel generators because of the trade-off between operational profit, achieved by reducing fuel costs, and initial costs of wind power generators. Moreover, an optimal quantity scale of wind power generators is difficult to establish due to the difference both in output of wind power generators and in initial costs of power generators of every scale.

Dorvo [2] uses the Weibull density function to express wind distribution and estimate the parameters. On the other hand, very few attempts have been made to create an evaluation model of hybrid generators, whereas several studies have proposed evaluation models of diesel generators.

Therefore we propose an evaluation model for wind-diesel hybrid generators. Comparing wind-diesel hybrid generators and diesel generators in terms of profit, we estimated the threshold in the mean hourly data of wind speed when the value of hybrid generators exceeds the value of diesel generators.

Moreover, an optimal quantity scale for wind power generators has been created for practical application.

The Weibull density function is applied to wind speed distribution. A logistic regression formula is proposed as an output function for wind power generators. These parameters are estimated by the least squares method.

We will provide a guideline for deciding whether power companies should implement wind-diesel hybrid generators or stay with diesel generators.

2 Model

2.1 Value function of power generators

Here we assume that electricity demand per hour D, fuel cost per hour F and electricity price per hour P are constant. On the other hand, wind speed W is an uncontrolled variable. Moreover, diesel generators have capacity equal to maximum electricity demand and compensate for the lack of electricity generated by wind power. Assuming that profit is represented per hour, the value of power plants V^x is as follows:

$$V^x = \sum_{n=1}^{N^x} \frac{R^x - C^x}{(1+\rho)^n} - I^x, \tag{1}$$

where x is a variable standing for hybrid generators ($x = w, d$) or diesel generators ($x = d$), N^x is the lifetime of power generators [hour], I^x is initial cost of power generators, R^x is the revenue per hour, C^x is operation cost, ρ is the discount rate.

2.2 Operation revenue of diesel generators

When power is generated by diesel generators alone, the operation revenue R^d is as follows:

$$R^d = (P - F)D. \tag{2}$$

Fuel, at cost F, is needed to generate electricity in diesel generators.

2.3 Operation revenue of wind-diesel hybrid generators

In this article the Weibull distribution is used to express wind distribution [2]D The probability when the wind speed is below $\tilde{W}$ is as follows:

$$G(W \leq \tilde{W}) = 1 - \exp\left(-\left(\frac{\tilde{W}}{c}\right)^k\right), \tag{3}$$

where W is wind speed [m/s], $G(W)$ is the wind distribution function, k is a shape parameter and c is a scale parameter.

Furthermore, we approximate an output function of wind power generators using a logistic regression formula due to the similar shape between output of wind power generators and logistic curves. The logistic regression formula is as follows:

$$J(W) = \frac{1}{\frac{1}{u} + b_0 b_1^W}, \tag{4}$$

where $J(W)$ is an output function of wind generators [kW], b_0 and b_1 are coefficients, and u is a scale of wind power generators. When u is larger than D, excess wind power makes no profit, therefore the revenue of hybrid generators $\mathrm{E}\left[R^{d,w}\right]$is as follows:

$$\mathrm{E}\left[R^{d,w}\right] = \mathrm{E}^w\left[\mathbf{1}_{\{J(W)>D\}} PD + \mathbf{1}_{\{J(W)\leq D\}}(PJ(W) + (P-F)(D-J(W))\right] \tag{5}$$

$$= (P-F)D + F\left(\int_{w_i}^{w^*} J(W)g(W)dW + \int_{w^*}^{w_o} Dg(W)dW\right). \tag{6}$$

When $u \leq D$ all output of wind generators generates profit, so $\mathrm{E}\left[R^{d,w}\right]$ is as follows:

$$\mathrm{E}\left[R^{d,w}\right] = \mathrm{E}^w\left[PD - FD + FJ(W)\right] \tag{7}$$

$$= (P-F)D + F\int_{w_i}^{w_o} J(W)g(W)dW, \tag{8}$$

where $g(W)$ is the wind probability density function, w^* is wind speed when $J(W) = D$, w_i is cut-in wind speed and w_o is cut-out wind speed.[4] The first term of the right-hand side $(P-F)D$ in equations (6) and (8) represents revenue per hour when all electricity demand is met by diesel generators alone, whereas the second term $F\left(\int_{w_i}^{w^*} J(W)g(W)dW + \int_{w^*}^{w_o} Dg(W)dW\right)$ or $F\int_{w_i}^{w_o} J(W)g(W)dW$ represents the output of wind power generators which reduces the fuel costs of diesel generators.

2.4 Value of implementing wind power generators

Assuming that the lifetime of wind-diesel hybrid generators is equal, the value of hybrid generators $V^{d,w}$ is as follows:

[4] Cut-in wind speed is the wind speed when wind power generators start to operate, whereas cut-out wind speed is the wind speed when wind power generators stop operation.

$$V^{d,w} = \sum_{n=1}^{N^{d,w}} \frac{\left(\mathrm{E}\left[R^{d,w}\right] - C^{d,w}\right)}{(1+\rho)^n} - I^{d,w} + S, \tag{9}$$

$$C^{d,w} = C^d + C^w, \qquad I^{d,w} = I^d + I^w. \tag{10}$$

S is the subsidy which the government pays for implementing wind power generators, then the value of implementing wind power generators $\Pi = V^{d,w} - V^d$ is as follows:

$$\Pi = \sum_{n=1}^{N^{d,w}} \frac{\left(\mathrm{E}\left[R^{d,w}\right] - R^d - C^w\right)}{(1+\rho)^n} - I^w + S. \tag{11}$$

If the right-hand side of equation (11) is positive, wind power generators should be implemented.

3 Results and discussion

3.1 Estimation of the parameters

The parameters of the Weibull distribution are estimated by the least squares method. We used mean hourly data of wind speed (Japan Meteorological Agency) in Hachijojima in 2002. The results of estimating the parameters are $k = 1.87$, $c = 6.53$. The coefficient of determination is 0.999.

The parameters of the output function of wind generators are also estimated by the least squares method. We use power curve data (NORDEX 1,500 [kW]). The parameters are estimated to be $b_0 = 0.54$ and $b_1 = 0.47$. The coefficient of determination is 0.998. Since coefficient of determination is close to 1, we can recognize that the logistic regression formula can fairly approximate the output function of wind power generators.

3.2 Value of implementing wind power generators

We examine the influence of parameter c and u on the power generators value V^x. $D = 6{,}477$ [kW] (mean electricity demand in Hachijojima from April to September in 2002), $\rho = 0.05$ [%/year], $P = 20$ [yen] (data from the web site of Okinawa Electric Power Company), $W_i = 3$ and $W_o = 25$ [m/s] (NORDEX 1,500 [kW]) are used in this experiment. We assume that I^w, S and C^w increase with the number of wind power generators. The other parameters are shown in Table 1.

Fig. 1 shows the value of wind-diesel hybrid generators and diesel generators when the scale parameter c varies. This figure indicates the threshold in mean hourly wind speed data when the value of wind-diesel hybrid generators exceeds the value of diesel generators.

From the Fig. 1, hybrid generators benefit when scale parameter c is over c^*, because initial cost I is constant. Since c represents the intensity of wind speed, wind-diesel hybrid generators should be implemented in isolated islands where average wind speed is high.

Table 1. Experimental data

	diesel generators	wind power generators
Capacity [kW]	11,000	1,500
Initial cost I [million yen]	1,540	375
Operation cost C [yen/hour]	8,790	645
Fuel cost F [yen/kWh]	10.4	—
Subidy S [million yen]	—	125
Lifetime of generators N [hours]	131,400	131,400

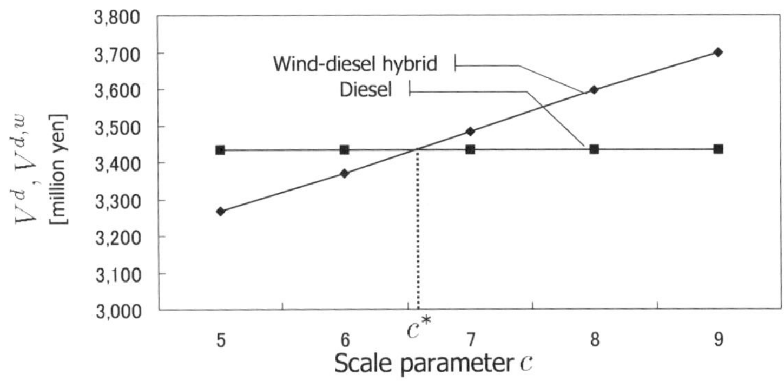

Fig. 1. The scale parameter c and the value of power generators (c^*=6.65)

3.3 Value of implementing wind power generators

The implementation value of wind power generators is shown in Fig. 2. When $c \leq c^*$, initial cost has strong effect on Π. Therefore Π is decreasing in everywhere. On the other hand, initial cost has little effect on Π when $c > c^*$. Therefore Π is increasing to optimal scale.

Fig. 2 shows that an optimal scale of wind power generators exists when $c > c^*$, because initial cost is increasing proportionally, whereas operational profit is decreasing in $u > D$ due to sunk electricity generators.

Moreover, an optimal scale varies in respect to wind abundance and speed because the ratio of decreasing operational profit becomes smaller as wind speed increases. Fig. 2 also indicates that sunk cost is generated when wind speed is not accurately estimated.

4 Conclusion

We have proposed an evaluation model of wind-diesel hybrid generators. By comparing the value of wind-diesel hybrid generators and that of diesel generators, we estimated the threshold in mean hourly wind speed data when the value of hybrid generators exceeds the value of diesel generators. Moreover, an

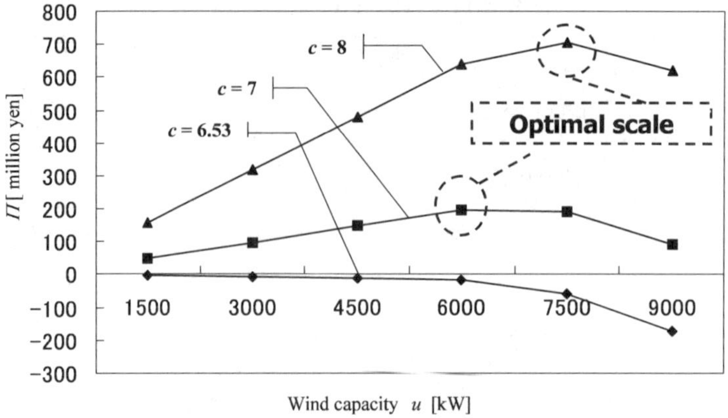

Fig. 2. The scale of wind power generators u and the value of implementing wind power generators (c^{*}=6.65)

optimal quantitity scale of wind power generators has been created for practical application. It was found from the results that an optimal scale varies in respect to the abundance of wind. This fact cannot be overemphasized because the wrong quantity of wind power generators may be constructed if the average wind speed is not measured precisely.

The proposed model has not been verified since acquisition of profit data is difficult. Furthermore, we have assumed that electricity demand is constant. Future research is needed to verify the proposed model and to allow for variable electricity demand.

Nevertheless, it is noteworthy that the evaluation model proposed here can at least estimate the optimal scale of wind power generators and clarify how an optimal quantity scale varies according to wind abundance and speed. Effective management is expected to result from the utilization of this model.

References

1. Dixit AK, Pindyck RS (1994) Investment under Uncertainty. Princeton University Press, Princeton
2. Dorvo ASS (2002) Estimating wind speed distribution.Energy Conversion and Management 17:2311-2318
3. NEDO (2000) An introduction guide book of Wind power generators. New Energy and Industrial Technology Development Organization

Part XIII

e-Business and Computer Sciences

Mobile Dienste zum Terminmanagement bei Geschäftsprozessen mit Kundenkontakt

Mario Hopp[1], Anastasia Meletiadou[2], J. Felix Hampe[3]

[1] Jan AG, Hans-Krämerstr. 31, 94469 Deggendorf, m.hopp@jan-ag.de
[2] Institut für Wirtschafts- und Verwaltungsinformatik, Universität Koblenz-Landau, Universitätsstr. 1, 56070 Koblenz, nancy@uni-koblenz.de
[3] School of Computing and Information Science, University of South Australia, City West Campus, 27-29 North Terrace, SM1-06,Adelaide SA 5000, Australia felix.hampe@unisa.edu.au

1 Einleitung

Laut einer Studie von EITO [5] lagen die mobilfunkbasierten MCommerce-Umsätze in Westeuropa 2003 bei rund 1,1 Mrd. Euro. Bis zum Jahr 2007 wird mit einem Anstieg der mCommerce Umsätze auf 24,7 Mrd. Euro gerechnet [6]. Es besteht aber breiter Konsens, dass es derzeit an vermarktbaren mobilen Mehrwertdiensten noch mangelt. Zu einem ähnlichen Ergebnis kommt auch die Studie "Mobile Solutions & Services" von Metagroup [13]. Weiterhin sieht die Studie den Schwerpunkt mobiler Unternehmensanwendungen im Segment Personal Information Management (PIM), also der Termin- und Kontaktorganisation, dem Senden und Empfangen von E-Mails sowie dem Abrufen von aktuellen Informationen und Nachrichten (Verkehr, Reise, Fahrplan, Stau etc.) im B2E-Bereich [16].

Mobile Dienste und Applikationen mit dem Schwerpunkt Terminmanagement stehen auch im Zentrum dieses Beitrags. Das Beispielszenario, welches durchgängig in diesem Beitrag betrachtet wird, gehört zu der Klasse der zeitabhängigen mobilen Anwendungen: Die Patienten einer Arztpraxis können über ein mobiles Endgerät nahezu jederzeit und ubiquitär einen Termin koordinieren. Sie betreten damit ein virtuelles Wartezimmer ohne real vor Ort sein zu müssen. Stattdessen werden sie so benachrichtigt, dass sie die Praxis termingerecht erreichen. Eventuelle Verspätungen oder Terminverschiebungen sind somit flexibler abzustimmen.

Der diesem Beitrag zugrunde liegende Forschungsansatz ist Design Research [17, 18], wobei es sich um eine frühe Phase handelt, da derzeit noch keine empirische Analyse zur Akzeptanz des abgeleiteten Artefakts vorliegt. Der Beitrag ordnet sich somit in die Kategorie ‚research in progress' ein.

2 Ein Modell für mobile Dienste zeitkritischer Geschäftsprozessen mit Kundenkontakt

Die folgende Abbildung 1 zeigt eine schematische Darstellung der vorgeschlagenen Infrastruktur zur Abwicklungsunterstützung eines zeitkritischen Kundenkontakts. An einem solchen Szenario sind je nach Anwendungsfall zwei oder drei Parteien beteiligt: der Kunde, der Anbieter der eigentlichen Dienstleistung (im folgenden Szenario der Arzt) und eventuell ein ASP. Der Kunde aktiviert den entsprechenden Dienst auf seinem mobilen Gerät und sowohl er als auch der Anbieter der Leistung können dann Interaktionen (z.B. zeitbezogene Anfragen, Bestätigungen oder Änderungsmitteilungen) initiieren.

Das primäre Ziel eines solchen Systems ist die Prozessoptimierung. Die Gestaltung der Abläufe auf Anbieter- und Kundenseite wird durch bestimmte Rahmenbedingungen beeinflusst. Dazu gehören etwa Prioritäten bei der Abarbeitung von Kundenanforderungen (zum Beispiel die bevorzugte Behandlung von Notfällen in einer Arztpraxis) oder betriebswirtschaftliche Ziele, wie die Anforderung, dass der Anbieter möglichst keine Leerlaufzeiten hat (z. B. leeres Wartezimmer in

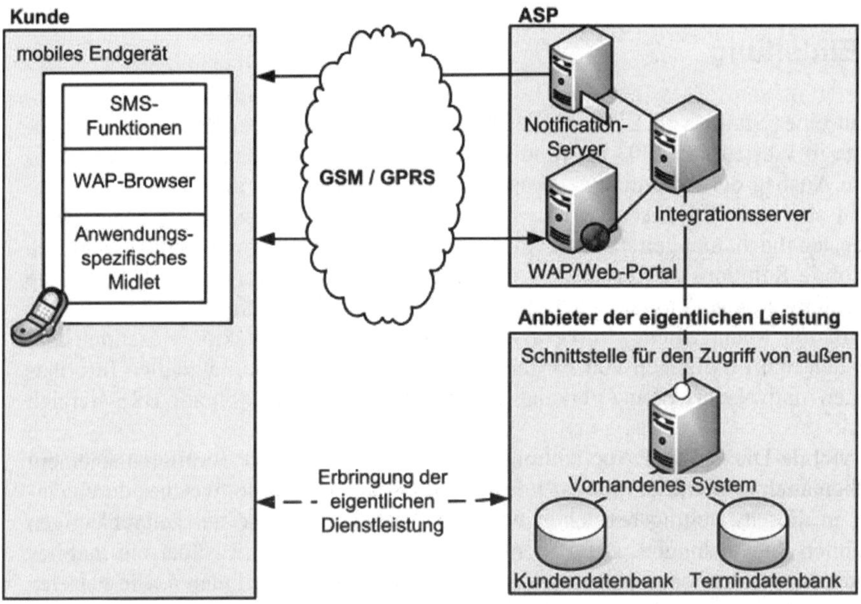

Fig. 1. Mobilnetz-Infrastuktur zur Terminkoordination zeitkritischer Geschäftsprozesse

der Arztpraxis). Bei traditionellen Ansätzen werden diese Ziele u.a. dadurch erreicht, dass immer ein gewisses Kundenkontigent „auf Vorrat" warten muss und sich deren Wartezeit beim Auftreten von priorisierten Sonderfällen (z.B. Unfallpatient) sogar noch unvorhergesehen verlängert. Durch intelligentes Management der demnächst zu bedienenden Kunden und die Verwendung von zeitnahen Benachrichtigungen lassen sich trotz der Beachtung der o.g. Rahmenbedingungen die Wartezeiten auf beiden Seiten minimieren. Dazu ist jedoch eine fortwährende Berücksichtigung der aktuellen Situation und eine gewisse Schätzung der aktuell ablaufenden Bedienvorgänge durchzuführen.

3 Ein generischer Ansatz: EQueue

EQueue (Electronic Queue) ist ein System, das ein solches virtuelles Wartezimmer (Warteschlange) simuliert. Kunden wird es ermöglicht, unter der Nutzung eines mobilen Endgerätes (Handy, PDA) und von jedem beliebigem Ort Termine zu vereinbaren und sich über deren Zustände zu informieren, z.B. wird er über etwaige Verspätungen und/oder Verschiebungen informiert.

Die Kommunikation zwischen Kunde und Dienstleister läuft folgendermaßen ab: – Der Kunde fragt einen Termin an – das EQueue-System gibt einen freien Termin zurück – der Kunde bestätigt diesen Terminvorschlag – das EQueue-System trägt den Termin ein und reiht den Kunden in die EQueue ein – das EQueue-System synchronisiert sich mit der Verwaltungssoftware (Terminvereinbarungssoftware) beim Dienstleister und informiert das zuständige Personal – das EQueue-System überwacht Termine bezüglich etwaiger Verspätungen und sendet dem Kunden Bestätigungen für seinen Termin – der Kunde nimmt den Termin wahr – das Bedienpersonal teilt dem EQueue-System das Eintreffen und Bedienen des Kunden mit – das EQueue-System überwacht etwaige Verspätungen und Verschiebungen – das Bedienpersonal informiert das EQueue-System über die abgeschlossene Bedienung.

3.1 Realisierungsansätze

Das EQueue-Gesamtsystem besteht aus zwei Hauptkomponenten (siehe Abbildung 2), dem EQueue-Applikationsserver (Anbieter) und dem EQueue-Client (Kunde). Es ist in Form eines J2EE-Systems [8] realisiert. Clients von EQueue (MIDLET) sind vorzugsweise mobile Endgeräte wie Handhelds (PDAs) und Mobiltelefone, die über ein GSM-Netz mit dem EQueue-Applikationsserver kommunizieren. Nachfolgend wird kurz auf die alternativen Entwicklungsansätze eingegangen und deren Vor- bzw. Nachteile im Abriss diskutiert.

- **Alternative I:** Service mit JSP/Servlet [11] / und EJB [4] - Bei diesem Lösungsansatz wird versucht, die Dienste des EQueue-Applikationsserver als Standardservices zu implementieren, die mit Hilfe von JSP (Java Server Pages) oder Servlets über das Internet zugänglich gemacht werden. Auf diese Art von

Diensten kann demzufolge über einen Web-Browser oder auch eine selbst entwickelte Software zugegriffen werden. Im Rahmen von EQueue war diese Variante nur bedingt geeignet, da die Kommunikation mit dem Server auf das HTTP-Protokoll beschränkt wäre.

- **Alternative II:** Getrennte Kommunikationskanäle - Die Limitierungen des vorherigen Ansatzes könnten durch die Etablierung eines zusätzlichen Kommunikationskanals überwunden werden. Dadurch wäre ein Kanal für diverse Endgeräte bezogen auf die Darstellungsdaten (HTTP, JSP/Servlet) reserviert, während ein zweiter (z.B. auf XML-Basis) Nutzdaten vom Client zum Server transportieren würde. Allerdings wäre die Verwendung zweier getrennter ein zu kleiner Gewinn an Flexibilität, als dass er den zusätzlichen Entwicklungsaufwand rechtfertigen würde.
- **Alternative III:** J2ME [9] und EJB [4] - Mit der Verwendung von J2ME könnte der EQueue-Client Daten direkt mit den EJBs im Applikationsserver austauschen oder deren Methoden aufrufen. Damit könnte auf JSPs verzichtet werden und ein zweiter Kommunikationskanal wäre ebenfalls überflüssig. Dieser Lösungsansatz schien für EQueue am besten geeignet und wurde deshalb für die Entwicklung von EQueue verfolgt.

3.2 EQueue-Architektur

EQueue baut auf dem JBoss Applikationsserver [10] auf und nutzt verschiedene Protokolle zur Kommunikation zwischen Server und Clients.

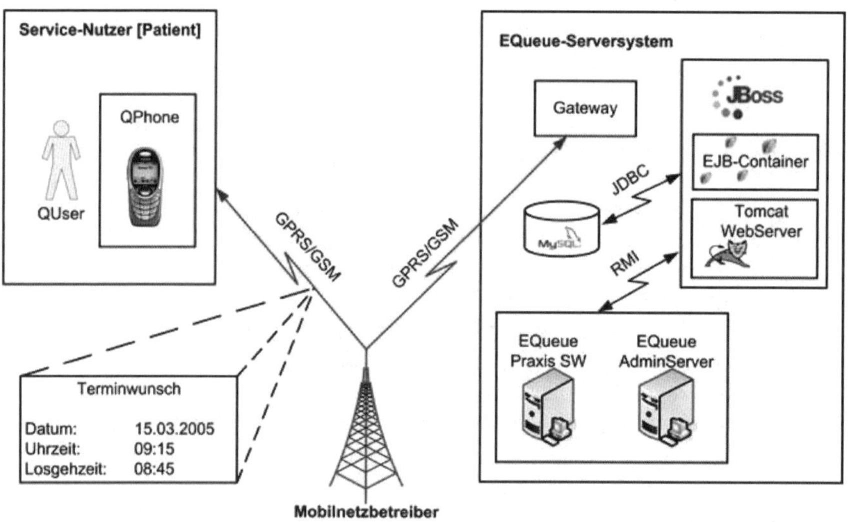

Fig. 2. EQueue Architektur

Zur Speicherung der Daten wurde die relationale Datenbank MySQL [15] verwendet. Die Software der EQueue-Clients ist in Java geschrieben. Der EQueue-Client, in Form einer J2ME-Applikation auf einem mobilen Endgerät (Handy), empfängt Daten vom Applikationsserver des Dienstanbieters und sendet die durch den Nutzer eingegebenen Informationen über GPRS an den EQueue-Applikationsserver (siehe Abbildung 2).

Um das System verwalten zu können, wurde eine Software mit dem Namen *EQueueAdminserver* implementiert, welche ausschließlich für den Betreiber von EQueue gedacht ist. Diese ist zuständig für die Verwaltung von systemtechnischen Daten, wie die Verwaltung der Queue usw. Für die Administration der relevanten Daten (Kundendaten, Termine) ist der Dienstleister (Artzpraxis) zuständig, entweder durch eine vorhandene ERP-Lösung (Praxissoftware) oder durch die *EQueue Praxis SW,* welche eine erweiterte Lösung der vorhandenen Praxissoftware darstellt (siehe Abbildung 2).

Die Software, welche auf dem Endgerät des Benutzers ausgeführt wird und ihm unter anderem eine Benutzeroberfläche zum Zugriff auf das Funktionsangebot des EQueue-Systems bietet, wird hier als EQueue-Client (*QPhone*) bezeichnet. Der Kunde hat die Möglichkeit, auf dem mobilen Endgerät diverse Profile zu hinterlegen. Einerseits kann er eigene Lokationen speichern, von denen er den Serviceleister (Arztpraxis) in einer bestimmten Zeit erreichen kann. Andererseits werden hier auch die unterschiedlichen Dienstleister (Ärzte), d.h. (Praxis-)Daten in Profilen hinterlegt.

4 Betriebswirtschaftliche Aspekte

Prinzipiell eignet sich der erstellte Prototyp als Basis für die Entwicklung eines Systems als mobiles Terminvereinbarungs- und Überwachungssystem für den privaten als auch geschäftstätigen Nutzer (z.B. für technische Wartungsdienste, Paket-Zustellservices, Anreiseunterstützung). Durch die Plattform- und Geräteunabhängigkeit lassen sich Clients für eine große Anzahl an verschiedenen Endgeräten implementieren. Der Mehrwert des EQueue-System zeigt sich konkret an folgenden Punkten:

- **Kosteneinsparungen:** Diese sind in personeller Hinsicht im Verwaltungsbereich durch die automatisierte und selbststeuernde Terminverwaltung möglich.
- **Zeiteinsparungen:** Eine gezielte und genauere Terminplanung (Melde- und Rückmeldemechanismen auf Kunden- und Dienstleisterseite) führt zu verkürzten Reaktionszeiten. Daraus ergeben sich optimierte Abwicklungszeiten, was wiederum zu besseren Durchlaufzeiten führt.
- **Verbesserung der Kundenzufriedenheit:** Die Optimierung im Zeitmanagement hat zur Folge, dass die Ablaufprozesse, d.h. die prozesstechnischen Behandlungsabläufe sowie die gegenwärtigen Bedingungen und Interaktionen im Behandlungsablauf genauer analysiert werden müssen [2, 3, 7]. Somit würden sich präzisere Behandlungszeiträume ergeben, welche die Grundlage für ein-

zelne Durchlaufzeiten bilden, die letztlich für eine optimierte Terminplanung unabdingbar sind. Ein weiterer Vorteil ist eine sich daraus ergebende Transparenz der behandlungstechnischen Maßnahmen auf Leistungsempfängerseite einerseits und auf Kostenträgerseite andererseits – somit könnte sich eine erhöhte Überprüfbarkeit der Leistungsentgelte könnte ergeben! – . Der Wert von aufgeklärten Patienten gegenüber ihren Gesundheits- und Behandlungsdaten ist in der Literatur schon mehrmals aufgezeigt worden [12, 14].

Referenzen

1. Amberg M, Daum M, Krcmar H (2002) Compass- Ein Kooperationsmodell für situationsabhängige mobile Dienste In: Hampe JF, Schwabe G (eds) Mobile and Collaborative Business 2002-Proceedings zur Teilkonferenz der Multikonferenz Wirtschaftsinformatik 2002, Nürnberg
2. Berger K (2004) Behandlungspfade als Managementinstrument im Krankenhaus In: Greiling M (ed) Pfade durch das Klinische Prozessmanagement: Methodik und aktuelle Diskussionen. Kohlhammer Stuttgart, pp 42-64
3. Eiff W, Ziegenbein R (2003) Entwicklung von Prozessmodellen im Krankenhaus. In: Eiff W, Ziegenbein R (eds) Geschäftsprozessmanagement, Methoden und Techniken für das Management von Leistungsprozessen im Krankenhaus, Bd.2. Bertelsmann Stiftung, Gütersloh
4. EJB Enterprise JavaBeans Technology, http://java.sun.com/products/ejb/
5. European Information Technology Observatory (2005), http://www.eito.com/start.html
6. Gatzke M, Heiders C (2004) Mobile Applications - Wo liegen die Chancen? , http://www.ecin.de/mobilebusinesscenter/applikationen/index-2.html
7. Greiling M, Hofstetter J (2002) Patientenbehandlungspfade optimieren, Prozessmanagement im Krankenhaus. Baumann Fachzeitschriften Verlag, Kulmbach
8. J2EE Java 2 Platform, Enterprise Edition, http://java.sun.com/j2ee/
9. J2ME Java 2 Platform, Micro Edition, http://java.sun.com/j2me/index.jsp
10. JBoss.com JBoss Application Server, http://www.jboss.org/products/jbossas
11. JSP JavaServer Pages Technology, http://java.sun.com/products/jsp/
12. Maysy D, Baker D, Butros A, Cowles KE (2002) Giving Patients Access To Their Medical Records Via The Internet: The PCASSO Experience Journal of the American Medical Association 9, pp 181-191
13. Meta Group Inc. (2005), http://www.metagroup.de/, http://www.gartner.com/
14. Munir S, Boaden R (2001) Patient Empowerment And The Electronic Health Record In: Patel V, Rogers R, Haux R (eds) MEDINFO 2001 Proceedings of the 10th World Congress on Medical Informatics, pp 663-665
15. MySQL MySQL, http://www.mysql.de/
16. o.V. (2004) Deutsche Unternehmen setzen auf mobile Lösungen, http://www.teletalk.de/teletalk-compact/Archiv/08.03.2004/newsletter_.htm
17. Purao S (2002) Truth or Dare: Design Research in Information Technologies, http://purao.ist.psu.edu/working-papers/dare-purao.pdf
18. Vaishnavi V, Kuechler B (2004) Design Research in Informations Systems, http://www.isworld.org/Researchdesign/drisISworld.htm

Biometrische Absicherung von Web-Applikationen mit BioW3

Götz Botterweck[1], J. Felix Hampe[2] und Sven Westenberg[1]

[1] Institut für Wirtschafts- und Verwaltungsinformatik, Universität Koblenz-Landau, Universitätsstr. 1, 56070 Koblenz, {botterwe,westi}@uni-koblenz.de

[2] School of Computing and Information Science, University of South Australia, City West Campus, 27-29 North Terrace, SM1-06, Adelaide SA 5000, Australia felix.hampe@unisa.edu.au

1 Anforderungen an biometrische Sicherheitsmechanismen für E-Commerce Applikationen

1.1 Biometrische Sicherheitssysteme

Für biometrische Sicherheitssysteme lassen sich drei grundlegende Aufgaben unterscheiden [4]: Beim **Enrollment** werden neue Benutzer in das System aufgenommen. Dabei werden erstmals biometrische Merkmale gemessen und daraus Kontrolldatensätze (Templates) erstellt, die im späteren Betrieb als Referenz herangezogen werden. Dann können Benutzer auf zwei Arten überprüft werden: Bei einer **Verifikation** wird der aktuelle Benutzer auf Übereinstimmung mit *einer* bestimmten Identität überprüft.[1] Bei einer **Identifikation** wird die Identität des aktuellen Benutzers durch Suche in einer Datenbank von *n* Identitäten und Matching mit den entsprechenden biometrischen Templates festgestellt oder die Ausgabe „Benutzer unbekannt" geliefert.

Ein biometrisches System kann als **binärer Klassifikator** modelliert werden, der *versucht*, eine Ja-Nein-Antwort auf die Frage zu liefern, ob der aktuelle Benutzer mit der vorgeblichen Identität übereinstimmt. Die Qualität eines solchen Verfahrens lässt sich mit Fehlerraten wie einer **False Accept Rate (FAR)** oder **False Reject Rate (FRR)** bewerten. Es ist zwar wünschenswert beide Fehlerraten zu minimieren, leider führt aber eine Verringerung der FAR in der Regel zu einer

[1] Die Verifikation entspricht ungefähr der Authentifikation in der allgemeinen IT-Sicherheit.

Erhöhung der FRR und umgekehrt. Ein **Receiver Operating Characteristic (ROC)** Diagramm bietet eine grafische Darstellung dieses Zusammenhangs [8].

1.3 Allgemeine Anforderungen

Beim Einsatz von biometrischen Sicherheitssystemen ist zunächst ein geeignetes biometrisches Merkmal zu wählen. Hier sind neben grundsätzlichen Anforderungen (Vorhandensein bei allen potentiellen Benutzern, Unterscheidungskraft, Langlebigkeit, Messbarkeit) Aspekte wie die erreichbare Erkennungsgenauigkeit, die dafür notwendigen Kosten, der Zeitaufwand für einen Erkennungsvorgang, die Akzeptanz durch Benutzer und die Möglichkeit einer Manipulation des Messvorganges zu berücksichtigen [4]. Zusätzlich müssen praktische Anforderungen wie die Integration mit existierenden Systemen bedacht werden.

Im Kontext von E-Commerce Anwendungen ergeben sich aus der räumlichen und organisatorischen Distanz zwischen Benutzer und Betreiber besondere Herausforderungen etwa beim Enrollment (Wie stellt man eine erste zuverlässige Identifikation sicher?) oder falls zusätzliche Hardware auf Nutzerseite benötigt wird (Fingerprint-Leser für alle Amazon-Kunden?). Für eine konkrete Anwendung können die Fehlerraten in Abhängigkeit von den Rahmenbedingungen optimiert werden (hohe FAR oder hohe FRR?). Hier sind u. a. die „Kosten“ für falsche Entscheidungen zu berücksichtigen (z. B. bei einem frustrierten Kunden) [8].

1.4 Ziele des BioW3-Ansatzes

Hauptziel des hier vorgestellten Ansatzes war die einfache Ergänzung eines Standard-Webservers durch eine biometrische Authentifikation. Dabei waren ein modularer Aufbau und die Verwendung von offenen Schnittstellen wichtig. So sollte es möglich sein, eine existierende Webserver-Installation durch Hinzufügen unserer Module zu ergänzen. Weiterhin sollte die Absicherung unabhängig von den zu schützenden Daten oder Anwendungen realisiert werden. Außerdem wurde eine möglichst einfache Konfiguration angestrebt. Der Mechanismus sollte einen Schutz gegen Replay-Attacken und „Session Hijacking” (Übernehmen einer bereits authentifizierten Benutzersitzung) bieten und offen für den Betrieb mit verschiedenen Telefonnetzen, insbesondere Voice-over-IP-Netzen sein.

2 Absicherung von Web-Applikationen mit BioW3

2.1 Ablauf einer Session aus Benutzersicht

Der Benutzer kommt zum ersten Mal in Kontakt mit BioW3, wenn er eine geschützte Ressource vom Webserver anfordert. Falls er noch nicht beim System angemeldet ist, wird er auf eine Login-Seite umgeleitet, die ihm die nötigen Informationen zum weiteren Vorgehen anzeigt. Dies sind die Telefonnummer und ein Text den er später bei der Abgabe seines Voiceprints vorlesen muss.

Der Benutzer ruft die angezeigte Telefonnummer (mit einem VoIP- oder einem Festnetzanschluss) an und befolgt die angesagten Hinweise. Dabei wird er zunächst durch Sprechen seines Namens oder seiner persönlichen Identifikationsnummer identifiziert und dann seine Identität durch Vorlesen des vorher angezeigten Textes verifiziert. Nach der Aufnahme aller Stimmproben entscheidet die Sprecher-Verifikation, ob dem Benutzer der Zugang erteilt oder verweigert wird. Bei erfolgreicher Verifikation kann der Benutzer über einen Link in der Login-Seite seine – jetzt autorisierte – Sitzung fortsetzten. Bei Misserfolg wird der Zugang verwehrt und der Grund in einer Fehlermeldung erklärt.

2.2 Entwurfsentscheidungen

Hauptgrund für die Entscheidung für Sprache als biometrisches Merkmal war die angestrebte Kombination mit Web-Anwendungen, bei denen eine einfache Anbindung entfernter Benutzer notwendig ist. Dabei ist eine Distribution/Installation zusätzlicher biometrischer Hardware problematisch. Wegen allgemeiner Verfügbarkeit erschien daher die Verifikation von Sprachcharakteristika über ein Telefon- oder VoIP-Netz gegenüber anderen biometrischen Merkmalen [4] vorteilhaft. Weiterhin spricht für dieses Verfahren die einfache Handhabung und relativ hohe Benutzerakzeptanz. Im beabsichtigten Anwendungskontext wird dafür das etwas niedrigere Sicherheitsniveau als akzeptabel angesehen.

Weitere Entwurfsentscheidungen treten bei der Festlegung der Vorgehensweise bei der Anmeldung auf. Dabei ist zwischen einer einfachen Bedienung und einem hohen Sicherheitsniveau abzuwägen. Neben den Parametern und Schwellwerten für den Verifikationsvorgang sind hier die zeitlichen Abläufe zu bedenken: Wie lange hat ein Benutzer Zeit, nach der Anzeige der Login-Informationen die angezeigte Telefonnummer anzurufen? Wie lange darf ein Benutzer seine Sitzung (ohne Request) ruhen lassen, bevor diese Sitzung automatisch beendet wird?

Bei der Konzeption sind auch potentielle Angriffsmöglichkeiten und entsprechende Gegenmaßnahmen zu bedenken (Tabelle 1).

Tabelle 1. Angriffsmöglichkeiten und entsprechende Gegenmaßnahmen

Angriffsmöglichkeit	Gegenmaßnahmen
Abhören der HTTP-Verbindung	Verwendung von HTTPS
Man-in-the-Middle-Attacke	Identifikation des echten Servers über Zertifikate
„Hijacken“ einer authentifizierten Session z. B. über IP- oder DNS-Spoofing	Kombination von Cookies und Plausibilitätskontrolle von Verbindungsmerkmalen
Replay-Attacke (Wiederverwenden eines bereits verwendeten Sicherheitsmerkmals)	Variation der vorzulesenden Texte

2.3 Architektur

Das BioW3-System besteht im Wesentlichen aus drei Komponenten (Fig. 1): Einem Webserver, der um das BioW3-Authentifikationsmodul erweitert wurde, einem Speech-Server zur Abwicklung der Sprecherverifikation über eine Telefon- oder VoIP-Schnittstelle und einer Datenbank zur zentralen Speicherung von benutzer- und sessionbezogenen Daten. Wenn eine komplexere Webanwendung gesichert wird, kann dazu noch ein Back-End, wie etwa ein ERP-System hinzukommen. Als Grundlage für die prototypische Realisierung wurden der Apache HTTP Server Version 2.0, Nuance Speech Verifier 3.5 sowie ein MySQL-Datenbankserver eingesetzt. Die Komponenten können je nach Anwendungskontext und Performance-Anforderungen auf mehrere Rechner verteilt werden.

2.4 Ablauf einer Session aus interner Sicht

Der Apache HTTP Server [1] ermöglicht über die Verwendung von sogenannten „Hooks" die Anmeldung von externen Funktionen, die an bestimmten Zeitpunkten während der Verarbeitung der Request-Response-Loop aufgerufen werden. Diese Loop ist in zahlreiche Phasen unterteilt, von denen hier vor allem die Security Phase interessant ist. Zur Realisierung des BioW3-Authentifikationsmoduls wurden u. a. zwei Funktionen als Handler für die Security Phase implementiert:

Der **check_id Handler** überprüft zunächst, ob der aktuelle Zugriff zu einer bereits authentifizierten Session gehört oder die Session wegen zu langer Nicht-Benutzung abgelaufen ist. Dann werden einige Plausibilitätstests zum Schutz gegen Manipulationen durchgeführt. Je nach Ergebnis wird der Benutzer dann auf ein Login-Formular oder eine Zugriff-Verweigert-Meldung umgeleitet. Dabei verwenden wir einen internen Redirect [2], um eine Verzögerung durch unnötige weitere HTTP-Requests zu vermeiden und um besondere Inhalte (Login-Formular)

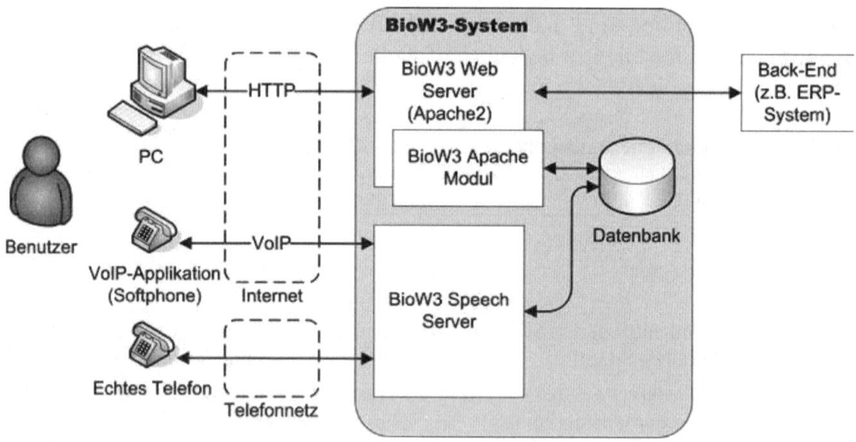

Fig. 1. Architektur des BioW3-Systems

nicht über eine eigene URL nach außen sichtbar zu machen.

Im **auth_checker Handler** werden die eigentlichen Entscheidungen über Zugriff oder Verweigerung gefällt. Einige Seiten (Login-Formular, Fehlermeldungen) sind jederzeit auch ohne Authentifikation zugreifbar. In allen anderen Fällen, wird überprüft, ob die aktuelle Ressource laut Konfiguration (.htaccess-Datei) mit dem BioW3-Modul geschützt werden soll und welche Benutzer Zugriffsrechte haben.

Während des Betriebs von BioW3 werden alle Informationen, die systemübergreifend verfügbar sein müssen in einer zentralen Datenbank gespeichert. Hierzu zählen insbesondere Angaben zur Verwaltung von gerade aktiven Benutzersitzungen sowie Informationen zur Plausibilitätskontrolle (IP-Adresse, verwendeter Webbrowser, Zeitpunkt des letzten Zugriffs, ...), die zur Abwehr von Session-Hijacking-Attacken dient.

2.5 Identifikation und Verifikation des Sprechers

Unter den verschiedenen Methoden, die der Nuance Speech Verifier zur Identifikation und Verifikation anbietet, haben wir uns für eine Methode entschieden, die es einerseits dem Benutzer erlaubt, sich ohne Vorabidentifikation, nur durch Sprechen seines Namens zu identifizieren und die andererseits für den Umgang mit größeren Benutzerdatenbanken geeignet ist („large scale identity claim“) [6]. Um die Erkennungsgenauigkeit zu erhöhen, können die Benutzer zusätzlich in Gruppen eingeteilt werden. Sie werden dann zu Beginn des Dialogs z. B. nach ihrer Abteilung oder ihrem Wohnort gefragt.

2.6 Verschiedene Telefonnetze

Die BioW3-Applikation ist für den Einsatz mit einem herkömmlichen Telefonnetz aber auch über VoIP-Verbindungen (z. B. via SIP) geeignet. Dabei ergeben sich je nach Telefonnetz Vor- und Nachteile. So kann bei Verwendung eines Mobilfunknetzes die personenbezogene Telefonnummer (Caller ID) als weiterer Faktor in die Authentifikation einbezogen werden. Bei VoIP ist eine Übertragung z. B. auch per Softphone-Applikation möglich, es wird außer einem Mikrofon oder Headset keine weitere Hardware benötigt. Andererseits wird hier die Authentifikation über ein zusätzliches getrenntes (Telefon-) Netz aufgegeben. Es müssen daher gesonderte Sicherheitsüberlegungen angestellt werden (z. B. Replay-Attacke durch Aufzeichnen und Wiederverwenden einer VoIP-Sprachverbindung).

3 Andere Arbeiten in diesem Themengebiet

Es gibt bereits zahlreiche Forschungsprojekte und kommerzielle Produkte, die Sprache als biometrisches Merkmal verwenden [7, 9]. Diese zielen aber häufig auf die Absicherung einer Call-Center- oder Telefonie-Anwendung. Weitere Ansätze kombinieren Sprache mit anderen biometrischen Merkmalen um eine „Multi-Faktor-Authentifikation" durchzuführen und damit eine höheres Sicherheitsniveau zu erreichen. Interessant erscheint hier Biobex [5] u. a. wegen der Integration mit der J2EE-Architektur. e-SentriNet [3] ist von der grundsätzlichen Idee mit unserem Ansatz vergleichbar (biometrische Absicherung von Web-Applikationen), verwendet jedoch Fingerabdrücke und erfordert daher besondere Hardware auf Nutzerseite.

4 Fazit und Ausblick

Nach einer kurzen Diskussion der Anforderungen an biometrische Verfahren im Kontext von E-Commerce haben wir den Ansatz **BioW3** vorgestellt, der es ermöglicht, eine Web-Applikation einfach und ohne die Distribution/Installation zusätzlicher Hardware auf Seiten des Nutzers durch biometrische Sicherheitsmechanismen abzusichern. Ein mögliches Einsatzgebiet wäre das Identitätsmanagement (z. B. für die Durchführung von Password Resets), denkbar ist aber generell jede Web-Anwendung mit Sicherheitsbedarf.

Referenzen

1. Apache Software Foundation (2005) The Apache HTTP Server Project, http://httpd.apache.org/
2. Apache Software Foundation (2005) Request Processing in Apache 2.0, http://httpd.apache.org/docs-2.0/developer/request.html
3. e-SentriNET Authentication for Web Servers, http://www.isl-biometrics.com/products/e-sentrinet.htm
4. Maltoni D, Maio D, Jain AK, Prabhakar S (2003) Handbook of Fingerprint Recognition. Springer, Berlin Heidelberg New York, http://bias.csr.unibo.it/maltoni/handbook/
5. Nagappan R, Lampinen T (2005) Building Biometric Authentication for J2EE, Web, and Enterprise Applications, http://developers.sun.com/prodtech/identserver/reference/techart/bioauthentication.html
6. Nuance Communications (2004) Nuance Verifier Developer's Guide Version 3.5, Menlo Park, CA, U.S.A.
7. saflink Corporation SAFLINK Network Security Products, http://www.saflink.com/network/
8. van Schalkwyk J (2001) The magnificent ROC (Receiver Operating Characteristic curve), http://www.anaesthetist.com/mnm/stats/roc/
9. Vocent Solutions Inc. (2003) Voice Authentication Solutions, http://www.vocent.com/products/

Performance-Measurement- und Analyse-Konzepte im Hochschulcontrolling

Jonas Rommelspacher[1], Lars Burmester[2], Matthias Goeken[3]

[1 2] Institut für Wirtschaftsinformatik Philipps-Universität Marburg
[3] HfB - Business School of Finance & Management

1 Einleitung

Die Universitäten und Hochschulen Deutschlands stehen vor gravierenden Veränderungen. Auslöser hierfür sind zum einen geänderte gesetzliche Rahmenbedingungen, die insbesondere die Finanzierung und die Organisations- bzw. Entscheidungsstrukturen betreffen. Andererseits bedingt die anhaltende Mittelknappheit, dass Hochschulen und deren Fakultäten intern wie extern im Wettbewerb um knappe Ressourcen stehen (Müller-Böling und Schreiterer 1999; Ziegele 2002).

In Hessen wurde im Zuge der Novellierung des Hochschulgesetzes (HHG) ein neues Steuerungsmodell für Hochschulen etabliert. Wesentliche Elemente sind der Abschluss von Zielvereinbarungen und die leistungsorientierte Mittelzuweisung.

Zielvereinbarungen werden zwischen der Hochschulleitung und dem Ministerium abgeschlossen und beinhalten Leistungs- und Entwicklungsziele der Hochschule für einen fest definierten Zeitraum (Weber 2003). Mit der leistungsorientierten Mittelzuweisung wurde ein zweiter Koordinationsmechanismus eingeführt, der einem Paradigmenwechsel in der Hochschulfinanzierung gleichkommt. Statt der bislang geltenden inputorientierten Finanzierung, werden Finanzmittel nun output- bzw. ergebnisorientiert auf Grundlage wohldefinierter Formeln als Globalbudget zugewiesen. So richtet sich bspw. das Grundbudget der Hochschule nach den Studierenden in der Regelstudienzeit, wobei für jeden Studierenden ein kostenorientierter Festbetrag zugewiesen wird. Ähnliche Budgetmodelle existieren für die Bereiche Forschung, wissenschaftliche Nachwuchsförderung und Internationalität. Über das zugewiesene Budget können die jeweiligen Leistungseinheiten eigenverantwortlich verfügen.

Insgesamt ergibt sich durch diese Maßnahmen ein erheblicher Aufgaben- und Kompetenzzuwachs für die Entscheidungsträger der Hochschule (Hochschul- bzw. Fakultätsleitung). Zur adäquaten Wahrnehmung dieser Aufgaben bedarf es diverser Führungsinstrumente und Entscheidungsgrundlagen bezüglich der Kernprozesse (Lehre, Forschung, Service) (Küpper und Zboril 1997). Zur Schaffung eines den Aufgaben gerecht werdenden Führungsinstrumentariums wurde 2002 am Fachbereich Wirtschaftswissenschaften der Philipps-Universität Marburg das Projekt Effizienz und innovative Steuerung von Fachbereichen (EiSFach) initiiert (Goeken und Burmester 2004).

Im Folgenden wird zunächst das im Rahmen des Projekts EiSFach erarbeitete Performance-Measurement-Konzept sowie dessen bisherige Implementierung vorgestellt (2). Darauf aufbauend werden die Erweiterungen um eine Balanced-Scorecard-Lösung, Effizienzmessung mittels Data-Envelopment-Analyse und Zeitreihenanalyse erörtert (3). Der Beitrag schließt mit Fazit und Ausblick (4).

2 Implementierte Lösung

Gegenstand des Pilotprojekts EiSFach ist die Entwicklung eines Controllingkonzepts und eines Kennzahlensystems sowie die informationstechnische Umsetzung mittels eines Data-Warehouse-Systems. In seiner ersten Ausbaustufe folgt das Controllingkonzept und dessen Umsetzung einer informationsorientierten Controllingkonzeption. Das Ziel besteht darin, Transparenz über die primären Leistungsprozesse Lehre und Forschung herzustellen. Die erwünschten Folgen wären bspw. die leichtere Identifikation von Problembereichen in den Prozessen oder die positive Beeinflussung von Willensbildungs- und Entscheidungsprozessen mit positiven Effekten für die Qualität von Studium und Lehre.

Als Datenbasis für die Implementierung dienen insbesondere die Stammdaten der Studierenden, Prüfungsdaten, Personaldaten sowie Daten der Lehrevaluation. Diese wurden zunächst aus den operativen Datenquellen in ein Data Warehouse geladen, konsolidiert und zur weiteren Verwendung multidimensional, in Form von OLAP-Hypercubes aufbereitet. Auf Grundlage dieses konsolidierten Datenbestands wurden einige grundlegende Berichte, Auswertungen und Analysen implementiert. Hierbei handelt es sich bspw. um Auslastungs- und Deckungsbeitragsrechnungen der einzelnen Abteilungen, Auswertungen der Studierendenstatistiken sowie Kohorten- bzw. Studienverlaufsanalysen (Goeken und Burmester 2004).

Die dabei realisierten Berichte sind im Wesentlichen vergangenheitsorientiert. Der Schwerpunkt liegt auf dem Tagesgeschäft, sodass nur wenig Handlungsempfehlungen für die Steuerung bzw. konkrete Entscheidungstatbestände generiert werden. Diese Defizite versucht die beschriebene Fortentwicklung zu überwinden.

3 Analytische Erweiterungen

3.1 Balanced Scorecard

Nachdem im bisherigen Projektverlauf mittels einer informationsorientierten Controllingkonzeption die informatorische Basis gelegt wurde, soll die Lösung gemäß einer planungs- und kontrollorientierten Controllingkonzeption (Horvath 2003) weiterentwickelt werden. Dafür bieten sich insbesondere Performance-Measurement-Systeme an. Im Gegensatz zu klassischen Finanzkennzahlensystemen verfügen sie über keine Spitzenkennzahl und scheinen geeigneter, den in der Realität existierenden Zielpluralismus abzubilden. Ein bekanntes System ist die von KAPLAN/NORTON entwickelte Balanced Scorecard (BSC) (Kaplan und Norton 1992). Dort wird die Unternehmensstrategie in monetäre und nichtmonetäre Kennzahlen überführt und es werden Ursache-Wirkungs-Zusammenhänge zwischen Kennzahlen unmittelbar verdeutlicht. Obgleich die BSC in ihrer originären Form für den privatwirtschaftlichen Sektor entwickelt wurde, wird sie aufgrund der Integration nichtmonetärer Kennzahlen dem Zielpluralismus und Sachzielcharakter einer Universität eher gerecht als eine rein finanzielle Rechnung (z.B. Return on Investment).

RÖBKEN empfiehlt aufgrund des zu erwartenden Widerstandes an großen deutschen Volluniversitäten die Einführung auf Fachbereichs- und Institutsebene und sieht hier die Möglichkeit von Motivationssteigerungen der Mitarbeiter durch größere Transparenz (Röbken 2003).

Basierend auf den guten Erfahrungen des Einsatzes in Hochschulen (vgl. Röbken 2003) wurde eine BSC zur Steuerung des Fachbereiches Wirtschaftswissenschaften der Philipps-Universität Marburg aufgebaut. Die Kennzahlen werden aus dem existierenden Data Warehouse befüllt. Zur Umsetzung wird die Software *SAS Strategic Performance Management* verwendet.

Aus der Grundkonzeption von KAPLAN/NORTON werden die Perspektiven *Finanzen*, *Potenziale* und *Kunden* übernommen. Statt der *Internen Prozessperspektive* wird die Perspektive *Öffentlicher Auftrag* implementiert. In Fig. 1 sind neben den genannten Perspektiven die strategischen Ziele und die mit ihnen verknüpften Kennzahlen dargestellt. Zur Verwendung der BSC sind zudem Zielvorgaben für die jeweiligen Kennzahlen festzulegen, um deren Leistungen vergleichen zu können. Vor dem Produktivbetrieb muss die BSC intensiv mit den beteiligten Entscheidungsträgern diskutiert werden, wobei der vorgestellte Prototyp als Stimulus für einen strategiebildenden Diskurs dient.

FB02BSC Scorecard: Key Metrics Table, Score 0.64

Measures	Status	Actual	Target	Perf
Angeworbene Drittmittel im Verhältnis zum Gesamtbudget	1,0160	0.1016	0.1	1,0160
Zitationen (SSCI) in den letzten 5 Jahren	1,0000	47	47	1,0000
ausländische Studierende im Verhältnis zu allen Studierenden	1,0260	0,1539	0,15	1,0260
Frauenquote	0,8080	0.404	0.5	0,8080
Durchschnittliche Fachstudiendauer	0,8700	11,5	10	0,87
Betreueungsquote Mitarbeiter je Student	1,3600	0,068	0,05	1,3600
Sachmittel ITAusstattung im Verhältnis zum Gesamtbudget	0,7200	0.0072	0.01	0,7200
Sachmittel Bibliothek in Verhältnis zum Gesamtbudget	0,5300	0,0106	0,02	0,5300
Personalausstattung nichtwissenschaftliches Personal	0,9947	14.92	15	0,9947
Studierende in der Regelstudienzeit	0,8826	1.111,00	1.258,8	0,8826
Deckungsbeitrag	3,0071	1.503.530	500.000	3,0071

Fig. 1. Performance, Traffic Lighting & Exception Reporting der BSC (Screenshot SAS)

Die Wahrnehmung der Leistung eines Kennzahlenbereichs kann dem Nutzer durch diverse Visualisierungsformen (Traffic-Lighting, Pfeile etc.) erleichtert werden. Auch andere Darstellungen, wie z.B. Spinnennetz- und Radardiagramme werden verwendet. Zudem werden aus Gründen der Transparenz dem Nutzer die Beziehungen zwischen den einzelnen Kennzahlen veranschaulicht.

3.2 Data-Envelopment-Analyse

Im bisherigen Projektverlauf wurde von den originären universitären Leistungsprozessen Forschung und Lehre ausgegangen. Bei bisherigen Berichten und Analysen wurden sie jedoch v. a. getrennt voneinander betrachtet und, daraus folgend, der Output nicht unmittelbar mit dem Input in Verbindung gebracht.

Sowohl zur Messung von Ineffizienzen als auch zur Integration nicht monetär bewertbarer Größen eignet sich die Data-Envelopment-Analyse (DEA) (Charnes et al. 1978). Die elementare Neuerung der DEA ist, dass multiple Inputs bzw. Outputs nicht exogen gewichtet werden müssen, sodass – anders als bei Produktionsfunktionen – keine Annahmen über funktionale Zusammenhänge zwischen Input- und Outputfaktoren erforderlich sind. Es werden für jede einzelne Entscheidungseinheit Effizienzwerte als Quotient der Summe der gewichteten Outputs im Verhältnis der Summe der gewichteten Inputs errechnet. Die effizientesten Entscheidungseinheiten liegen auf der Effizienzgrenze und markieren somit die zu erreichende Benchmark.

Ziel der DEA ist es, die Effizienz der untergeordneten Entscheidungseinheiten (Lehrstühle, Institute etc.) zu messen. Als Proxy für den Input dient die Anzahl der Mitarbeiter. Für die o. g. Prozesse sollen die Anzahl der abgenommenen Prüfungen und die Anzahl der Publikationen als Outputkennzahlen fungieren. Da der Input mittelfristig konstant ist, bietet sich ein outputorientiertes DEA-Modell an. Die Effizienzberechnung wird mit der Software *SAS Enterprise Guide* durchgeführt. Untenstehend (Fig. 2) sind für ausgewählte Lehrstühle die Inputs und Outputs sowie die errechneten Ergebnisse für eine Periode dargestellt.

DMU	Input	Output-Forschung	Output-Lehre	Effizienz	Gewichteter Output Forschung	Gewichteter Output Lehre	Slack Forschung	Slack Lehre
Lehrstuhl 1	3	3	75.5	1.00	0.075	0.925	0.000	0.000
Lehrstuhl 2	3	0	50.5	0.67	0.000	1.000	3.000	0.000
Lehrstuhl 3	4	6	54.5	0.61	0.182	0.818	0.000	0.000
Lehrstuhl 4	5.67	6	82	0.61	0.129	0.871	0.000	0.000
Lehrstuhl 5	5.67	40	74.5	1.00	0.521	0.479	0.000	0.000
Lehrstuhl 6	3.5	11	49.5	0.75	0.310	0.690	0.000	0.000

Fig. 2. Inputs, Outputs und Effizienzmaße (Beispieldaten ausgewählter Lehrstühle)

In der zweiten Spalte ist die berechnete Effizienz dargestellt. Lehrstuhl 1 und Lehrstuhl 5 sind effizient. Die Lehrstühle 3, 4 und 6 müssen ihre Outputs proportional um 64 %, 64 % bzw. 33 % steigern, um effizient zu sein. Etwas anders liegt der Fall bei Lehrstuhl 2. Dort muss, ehe mittels einer proportionalen Steigerung die Effizienzgrenze erreicht werden kann, der Forschungsoutput erhöht werden, da die Schlupfvariable *Slack Forschung* ungleich null ist.

Positiv hervorzuheben an der vorgestellten Effizienzmessung ist die endogene Ermittlung der Gewichte und die konkret ableitbaren Handlungsempfehlungen. Problematisch ist insbesondere der verwendete Indikator für den Forschungsoutput, welcher die Leistung lediglich in einer quantitativen Dimension erfasst. LUPTÁCIK schlägt daher eine Erweiterung auf (qualitative) Publikationsklassen und Gewichtsbeschränkungen der erstellten Klassen vor (Luptácik 2003). Diesem Ansatz wurde nicht gefolgt, um nicht wertend in das Modell einzugreifen.

3.3 Zeitreihenanalyse

Aufgrund des funktionalen Zusammenhangs zwischen Leistung und Mittelzuweisung an Hochschulen sind zur Entscheidungsunterstützung Zukunftsprognosen der budgetrelevanten Kennzahlen notwendig. Zur Vorhersage werden modellgestützte Prognoseverfahren, wie etwa ARIMA(p,d,q)-Prozesse (Box und Jenkins 1976) eingesetzt. Diese erfreuen sich zunehmender Beliebtheit, da die mit ihnen erstellten Kurzfristprognosen den Ergebnissen naiver Prognoseverfahren (z.B. exponentielles Glätten) meist überlegen sind. Zur Umsetzung wird der *SAS Enterprise Guide* eingesetzt.

Basierend auf den vorhandenen Daten wird nach einer saisonalen Bereinigung eine ARIMA-Prognose durchgeführt. Vor der Durchführung werden die Ordnungen des ARIMA-Prozesses mit dem Informationskriterium von BAYES (BIC) bestimmt. Untenstehend ist die Prognose der Kennzahl „Anzahl der Studierenden" dargestellt: Es ist deutlich zu erkennen, dass sich der stochastische Prozess (durchgezogene Linie) den saisonbereinigten Daten (gestrichelte Linie) anpasst. Die implementierten Prognoseverfahren lassen sich auf alle in der Data-Warehouse-Datenbank enthaltenen Zeitreihen anwenden und könnten etwa für Erlösprognosen verwendet werden.

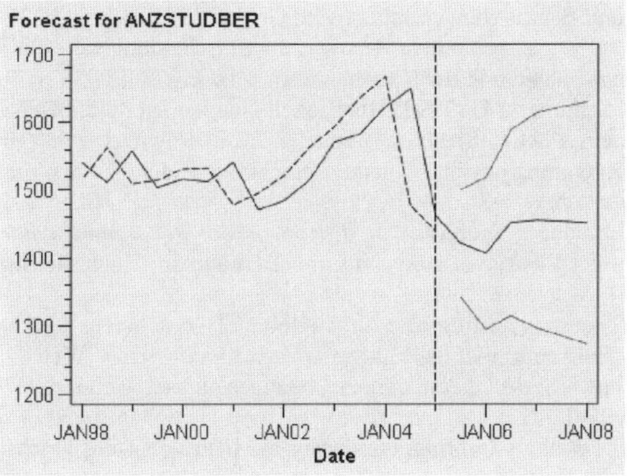

Fig. 3. Prognose mit ARIMA (4,1,0)-Prozess und 95 % Konfidenzintervall).

4 Fazit und Ausblick

Performance-Measurement- und Analyse-Konzepte können eine bestehende, im Wesentlichen auf die Informationsbereitstellung ausgerichtete Data-Warehouse-Lösung erweitern, sodass ein höheres Maß an Entscheidungsunterstützung erzielt werden kann. In diesem Beitrag wurde die Erweiterung eines bestehenden Data-Warehouse-Systems um eine BSC, eine Effizienzmessung mittels DEA sowie eine Zeitreihenanalyse für das Hochschulcontrolling vorgestellt. Die genannten Konzepte erweisen sich als grundsätzlich anwendbar in diesem Einsatzgebiet. Da auf einem vorhandenen Data Warehouse aufgebaut werde konnte, war die vorgestellte Erweiterung mit einem überschaubaren Aufwand zu realisieren.

Zur Lösung zukünftiger Herausforderungen an Hochschulen ist jedoch nicht alleine die Verfügbarkeit des Instrumentariums von Relevanz, sondern vielmehr seine Akzeptanz und aktive Nutzung zur Entscheidungsfindung.

Literaturverzeichnis

Box G, Jenkins G (1976) Time Series Analysis-forecasting and control. Revised Edition, Holden-Day-Verlag, San Francisco et al.

Charnes A et al. (1978) Measuring the efficiency of decision-making units. European Journal of Operational Research, Vol. 2 (1978), Issue 6, S. 429-444

Goeken M, Burmester L (2004) Entwurf und Umsetzung einer Business-Intelligence-Lösung für ein Fakultätscontrolling. In: Chamoni P et al. (Hrsg) Multikonferenz Wirtschaftsinformatik (MKWI) 2004, Band 2, INFIX, Berlin, S. 137-152

Horváth P (2003) Controlling. 9. Auflage, Vahlen-Verlag, München

Kaplan R, Norton D (1992) The Balanced Scorecard-Measures That Drive Performance. In: Harvard Business Review, Vol. 70 (1992), Issue 1, S. 71-79

Küpper HU, Zboril NA (1997) Rechnungszwecke und Struktur einer Kosten-, Leistungs- und Kennzahlenrechnung für Fakultäten. In: Becker W, Weber J (Hrsg.) Kostenrechnung. Stand und Entwicklungsperspektiven, Gabler, Wiesbaden, S. 337-365

Luptácik M (2003) Data Envelopment Analysis als Entscheidungshilfe für die Evaluierung von Forschungseinheiten in der Universität. In: ZfB Erg. 3 (2003), S 59-73

Müller-Böling D, Schreiterer U (1999) Hochschulmanagement durch Zielvereinbarungen – Perspektiven eines neuen Steuerungsmodells. In: Fedrowitz J et al. (Hrsg.) Hochschulen und Zielvereinbarungen – neue Perspektiven der Autonomie, Bertelsmann-Stiftung, Gütersloh, S. 9-25

Röbken H (2003) Balanced Scorecard als Instrument der Hochschulentwicklung – Projektergebnisse an der Reykjavik University. In: Beiträge zur Hochschulforschung. 25. Jg. (2003), Heft 1, S. 102-121

Weber H (2003) Steuerungsinstrumente für Autonome Hochschulen. Zielvereinbarungen Land Hochschule einschließlich Budgetierung. „Der hessische Weg". http://www.hmwk.hessen.de/md/content/sonstiges/aufsatz_weber.pdf Abruf am: 2005-07-21

Zboril NA (1998) Fakultäts-Informationssystem als Instrument des Hochschulcontrolling. Dissertation, Stuttgart

Ziegele F (2002) Reformansätze und Perspektiven der Hochschulsteuerung in Deutschland. Beiträge zur Hochschulforschung 24 (3), S. 106-121

Risikoanalyse und Auswahl von Maßnahmen zur Gewährleistung der IT-Sicherheit

Brigitte Werners und Philipp Klempt

Institut für Sicherheit im E-Business (ISEB), Fakultät für Wirtschaftswissenschaft, Ruhr-Universität Bochum, 44780 Bochum

Abstract. Zur Gewährleistung der IT-Sicherheit in einem Unternehmen sind zunächst die Anforderungen zu spezifizieren, der erreichte Sicherheitsstand zu beurteilen und anschließend diejenigen Sicherheitsmaßnahmen zu ermitteln, deren Umsetzung eine Optimierung des Sicherheitsniveaus bewirkt. Wurden im Rahmen der Schutzbedarfsfeststellung IT-Komponenten mit hohem oder sehr hohem Schutzbedarf identifiziert, so sind zusätzliche Analysen zur Einschätzung der Risiken erforderlich. Mit diesem Beitrag wird eine Methode vorgestellt, wie zum einen das Risiko einzelner IT-Komponenten evaluiert werden kann und zum anderen konkrete Maßnahmen hinsichtlich ihres Beitrags zur Steigerung des Sicherheitsniveaus eines Unternehmens bewertet werden können. Dies schafft die Basis für die Auswahl optimaler Maßnahmen zur Gewährleistung der IT-Sicherheit. Die Ermittlung des Risikos und die Auswahl der Maßnahmen unter gegebenen Restriktionen werden mit einem auf der Fuzzy-Sets-Theorie basierenden Ansatz durchgeführt.

1 IT-Sicherheit und Grundschutzhandbuch (GSHB)

Die rasante Zunahme an Informationssystemen in den Unternehmen steigert den Bedarf an praxisnahen Verfahren zur Gesamtbeurteilung der IT-Sicherheit. Nicht nur die Finanz- und Versicherungsindustrie zeigen Interesse an einer Quantifizierung von IT-Sicherheit [5], sondern auch die Unternehmen selbst, deren IT-Sicherheit wichtiger Wettbewerbsfaktor ist [1]. Bislang mangelt es an anerkannten und praktikablen Methoden, mit denen sich die IT-Sicherheit ganzheitlich in einem Unternehmen bewerten lässt [5].

In diesem Beitrag wird ein neu entwickeltes, praxisnahes Verfahren zur Evaluation der IT-Sicherheit in einem Unternehmen vorgestellt, das auf den Maßnahmenempfehlungen des IT-Grundschutzhandbuchs (GSHB) des Bundesamtes für Sicherheit in der Informationstechnik (BSI) basiert. Im GSHB werden Standardsicherheitsmaßnahmen für typische IT-Anwendungen und IT-Systeme empfohlen. Das Ziel dieser IT-Grundschutz-Empfehlungen besteht darin, durch die Umsetzung von organisatorischen, personellen, infrastrukturellen und technischen Sicherheitsmaßnahmen ein Sicherheitsniveau zu erreichen, das für den normalen Schutzbedarf angemessen und ausreichend ist und als Grundlage für hochschutzbedürftige IT-Systeme und IT-Anwendungen dienen kann [2]. Die Bewertungen der einzelnen sicherheitsrelevanten Aspekte werden in dem vorgestellten Verfahren mit einem auf der Fuzzy-Sets-Theorie basierenden Ansatz durchgeführt.

Das Verfahren setzt auf der im GSHB beschriebenen IT-Grundschutzanalyse auf. In diesem Rahmen wird der betrachtete IT-Verbund[1] mit den Bausteinen des Handbuches nachgebildet, wobei jeder Baustein einer von fünf Schichten angehört (Übergreifende Aspekte, Infrastruktur, IT-Systeme, Netze, IT-Anwendungen) [7]. Die Bausteine befassen sich mit typischen IT-Komponenten wie z.B. „IT-Sicherheitsmanagement", „Verkabelung", „Windows 2000 Client", „Firewall" oder „WWW-Server". Für jeden Baustein sieht das GSHB Maßnahmenempfehlungen vor, die sich hinsichtlich ihrer Wichtigkeit unterscheiden:

- Die Maßnahmen vom Typ A stellen die *unabdingbaren* Sicherheitsmaßnahmen dar, die für alle drei Stufen der IT-Grundschutz-Qualifizierung umzusetzen sind.
- Die Maßnahmen vom Typ B werden als *wichtigst* bezeichnet und sind für die Aufbaustufe und das Zertifikat umzusetzen.
- Die Umsetzung der Maßnahmen vom Typ C ist nur für das IT-Grundschutz-Zertifikat notwendig.

Für jede aufgeführte Maßnahme ist zu entscheiden, ob deren Umsetzung erforderlich oder entbehrlich ist. Folgende Abbildung veranschaulicht den strukturellen Aufbau eines IT-Verbundes und den Entscheidungsprozess:

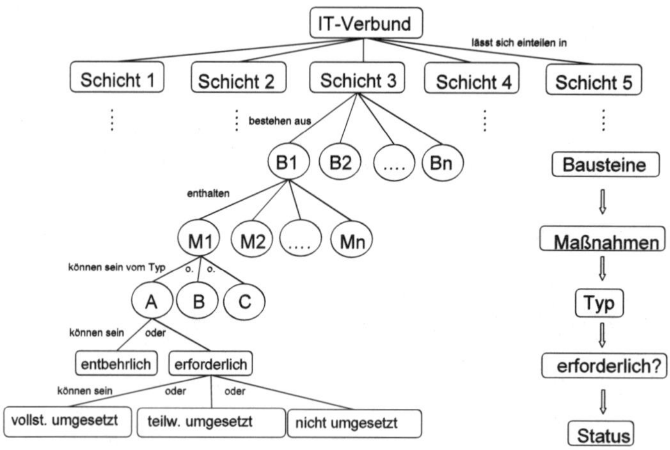

Abb. 1. Struktureller Aufbau eines IT-Verbundes

[1] Ein IT-Verbund bezeichnet die Gesamtheit der infrastrukturellen, organisatorischen und technischen Komponenten, die der Aufgabenerfüllung in einem definierten Anwendungsbereich der Informationsverarbeitung dienen, vgl. [3].

2 Evaluationsverfahren

Mit dem nachfolgend beschriebenen Evaluationsverfahren lässt sich für jeden Baustein, jede Schicht und den gesamten IT-Verbund eine Kennzahl für den Grad der IT-Sicherheit ermitteln, der sich aus der Umsetzung der erforderlichen Maßnahmen bestimmt. Der Bewertungsansatz auf Grundlage der Fuzzy-Sets-Theorie stellt hierbei sicher, dass die Wichtigkeits-Ausprägungen der verschiedenen Maßnahmentypen (s.o.) ebenso berücksichtigt werden wie die Interdependenzen der Bausteine untereinander, sodass auch das schwächste Glied in der Kette [6] angemessen berücksichtigt wird. So ergeben bspw. eine starke Sicherheitstechnik und eine schwache Organisation nicht automatisch ein Sicherheitsniveau mittlerer Güte, wie folgende überspitzt formulierte Frage verdeutlichen möge: Was nützt das stärkste Authentifizierungsverfahren in einem Unternehmen, wenn die Mitarbeiter für IT-Sicherheit nicht sensibilisiert sind und ihr Zugangspasswort unter der Tastatur ihres Rechners deponieren oder auf Anfrage am Telefon herausgeben?

2.1 Modellierung der Umsetzung erforderlicher Maßnahmen

Im ersten Schritt des Evaluationsverfahrens wird für jede Maßnahme geprüft, ob eine Umsetzung für das betrachtete Unternehmen u erforderlich oder entbehrlich ist:

U steht für die Grundmenge aller Unternehmen und M für die Menge aller Maßnahmen im GSHB. Auf der Menge $M \times U$ wird die unscharfe Menge $\tilde{E}$ (für erforderlich) definiert und durch folgende Zugehörigkeitsfunktion bestimmt:

$$\mu_{\tilde{E}}(M_{ij}, u) = \begin{cases} 1 & \text{falls } M_{ij} \text{ für } u \text{ erforderlich} \\ 0 & \text{falls } M_{ij} \text{ für } u \text{ entbehrlich} \end{cases}$$

Diese Zugehörigkeitsfunktion stuft jede Maßnahme $M_{ij} \in M$ in Abhängigkeit vom betrachteten Unternehmen $u \in U$ als erforderlich oder entbehrlich ein. M_{ij} bezeichnet dabei die *i-te* Maßnahme ($i=1,\ldots,N_j$) im Baustein B_j ($j \in J$). Hier wird also die charakteristische Funktion als Zugehörigkeitsfunktion verwendet, um die für eine spezielle Unternehmung notwendigen Maßnahmen zu erfassen.

Im nächsten Schritt wird der Beitrag der erforderlichen Maßnahmen zur IT-Sicherheit in dem betrachteten Unternehmen ermittelt. Hierzu wird zunächst für jede Maßnahme untersucht, ob sie komplett, teilweise oder nicht umgesetzt wurde, unabhängig davon, ob die Umsetzung der jeweiligen Maßnahme für das spezielle Unternehmen erforderlich oder entbehrlich ist:

Auf der Menge aller Maßnahmen M wird die unscharfe Menge $\tilde{S}$ (für Status) durch die folgende Zugehörigkeitsfunktion definiert:

$$\mu_{\tilde{S}}(M_{ij}, u) = \begin{cases} 1 & \text{falls } M_{ij} \text{ in } u \text{ umgesetzt} \\ 0{,}5 & \text{falls } M_{ij} \text{ teilweise in } u \text{ umgesetzt} \\ 0 & \text{sonst} \end{cases}$$

Diese Zugehörigkeitsfunktion ordnet jeder Maßnahme $M_{ij} \in M$ den Zugehörigkeitsgrad zu, der als Maß für die Umsetzung in u interpretiert werden kann. Da wie oben erwähnt nur die erforderlichen Maßnahmen einen Beitrag zur IT-Sicherheit leisten, wird nachfolgend die unscharfe Menge $\widetilde{MS}$ (für Maßnahmenstatus) definiert, die sich als Durchschnitt der unscharfen Mengen $\tilde{E}$ und $\tilde{S}$ ergibt:

$$\mu_{\widetilde{MS}}(M_{ij},u) := \min\left\{\mu_{\tilde{E}}(M_{ij},u); \mu_{\tilde{S}}(M_{ij},u)\right\} \qquad \forall u \in U,\ M_{ij} \in M$$

Die als erforderlich identifizierten Maßnahmen können demnach einen Zugehörigkeitsgrad von null, 0,5 oder eins zu dieser Menge annehmen, während die entbehrlichen Maßnahmen einen Zugehörigkeitsgrad von null haben und somit keinen Beitrag zur IT-Sicherheit leisten.

Im nächsten Schritt des Evaluationsverfahrens wird für jeden Baustein und für jede Schicht der Umsetzungsgrad der Maßnahmen getrennt nach Maßnahmen-Typ ermittelt. Hierzu werden zunächst die charakteristischen Funktionen als Zugehörigkeitsfunktionen der unscharfen Mengen $\tilde{A}$, $\tilde{B}$ und $\tilde{C}$ definiert, die allen Maßnahmen M_{ij} den Grad der Zugehörigkeit zum jeweiligen Typ zuordnen. Nachfolgend ist die Zugehörigkeitsfunktion der unscharfen Menge $\tilde{A}$ angegeben:

$$\mu_{\tilde{A}}(M_{ij}) = \begin{cases} 1 & \text{falls } M_{ij} \text{ vom Typ A} \\ 0 & \text{falls } M_{ij} \text{ nicht vom Typ A} \end{cases}$$

Die Zugehörigkeitsfunktionen von $\tilde{B}$ und $\tilde{C}$ werden dementsprechend definiert.

Zur Berechnung des Umsetzungsgrads des Maßnahmen-Typs A in einem Baustein B_j wird die unscharfe Menge $\widetilde{UB_jA}$ (für Umsetzungsgrad Baustein B_j Maßnahmen-Typ A) eingeführt, die über folgende Zugehörigkeitsfunktion definiert ist als:

$$\mu_{\widetilde{UB_jA}}(u) = \frac{\sum_{i=1}^{N_j} \min\left\{\mu_{\tilde{A}}(M_{ij}); \mu_{\widetilde{MS}}(M_{ij},u)\right\}}{\sum_{i=1}^{N_j} \min\left\{\mu_{\tilde{A}}(M_{ij}); \mu_{\tilde{E}}(M_{ij},u)\right\}} \qquad \forall\ j \in J$$

Der Zugehörigkeitsgrad zu dieser Menge kann als Umsetzungsgrad des Maßnahmenbündels vom Typ A im Baustein B_j interpretiert werden. Im Zähler dieser Funktion werden die Status-Beiträge der Maßnahmen zur IT-Sicherheit aggregiert, sofern die betrachtete Maßnahme erforderlich und vom Typ A ist. Im Nenner der Funktion werden die Zugehörigkeitsgrade aller Maßnahmen, die erforderlich und vom Typ A sind, aufsummiert. Die Konstante N_j gibt die Anzahl der im GSHB aufgeführten Maßnahmen im Baustein B_j an. Die Zugehörigkeitsfunktionen für $\widetilde{UB_jB}$ und $\widetilde{UB_jC}$ werden analog definiert.

Nachdem für jeden Baustein und jede Schicht der Umsetzungsgrad der Maßnahmenbündel vom Typ A, B und C definiert wurde, wird aus diesen Informationen über ein Regelsystem die Sicherheit der Bausteine bzw. der Schichten bestimmt, die als Zugehörigkeitsgrad $\mu_{\widetilde{SB_j}}(u)$ zu den unscharfen Mengen $\widetilde{SB_j}$ (für Sicherheit

Baustein B_j) bzw. $\mu_{\widetilde{SIS}_l}(u)$ zu den unscharfen Mengen $\widetilde{SIS}_l$ (für Sicherheit Schicht S_l) definiert ist.

Aus diesen Größen wird die Sicherheit eines IT-Verbundes abgeleitet, die von der Sicherheit der einzelnen Schichten abhängt. Eine Aggregation der Zugehörigkeitsgrade $\mu_{\widetilde{SIS}_l}(u)$ muss den Aspekt „Sicherheit ist nur so stark wie ihr schwächstes Glied“ angemessen berücksichtigen. Daher werden die unscharfen Mengen $\widetilde{SIS}_l$ im vorgestellten Evaluationsverfahren mit dem *γ-Operator* ohne Kompensation verknüpft [9].

2.2 Beispiel

Das folgende Beispiel demonstriert die Vorgehensweise und die Vorteile des vorgestellten Verfahrens an einem konkreten Anwendungsfall. Die Evaluation der IT-Sicherheit für den untersuchten IT-Verbund hat eine Gesamtsicherheit von 0,59 auf einer Skala von 0 bis 1 ergeben. Dieses unbefriedigende Untersuchungsergebnis ist darauf zurückzuführen, dass in der Schicht 1 nur eine Sicherheit von 0,38 vorliegt. Zahlreiche unabdingbare Maßnahmen des Typs A wurden in dieser Schicht bzw. in den Bausteinen dieser Schicht nicht umgesetzt, wodurch die Gesamtsicherheit deutlich gemindert wird. Eine einfache Mittelung der Ergebnisse aus den einzelnen Schichten hätte zu einer Sicherheit von 0,70 geführt, wodurch fälschlicherweise der Eindruck eines zufrieden stellenden Ergebnisses entstehen könnte.

Um den zur Ermittlung der IT-Sicherheit erforderlichen rechentechnischen Aufwand zu reduzieren, wurde ein Tool entwickelt und eingesetzt, das die Berechnungen übernimmt [8]. Neben der Feststellung der Sicherheit in den einzelnen IT-Komponenten kann so simuliert werden, wie sich die Umsetzung einer Maßnahme auf die Sicherheit des IT-Verbundes auswirkt.

Nach einer Weiterentwicklung des Verfahrens wird das Tool in der Lage sein konkrete Anweisungen zu geben, welche Maßnahmen bei gegebenem Sicherheitsbudget umzusetzen sind, sodass das Gesamtsicherheitsniveau des IT-Verbundes maximiert wird. In dem aufgeführten Beispiel könnten dies die Maßnahmen 3, 7 und 15 aus dem Baustein 4 in der Schicht 1 sein.

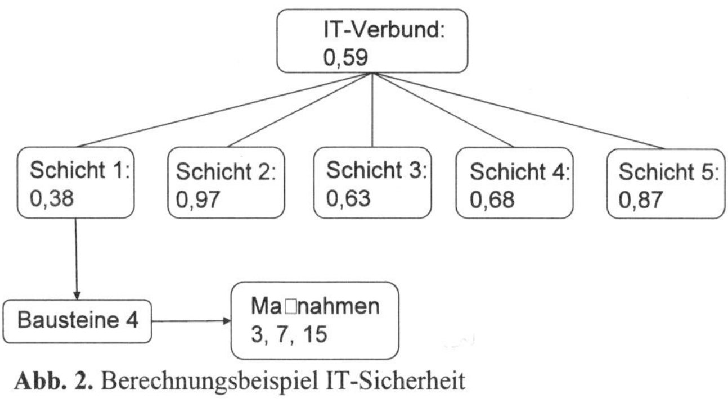

Abb. 2. Berechnungsbeispiel IT-Sicherheit

3 Ergebnisse

Da das vorgestellte Verfahren auf den Maßnahmenempfehlungen des GSHB basiert, ist sichergestellt, dass bei der Evaluation der IT-Sicherheit nicht nur technische, sondern ebenso organisatorische, personelle und infrastrukturelle Aspekte berücksichtigt werden. Der Bewertungsansatz auf Grundlage der Fuzzy-Sets-Theorie berücksichtigt auch schwache Glieder in dem zu bewertenden IT-Verbund angemessen. Mit dem vorgestellten Verfahren lassen sich für jede einzelne IT-Komponente und den gesamten IT-Verbund Kennzahlen ermitteln, sodass die Gesamtsicherheit in einem Unternehmen quantifiziert werden kann.

Dieses Verfahren wird dahingehend weiterentwickelt, dass die bei der Umsetzung von IT-Sicherheitsmaßnahmen erforderlichen Ressourcen erfasst werden. So lassen sich mit einem vorgegebenen, begrenzten IT-Sicherheitsbudget diejenigen Maßnahmen ermitteln, deren Umsetzung eine Optimierung des Sicherheitsniveaus bewirkt.

Weiter wird an einem Risikoanalyseverfahren für einzelne IT-Komponenten gearbeitet, das ebenfalls auf der Fuzzy-Logik basiert. Dies hat den Vorteil, dass nicht exakt bewertbare Größen wie Gefährdungen und Schadenshöhen dargestellt und in den Bewertungsprozess mit einbezogen werden können [4]. Mit einem derartigen System kann das Management in seinen Entscheidungen hinsichtlich Sicherheitsinvestitionen unterstützt werden.

Literatur

1. Böhmer, Wolfgang (2002) VPN – Virtual private networks, München 2002
2. BSI (2003a) IT-Grundschutzhandbuch, Stand Oktober 2003, http://www.bsi.de/gshb/deutsch/download/GSHB2003.pdf, abgerufen Dezember 2003
3. BSI (2003b) BSI Schulung IT-Grundschutz, http://www. bsi.de/gshb/webkurs/itgswbt.zip, abgerufen Januar 2004
4. De Ru, W.G.; Eloff, J.H.P. (1996) Risk analysis modelling with the use of fuzzy logic, in: Computers & Security, Band 15, Nr. 3, 1996, S.239 – 248.
5. Matousek, Michael; Schlienger, Thomas; Teufel, Stephanie (2004) Metriken und Konzepte zur Messung der Informationssicherheit, in: HMD - Praxis der Wirtschaftsinformatik, Jg. 41, Heft 236, 2004, S. 33 - 41
6. Schneier, Bruce (2001): Secrets & lies, Heidelberg u.a. 2001.
7. Werners, Brigitte; Klempt, Philipp (2005a) Standards und Kriterienwerke zur Zertifizierung von IT-Sicherheit, Arbeitsbericht Nr. 9 des Instituts für Sicherheit im E-Business, Bochum 2005
8. Werners, Brigitte; Klempt, Philipp (2005b) Verfahren zur Evaluation der IT-Sicherheit in Unternehmen, Arbeitsbericht Nr. 12 des Instituts für Sicherheit im E-Business, Bochum 2005
9. Zimmermann, Hans-Jürgen (2001) Fuzzy Set Theorie – and Its Applications, 4., überarb. Auflage, Boston u.a. 2001

m-Parking - Mobile Parking Payment Systems in Europa

Christine Strauß[1], Melitta Urbanek[1], and Gernot Wörther[2]

[1] Institut für Betriebwirtschaftslehre, Produktion und Logistik, Universität Wien
A - 1210 Wien, Brünner Straße 72
christine.strauss@univie.ac.at, mountietta@a1.net

[2] Digital Business Research and Development, eCommerce Competence Center
A - 1220 Wien, Donau-City-Strasse 1
gernot.woerther@ec3.at

Zusammenfassung Mobile Bezahlsysteme im Rahmen urbaner Parkraumbewirtschaftung stellen eine alternative Anwendung des eBusiness dar und finden zunehmend Verbreitung und Akzeptanz in Europa. Mobile Parkgebührenbezahlung über In-Vehicle devices, anwählbare Dialling a Pay and Display Maschinen und Handyparksysteme werden im vorliegenden Beitrag hinsichtlich ihres Einsatzes in Europa im Überblick präsentiert; ferner werden die Vor- und Nachteile der mobilen Bezahlsysteme jeweils aus der Sicht der Anwender und aus der Sicht der Betreiber gegenübergestellt. Der Fokus des Beitrags liegt auf einer detaillierten Analyse unterschiedlicher Handyparksysteme in Europa. Anhand geeigneter Kriterien, wie Technologieoptionen, Registrierungsmöglichkeiten, Abrechnungsalternativen, Kontoführung, Fahrzeugerkennung, Kontrollmechanismen, Tourismusverträglichkeit und Zusatzdienste, wird ein systematischer Vergleich von Handyparksystemen durchgeführt.

1 Einleitung

Parkraum stellt im innerstädtischen Bereich ein knappes Gut dar. Zur Anwendung des Verursacherprinzips, Attraktivitätssteigerung des öffentlichen Verkehrs und Rückgewinnung von Flächen für parkfremde Zwecke wird in den meisten Kommunen eine systematische Parkraumbewirtschaftung angestrebt [5]. Neben konventionellen Systemen wie etwa der Gebührenentrichtung mittels Parkschein, Parkuhr oder Parkscheibe sind in den letzten Jahren auch mobile, IKT-basierte Systeme der Parkraumbewirtschaftung entstanden. Diese neuen Systeme finden in Europa zunehmend Verbreitung und befinden sich in vielen Städten bereits erfolgreich im Einsatz. Beispielsweise werden in Wien nach einem dreimonatigen m-parking Pilotversuch [14] bereits mehr als 160.000 Parkaktivitäten monatlich über Handyparken abgewickelt (Stand: Mai 2005) [7]. In dem vorliegenden Artikel werden die verschiedenen Mobile Parking Payment Systems (MPPS) klassifiziert und ihre Vor-

und Nachteile aus Betreiber- und Kundenperspektive analysiert. Ferner erfolgt ein Funktionsvergleich ausgewählter Systeme mobiler Parklösungen in Europa.

2 Typen und Funktionen mobiler Parking Payment Systeme

Bei Parkraumüberwachungssystemen können konventionelle und alternative Systeme unterschieden werden, wobei alternative Systeme allgemein in nutzerbediente monofunktionale sowie nutzerbediente multifunktionale Systeme eingeteilt werden [3]. Während monofunktionale Systeme ausschließlich zur Abwicklung von Parkvorgängen genutzt werden, ermöglichen multifunktionale Systeme mehrere, völlig unterschiedliche Anwendungen. Gegenstand der vorliegenden Arbeit sind alternative Parkraumüberwachungssysteme, sogenannte Mobile Parking Payment Systems (MPPS), die ferner gemäß einer Klassifikation der European Parking Association in In-Vehicle Device, Dialling a Pay and Display Machine (DPDM) sowie Handyparksysteme eingeteilt werden [1].

2.1 In-Vehicle Device Systeme

Bei In-Vehicle Device Systemen wird ein von außen sichtbares Gerät im Fahrzeug platziert, über das die Parkgebühr entrichtet wird und das vom Parkraumüberwachungspersonal kontrolliert werden kann. Dabei wird zwischen Geräten mit und ohne Datenübertragungsfunktion unterschieden. Bei Geräten ohne Datentransfer handelt es sich um Systeme, die meist durch Einschalten aktiviert und durch Sichtkontrolle überprüft werden. Hierzu zählen auch die beispielsweise in Wien üblichen Parkschecks ("Parkscheine") [14].

2.2 Dialling a Pay and Display Machine

Bei allen Dialling a Pay and Display Machine Systemen (DPDM) löst der Parkkunde durch einen Anruf bei einer kostenpflichtigen Telefonnummer die Ausstellung eines Parktickets über den Parkautomaten aus. Dabei ist zwischen zwei Varianten zu unterscheiden: Bei einer Variante von Dialling a Pay ruft der Parkkunde eine zentrale, kostenpflichtige Nummer an um anzugegben, auf welchem Parkscheinautomaten das Parkticket ausgegeben werden soll. Im anderen Fall ruft der Parkkunde eine dem Parkautomaten spezifisch zugeordnete, kostenpflichtige Nummer an.

2.3 Handyparken

Ein bereits derzeit weit verbreitetes MPPS ist das Handyparken. Einer der Gründe für die Akzeptanz ist der hohe Verbreitungsgrad der Mobiltelefonie in der Bevölkerung [15]. Beim Handyparken bestellt der Parkkunde üblicherweise über sein Mobiltelefon ein Parkticket und bekommt dieses auch über das Mobiltelefon, beispielsweise mittels SMS, ausgestellt. Es gibt verschiedene Varianten der technischen Umsetzung des Handyparkens, wie beispielsweise Voice, SMS, MMS oder WAP.

2.4 Funktionen von MPPS

Es gibt acht generische Prozesse, die sämtliche Funktionen mobiler Parking Payment Systeme beschreiben, die im folgenden kurz dargestellt werden.

1. Damit Kunden ein MPPS nutzen können, müssen sie die *Registrierung* durchführen. Dabei werden üblicherweise folgende Informationen übermittelt: Personendaten, Mobilfunknummer, Kennzeichen des Fahrzeugs, sowie Zahlungsinformationen. Die Registrierung kann über ein Callcenter, das Internet, ein Formular oder mittels Mobiltelefon beispielsweise via WAP-Formular, SMS oder MMS erfolgen.
2. Die *Auslösung des Parkvorganges* wird durch den Kunden vorgenommen. Dafür stehen je nach Implementierung des Parksystems einer oder mehrere der folgenden Kommunikationskanäle zur Verfügung: Ein Sprachanruf in einem Callcenter bzw. bei einem IVR (Interactive Voice Response) Server, eine SMS, MMS, WAP-Applikation oder das Internet. Zur Verhinderung von Missbrauch kann im Bedarfsfall ein Pin-Code-Mechanismus verwendet werden, der allerdings die Benutzerfreundlichkeit reduziert. Bezüglich der Auswahl der Parkregion gibt es drei mögliche Lösungsansätze: i) Eine einzige Nummer bzw. WAP-Adresse gilt für die gesamte Kommune, und der Kunde gibt aktiv die betreffende Parkzone an; ii) jede Parkzone besitzt ihre eigene Nummer bzw WAP-Adresse; iii) mittels GPS wird die Position des Fahrzeugs und somit die zu verrechnende Zone vom MPPS ermittelt.
3. Das *Beenden des Parkvorganges* kann entweder durch den Kunden aktiv durchgeführt werden oder findet durch Zeitablauf statt. In-Vehicle Devices werden einfach deaktivert, während Handyparksysteme oft über eine Reminder-Funktion verfügen, welche den Kunden an den Ablauf der ursprünglich gewählten Parkdauer erinnert. Aus technischer Sicht bietet das Handyparken die Möglichkeit der minutengenauen Abrechung, allerdings ist dieser Vorteil aufgrund legislativer Vorgaben zurzeit noch nicht überall realisierbar.
4. Von einigen Systemen werden *Web-Funktionen* angeboten. Sie dienen in erster Linie zur Erhöhung der Benutzerfreundlichkeit. Typische Features, die angeboten werden, sind Übersichten über vergangene Parkaktivitäten, die Erstellung von Parkreports, die Verwaltung des Benutzerprofils, sowie das Auslösen bzw. Beenden des Parkprozesses.
5. *Customer Care* spielt bei MPPS eine zentrale Rolle zur Unterstützung des Kunden und Behandlung von Problemfällen sowie Rückfragen, die besonders bei Neukunden auftreten können.
6. Der *Billing-Process* ist ein integraler Bestandteil des Handyparkens. Dabei gibt es sowohl beim Zahlungszeitpunkt als auch bei den zur Verfügung stehenden Zahlungsmitteln verschiedene Möglichkeiten. Allgemein kann zwischen Pre- und Post-Paid Verfahren unterschieden werden. Bei ersterem muss ein Guthaben erworben werden, bevor ein Parkticket ausgestellt wird, bei letzterem Verfahren wird die erbrachte Leistung im Nachhinein abgerechnet, wobei die Zahlung wöchentlich, monatlich oder bei Erreichung eines bestimmten Betrages fällig wird. Typische Verrechnungsmodi sind hier die Abrechnung über Kreditkarte, Mobile Payment Provider, wie beispielsweise die paybox [11], oder über die Telefonrechnung.
7. Die *Parkraumkontrolle* wird – wie bei konventionellem Parken – vom Parkraumüberwachungspersonal durchgeführt. Dabei muss das Personal die Möglichkeit

haben, die Gültigkeit eines Parktickets zu überprüfen und gegebenenfalls eine Strafverfügung auszustellen. In der Praxis bedeutet dies, dass diese Personen für das Handyparksystem ausgebildet und gegebenenfalls mit entsprechenden Kontrollgeräten ausgestattet werden müssen.

8. Das *Parkbetreiber-Managementsystem* erlaubt dem Betreiber das gesamte System zu verwalten, Änderungen im System vorzunehmen, die Aktivitäten im Parkraum (bezüglich der elektronisch ausgestellten Parkscheine) festzustellen sowie die Einhebung von Parkstrafen zu administrieren.

2.5 Vor- und Nachteile von MPPS

Nachdem allgemeine Klassen IKT-gestützer Parksysteme vorgestellt wurden, werden nun in aggregierter Form die wichtigsten Vor- und Nachteile der einzelnen MPPS getrennt nach Kundensicht (Tab. 1) und Betreibersicht (Tab. 2) dargestellt.

Tabelle 1. Vor- und Nachteile von MPPS aus Kundensicht

System	Vorteile	Nachteile
In-Vehicle Device	+ minutengenaue Abrechnung + einfache Bedienbarkeit + bargeldlose Bezahlung	- Kosten für das Gerät - für Touristen nicht geeignet
DPDM	+ Online-Kostenkontrolle + Erinnerungs-SMS + Abrechnung über verschiedene Bezahlsysteme + tourismustauglich + bargeldlose Bezahlung	- ortsgebundene Verlängerung - keine minutengenaue Abrechnung - handygebunden - Mobilfunkkosten
Handyparken	+ Online-Kostenkontrolle + Erinnerungs-SMS + Abrechnung über verschiedene Bezahlsysteme + tourismustauglich + ortsunabhängige Verlängerung + bargeldlose Bezahlung	- handygebunden - Mobilfunkkosten

3 Vergleich von MPPS in Europa

In diesem Abschnitt werden fünf MPPS verglichen, die in Europa eingesetzt werden. Es werden zu den in Kapitel 2 vorgestellten Systemtypen jeweils ein bzw. zwei in der Praxis verwendete Systeme exemplarisch herausgegriffen. Als Beispiele für In-Vehicle Device Anwendungen wurden park-line [10] und Park-O-Pin [12] untersucht. Als Vertreter einer DPDM-Lösung wurde m-park [9] ausgewählt, das unter anderem in Bremen angeboten wird; für das Handyparken wurden VIP.parking [13] [4] sowie das in Wien eingesetzte m-parking [8] analysiert. Tabelle 3 zeigt für die genannten, ausgewählten Praxislösungen im Überblick einen Funktionsvergleich anhand der Kritierien Registrierung-, Konto-, Park-, Zahlungs- und Kontrolloptionen sowie der Möglichkeiten zum Starten und Beenden der Parktransaktion (vgl. [6]).

Tabelle 2. Vor- und Nachteile von MPPS aus Betreibersicht

System	Vorteile	Nachteile
In-Vehicle Device	+ geringe Investitionskosten + unveränderte Kontrollprozesse + Datenbasis für strategische Entscheidungen + bargeldloser Zahlungsverkehr	- für Touristen nicht geeignet
DPDM	+ Verlängerung der Serviceintervalle + flexible Tarifgestaltung + Beibehaltung bestehender Ticket-Typen + Beibehaltung vorhandener Kontrollgeräte + Datenbasis für strategische Entscheidungen + bargeldloser Zahlungsverkehr	- Investitionskosten - Umrüstung der Ticket-Automaten
Handyparken	+ Datenbasis für strategische Entscheidungen + flexible Tarifgestaltung + effiziente Kontrolle + bargeldloser Zahlungsverkehr	- Investitions-, Betriebs- und Instandhaltungskosten

4 Zusammenfassung und Ausblick

Dieser Beitrag gibt einen Überblick über verschiedene Möglichkeiten der Parkraumbewirtschaftung mithilfe neuer Technologien. Dabei können drei Systemtypen unterschieden werden (In-Vehicle Device, DPDM und Handyparken), die unterschiedliche Vor- und Nachteile jeweils für den Benutzer und den Betreiber aufweisen. Anhand ausgewählter Praxislösungen von Mobile Parking Payment Systems in Europa wird ein detaillierter Überblick über unterschiedliche Systeme und deren Ausprägungen gegeben.

Diese Analyse kann als grobe Richtlinie für Entscheidungen bei zukünftigen Systemeinführungen genutzt werden. Letztlich werden neben technologischen Kriterien vor allem tiefergehende Analysen der Kundenwünsche sowie Überlegungen zur Benutzerfreundlichkeit über erfolgreiche, flächendeckende Implemetierungen entscheiden.

Literatur

1. Dahlström E (2002) Report on Mobile Parking Payments, European Parking Association, http://www.europeanparking.com, letzter Abruf am 25.07.05
2. Haimböck J (2004) Elektronische Parkraumbewirtschaftung in Wien, Magistrat der Stadt Wien, Magistratsabteilung 4
3. Schäfer PK (2004) Alternative Methoden zur Überwachung der Parkdauer sowie zur Zahlung der Parkgebühren, Dissertation, Technische Universität Darmstadt
4. Soudi A (2003) Erfolgreiche Mobile Services und Anwendungen für den mobilen User im Privatbereich, Diplomarbeit, Wirtschaftsuniversität Wien
5. Taupe M (2001) Parkraumbewirtschaftung und die Konkurrenzbeziehung zwischen Innenstadthandel und Einkaufszentren, Diplomarbeit, Wirtschaftsuniversität Wien

Tabelle 3. Funktionsvergleich ausgewählter MPPS-Typen

Funktion \ Typ	In-Vehicle Device		DPDM	Handyparken	
	park-line	Park-O-Pin	m-park	VIP.parking	m-parking
alle Mobilfunkbetreiber im Land	X	na	X	X	X
Registrierungsoptionen					
Keine Registrierung	-	X	-	X	-
Firmenregistrierung	X	-	X	-	-
Call Center	-	-	X	-	-
Internet	X	-	X	-	X
SMS	-	-	-	-	X
Konto					
kein Konto	-	-	-	-	X
Pre-paid Telefonnummern	-	-	-	X	-
Service mit jedem Kfz möglich	-	X	X	X	X
Optionen für Starten/Beenden der Parktransaktion					
Anruf	X	-	-	-	-
Parktransaktionen mittels SMS	-	-	-	X	X
Online Parken mit WAP	X	-	-	-	-
manuelles Ein- & Ausschalten	-	X	-	-	-
Ablauf des Parkscheins	-	-	X	-	-
Parkoptionen					
minutengenaue Parkzeit	X	X	-	X	-
Einstellung einer bestimmten Parkzeit	-	-	X	X	X
Bestätigung via SMS oder Anruf oder IVR	-	-	-	X	X
Erinnerungs-SMS bei vorbestimmter Parkzeit	-	-	X (optional)	X	X
Tourismustauglichkeit	-	-	-	X	X
Zahlung					
kein Aufladen eines virtuellen Parkkontos notwendig	-	-	X	-	X
Rechnung	X	-	-	-	-
Bankeinzug, Überweisung	X	-	X	-	-
Kreditkarte	-	-	X	-	X
Mobiltelefonrechnung	-	-	-	X	X (nur A1)
M-Payment Provider (z.B. Paybox)	-	-	-	-	X
sofortige Abbuchung vom Guthaben	-	X	-	-	X
Kauf einer Parkkarte	X	X	-	-	-
Enforcement - Fahrzeugerkennung					
In-vehicle unit	X	X	X (Parkschein)	-	-
Kfz-Nummernerkennung	-	-	-	X	X
Transponder card	X	-	-	-	-
Parkplatzerkennung (keine Fahrzeugerkennung)	X	-	-	-	-
kein Barcodeleser notwendig	X	X	na	na	na

6. Urbanek M (2005) Mobile Business und Parkraumbewirtschaftung, Diplomarbeit, Universität Wien, in Arbeit
7. http://www.derstandard.at, letzter Abruf am 02.07.05
8. http://www.m-parking.at, letzter Abruf am 25.07.05
9. http://www.mpark.de/mpark/index.jspx, letzter Abruf am 25.07.05
10. http://www.park-line.nl/en/default.htm, letzter Abruf am 25.07.05
11. http://www.paybox.at, letzter Abruf am 12.07.05
12. http://www.pin-gmbh.de, letzter Abruf am 22.07.05
13. http://www.vipnet.hr/cw/d-show?idc=3913842, letzter Abruf am 22.07.05
14. Parkraumbewirtschaftung in Wien (1997), Stadtplanung Wien, Magistratsabteilung 18
15. http://www.itu.int/ITU-D/ict/statistics/at-glance/cellular04.pdf, letzter Abruf am 12.07.05

Part XIV

Sustainable Systems

Energieorientierte Maschinenbelegungsplanung auf Basis evolutionärer Algorithmen

Markus Rager, Axel Tuma, Jürgen Friedl

Lehrstuhl für Produktions- und Umweltmanagement, Universität Augsburg, 86135 Augsburg

Zusammenfassung Ziel des vorliegenden Beitrages ist es, die ökonomischen Einsparpotentiale einer *energieorientierten Maschinenbelegungsplanung* bei diskontinuierlichen Produktionsprozessen anhand eines Beispiels aus der Textilindustrie aufzuzeigen. Dazu wird das vorliegende Planungsproblem als *resource levelling problem* abgebildet. Der entwickelte, heuristische Lösungsansatz beruht auf einem *evolutionären Algorithmus*.

1 Energieorientierte Maschinenbelegungsplanung

Analysiert man die Rahmenbedingungen moderner industrieller Produktionsprozesse, so sind diese aufgrund einer verschärften Wettbewerbssituation durch einen starken Druck zur Kostenreduktion geprägt. Demgegenüber steht ein stetiger Anstieg der Energiepreise, der sich gerade in energieintensiven Branchen auf den Unternehmenserfolg auswirkt. Die durch technische Lösungen realisierbaren Einsparpotentiale an eingesetzten Energieträgern sind in vielen Fällen bereits ausgeschöpft oder nur durch hohe Investitionen realisierbar. Dementsprechend liegt der Fokus, neben technischen Ansätzen, zusätzlich auf organisatorischen Maßnahmen zur Umsetzung einer energieeffizienten Produktion. Vor diesem Hintergrund werden (diskontinuierliche) Produktionsprozesse auf mögliche Einsparpotentiale an eingesetzten Energieträgern untersucht, bei denen der Hauptanteil der eingesetzten Energieträger in zentralen Umwandlungsanlagen (z.B. Dampferzeuger) in Nutzenergie (z.B. Prozessdampf) transformiert und dann in dezentralen Produktionsanlagen (z.B. dampfbeheizte Färbeaggregate) eingesetzt wird (vgl. Abb. 1). Derartige Produktionssysteme finden sich beispielsweise in der Chemischen Industrie (z.B. Produktion von Feinchemikalien), der Lebensmittelindustrie (z.B. Getränkeherstellung) sowie bei Veredlungsprozessen in der Textilindustrie (z.B. Färbereien) wieder.

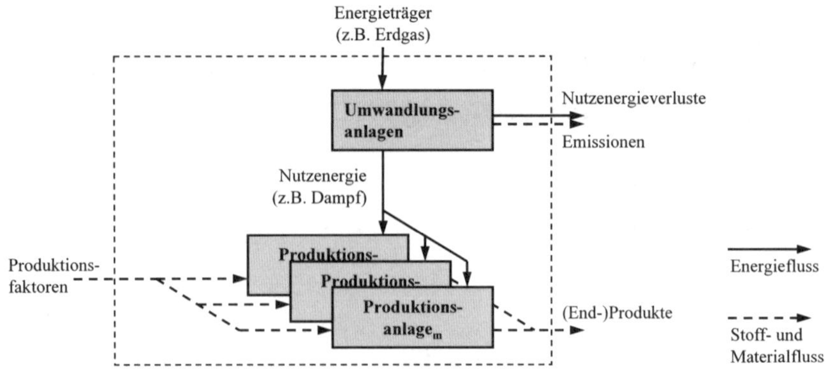

Abb. 1. Produktionssystem mit dezentralen Energieumwandlungsanlagen

In solchen Produktionssystemen ist die Steuerung der Umwandlungsanlagen (z.B. Steuerung der Brennstoffzufuhr) eng an die von den einzelnen Produktionsanlagen gleichzeitig nachgefragte Menge an Nutzenergie gekoppelt. Dabei kann es durch ungünstige Überlagerungen der einzelnen Nachfragemengen zu starken Schwankungen in der abgerufenen Gesamtmenge an Nutzenergie (Gesamtlast) kommen. Durch diese kurzfristigen Schwankungen (Minutenbasis) kann der eingesetzte Energieträger nicht effizient genutzt werden und es entstehen Nutzenergieverluste und vermeidbare Emissionen. Die Ursachen dafür werden exemplarisch anhand des Betriebs eines Dampfkessels aufgezeigt:[1]

- *abgerufene Gesamtlast fällt unter die Ausschaltgrenze*
  Wird eine (Grund-)Lastgrenze unterschritten, ist eine Drosselung des Brenners nicht mehr möglich und die Brennstoffzufuhr wird ganz unterbrochen. Beim erneuten Anfahren müssen die Rauchgaszüge durchgelüftet werden, wodurch Nutzenergie verloren geht.
- *abgerufene Gesamtlast ist oberhalb der Ausschaltgrenze, aber schwankt stark*
  Einerseits kann die Anlage nicht bezüglich ihres optimalen technischen Wirkungsgrades betrieben werden, wodurch sich der Brennstoffeinsatz erhöht. Anderseits sind häufige Schaltvorgänge der regelungstechnischen Einrichtungen notwendig, die deren Lebensdauer verringern.
- *abgerufene Gesamtlast führt zu Spitzenlasten:*
  Überschreitet die abgerufene Gesamtlast eine (Spitzen-)Lastgrenze, können spezielle Spitzenlasteinrichtungen (z.B. Spitzenlastbrenner) zugeschaltet werden. Diese weisen allerdings häufig einen schlechteren Wirkungsgrad auf.

Demnach kann die angestrebte Reduktion des Energieträgereinsatzes bei gegebener Ausbringungsmenge an (End-)Produkten durch die *Glättung der auftretenden Schwankungen der Gesamtlast* als entsprechendes Ersatzziel bei der Planung

[1] Ähnliche Ursachen für eine ineffiziente Energieausnutzung lassen sich auch beim Betrieb von anderen technischen Konfigurationen von Umwandlungsanlagen (z.B. Blockheizkraftwerke, Gasturbinen-Kraftwärmekopplungen) beobachten. (vgl. [6])

der Produktionsprozesse operationalisiert werden. Da für diese Glättung eine hohe zeitliche Auflösung erforderlich ist, stellt die *Maschinenbelegungsplanung* im Rahmen der operativen Produktionsplanung und -steuerung einen geeigneten Ansatzpunkt dar. Bei der Maschinenbelegungsplanung werden die dem Produktionssystem für einen bestimmten Zeitraum (z.B. ein Tag) zugeteilten Produktionsaufträge mit einer stunden- bis minutengenauen Zeiteinteilung auf den einzelnen Produktionsanlagen eingeplant [2].

2 Problemformulierung

Nachfolgend wird die Problemformulierung eines energieorientierten Maschinenbelegungsproblems anhand einer Garnfärberei aufgezeigt. Dort werden in identischen, dampfbeheizten Färbeaggregaten (Produktionsanlagen), die über ein zentrales Kesselhaus mit Prozessdampf (Umwandlungsanlage) versorgt werden, Garne nach fest vorgegebenen Rezepturen gefärbt. Dabei können die Produktionsaufträge in beliebiger Reihenfolge auf einer Produktionsanlage abgearbeitet werden, sind aber innerhalb des Planungshorizonts fertig zu stellen. Die Färberezeptur unterteilt jeden Färbeauftrag (Produktionsauftrag) in sich abwechselnde Färbe- und Spülphasen (Operationen) unterschiedlicher Länge. Während der Färbephasen wird das wässrige Färbebad, in dem die Garne gefärbt werden, mit einer konstanten Dampfzufuhr aufgeheizt. Hingegen erfolgt in den Spülphasen keine Dampfbeheizung. Die von einem Produktionsauftrag nachgefragte Menge an Nutzenergie kann somit durch ein in Abb. 2 dargestelltes Lastprofil abgebildet werden, wodurch sich die Gesamtlast an den Umwandlungsanlagen berechnet lässt.

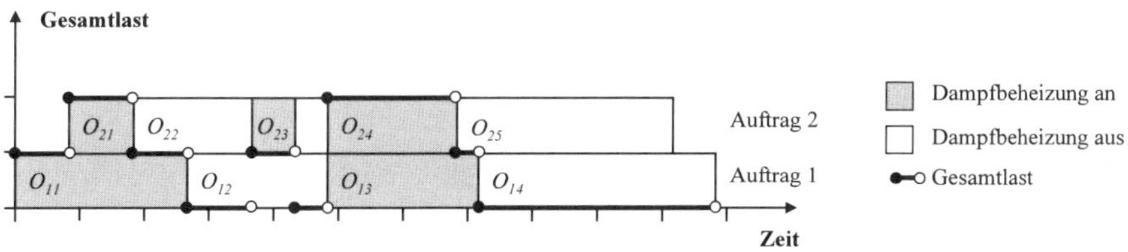

Abb. 2. Lastprofile von Färbeaufträgen und resultierende Gesamtlast

Für das beschriebene Problem besteht ein Maschinenbelegungsplan aus einer Zuordnung für jeden Produktionsauftrag $i = 1,...,n$ von einem Zeitintervall auf einer der $k = 1,...,m$ identischen Produktionsanlagen. Jeder Produktionsauftrag i besteht aus einer Abfolge von Operationen $O_{ij} := (i, j)$, $j = 1,...,n_i$ mit einer Bearbeitungszeit p_{ij} und einer konstanten, nachgefragten Menge an Nutzenergie pro Zeiteinheit e_{ij} . Die Menge aller Operationen sei Ω . Nach Start des Produktionsauftrages i müssen dessen Operationen der Reihenfolge $O_{i1} \to O_{i2} \to ... \to O_{in_i}$ ohne Unterbrechung auf einer Produktionsanlage k nacheinander bearbeitet werden, wodurch $\sum_{j=1}^{n_i} p_{ij}$ die Gesamtdauer eines Auftrages determiniert. Jede Produktionsanlage kann zu einem Zeitpunkt $t \in \{0,...,T\}$ nur einen Produktionsauftrag gleichzeitig ausführen und jeder Produktionsauftrag darf in jedem Zeitpunkt t nur von einer Produktionsanlage ausgeführt werden. Alle Produktionsaufträge und Produktionsanlagen sind ab $t = 0$ verfügbar und alle Produktionsaufträge müssen innerhalb des festen Zeithorizonts T fertig gestellt werden.

Ein Maschinenbelegungsplan $\boldsymbol{S}$ kann somit durch eine Abbildung $\boldsymbol{S}$*:* $i \mapsto \left(M_i^S, S_i^S\right)$ dargestellt werden, wobei $M_i^S \in \{1,...,m\}$ die Produktionsanlage, auf der ein Produktionsauftrag i ausgeführt wird und $S_i^S \in \{0,...,T\}$ seinen Startzeitpunkt bezeichnet. Die Startzeit S_{ij}^S der Operation O_{ij} ist somit definiert als $S_{ij}^S = \begin{cases} S_i^S & j = 1 \\ S_i^S + \sum_{j'=1}^{j-1} p_{ij'} & j = 2,...,n_i \end{cases}$.

Die Menge der gleichzeitig zu einer Zeit t in Bearbeitung befindlichen Produktionsaufträge ist dann $P_t^S := \left\{O_{ij} \in \Omega \middle| S_{ij}^S \le t \le S_{ij}^S + p_{ij}\right\}$ und die abgerufene Gesamtlast im Zeitpunkt t ist $e_t^S := \sum_{O_{ij} \in P_t^S} e_{ij}$.

Eine Maschinenbelegung ist zulässig, wenn für alle Produktionsaufträge i gilt, $S_i^S + \sum_{j=1}^{n_i} p_{ij} \le T$ und wenn für jedes Paar von Produktionsaufträgen (i,i'), das auf der gleichen Produktionsanlage ($M_i^S = M_{i'}^S$) ausgeführt wird, gilt, dass $S_i^S + \sum_{j=1}^{n_i} p_{ij} \le S_{i'}^S$ oder $S_{i'}^S + \sum_{j=1}^{n_{i'}} p_{i'j} \le S_i^S$.

Berücksichtigt man die durchschnittlich nachgefragte Gesamtlast $\overline{E} = \frac{1}{T}\sum_{i=1}^{n}\sum_{j=1}^{n_i} e_{ij}$, so stellt die Zielvorstellung *min* $\sum_{t=0}^{T}(\overline{E} - e_t^S)^2$ eine adäquate Umsetzung der Zielsetzung der angestrebten Glättung der nachgefragten Gesamtlast an der Umwandlungsanlage dar.

Das vorliegende Problem der Glättung der nachgefragten Gesamtlast kann somit als Spezialfall des in der Literatur beschriebenen *resource levelling problem* [3] interpretiert werden. Dabei kann aufgrund der Komplexität des vorliegenden Problems im Allgemeinen keine optimale Lösung bestimmt werden. Weiterhin sind auch Ansätze des Project Scheduling (vgl. z.B. [3], [4]) aufgrund der fehlenden Reihenfolgebeziehungen zwischen den Produktionsaufträgen nur ungenügend übertragbar. Daher wird ein heuristischer Lösungsansatz auf Basis von evolutionären Algorithmen für die vorliegende energieorientierte Maschinenbelegung vorgeschlagen und umgesetzt.

3 Heuristischer Lösungsansatz auf Basis evolutionärer Algorithmen

Prinzipiell besteht das beschriebene energieorientierte Maschinenbelegungsproblem aus einem Zuordnungsproblem (Zuordnung der Produktionsaufträge zu Produktionsanlagen) und der Bestimmung der Startzeiten der einzelnen Produktionsaufträge. Somit werden nachfolgend, ausgehend von einer zulässigen Maschinenbelegung[2], mittels eines evolutionären Verfahrens die Startzeiten der einzelnen Produktionsaufträge ermittelt.

Evolutionäre Algorithmen sind stochastische Suchverfahren, die sich an den Prinzipien der biologischen Evolution orientieren, wobei in jeder Iteration mit mehreren potentiellen Lösungen (Individuen) gearbeitet wird [5]. Die Spezifikation eines evolutionären Algorithmus umfasst dabei (i) die Form der Repräsentation des Problems, (ii) die Ausgestaltung der Selektion, der Rekombination und der Mutation sowie (iii) die Bestimmung einer Startlösung, einer Bewertungsfunktion und eines Abbruchkriteriums.

In der Anwendung von evolutionären Verfahren auf Maschinenbelegungsprobleme wird die Repräsentation eines Lösungsvektors oftmals in Form einer *Permutationscodierung* (Liste nacheinander auszuführender Produktionsaufträge) umgesetzt. Der wesentliche Vorteil dabei ist, dass bei fester Zuordnung der Produktionsaufträge zu Produktionsanlangen durch die Auflösung der (Permutations-) Liste die Zulässigkeit der (Zwischen-) Lösungen sichergestellt wird. Weiterhin können dadurch auch gewollte Pausen[3] zwischen den Produktionsaufträgen in Form von Pausenaufträgen realisiert werden. Insgesamt kann ein kompletter Maschinenbelegungsplan und somit die Startzeiten der einzelnen Produktionsaufträge durch die feste Zuordnung der Produktionsaufträge und die (Permutations-)Liste eindeutig bestimmt werden.

Die weitere Spezifikation des vorgeschlagenen evolutionären Algorithmus basiert auf Standardoperatoren:[4] Die Selektion der Eltern für die Individuen der nächsten Generation innerhalb des vorgeschlagenen evolutionären Algorithmus erfolgt auf Basis der *roulette wheel selection*. Die Rekombination ist in Form eines *partially matched crossover*, das speziell für die Permutationscodierung entwickelt wurde, ausgestaltet. Die Mutation erfolgt durch eine *inversion mutation*. Die Startlösung stellen zufällig erzeugte zulässige Maschinenbelegungen, die um die Pausenaufträge erweitert sind, dar. Als Bewertungsfunktion dient die beschriebene Zielvorstellung zur Glättung der Gesamtlast und als Abbruchkriterium die Anzahl der Iterationen.

---

[2] Da im vorliegenden Fall die zur Verfügung stehende Bearbeitungskapazität größer als die Summe der Bearbeitungszeiten der Produktionsaufträge ist und sich daher genügend Freiheitsgrade bezüglich der Zuordnung der Aufträge ergeben, kann die Ermittlung eines zulässigen Maschinenbelegungsplans mit einfachen Algorithmen erreicht werden.

[3] Diese Pausen entstehen, wenn die Summe der Bearbeitungszeiten der Produktionsaufträge auf einem Aggregat kleiner ist als der vorgegebene Zeithorizont.

[4] zur genauen Beschreibung der einzelnen Verfahren siehe [1] bzw. [5]

4 Numerische Ergebnisse

Nachfolgend werden die Ergebnisse des untersuchten Szenarios (32 Färbeaggregate, 126 Färbeaufträge, Zeithorizont 24 Stunden bei 3-Schicht-Betrieb, Zeitauflösung 5 Minuten) dargestellt. Durch die Formulierung des energieorientierten Maschinenbelegungsproblems als resource levelling problem und die vorgeschlagene heuristische Lösung auf Basis eines evolutionären Algorithmus konnte eine signifikante Glättung der nachgefragten Gesamtlast im vorliegenden Beispiel erreicht werden (vgl. Abb. 3). Darauf aufbauend kann durch eine geeignete Simulation des Betriebs der Umwandlungsanlage im Kesselhaus der betrachteten Garnfärberei die Einsparung an eingesetztem Energieträger quantifiziert werden. Das so ermittelte Einsparpotential an eingesetztem Erdgas beträgt im Durchschnitt 6 Prozent.

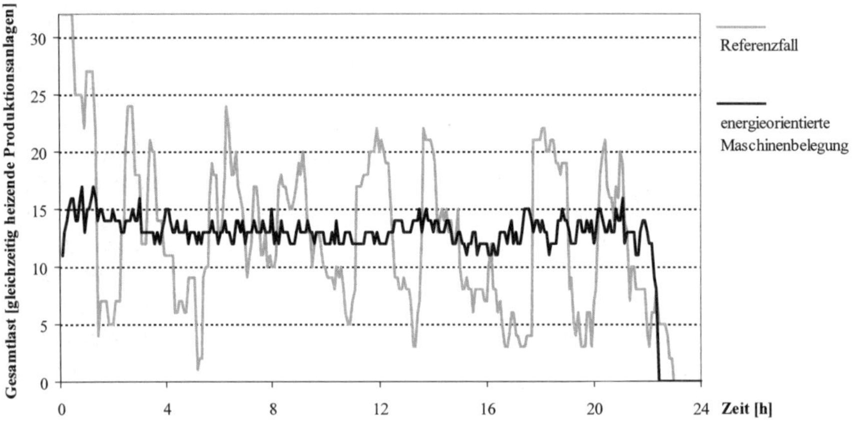

Abb. 3. Glättung der Lastgangkurve

Literatur

1. Eiben AE, Smith JE (2003) Introduction to evolutionary computation. Springer, Berlin Heidelberg New York.
2. Günther H-O, Tempelmeier H (2000) Produktion und Logistik. 4. Aufl., Springer, Berlin Heidelberg New York.
3. Neumann K, Zimmermann J (1999) Resource levelling for projects with schedule-dependent time windows. European Journal of Operational Research 117, 591-605.
4. Neumann K, Zimmermann J (2000) Procedures for resource levelling and net present value problems in project scheduling with general temporal and resource constraints. European Journal of Operational Research 127, 425-443.
5. Pohlheim H (2000) Evolutionäre Algorithmen: Verfahren, Operatoren und Hinweise. Springer, Berlin Heidelberg New York.
6. Schmitz KW, Schaumann G (2005) Kraft-Wärme-Kopplung. 3. Aufl., Springer, Berlin Heidelberg.

Multi Objective Pinch Analysis (MOPA) Using PROMETHEE to Evaluate Resource Efficiency

Hannes Schollenberger, Martin Treitz, Jutta Geldermann, Otto Rentz

French-German Institute for Environmental Research, University of Karlsruhe, Hertzstr. 16, 76187 Karlsruhe, Germany
hannes.schollenberger@wiwi.uni-karlsruhe.de

Summary. Process optimisation on the basis of detailed process characteristics requires the simultaneous consideration of different mass and energy flows and leads to a multi-criteria problem. Different technological options can be compared based on targets calculated by pinch analyses and further criteria. The approach is applied to a case study on bicycle coating.

1 Introduction

The problem of defining resource efficiency is discussed in this paper as a goal for process improvement efforts. For this purpose the outranking multi-criteria approach PROMETHEE [Brans et al., 1986] is proposed for evaluating the strengths and weaknesses of different technological options which are assessed using pinch analysis inter alia.

The methodology is employed in a case study on bicycle coating and addresses the best way to react to legislative changes that stipulate reduced solvent emission concentrations in the waste air of the spray booth and the drying oven.

2 Multi Objective Pinch Analysis (MOPA)

Multi Objective Pinch Analysis (MOPA) [Geldermann et al., 2005] is a combination of pinch analyses with different targets (energy, wastewater, solvents) and a subsequent multi-criteria analysis. The starting point of the analysis is a process characterisation based on thermodynamic, chemical and engineering principles focussing on the entire system's performance rather than the improvement of single process units while considering the catalogue of *Best Available Techniques* (BAT) and a screening of emerging technologies.

The implemented pinch analysis approach (e.g. [Linnhoff and Hindmarsh, 1983] is used to calculate the theoretical minimal target values. These are used together with the characterising data of the different technological options to identify economically reasonable technologies in a multi-criteria analysis incorporating further criteria as for example operating and investment costs.

3 Evaluation of Resource Efficiency using PROMETHEE

Process data and value judgements are the key input parameters to the multi-criteria analysis in MOPA. Therefore, modelling the decision maker's preferences is addressed here. The discussion implies that it is not only necessary to analyse the influence of the uncertainty in the process data, but also in the modelling of the decision maker's value judgements.

Preference Functions and Parameters

Once discrete alternatives, criteria and attribute values have formally been identified the next fundamental modelling step in PROMETHEE is the characterisation of a preference function $P_j(.,.) : A \times A \rightarrow [0,1]$ for each criterion. [Belton and Stewart, 2002] define preferences as "models describing the relative importance or desirability of achieving different levels of performance for each identified criterion" by incorporating value judgements about the decision makers's preferences.

The relation $P_j(.,.)$ is based on the difference in the attribute values of the criteria. [Brans et al., 1986] propose six different generalised preference functions defined by threshold values of indifference (q) and strict preference (p), or an inflection point s in case of the Gaussian distribution (Figure 1).

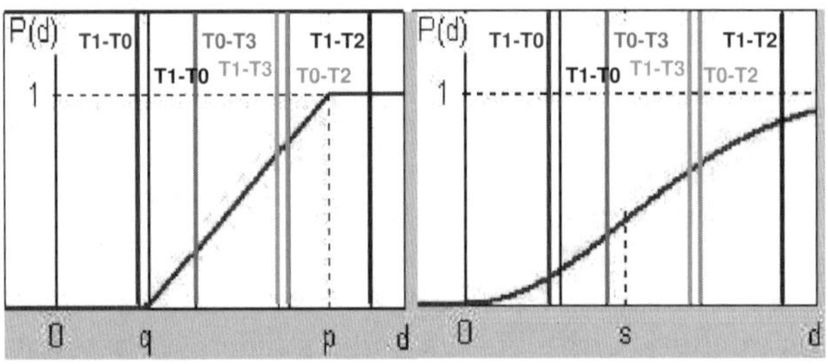

Fig. 1. Preference Function Type V (left) and VI (right) of the criterion *Energy*

If the difference is smaller than q the decision maker has no preference between alternatives a and b with respect to this one criterion, whereas there is a strict preference for a over b if $d_j(a,b) > p$. Since no objective preferences exist and value judgements are, at least partially, formed only during the modelling process, the multi-criteria method should enable the decision maker to better comprehend his underlying assumptions and model his own preferences [Belton and Stewart, 2002, Basson, 2004].

Sensitivity Analysis

Considering the above mentioned aspects it is important to carry out sensitivity analyses that iteratively re-model the decision problem and increase knowledge about it. Besides investigating the robustness of the results against the parameters p, q, and s of PROMETHEE by carrying out a *Monte Carlo Simulation* using certain uncertainty levels (e.g. $\pm 10\%$) and evaluating all possible preference combinations, the *principal component analysis* (PCA) [Timm, 2002], based on the matrix of the *single criterion net flows* in which the strength and weaknesses of an alternative is analysed, can also be employed. By rotating the original axes into the new principal axes the cloud of alternatives is projected onto the so-called GAIA plane (cf. Figure 2) spanned by the two eigenvectors with the largest eigenvalues (henceforth called 1st and 2nd principal component) [Brans and Mareschal, 2005].

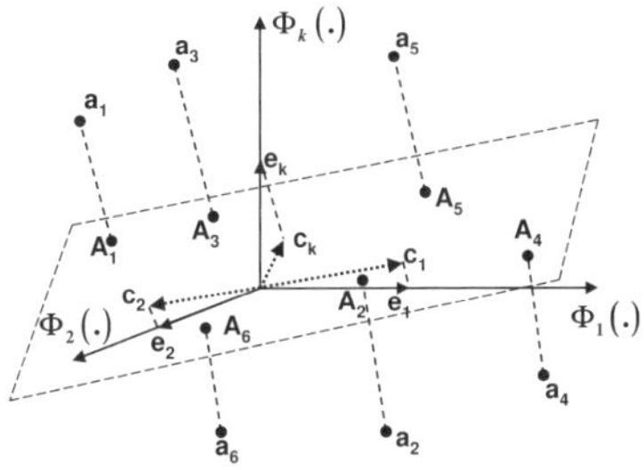

Fig. 2. Projections on the GAIA Plane [Brans and Mareschal, 2005]

Besides alternatives and criteria axes, also the weighting vector (the *PROMETHEE decision stick* π) can be projected on the GAIA plane. The weights describe the relative importance of the different criteria. In PROMETHEE, the weighting factors represent a kind of "voting power" for each criterion, rather than trade-offs or compensation ratios. In the case of valuating environmental impacts, specialized weighting methods (e.g. eco-indicator, shadow prices, eco-taxes) are discussed. A simultaneous valuation of both environmental and economic indicators even adds complexity to the discussion of how to derive weighting factors.

Furthermore, behavioural aspects must be considered that reflect different perceptions by wording a criterion in terms of "losses" to one reference point or in terms of "gains" to a different one [Kahneman et al., 1982]. Additionally, biases due to the number of sub-criteria or the hierarchical level (see [Belton and Stewart, 2002, Basson, 2004]) must also be taken into account.

By defining upper and lower bounds for each weighting factor the convex hull of all valid weighting combinations can be projected on the GAIA plane, visualising the range of π, i.e. the *PROMETHEE VI area*. If attribute values are furthermore characterised by a specific uncertainty level, a scatter plot based on a *Monte Carlo Simulation* can be displayed, visualising the distinguishability between all alternatives [Basson, 2004]. Consequently, these analyses allow the simultaneous change of all weighting factors.

The presented approach allows to consider the uncertainty in the process data as well as in the value judgements of the decision maker. The results of the different analyses that can be carried out lead to a deeper understanding of the decision problem itself. Hence, the decision maker is able to evaluate the influence of the chosen parametric description of his preferences and to investigate the sensitivity to its uncertain framework.

4 Case Study

The basic question and data of this case study were originally presented in [Geldermann et al., 2006] aimed at finding the most effective utilisation of solvents in waste air from the serial coating of bicycles: Recovery of solvents or thermal utilisation. However, this paper concentrates on the question of how to fulfill legal obligations pertaining to the maximum solvent concentration allowable in waste air. Therefore, not only are the additive measures thermal incineration (T_2) and solvent condensation (T_3) considered, but also the switch to waterborne coatings (T_1) as an integrated measure for emission reduction. These alternatives are compared to the status quo (T_0), which is still legal for two more years (cf. Table 1). The minimum values for each criterion (S_{pinch}) are calculated using among others pinch analyses. They are used as targets and only the differences between the attribute values and them are used in the multi-criteria analysis. The objective of the case study is to find out if it is worthwhile to modify the process earlier than legally required.

Table 1. Data of the Case Study

Parameter	w_i	S_{pinch}	T_0	T_1	T_2	T_3
Energy $[kWh/h]$	0.15	81.8	1 406	1 381	1 477	1 448
Fresh water $[m^3/h]$	0.15	0.5	2.1	2.1	2.0	2.2
Solvents Conc. $[mgC/m^3]$	0.20	0	694	66	7	139
Investment [€/a]	0.20	0	78 500	88 250	140 250	133 000
Purchased solvents [€/a]	0.30	0	19 340	12 918	19 340	4 835

In addition to energy and water consumption, solvent concentration in the waste air, operating costs (represented by costs of purchased solvents) and investment costs complete the list of criteria. In this case specific economic life-times and interest rates have been assumed and therefore the status quo costs depend on the residual value of the current plant.

The solvent concentration can be reduced most effectively with thermal incineration, which on the other hand has the highest investment costs and induces no savings in the amount of purchased solvents per year, while water consumption is not affected significantly.

Figure 3 shows the sensitivity of the criterion specific net flow of *Energy* to the uniformly distributed parameter s. The mean of s is half of the difference between the maximal and the minimal attribute value, the uncertainty level is assumed to be $\pm 10\,\%$. The overall contribution does not significantly change by varying s. Further calculations show that in the case of the permutation of all preference types I through VI for all criteria, alternative T_1 is the best alternative for all possible combinations.

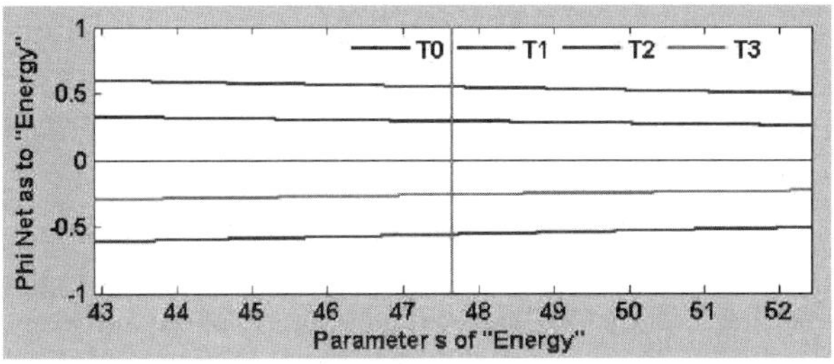

Fig. 3. Sensitivity of the Net Flow of *Energy* to s

The GAIA plane (Figure 4) is displayed for all criteria using preference type VI (parameter s is half the difference between the max. and the min. attribute value of each criterion). Roughly 87 % of the total information is illustrated in the diagram. The grey area visualises the PROMETHEE VI area by assuming an absolut variation of $\pm 5\,\%$ in the weighting of each criterion simultaneously. Additionally, the scatterplot of the projections of the alternatives T_0 to T_3 shows the variation of the results as a consequence of the uncertainty of the data which is modelled using a Monte Carlo Simulation with an uncertainty level of 10 % and a normal distribution.

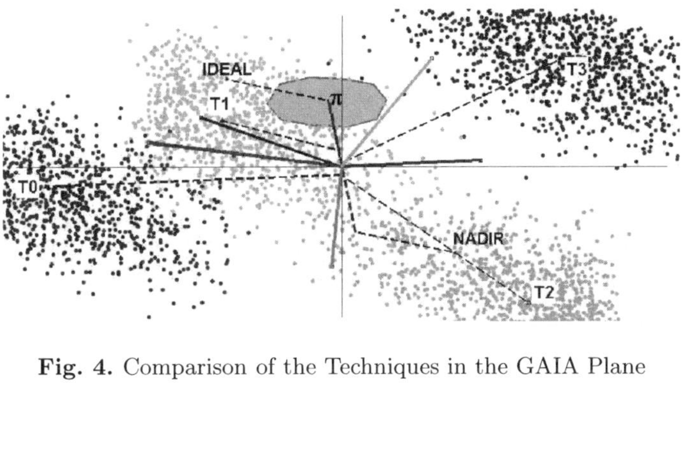

Fig. 4. Comparison of the Techniques in the GAIA Plane

5 Conclusions

This paper shows the application of a multi-criteria analysis to the process of industrial bicycle coating. The application of several sensitivity analyses proves that it can provide a good understanding of the effects of the different parameters. Therein, the use of a Monte Carlo Simulation offers the possibility of an analysis comprising several parameters simultaneously.

Consequently, the application of the Monte Carlo Simulation in combination with a principal component analysis can help to understand the impact of the uncertainties in the value judgements on the overall results [Basson, 2004]. It may be deeper than the one of the uncertainty in the weighting factors only. This allows an evaluation of the robustness of the decision. Hence, the aim of further experiments will be the calculation of probabilities and stability intervals of the ranking of the alternatives considering different uncertainty levels and parameter combinations.

Acknowledgement. This research is funded by a grant from the *VolkswagenStiftung*. We would like to thank them for the excellent support of our research, as well as Dr. A. Berg, J. Neugebauer, M. Zacarías and F. Saavedra from the Unidad de Desarrollo Tecnológico (UDT), Concepción, Chile, for the close cooperation.

References

L. Basson. *Context, Compensation and Uncertainty in Environmental Decision Making.* PhD thesis, Dep. of Chemical Engineering, University of Sydney, 2004.

V. Belton and T. Stewart. *Multiple Criteria Decision Analysis - An integrated approach.* Kluwer Academic Press, Boston, 2002.

J.-P. Brans and B. Mareschal. PROMETHEE Methods. In J. Figueira, S. Greco, and M. Ehrgott, editors, *Multiple Criteria Decision Analysis - State of the Art Surveys*, pages 163–195. Springer, 2005.

J.-P. Brans, P. Vincke, and B. Mareschal. How to select and how to rank projects: The PROMETHEE method. *European Journal of Operational Research*, 24: 228–238, 1986.

J. Geldermann, H. Schollenberger, M. Treitz, and O. Rentz. Multi Objective Pinch Analysis (MOPA) for Integrated Process Design. In H. A. Fleuren, D. Hertog, and P. M. Kort, editors, *Operations Research Proceedings 2004*, pages 461–469. Springer, 2005.

J. Geldermann, M. Treitz, H. Schollenberger, and O. Rentz. Evaluation of VOC recovery strategies: Multi Objective Pinch Analysis (MOPA) for the evaluation of VOC recovery strategies. *OR Spectrum*, 28(1):1–18, 2006.

D. Kahneman, P. Slovic, and A. Tversky. *Judgment under uncertainty: heuristics and biases.* Cambridge University Press, Cambridge, 1982.

B. Linnhoff and E. Hindmarsh. The pinch design method for heat exchanger networks. *Chemical Engineering Science*, 38(5):745–763, 1983.

N. H. Timm. *Applied multivariate analysis.* Springer, New York, 2002.

An Emission Analysis on Toxic Substances (SPM and NO_x) from Transportation Network System in Tokyo of Japan

Kiyoshi DOWAKI*, Kouichiro YOSHIYA and Shunsuke MORI****

*Tokyo University of Science, 2641 Yamazaki, Noda, Chiba, Japan 278-8510
Current Affiliation: Research Institute of Innovative Technology for the Earth (RITE)
9th Floor, No. 3 Toyokaiji Bldg., 2-23-1, Nishi-shimbashi, Minato-ku, Tokyo, 105-0003, Japan (e-mail: dowaki@rite.or.jp)
**Tokyo University of Science, 2641 Yamazaki, Noda, Chiba, Japan 278-8510

Abstract. Recent years, the increases in toxic substances of NO_x and/or SPM (Suspended Particulate Matter) from vehicles come to be serious problem. In this paper, we estimated the concentrations of SPM and NO_x using a traffic flow model. This model has a characteristic of which the traffic volume in each link and the average speed can be decided by solving the problem on the travel time minimization. We also considered the traffic congestions and/or the toll roads in the model, and these effects were reflected to it by converting the time functions. About the correlation between the traffic volume in our model and the real volume, we obtained the correlation coefficient of 0.74. Simultaneously, we got the result that the concentration of NO_x was approximately 70 to 230 ppb. That of SPM was approximately 40 to 100 $\mu g/m^3$.

Key words: Traffic flow model, SPM, NO_x, Tokyo area

1 Introduction

Nowadays, in Tokyo area, the increases in toxic substances of SPM (Suspended Particulate Matter) and/or NO_x come to be serious problem. SPM which was emitted from mainly freight car might cause the serious disease such as a cancer. NO_x might cause air pollution. In summer time of Japan, photo-chemical smog often takes place. In Tokyo, for recent 20 years, the number of freight car has increased rapidly with the emissions of SPM and NO_x. Also, this area is suffering from chronic traffic congestion. After all, some residents living in the vicinity of Tokyo area might have increased the risk of serious disease such as a cancer.

Diesel engines have many advantages over gasoline engines, including better fuel efficiency and lower emissions of some air pollutants (carbon monoxide and hydrocarbons). That is, diesel engines emit less carbon dioxide (CO_2, a greenhouse gas) per unit of work. However, uncontrolled diesel engines emit high concentrations of particles, NO_x and aldehydes and low concentrations of CO and hydrocarbons. Due to these compounds, the relationship between diesel exhaust exposure and risk of lung cancer has been a public health concern for several decades. In urban areas, direct emissions of SPM from diesel engines represent about 10% of the mass of ambient particles [1].

In this paper, using the data on the traffic volume and on the emissions on SPM and NO_x, we established the traffic flow model and estimated the concentration of their substances in Tokyo area.

So far, some studies have been made to estimate environmental emissions by motor vehicles within the Tokyo metropolitan area taking traffic flow into account [2, 3, 7]. For instance, Kudoh et al. estimated the emissions of NO_x and CO_2 in Tokyo area using the traffic network model a logistic algorithm [7]. However, they were not referred to the emissions of SPM with the risk of serious disease. The concentrations rather than the emissions would be more significant in assessing the risk on diseases [1]. Another studies in traffic engineering treat traffic flow dynamically [4, 5]. In addition, the structure of traffic flow model taking the traffic congestion into consideration has already established (cf. Akamatu et al. [6]). These studies only focus on simulating real traffic flow in not wider areas but limited areas. Few studies offer much towards environmental concerns.

Thus, we focused on their concentrations in Tokyo area (ex. the inner area of Kanjou No. 7 line) where toxic substances were emitted too much. Based on OD data, we established the traffic flow model considering the toll road and the traffic congestion. We allocated the traffic volume in each link and estimated its average speed using the travel time minimization methodology. Finally, we estimated the concentration intensities using the regression equations on the concentrations of SPM and NO_x every specific term.

2 Computation methodology

2.1 Structure of traffic network

There are mainly two methods to allocate cars in each link. One is due to the Warshall-Floyd algorithm based on the logistic theorem [8]. In this method, the path choice is conducted based on the probability distribution on travel time. The other one, which we used in this model, is due to the system optimization method. This means that the allocation of cars in each link is conducted so that the total travel time of whole network is minimum. In this method, the formulation of problem is relatively easier. However, more paths have to be described in the formulation. That is, it is necessary to increase the possibilities of the path choice due to cars. Likewise, since the time (or the cost) function is employed, it would be easy to deal with the introduction of road pricing for mitigating toxic substances.

In order to formulate, we prepared the traffic network model shown in Figure 1. According to Fig.1, Kanjou No.7 line is represented as dotted line and the numbers of link (road) and of node (intersection) are 262 and 71, respectively.

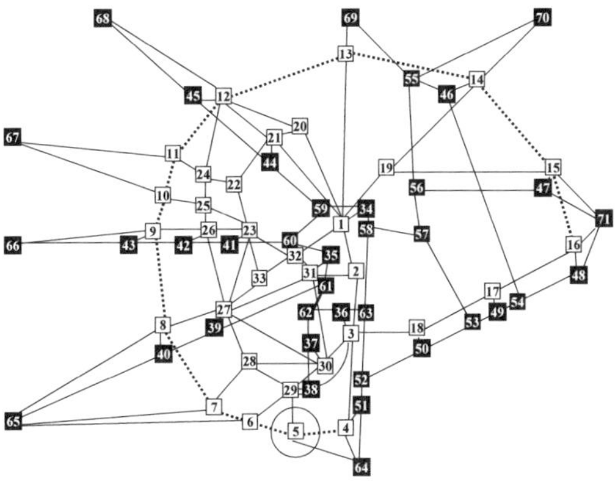

Fig.1 Traffic network[1]

OD data [9, 10] includes traffic data for the following eight types of vehicles (All vehicles are classified within these eight categories as stated by Japanese law): light cars, motorcars, buses, light freight cars, small freight cars, light vans, freight cars and special cars, e.g. emergency vehicles. However, since our aims are to estimate the concentrations of SPM and NO_x, we focused on motorcars and freight car whose traffic volumes are larger relatively. Traffic outside the area is assumed to arrive and depart from the edge of the nodes (see Fig. 1). Some of this traffic is allocated to the edge in order that the 24 hours traffic data given by the census [10] can become consistent with the OD data. We also considered the passed traffic volume in this network. Since OD data has the traffic volume of 24 hours, we have to distribute the data every 3 hours in order to examine the concentrations dynamically. The traffic volume of 1 day was divided into the data of every 3 hours using the traffic ratio every time [11].

2.2 Conversion equation on concentrations

About the concentrations of SPM and NO_x, we assumed that the concentration equations should be represented by the emissions of SPM and NO_x, the wind data (speed and direction) and the distance to a measurement point [14].

[1] *: Point measured SPM and NO_x emissions, Dotted line: Kanjou No. 7 line (road), Circle: the intersection on which we focused in our paper. Number in the square box indicates intersection and it is corresponding to the point of an origin or a destination in the OD table.

$$E_{SPM} = 0.756x_1 - 0.296x_2 - 0.966x_3 + 1.72\times10^{-3}x_4 + 9.866 \quad R^2 = 0.752 \tag{1}$$

$$E_{NO_x} = 0.760x_1 - 0.294x_2 - 2.30x_3 + 3.65\times10^{-4}x_4 + 26.03 \quad R^2 = 0.729 \tag{2}$$

where E_j [$\mu g/m^3$ or ppb], x_1 [$\mu g/m^3$ or ppb], x_2 [m/s], x_3 [m], x_4 [g] and R^2 are the concentration in j-kind, the concentration before 1 period, the value which multiplied a direction by a speed, the minimum distance from a center of intersection to a detected point, the emissions at current period and a coefficient of determination, respectively. We analyzed the conversion equations using 672 measured data.

2.3 Formulation

Assuming that the traffic volume of OD pair given from OD data between i-node and j-node is shown as OD_{ij}, OD_{ij} is balanced to the summation of the traffic volume in each link. That is,

$$OD_{ij} = q_{ij1} - q_{ij3} - q_{ij4} + q_{ij7} + q_{ij9} + \cdots + q_{ijk} = \sum_k \delta_{ijk} q_{ijk} \tag{3}$$

where q_{ijk} is the traffic volume from i-node to j-node in k-link. Commonly, we select a few paths between i-node and j-node so as to find out optimal solutions. Also, δ_{ijk} is a variable of 0 or 1. This means that all paths are not always selected between i-node and j-node.

Since the total traffic volume q_k in k-link is

$$q_k = \sum_i \sum_j \delta_{ijk} q_{ijk} \tag{4},$$

the travel time t_k in k-link is represented as Eq. 5 using the Bureau of Public Roads (BPR) formula [12].

$$t_k\left(=\frac{d_k}{v_k}\right) = \begin{cases} \dfrac{d_k}{v_0}\left\{\dfrac{1+\alpha(q_k/c_k)^\beta}{1+\alpha}\right\} & v_k \le v_o \\ \dfrac{d_k}{v_0} & v_k > v_o \end{cases} \tag{5}$$

where v_k, v_0, d_k, c_k, α and β are the average traffic velocity in k-link, the speed limit in k-link, the length of k-link, the traffic capacity in k-link and parameters, respectively. d_k is also provided from the OD data [9,13]. c_k is the traffic capacity ,of which general road is 30000, toll road with two lanes is 36000 and toll road with three lanes is 54000. In addition, the parameters of α and β are used for $\alpha = 1$ and $\beta = 2$ commonly. Hence, it is necessary to minimize the following equation including in the benefit toll road in order to allocate cars in each link.

$$\min \sum_k (t_k + p_k/3000) \tag{6}$$

where p_k is the toll in k-link. Usually, the toll fee is 700 or 1400 yen/vehicle.

Using GAMS (General Algebraic Modeling System) developed by World Bank, we solved the non-linear equations which consist of Eqs. 4-6. Simultaneously, we predicted the concentrations using the concentration equations (see Eqs. 1 and 2) at every intersection.

3 Computation results

First, we examined the correlation coefficient on the allocation of cars due to our computation by comparing with the census data (real data). Although the correlation coefficient of motorcars was 0.53, that of freight car was 0.74. Since we know that the concentrations are strongly affected by the emissions from freight cars, our result which is the coefficient of 0.74 in freight cars has relevance to some extent.

Next, using Eqs. 1 and 2, we compared the calculated concentrations on SPM and NO_x with the measured data at every intersection. Figures 2 and 3 show the annual average concentrations of SPM and NO_x. The gray bar is represented as a deviation between the measured data and the calculated results. Although the maximum values of error bar are approximately 30% of SPM and 50 % of NO_x, we could see the tendency of concentrations in time series. These results are dependent upon the traffic volume. If the matching precision for real traffic volume is over 0.7, these in the model are reliable. There were some intersections with the smaller error bar.

Finally, we examined the spreading aspects of SPM and NO_x at Matsubarabashi intersection using the results of traffic flow simulation (see Fig. 1). About the area of 130 m×130 m, we plotted the concentration intensities of SPM and NO_x in Figures 4 and 5 due to the following equation.

$$v l \left(\frac{\partial^2 E_i}{\partial x^2} + \frac{\partial^2 E_i}{\partial y^2} \right) = \frac{\partial E_i}{\partial t} \tag{7}$$

where v [m/s] and l [m] are the wind speed and the grid width (=5 [m]). These results would indicate the impacts of toxic substances due to wind and/or traffic volume.

4 Conclusion

We were able to estimate the concentrations of SPM and NO_x using the statistical equations on the toxic substances even if we do not solve the diffusion equation. Likewise, we would obtain the significant information on the substances.

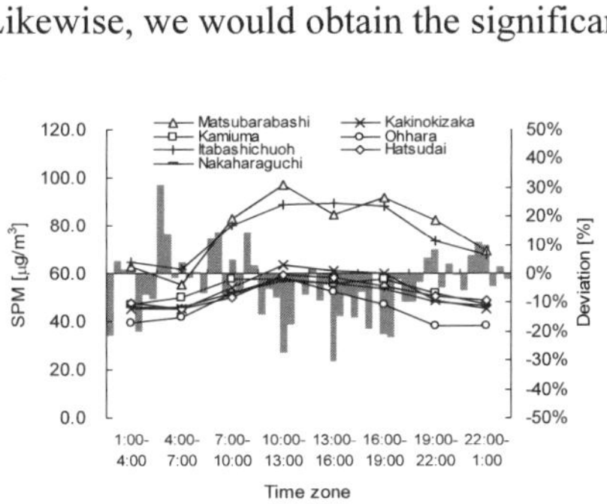

Fig. 2 SPM concentration at intersection

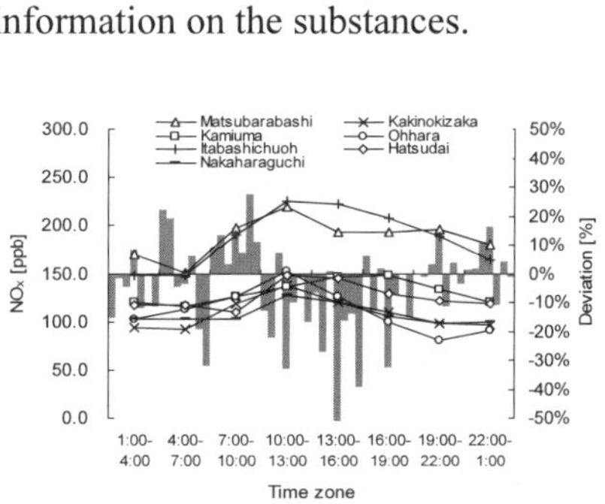

Fig. 3 NO_x concentration at intersection

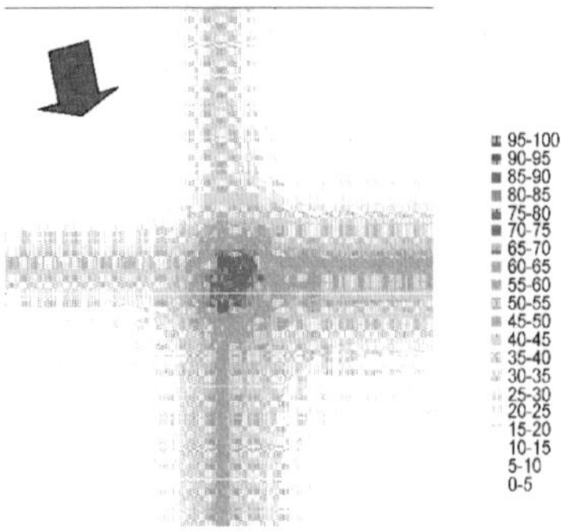

Fig. 4 The spreading aspect of SPM

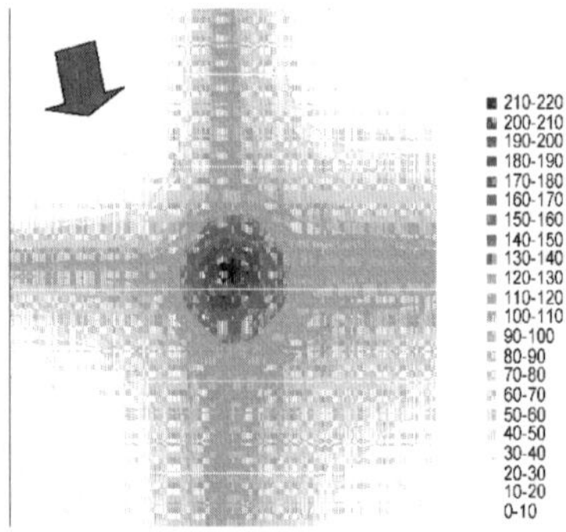

Fig. 5 The spreading aspect of NO_x

Reference

1. Health Effect Institute (2002) Research Directions to Improve Estimates of Human Exposure and Risk from Diesel Exhaust
2. Moriguchi Y, et al. (1993) Development of estimation system of environmental improvement by traffic pollution measures in wide area (in Japanese). Infrastructure Planning Review 11
3. Akisawa A, et al. (1995) Evaluation of optimal land-use structure minimizing the energy consumption in cities (in Japanese). In: Proceedings of the 1998 Conference on Energy Systems and Economics: 21–6
4. Yoshii T, et al. (1995) A network simulation model for oversaturated flow on urban motorways (in Japanese). Traffic Engineering; 30 1:33–41
5. Okamura H, et al. (1996) Development and verification of simulation model for urban road network (in Japanese). Proceedings of the 16th Conference on Traffic Engineering: 93–96.
6. Akamatsu T, Makino Y, Takahashi E. (1998) Semi-dynamic Traffic Assignment Models with Queue Evolution and Elastic OD Demands (in Japanese). Infrastructure Planning Review 15: 535-545.
7. Kudoh Y, et al. (2001) Environmental evaluation of introducing electric vehicles using a dynamic traffic-flow model. Applied Energy; 69: 145-159
8. Iri M, et al. (1976) Network Theory (in Japanese). Niikagiren, Tokyo
9. Kanto Regional Construction Bureau (1998) Ministry of Construction The OD data in Kanto area of Japan.
10. Road Bureau Road of Construction Ministry (1994) Traffic census
11. Traffic Bureau of Tokyo Metropolitan Police (2000) Statistical table on the traffic in Tokyo
12. U.S. Department of Commerce, Urban Planning Division Bureau of Public Roads (1964) Traffic Assignment Manual. Washington D.C.
13. Metropolitan Expressway Public Corporation (2001) The 24th report of OD investigation
14. Environmental Bureau of the Tokyo Metropolitan Government (1996) Investment report of traffic volume and exhaust gases from vehicles (outline) (in Japanese)

Planning and Evaluation of Sustainable Reverse Logistics Systems

Grit Walther, Eberhard Schmid, Sanne Kramer, Thomas Spengler [1]

Department of Production & Logistics Management, Braunschweig University of Technology, Katharinenstr. 3, 38106, Braunschweig, g.walther@tu-bs.de

Abstract: Different alternatives exist for installation of scrap treatment systems in order to comply with the German law on discarded electronic equipment. Thus, the aim of this contribution is to compare various treatment alternatives considering economic but also ecological, technical, infrastructural, and social objectives. The MADM-method PROMETHEE is used for the evaluation of these systems. Since many relevant evaluation attributes depend on short-term decisions within a given infrastructure, these decisions are anticipated based on activity based optimization models. Depending on the operation of the treatment systems, either an economic efficient solution applying a LP-model or a sustainable solution applying Weighted Goal Programming (WGP) is calculated.

Introduction

As result of the adoption of the European directive on waste electrical and electronic equipment (WEEE) and the implementation of the German electronic law (ElektroG), systems for take-back and treatment of WEEE will have to be implemented; existing systems will have to be improved.

Since different alternatives exist for installation of scrap treatment systems, the aim of this contribution is to compare various treatment alternatives. To meet concerns of sustainable development, economic but also ecological, technical, infrastructural, and social objectives have to be taken into account.

In chapter 2, alternatives as well as relevant attributes and targets for evaluation will be presented, and the application of the MADM-method PROMETHEE for evaluation of the discrete infrastructural alternatives is described. Since many relevant evaluation attributes, e.g. degree of recycling achieved, emissions produced etc., depend on short-term decisions within a given infrastructure, these decisions are to be anticipated. This is done based on an activity based material flow model applying either an economic objective function or a sustainable multi-objective function presented in chapter 3. We will close with an outlook.

---

[1] This work has been promoted by the German "Deutsche Forschungsgemeinschaft" (DFG). The authors would like to thank for the support.

Evaluation of infrastructural alternatives

Different interplant as well as intraplant alternatives exist for future treatment of electronic scrap. As infrastructural interplant network alternatives, decentralized systems with various small and medium sized companies of inferior capacity as well as more centralized systems with companies having higher capacities are feasible. As intraplant treatment systems, unchained working stations are currently installed. However, line-chained working stations are also possible. Automated disassembly is only feasible for goods with low variability like washing machines or TV sets. Therefore, partly automated systems using automated disassembly cells in combination with line-chained disassembly working stations may be installed in the future. Fully automated systems can be considered in order to anticipate technological progress.

Table 1. Overview of inter- and intra-plant infrastructural system alternatives

Interplant infrastructure	Decentralized alternative Advanced centralization Complete centralization
Intraplant infrastructure	Unchained working stations Chained working stations Partial automation Complete automation

When implementing and improving treatment systems, the disassembly and recycling companies themselves will mainly pursue economic and technical targets. However, ecological and social targets are important from the macro-level point of view, e.g. for political decision makers being responsible for the national implementation of the WEEE-directive, for manufacturers being responsible for the assignment of disassembly contracts, for ecologically oriented users of electrical devices, and for other sustainability-oriented stakeholders. Relevant targets, subtargets, and attributes for evaluation of treatment systems are presented in table 2.

Table 2. Targets and attributes of the multi-attributive evaluation of treatment systems (* attributes depending on short-term decisions of the system)

Target and Sub-targets	Attributes
Economic targets	
annual profit	fixed costs; *contribution margin**
Technical targets	
utility of technique	automisation; technical state of the art; work security
system flexibility	flexibility of capacity; know-how; specialization
Ecological targets	
recycling targets	*reuse*; recycling*; energy recovery**
ecological influences	*CO_2 emissions**
Social targets	
created jobs	Nr. of jobs: for low-skilled/disabled/ in economically weak areas/other; job quality

Since a discrete number of infrastructural alternatives is to be evaluated, MADM-methods are applied. As attributes may be measured in quantitative (costs, CO_2-emissions) as well as in qualitative (high/low technological standard) units, and as it is easier for decision makers to compare alternatives for every single attribute than it is to assign values to all alternatives and attributes at once, an outranking method is applied. We use the method PROMETHEE (Preference Ranking Organisation METHod for Enrichment Evaluations) since it has advantages such as simplicity, clarity, efficiency, and low information requirements (Brans et al. 1986; Zimmermann/Gutsche 1991).
The evaluation attributes in table 2, which are italicized and marked by *, strongly depend on short-term decisions within a given infrastructure. Thus, these decisions are to be anticipated in order to evaluate the treatment systems.

Anticipation of short-term decisions and resulting objectives

The operational decisions, resembling the degrees of freedom within a given infrastructure, are mainly related to material flows and processes within the systems. Short-term decisions concern the quantities of products transported from collection points to disassembly companies and between disassembly companies, but also the execution of disassembly activities at the disassembly companies as well as the delivery of material fractions and components to land-filling and recovery sites. An overview of material flows within a treatment system is given in figure 1.

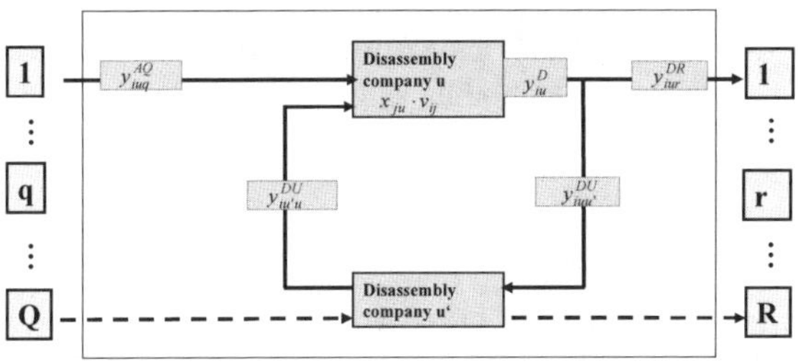

Fig. 1. Material flow model for disassembly company u

The input of a disassembly company is made up by appliances from all sources of the network (y_{iuq}^{AQ}) as well as by appliances and prefabricated parts from other disassembly companies ($y_{iu'u}^{DU}$). This input is then transformed by disassembly, which is modelled as the number of executions of a disassembly activity (x_{ju}) multiplied with an input-output-coefficient (v_{ij}) representing the input-output-relationship of the disassembly activity *j*. This is resulting in the output of the company (y_{iu}^{D}).

$$\sum_{q=1}^{Q} y_{iuq}^{AQ} + \sum_{\substack{u'=1 \\ u' \neq u}}^{U} y_{iu'u}^{DU} + (\sum_{j=1}^{J} x_{ju} \cdot v_{ij}) = y_{iu}^{D} \qquad i = 1..I; u = 1..U \qquad (1)$$

The output (y_{iu}^{D}) is further delivered to either recycling companies and disposal sites (y_{iur}^{DR}) or for further disassembly to other disassembly companies ($y_{iuu'}^{DU}$).

$$y_{iu}^{D} = \sum_{\substack{u'=1 \\ u' \neq u}}^{U} y_{iuu'}^{DU} + \sum_{r=1}^{R} y_{iur}^{DR} \qquad i = 1..I; u = 1..U \qquad (2)$$

Different restrictions exist regarding the quantity of available products at sources, disassembly capacities at companies as well as capacities at recycling and disposal sites. There are also non-negativity constraints. For a complete presentation of the model we refer to (Spengler/Walther 2004; Walther, 2005).

The objective function depends on the overall target of the treatment system. The system can either be aiming at the most efficient economic solution, thus neglecting other ecological and social criteria. For such cases, contribution margin as result of product (e_i^A) and recycling (e_{ir}^V) revenues minus disassembly (c_{ju}^{Z}), transportation ($c_{qu}^{TQ}, c_{uu'}^{TU}, c_{ur}^{TR}$), and sorting costs ($c_i^S$) is maximized. Since this is a single objective only, Linear Programming (LP) can be applied (Walther/Spengler 2005).

$$f(economy) = Max \sum_{u=1}^{U} \left(\sum_{i=1}^{I} \left(\sum_{q=1}^{Q} (e_i^A - c_{qu}^{TQ} - c_i^S) \cdot y_{iuq}^{AQ} + \sum_{\substack{u'=1 \\ u' \neq u}}^{U} (-c_{uu'}^{TU}) \cdot y_{iu'u}^{DU} + \sum_{r=1}^{R} (e_{ir}^V - c_{ur}^{TR}) \cdot y_{iur}^{DR} \right) - \sum_{j=1}^{J} x_{ju} \cdot c_{ju}^{Z} \right) \qquad (3)$$

However, if sustainability oriented OEMs or municipal authorities operate the system, ecological and social objectives may play an important role too.

Ecological influences are represented by CO_2-emissions. These emissions are calculated minimizing total transportation distances (D_{qu}, $D_{uu'}$, D_{ur}) from communities to disassembly companies, between disassembly companies, and from disassembly companies to recycling and land-filling facilities multiplied with a CO_2-emission factor (α^{TQ}, α^{TU}, α^{TR}) depending on the vehicle used at the different transportation levels.

$$f(CO_2) = Min \sum_{i=1}^{I} \sum_{u=1}^{U} \left(\sum_{q=1}^{Q} \alpha^{TQ} \cdot D_{qu} \cdot y_{iuq}^{AQ} \right) + \left(\sum_{\substack{u'=1 \\ u \neq u'}}^{U} \alpha^{TU} \cdot D_{uu'} \cdot y_{iu'u}^{DU} \right) + \left(\sum_{r=1}^{R} \alpha^{TR} \cdot D_{ur} \cdot y_{iur}^{DR} \right) \qquad (4)$$

Recycling targets are calculated based on the material flow model, maximizing the amount of scrap that is send to re-use, recycling, and energy-recovery facilities. This is done multiplying the amount of a material fraction send to a special recycling facility (y_{iur}^{DR}) with fraction and facility specific factors (ρ_{ir}, γ_{ir}, θ_{ir}) representing the amount of scrap approved to be reused, recycled, or recovered. Thereby, the amount of scrap that is send to land-fills is minimized.

$$f(reuse) = Max \sum_{i=1}^{I} \sum_{u=1}^{U} \sum_{r=1}^{R} y_{iur}^{DR} \cdot \rho_{ir}\,; \qquad f(recycling) = Max \sum_{i=1}^{I} \sum_{u=1}^{U} \sum_{r=1}^{R} y_{iur}^{DR} \cdot \chi_{ir}\,;$$

$$f(energy) = Max \sum_{i=1}^{I} \sum_{u=1}^{U} \sum_{r=1}^{R} y_{iur}^{DR} \cdot \theta_{ir} \qquad (5)$$

If a simultaneous consideration of all short-term objectives is necessary, vector optimization methods are to be applied. In our problem, no cardinal information regarding a utility function is available. In addition, all objective functions are continuous and none of the objective functions is infinitely more important than others. Therefore, we apply Weighted Goal Programming (WGP) in the following (Tamiz et al, 1998).

For setting the goals, the maximum or minimum value of each alternative is calculated applying each objective function individually. Afterwards, all functions are considered simultaneously minimizing the weighted and normalized distances to these calculated goals. This results in a Pareto efficient achievement function. For functions to be maximized (recycling, energy recovery, reuse, contribution margin), underachievement of the goal is calculated, while overachievement is calculated for the function to be minimized (CO_2-emissions). To ensure commensurability, we normalize the deviations applying percentage normalization (Tamiz et al., 1998).

Weights for every criteria c (w_c) are determined based on questionnaires answered by politicians, OEMs, and other stakeholders. If weightings vary with regard to the different groups, evaluations may be carried out for every individual stakeholder group. If decision makers at operational level are assumed to have the same preferences as for evaluation at infrastructural level, the weights determined at the strategic level can be used.

Since it is hard to assess a priori, whether the overall sum of the deviations or the maximum deviation of all deviational variables is to be minimized, a combination of the Chebyshev (MINIMAX) and the MINSUM WGP is used (Romero/Linares, 2002). A sensitivity analysis is applied varying the parameter λ. Since (6) is not a smooth function, an equivalent problem is solved by substituting the first part of the function by an equivalent LP-formulation (compare Tamiz et al., 1998; Romero/Linares, 2002).

$$Min\left[(1-\lambda)\left(\sum_{c=1}^{5}(w_c \cdot \frac{d_c}{N_c})^{\infty}\right)^{1/\infty} + \lambda\sum_{c=1}^{5}(w_c \cdot \frac{d_c}{N_c})\right] \tag{6}$$

$$\text{s.t. } \lambda = (0;1)$$

Linking infrastructural and operational decisions

The linking of the infrastructural evaluation and the anticipation of short-term decisions is presented in figure 2.

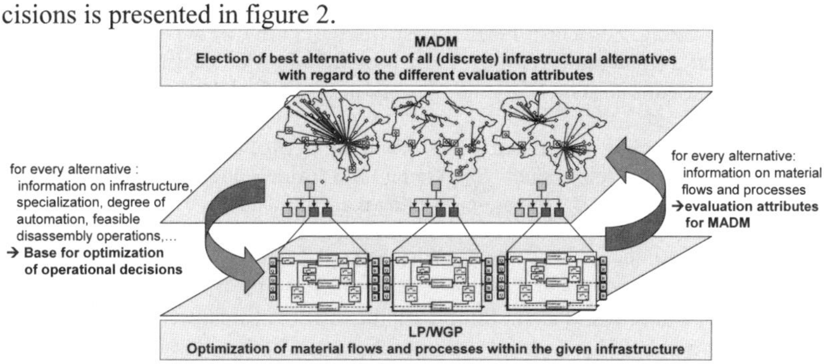

Fig. 2. Linking infrastructural evaluation and anticipation of short-term decisions

When integrating the methods as described above, alternatives and infrastructural layouts are to be determined first, since these long-term decisions strongly influence short-term decisions. Afterwards, operational performance is anticipated applying either Linear Programming or Weighted Goal Programming for every strategic alternative. Whether LP or WGP is to be used depends on the predicted targets of the decision maker operating the recycling system. Doing so, future material flows and thus attributes needed for evaluation of the infrastructure can be deduced. Based on these attribute values, the MADM-evaluation for selection of the best strategic alternative can be applied.

Perspective

Core element of the presented approach is a two-step evaluation method. Within this method, short-term decisions within given infrastructures are calculated using LP or WGP. The calculated results are used as a-priori information for multi-criteria decisions at strategic level. Next steps to be taken are to feed all required data into the model and to evaluate different recycling systems. Based on the results generated by this procedure, recommendations and proposals on how to establish and adjust treatment systems can be derived for political decision makers, for OEMs oriented towards sustainability, and for other stakeholders with ecological and/or social interests. Additionally, the effect of economic or sustainable oriented operation of the treatment system within a given infrastructure can be predicted.

References

Brans, J.P.; Vincke, Ph.; Mareschal, B. (1986): How to select and how to rank projects – the PROMETHEE method. European Journal of Operational Research 24, 228-238.

Romero, C.; Linares, P. (2002): Aggregation of preferences in an environmental economics context: A goal programming approach. Omega, 30, 89-95.

Spengler, Th.; Walther, G. (2005): Strategische Planung von Wertschöpfungsnetzwerken zum Produktrecycling, Zeitschrift für Betriebswirtschaft (ZfB), 3, 247-275.

Tamiz, M.; Jones, D.; Romero, C. (1998): Goal programming for decision making: An overview on the current state-of-the-art, European Journal of Operational Research 111, pp. 569-581.

Walther, G. (2005) Recycling von Elektro- und Elektronik-Altgeräten – Strategische Planung von Stoffstrom-Netzwerken für kleine und mittelständische Unternehmen, Gabler Edition Wissenschaft, Universitas-Verlag u.a., Wiesbaden

Walther, G.; Spengler, T. (2005): Impact of WEEE-directive on Reverse Logistics in Germany. International Journal of Physical Distribution and Logistics Management (IJPDLM), 35(5), 337-361.

Zimmermann, H.-J.; Gutsche, L. (1991) Multi-Criteria Analyse. Einführung in die Theorie der Entscheidungen bei Mehrfachzielsetzungen, Springer Verlag, Berlin/Heidelberg

Part XV

Revenue Management

Simultaneous Dynamic Pricing and Lot-sizing Decision for a Discrete Number of Price Variations

Sandra Transchel, Stefan Minner

University of Mannheim, Mannheim Business School, Schloss, 68131 Mannheim, Germany.
{sandra.transchel;minner}@bwl.uni-mannheim.de

Abstract: We investigate the impact of a dynamic pricing strategy on the economic ordering decision where a discrete number of price changes within each order cycle is allowed. Customer reaction to prices is modelled by a linear price response function and the ordering process is subject to variable procurement cost and setup cost. Inventories are subject to holding cost. The objective is to maximize average profit by choosing the optimal lot-size and pricing strategy.

1 Introduction

The application of innovative pricing techniques to improve supply chain performance is receiving growing attention in many industries. In particular, the coordination of dynamic pricing and other operations decisions is still at an early stage and offers significant opportunities for improving supply chain performance. In this paper we investigate the coordination of pricing and lot-sizing decisions in an economic order quantity (EOQ) and monopolist pricing environment where a discrete number of price changes is allowed over an order cycle. For a linear price response function, closed-form solutions for pricing and timing strategies are obtained. We present an example to illustrate the properties of optimal pricing and timing policies.

In the research of simultaneous pricing and inventory there is a large stream of literature. Whitin (1955) and Kunreuther and Richard (1971) have shown that in a centralized organization where pricing and ordering decisions are made simultaneously the profit increases significantly compared to sequential decision making. Rajan et al. (1992) derive simultaneous pricing and ordering policies for a retailer under standard EOQ assumptions similar to our model. They consider a deterministic problem where continuous price adjustments are allowed and the product is subject to physical decay and value drop. Gupta et al. (2002) consider a discrete-time model with deterministic demand, time-dependent reservation prices, and a given number of price changes. Netessine (2005) generalizes these results for a more general type of customer arrival rate. The focus of his paper is the optimal choice of prices and timing when there is a capacity/inventory constraint.

2 Model description

Consider a monopolistic retailer who is selling a product on a single market without competition over an infinite planning horizon. We assume that customer demand is a continuous function of the sales price P and arrives dynamically at rate $D(P)$ which is a differentiable and non-increasing function in P with $D(P) \geq 0$ and $\frac{\partial D}{\partial P} \leq 0$. At every point in time the demand rate depends solely on the current price, that means, the customers are willing to buy as soon as the price is below their reservation price. We do not include forward buying or postponement in the case of dynamically changing prices.

Following the assumptions of the classical EOQ model, the retailer has to place replenishment orders in batches of size Q every T periods during the infinite planning horizon. With the release of any single batch there is an associated setup cost F and a variable procurement cost c per unit. Furthermore, the supplier has no capacity constraints and the overall order quantity is delivered in one shipment without any delay. Products delivered but not yet sold are kept in inventory subject to holding cost h per unit and unit of time. Backorders are not permitted.

The retailer establishes a number N of different prices per order cycle. In problems where both pricing and inventory decisions need to be made simultaneously, it might be appropriate that the selling price increases over the order cycle. By such pricing strategy the retailer gives incentives that the demand rate is higher at the beginning of the order cycle what results in a reduced average inventory and therefore in reduced average holding costs. The administrative cost associated with price setting is denoted by $\kappa(N)$ and is a non-decreasing function of N. For given N the retailer has to establish the time intervals $[t_{i-1}, t_i)$ and the associated price P_i. The time t_N corresponds to the cycle length T and $t_0 = 0$. We denote the optimal decision variable by a superscript "$*$". Due to the price increases the demand rate decreases from D_i to D_{i+1} where $D_i = D(P_i)$. The decision making will be separated into two stages. At the first stage the number of price settings N will be optimized anticipating the optimal timing and sizing of prices for any given N. At the second stage the retailer optimizes the prices for a given number of prices N, the points in time where the price is adjusted, and the optimal cycle length. The objective is to maximize the average profit per unit of time and can be formulated sequentially as follows:

$$\Pi^* = \max_N \left[\max_{P_1,\cdots,P_N,t_1,\cdots,t_N} \left\{ \Pi^{(N)} - \kappa(N) \right\} \right], \tag{1}$$

with

$$\begin{aligned} \Pi^{(N)} =& \frac{1}{t_N} \left[\sum_{i=1}^{N} (P_i - c) D_i(P_i) \left(t_i - t_{i-1}\right) - \frac{h}{2} \sum_{i=1}^{N} D_i(P_i) \left(t_i - t_{i-1}\right)^2 \right. \\ & \left. - h \sum_{i=1}^{N-1} \left(\left(t_i - t_{i-1}\right) \sum_{j=i+1}^{N} D_j(P_j) \left(t_j - t_{j-1}\right) \right) - F \right]. \end{aligned} \tag{2}$$

Equation (2) represents an optimization problem where we determine the optimal prices $P^* = (P_1^*, \cdots, P_N^*)$ and the associated timing of price changes $t^* = (t_1^*, \cdots, t_N^*)$ simultaneously for a given number of N. Equation (1) calculates N^* that maximizes the average profit. The average retailer profit per unit of time is given by the revenue over the cycle minus the purchasing cost, the inventory holding cost, and the setup cost over the order cycle, divided by the cycle length.

In order to maximize the average profit, we differentiate (2) with respect to P_i and t_i, for $i = 1, \cdots, N$. The first-order necessary conditions for the optimal prices and the optimal timings are characterized by

$$P_i^* = c - \frac{D_i(P_i^*)}{D_i'(P_i^*)} + \frac{h}{2} \left(t_i^* + t_{i-1}^*\right), \tag{3}$$

$$t_i^* = \frac{(P_i^* - c) D_i(P_i^*) - (P_{i+1}^* - c) D_{i+1}(P_{i+1}^*)}{h(D_i(P_i^*) - D_{i+1}(P_{i+1}^*))}, \tag{4}$$

where $D_i'(P_i) := \frac{\partial D_i(P_i)}{\partial P_i}$. The first order condition for the optimal cycle length provides:

$$t_N^* = \sqrt{\frac{2F}{hD_N(P_N^*)} - \frac{1}{D_N(P_N^*)} \sum_{i=1}^{N-1} (t_i^*)^2 (D_i(P_i^*) - D_{i+1}(P_{i+1}^*))}. \tag{5}$$

To gain more structural insights, we analyze the model for a linear price response function.

3 Linear price response

Assume that at any point in time t the market potential is denoted by a and an amount of bP customers decide that the price is too high and do not buy. We assume $a > bc$. In particular, there is a price $P = \frac{a}{b}$ (*reservation price*) where the demand rate drops to zero. The demand rate is as follows

$$D(P) = \begin{cases} a - bP & : \quad 0 \leq P \leq \frac{a}{b} \\ 0 & : \quad P > \frac{a}{b} \end{cases} . \tag{6}$$

In the case that price changes are limited to a fixed number N, the following propositions provide important implications on the optimal cycle length, the optimal points in time where the price will be adjusted, and the behaviour of both depending on variation of N.

Proposition 1 *In an order cycle where the retailer is allowed to charge N different sales prices and the demand response is linear function in sales price, the time intervals $[t_{i-1}, t_i)$ are equidistant with*

$$\delta^* := t_i^* - t_{i-1}^* = t_{i+1}^* - t_i^*$$

for all $i = 1, \cdots, N-1$.

Proof. From (3) and (4) we find

$$P_i^* = \frac{1}{2}\left(\frac{a}{b} + c + \frac{h}{2}\left(t_i^* + t_{i-1}^*\right)\right) \quad \text{and} \quad t_i^* = \frac{1}{h}\left(P_i^* + P_{i+1}^* - \left(\frac{a}{b} + c\right)\right).$$

Inserting P_i^* and P_{i+1}^* into the equation for t_i^* leads to the condition $t_i^* = \frac{t_{i-1}^* + t_{i+1}^*}{2} \Longleftrightarrow t_i^* - t_{i-1}^* = t_{i+1}^* - t_i^* = \delta^*$. □

According to Proposition 1,

$$t_i^* = i\frac{t_N^*}{N} \quad \text{and} \quad P_i^* = \frac{1}{2}\left(\frac{a}{b} + c + \frac{h}{2}\frac{(2i-1)}{N}t_N^*\right). \tag{7}$$

Equation (7) indicates that the optimal price will increase over the cycle and at every time t_i the retailer increases the price by a constant $\frac{h}{2}\frac{t_N^*}{N}$. Using (7) in (5) the optimal cycle length results from

$$\frac{4N^2 - 1}{N^2}t_N^3 - \frac{6(a - bc)}{hb}t_N^2 + \frac{24F}{h^2 b} = 0. \tag{8}$$

The solution of (8) requires to find the roots of a cubic polynomial, e.g., see Bronshtein et.al. (2004).

Proposition 2 *If the response function is linear and N price changes are allowed in an order cycle, there exists a unique optimal cycle length $t_N^* > 0$ if*

$$F \leq \frac{4}{3}\frac{(a - bc)^3}{b^2 h}\frac{N^4}{(4N^2 - 1)^2} =: F_{max} \tag{9}$$

and the optimal cycle length t_N^* is

$$t_N^* = \begin{cases} -2\frac{(a-bc)}{hb}\frac{N^2}{(4N^2-1)}\left(2\cos(\frac{\pi}{3}+\frac{\phi}{3})-1\right) & : F \leq \frac{2}{3}\frac{(a-bc)^3}{hb^2}\frac{N^4}{(4N^2-1)^2} \\ \phi = arccos\left(1-\frac{3F}{2}\frac{hb^2}{(a-bc)^3}\frac{(4N^2-1)^2}{N^4}\right) & \\ & \\ 2\frac{(a-bc)}{hb}\frac{N^2}{(4N^2-1)}\left(2\cos(\frac{\pi}{3}+\frac{\phi}{3})+1\right) & : F > \frac{2}{3}\frac{(a-bc)^3}{hb^2}\frac{N^4}{(4N^2-1)^2} \\ \phi = arccos\left(\frac{3F}{2}\frac{hb^2}{(a-bc)^3}\frac{(4N^2-1)^2}{N^4}-1\right) & \end{cases}.$$

Furthermore, t_N^* is increasing N.

The proof is illustrated in [8]. Thus, the order frequency of the retailer is lower the more price adjustments are allowed over an order cycle. Compared to static pricing, a dynamic pricing strategy has two effects. First, a lower price at the beginning of the cycle results in a higher demand rate. The second effect is similar to a reduction of holding costs in the EOQ model. The lower the average holding cost the lower is the order frequency. For the dynamic pricing strategy the retailer charges a lower price at the beginning of the cycle and reduces the stock level and the average holding costs over the entire order cycle. The managerial intuition behind these effects is as follows. The retailer places the orders in lots, that means, the later an item is sold the higher the inventory holding cost for this item. For this reason, the retailer has an incentive to reduce inventories at the beginning of the order cycle.

4 Numerical results

Consider an example of a retailer with a linear demand rate $D(P) = a - bP$, with the market potential $a = 500$, and sensitivity parameter $b = 20.4$. The setup cost is $F = 900$, purchasing cost $c = 15$ per unit, and inventory holding cost $h = 1.5$ per unit and time unit. Furthermore, we set the menu cost $\kappa(N) = 1$. The results of Table 1 indicate that there exists potential for improvement of the optimal profit per time unit with only a few adjustments. It is optimal to adjust the selling price 4 times over an order cycle.

N	Π^*	Q^*	t_N^*	N	Π^*	Q^*	t_N^*
1	-8.84	277	4.33	6	6.55	297	5.20
2	5.17	291	4.88	7	5.75	298	5.22
3	7.45	295	5.06	8	4.87	298	5.23
4	7.66	297	5.14	9	3.96	298	5.23
5	7.24	297	5.18	10	3.02	299	5.24

Table 1. Optimal profit, order quantity, and cycle length for different number of price changes

In this example the results indicate that when the retailer optimizes the profit on the basis of constant pricing, the product is not profitable and the loss per time unit is 8.84. With a single price adjustment, the product generates a positive profit of 5.17 per time unit. The additional benefit is decreasing with the number of price changes. As shown in previous sections, the order cycle and the order quantity increase with increasing N.

5 Conclusion

This paper analyzes a problem of jointly determining the profit-maximizing pricing strategy and lot-sizing policy in terms of intertemporal price discrimination where a discrete number of price changes are allowed. For the linear model, we found analytical solutions for the optimal prices, the optimal times where the price is adjusted, and the optimal cycle length and we have proven that the length of the time intervals where a particular price is charged is equidistant. From numerical investigations we have indicated that the optimal profit is increasing and the marginal revenue is decreasing with the number of price changes. Furthermore, it was shown that the optimal cycle length and the order quantity is increasing in the number of price variations. Thus, the order frequency of the retailer is lower the more price adjustments are allowed over an order cycle. Natural extensions of the model would be to introduce competition, forward buying, and postponement. Another extension would be the consideration of multiple products and capacity constraints where at least two different products require the same capacity.

References

1. Bronshtein I N, Semendyayev K A, Musiol G, Muehlig H (2004). Handbook of Mathematics. 3rd edn. Springer, Berlin et.al.
2. Eliashberg J, Steinberg R (1993). Marketing-production joint decision making. In: Eliashberg J, Lilien J D (eds) Management Science in Marketing, Volume 5 of Handbooks of Operations Research and Management Science. North Holland, Amsterdam et al.
3. Whitin T M (1955). Inventory Control and Price Theory. Management Science 2:61-68
4. Kunreuther H, Richard J F (1971). Optimal pricing and inventory decisions for non-seasonal items. Econometrica 39:173-175
5. Rajan A, Steinberg Ra, Steinberg Ri (1992). Dynamic Pricing and Ordering Decisions by a Monopolist. Management Science 38:240-262
6. Netessine S (2005). Dynamic pricing of inventory/capacity with infrequent price changes. Forthcoming in European Journal of Operational Research
7. Gupta D, Hill A V, Bouzdine-Chameeva T (2002). A pricing model for clearing end of season retail inventory. Working Paper, University of Minnesota
8. Transchel S, Minner S (2005). The impact of dynamic pricing on the economic order decision. Working Paper No.2/2005, University of Mannheim

Optimal Fares for Public Transport*

Ralf Borndörfer, Marika Neumann, and Marc E. Pfetsch

Konrad-Zuse-Zentrum für Informationstechnik Berlin, Takustr. 7, 14195 Berlin, Germany; Email:{`borndoerfer, marika.neumann, pfetsch`}`@zib.de`

Summary. The *fare planning problem* for public transport is to design a system of fares that maximize the revenue. We introduce a nonlinear optimization model to approach this problem. It is based on a discrete choice logit model that expresses demand as a function of the fares. We illustrate our approach by computing and comparing two different fare systems for the intercity network of the Netherlands.

1 The Fare Planning Problem

The influence of fares on passenger behavior and revenues is traditionally studied from a macroscopic point of view. Classical topics are the analytic study of equilibria [7], price elasticities [3], and the prediction of passenger behavior [1]. The only approaches to fare optimization on a more detailed level that we are aware of are the work of Hamacher and Schöbel [5] on the optimal design of fare zones and the work of Kocur and Hendrikson [6] and De Borger et al. [4] who introduced a model for maximizing the revenue and the social welfare, respectively. In contrast to these approaches, our model for fare optimization takes different origins and destinations of travel into account, i.e., we consider a "network effect". Our aim in this article is to show that such a model can be a versatile tool for optimizing fare systems. While the general model has been introduced in [2], we focus here on the comparison of two examples.

Consider a traffic network with nodes (stations) V, *origin-destination pairs* (OD-pairs) $D \subseteq V \times V$, and a finite set $\mathcal{C}$ of *travel choices*; for examples see Section 2.1. Let $\boldsymbol{x} \in \mathbb{R}^n_+$ be a vector of fare variables $x_1, \ldots, x_n$, which we call *fares* in the following. Fares can be restricted to a polyhedron $P \subseteq \mathbb{R}^n_+$. Further, let $p^i_{st} : \mathbb{R}^n \to \mathbb{R}_+$, $\boldsymbol{x} \mapsto p^i_{st}(\boldsymbol{x})$ be the *price* for traveling from s to t and travel choice $i \in \mathcal{C}$. Similarly, let $d^i_{st}(\boldsymbol{x})$ determine the *demand* of

* Supported by the DFG Research Center MATHEON "Mathematics for key technologies" in Berlin

passengers for this combination. In our examples, demand functions and price functions are differentiable and $P = \mathbb{R}^n_+$. Given fares $\boldsymbol{x}$, the *revenue* $r(\boldsymbol{x})$ is:

$$r(\boldsymbol{x}) := \sum_{i \in C} \sum_{(s,t) \in D} p^i_{st}(\boldsymbol{x}) \cdot d^i_{st}(\boldsymbol{x}) .$$

Our general model for the fare planning problem reads:

$$\text{(FPP)} \qquad \begin{array}{l} \max r(\boldsymbol{x}) \\ \;s.t.\; \boldsymbol{x} \in P . \end{array}$$

(FPP) is a nonlinear program that may be quite hard to solve in general.

2 Discrete Choice Demand Functions

We use a discrete choice logit model to obtain realistic demand functions d^i_{st}. Our exposition assumes that the reader is familiar with such a construction. We refer to Ben-Akiva and Lerman [1] for a thorough exposition.

The model is as follows. A passenger traveling from s to t performs a random number of trips $X_{st} \in \mathbb{Z}_+$ during a time horizon T, i.e., X_{st} is a discrete random variable. We assume that $X_{st} \leq N$ and that the same travel alternative is chosen for all trips, i.e., passengers do not mix alternatives. For these trips, a passenger chooses among a finite set A of *alternatives* for the travel mode, e.g., single ticket, monthly ticket, bike, car travel, etc.

Associated with each alternative $a \in A$ and OD-pair $(s,t) \in D$ is a random *utility* variable U^a_{st} which may depend on the passenger. Each utility is the sum of an observable part, the *deterministic utility* V^a_{st}, and a random *disturbance term* ν^a_{st}. We consider the utility $U^a_{st}(\boldsymbol{x}, k) = V^a_{st}(\boldsymbol{x}, k) + \nu^a_{st}$, which depends on the fare system $\boldsymbol{x}$ and the number of trips k. Assuming that each passenger chooses the alternative with the highest utility, the probability of choosing alternative $a \in A$ (for given $\boldsymbol{x}$ and k) is

$$P^a_{st}(\boldsymbol{x}, k) := \mathbb{P}\big[V^a_{st}(\boldsymbol{x}, k) + \nu^a_{st} = \max_{b \in A}(V^b_{st}(\boldsymbol{x}, k) + \nu^b_{st})\big].$$

In a *logit model*, the ν^a_{st} are Gumbel distributed and the probability for choosing alternative a for $(s,t) \in D$ can explicitly computed by the formula (see [1]):

$$P^a_{st}(\boldsymbol{x}, k) = \frac{e^{\mu V^a_{st}(\boldsymbol{x},k)}}{\sum\limits_{b \in A} e^{\mu V^b_{st}(\boldsymbol{x},k)}} .$$

Here $\mu > 0$ is a scaling parameter for the disturbance terms ν^a_{st}.

We derive demand functions for (FPP) from this discrete choice model by defining the travel choices as $C = A \times \{1, \ldots, N\}$ and setting

$$d^{a,k}_{st}(\boldsymbol{x}) \;=\; d_{st} \cdot P^a_{st}(\boldsymbol{x}, k) \cdot \mathbb{P}[X_{st} = k], \tag{1}$$

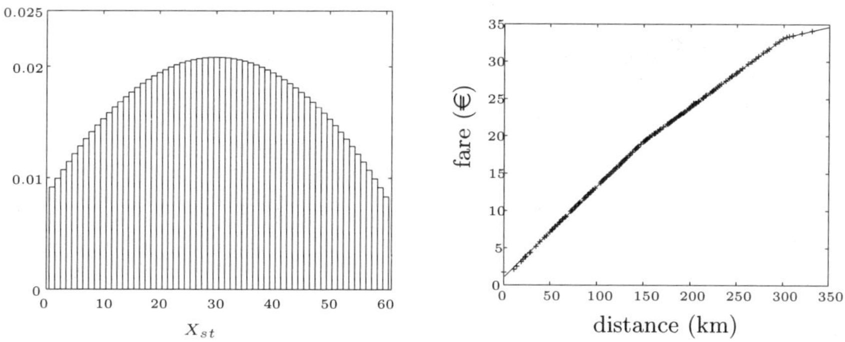

Fig. 1. *Left:* Probabilities for the discrete random variables X_{st} representing the number of trips. *Right:* Samples for the fare system of NS Reizigers and corresponding piecewise linear approximation (with three pieces).

where d_{st} is the number of passengers that travel from s to t. The *expected* revenue (over the probability spaces for X_{st} and ν_{st}^{a}) can then be written as:

$$r(\boldsymbol{x}) = \sum_{a \in A'} \sum_{k=1}^{N} \sum_{(s,t) \in D} p_{st}^{a,k}(\boldsymbol{x}) \cdot d_{st}^{a,k}(\boldsymbol{x}),$$

where A' is the set of public transport alternatives and $p_{st}^{a,k}(\boldsymbol{x})$ is the price function. Note that $r(\boldsymbol{x})$ is differentiable if V_{st}^{a} and $p_{st}^{a,k}$ have this property; compare the examples in the next section.

2.1 Two Examples

We will demonstrate our approach by two examples. In the first example we work with alternatives "standard ticket" (S), "reduced ticket" (R), and "car" (C), i.e., $A = \{S, R, C\}$. In the second example we work with alternatives "monthly ticket" (M), "single ticket" (S), and "car" (C), i.e., $A = \{M, S, C\}$.

Both examples use a time horizon T of one month. We set the scaling parameter to $\mu = 0.01$. The (discrete) probabilities for the number of trips X_{st} are defined using the function $1 - \frac{1}{1500} \cdot (k - 30)^2$ and normalizing. The resulting probabilities are independent of the OD-pair $(s,t) \in D$ and are centered around 30 in an interval from 1 to $N := 60$, see Figure 1.

Our data for the intercity network of the Netherlands is taken from a publicly available GAMS model by Bussieck (`www.gams.com/modlib/libhtml/lop.htm`). It consists of a network containing 23 nodes (stations) and a corresponding upper-triangular origin-destination matrix (d_{st}^{0}) with 210 nonzero entries that account for a symmetric bidirectional traffic. We added to this data the currently valid fares, distances, and travel times taken from the internet site of the railway company NS Reizigers (`www.ns.nl`). It turns out that these fares are determined by a piecewise linear function with three pieces

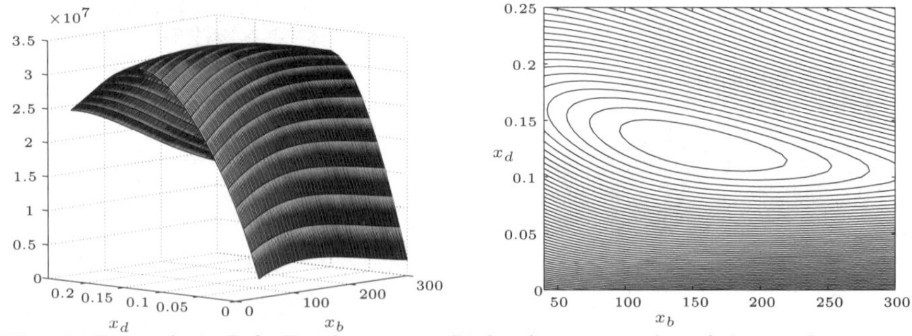

Fig. 2. Example 1. *Left:* Total revenue. *Right:* A contour plot of the total revenue. The optimum is at $x_b \approx 153.31$ and $x_d \approx 0.13$.

depending on distance, see Figure 1. With these data, the total demand is 91,791 and the current total revenue is 860,991 € per day.

Distances and travel times for alternative "car" were obtained from the routing planer Map24 (`www.map24.com`); we used the quickest route between the corresponding train stations. The price for alternative "car" is the sum of a fixed cost Q and distance dependent operating costs q, i.e., $p_{st}^{C,k}(\boldsymbol{x}) = Q + q \cdot \ell_{st}^c \cdot k$, which is constant; here, ℓ_{st}^c denotes the distance between s and t in kilometers for a car. We set $Q = 100$ € and $q = 0.1$ €.

We extrapolated the OD-matrix (d_{st}^0) in order to also include car traffic as follows. Using alternatives "car" as above and alternative "NS Reizigers" (with the current fares and travel times), we estimated for each OD-pair the percentage q_{st} of passengers using public transport applying (1) with $k = 30$. The total number of travelers between s and t is then $d_{st} = 100 \cdot d_{st}^0 / q_{st}$. The total number of passengers in (d_{st}) is 184,016.

Example 1: Standard Ticket, Reduced Ticket, and Car

We consider two fares x_d and x_b (hence $n = 2$). Namely, x_d is a distance fare per kilometer for standard tickets, and x_b is a basic fare that has to be paid once a month in order to buy reduced tickets that provide a 50% discount in comparison to standard tickets. We write $\boldsymbol{x} = (x_b, x_d)$ and set the prices for alternatives standard and reduced ticket to $p_{st}^{S,k}(\boldsymbol{x}) = x_d \cdot \ell_{st} \cdot k$ and $p_{st}^{R,k}(\boldsymbol{x}) = x_b + \frac{1}{2}\, x_d \cdot \ell_{st} \cdot k$, respectively, where ℓ_{st} denotes the shortest distance in the public transport network between s and t in kilometers.

We assume that the utilities are affine functions of prices and travel times t_{st}^a between s to t with alternative a. The utilities depend on the number of trips k. More precisely, we set:

$$\begin{aligned} U_{st}^S(x_b, x_d, k) &= -\delta_1 \cdot x_d \cdot \ell_{st} \cdot k - \delta_2 \cdot t_{st}^S \cdot k + \nu_{st}^S && \text{"standard ticket"} \\ U_{st}^R(x_b, x_d, k) &= -\delta_1 \left(x_b + \tfrac{1}{2}\, x_d \cdot \ell_{st} \cdot k\right) - \delta_2 \cdot t_{st}^R \cdot k + \nu_{st}^R && \text{"reduced ticket"} \\ U_{st}^C(x_b, x_d, k) &= -\delta_1 \left(Q + q \cdot \ell_{st}^c \cdot k\right) - \delta_2 \cdot t_{st}^C \cdot k + \nu_{st}^C && \text{"car".} \end{aligned}$$

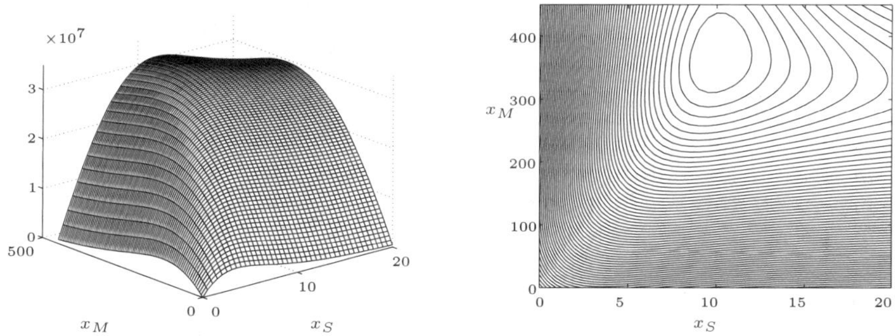

Fig. 3. Example 2. *Left:* Total revenue. *Right:* A contour plot of the total revenue. The optimum is at $x_S \approx 10.54$ and $x_M \approx 368.85$.

Here, δ_1 and δ_2 are weight parameters; we use $\delta_1 = 1$ and $\delta_2 = 0.1$, i.e., 10 minutes of travel time are worth 1 €.

Altogether, the fare planning problem we want to consider has the form:

$$\begin{aligned} \max \sum_{k=1}^{N} \sum_{(s,t)\in D} d_{st} \cdot \frac{\mathbb{P}[X_{st} = k]}{\sum_{b\in A} e^{\mu V_{st}^b(\boldsymbol{x},k)}} \cdot \Big[(x_d \cdot \ell_{st} \cdot k) \cdot e^{\mu V_{st}^S(\boldsymbol{x},k)} + \\ (x_b + \tfrac{1}{2}\, x_d \cdot \ell_{st} \cdot k) \cdot e^{\mu V_{st}^R(\boldsymbol{x},k)} \Big] \\ s.t. \quad \boldsymbol{x} \geq \boldsymbol{0}. \end{aligned} \tag{2}$$

Note that the revenue function is differentiable.

Example 2: Single Ticket, Monthly Ticket, and Car

We consider the fares x_M (for the monthly ticket) and x_S (for the single ticket) and write $\boldsymbol{x} = (x_M, x_S)$. We set the cost for alternative "monthly ticket" and "single ticket" to $p_{st}^{M,k}(\boldsymbol{x}) = x_M$ and $p_{st}^{S,k}(\boldsymbol{x}) = x_S \cdot k$, respectively.

Analogously to the previous example we set the utility function as follows:

$$\begin{aligned} U_{st}^M(x_M, x_S, k) &= -\delta_1 \cdot x_M - \delta_2 \cdot t_{st}^M \cdot k + \nu_{st}^M && \text{“monthly ticket”} \\ U_{st}^S(x_M, x_S, k) &= -\delta_1\, (x_S \cdot k) - \delta_2 \cdot t_{st}^S \cdot k + \nu_{st}^S && \text{“single ticket”} \\ U_{st}^C(x_M, x_S, k) &= -\delta_1\, (Q + q \cdot \ell_{st}^c \cdot k) - \delta_2 \cdot t_{st}^C \cdot k + \nu_{st}^C && \text{“car”.} \end{aligned}$$

Here again, we use $\delta_1 = 1$ and $\delta_2 = 0.1$.

Altogether the fare planning program for this example is

$$\begin{aligned} \max \sum_{k=1}^{N} \sum_{(s,t)\in D} d_{st} \cdot \frac{x_M \cdot e^{\mu V_{st}^M(\boldsymbol{x},k)} + x_S \cdot k \cdot e^{\mu V_{st}^S(\boldsymbol{x},k)}}{\sum_{b\in A} e^{\mu V_{st}^b(\boldsymbol{x},k)}} \cdot \mathbb{P}[X_{st} = k] \\ \text{s.t.} \quad \boldsymbol{x} \geq \boldsymbol{0}. \end{aligned} \tag{3}$$

Table 1. Comparison of the results of Example 1 ("standard ticket, reduced ticket, and car") and Example 2 ("single ticket, monthly ticket, and car").

	revenue	demand	modal split				
Status quo	25,829,730.0	91,791	50.1%				
Example 1	34,201,767.8	126,786	68.9%	standard:	37.1%	reduced:	31.8%
Example 2	31,813,156.4	110,999	60.3%	single:	35.4%	monthly:	24.9%

Results

We solved models (2) and (3) using a Newton-type method in Matlab 7 and confirmed the results by the Nelder-Mead method. The optimal fares for Example 1 are $x_b \approx 153.31$€ and $x_d \approx 0.13$€; see also Figure 2. The optimal fares for Example 2 are $x_S \approx 10.54$€ for the single ticket and $x_M \approx 368.85$€ for the monthly ticket; see also Figure 3. Table 1 compares revenue (per month), demand (per day), and modal split (percentage of passengers using public transport and the corresponding alternatives, respectively).

In Example 1, alternatives "standard" and "reduced ticket" attract more passengers for every OD-pair than the current fare system. The "reduced ticket", in particular, is used by passengers who often travel long distances. Similarly, in Example 2, passengers traveling long distances buy a "single ticket" if the number of trips is small and a "monthly ticket" if the number is high. In both examples, optimized fares result in a higher revenue and larger demand than in the status quo, that is, we have managed to attract additional passengers to public transport and at the same time improved revenue.

References

1. M. Ben-Akiva and S. R. Lerman, *Discrete Choice Analysis: Theory and Application to Travel Demand*, MIT-Press, Cambridge, 1985.
2. R. Borndörfer, M. Neumann, and M. E. Pfetsch, *Fare planning for public transport*, Report 05-20, ZIB, 2005. `http://www.zib.de/Publications/abstracts/ZR-05-20`.
3. J. F. Curtin, *Effect of fares on transit riding*, Highway Research Record 213, Washington D.C., 1968.
4. B. De Borger, I. Mayeres, S. Proost, and S. Wouters, *Optimal pricing of urban passenger transport, A simulation exercise for Belgium*, J. Transport Economics and Policy, 30 (1996), pp. 31–54.
5. H. W. Hamacher and A. Schöbel, *On fair zone designs in public transportation*, in Proc. of CASPT, vol. 430 of LNEMS, Springer-Verlag, 1995, pp. 8–22.
6. G. Kocur and C. Hendrickson, *Design of local bus service with demand equilibrium*, Transportation Sci., 16 (1982), pp. 149–170.
7. P. A. Pedersen, *On the optimal fare policies in urban transportation*, Transportation Res. Part B, 37 (2003), pp. 423–435.

Auswirkungen eines kontinuierlichen Fleet Assignment Prozesses

Michael Frank, Martin Friedemann, Michael Mederer und Anika Schröder

Technische Universität Clausthal

1 Einleitung

Die Verteilung der im Flugzeug vorhandenen Sitzplätze auf Basis von Passagierprognosen ist die zentrale Aufgabe des Revenue Management. Das Fleet Assignment garantiert hierbei eine optimale Zuordnung von angebotener Kapazität und erwarteter Nachfrage und erfolgt erstmals mit Erstellung des Flugplanes. Die Qualität der Prognose nimmt zum Abflug hin zu, wodurch ein neues Fleet Assignment höhere Erträge erzielen kann. Mit dieser Idee konnten bereits Berge and Hopperstad (1993) einen positiven Revenue Effekt nachweisen, dessen Höhe sich im Rahmen von 1,2 bis 4,9 Prozent bewegt. Der Schwerpunkt dieser Arbeit liegt im Gegensatz zu Berge and Hopperstad (1993) in der Untersuchung der zeitlichen Entwicklung des Potenzials.

2 Vorgehensweise

Eine ex-post Abschätzung des Revenue Potenzials weist viele methodische Schwächen auf, weshalb eine ereignisorientierte stochastische Simulation zum Einsatz kommt. Dabei können durch die Erzeugung von Nachfrageströmen der zeitliche Verlauf der Anfragen, das Volumen der Nachfrage sowie die Up-Sell Effekte Berücksichtigung finden. Die Verwendung eines Prognose Moduls garantiert die Abbildbarkeit von Ungenauigkeiten der Prognose in der Realität. Unter Einbezug der zeitlichen Komponente in der Simulation können auch Wechselwirkungen zwischen Fleet Assignment und Buchungssteuerung dargestellt werden. Um die Ergebnisse dieser Untersuchung auf das gesamte Lufthansa Netzwerk übertragen zu können, ist die Verwendung eines möglichst realistischen Teilnetzes an Stationen, Flügen und den damit verbundenen Umsteigeverbindungen erforderlich. Um den Rechenaufwand, sowie den Umfang von Inputdaten in Grenzen zu halten, basiert die Untersuchung auf einem Knoten am Hub Frankfurt. Bei diesen so genannten Knoten handelt es sich um zeitliche Häufungen von ankommenden gefolgt von abgehenden

Flügen mit dem Zweck einer guten Verbindungsqualität. Dieser Ausschnitt gewährleistet eine repräsentative Anzahl kontinentaler (ca. 75%) zu interkontinentalen (ca. 25%) Flügen. Konkret ergeben sich für vorliegende Untersuchung 184 Flüge mit 90 verschiedenen Zielflughäfen an einem Montagmorgen. Unter Berücksichtigung der wesentlichen Verkehrsströme resultieren daraus 880 Reisewege.

Der durchschnittliche Sitzladefaktor des Teilnetzes ist mit ca. 77% relativ hoch, was auf das nachfrageintensive Zeitfenster zurück zu führen ist. Die Flugzeugkapazität reicht von 44 Sitzen in einem Regionaljet bis zu 390 Sitzen in der Boeing 747. Der Flugzeug Mix ist für die Flotte der Lufthansa als repräsentativ anzusehen. Das für die Auswertung erforderliche Simulationsmodell wird im folgenden Abschnitt detailliert erläutert.

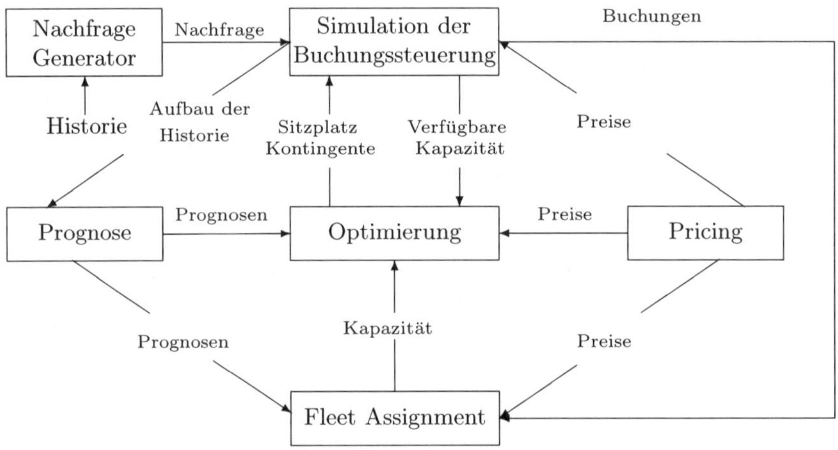

Abb. 1. Aufbau der Simulationsumgebung

2.1 Nachfrage Generator

Die Generierung der Nachfrage erfolgt auf Basis historischer Daten aus dem Revenue Management System der Lufthansa. Im Weiteren bezieht sich der Begriff der Nachfrage immer auf die Originärnachfrage, die sich aus den historischen eingegangenen Buchungen und dem vermeintlichen Spill während des Buchungsprozesses ergibt.

Es wird davon ausgegangen, dass nicht jede Buchungsklasse als eigenes Produkt von den Passagieren wahrgenommen wird, sondern dass die Buchungsklassen zu Gruppen zusammengefasst werden können, die jeweils ein Nachfragesegment bedienen. Hier werden drei Nachfragesegmente angenommen, wobei dem Ersten zwei und den Folgenden drei bzw. vier Buchungsklassen zugeordnet sind.

Aus historischen Daten wird für jedes Segment und mögliche Verbindung neben dem Mittelwert und der Varianz der Nachfrage deren zeitlicher Verlauf bestimmt. Basierend auf diesen Daten und der Annahme einer Gammaverteilung für das Volumen der Nachfrage wird ein Strom von Anfragen generiert (vgl. Talluri and van Ryzin, 2004b, S.613). Zur Vereinfachung werden in diesem Nachfragemodell No-Shows und Gruppenbuchungen vernachlässigt.

Um die bestehenden Konditionsunterschiede der Buchungsklassen innerhalb eines Segmentes zu berücksichtigen, werden Anfragen aus den jeweiligen Nachfrageklassen mit verschiedenen Wahrscheinlichkeiten in die Buchungsklassen eingestuft. Dabei bezeichnet q_i die Wahrscheinlichkeit, dass ein Kunde die Buchungsklasse i anfragt. Mit der höchsten Wahrscheinlichkeit erfolgt die Anfrage in der Buchungsklasse mit der geringsten Ertragswertigkeit des zugehörigen Segments. Stehen in dieser Klasse i keine Sitzplätze zur Verfügung, so wird die Anfrage abgelehnt und mit einer Up-Sell Wahrscheinlichkeit von p_i erneut in der nächst höheren Buchungsklasse $i-1$ gestellt. Dieser Up-Sell Algorithmus wird solange fortgesetzt, bis der Passagier sich für eine Buchungsklasse oder gegen die Beförderung entschieden hat (vgl. Abbildung 2). Um die Zahl der Inputparameter zu reduzieren, wird angenommen, dass die Wahrscheinlichkeit eines Up-Sell unabhängig von der ursprünglichen Buchungsklasse konstant ist, d.h. $q_i = q_j$ für alle i, j.

2.2 Fleet Assignment

Im Prozess der Flugplanerstellung wird zunächst basierend auf Erfahrungswerten eine Kapazität eingestellt. Diese Kapazitätszuordnung wird etwa neun Wochen vor Abflugsdatum optimiert, indem die vorhandenen Flugzeuge dann anhand der aktuellen Nachfrageprognose den zu fliegenden Strecken zugeordnet werden. In der Simulation werden für den betrachteten Netzausschnitt als Initialzuordnung die am häufigsten eingesetzten Kapazitäten ermittelt. Daraus resultiert ein Pool an verfügbaren Kapazitäten, zwischen denen unter Berücksichtigung technischer Realisierbarkeit Täusche durchgeführt werden können. Während des Buchungsprozesses unterliegen die Prognosen der Passagierzahlen Veränderungen, wodurch ein wiederholtes Fleet Assignment zusätzlichen generieren kann.

Ziel des Fleet Assignment ist es, eine umsatzoptimale Zuordnung aller Flugzeuge zu den entsprechenden Strecken zu erreichen. Der Erlös ergibt sich aus der prognostizierten Passagieranzahl begrenzt durch die Flugzeugkapazität multipliziert mit dem induzierten Durchschnittsyield. Letzter wird aus den Passagieren errechnet, die ihrer Wertigkeit entsprechend bis zum Erreichen der Kapazität angenommen werden. Randbedingungen garantieren, dass einer Strecke keine geringere Kapazität zugewiesen wird, als die bereits erfolgten Buchungen benötigen. Aufgrund des verwendeten Zeitfenster können rotationelle Aspekte vernachlässigt werden. Aufgrund von Crew Restriktionen erfolgen in der Realität kaum Täusche von Flugzeugen auf Interkontinental Strecken, weshalb diese in der Simulation ausgeschlossen sind. Zur Lösung

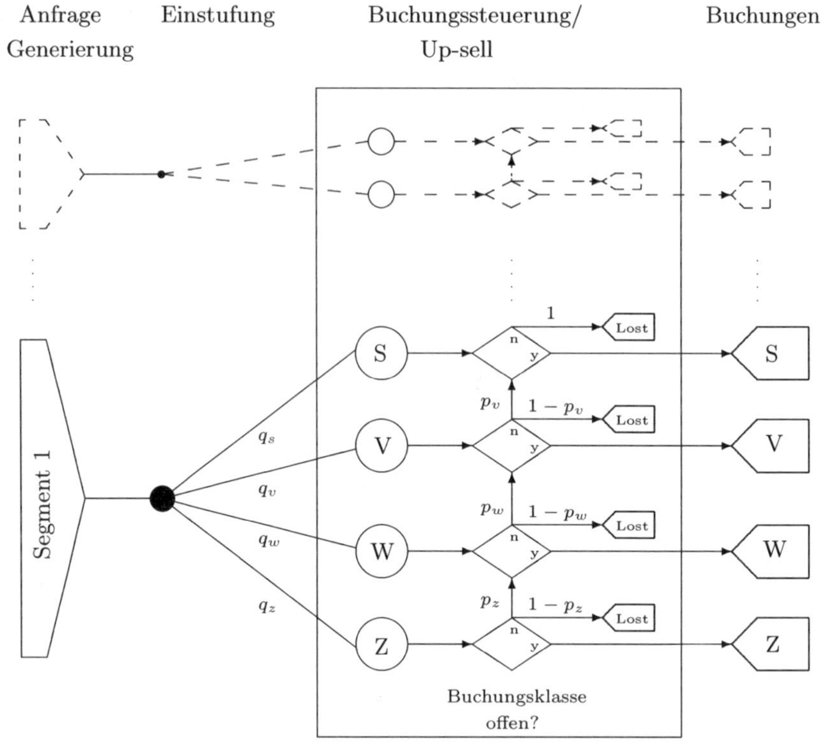

Abb. 2. Abarbeitung von Anfragen im Simulationsmodell

des beschriebenen Fleet Assignment Problems werden heuristische Methoden eingesetzt.

2.3 Prognose und Optimierung

Ein Überblick über bestehende Lösungverfahren für das Seat Inventory Control Problem findet sich in Talluri and van Ryzin (2004b). Zeni (2001) stellt Prognoseverfahren vor, welche häufig bei Fluggesellschaften zum Einsatz kommen. Die Mehrzahl an Methoden nehmen die Unabhängigkeit der Buchungsklassen und den Einsatz fixer Kapazitäten an.

Realistischeres Passagierverhalten wird beispielsweise von Brumelle and McGill (1993); Talluri and van Ryzin (2004a); Bodily and Weatherford (1995) untersucht. De Boer (2003) hat ein Verfahren entwickelt, welches wechselnde Kapazitäten in den Optimierungprozess mit einbezieht. Diese Methoden sind eher theoretisch motiviert und vernachlässigen wichtige Aspekte der Praxis, weshalb sie für diese Untersuchung nicht angewendet werden.

Um das ermittelte Potenzial in die Realität übertragen zu können, kommen bewährte Verfahren der Praxis von Fluggesellschaften zum Einsatz, das Exponential Smoothing zur Prognose und ein auf heuristischen Bid Preisen (vgl. Talluri and van Ryzin, 2004b) beruhendes Verfahren zur Optimierung der Sitzplatzkontingente.

3 Ergebnisse

Die Simulation besteht aus zwei Phasen. In der Kalibrierungsphase werden die Einstufungswahrscheinlichkeiten zu den gewählten Up-Sell so bestimmt, dass der Output den in der Realität beobachteten Daten entspricht. Der für die Kalibrierung genutzte Ausgangsfall bezieht sich auf ein einmaliges Fleet Assignment neun Wochen vor Abflug. Mit den so ermittelten Werten für die Einstufung werden während der Auswertungsphase die Erlös Auswirkungen eines Fleet Assignment zu jedem DCP bis drei Tage vor Abflug ermittelt.

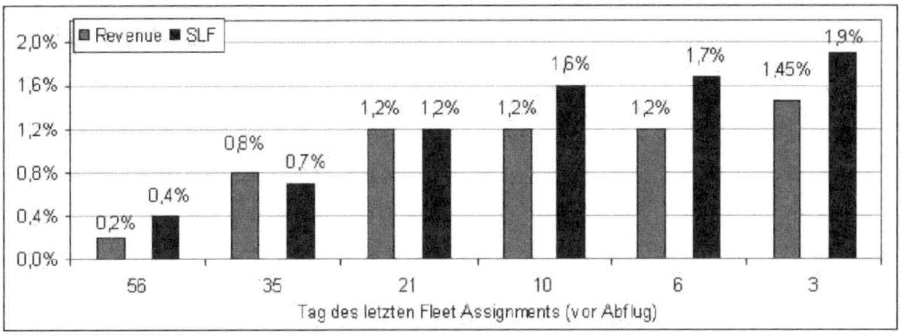

Abb. 3. Erlös- und Auslastungs-Verbesserung durch einen kontinuierlichen Fleet Assignment Prozess

Mit Hilfe eines kontinierlichen Fleet Assignment Prozesses kann der Sitzladefaktor um etwa 1,9% gesteigert werden, was mit einer Erhöhung des Erlöses um 1,45% einhergeht (vgl. Abbildung 3). Ein Großteil dessen wird bis etwa drei Wochen vor Abflug realisiert. Die Ursache liegt in der zeitlichen Verteilung der Nachfrage. Da bis zu diesem Zeitpunkt ein Großteil der Buchungen eingegangen ist, gewinnt die Nebenbedingung nicht unter die Buchungszahlen zu tauschen stark an Bedeutung. Das Potenzial wird dabei von nur wenigen Abtäuschen getragen. Etwa zwei Drittel des Potenzials werden bereits mit einem Drittel der möglichen Abtäusche umgesetzt.

Wie zuvor erwähnt werden im Rahmen dieser Arbeit die Ergebnisse einiger Sensitivitätsanalysen aufgezeigt, wobei der in Abbildung 3 vorgestellte Fall als Referenz genutzt wird. Wichtige Einflussfaktoren bilden das Volumen der Nachfrage und deren Varianz. Es werden sowohl Szenarien mit einem um 20% erhöhten bzw. abgesenkten Nachfragevolumen als auch solche mit einer um 50% erhöhten bzw. gesenkten Schwankung im Volumen untersucht.

Eine Änderung des Volumens wirkt sich unterschiedlich auf das Potenzial aus. Wie Abbildung 4 zeigt, hat eine Verringerung einen wesentlich größeren

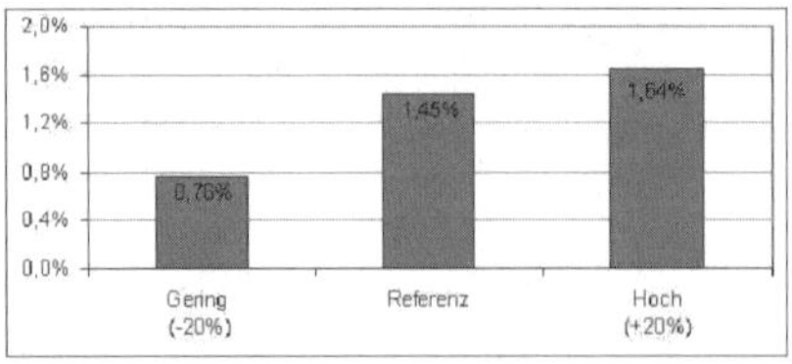

Abb. 4. Sensitivität bezüglich des Nachfrage Volumens

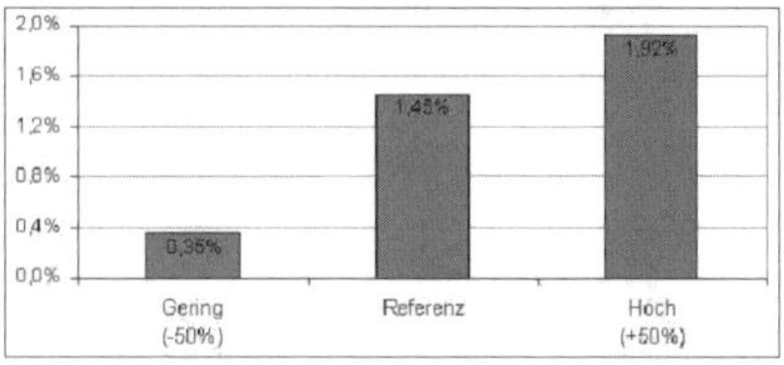

Abb. 5. Sensitivität bezüglich der Schwankung der Nachfrage

Einfluss als eine Erhöhung. Dieser Tatbestand erklärt sich dadurch, dass die verringerte Nachfrage dazu führt, dass sie auf vielen Flugstrecken niedriger ist als das Angebot und somit kaum Potenzial für Abtäusche besteht.

Bei einer veränderten Schwankung der Nachfrage ist dieser Effekt noch stärker ausgeprägt. Mit einer um 50% erhöhten Schwankung wird ein Potenzial von 1,9% erzielt. Wird die Schwankung um 50% reduziert, ist die Prognose so gut, dass durch einen kontinuierlichen Fleet Assignment Prozess kaum erkennbare Steigerungen im Revenue erzielt werden können.

Um den Einfluss des Up-Sell zu untersuchen, wurde die Simulation in zwei Szenarien neu kalibriert und mit diesen Einstellungen die Potenziale erneut ermittelt. Es wird hierbei von einem minimalen Up-Sell von 20% und einem maximalen Up-Sell von 80% ausgegangen, was zu einem Potenzial von 1,15% bis 2,45% führt.

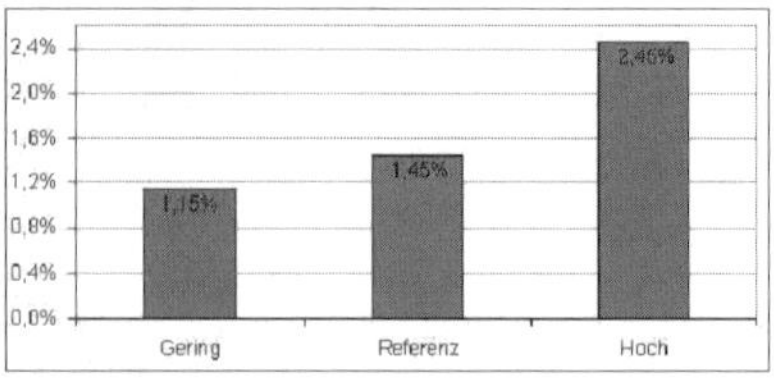

Abb. 6. Sensitivität: Up-Sell Parameter

4 Zusammenfassung und Ausblick

Durch die Simulation eines Revenue Management Systems wie es bei Fluggesellschaften in der Praxis eingesetzt wird, konnte ein positiver Erlöseffekt für einen kontinuierlichen Fleet Assignment Prozess nachgewiesen werden. Die Realisierung eines Großteiles des Potenzials bis drei Wochen vor Abflug ist operationell umsetzbar.

Die Vernachlässigung von Gruppenbuchungen, No-Shows und Stornierungen in den Nachfragedaten stellt eine wesentliche Vereinfachung der Realität dar. Deren Abbildung würde jedoch den Prognosefehler erhöhen und sich damit eher positiv auf das Potenzial des Fleet Assignment auswirken. Die Ausdehnung des betrachteten Netzwerkes auf einen gesamten Tag mit den rotationellen Verknüpfungen der Flugzeuge stellt eine weitere Herausforderung dar.

Literaturverzeichnis

Berge, M. E. and Hopperstad, Craig, A. (1993). Demand driven dispatch: A method for dynamic aircraft capacity assignment, models and algorithms. *Operations Research*, 41(1):153–168.

Bodily, S. E. and Weatherford, L. R. (1995). Perishable-asset revenue management: Generic and multiple-price yield management with diversion. *Omega*, 23(2):173–185.

Brumelle, S. and McGill, J. (1993). Airline seat allocation with multiple nested fare classes. *Operations Research*, 41(1):127–137.

De Boer, S. V. (2003). *Advances in Airline Revenue Management and Pricing.* PhD thesis, Massachusetts Institute of Technology.

Talluri, K. and van Ryzin, G. (2004a). Revenue management under a general discrete choice model of consumer behavior. *Management Science*, 50(1):15–33.

Talluri, K. T. and van Ryzin, G. J. (2004b). *The theory and practice of Revenue Management.* Kluwer Academic Publishers.

Zeni, R. H. (2001). *Improved forecast accuracy in Revenue Management by uncontraining demand estimates from censored data.* PhD thesis, The State University of New Jersey.

Part XVI

Marketing

Monotonic Spline Regression to Estimate Promotional Price Effects: A Comparison to Benchmark Parametric Models

Andreas Brezger[1] and Winfried J. Steiner[2]

[1] Department of Statistics, University of Munich, Ludwigstr. 33, 80539 Munich, Germany andreas.brezger@stat.uni-muenchen.de
[2] Department of Marketing, University of Regensburg, Universitätsstr. 31, 93053 Regensburg, Germany winfried.steiner@wiwi.uni-regensburg.de

[7] and [10] have suggested nonparametric regression techniques to estimate promotional price effects flexibly, and both obtained superior performance for their semiparametric sales response models compared to strictly parametric functions. Like these researchers, we also suggest a semiparametric approach to estimate price promotion effects: we model a brand's unit sales as (1) a nonparametric function of own- and cross-item price variables using Bayesian P-splines and (2) a parametric function of other promotional instruments. Unlike these researchers, we impose monotonicity constraints to avoid too much flexibility of our nonparametric estimator which may otherwise lead to economically implausible results (i.e., nonmonotonic price response curves). Our results from an empirical application based on weekly store-level scanner data show that the constrained semiparametric model clearly outperforms two widely used parametric specifications in validation samples.

1 Introduction

It is well documented that retail price promotions substantially increase sales of brands (e.g., [2]). There is also empirical evidence that a temporary price cut by a brand under promotion may decrease sales of competitive items significantly (e.g., [3]). However, little was known about the shape of own- and cross-item promotional price response curves until recently. Most studies addressing this issue employed strictly parametric functions, and came to different results from model comparisons (e.g., see [10] for an overview). Today, multiplicative, semilog and log-reciprocal functional forms are the most widely used parametric specifications to represent nonlinearities in sales response to promotional price cuts (e.g., [1], [9], [11]). It is important to note that these parametric functional forms are inherently monotonic, i.e., decreasing for own-price effects and increasing for cross-price effects, which is in accordance with economic theory (e.g., [6]).

In order to explore the shape of promotional price response curves more flexibly, [7] and [10] suggested nonparametric regression techniques. Specifically, [7] proposed a stochastic spline regression approach and [10] a Kernel regression approach, and both obtained superior performance for their models compared to strictly parametric models. [7], however, found strong irregularities in own-price response for some of the brands examined, resulting in less smooth and nonmonotonic shapes. The authors themselves pointed out that in case of an insufficient number of data points, the estimated functions may show irregularities where none exist. The promotional price response curves estimated by [10] were indeed much more smooth though not untroubled by nonmonotonicities. Especially, one brand's own-item price response curve indicated a decrease in unit sales as price cuts become very deep which is difficult to interpret from an economic point of view. [10] noted that such nonmonotonic effects might be due to chance. Anyway, the large improvement in predictive validities in favor of the Kernel estimator reported by [10] strongly support the use of nonparametric regression techniques to estimate promotional price effects. In the following, we show how the problem of nonmonotonicity can be dealt with.

We follow [7] and [10] and suggest a semiparametric approach to estimate price promotion effects: we model a brand's unit sales as (1) a nonparametric function of own- and cross-item price variables using Bayesian P-splines (e.g., [8]) and (2) a parametric function of other promotional instruments. Unlike [7] and [10], we additionally impose monotonicity constraints to avoid too much flexibility of our nonparametric estimator which may otherwise lead to economically implausible results, as discussed above. Importantly, imposing monotonicity constraints does not preclude the estimation of particular pricing effects like steps and kinks at certain price points or threshold and saturation effects at the extremes of the observed price ranges. Our results from an empirical application based on weekly store-level scanner data show that the constrained semiparametric model clearly outperforms two parametric benchmark models in validation samples (see section 3).

2 Model formulation and estimation

We use the following additive model:

$$\ln(Q_{is,t}) = \alpha_{is} + \sum_{j=1}^{J} f_{ij}(P_{js,t}) + \sum_{j=1}^{J} \gamma_{ij} D_{js,t} + \sum_{q=2}^{4} \delta_{iq} T_{q,t} + \epsilon_{is,t}, \quad \epsilon_{is,t} \sim N(0, \sigma^2) \tag{1}$$

$Q_{is,t}$ denotes unit sales of brand i in store s, $s = 1, \ldots, S$, and week t, $t = 1, \ldots, T$. f_{ij}, $j = 1, \ldots, J$, are unknown smooth functions of prices $P_{js,t}$ of the available J brands in store s and week t, referring to own price of brand i ($j = i$) and prices of competing brands ($j \neq i$). $D_{js,t}$ is a dummy variable capturing usage ($= 1$) or nonusage ($= 0$) of a display for brand j in store s and week t, and $T_{q,t}$ is a seasonal dummy variable indicating if week t belongs to the q-th quarter, $q = 2, 3, 4$. The seasons considered as variables are summer (q=2), autumn (q=3) and winter (q=4), with spring representing the reference season. α_{is} is a random store effect for brand i accounting for heterogeneity in baseline sales between different stores, γ_{ij} are parameters measuring own display ($i = j$) and cross display ($i \neq j$) effects, and δ_{iq} are parameters capturing seasonal effects in the consumption of products. $\epsilon_{is,t}$ is a disturbance term for brand i in store s and week t.

For modeling the unknown functions f_{ij}, $j = 1, \ldots, J$, we follow [8] who proposed a Bayesian version of the P-splines approach introduced in a frequentist setting by [5]. Accordingly, we assume that the unknown functions can be approximated by a linear combination of M cubic B-spline basis functions. Denoting the m-th basis function by B_{jm}, we obtain (suppressing index i for the brand under consideration)

$$f_j(x_j) = \sum_{m=1}^{M} \beta_{jm} B_{jm}(x_j).$$

A relatively large number of knots (usually between 20 and 40) is suggested to ensure sufficient flexibility, and a roughness penalty on adjacent regression coefficients β_{jm} is introduced to avoid overfitting. This penalization is accomplished by a second order random walk defined by

$$\beta_{jm} = 2\beta_{j,m-1} - \beta_{j,m-2} + u_{jm} \tag{2}$$

with normally distributed errors $u_{jm} \sim N(0, \tau_j^2)$ and diffuse priors β_{j1} and $\beta_{j2} \propto const$, for initial values. The second order random walk is the stochastic analogue to the second order difference penalty suggested by [5]. In our Bayesian approach, the amount of smoothness is controlled by the variance parameter τ_j^2 and can be estimated simultaneously with the regression coefficients by defining an additional hyperprior for the variance parameters τ_j^2. We assign inverse Gamma $IG(a_j, b_j)$ distributions on the variance parameters τ_j^2 and the scale parameter σ^2. We choose $a_j = b_j = 0.001$ leading to almost diffuse priors.

To obtain monotonicity, i.e., $f_j'(x) \leq 0$ for own-price response or $f_j'(x) \geq 0$ for cross-price response, it is sufficient to guarantee that subsequent parameters are ordered, such that

$$\beta_{j1} \geq \cdots \geq \beta_{jM} \quad \text{or} \quad \beta_{j1} \leq \cdots \leq \beta_{jM}, \tag{3}$$

respectively. In our approach, these constraints are imposed by introducing indicator functions to truncate the prior appropriately to obtain the desired support. For the fixed effects, i.e., display and seasonality effects, we assume diffuse priors, and for brand i's random store effect we assume $\alpha_{is} \sim N(0, \tau_\alpha^2)$, $s = 1, \ldots, S$. Since the posterior distribution of all the parameters given the data of this model is analytically intractable, we employ Markov Chain Monte Carlo (MCMC) techniques to obtain estimates for the parameters of interest. More specifically, we subsequently draw from the full conditionals $p(\beta_j|\cdot)$, $j = 1, \ldots, J$, $p(\gamma|\cdot)$ and $p(\alpha|\cdot)$. Technical details on the full conditionals, especially that of the smooth functions, the employed sampling scheme and efficient implementation, are available from the authors upon request.

We compare our semiparametric model (1) to the following two parametric models

$$\ln(Q_{is,t}) = \alpha'_{is} + \sum_{j=1}^{J} \beta'_{ij} ln(P_{js,t}) + \sum_{j=1}^{J} \gamma'_{ij} D_{js,t} + \sum_{q=2}^{4} \delta'_{iq} T_{q,t} + \epsilon'_{is,t} \tag{4}$$

$$\ln(Q_{is,t}) = \alpha''_{is} + \beta''_{ii} P_{is,t} + \sum_{j \neq i}^{J} \beta''_{ij}(1/P_{js,t}) + \sum_{j=1}^{J} \gamma''_{ij} D_{js,t} + \sum_{q=2}^{4} \delta''_{iq} T_{q,t} + \epsilon''_{is,t} \tag{5}$$

Models (2) and (3) differ from the semiparametric model (1) only with regard to the specification of price effects. Model (2) represents a multiplicative (log-log) functional form like the well-known SCAN*PRO model in its parametric versions ([11]), and β'_{ij} represents the (constant) elasticity of unit sales of brand i with respect to the price of brand j ($j = i$ or $j \neq i$). Model (3) follows [3] and is semilog in own price and log-reciprocal in competitive prices. Accordingly, β''_{ii} corresponds to the own-item price effect of brand i and β''_{ij} to the cross-item price effects ($j \neq i$). All involved full conditionals in models (2) and (3) are fully known and can therefore be easily updated by Gibbs sampling steps.

3 Empirical Study

We use weekly store-level scanner data from a major supermarket chain for eight brands of refrigerated orange juice. The data include unit sales, retail prices and display activities for these brands in 81 stores of the chain over a time horizon of 89 weeks. The data were provided by the James M. Kilts Center, GSB, University of Chicago.

Among the brands are 2 *premium* brands, 5 *national* brands and the supermarket's own *private label* brand. In the following, we illustrate our methodology for one of the national brands. To account for multicollinearity and for the fact that cross-item price effects are usually much lower than own-item price effects (see, e.g., [6]), we capture cross-promotional effects at the tier level rather than the individual brand level: we define $price\_premium_{st}$ ($price\_national_{st}$) as the minimum price for a premium brand (national brand) in store s and week t, and dummy variables $display\_premium_{st}$

($display\_national_{st}$) which indicate whether at least one of the premium brands (national brands) was on display (= 1) or not (= 0) in store s and week t. It is important to note that price and display activities of the national brand under consideration are excluded from the computation of $price\_national_{st}$ and $display\_national_{st}$.

We compare the performance of models (1)-(3) in terms of the Average Mean Squared Error (AMSE) in validation samples (also compare [10]). Specifically, we randomly split the data into nine equally-sized subsets and performed nine-fold cross-validation. For each subset, we fitted the respective model to the remaining eight subsets making up the estimation sample and calculated the Mean Squared Prediction Error (MSE) of the fitted model when applied to the observations in this holdout subset ([4]). Finally, the AMSE measure is calculated by averaging the individual MSE values across the nine holdout subsets.

The validation results are displayed in table 1 and indicate that the monotonic semiparametric model (1) clearly outperforms the inherently monotonic parametric models (2) and (3). Importantly, the AMSE is 48,65 percent lower for the semiparametric model (1) than for the best parametric model, the multiplicative model (2). Figure 1 depicts own- and cross price effects on unit sales of the national brand analyzed and reveals why model (1) performs so much better (due to space limitations, we do not show the estimated curves for model (3) which are quite similar to those of the multiplicative model). The nonparametrically estimated own price response curve shows a reverse s-shape with an additional increase in sales for extremely low prices. This strong sales spike can be attributed to an odd pricing effect at 99 cents, the lowest observed price for the national brand. The cross-price response curve with respect to the premium brands shows an s-shape and a strong kink at a price of two dollars, below which the unit sales of the national brand rapidly decrease. It is obvious that both parametric models cannot capture such complex own- and cross-price response patterns. The estimated cross-price effects with respect to the national brand tier and the private label brand show less dramatic differences between the models, although the nonparametric curves reveal slightly convex shapes, as compared to rather concave shapes of the parametric models. All estimated price effects of the parametric models (2) and (3) were significant at 5%. The cross-display effect of the private label brand was not significant at 5%, while all other display effects were significant at 5% and show the expected sign.

Table 1. Evaluation of models in terms of AMSE.

Model specification	AMSE
semiparametric model (1)	52139.7
parametric model (2)	101536.1
parametric model (3)	110245.5

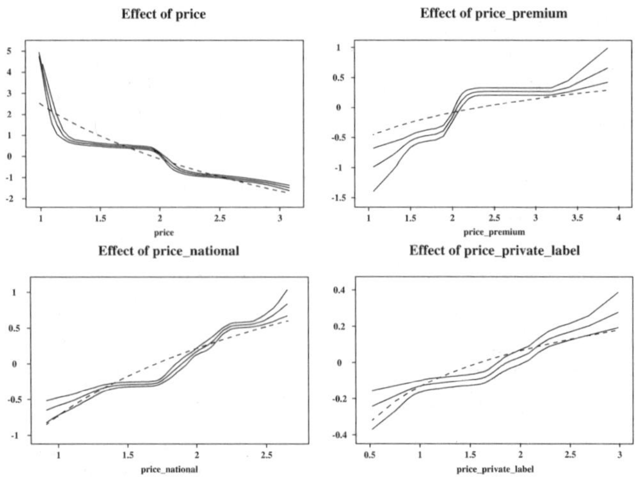

Fig. 1. Own-price (*price*) and cross-price (*price_premium*, *price_national*, *price_private_label*) effects estimated by the semiparametric model (solid lines) and model (2), the best parametric model (dashed lines). Also shown are the 95% pointwise credible intervals for the semiparametric model.

References

1. Blattberg RC and George EI (1991) Shrinkage Estimation of Price and Promotional Elasticities, Journal of the American Statistical Association, 86(414):304–315
2. Blattberg RC, Briesch R and Fox EJ (1995) How Promotions Work, Marketing Science, 14(3)(Part 2):G122–G132
3. Blattberg RC and Wisniewski KJ (1989) Price-Induced Patterns of Competition, Marketing Science, 8(4): 291–309
4. Efron B and Tibshirani RJ (1998) An Introduction to the Bootstrap, Chapman and Hall/CRC, Boca Raton
5. Eilers PHC and Marx BD (1996) Flexible smoothing using B-splines and penalized likelihood (with comments and rejoinder) Statistical Science, 11(2):89–121
6. Hanssens DM, Parsons LJ and Schultz RL (2001) Market Response Models: Econometric and Time Series Analysis, Chapman and Hall, London
7. Kalyanam K, Shively TS (1998) Estimating Irregular Pricing Effects: A Stochastic Spline Regression Approach, Journal of Marketing Research, 35(1):16–29
8. Lang S and Brezger A (2004) Bayesian P-splines, Journal of Computational and Graphical Statistics, 13:183–212
9. Montgomery AL (1997) Creating Micro-Marketing Pricing Strategies Using Supermarket Scanner Data, Marketing Science, 16(4):315–337
10. van Heerde HJ, Leeflang PSH and Wittink DR (2001) Semiparametric Analysis to Estimate the Deal Effect Curve, Journal of Marketing Research, 38(2):197–215
11. —— (2002) How Promotions Work: SCAN*PRO-Based Evolutionary Model Building, Schmalenbach Business Review, 54:198–220

Robust Preference Measurement

A Simulation Study of Erroneous and Ambiguous Judgement's Impact on AHP and Conjoint Analysis

Sören W. Scholz, Martin Meißner, and Ralf Wagner

Business Administration and Marketing
Bielefeld University, D-33615 Bielefeld, Germany

Summary. Despite the recent methodological progress to unburden respondents in preference analysis the quality of consumers' judgements is fundamental for marketing research results. Surprisingly, the impact of ambiguous and erroneous judgments given by the respondents is widely neglected in the marketing literature. In this paper we compare the Analytic Hierarchy Process and Conjoint Analysis with respect to the impact of random errors as well as ambiguities in preference statements by means of Monte Carlo simulation studies. Referring to Thurstone's law of comparative judgements, we demonstrate the superior robustness of the Analytic Hierarchy Process in dealing with these kinds of perturbing effects.

1 Introduction

Conjoint Analysis (CA) is marketing researchers' favorite in consumer preference measurement and has been widely discussed in recent OR-related publications. The Analytic Hierarchy Process (AHP) is an alternative methodology and is well-established in multi-attributive utility measurement for supporting expert decision making. The latter methodology is widely used in managerial decision making but rather uncommon in consumer research. Recent empirical studies that compare these two methods in the application to consumer preference measurement claim that the AHP has some appealing features and is at least en par with respect to accuracy, sacrifice of time, and motivational aspects involved in preference measurement (Scholl *et al.* (2005)).

Both methods are based on the assumption that preference statements are deterministic. Psychological research has proven that this basic assumption is rarely met. The effect of errors in human judgment is well known in psychometric measurement literature. Fischhoff (1991) refers to the substantial instability of human preferences: Preferences appear to be remarkably vulnerable in the light of framing effects and the response mode in which preferences are induced. Increased motivational and cognitive efforts on the part of the decision maker hardly will compensate for these effects. Pommerehne *et al.*

(1982, 573) pinpoint this in the following way: "Even when the subjects are exposed to strong incentives for making motivated, rational decisions, the phenomenon of preference reversal does not vanish." Schmidt and Hunter (1999, 183) conclude in the following statement: "There is no such thing as errorless measurement". Facing these discrepancies of concerns in the psychometric literature on the one hand and the contemporary marketing research practice on the other hand, this study aims at

- relating AHP and CA to errors discussed in psychometric literature and
- exploring the robustness of AHP and CA in a controlled setting of a Monte Carlo simulation.

This paper is structured as follows: First, we discuss reasons for erroneous statements in both AHP and CA measurement scales. Subsequently, we outline the methodology of preference measurement by means of AHP. Then, we describe the Monte Carlo simulations evaluating the proneness of AHP and traditional CA to erroneous measurement as well as inconsistencies, fuzziness and uncertainty in the consumer preference statements. Finally, we conclude with a brief discussion and outlook to future research issues.

2 Errors in Preference Measurement

Random response errors appear on many occasions, driven by variations in attention, mood, feelings, mental efficiency, or general mental state during the preference elicitation (Schmidt and Hunter (1999)). Cognitive psychology and human information processing teaches us that there is considerable noise in the human central nervous system at any given moment influencing the accuracy of preference statements. Therefore, Fujii and Gärling (2003) assume that preferences are made up of two parts: (1) a preference component which is determined by an invariant utility function and (2) a component which is based on the context.

The basic theory dealing with these inaccuracies is given by the law of comparative judgment. The measurement model of Thurstone (1927) and the preference measurement model in the AHP (Saaty (1980)) resemble each other in that both are estimated on a unidimensional scale which expresses the decision maker's preference for alternatives or stimuli in a pair-wise comparison. The comparative judgment model assumes that stimuli are represented along a psychological continuum by real-valued random variables ($S_k \, \forall k \in K$). Thurstone defines each psychological magnitude to be mediated as a *discriminal process* in which the organism identifies, distinguishes, or reacts to stimuli. Each stimulus–when presented to an observer–gives rise to a discriminal process. Due to momentary fluctuations a given stimulus does not always excite the same discriminal process, but may excite one with a higher or lower value on the psychological continuum. Thus, the value x_{kl} which denotes the preference of stimulus k over stimulus l is an occurrence of the value of the random variable X_{kl} representing this psychological continuum. In line with Thurstone (1927) we suggest that the decision maker's preference judgments x_{kl} are independently sampled from a normal distribution. The standard deviation of the distribution of the preference judgments (when presented repeatedly) is called the *discriminal dispersion* of the compared pair of stimuli. Additional to these effects of ambiguity and uncertainty in the preference judgments, which make up the core of the AHP (see section 3 for details), further effects can perturb the real preference structure of the decision maker. First, the decision maker can produce erroneous statements when quoting his preferences. These errors can influence the direction of the preference structure (i.e. alternative k is preferred to l instead of l is preferred to k). Second, the real strength of a preference can be over- or underestimated because of an inaccurate statement of the decision maker's preferences. Of course, these two effects can also stem from an erroneous input of data (e.g., by typing the results from paper forms into spreadsheet tables).

In a similar manner Saaty as well as Thurstone propose to measure the preferences of the decision makers by means of real values. He or she is asked to answer on a verbal scale. Consequently, the measurement process is prone to effects of ambiguity. Thurstone's law of comparative judgment relates to the proportion of times any given stimulus i is judged greater on a given attribute than any other stimulus j to the psychological scale values and discriminal dispersions of the two stimuli on the psychological continuum. The scale difference between this discriminal process $(S_k - S_l)$ which is denoted as the discriminal difference between two stimuli can be mathematically expressed in the following way:

$$S_k - S_l = x_{kl} \cdot \sqrt{\sigma_k^2 + \sigma_l^2 - 2\rho\sigma_k\sigma_l} \tag{1}$$

Referring to Thurstone's (1927) case V in the law of comparative judgment, we assume the discriminal dispersions for pairs of stimuli to be constant on the psychological scale which implies that $\sigma_k = \sigma \quad \forall\ k \in K$ and that the covariance term for all pairs of stimuli is zero ($\rho = 0$).

While traditional CA is based upon the evaluation of single separate stimuli in terms of alternative profiles on a classical rating scale, this method involves no comparative judgments as described above. In this case, we also assume that the preference statements include errors as described in the law of categorical judgment which is rather similar to the aforementioned law of comparative judgment (Torgerson (1958)).

3 Outline of the AHP Method

Saaty (1980) developed the AHP as one of the supporting systems for multi-criteria decision making and as a tool for analyzing the decision making process. Primarily, the AHP is an additive weight aggregation of priority scores that have been derived from subjective scores for pair-wise comparisons of the lowest level criteria. That is to say, the AHP utilizes the subjective judgments of each decision maker on the input side and delivers quantified weights for each alternative as an output.

The AHP proceeds in three steps to derive this final output: First, the decision is broken down into a hierarchical structure of elements such as goal, criteria, sub-criteria, and alternatives that make up the complete decision problem at hand. These are translated into a hierarchical decision tree. This structure enables us to divide and conquer complex decision problems into manageable sub-problems. The hierarchy indicates a relationship between elements on one level with those of the level immediately below them. Second, pair-wise comparisons of the elements are derived and third, the weight of each alternative is calculated (Saaty (1980)). The pair-wise comparison process improves the accuracy of these weights as it allows the decision maker to focus on a series of $(n-1)/2$ simple, straightforward questions, with n denoting the number of (sub-) criteria or alternatives being involved on one sub-problem.

Usually a nine-point-scale is used to measure the decision maker's preference towards an attribute-level i versus j. The scale ranges from "i and j are equal" (scale point 1) to "i is absolutely preferred to attribute-level j" (scale point 9). The scale simply transforms verbal judgments of the decision maker into priority ratios a_{ij}. Larger values on the scale express stronger preferences for attribute-level i. The reciprocal values of the priority ratio a_{ji} with $a_{ji} = 1/a_{ij}$ provide us with evidence on how much j is preferred over i. Accordingly, all pair-wise comparisons that are measured with respect to a higher level element of the hierarchy can be subsumed in reciprocal matrix $\mathbf{A} = (a_{ij})_{n \times n}, \forall\, i,j = 1,\ldots,n$. For consistency the pair-wise comparisons a_{ij} must have the form $a_{ij} = (w_i/w_j), \forall\, i,j = 1,\ldots,n$. From eigenvalue theory it is known that up to small perturbations this is the classical eigenvalue problem of the form $\mathbf{Aw} = \lambda_{max}\mathbf{w}$, where λ_{max} is the maximum principal eigenvalue (Perron root) of matrix $\mathbf{A}$. The corresponding principal right eigenvector $\mathbf{w}$ includes the relative utility u_{hi} for each attribute-level with respect to an element h of the immediately higher level in the hierarchy.

4 Simulation Study

The focus of this simulation study is to investigate how the traditional CA (using the full-profile approach by means of the rating method on a nine-point scale) and the AHP (using classical eigenvector method for weights estimation as outlined in section 3) handle the aforementioned disturbance in preference statements. We assume that the statements of the decision makers are probabilistic and therefore prone to uncertainty and ambiguity rather than deterministic values of the real preferences of the decision maker.

In order to ensure a suitable comparison of the CA and the AHP, we designed a decision making situation with approximately the same complexity for both, the AHP as well as CA. As CA is not capable of representing multi-level hierarchies, we use a simple two-level design including four attributes with three levels each which is identical to the design used by Scholl *et al.* (2005). Adopting a fractional factorial design leads to 9 profiles (Kuhfeld (2004)). The AHP method includes $4 \times 3 = 12$ pair-wise comparisons on the bottom level and 6 pair-wise comparisons on the higher level. Consequently, the decision maker has to answer twice as many Likert-type ratio scale questions in the AHP setting than in CA. Noteworthy, Scholl *et al.* (2005) found that decision makers need only half the time to answer the required pair-wise comparisons compared to answer the corresponding statements in CA.

On the basis of these fairly comparable initial preference statements we conducted Monte Carlo simulations by perturbing preference statements on the basis of the theoretical underpinnings outlined in section 2. We assume each judgement to be normally distributed on the psychological continuum applying the law of comparative judgement for the AHP and law of categorical judgement for CA. We simulate 300 perturbations for each set of the 15 corresponding AHP and CA judgements by adding a normally distributed error ϵ (perturbing factor) with deviation $s(\epsilon)$ and mean $\overline{\epsilon} = 0$ to each of the initial nine-point Likert-scale statements. In order to measure the impact of the perturbation of the initial preference statements we compute Pearson's r and mean deviation (d) of the perturbed resulting preference weights in comparison to the initial weights as well as the sum of squared errors (SSE). The results–subject to varying $s(\epsilon)$–are shown in Table 1.

	AHP			CA		
$s(\epsilon)$	r	d	SSE	r	d	SSE
.5	.9925	.0293	.1711	.9424	.0621	.2205
1.0	.9857	.0500	.2775	.8532	.1121	.5615
2.0	.9581	.0760	.4131	.6947	.1785	1.1643
3.0	.9165	.1003	.5658	.5272	.2330	1.8006

Table 1. Accuracy of AHP and CA with respect to different levels of perturbed preference statements ($s(\epsilon)$)

The AHP shows a highly accurate projection of the initial preference structure up to a critical level of perturbations of the initial statements. Even in the worst case with a standard error $s(\epsilon) = 3.0$, the AHP weights have a correlation higher than .9 with the true weights, but the correlation of the part-worth of the CA drops down to less than .6. Thus, CA provides only an accurate computation of the weights in the case of small distortion of the true preferences. Contrastingly, the AHP is robust to fuzzy and ambiguous preference judgments.

5 Discussion and Conclusions

The goal of the study is twofold. First, we systematize errors related to formal preference analysis and link them to underlying psychometric theory. Starting with case V of Thurstone's law of comparative judgment we investigate the implications of additive errors described by a normal distribution. Second, an explorative comparison of CA and AHP with respect to the robustness against errors in respondents' judgments is conducted. The results show a significant superiority of the AHP in comparison to traditional CA in the accuracy when dealing with the phenomena of inconsistencies as well as judgmental uncertainty or errors. CA only compensates for small derivations in stated and actual preferences. The AHP proves to be trustworthy, essentially because the huge amount of redundancy in the pair-wise comparisons makes the process fairly insensitive to judgmental errors. Thus, we claim the AHP to be a promising approach to measure consumer preferences in typical marketing research settings where uncertainty is an indispensable feature due to information asymmetry when considering various decision alternatives.

Regarding the negligible impact of ambiguous or erroneous judgments on the accuracy of preference elicitation via AHP, it is easy to question whether more sophisticated AHP methods, particularly the fuzzy AHP approach, really lead to substantial improvements in preference measurement. While this issue requires further research, the experimental design applied in this study provides a suitable starting point for this investigation.

References

Fischhoff, B. (1991). Value elicitation: Is there anything in there? *American Psychologist*, **46**, 837–847.

Fujii, S. and Gärling, T. (2003). Application of attitude theory for improved predictive accuracy of stated preference methods in travel demand analysis. *Transport Research A*, **37**(4), 289–402.

Kuhfeld, W. F. (2004). *Marketing Research Methods in SAS: Experimental Design, Choice, Conjoint, and Graphical Techniques*. SAS, Carry.

Pommerehne, W. W., Schneider, F., and Zweifel, P. (1982). Economic theory of choice and the preference reversal phenomenon: A reexamination. *The American Economic Review*, **72**(3), 569–574.

Saaty, T. L. (1980). *The analytic hierarchy process*. McGraw-Hill.

Schmidt, F. L. and Hunter, J. E. (1999). Theory testing and measurement error. *Intelligence*, **27**(3), 138–198.

Scholl, A., Manthey, L., Helm, R., and Steiner, M. (2005). Solving multiattribute design problems with analytic hierarchy process and conjoint analysis: An empirical comparison. *European Journal of Operational Research*, **164**, 760–777.

Thurstone, L. (1927). A law of comparative judgment. *Psychological Review*, **34**, 273–286.

Torgerson, W. (1958). *Theory and methods of scaling*. Wiley, New York.

Improving the Predictive Validity of Quality Function Deployment by Conjoint Analysis: A Monte Carlo Comparison

Daniel Baier and Michael Brusch

Chair of Marketing and Innovation Management, Brandenburg University of Technology, Konrad-Wachsmann-Allee 1, D-03046 Cottbus, Germany
daniel.baier@tu-cottbus.de and m.brusch@tu-cottbus.de

Recently, for quality function deployment (QFD), new approaches have been developed to improve the estimation of the influence of technical features and engineering characteristics on relevant product attributes and of the importance of these attributes for the customer. Instead of group discussions with qualitative consensus outcomes these new approaches make use of regression analysis or of conjoint analysis. Applications indicate superior predictive validity of these new approaches. In this paper, this superior validity is analyzed in more detail. A factorial design with synthetically generated data is used for comparisons.

1 Introduction

Quality function deployment (QFD) is a systematic process for new product development which was developed and first used in Japan in the mid 1960s, introduced to the western world in the 1980s and has been consequently improved and modified since then (see [1]). QFD has two main advantages compared to other new product development techniques: (1) better and earlier determination of product attributes and the product quality searched by the customer and (2) better and earlier determination of key manufacturing requirements in advance. Recent surveys ([7] on QFD usage in Japan and the U.S.) and literature reviews ([8] with 650 reviewed QFD articles) document the wide spread usage of this tool in research and practice.

Characterized briefly, the QFD methodology tries to transform the customer's requirements along a chain of tables and graphs into technical features and engineering characteristics that lead – finally – to certain demands on production and assembly. So, e.g., in a first step the relation between

- customer's requirements described by product attributes (PAs) like "small", "easy to handle" in the voice of the customer and

- technical features described by engineering or product characteristics (PCs) like "size in mm" or "number of switches" in the voice of the engineer

is formalized and documented in a key document, the so-called house of quality. For this parametrization, objective measurement has often been demanded in the literature (i.e. with the help of controlled experiments). However, instead, often group discussions are used with a high level of subjectivity.

To eliminate this potential source of error, various authors proposed the usage of regression analysis (e.g., [10], [13], [2]) for measuring the PC influence on the PAs or the integration of conjoint analysis (CA) into the approach (see, e.g., [12], [9], [11]). Recently (see [3]), a CA based approach has been developed which showed superiority compared to the traditional QFD approach in empirical applications. The present article analyzes this superiority in more detail. The "new" CA based approach for QFD is briefly discussed (Sect. 2). Then, a Monte Carlo comparison of the traditional and the "new" CA approach follows (Sect. 3). A discussion closes the article (Sect. 4).

2 Linking QFD and CA

The essential advantage of CA compared to direct measurement is the "conjoint" evaluation task (see, e.g., [4], [5], [6]). The influence or importance of product features is measured via evaluations of synthetic feature level combinations, what forces the respondents to make trade-offs between different features and feature levels. We make use of this in our new CA based QFD approach when PA importances in the eyes of the customer (step 2 of the new approach) and when PC influences on the PAs are measured (step 4). Table 1 summarizes the five steps of a "new" (see [3]) CA based approach in comparison to the traditional approach. For the application of adaptive conjoint analysis (ACA, the most popular CA technique (e.g., [5]), more hints are given in steps 2 and 4.

First, PAs are selected. In both approaches (also true for the third step) known techniques can be used for this purpose (e.g., focus groups, rep-tests, explorative interviews). However, the CA based approach requires an attribute and level number limitation. The second step is the evaluation of the PA importance. In the traditional approach this is done on the basis of direct questioning of customers or experts, e.g., the so-called QFD team. In the CA based approach a conjoint study is used for this purpose. For the evaluation of the PC influence on the PAs in step 4, the CA based approach makes again use of conjoint studies. For each PA, members of the QFD team are questioned. The PCs are evaluated on how strongly they contribute to the fulfilment of the respective PA. The determined influences are standardized (similar to step 2) so that the influences of the PC on each PA add up to 1. A similar standardization should also be executed in the traditional approach

Table 1. Traditional vs. CA based approach for measuring the importance of product characteristics in the eyes of the customer (ACA=Adaptive Conjoint Analysis, OLS=Ordinary Least Squares)

	Traditional approach	CA based approach
Step 1: Selecting product attributes (PAs)		
No. of PAs	arbitrary	max. 30 (ACA)
No. of PA levels	no levels specified	2 up to 9 for each PA (ACA)
Step 2: Evaluating PA importances		
Respondents	customers, QFD Team	customers, QFD Team
Data collection and analysis	direct ratings of PAs, e.g., 0(unimp.), ..., 4(important)	ratings of stimuli, analyzed by OLS imp. estimation (ACA)
Results	PA (importance) shares	PA (importance) shares
Step 3: Selecting product characteristics (PCs)		
No. of PCs	arbitrary	max. 30 (ACA)
No. of PC levels	no levels specified	2 up to 9 for each PC (ACA)
Step 4: Evaluating the influence of PCs on PAs		
Respondents	experts, QFD Team	experts, QFD Team
Data collection and analysis	group discussion w.r.t. each PA resulting in 0, 1(=△), 3(=o) or 9(=⊙) as weights for each PC	for each PA: ratings of stimuli, analyzed by OLS importance estimation (ACA)
Results	PC (imp.) shares per PA	PC (imp.) shares per PA
Step 5: Computing PC importances in the eyes of the customer		
Calculation	sum of PC shares per PA weighted by PA shares	sum of PC shares per PA weighted by PA shares

where weights are determined in QFD team discussions. The calculation of the PC importance in view of the customer is implemented in a fifth step.

This "new" CA based approach already showed superiority to the traditional QFD approach in empirical applications (see [3]). Here, this superiority is analyzed in the following section in more detail using a factorial design with synthetically generated and disturbed data.

3 A Monte Carlo Comparison

In order to compare the traditional and the new CA approach using synthetic data, available applications of QFD for new product development were checked for characteristics. As a result, a factorial design with eight factors – each with three levels – varies

- the structure of the relationship matrix (e.g., equal numbers of PAs and PCs, much more PAs than PCs, much more PCs than PAs which results in square or rectangular relationship matrices),

- the dimension of the relationship matrix (e.g., small, medium, or large relationship matrices),
- the percentage of "active" influences of PCs on PAs (e.g., small, medium, or large number of positive values in the relationship matrix),
- the structure of "active" influences of PCs on PAs (e.g., small, medium, or large blocks of positive values in the relationship matrix),
- the distribution of the influence values of PCs on PAs (positive values uniformly distributed vs. skew distributions with more extreme values),
- the error in measuring influence values of PCs on PAs (using normal distributions with small, medium, and large standard deviations for generating additive error),
- the distribution of the PA importance values (values uniformly distributed vs. skew distributions with more extreme values),
- the error in measuring PA importance values (using normal distributions with small, medium, and large standard deviations for generating additive error).

Basing on this factorial design synthetic applications of the traditional and the CA based approach were generated:

- For the traditional approach and each dataset, the generated values according to the factorial design were rounded to 0-, 1-, 3-, and 9-values (for the influence) resp. 0-, 1-, ..., 9-values (PA importance). The superimposed measurement error reflects the fuzziness of the qualitative approach with divergent ratings.
- For the CA based approach and each dataset, conjoint ratings were generated using the "true" influence values of PCs on PAs as well as PA importance values. The resulting conjoint ratings were superimposed by measurement error according to the factorial design and again rounded to 0-, 1-, ..., 9-values. Then, the resulting total of number of PAs plus one generated sets of conjoint data could then be used to estimate influence values of PCs on PAs as well as PA importance values using OLS (see Table 1 for the CA based approach).
- Additionally, for both approaches, a holdout set of 20 randomly generated combinations of PA values were used to test the validity of both approaches by calculating ratings of this stimuli based on the "true" (without measurement, regression and rounding errors) and the "estimated" influence and importance values.

With three replications and a full factorial approach a total of 3^9=19,683 synthetic datasets were generated and analyzed. For each dataset, R^2 values between "true" and "estimated" holdout ratings were calculated. Table 2 shows the results of this calculations giving mean R^2 values with respect to (w.r.t.) each approach as well as each factor level. t- and F-tests indicate significance of the differences w.r.t. methods and factors.

The results show a clear superiority of the "new" approach over the traditional approach across a huge variety of factor levels. Even though this superiority is less striking in case of a high error in measuring importance values as well as in case of uniformly distributed influence values, the Monte Carlo comparison shows that the "new" approach with the OLS estimation of real influence and importance values (instead of the integer restriction in the traditional approach) should be preferred.

Table 2. Monte Carlo comparison of the validity of the traditional and the "new" CA based approach using mean R^2 values w.r.t. the holdout set of stimuli (no. = number, PA = product attributes, PC = product characteristics, n=19,683 datasets)

Factor	Level	R^2 trad. approach	R^2 "new" approach	R^2 both approaches
Structure of the relationship matrix	equal PA and PC nos.	.653	.726***	.689
	PC no.=2x PA no.	.599	.701***	.650
	PA no.=2x PC no.	.713	.765***	.739***
Dimension of the relationship matrix	36 cells	.732	.816***	.774***
	144 cells	.686	.750***	.718
	1800 cells	.547	.626***	.586
Percentage of active influences in the relationship matrix	11.11 %	.550	.649***	.599
	22.22 %	.661	.743***	.702
	33.33 %	.754	.800***	.777***
Structure of active influences in the relationship matrix	1 block	.664	.738***	.701**
	2 of 4 blocks are active	.652	.727***	.689
	3 of 9 blocks are active	.649	.728***	.688
Distribution of the influence values	uniform distribution	.684	.691^{ns}	.687
	triangle distribution	.698	.738***	.718***
	half-triangle distribution	.583	.763***	.673
Measurement error influences in the relationship matrix	σ=.1	.845	.898***	.872***
	σ=.3	.775	.832***	.803
	σ=.9	.345	.462***	.403
Distribution of the importance values	rectangular distribution	.645	.708***	.676
	triangle distribution	.659	.729***	.694
	half-triangle distribution	.661	.755***	.708***
Measurement error importance values	σ=.1	.660	.792***	.726***
	σ=.3	.657	.765***	.711
	σ=.9	.648	.635*	.641
Overall		.655	.731***	

****: significant differences within rows (t-Test) and columns (F-test) at the p<.001 level; **: at the p<.01 level; *: at the p<.1 level; ns: not significant*

4 Conclusion and outlook

The "new" CA based approach for QFD shows a number of advantages in comparison to the traditional approach. PA importances as well as PC influences on PAs are measured "conjoint" resp. simultaneously. Furthermore, the calculated weights are more precise (real valued instead of 0-, 1-, 3-, or 9-values) which resulted in a higher predictive validity. The Monte Carlo comparison has shown a clear superiority in a huge variety of simulated empirical settings.

References

1. Akao Y (1990) QFD, Integrating customer requirements into product design. Productivity Press, Cambridge, MA
2. Askin RG, Dawson D (2000) Maximizing customer satisfaction by optimal specification of engineering Characteristics. IIE Transactions 32:9–20
3. Baier D, Brusch M (2005) Linking quality fuction deployment and conjoint analysis for new product design. In: Baier, D, Decker, R, Schmidt-Thieme, L (eds) Data analysis and decision support. Springer, Berlin, 189–198
4. Baier D, Gaul W (1999) Optimal product positioning based on paired comparison data. Journal of Econometrics 89:365–392
5. Baier D, Gaul W (2003) Market simulation using a probabilistic ideal vector model for conjoint data. In: Gustafsson A, Herrmann A, Huber F (eds) Conjoint measurement - methods and applications. 3rd ed., Springer, Berlin, 97–120
6. Brusch M, Baier D, Treppa A (2002) Conjoint analysis and stimulus presentation: a comparison of alternative methods. In: Jajuga K, Sokolowski A, Bock HH (eds) Classification, clustering, and analysis. Springer, Berlin, 203–210
7. Cristiano JJ, Liker JK, White CC (2000) Customer-driven product development through Quality Function Deployment in the U.S. and Japan. Journal of Product Innovation Management 17:286–308
8. Chan LK, Wu ML (2002) Quality Function Deployment: a literature review. European Journal of Operational Research 143:463–497
9. Gustafsson A (1996) Customer focused product development by conjoint analysis and Quality Function Deployment. Linköping University Press, Linköping
10. Hauser JR, Simmie P (1981) Profit maximizing perceptual positions: an integrated theory for the selection of product features and price. Management Science 27:33–56
11. Pullman ME, Moore WL, Wardell DG (2002) A comparison of Quality Function Deployment and conjoint analysis in new product design. Journal of Product Innovation Management 19:354–364
12. Urban GL, Hauser JR (1993) Design and marketing of new products. Prentice Hall, Englewood Cliffs, NJ
13. Yoder B, Mason D (1995) Evaluating QFD relationships through the use of regression analysis. In: Proceedings of the Seventh Symposium on Quality Function Deployment, ASI&GOAL/QPC. American Supplier Institute, Livonia, MI, 239–249

System Dynamics Based Prediction of New Product Diffusion: An Evaluation

Sabine Schmidt[1] and Daniel Baier[2]

[1] Chair of Planning and Innovation Management, Brandenburg University of Technology, Konrad-Wachsmann-Allee 1, 03046 Cottbus, Germany
[2] Chair of Marketing and Innovation Management, Brandenburg University of Technology, Konrad-Wachsmann-Allee 1, 03046 Cottbus, Germany

Abstract: System Dynamics (SD) is a methodology that can be used for analysing and understanding complex feedback systems. Influencing factors, time delays as well as dynamic relations between factors and effects can be assumed and used for simulations and the development of strategies. Whereas the aim of SD models is the better understanding of the relationship between underlying structure and behaviour of the feedback system, it can also – at least in principal – be used for forecasting. This paper analyses this application field by using real and generated data on new product diffusions in a calibration – validation – setting.

1 Introduction

The use of assumptions and forecasts about the launch of new products is essential and inevitable for business decisions. It can be used to increase competitive advantage of a company: The pre-launch activities and decisions about time to market, a balanced product portfolio, pricing, advertising, relevant customer segments, offensive and defensive strategies are incorporated. However, the basis for decision making is extremely hazardous because no or only few data is available. In this context, diffusion models can be used to reduce the uncertainty since they try to predict achievable market share, the speed of the diffusion, and which key variables influence and accelerate the diffusion process even.

One possibility to analyse the diffusion of new products could be the approach of System Dynamics (SD), founded by J. W. Forrester in the 1950s. This methodology offers the possibility of building calibrated models which support the decision and policy making [6, 17] and could be used – at least in principal – for forecasting [9]. This article tries to investigate the validity of such an SD usage in comparison to a regression approach, well aware of the fact that the SD community in general is suspicious in using SD for this purpose.

2 New product diffusion models – The Bass-Model

The initial point of the diffusion research in marketing is seen since the fundamental models in the 1960s by Fourt, Woodlock and Mansfield were combined by Bass to a mixed-influence model [1]. After that, a variety of further refinements were formed that incorporate marketing-mix influence, replacement and multiple purchases as well as spatial, dynamic and disaggregate-level aspects [3, 12, 14]. In general their purpose is to characterise the development of the potential adopters (market potential) to the adopters over a time horizon and to capture the life-cycle dynamics of new products [12]. The investigation of this paper will concentrate on the fundamental model by Bass. This model

$$q_t = \alpha(Q_{max} - Q_{t-1}) + \beta(Q_{t-1} / Q_{max})(Q_{max} - Q_{t-1}) \tag{2.1}$$

with

q_t	sales in period t $(t=0,...,T)$,
α	coefficient of innovation $(\alpha>0)$,
β	coefficient of imitation $(\beta>0)$,
Q_{max}	market potential or maximum cumulated sales,
Q_t	cumulated sales up to period t $(Q_0=0)$

calculates the sales in a period as the sum of sales coming from innovation demand $(\alpha(Q_{max}-Q_{t-1}))$ and from imitative demand $(\beta(Q_{t-1}/Q_{max})(Q_{max}-Q_{t-1}))$. α and β are constant rates of innovation and imitation. Under the assumption that $\alpha<\beta$ the q_t grows up to a peak and after that declines. Otherwise the sales curve falls continuously [2]. The rate parameters have to be estimated using experts or – if available – observed time series data.

Bass transformed equation (2.1) into a second-degree polynomial

$$q_t = a_0 + a_1 Q_{t-1} + a_2 Q_{t-1}^2 \tag{2.2}$$

with

$$a_0 = \alpha Q_{max}, \quad a_1 = \beta - \alpha, \quad a_2 = -\beta / Q_{max}$$

which can be easily used for estimation using observed times series data for Q_t $(t=1,...,T)$ and ordinary least squares (OLS) [1, 8, 15, 16].

On the basis of a meta-analysis of empirical studies the Bass Model was confirmed as an empirical generalization. Bass considered widely the demand as a homogeneous mass. The typical left scarped trend of the diffusion process utilises fewer than 50% of the market potential in the peak of the life cycle curve. In spite of that the uncertainty about the speed of growth and the decline is not eliminated [2].

3 System Dynamics methodology in diffusion research

System dynamics (SD) has been used for studying and managing complex systems with feedback loops and time delays [17]. The system structure allows hypothesizing approximate rules of behaviour based on sensitivity analysis and scenarios. But noise limits the quality and ability to predict accurately [6]. A well-calibrated SD model is used to more reliable forecasts in particular of short- and mid-term trends than statistical models. Additionally it can be considered as a part of an early-warning or on-going learning system [9].

In System Dynamics research various diffusion models capture different market structures, components of the marketing-mix, general aspects of the innovation process as network effects and the process of innovation diffusion itself [see, e.g., 10, 11, 13]. Younger approaches focus on agent based and network externalities modelling [see, e.g., 2, 7, 17].
The above-mentioned Bass Model can be modified to System Dynamics methodology (see Fig. 1).

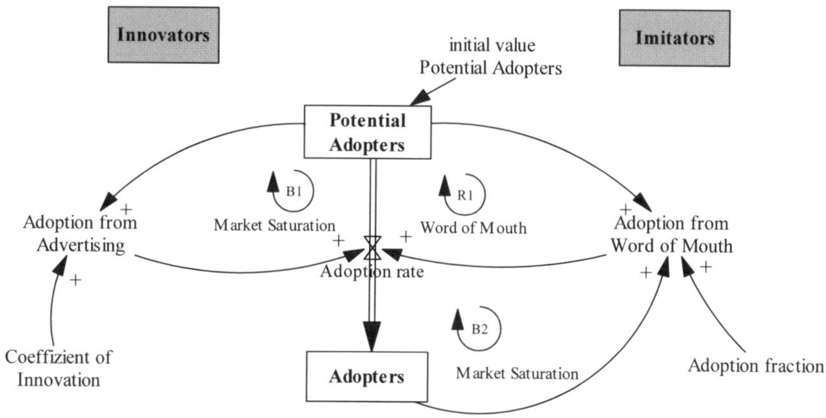

Fig. 1. System Dynamics Structure of the Bass Diffusion Model

The *adoption rate* changes the *levels* (number of potential adopters and number of adopters) and is influenced by two means of communication: mass media or advertising and word of mouth. In the first periods innovators have a great influence on the sales, later on the adopters.

4 Empirical comparison of SD and regression analysis

In the empirical investigation we use observed (and additionally disturbed) diffusion data and compare the modelling results using the above mentioned regression analysis approach and the SD approach both with respect to mean R^2 measures.
The basic data are real annual time series of five single products: camcorder, CD-Player, extractor hood, colour television set, freezer [4]. These data series show the sales over different time periods only of the first purchases. Repeat purchases are not considered.
Object of the investigation are disturbed real data using the following four factor design: (1) the disturbance by measurement errors (additive normally distributed error with no, low, or medium standard deviation), (2) the disturbance by systematic errors (no, one, or two outliers) and (3) the length of the calibration period (all data or only 75 % of the available data are used for calibration) as well as (4) the underlying times series (time series 1, 2, 3, 4, and 5). Each cell of the factorial design was repeated three times resulting in 270 datasets. The results of the comparison are presented in Table 1.

Table 1. Comparison of validity using mean R^2 measures in the calibration and the validation period w.r.t. to OLS and SD estimates

		Calibration period			**Validation period**		
Factor	**Level**	**Regression**	**SD**	**Mean**	**Regression**	**SD**	**Mean**
Measurement error	No	0.605	0.638	0.622	0.516	0.578	0.547
	Low	0.577	0.625	0.601	0.501	0.561	0.531
	Medium	0.560	0.620	0.590	0.480	0.531	0.505
Systematic error	No outlier	0.571	0.631	0.601	0.491	0.568	0.529
	1 outlier	0.590	0.605	0.597	0.500	0.52	0.510
	2 outliers	0.581	0.647	0.614	0.506	0.583*	0.544
Calibration period	All data	0.554	0.606	0.580*	0.554	0.606	0.580***
	75% data	0.608	0.649	0.629*	0.444	0.507	0.476
Time series	1	0.617	0.715	0.666**	0.573	0.673	0.623***
	2	0.591	0.742*	0.666	0.496	0.679**	0.587
	3	0.532	0.521	0.526	0.439	0.442	0.441
	4	0.590	0.592	0.591	0.408	0.413	0.410
	5	0.573	0.569	0.571	0.579	0.577	0.578
Mean		0.581	0.628		0.499	0.557*	

OLS Ordinary Least Squares, *SD* System Dynamics.
***: significant differences in rows (t-test) and columns (F-test) at the p<.001 level, **: at the p<.01 level, *: at the p<.1 level.

Each dataset was used for calibrating the Bass Model using Ordinary Least Squares (OLS) in case of regression analysis and a modified Powell search nonlinear optimiser in the System Dynamics system. An R^2 measure was used to compare sales estimates with the corresponding undisturbed time series in the calibration period and over the whole period for controlling predictive validity.

The comparison showed significant differences and in the most cases the System Dynamics model had an improved fit. The traditional OLS method calculates biased estimators and was outperformed by the quadratically convergent Powell's method [1, 5, 8].

5 Conclusion and Outlook

In this paper the empirical comparison of the System Dynamics approach and the traditional regression analysis shows that the System Dynamics model has predominantly a better estimation. Further research should deal with extensions of diffusion model, specific network effects influencing diffusion, the agent-based view, prediction of the important turning points in the S-shaped curve and using more real data series of different products.

References

1. Bass FM (2004) A New Product Growth Model for Consumer Durables. Management Science 50 (12 Supplement): 1825-1832
2. Bass FM (2004) Comments on "A New Product Growth for Model Consumer Durables". Management Science 50 (12 Supplement): 1833-1840
3. Bass FM, Krishnan T, Jain D (1994) Why the Bass model fits without decision variables. Marketing Science 13 (3): 203-223
4. Bähr-Seppelfricke U (1999) Diffusion neuer Produkte – Der Einfluss von Produkteigenschaften. Deutscher Universitäts-Verlag GmbH Wiesbaden
5. Barlas Y (1996) Formal aspects of model validity and validation in System Dynamics. System Dynamics Review 12 (3): 183-210.
6. Forrester JW (1961) Industrial Dynamics. Cambridge MIT Press
7. Goldenberg J, Libai E, Muller E (2004) From density to destiny: Using spatial dimension of sales data for early prediction of new product success. Marketing Science 23 (3): 419-428
8. Lilien G, Kotler P, Moorthy KS (1992) Marketing Models. Prentice-Hall International, INC.
9. Lyneis JM (2000) System Dynamics for market forecasting and structural analysis. System Dynamics Review 16 (1): 3-25
10. Maier FH (1998) New product diffusion models in innovation management – a system dynamics perspective. System Dynamics Review 14 (4): 285-308

11. Maier FH (1995) Die Integration wissens- und modellbasierter Konzepte zur Entscheidungsunterstützung im Innovationsmanagement. Duncker & Humblot Berlin
12. Mahajan V, Peterson RA (1985) Models for Innovation diffusion. SAGE Publications, Beverly Hills London New Delhi
13. Milling P, Maier F (1996) Invention, Innovation und Diffusion, Eine Simulationsanalyse des Managements neuer Produkte. Duncker & Humblot Berlin
14. Parker P (1994) Aggregate diffusion forecasting models in marketing: A critical review. International Journal of Forecasting 10 (2): 353-380
15. Putsis JR W P, Srinivasan V (2000) Estimation Techniques for Macro Diffusion Models. In: Mahajan V, Muller E, Wind Y (eds) New Product Diffusion Models. Kluwer, Boston, pp 263-291
16. Schmalen H, (1989) Das Bass-Modell zur Diffusionsforschung. Zeitschrift für betriebswirtschaftliche Forschung 41 (3): 210-226
17. Sterman JD (2000) Business Dynamics – Systems Thinking and Modeling for a Complex World. Irwin Mc Graw Hill

Part XVII

Managerial Accounting

Portfolio Optimization as a Tool for Knowledge Management

Hennie A.M. Daniels, Martin T. Smits

Center for Research on Information Systems and Management (CRISM). School of Economics and Business Administration. Tilburg University(www.uvt.nl), the Netherlands. Corresponding author: M.T.Smits@uvt.nl.

Introduction

In today's business environment managers are trying to find new approaches to improve their organization's performance. Therefore they need to manage their sources of competitiveness effectively. In knowledge intensive organizations, these sources rely more and more on the intangible parts of the organization; the knowledge and know-how of employees, the relationships of the organization with its stakeholders, its trademarks, patents, etc. Management of these resources –also known as intellectual capital- should enable the organization to sustain viability, success, and basis for innovation (Wiig 1997, Davenport and Prusak, 2000).

How exactly the knowledge resources and knowledge management (KM) processes tie to strategic, tactical, and operational business objectives and work-flow is often left implicit or not addressed at all in business practice (Nahapiet and Goshal, 1998). To specify these relationships, Smits and de Moor (2004) developed the 'knowledge governance framework', linking operational KM to long-term KM to organizational objectives. Other frameworks (Holsapple 2001) focus on specific knowledge linked to one business objective.

Figure 1 presents the knowledge governance framework, showing the knowledge resources, either as 'available resources' (lower left side) or 'in use', i.e., assigned to projects or business processes (lower right side). The central part of Figure 1 shows the links between operational KM, long-term KM, and business strategy. Operational KM performs activities such as assigning knowledge resources to projects, forming project teams, based on customer needs, available knowledge resources and long-term KM objectives (Smits and de Moor, 2004).

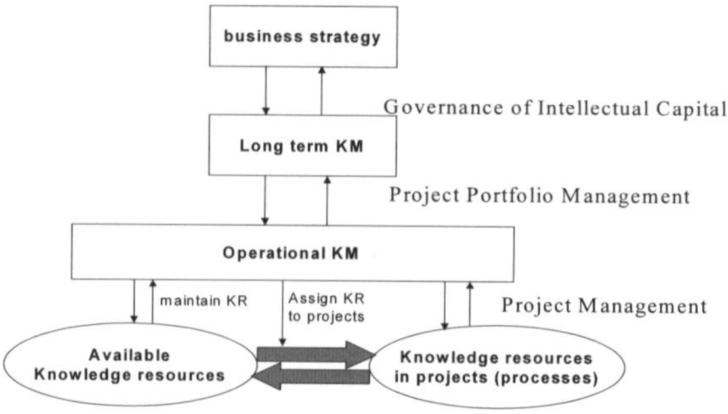

Figure 1 The Knowledge Governance Framework: positioning Intellectual Capital, Knowledge Management (KM), Project portfolio Management, and Project Management (based on Smits and De Moor, 2004)

Knowledge development typically occurs in projects and other operational activities (Blackler 2002). In projects, various experts work together to create a product or service that meets a business requirement. After the project the knowledge workers return to their 'home base' adding knowledge acquired in the project to the knowledge resources. How to effectively manage knowledge in a project environment is an open question. (Millen, 2002; Wenger, 2002).

Management can also focus on the overall strategic objectives and directions set as part of governance of intellectual capital (Figure 1). The knowledge resources or intellectual capital and the targets set by the organization's strategy need to be matched (Roos 1997; Zou 2003). The establishment of this relationship is the purpose of our model, in which the contribution of each individual project (on the operational level) is estimated on the program (or long-term) level, for each of the objectives.

Portfolio management

Projects are usually proposed by technical staff, or by marketing and production staff. The focus of the projects can range from technical problems, product improvement demanded by customers, process improvement for production or the development of new knowledge. However, usually only limited resources are available for innovative projects. The optimal composition of the portfolio is a set of projects, where capacity constraints, strategic objectives and cash flow are carefully balanced. In portfolio optimization there are three main goals:

- Value Maximization: Resources should be allocated to projects in such way that the total value of the project portfolio is maximized.

- Strategic direction: The portfolio should be a derivative of the strategy as formulated by the organization.
- Balance: A good portfolio should be balanced, based on several parameters, which vary among departments. Parameters are e.g.: long-term versus short-term projects, high versus low risk projects, strategic fit versus economic value.

Method for portfolio selection

Our approach of calculating a set of optimal portfolios based on three dimensions: financial revenues, strategic fit, and risk (Daniels and Noordhuis, 2002). Managers should be able to select a portfolio from this set. To find optimal project portfolios, we apply a multi-objective non-linear integer optimization model.

Input variables

- Intellectual Capital Indicators. These indicators (I_1, I_2... I_n) are chosen by managers and considered to be the most relevant for project selection in an organization. If an organization does not have multiple indicators, the total intellectual value of a project is estimated on one indicator, like in our tool.
- Weights. If multiple intellectual capital indicators are used, then we assign a weight to each indicator indicating it's relative importance. Since weights are just mere numeric preferences, a 5-point scale suits perfectly. The final weight of indicator I_i is denoted by α_i.
- Project Contributions. For each project, management estimates the expected contribution of the project to the intellectual capital indicators I_n. In our tool we only use one indicator, so, managers only estimate the (strategic) contribution of project i to the intellectual capital, using a five point Likert scale.
- Capacities. Each project requires knowledge resources in the form of available expertise. We assume that for each area of expertise a number of hours TC_k are available during the time period considered. For each individual project i we need an estimate of the capacity-requirements. This results in a capacity matrix C_{ik}, representing the capacity required for project i of type k. In the tool we do not (yet) distinguish between different areas of expertise, and assume that all available expertise can be used in any project.
- Financial information. For each project the model requires financial data. In our tool managers estimate the profits of each project by simply subtracting the estimated total project costs from the estimated total project revenues.
- Project Risk. We assume that each project has a certain probability p_i to fail. The probability can be estimated using historical data or decomposing the probability in several sub probabilities. In our tool, managers estimate the total risk of each project on a five point Likert scale (1 = low risk, 5 = high risk project), corresponding to probabilities of failure p_i of 0.1, 0.3, 0.5, 0.7, and 0.9.

Constraints

- Capacity constraints. The number of constraints depends on the number of capacities. Capacities can be, in the case of an R&D department, teams of knowledge workers, experts in their field. Every constraint states that the amount of capacity used should be less or equal than the total amount available.
- Cash flow constraints. Cash flow constraints are optional in this model. One might include cashflow constraints in each period, for example per month.
- Risk constraints. A natural risk constraint is that the probability that k or more projects fail is less then a certain threshold. The overall risk score of the project portfolio is computed from the individual risk factors of the projects.

Objective function and model output

The model's decision variables are binary for each project, meaning that a project will be 'part' or 'no part' of a portfolio. The model optimizes the portfolio simultaneously on two objectives: total strategic fit (based on the individual project intellectual indicators) and total risk (based on the individual project risks).

Risk can be easily exchanged by a financial indicator. We did not choose for all three objectives to be part of the optimization objective function. In this way managers can balance their portfolio more easily.

The nature of multi-objective programming is that there is no unique optimal solution but, most likely, a set of possible solutions. This set is called the non-dominated set, or Pareto optimal set. A portfolio is non-dominated if there is no other solution which would improve at least one objective function and not worsen any other. It is useful to express non-dominance in terms of a simple vector comparison. Essentially, a vector $x^* \in C$ is said to be Pareto optimal if all other vectors have a higher value for at least one of the objective functions, or else have the same value for all objectives.

Applying the method in a case

The method was used to develop a Portfolio Selection Tool (PST). The tool was applied in order to evaluate its usability in a project-based, medium sized (400 fte), knowledge intensive company in the insurance industry. This company is well experienced in project management and uses a rich set of criteria to evaluate individual project proposals, and runs a portfolio of about 40 projects.

Table 1. Examples of project data (three projects) used as input in the project selection tool

Project	Resources (fte)	Strategic value	Risk	Revenues (euro * 1000)
P1	15	1	0.5	80
P2	14	5	0.2	100
P3	10	3	0.7	100

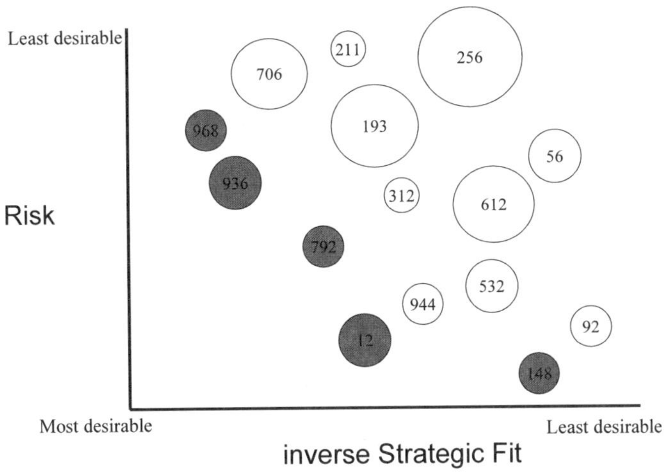

Figure 2. Non-dominated solutions (grey) resulting from the optimization model. Other (dominated) portfolios are represented in white circles. (Each circle has a number for identification). The diameter of a circle indicates total profits.

First organizational data are entered, being the available personnel resources (fte) and the average costs per fte. Project data are entered (Table 1): project name, resources (fte) used by the project, strategic value, project risk, and estimated project financial revenues.

After entering all data for all projects, the tool calculates all feasible portfolios that fit within the available resources. The output of the tool consists of the following parts:

- A (long) table of feasible portfolio's (i.e. the portfolios satisfying the capacity constraints) with total costs, total revenues, total strategic value, and total risk per portfolio. Note that for 15 projects, 2^{15} possible portfolio's are evaluated;
- A table showing for each project how often the tool has selected the project in a feasible portfolio (projects can occur in all portfolio's, or in none, being an indicator for project attractiveness);
- A table showing the minimum total risk, minimum total costs, maximum total revenues, maximum total profits, and maximum total strategic value of the feasible project portfolios;
- A table of non-dominated (Pareto-optimal) solutions, created from the list of feasible portfolios;

The final result of the optimization, the Pareto-optimal set, is plotted as a bubble diagram (Figure 2). The x-axis displays the strategic fit, the y-axis the risk of the portfolio. The size of the bubble indicates the financial value (total profits) of the portfolio. Figure 2 shows five non-dominated solutions to the optimization problem. The most left portfolio has the highest strategic fit of all the portfolios and has a somewhat small total profits.

Portfolio management can be simulated in the PST by adding new projects to the project list, by removing projects from the list, by changing project parameters, by changing the costs per fte, and by changing the available knowledge resources (fte). When the required changes have been implemented, a new output of the PST is created and effects of the changes can be used to determine long-term and short-term knowledge management interventions.

Conclusions and future research

The method and the tool give managers a good overview of how an optimal portfolio could be composed given the projects they gave as input. The tool is a decision support system and enables a solid starting point of reasoning and discussion about the optimal portfolios in a given situation. Many questions are still left open for future research: The tool was applied in a medium sized organization: can it be used in larger companies running multiple portfolios? Which people should be included to fully utilize the potential of the tool?

Acknowledgement: This research was part of the METIS project sponsored by the Telematics Institute in the Netherlands (www.telin.nl).

References

Blackler F. (2002): Knowledge, knowledge work, and Organizations. In: Chun Wei Choo and N. Bontis (2002), Oxford University Press., pp 47-64.

Daniels, H.A.M., Noordhuis, H. (2002), "Management of intellectual capital by optimal portfolio selection" In: Proceedings of the 4th international conference on practical aspects of knowledge management, Vienna.

Davenport, Th.H., Prusak, L. (2000), "Working knowledge: how organizations manage what they know". Harvard Business School Press, Boston, USA.

Millen, D.R., Fontaine, M.A., Muller, M.J. (2000): "Understanding the Benefits and Costs of Communities of Practice". Communications of the ACM, (45) 69-73.

Nahapiet J, Ghoshal S (1998): Social Capital, Intellectual Capital, and the Organizational Advantage. Ac. Of Man. Rev. (23) 2: 242-266.

Nonaka, I, Toyama R & Konno N (2000). SECI, Ba and Leadership: a Unified Model of Dynamic Knowledge Creation. Long Range Planning 33, 5-34.

Roos, J., Roos, G., Dragonetti, N.C., and Edvinsson, L. (1997). "Intellectual Capital: Navigating in the New Business Landscape". Macmillan: London.

Smits, M., De Moor, A. (2004), "Measuring knowledge management effectiveness in communities of practice. In: Proceedings of HICSS, (Ed: Sprague) IEEE, Cal.

Wenger, E., McDermott, R., Snyder, W. (2002): Cultivating Communities of practice, Harvard Business School Press, Boston, USA.

Wiig, K.M. (1997), "Integrating Intellectual Capital and Knowledge Management". Long Range Planning Vol 30 No. 3 pp 399-405

Zou, A.Z., Fink, D. (2003), "The intellectual capital web; A systematic linking of intellectual capital and knowledge management". J. of Intellectual Capital (4) 1:34-48.

Berücksichtigung nicht-finanzieller Aspekte im Rahmen eines Entscheidungsmodells für Zwecke der Unternehmenssteuerung

Dirk Heyne

KPMG Deutsche Treuhand-Gesellschaft AG
Ganghoferstraße 29, 80339 München, e-mail: DHeyne@kpmg.com
Technische Universität Ilmenau, Institut für Wirtschaftsinformatik,
PF 100565, 98684 Ilmenau, e-mail: Dirk.Heyne@tu-ilmenau.de
Germany

Abstract

Es wird eine Methode vorgestellt, die es ermöglicht nicht-finanzielle Zusammenhänge bei der Entwicklung von Entscheidungsmodellen zur unternehmensübergreifenden Koordination zu berücksichtigen. Das untersuchte Problem entsteht durch die Forderung, auch nicht-finanzielle Ziele im Rahmen von quantitativen Zielsystemen zu berücksichtigen. Die vorgeschlagene Methode wird auf den Bereich Absatz exemplarisch angewandt.

1. Einleitung

Der Schwerpunkt dieses Beitrags soll auf Herausforderungen im Zusammenhang mit der Festlegung der Unternehmensziele liegen, wie sie im Rahmen der Planung auftreten. Im Schritt der Zielfestlegung ist es erforderlich, das in der Regel finanzielle Gesamtziel in konkrete operative Ziele umzusetzen. Das Balanced Scorecard (BSC) Konzept propagiert, bei der Operationalisierung insbesondere auch nicht-finanzielle Ziele zu berücksichtigen [9]. Wesentlich ist die Abbildung der Zusammenhänge zwischen den Zielen. Diese werden bislang durch qualitative Ursache-Wirkungsschaubilder visualisiert. Die wesentliche Kritik aus Wissenschaft und Praxis an diesem und ähnlichen Konzepten liegt darin begründet, dass die qualitativen Ursache-Wirkungsbeziehungen es nicht erlauben die Kohärenz der Unternehmensziele hinsichtlich der Gesamtzielsetzung zu gewährleisten. Außerdem stehen die Kausalitäten grundsätzlich in Frage [13].

Das Kausalitätsproblem wurde durch Hillbrand et al. [4] untersucht. Sie schlagen einen generischen Decision-Support-Ansatz vor, um den Zusammenhang, der durch qualitative Ursache Wirkungsbeziehungen dokumentiert wird, zu verifizieren. Dieser Ansatz eignet sich das Teilproblem mangelnder objektivierter Zusammenhänge zu adressieren, hilft jedoch nicht bei der Zieloperationalisierung bzw. Koordination, da der funktionale Zusammenhang unbekannt bleibt.

Die Untersuchung bestehender Arbeiten des Operations Research insbesondere im Rahmen der Unternehmensplanung [14], sowie bestehender Management Konzepte [15], Ansätze zum Performance Management (PM) [9], Controlling Konzeptionen [6][10], Arbeiten zu Entscheidungsunterstützungssystemen [16] liefert das Ergebnis, dass bislang kein integrierter Ansatz vorliegt, der geeignet ist die einleitend formulierte Problemstellung abzudecken.

In diesem Beitrag wird deshalb eine Methode vorgestellt, nicht-finanzielle Faktoren in quantitative Optimierungsmodelle mittels nicht-linearer Approximation der Zusammenhänge zu integrieren[1].

2. Ansatz

Um die angesprochenen „weichen“ Faktoren wie z.B. Kundenzufriedenheit berücksichtigt zu können, ist die Abbildung der ex-ante unbekannten funktionalen Zusammenhänge zwischen den Handlungs- und Zustandsvariablen erforderlich.

Einen flexiblen Ansatz zur Lösung des damit entstehenden Approximationsproblems stellen Neuronale Netze (NN) dar [7]. Die Netzwerkfunktion eines für die vorliegende Approximation verwendeten Multi-Layer-Perceptron (MLP) Netzwerkes mit drei Schichten kann in allgemeiner Form angegeben werden mit [2]:

$$f(X,w)=g_o\left(\sum_{i=0}^{I}\alpha_i x_i+\sum_{h=1}^{H}\beta_h\, g_h\left(\sum_{i=0}^{I}\gamma_{hi}x_i\right)\right), \tag{2.1}$$

wobei die folgenden Bezeichnungen verwendet werden:

g_o (·) Aktivierungsfunktion des Ausgabeneurons,
g_h (·) Aktivierungsfunktion der verdeckten Neuronen,
x exogene Variable (Eingangsgrößen),
α, β, γ Netzwerkgewichte w,
I Anzahl der Eingangsneuronen,
H Anzahl der verdeckten Neuronen.

[1] Auf die grundsätzliche Bedeutung des OR für PM Aufgabenstellungen weisen z.B. auch Dyson [3] und Pidd [12] hin.

Nach erfolgter Modellierung der funktionalen Zusammenhänge kann das resultierende nichtlineare Optimierungsmodell (vgl. Eq. 3.1 i.V.m. Eq. 2.1) mittels Newton-SQP-Verfahren gelöst werden [1].

3. Modellbeispiel

Nachfolgend soll der vorgestellte Ansatz exemplarisch auf ein Beispiel aus dem Absatzbereich angewandt werden.

Für das Absatz-Kriterium soll gelten[2]:

$$C_t^A\left(a_t^A\right)=\left(\left(P_t-C_t^H\right)\cdot Y_t\left(am_t\right)-\left(C_t^A\left(am_t\right)\right)\right)\rightarrow \max, \tag{3.1}$$

wobei die folgenden Bezeichnungen verwendet werden:

P_t durchschnittlicher Preis der Produkte in der Periode,

C_t^A Periodenkosten der absatzfördernden Instrumente,

C_t^H durchschnittliche Stückkosten der Herstellung,

Y_t Periodenabsatz,

am_t Vektor der möglichen absatzfördernden Instrumente,

t Periode.

Das Kriterium zielt auf die Maximierung des Deckungsbeitrages durch optimale Preisfestsetzung bei gleichzeitig kostenoptimalem Einsatz absatzfördernder Instrumente zur Erreichung des damit verbundenen Absatzvolumens ab. Das Augenmerk liegt auf den absatzfördernden Maßnahmen, die sowohl den Absatz Y als auch die Kosten C^A beeinflussen. Hierfür stehen verschiedene Instrumente zur Verfügung, wie sie aus den Marketing-Mix (Teil-) Modellen bekannt sind [5]. Die Instrumente sind (partiell) gegeneinander substituierbar. Diese Interaktionsproblematik wurde für die Frage, wie die verschiedenen Instrumente optimal eingesetzt werden können um eine angestrebte Absatzmenge zu erreichen, in bestehenden Arbeiten so noch nicht adressiert [17]. Gleiches gilt für die Carry-Over-Effekte zwischen den Variablen. In der Tabelle 3.1 sind die Einflussfaktoren zusammengefasst die in dem vorliegenden Papier für die Abschätzung der Absatzmenge berücksichtigt werden sollen.

---

[2] Aus Gründen der Übersichtlichkeit wird hier zunächst der Einprodukt Fall diskutiert. Durch Einführung einer Koordinationsebene könnte auch der Mehrproduktfall berücksichtigt werden [14].

Table 3.1. Absatzmengen relevante Parameter/ Variable der Absatzsituation sowie für das Szenario unterstellte Verzögerung

Parameter und Entscheidungsvariable	Delay [3]
Marktanteil (soweit durch Werbung verursacht) [%]	0
Preis [GE]	0
Absatzmenge [Stück][4]	1
Kundenzufriedenheit [0, 100]	0

Um eine Entscheidung zwischen den verschiedenen Variablen (durch Werbung beeinflusster Marktanteil, Kundenzufriedenheit) treffen zu können, müssen diese bewertet werden. Es wird deshalb ein Kosten-Koeffizient ermittelt der den Budgetverbrauch abbildet.

4. Versuchsdurchführung

Es kommt eine Drei-Ebenen Architektur aus Simulator, NN und Optimierungspaket zum Einsatz. Da keine Realdaten zur Verfügung standen, wurde ein für dieses Problem geeigneter Simulator entwickelt, um die erforderlichen Trainingsdaten zu generieren. Zur Approximation von Y(am) bzw. C^A(am) wird ein dynamisches NN[5] mit vier Eingangsgrößen (Marktanteil, Kundenzufriedenheit, Preis, Absatzmenge der Vorperiode) herangezogen[6]. Im nächsten Schritt ist mit Hilfe der Netzparameter die Formulierung des Optimierungsproblems (Eq. 3.1) möglich.

Für dieses gilt: $A \cdot x \leq b$

Die Koeffizientenmatrix A die die Kostensätze für den Einsatz eines Instruments angibt und das Budget b je Instrument sind gegeben. x betrifft die Entscheidungsvariablen – hier die erforderliche Kundenzufriedenheit bzw. Marktanteil und den geeigneten Preis. Die Herstellungskosten betragen 5 GE. Für die Eingangsgrößen wurden außerdem Ober- und Untergrenzen definiert: Die Kundenzufriedenheit kann zwischen 10% und 60% betragen. Der Marktanteil darf ebenfalls zwischen 10% und 60% schwanken. Der Preis kann zwischen 10 und 15 GE liegen. Die Ergebnisse sind in Tabelle 4.1 dargestellt:

---

[3] Delay = Anzahl der dem Betrachtungszeitpunkt vorangehenden rel. Zeitschritte

[4] Es wird das Ergebnis der vorangegangenen Periode berücksichtigt.

[5] Feedforward-Multilayer-Perceptron-Netz mit drei Schichten und sigmoider Aktivierungsfunktion in der verdeckten Schicht.

[6] Dies sind in der Regel nicht die einzigen Einflussfaktoren auf die Absatzmenge.

Table 4.1. Ergebnisse Szenario

Variable/ Parameter	Wert
Kosten des Faktorverbrauchs zur Sicherstellung der Kundenzufriedenheit/ des Marktanteils in Höhe von je 1%	5.000/ 3.000
Absatz	9.504.500
Absatz Vorperiode	8.200.000
Erforderliche Kundenzufriedenheit	60%
Erforderlicher Marktanteil	60%
Preis	15
Resultierender Deckungsbeitrag	94.565.000

Es zeigt sich, dass die Instrumente in dem maximalen Umfang eingesetzt werden müssen um, unter Berücksichtigung der Absatzmenge der Vorperiode, das maximal erwünschte Absatzziel in Höhe von 10 Mio. Stück knapp erreichen zu können. Im Vergleich zu der Kundenzufriedenheit (58,6%), dem Marktanteil (58,6%) und dem gewählten Preis (12,9 GE) der Vorperiode wird deutlich, dass der verstärkte Einsatz der drei Instrumente hier zu einem Absatzwachstum von 1,3 Mio. Stück führt. In einem weiteren Experiment wurde die maximale Absatzmenge auf 9 Mio. Stück beschränkt. Hier zeigte sich, dass es hier nicht notwendig ist, den Rahmen der sich bei der Kundenzufriedenheit und dem Marktanteil bietet, auszuschöpfen. Die beiden Instrumente lagen in einem Bereich um 10% bzw. 30%. Das kann auf die verschiedenen Faktorkosten zurückgeführt werden.

5. Zusammenfassung und Ausblick

Mit dieser Methode wurde die funktionale Verknüpfung zwischen verschiedenen Zielen (hier: Absatzmenge, Kundenzufriedenheit, Marktanteil) hergestellt und in ein umfassenderes Optimierungsmodell eingebunden. Dies ermöglichte die Entscheidung über den Umfang der Einbindung der verschiedenen möglichen Instrumente.

Speziell bei der hier gewählten Vorgehensweise, die zu berücksichtigenden Einflussfaktoren auf Basis vorhandener Teilmodelle abzuleiten, wird zwar die ökonomische Kausalität sichergestellt, es ist jedoch notwendig für den Realfall zu verifizieren, ob ggf. weitere Erklärungsvariablen einen Erklärungsbeitrag leisten können.

Weitere Forschungsaktivitäten beschäftigen sich mit der Entwicklung und Verknüpfung einzelner Modelle in einen umfassenderen Unternehmenskontext.

Literatur

[1] Alt W (2002) Nichtlineare Optimierung, Vieweg
[2] Anders U (1997) Statistische neuronale Netze, Vahlen
[3] Dyson RG (2000) Strategy, performance and operational research, Journal of the Operational Research Society, pp 5-11
[4] Hillbrand C, Karagiannis D (2002) Using Artificial Neural Networks to prove hypothetic cause-and-effect relations: A Metamodel-based approach to support decisions, in Piattini, M., Filipe, J. and Braz, J. (Ed.): Proceedings of ICEIS 2002 – the fourth Conference on Enterprise Information Systems, vol.1, pp 367-373, Spain
[5] Homburg Ch (2000) Quantitative Betriebswirtschaftslehre, 3.Auflage, Gabler
[6] Homburg C (2001) Hierarchische Controllingkonzeption, Physica
[7] Hornik K, Stinchcombe M, White, H (1990) Universal Approximation of an Unknown Mapping and Its Derivatives Using Multilayer Feedforward Networks, Neural Networks, 3, pp 551-560
[8] Janetzke P, Falk J (2005) Der Beitrag der künstlichen Intelligenz zur betrieblichen Prognose, in: Mertens, P. et al.: Prognoserechnung, 6.Auflage, Physica
[9] Kaplan RS, Norton DP (1996) Balanced Scorecard: Turning Strategy into Action, Harvard Business School Press
[10] Küpper H-U (1997) Controlling, 2.Auflage, Schäffer-Poeschel
[11] Melody JW (1999) On universal Approximation using Neural Networks, ECE 480 Project
[12] Pidd M (2004) Contemporary OR/MS in strategy development and policy making: some reflections, Journal of the Operational Research Society 55, pp 791-800
[13] Ruhtz V (2001) Balanced Scorecard im Praxistest: wie zufrieden sind die Anwender? Frankfurt am Main
[14] Schneeweiss C (2003) Distributed Decision Making, 2.Auflage, Springer
[15] Steinmann H, Schreyögg G (2000) Management: Grundlagen der Unternehmensführung: Konzepte-Funktionen-Fallstudien, 5.Auflage, Gabler
[16] Turban E, Aronson J, Liang T-P (2005) Decision Support Systems, 7.ed., Prentice Hall
[17] Wiedmann K-P, Buckler F (2001) Neuronale Netze im Marketing Management, Gabler

Wirtschaftliche Folgen von Verträgen – eine Simulationsstudie

Markus Spiekermann

Universität Paderborn – Schwerpunkt Wirtschaftsinformatik 1 – Betriebswirtschaftliche Informationssysteme (BIS)

In Verträgen werden die Rahmenbedingungen für die zukünftige Informations-, Güter- und Geldlogistik von Unternehmen fixiert. Neben den Preisen und Mengen sind insbesondere zeitliche Aspekte wie Bindungsdauern, Kündigungsfristen sowie Dispositions- und Zahlungstermine relevant. Die Unternehmen unterscheiden sich in Ihrer Vermögens- und Aufwandsstruktur. Diese Unterschiede lassen sich zur Typenbildung nutzen, beispielsweise um personalintensive von anlagenintensiven Unternehmen abzugrenzen. Verträge lassen sich Bilanz- und GuV Positionen zuordnen. In einer Studie am Beispiel von typischen Zuliefererunternehmen der Automobilbranche soll gezeigt werden, wie sich die Vertragsgestaltung auf die Kennzahlen eines Unternehmens auswirkt. Dazu werden die Zahlungsströme von Unternehmen in verschiedenen Konjunkturlagen und mit abweichenden Verträgen simuliert. Die Ergebnisse sollen realen Unternehmen als Orientierungshilfe für die Vertragsgestaltung dienen. (*Verträge, Zahlungen, Zahlungsstatus, Simulation*)

Einführung

Unternehmen interagieren mit Ihrer Umwelt auf der Basis von Verträgen. Mit der Entscheidung für oder wider einen Vertrag werden die Rahmenbedingungen für die zukünftige Informations-, Güter- und Geldlogistik festgelegt. Verträge fungieren dabei als Brücke zwischen den Interessen der Beteiligten. Zahlungen stellen die Gegenleistungen für die Lieferung von Produktionsfaktoren und Produkten, den Aufbau von Kapazitäten, den Verzicht auf Freizeit etc. dar. Einzahlungen resultieren aus Verträgen mit Kunden, Auszahlungen werden durch die Verträge mit Faktoren liefernden Stakeholdern bestimmt. Daher sollten sie als Basis für Informationssysteme genutzt werden, um den Verantwortlichen im Unternehmen die Wirkungen von Entscheidungen frühzeitig zu verdeutlichen.

Analysen von Verträgen in der Literatur

Verträge haben in den letzten Jahren verstärktes Interesse in den Theorien der Wirtschaftswissenschaften gefunden.

(Wielenberg 1999) und andere untersuchten, unter welchen ökonomischen Bedingungen Verträge zustande kommen. Ein Beispiel sind „Pfänder" die eine Vertragspartei bei der anderen hinterlegt, bevor eine Seite spezifische Investitionen tätigt. Bei Verträgen und den vorherigen Verhandlungen gibt es oft einen „Trade off" zwischen den Beteiligten, dessen Details stark von der Stellung der Vertragspartner abhängen. Die Machtverhältnisse und Informationsasymmetrien zwischen den Beteiligten sollen nicht Gegenstand dieser Untersuchung sein. Eine Einführung in die ökonomische Vertragstheorie gibt (Richter 2000).

Im englischsprachigen Raum existieren Untersuchungen zu einzelnen Vertragstypen. (Tsay und Lovejoy 1999) untersuchten wie sich Lagerbestände entlang einer Supply Chain entwickeln. Dabei bestehen zwischen den modellierten Unternehmen langfristige Verträge, die flexible Mengenanpassungen innerhalb gewisser Bandbreiten („Quantity Flexibility Contracts" / „Rolling Horizon Contracts") erlauben. Die Abnehmer geben den Lieferanten Dispositionen („Forecast"), über zukünftige Mengen an. Die Autoren fokussieren in der Untersuchung insbesondere auf die Entwicklung von Lagerbeständen bei den Unternehmen, die durch Zeitverzögerungen im Informationsfluss („Time Lags") und Unterschiede in der Flexibilität („Flexibility Constraints") der Verträge entstehen („Bullwhip-Effects"). Andere Typen von Verträgen untersuchten u. a. (Corbett C. J. / Zhou, D. / Tang, C. S., 2001) und (Barnes-Schuster, D. / Bassok, Y. / Anupindi, R., 2000). Die genannten Autoren führten ihre Untersuchungen mittels analytischer Methoden durch. Für einfache Problemtypen konnten so allgemeine Aussagen abgeleitet werden. Der Untersuchungsrahmen war allerdings oft auf einen Vertrag eines bestimmten Typs zwischen zwei Akteuren und zwei Perioden beschränkt. Im Bereich des Rechnungswesens prägte Riebel in seinen Arbeiten den entscheidungsorientierten Kosten- und Erlösbegriff. Er war es, der zeitliche Dimensionen von Entgelten und disponiblen Zahlungen berücksichtige, und eine darauf ausgerichtete Unternehmensrechnung erdachte (Riebel 1994).

Analyse von Verträgen

Aus betriebswirtschaftlicher Sicht können die Merkmale eines Vertrags in drei Gruppen eingeteilt werden:

- Die Bindungsmerkmale beschreiben, für welchen Zeitraum eine Vereinbarung zwischen den Vertragspartnern besteht und wie diese zu reversieren sind. Die Entscheidung über die Bindung durch einen Vertrag wird durch Manager getroffen und als Reversion bezeichnet. Von der Bindung hängen Leistungen und Zahlungen ab.

- Leistungsmerkmale definieren die Produktions- und Liefermengen für einen Zeitraum. Sie enthalten neben Zeiten auch quantitative und qualitative Angaben. Konkrete Spezifikationen des Abnehmers für einen Zeitraum, z. B. die Liefermenge für einen Monat, werden als Leistungsdisposition bezeichnet. Für Leistungsdispositionen existieren bereits diverse Lösungsansätze im Rahmen der Produktionsplanung. Für die Simulationsstudie sind insbesondere die zeitlichen Merkmale relevant, von denen sich Zahlungen ableiten lassen. Die vertraglich vereinbarten Leistungen werden gemäß den traditionellen Sichten der Betriebswirtschaftslehre in Produkte und Faktoren unterteilt. Die Faktoren wiederum lassen sich in Potential- und Repetierfaktoren unterteilen. Bei Potentialen, die zum Eigentum des Unternehmens gehören, ist es z. B. wichtig, ob die Totalkapazität definiert oder ob die Nutzungsdauer variabel ist. Bei einigen (Fremdleistungs-) Potentialen wie z. B. Arbeitsverträgen ist die Kapazität pro Periode beschränkt. Bei Repetierfaktoren kann z. B. bei langfristigen Verträgen eine Mindestabnahmemenge vereinbart sein.
- Zahlungsmerkmale beschreiben die Termine und Höhe der finanziellen Gegenleistungen und wie diese zu disponieren sind. Eine Zahlung kann durch eine Bindungsentscheidung, eine Leistung oder durch eine andere Zahlung begründet sein. Zahlungen können daher ihrem Grund nach gekennzeichnet werden und zusätzlich danach, ob der Zahlungstermin oder -weg beeinflusst werden kann. Im Rechnungswesen wird auf der Basis von Belegen gebucht. Dadurch werden Zahlungen nur im Nachhinein erfasst. Dieses Vorgehen soll durch das Paradigma „Keine Buchung ohne Vertrag" abgelöst werden, denn Verträge enthalten Informationen über zukünftige Zahlungen. Die Status von Zahlungen setzen sich aus dem Status der zugrunde liegenden Bindung und ihrer Beeinflussbarkeit durch Disposition von Leistungen und ggf. Zahlungsterminen zusammen. Eine „irreversibel vordisponierte" Zahlung ist eine Zahlung, bei der die Bindung nicht mehr aufgehoben werden kann und die auch in der Höhe feststeht. Beispiele dafür sind Zahlungen für eine vertraglich vereinbarte Grundgebühr oder eine Mindestabnahmemenge. Kann der Abnehmer der Leistung am Betrachtungszeitpunkt die Höhe der Zahlung noch durch Dispositionen beeinflussen, so ist diese Zahlung „irreversibel disponibel". Der Betrachter kann somit die Einflussmöglichkeiten auf zukünftige Zahlungen beurteilen. Fristablauf und Entscheidungen verändern die Zahlungsstatus im Zeitablauf bis sie den Status „erfolgt" erreichen und damit den Werten des herkömmlichen, belegbasierten Rechnungswesens entsprechen.

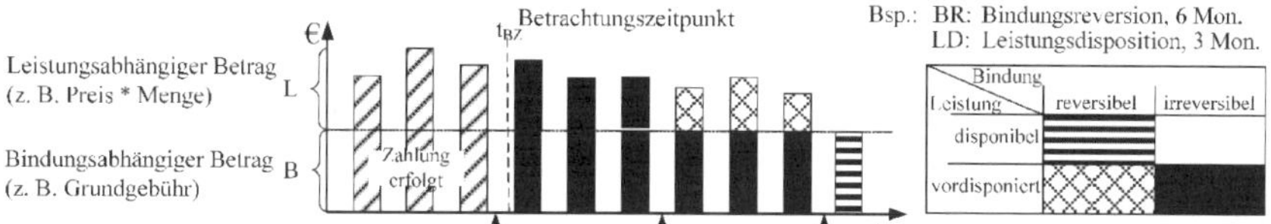

Fig. 1. Zahlungsstatus im Zeitablauf

Simulationsstudie

Das Ziel dieser Untersuchung ist es, auf Basis von numerischen Experimenten die Günstigkeit von Vertragsstrukturen für typische Unternehmen zu untersuchen. Dazu wurde ein deterministisches, diskretes Simulationsmodell erstellt.

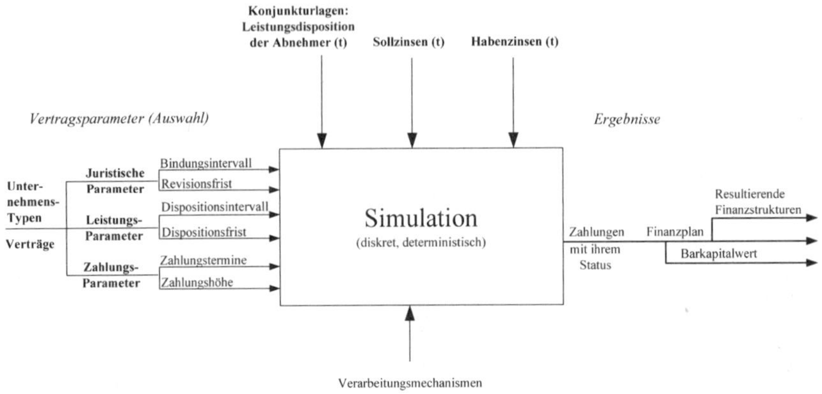

Fig. 2. Aufbau der Simulationsstudie

Verträge werden über Leistungen abgeschlossen und lassen sich daher analog zu einer Bilanz oder Gewinn- und Verlustrechnung gliedern. Ein Kaufvertrag über Potentialfaktoren wird demnach dem Anlagevermögen zugeordnet, während Arbeitsverträge den Personalaufwand in der Gewinn- und Verlustrechnung berühren. Die Bewertung von Vertragsgegenständen in Jahresabschlüssen soll nicht näher betrachtet werden, jedoch soll deren Gliederungssystematik übernommen werden. Basis sind die Jahresabschlussdaten von vier Automobilzulieferertypen (IKB 2002). Danach lassen sich z. B. personalintensive von anlagenintensiven Unternehmen unterscheiden. Die Zahlungsparameter für die Personalverträge, Potentialfaktorverträge, etc. wurden angepasst, damit diese bei Beginn der Simulation den Verhältnissen in den Jahresabschlüssen der Unternehmenstypen entsprechen.

Um die Realitätsnähe zu gewährleisten wurde ein einfaches Produktionsverfahren (Kunststoffspritzguss von Schaltknäufen) eines realen Automobilzulieferers gewählt. Es ist nicht das Ziel, mit Produktionsplanungssystemen zu konkurrieren, sondern Aussagen über die resultierenden Zahlungs- und Jahresabschlussstrukturen zu erhalten. Daher sind die Dispositionen über den Faktoreinsatz und andere Verarbeitungsmechanismen nicht immer optimal, jedoch nachvollziehbar.

Die Vertragsparameter definieren, in welchen Intervallen die Kunden und das eigene Management über die Bindung, Leistung und Zahlung entscheiden können. Zudem wird festgelegt, nach welcher Zahlungsfunktion sich die Preise errechnen lassen. Die Vertragsparameter werden systematisch variiert indem z. B. Reversions- und Dispositionsintervalle in ihrer Länge und Lage zueinander verändert werden.

Die variierten Vertragsparameter gehen mit exogen vorgegebenen Konjunkturlagen in Form von monatlichen Abnahmemengen und Zinssätzen für die Geldanlage und –aufnahme in das Simulationssystem ein. Dieses kann nur entsprechend der eingestellten Vertragsparameter reagieren.

Die resultierenden Zahlungsströme werden in einem vollständigen Finanzplan aufgezeichnet und ausgewertet. Um den Vergleich zwischen den Simulationsläufen zu erleichtern, werden Kennzahlen (z. B. Kapitalwert) verglichen.

Ausblick

Aussagen, wie ein Unternehmen in bestimmten Konjunkturlagen seine Verträge gestalten sollte, sind das Ziel dieser Studie. Fragestellungen wie: „Ist es günstig niedrige aber langfristig irreversible Zahlungen für seine Produkte zu verlangen oder ist es besser kurzfristig abgesicherte, hohe Zahlungen zu vereinbaren?" sollen beantwortet werden. Auch diese Ergebnisse müssen im jeweiligen Kontext betrachtet werden und oft wird das Übertragen auf die Situation eines realen Unternehmens nicht vollständig gelingen. Verträge haben ein großes Potential für das Management, wenn geeignete Informationsinstrumente bereitgestellt werden. Die wirtschaftlichen Folgen der oft unbeachteten juristischen Merkmale sind dazu in diesen Systemen abzubilden. Mögliche Weiterentwicklungen wären z. B. Systeme, welche die Entscheidungen über Verträge optimieren z. B. auf Basis von Realoptionen. Dazu sind jedoch noch umfangreiche Forschungsarbeiten z. B. über die notwendigen Regelwerke für das Finden von Entscheidungen erforderlich.

Quellen

Allen, J.F.: Maintaining Knowledge about Temporal Intervals. In Communications of the ACM 26 (1983), Nr. 11, S. 832-843.

Barnes-Schuster, D. / Bassok, Y. / Anupindi, R.: Coordination and Flexibility in Supply Contracts, 2000

Bassok, Y. / Anupindi, R.: Analysis of Supply Contracts with Commitments and Flexibility, Northwestern University, 1998

Brandt, C.: Wirtschaftliche Vertragsfolgen im IT-Umfeld eines Automobilzulieferers - Eine Simulationsstudie auf Basis einer Vertragsdatenbank, Diplomarbeit, Paderborn, 2005

Corbett C.J. / Zhou, D. / Tang, C.S.: Designing Supply Contracts: Contract Type and Information Asymetry, University of California, Los Angeles, 2001

Dresing, H.: Einsatz von zeitorientierten Datenbanken für Verträge im Rechnungswesen, Diss. Univ. Paderborn 1998

Ester, B. / Baumgart, G.: Cash-Flow Aspekte bei der Supply Chain-Gestaltung, in: Pfohl H.C.: Supply Chain Management: Logistik plus? Logistikkette – Marketingkette – Finanzkette, Berlin 2000, S.141 – 159

Fischer, J.: Einsatzmöglichkeiten zeitorientierter Vertragsdatenbanken im Controlling, in: Wirtschaftsinformatik (1997), S. 55 - 63

Fischer, J.: Aktive Datenbankmanagementsysteme, in: Wirtschaftsinformatik (1996), S. 435 - 438

Fischer, J. / Hoos, J.: Vertragsbasierte, datenbankgestützte Buchhaltung in Lieferketten, in: 46th Scientific Colloquium Ilmenau Technical University, 2001

Fischer, J. / Hoos, J.: Vertragscontrolling als Schlüsselelement des Supply Chain Managements, in: Dangelmaier, W. / Emmrich, A. / Kaschula, D. (Hrsg): Modelle im E-Business, Paderborn 2002, S. 255 - 272

Fischer, J. / Spiekermann, M / Wüst, A.: Vertragsmanagement in Lieferketten, 9. Magdeburger Logistik Tagung, Universität Magdeburg, 2003

Fischer, J. / Walter, A. / Dresing, H.: Datenbankgestützte, vertragsbasierte Buchhaltung, in: König, W. (Hrsg.): Wirtschaftsinformatik ′95′, Heidelberg, 1995, S. 429 – 441

Deutsche Industriekreditbank, Automobilzulieferer – Bericht zur Branche, Düsseldorf, 2002

Knolmayer, G. / Bötzel, S. / Disterer, G.: Zeitbezogene Daten in betrieblichen Informationssystemen, in: Rückle, D. (Hrsg.): Aktuelle Fragen der Finanzwirtschaft und der Unternehmensbesteuerung, Wien, 1991, S. 287-319

Meyer, A.: Finanzinnovationen: Optimale Verträge im Rahmen von Ein- und Mehr-Agenten-Ansätzen der Prinzipal-Agent-Theorie. In: Mathematical systems in economics. Frankfurt am Main, 129, 1992.

Richter, R.: Verträge aus wirtschaftstheoretischer Sicht. In: Franz, W. / Hesse, H. / Ramser, H.J. / Stadler, M. (Hrsg.): Ökonomische Analyse von Verträgen. Tübingen, 29, 2000; S. 1-24.

Riebel, P.: Einzelkosten- und Deckungsbeitragsrechnung, 7. Auflage, Wiesbaden, 1994

Sinzig, W.: Datenbankorientiertes Rechnungswesen, 3. Auflage, Berlin, Heidelberg, New York, Tokio, 1990

Schellhas, C.: Bilanzielle und finanzielle Wirkungen von Vertragsfolgen eines Automobilzulieferers - eine Simulationsstudie auf Basis einer Vertragsdatenbank, Diplomarbeit, Paderborn, 2005

Tsay, A.A. / Lovejoy, W.S.: Quantity Flexibility Contracts and Supply Chain Performance in: Manufacturing & Service Operations Management Vol. 1, No. 2, 1999, S. 89-111

Wielenberg, S.: Investitionen in Outsourcing – Beziehungen. Diss.; Wiesbaden, 1999

Wüst, A: Kapazität- und finanzorientiertes Vertragsmanagement in Supply Chains bei Automobilzulieferern – ein prototypischer, datenbankgestützter Ansatz, Diplomarbeit, Paderborn, 2002

Part XVIII

Tourism, Entertainment and Sports

Identifying Segments of a Domestic Tourism Market by Means of Data Mining

Gül Gökay Emel and Çağatan Taşkın

Department of Business Administration, Uludağ University, Görükle Campus, 16059 Bursa, Turkey. ggokay@uludag.edu.tr, ctaskin@uludag.edu.tr

Abstract. This paper helps to analyse a typical problem seen in the marketing systems of firms in tourism industry. The problem here is the difficulty in determining the market segments for an optimal customer management. In this work, data mining is used as a decision support tool in order to extract previously unknown patterns and ultimately comprehensible information from large databases which traditional statistical tools can not extract. The research is conducted in Bursa, the fourth biggest city of Turkey. The multi-dimensional analysis of this domestic market is very important for foreign hotel investors, tour operators and travel agencies in their investment, marketing and management strategies. For this multi-dimensional analysis, visual and robust data mining software Clementine 8.1 is used for the classification task of data mining in order to determine the market segments for optimal customer management.

1 Introduction

The tourism has become a key industry for many countries all over the world. There is an intense competition in the tourism market where consumers' needs and wants are very diverse. Thus, consumers' heterogeneous attitudes, behaviours and demands should be homogenized in order to gain competitive advantage. Accurate segments can be identified with the help of knowledge extraction from huge amounts of consumer data. One of the disciplines which is used for knowledge extraction is data mining. Data mining can be defined as knowledge discovery from databases as the process of nontrivial extraction of implicit, previously unknown and potentially useful knowledge from large databases for creating effective strategies (Tsay and Chiang 2005; Bose and Mahapatra 2001). This technology is motivated by the need of new techniques to help analyse, understand or even visualize the huge amounts of stored data gathered from business and scientific applications (Liao and Chen 2004; Akat et al. 2005).

This paper is concerned with the classification task of data mining in order to determine the market segments in a domestic tourism market. Data mining is used as a decision support tool for optimal customer management. C5.0 which is a classification algorithm is implemented to identify market segments according to accommodation time, monthly income and age; all of which are important attributes in tourism marketing. The aim of the paper is to identify different segments of domestic tourists which are of high interest in terms of mentioned attributes. With this information, some of the "valued customers" can be targeted for special treatment according to the strategic marketing plans and niche markets can be found.

2 Classification Task and Market Segmentation

Classification can be defined as finding a classifier that results from training datasets with predetermined targets, fine-tuning it with test datasets, and using it to classify other datasets of interest (Chen et al. 2005; Kelly et al. 1999). From a tourism marketing perspective, classification can be used to arrange customers/tourists into pre-defined segments that allow the size and structure of market groups to be monitored. Classification uses the information contained in sets of predictor variables, such as demographic and lifestyle data, to assign customers to segments (Magnini et al. 2003). There exists various ways of constructing classifiers in the form of, for example, rules, decision trees, Bayesian networks, support vectors machine, etc. Decision trees classifiers, such as Quinlan's C4.5/5.0 classifier and its extensions, have received considerable attention due to its speed and clarity (Chen et al. 2005).

There are studies which mention the importance of market segmentation for the tourism industry (Davies 2003; Dibb and Simkin 2001). There is also a research including the overview of empirical international market segmentation studies and segmentation methods used (Steenkamp and Hofstede 2002). Most of the researches mentioned above include the use of traditional statistical methods. Recently, data mining methods have become very popular in segmentation. These studies include the use of C5.0 algorithm (Cardaso and Moutinho 2003), artificial neural networks (Kim et al. 2003; Bloom 2005), self-organizing feature map and K-means algorithm (Kuo et al. 2002) for segmentation in tourism markets. There is also a research including the comparison of linear and non-linear techniques used for segmentation for a tourism market (Bloom 2004).

3 Research

3.1 Population and Sampling Size

The raw data are collected by a questionnaire. The population of the study is the customers of travel agencies in Bursa. Random sampling is chosen as the sampling method. 204 questionnaire forms are filled out of 1000 distributed. In addition to 17 different questions related to the domestic customers' profiles, there are 32 sentences concerning the degree of importance about the accommodation type selection behaviour of domestic customers on a five-point Likert scale. In Likert questions, answers 1, 2, 3, 4 and 5 refer to "not important at all", "not important", "important", "very important" and "extremely important" respectively.

3.2 Steps of the Analysis

Clementine 8.1 which is used here for the analysis, works with data streams. Every operation on this stream is exploited by the nodes that are connected to each other with arrows. A basic stream contains a source node, a field/record node and an output node. The model of the research can be seen in Figure 1.

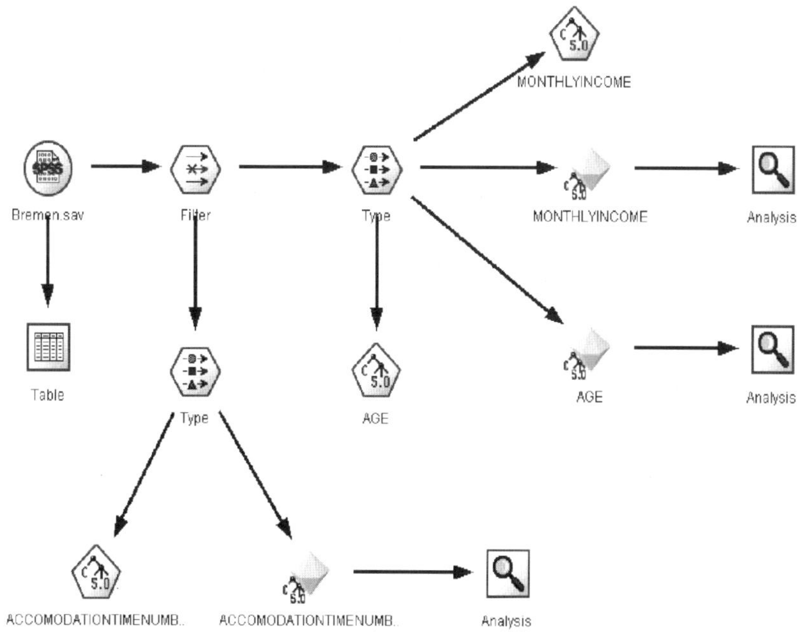

Fig. 1. The Model

As it is seen from Figure 1, the process of the model starts from the database node and goes on with the filter node where unnecessary fields are removed. Then, a type node is connected to the data stream in order to specify the field types. After specifying the field types, C5.0 modelling node is added to the type node. The target field, input fields and the output type, which are chosen as rule set, are specified in C5.0 node. The target field of the first C5.0 node is specified as accommodation time. Then C5.0 node is executed. Extracted classification rules are shown in Table 1 and Table 2.

Table 1: Classification Rule 1 For Accommodation Time

IF Accommodation Type = Holiday Village
AND Preference of Bed Type = Twin Beds
AND L_Price of Extra Services = "important"
THEN Accommodation Time = 6 and more nights

Table 2: Classification Rule 2 For Accommodation Time

IF Accommodation Type = Hotel
AND L_Comfort of Bed & Pillows = "important"
AND L_Quality&Diversity of Food = "extremely important"
THEN Accommodation Time = 6 and more nights

According to the extracted classification rules in Table 1 and Table 2, an important segment is defined which includes domestic tourists who prefer long stays (6 and more nights). "Long-stay" segment has two sub-segments. One of these sub-segments has a profile of tourists who prefer holiday village and twin beds. The price of extra services is also important for this sub-segment. The second sub-segment is composed of tourists who prefer hotels. For the second sub-segment, comfort of beds is important and quality and diversity of food is extremely important.

Table 3: Classification Rules For Monthly Income

Rules : 3000 NTL <Monthly Income < 5000 NTL
IF Age = 40-49
AND L_Room Size = "very important"
AND L_Sea Activities = "very important"
THEN 3000 NTL<Monthly Income <5000 NTL
IF Age = 40-49
AND L_Security = "important"
AND L_Sea Activities = "extremely important"
THEN 3000 NTL<Monthly Income <5000 NTL
IF Purpose of Accommodation = Business
AND L_Security = "important"
AND L_Room Cleanness = "extremely important"
THEN 3000 NTL<Monthly Income <5000 NTL

Classification rules in which the target variable is monthly income, is shown in Table 3 above. The most important segment is the domestic tourists who have high income. Classification rules that belong to domestic tourists whose monthly income is between 3000 New Turkish Liras (1 Euro = 1,62 NTL) and 5000 New Turkish Liras, includes two main segments. First segment is composed of tourists, aged between 40 and 49 years and who like sea activities very much. They also give importance to room size and security. The second segment consists of domestic tourists whose purpose of accommodation is business. They also give importance to security and room cleanness. In Table 4, classification rules where the target variable is age, can be seen. In this table, there is a segment consisting of domestic tourists aged between 40 and 49 years, married, whose income is between 3000 NTL and 5000 NTL and who like mountain activities very much.

Table 4: Classification Rules For Age

Rules : Age = 40-49
IF Marital Status = Married
AND 3000 NTL<Monthly Income<5000 NTL
AND L_Mountain Activities = "very important"
THEN Age = 40-49
IF 3000 NTL<Monthly Income<5000 NTL
AND L_Mountain Activities = "extremely important"
THEN Age = 40-49

Rule accuracy is the most important measure of rule quality and it is also called precision in information retrieval (Flach and Lavrac 2003). As seen from the model in Figure 1, the analysis nodes are connected to each C5.0 output node in order to calculate the analysis accuracy and standard error. The accuracy results are given in Table 5 below. Rule accuracy results show that the fraction of predicted positives that are true positives is %70,588, %75,490 and %71,569 for accommodation time, monthly income and age target fields, respectively.

Table 5: Analysis Accuracy & Standard Error

Target Variable	Analysis Accuracy (%)	Standard Error (%)
Accomodation Time	70,588	3,2
Monthly Income	75,490	3,3
Age	71,569	3

4 Conclusions

Data mining is a powerful tool for multi-dimensional data analysis. It has the ability of extracting previously unknown patterns that traditional statistical tools can not. In optimal customer management, identifying accurate market segments is up to well extracted knowledge. Thus, valued segments can be targeted and also niche segments can be found with the use of this knowledge.

In this study, C5.0 classification algorithm is used for prediction in identifying segments of a domestic tourism market. Classification rules are extracted according to the pre-defined attributes such as accommodation time, monthly income and age, in order to recognize the behavioural patterns of domestic tourists. C5.0 classification algorithm is preferred for the induction of propositional rules as it yields results that are easy to understand. The main findings of this study can be a useful source of information for hotel investors, tour operators and travel agencies in their investment, management and marketing strategies.

References

Akat Ö, Emel GG, Taşkın Ç (2005) The Use of Association Rule Mining for Hotel Customers Profiling: An Application in Bursa. 1st International Conference on Business Management & Economics in a Changing World, 16-19 June, Çeşme, Turkey.

Bloom JZ (2004) Tourist Market Segmentation with Linear and Non-Linear Techniques. Tourism Management **25**: 723-733.

Bloom JZ (2005) Market Segmentation: A Neural Network Application. Annals of Tourism Research **32**: 93-111.

Bose I, Mahapatra RK (2001) Business Data: A Machine Learning Perspective. Information & Management **39**: 211-225.

Cardaso MGMS, Moutinho L (2003) A Logical Type Discriminant Model for Profiling a Segment Structure. Journal of Targeting, Measurement and Analysis for Marketing **12**: 27-41.

Chen G, Liu H, Yu L, Wei Q, Zhang X (2005) A New Approach to Classification Based on Association Rule Mining. Decision Support Systems, Article In Press.

Davies B (2003) The Role of Quantitative and Qualitative Research in Industrial Studies of Tourism. International Journal of Tourism Research **5**: 97-111.

Dibb S, Simkin L (2001) Market Segmentation: Diagnosing and Treating the Barriers. Industrial Marketing Management **30**: 609-625.

Flach P, Lavrac N (2003) Rule Induction. In: Berthold M, Hand DJ (eds) Intelligent Data Analysis. Springer-Verlag, Berlin Heidelberg, pp. 229-267.

Kelly MG, Hand DJ, Adams NM (1999) Supervised Classification Problems: How to Be Both Judge and Jury. In: Hand DJ, Kok JN, Berthold MR (eds) IDA'99. Springer-Verlag, Berlin Heidelberg, pp. 235-244.

Kim J, Wei S, Ruys H (2003) Segmenting the Market of West Australian Senior Tourists Using an Artificial Neural Network. Tourism Management **24**: 25-34.

Kuo RJ, Ho LM, Hu CM (2002) Integration of Self-Organizing Feature Map and K-means Algorithm for Market Segmentation. Computers&Operations Research **29**: 1475-1493.

Liao SH, Chen YJ (2004) Mining Customer Knowledge For Electronic Catalog Marketing. Expert Systems with Applications **27**: 521-532.

Magnini VP, Honeycutt JR ED, Hodge SK (2003) Data Mining for Hotel Firms: Use and Limitations. Cornell Hotel and Restaurant Administration Quarterly **44**: 94-105.

Steenkamp JBEM, Hofstede FT (2002) International Market Segmentation: Issues and Perspectives. International Journal of Research in Marketing **19**: 185-213.

Tsay YJ, Chiang JY (2005) CBAR: An Efficient Method for Mining Association Rules. Knowledge-Based Systems **18**: 99–105.

Part XIX

Scheduling and Project Management

Scheduling Tests in Automotive R&D Projects

Jan-Hendrik Bartels, Jürgen Zimmermann

Clausthal University of Technology, Institute of Management and Economics, Julius-Albert-Str. 2, D-38678 Clausthal, Germany, e-mail: bartels@tibasc.de

Summary. In order to reduce testing costs in automotive R&D projects, we introduce an approach to schedule necessary tests such that the number of used experimental vehicles is minimized. Based on a multi-mode resource-constrained project scheduling model with cumulative resources, we propose a MIP formulation, which is solvable for small problem instances. A priority-rule based method serves to solve large problem instances. Finally, we present preliminary computational results.

1 Motivation

During the last decades the automotive industry has been confronted with a shortening product life cycle. Consequently, the time-to-market in the automotive industry has been reduced for being able to develop cars corresponding to the current needs of target customers. Moreover, a decreasing number of cars produced throughout the life-span of a model cycle has led to an increase in the portion of indirect costs. Due to the interaction with a decreasing time-to-market, a necessary reduction of development costs is challenging (cf. [3]).

The automotive product development process generally consists of two alternating stages. At first, new components are constructed by computer aided engineering. Afterwards, these components are tested with the help of experimental vehicles that have to be built by the prototype section. Tests are necessary to ensure the level of quality customers expect and to verify certain product attributes that are prescribed by law. As the construction of one experimental vehicle costs up to one million Euros, it is evident that the prototype section causes the majority of testing costs. Thus, the objective criterion in scheduling the tests is to reduce the demand for vehicles.

2 Problem Description and Mathematical Formulation

The problem of scheduling tests can be represented by a multi-mode resource-constrained project scheduling problem. Each test stage is considered as a single project containing n individual tests, which must be scheduled such that an objective criterion is optimized and the following constraints are met.

Temporal constraints between the start times of tests are given by an activity-on-node network (cf. e.g. [2]). This network contains a set V of nodes, each corresponding to a test. Two additional, fictitious tests *0* and n+*1* with duration zero represent the project's start and end. Moreover, for each minimum time lag, claiming that test j has to start at least d_{ij}^{min} time units after the start of test i, we add a forward arc with weight $\delta_{ij}=d_{ij}^{min}$ between nodes i and j to the network. Likewise, for each maximum time lag, claiming that j starts at most d_{ij}^{max} time units after the start of i, a backward arc with weight $\delta_{ji}=-d_{ij}^{max}$ is introduced. Let S_i be the start time of a test i. Then minimum as well as maximum time lags lead to restrictions $S_j-S_i \geq \delta_{ij}$. By a maximum time lag between tests *0* and n+*1* a maximum project duration $\bar{d}$ is prescribed. Precedence relations between some tests can be ensured by respective minimum time lags. Besides, we regard dependencies between engineering tasks and tests by release and due dates, which are represented by minimum time lags between tests *0* and i, and between tests i and n+*1*, respectively.

Since different variants of a model-line are developed simultaneously, different variants of experimental vehicles have to be distinguished. These variants differ by their engine, chassis, or body. As the results of several tests are independent of certain properties of the used vehicle, there may exist a set of alternative *modes* $M_i \subseteq M$ to perform a test i. Each mode corresponds to an experimental vehicle. Note that only resource allocations, but not test durations and temporal constraints, depend on the selected modes.

A *destructive test* $i \in D$, e.g. a crash test, is taken into account by occupying its used vehicle from the start of i to the end of the project. Moreover, a *partial ordered destructive relation* $(i,j) \in P$ implies that a test i disables the used vehicle to perform a certain test j afterwards. Thus, we claim that either tests i and j are performed in different modes or test j ends before test i starts.

Finally, as a vehicle has to be built before it can be used to perform any test, so-called *prototyping activities* have to be scheduled. Due to the limited capacity ρ_t of the prototype section, a renewable resource is introduced that limits the number $r_b(t)$ of prototyping activities b executed at the same point in time t. Then, since the vehicles are provided successively, the determination of the sequence in which the vehicles are built gets part of the optimization problem. Unfortunately, the modes prevent us from linking prototyping and testing activities by temporal constraints. Instead, we model each vehicle as a cumulative resource, which is usually used to represent a storage facility (cf. [2]). Initially, the inventory R_{v_k} of each cumulative resource comprises zero units in order to indicate that the respective vehicle v_k is not available for testing. At the end of a prototyping activity the inventory of the respective cumulative resource is incremented by one unit, indicating an additional vehicle that is available. A test depletes the inventory of the used vehicle by one unit at its start and, provided that the test is not destructive, the inventory is replenished by one unit at the test's end. Thus, all activities must be scheduled such that the inventory of no vehicle v_k falls below a minimum inventory $\underline{R_{v_k}} = 0$ at any point in time. Figure 1 illustrates the concepts we have introduced.

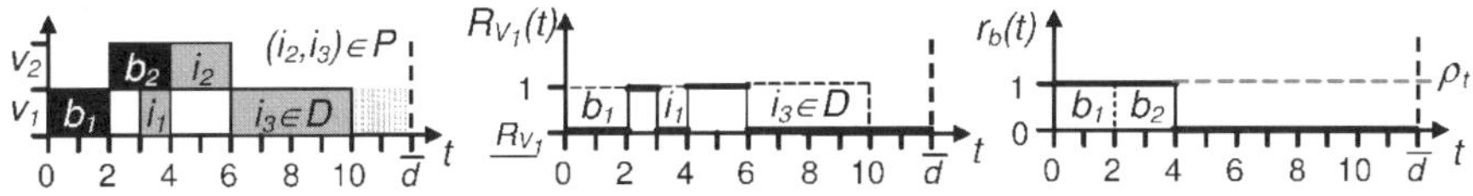

Fig. 1. Gantt chart representing a feasible solution; inventory-profile $R_{v_1}(t)$ of vehicle v_1; and (renewable) resource-profile $r_b(t)$ of the prototype section (example)

The considered scheduling problem can be described by the following mixed-integer program formulation, which serves to solve small problem instances occuring within R&D projects of model facelifts. Let x_{itv_k} be 1 if test i starts at time t on vehicle k of variant v, and 0 otherwise. If prototyping activity of vehicle v_k starts at time t, y_{tv_k} equals 1, and 0 otherwise. Finally, p_b and p_i denote the durations of prototyping and testing activities, respectively.

$$Min! \quad \sum_{t=0}^{\bar{d}} \sum_{v_k \in M} y_{tv_k} \tag{1}$$

w.r.t.

$$\sum_{v_k \in M_j} \sum_{t=0}^{\bar{d}} x_{jtv_k} \times t - \sum_{v_k \in M_i} \sum_{t=0}^{\bar{d}} x_{itv_k} \times t \geq \delta_{ij} \quad \forall \langle i,j \rangle \in E \tag{2}$$

$$\sum_{\tau=0}^{t-p_b} y_{\tau v_k} - \sum_{i \in D} \sum_{\tau=0}^{t} x_{i\tau v_k} - \sum_{i \in V \setminus D} \sum_{\tau=t-p_i+1}^{t} x_{i\tau v_k} \geq 0 \quad \forall v_k \in M, t \in \left[0,\bar{d}\right] \tag{3}$$

$$\sum_{v_k \in M} \sum_{\tau=t-p_b+1}^{t} y_{\tau v_k} \leq \rho_t \quad \forall t \in \left[0,\bar{d}\right] \tag{4}$$

$$\sum_{t=0}^{\bar{d}} x_{jtv_k} \times t - \sum_{t=0}^{\bar{d}} x_{itv_k} \times \left(t - \bar{d}\right) \leq \bar{d} \quad \forall (i,j) \in P, v_k \in M \tag{5}$$

$$\sum_{v_k \in M_i} \sum_{t=0}^{\bar{d}} x_{itv_k} = 1 \quad \forall i \in V \tag{6}$$

$$x_{itv_k} \in \{0,1\} \qquad y_{tv_k} \in \{0,1\} \qquad \forall i \in V, t \in \left[0,\bar{d}\right], v_k \in M \tag{7}$$

Objective function (1) serves to minimize the number of built vehicles. Restrictions (2) ensure that all temporal constraints $\langle i,j \rangle \in E$ are met. Due to (3), a test cannot occupy a vehicle v_k before the respective prototyping activity has been finished. As only one prototyping activity for vehicle v_k exists, the first term of (3) is smaller than or equal to one. Thus, (3) ensure for every point in time that either at most one destructive test has been started (second term) or at most one test is performed (third term) on each vehicle. By (4) we limit the number of prototyping activities that are executed at the same time according to the capacity ρ_t of the prototype section. The partial ordered destructive relations $(i,j) \in P$ are considered in (5). For $x_{itv_k} = 1$ the two $\bar{d}$ neutralize each other, such that, if test i and j are performed in the same mode v_k, j must start before i. Otherwise, (5) hold always true. Due to (6) each test must be executed, and (7) define the domains of the decision variables.

3 A Priority-Rule Based Method

Since problem (1)–(7) is NP-hard and – considering the development of new car models – problem instances consist of up to 600 tests and 25 variants of vehicles, we propose a priority-rule based heuristic as a solution procedure.

The proposed heuristic is based on a priority-rule method for problems with nonregular objective functions devised by Neumann and Zimmermann [1]. This method successively schedules the activities, such that the increase in the objective function value of the extended partial schedule S^C is minimum.

First of all, we have to explain how to represent the prototyping activities as they have an important impact on the solution procedure. As all these activities b have the same duration p_b and should start as early as possible, we can predetermine their completion times β_b and allocate them to vehicles within the procedure. Moreover, we set $\beta_{min}:=\min\{\beta_b \in B\}$ where B denotes the set of completion times from prototyping activities that have to be allocated.

In each main step of the procedure (see Fig. 3a) the test $i \in V \setminus C$ with highest priority value $\pi(i)$ is determined to be scheduled next, where set C contains all scheduled tests. This is repeated until all tests have been scheduled.

Test i must start between its earliest start time ES_i and its latest start time LS_i (see Fig. 2). Given a partial schedule S^C, $\tau^C_{iv_k} \subseteq [ES_i, LS_i]$ is the time domain in which test i can start such that neither the described minimum inventory $\underline{R_{v_k}}$ of vehicle v_k nor any partial ordered destructive relation is violated. By setting $\tau^C_{iv_k} := \tau^C_{iv_k} \setminus [0,\beta_{min}]$ for all vehicles v_k for which no prototyping activity b has been allocated so far, we ensure that such an activity b can be finished before test i starts. Since for fixed modes objective function (1) is locally regular, it is sufficient to consider only a set D^C_i of appropriate start times $\delta^C_{iv_k}$ for test i (cf. [2]). That means, considering all vehicles $v_k \in M_i$ being part of the current solution, D^C_i contains only those points in time $\delta^C_{iv_k} \in \tau^C_{iv_k}$, at which a scheduled test ends on a vehicle v_k, and $\delta^C_{iv_k} := \min\{t \in \tau^C_{iv_k}\}$.

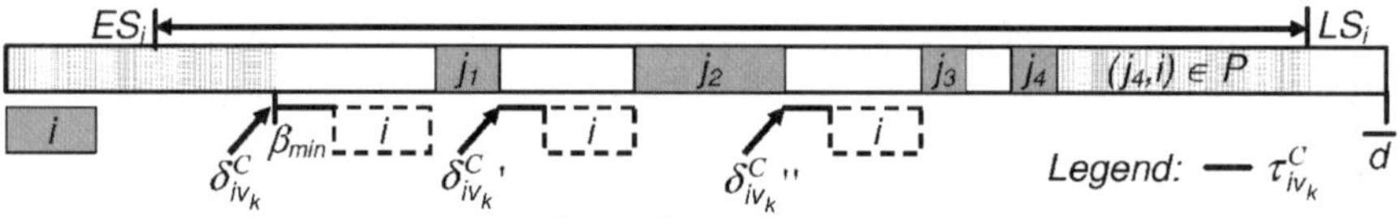

Fig. 2. Appropriate start times $\delta^C_{iv_k} \in D^C_i$ to schedule test i on a vehicle v_k (example)

Set D^C_i is computed by procedure "Schedule(i)" (see Fig. 3c). If $D^C_i \neq \emptyset$ holds true, test i can be performed on a vehicle contained in the partial solution and the smallest $\delta^C_{iv_k}$ is selected to schedule i. Otherwise, the possibility to execute test i on an additional vehicle is examined. For this purpose we set $S_i := \max\{ES_i, \beta_{min}\}$ in order to ensure that at least the prototyping activity with the smallest completion time β_{min} can be finished before i starts. If we obtain $S_i > LS_i$, it is not sufficient to add a vehicle and the procedure Unschedule(i) is called. But, if $S_i \leq LS_i$ holds true, we generate a prototyping activity and another vehicle becomes part of the solution. The variant of the vehicle is chosen such that test i can be performed on it (i.e. $v. \in M_i$) and the unsatisfied demand of the variant $\upsilon(v) = \sum_{\{i \in V | v. \in M_i\}} p_i$ is maximum. Finally, a prototyping activity is scheduled on each vehicle v_k for which no such activity has been scheduled yet and that is occupied by a test i before the maximum completion time of a prototyping activity β_{max}.

Having fixed the start time S_i of test i, earliest and latest start times of all tests $j \in V \setminus C$ must be updated, since there may exist time lags between tests i and j. Each test becoming critical (i.e. $ES_j = LS_j$) is scheduled subsequently.

An unschedule step (see Fig. 3b) is executed if a test i must be scheduled on an additional vehicle, but due to the small latest start time of i no prototyping activity can be allocated to this vehicle. Thus, we try to enlarge LS_i by removing a set U of tests j from the partial solution which restrict LS_i to a value smaller than β_{min}. However, if the initial latest start time $-d_{i0}$ is smaller than β_{min}, this approach is not expedient and we search for a test $i^* \in C$ that can be replaced by test i. If no i^* is found, the procedure terminates without finding a feasible schedule. By restricting the number of times a test i^* can be replaced, we prevent the procedure from cycling.

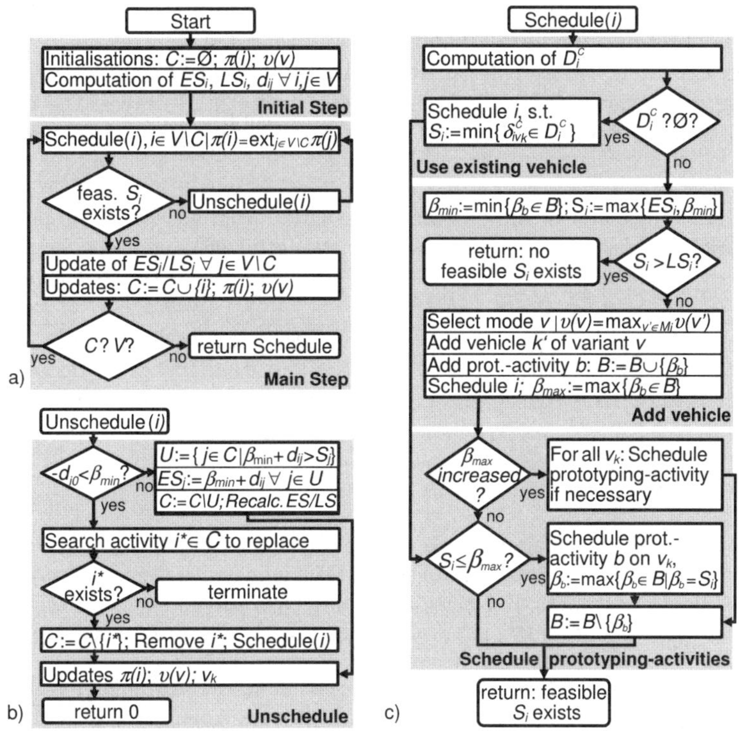

Fig. 3. Procedures: **(a)** Main function, **(b)** Unschedule(i), and **(c)** Schedule(i)

4 Preliminary Computational Experience

By means of the project generator *ProGen/max* (cf. [4]), we randomly generated 120 small and 120 large problem instances, each containing 20 or 600 tests as well as 4 or 25 variants of vehicles, respectively. The maximal project duration $\bar{d}$ was set to 120% of the length of the critical path ES_{n+1}.

The large problem instances were used to compare several priority rules by which the described priority values $\pi(i)$ are calculated. We tested "smallest latest start time first" (LST), "minimum number of modes first" (MNM), "most total successors first" (MTS) and "least unscheduled total successors first" ($LUTS$) (cf. [2]). Comparing these rules by their relative error Δ_{best} (see Table 1), we obtain that MTS and $LUTS$ behave quite similarly as they provide nearly the same scheduling sequence, whereas LST and MNM perform a little worse. The ratio p_{inf} of problem instances that no resource and time feasible solution had been computed for is acceptable for all rules.

The small problem instances served to examine the relative deviation Δ_{opt} between results from the proposed heuristic and exact solutions to the mixed-integer program (see Table 2). Besides, we examined the dependence of Δ_{opt} on the mode factor $MF = \sum_{i \in V} \frac{|M_i|}{|M|}$ and on the order strength $OS \in [0,1]$, which is minimum for parallel and maximum for serial project networks (cf. [4]). As expected, the deviation from an optimal solution increases with increasing mode factor and decreasing order strength. However, the obtained average relative deviation of 13.5% is not scalable to large problem instances.

Table 1. Comparison of different priority rules

	$LUTS$	MTS	LST	MNM
Δ_{best}	0.2%	0.4%	6.4%	7.0%
p_{inf}	-	-	-	1.7%

Table 2. Deviation from an exact solution depending on order strength (OS) and mode factor (MF)

OS/MF	0.75/0.5	0.75/0.75	0.25/0.5	0.25/0.75
Δ_{opt}	12.9%	13.2%	13.9%	14.1%

References

1. Neumann, K., Zimmermann, J. (2000) Procedures for Resource Leveling and Net Present Value Problems in Project Scheduling with General Temporal and Resource Constraints. European Journal of Operational Research 127: 425–443
2. Neumann, K., Schwindt, C., Zimmermann, J. (2003) Project Scheduling with Time Windows and Scarce Resources. Springer, Berlin
3. Risse, J. (2002) Time-to-Market-Management in der Automobilproduktion – Ein Gestaltungsrahmen für ein logistikorientiertes Anlaufmanagement. University of Berlin
4. Schwindt, C. (1998) Generation of Resource-Constrained Project Scheduling Problems Subject to Temporal Constraints. Report WIOR-543, University of Karlsruhe

Cyclic Scheduling Problems with Linear Precedences and Resource Constraints

Peter Brucker and Thomas Kampmeyer

Universität Osnabrück, Fachbereich Mathematik/Informatik
49069 Osnabrück
{Peter.Brucker, Thomas.Kampmeyer}@mathematik.uni-osnabrueck.de

1 Introduction

Cyclic scheduling is concerned with planing of operations that have to be repeated infinitely often. In [1, 2] the following problem is considered: Given is a set $T = \{1, \ldots, n\}$ of operations. The aim is to find a periodic schedule which assigns to each occurrence $< i; k >$ of an operation i a starting time $t(i; k)$ with minimal cycle time α. The schedule is called periodic if $t(i; k) = t(i; 0) + \alpha k$ for all $k \in \mathbb{Z}$ holds. Associated with each operation $i \in T$ is a dedicated machine $M(i) \in M = \{1, \ldots, m\}$ on which each occurrence $< i; k >$ of operation i must be processed. Occurrences of different operations that have to be processed on the same machine cannot overlap. Also there are given uniform precedence constraints between some operations i and j. These constraints are described by a double weighted graph $G = (T, E)$.

Hanen & Munier-Kordon [3] considered the problem with unlimited resources and linear precedence constraints. In the presence of linear precedence constraints the aim is to find a periodic schedule with minimal cycle time $w_i > 0$ for all $i \in T$, whereas the starting of the occurrences of different iterations can be computed by $t(i; k) = t(i; 0) + kw_i$ with $k \in \mathbb{Z}$. They show that this more general problem can also be reduced to a minimum cost-to-time ratio problem if the graph given by the linear precedence constraints is unitary. In [4] Cavory et al. developed a genetic algorithm for solving the same problem with limited resources, but they only considered non-periodic schedules.

Our main result is that the problem of finding a periodic schedule for a given unitary graph of linear precedence constraints and limited resources can be formulated by a mixed integer linear program. This allows a similar tabu-search procedure as the one developed in [1]. A more extended version of this paper can be found in [5]. The paper is organized as follows. In the next section we recall the results of Hanen & Munier-Kordon [3]. In the third section we derive the mixed integer linear programming formulation. Section 4 contains some concluding remarks.

2 A Basic Cyclic Scheduling Problem with linear precedence constraints (BLCP)

Let $T = \{1, \ldots, n\}$ be a set of generic tasks or operations (in connection with shop scheduling problems). Task (operation) i has processing time $p_i > 0$ and must be performed infinitely often. We denote by $< i; k >$ the k-th occurrence of the generic task i. A schedule is called **periodic** if $t(i;k) = t(i;0) + kw_i$ for all $i \in T$ and all $k \in \mathbb{Z}$. Here $w_i > 0$ is the cycle time of operation i. We also assume that $t_i := t(i;0) \geq 0$ for all $i \in T$. A periodic schedule is defined by the vectors $(t_i)_{i \in T}$ and $(w_i)_{i \in T}$.

Linear precedence constraints of the form

$$t(i; p_{ij}k + q_{ij}) + L_{ij} \leq t(j; p'_{ij}k + q'_{ij}) \tag{1}$$

for all $k \in \mathbb{Z}$ may be given for all arcs $(i,j) \in E$ of a directed graph $G = (T, E)$ with vertex set T. L_{ij} is called (start-start) **delay**. Delays are assumed to be positive integers. Also the following properties are satisfied: p_{ij}, p'_{ij} are positive integers and q_{ij}, q'_{ij} are integer values.

We also assume that the precedence constraints are **unitary**. That means that the graph G is strongly connected and all cycles c of G have the weight $\pi(c) = 1$, whereas the weight of an arc (i,j) is $\pi_{ij} = \frac{p'_{ij}}{p_{ij}}$ and the weight of a path μ is $\pi(\mu) = \Pi_{(i,j)\in\mu}\pi_{ij}$.

The aim is to minimize simultaneously the cycle time w_i for all operation $i \in T$.

In this section we show that $w_i = \alpha W_i$ with $\alpha = \frac{w_1}{\beta}$ for $i \in T$ with constants W_i and β depending on the values π_{1i} hold and that (1) can be written in the form

$$t_j - t_i \geq L_{ij} - \alpha H_{ij}$$

where the values H_{ij} are integer constants depending on the values π_{1i}.

Thus, the problem of finding a periodic schedule that minimizes the w_i-values subject to the constraints (1) can be reduced to the following problem which is usually called Basic Cyclic Scheduling Problem (BCSP)

$$\min \alpha \tag{2}$$

s.t.

$$t_j - t_i \geq L_{ij} - \alpha H_{ij} \; \forall \, (i,j) \in E \tag{3}$$

Problem (2) to (3) is equivalent to a special minimum cost-to-time ratio problem, which can be solved in polynomial time.

The reduction to the BCSP is due to Hanen & Munier-Kordon [3]. Here we recall the main results of their paper.

As one is looking for a periodic schedule, one can prove the following Lemma.

Lemma 1. *A periodic schedule meets the linear precedence constraints, if and only if the following inequality holds for any arc* $(i,j) \in E$*:*

$$t_j - t_i \geq L_{ij} + (w_i p_{ij} - w_j p'_{ij})k + w_i q_{ij} - w_j q'_{ij} \; \forall \, k \in \mathbb{Z}. \tag{4}$$

A direct conclusion of this Lemma is that $w_i p_{ij} - w_j p'_{ij} = 0$, because the inequality (4) must be true for all $k \in \mathbb{Z}$. Thus, condition (4) is equivalent to the following two conditions

$$t_j - t_i \geq L_{ij} + w_i q_{ij} - w_j q'_{ij} \tag{5}$$

and

$$w_i p_{ij} - w_j p'_{ij} = 0. \tag{6}$$

Note, that (6) is equivalent to

$$\pi_{ij} := \frac{w_i}{w_j} = \frac{p'_{ij}}{p_{ij}}. \tag{7}$$

Lemma 2. *The linear precedence constraints* $t_j - t_i \geq L_{ij} + w_i q_{ij} - w_j q'_{ij}$ *for any arc* $(i,j) \in E$ *can be rewritten to* $t_j - t_i \geq L_{ij} - \alpha H_{ij}$ *with integer* H_{ij}*.* H_{ij} *is called* ***height*** *of the arc* (i,j)*.*

A consequence of the proof of Lemma 2 is that one can rewrite the cycle time w_i for all operations $i \in T$ by

$$w_i = \frac{p'_{1i}}{p_{1i}} = \alpha \frac{\beta}{\rho_i} = \alpha W_i$$

with $W_i = \frac{\beta}{\rho_i} = \beta \cdot \frac{\gamma_i}{\beta_i} \in \mathbb{N}$, whereas $\rho_i = \frac{w_1}{w_i} = \frac{\beta_i}{\gamma_i}$ with $GCD(\beta_i, \gamma_i) = 1$ and $\beta := LCM(\beta_1, \ldots, \beta_n)$.

So one gets the following value for the height of an arc (i,j):

$$H_{ij} = W_j q'_{ij} - W_i q_{ij} \tag{8}$$

Therefore we consider only problems with $W_i \in \mathbb{N} \, \forall \, i \in T$.

Furthermore by minimizing α, one minimizes simultaneously the cycle time for all operations $i \in T$ because the values W_i depend on the values p_{1i} and p'_{1i} only.

The BLCP, which is described by (2) to (3) can be solved in polynomial time by using the same method which can be used for solving the BCSP, see e.g. [1]. Furthermore, there exists only a solution for the problem, if all cycles have a positive height.

3 A Cyclic Scheduling Problem with linear precedence constraints and resource constraints (CLSP)

In this section the previous problem is extended by resource constraints.

Associated with each operation i there is a dedicated machine $M(i) \in M = \{1, \ldots, m\}$, on which each occurrence $< i; k >$ of i must be processed. Occurrences of different operations to be processed on the same machine cannot overlap.

Therefore for all occurrences of pairs of operations i and j which are processed on the same machine one has to add the following constraints :

$$t(i;k) + p_i \leq t(j;k) \vee t(j:l) + p_j \leq t(i;k). \tag{9}$$

Thus, the general cycle time minimization problem for linear precedence constraints can be written as:

$$min\ \alpha \tag{10}$$

s.t.

$$t(i;k) = t(i;0) + \alpha k W_i \qquad i \in T, k \in \mathbb{Z} \tag{11}$$

$$t(i; p_{ij}k + q_{ij}) + L_{ij} \leq t(j; p'_{ij}k + q'_{ij}) \qquad (i,j) \in E, k \in \mathbb{Z} \tag{12}$$

$$t(i;k) + p_i \leq t(j;l) \vee t(j;l) + p_j \leq t(i;k) \; i,j \in T \text{ with } i \neq j \text{ and } M(i) = M(j), k,l \in \mathbb{Z} \tag{13}$$

Now we want to show that this problem is equivalent to a mixed integer linear program. In the proof of this equivalence, we use results which are based on the *Extended Euclidean Algorithm.* The Extended Euclidean Algorithm is presented in Theorem 1. A proof can be found e.g. in [6].

Theorem 1. ***Extended Euclidean Algorithm*** *Let* $a \in \mathbb{N}_0, b \in \mathbb{N}$. *The greatest common divisor GCD(a,b) can be written as linear combination of a and b.*

$$GCD(a,b) = u \cdot a + v \cdot b, \text{ with } u, v \in \mathbb{Z}$$

Now we will describe the main result of this section:

Theorem 2. *The problem* (10) *to* (13) *is equivalent to the following mixed integer linear program* (14) *to* (18).

$$\min \alpha \tag{14}$$

s.t.

$$t_j - t_i \geq L_{ij} - \alpha H_{ij} \qquad (i,j) \in E \tag{15}$$

$$t_j - t_i \geq p_i - \alpha K_{ij} \cdot GCD(W_i, W_j) \; i,j \in T \text{ with } i \neq j \text{ and } M(i) = M(j) \tag{16}$$

$$K_{ij} + K_{ji} = 1 \qquad i,j \in T \text{ with } i \neq j \text{ and } M(i) = M(j) \tag{17}$$

$$K_{ij} \in \mathbb{Z} \qquad i,j \in T \text{ with } i \neq j \text{ and } M(i) = M(j) \tag{18}$$

Proof. By substituting (11) into (12) we get

$$t_i + \alpha W_i(p_{ij}k + q_{ij}) + L_{ij} \leq t_j + \alpha W_j({p'}_{ij}k + {q'}_{ij})$$

$$\Leftrightarrow t_j - t_i \geq L_{ij} + k(p_{ij}\alpha W_i - {p'}_{ij}\alpha W_j) + \alpha(W_i q_{ij} - W_j {q'}_{ij})$$

With $\alpha W_i = w_i = w_j \pi_{ij} = w_j \frac{p'_{ij}}{p_{ij}}$ and $\alpha W_j = w_j$ we have $p_{ij}\alpha W_i - {p'}_{ij}\alpha W_j = 0$ and thus with (8)

$$t_j - t_i \geq L_{ij} + \alpha(W_i q_{ij} - W_j {q'}_{ij}) = L_{ij} - \alpha H_{ij}.$$

Now consider two tasks $< i; k >$ and $< j; l >$ to be processed on the same machine. Again (13) with (11) is equivalent to

$$t_i + \alpha k W_i + p_i \leq t_j + \alpha l W_j \vee t_j + \alpha l W_j + p_j \leq t_i + \alpha k W_i$$

or

$$p_i + \alpha(kW_i - lW_j) \leq t_j - t_i \vee t_j - t_i \leq -p_j + \alpha(kW_i - lW_j).$$

We have $\{-W_i k + W_j l | k, l \in \mathbb{Z}\} = \{m * GCD(W_i, W_j) | m \in \mathbb{Z}\}$, because of Theorem 1. So we get

$$p_i - \alpha m * GCD(W_i, W_j) \leq t_j - t_i \vee t_j - t_i \leq -p_j - \alpha m * GCD(W_i, W_j)$$

Therefore the numbers $t_j - t_i$ cannot be contained in the intervals

$$\ldots,] - p_j - \alpha m * GCD(W_i, W_j), p_i - \alpha m * GCD(W_i, W_j)[,$$

$$] - p_j - \alpha(m-1) * GCD(W_i, W_j), p_i - \alpha(m-1) * GCD(W_i, W_j)[, \ldots$$

Thus $t_j - t_i$ must be contained in one of the intervals

$$\ldots, [p_i - \alpha m * GCD(W_i, W_j), -p_j - \alpha(m-1) * GCD(W_i, W_j)], \ldots \quad (19)$$

which implies that for some integer K_{ij} we must have

$$p_i - \alpha K_{ij} * GCD(W_i, W_j) \leq t_j - t_i \leq -p_j + \alpha(1 - K_{ij}) * GCD(W_i, W_j)$$

With $K_{ji} := 1 - K_{ij}$ conditions (16) to (18) are satisfied. On the other hand if (15) to (18) are satisfied then conditions (11) to (13) hold if we set $t(i; k) := t_i + \alpha k W_i$ for $i \in T$ and $k \in \mathbb{Z}$. □

The next theorem shows that the intervals (19) cannot be empty.

Theorem 3. *Given is an feasible instance of a cyclic machine scheduling problem with linear precedence constraints. Then*

$$\alpha \geq \frac{p_i + p_j}{GCD(W_i, W_j)} \quad (20)$$

holds for every $\alpha \geq \alpha^*$ *and* $i, j \in T$ *with* $i \neq j$ *and* $M(i) = M(j)$*, whereas* α^* *is the optimal cycle time for the given instance.*

4 Concluding Remarks

A model for cyclic scheduling problems with linear precedences and resource constraints has been developed. The next step would be to adapt our tabu-search approach, which we developed for the problem with uniform precedence constraints in [1] to this problem. As problems with linear precedence constraints occur in production manufacturing, it would be of interest to see how the local search approach deals with these problems.

References

1. Brucker P, Kampmeyer T (2005) Tabu search algorithms for cyclic machine scheduling problems. Journal of Scheduling 8:303–322
2. Hanen C (1994) Study of a NP-hard cyclic scheduling problem: The recurrent job-shop. European Journal of Operational Research 72:82–101
3. Hanen C, Munier-Kordon A (2004) Periodic schedules for linear precedence constraints. Technical report, Laboratoire LIP6, Paris
4. Cavory G, Dupas R, Goncalves G (2005) A genetic approach to solving the problem of cyclic job shop scheduling with linear constraints. European Journal of Operational Research 161:73–85
5. Brucker P, Kampmeyer T (2005) Cyclic scheduling problems with linear precedences and resource constraints. OSM Reihe P, 261, Universität Osnabrück, Fachbereich Mathematik/Informatik
6. Scheid H (1994) Zahlentheorie, 2nd edition. BI Wissenschaftsverlag

Ein System zur Lösung multikriterieller Probleme der Ablaufplanung

Martin Josef Geiger

Lehrstuhl für Industriebetriebslehre (510A), Universität Hohenheim, D-70593 Stuttgart, mjgeiger@uni-hohenheim.de

1 Einleitung

Die Lösung multikriterieller Optimierungsprobleme umfasst zwei grundsätzliche Teilbereiche. Zum einen besteht die Problemlösung, analog zu monokriteriellen Problemen, in der Identifikation optimaler Alternativen, zum anderen ist eine Auswahl einer von einem Entscheidungsträger meistpräferierten Lösung zu treffen.

Beide Aspekte mehrkriterieller Problemstellungen werden entsprechend durch verschiedene Lösungskonzepte unterstützt. Auf der Seite der Identifikation optimaler Alternativen finden sich in aller Regel Adaptionen monokriterieller Optimierungskonzepte, beispielsweise in Form von Gewichtungsansätzen, welche durch Einsatz von Skalarisierungsfunktionen die Problemstellung in ein Problem mit einer einzigen Optimallösung überführen. Die entsprechende Auswahl einer Lösung wird in aller Regel durch interaktive Suchkonzepte realisiert, welche dem Entscheidungsträger einen Vergleich der effizienten Ergebnisse ermöglichen.

Der vorliegende Artikel beschreibt ein System zur Lösung multikriterieller Probleme der Ablaufplanung, welches beide Teilaspekte mehrkriterieller Problemstellung in ein Gesamtsystem integriert. Die Identifikation effizienter Ergebnisse wird hierbei durch eine Methodendatenbank unterstützt, welche auf die jeweilige Problemstellung adaptierbare metaheuristische Suchverfahren bereit stellt. Die Identifikation einer meistpräferierten Lösung wird im folgenden durch den Einsatz der Aspiration Interactive Method (AIM) [5] ermöglicht, und eine für eine Interaktion mit einem menschlichen Entscheidungsträger bedeutende Visualisierung der Ergebnisse ist verfügbar.

Nach erfolgreicher Teilnahme am Finale des *European Academic Software Award 2002* in Ronneby (Schweden) wurde das System mit einem der dort verliehenen Preise ausgezeichnet (http://www.easa-award.net/, http://www.bth.se/llab/easa_2002.nsf).

2 Problembeschreibung und Systemkonzeption

2.1 Multikriterielle Probleme der Ablaufplanung

Probleme der Ablaufplanung befassen sich allgemein mit der zeitlichen Planung einer Menge an Fertigungsaufträgen $\mathcal{J} = \{J_1, \ldots, J_n\}$, jeweils bestehend aus einer Menge an Arbeitsgängen $J_j = \{O_{j1}, \ldots, O_{jo_j}\}$, welche auf einer Menge an zur Durchführung der Bearbeitung bereit stehenden Maschinen $\mathcal{M} = \{M_1, \ldots, M_m\}$ eingeplant werden müssen [3]. Die zu einem Auftrag J_j gehörenden Operationen sind oftmals technologisch bedingten Präzedenzbeziehungen der Form $O_{jk} \rhd O_{jk+1} \, \forall \, k = \{1, \ldots, o_j - 1\}$ unterworfen, so dass die Beendigung der Bearbeitung von O_{jk} eine notwendige Voraussetzung für den Beginn von O_{jk+1} ist.

Eine Alternative der Problemstellung, ein so genannter Ablaufplan $x \in X$ legt unter Berücksichtigung aller Nebenbedingungen Starttermine der Operationen fest. Die Optimalität eines Plans x wird hierbei unter Einbezug einer Menge an Kriterien $G(x) = (g_1(x), \ldots, g_k(x))$ bestimmt, wobei im folgenden von Minimierungszielen ausgegangen wird.

$$\text{„min"} \; G(x) \tag{1}$$

$$x \in X \tag{2}$$

Da die relevanten Ziele oftmals konfliktärer Natur sind, existiert jedoch keine einzelne Alternative, welche alle Komponenten des Zielfunktionsvektors gleichermaßen minimiert. Die Lösung der Problemstellung besteht für diesen Fall in der Bestimmung aller Alternativen, für die die Ergebnisvektoren gemäß den Definitionen 1 und 2 Pareto-optimal sind. Die Menge aller Alternativen, die diese Eigenschaft besitzen, heißt Pareto Menge P.

Definition 1 (Pareto-Dominanz). *Ein Ergebnisvektor $y = G(x)$ dominiert einen Ergebnisvektor $y' = G(x')$, falls $\forall i \in \{1, \ldots, K\} : g_i(x) \leqslant g_i(x') \wedge y \neq y'$. Die Pareto-Dominanz von y gegenüber y' wird mit $y \preccurlyeq y'$ notiert, und die entsprechende Relation gilt für den Alternativenraum, d. h. $x \preccurlyeq x'$ genau dann wenn $G(x) \preccurlyeq G(x')$.*

Definition 2 (Pareto-Optimalität, Effizienz). *Eine Alternative $x \in \mathcal{X}$ heißt* Pareto-optimal, *falls $\nexists x' \in \mathcal{X} \mid G(x') \preccurlyeq G(x)$. Der zu einer Pareto-optimalen Alternative $x \in \mathcal{X}$ zugehörige Ergebnisvektor $y = G(x)$ heißt* effizient.

2.2 Das Lösungssystem MOOPPS

Das System MOOPPS (*A Multi Objective Optimization System for Production and Project Scheduling*) verfolgt einen zweistufigen Ansatz zur Lösung

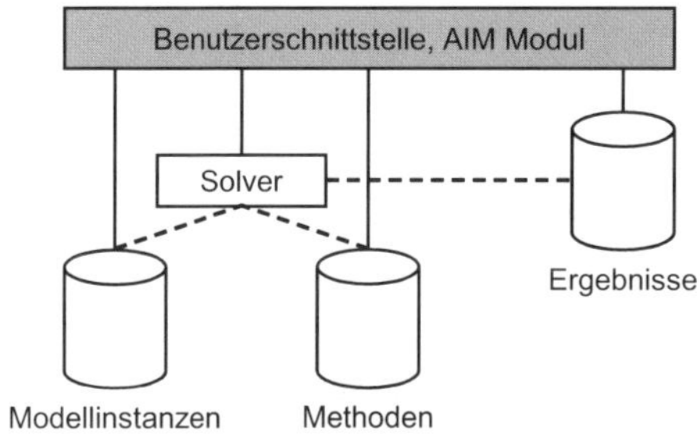

Abb. 1. Struktur des Systems

multikriterieller Probleme der Ablaufplanung. Abbildung 1 gibt die Hauptkomponenten an und verdeutlicht deren Zusammenwirken.

Ausgehend von der Definition metaheuristischer Suchverfahren auf der Grundlage einer Methodendatenbank können Modellinstanzen verwaltet und gelöst werden. Ein Solver verbindet beide Elemente und erzeugt nach der Durchführung von Testläufen Ergebnisse, welche über eine Benutzerschnittstelle ausgewertet werden können. In diese Komponente ist zudem ein Entscheidungsunterstützungsmodul integriert, welches die Auswahl einer meistpräferierten Alternative $x^* \in P$ unterstützt.

Die implementierten Methoden umfassen lokale Suchheuristiken auf der Grundlage eines Hillclimbing-Verfahrens [6], Evolutionäre Algorithmen (EA) [1] sowie ein mehrkriterieller Simulated Annealing Algorithmus [8].

Mögliche Zielkriterien [7] der Problemstellung sind die Minimierung des maximalen Fertigstellungszeitpunktes aller Aufträge C_{max} und die Minimierung der Summe der Fertigstellungszeitpunkte C_{sum}. Daneben können an Lieferterminen d_j der Aufträge J_j orientierte Kriterien wie die maximale Verspätung T_{max}, die Summe der Lieferterminüberschreitungen T_{sum} sowie die Anzahl der verspäteten Aufträge U optimiert werden. Eine maschinenorientierte Betrachtung der Güte einzelner Pläne ist unter dem Aspekt der Minimierung von Maschinenleerzeiten möglich.

3 Ergebnisse

3.1 Approximation effizienter Ergebnisse

Ein Reihe der Literatur entnommenen Modellinstanzen [1, 2] multikriterieller Flow Shop Scheduling Probleme wurden unter Einsatz verschiedener EA gelöst und mit den Ergebnissen einer multikriteriellen variablen Nachbarschaftssuche (MOVNS) verglichen. Die EA verwenden zur Erzeugung einer Approximation der Pareto Menge P Rekombinations-Nachbarschaften der Form Partially Mapped Crossover (PMX), Order Based Crossover (OBX), Uniform Order Based Crossover (UOBX) und Two-Point Crossover (TPOX). Die genaue Vorgehensweise ist hierbei in Algorithmus 1 als Pseudo-Code gegeben.

Algorithmus 1 MOEA

Setze $i = 1$
Setze $P^{approx} = \emptyset$
Generiere Population POP_i an n_{pop} zufälligen Ausgangslösungen
Aktualisiere P^{approx} mit POP_i
Wiederhole
  Wiederhole
    Selektiere Alternativen $x_1, x_2 \in POP_i \cup P^{approx}$
    Erzeuge $x'_1, x'_2 \in \mathbf{N}_c(x_1, x_2, z)$ mittels Zufallszahl z
    Wende mit p_{mut} Wahrscheinlichkeit $\mathbf{N}_m(x'_1, z)$ und $\mathbf{N}_m(x'_2, z)$ an
    Prüfe neue Alternativen auf Akzeptanz in POP_{i+1}
  Bis Anzahl Elemente in $POP_{i+1} = n_{pop}$
  Aktualisiere P^{approx} mit POP_{i+1}
  Setze $i = i + 1$
Bis Abbruchkriterium erfüllt

Die multikriterielle variable Nachbarschaftssuche setzt alternativ in jedem Schritt der Optimierungsläufe einen Operator aus der Menge der Exchange-, Backward-Shift- oder Forward-Shift-Nachbarschaft ein [6].

Algorithmus 2 MOVNS

Initialisiere Steuerparameter: Lege Nachbarschaften $\mathbf{N}_1, \ldots, \mathbf{N}_k$ fest
Generiere eine Ausgangslösung x
Setze $P^{approx} = \{x\}$
Wiederhole
  Wähle $x \in P^{approx}$ mit noch nicht untersuchter Nachbarschaft
  Wähle zufällig eine Nachbarschaft $\mathbf{N}_i$ aus $\mathbf{N}_1, \ldots, \mathbf{N}_k$
  Erzeuge $\mathbf{N}_i(x)$
  Aktualisiere P^{approx} mit allen $x' \in \mathbf{N}_i(x)$
  Wenn $x \in P^{approx}$ **dann**
    Markiere Nachbarschaft von x als untersucht
  Ende Wenn
Bis $\nexists x \in P^{approx}$ mit noch nicht untersuchter Nachbarschaft

Beiden Algorithmen ist gemein, dass Sie während der Suche ein Archiv P^{approx} an nichtdominierten Alternativen erzeugen, welches nach Terminierung der Testläufe die Approximation von P darstellt. Im Kontext des EA wird somit implizit eine Elitismusstrategie realisiert.

Probleme unterschiedlicher Größe von 20 Aufträge auf 5 Maschinen bis hin zu 100 Aufträge auf 20 Maschinen wurden erfolgreich gelöst. Die aus [2] entnommenen Instanzen sind hierbei mit „Ta $n \times m$“, die aus [1] entnommene als „Ba 49×15“ bezeichnet.

Tabelle 1 gibt die durchschnittlich erzielten Werte der Approximationsgüte auf der Grundlage der D_1 Metrik von Czyżak und Jaszkiewicz [4] wieder. Es wird deutlich, dass die erzielten Ergebnisse unter Einsatz der variablen Nachbarschaftssuche den EA insbesondere bei den kleineren Testinstanzen signifikant besser sind. Eine zusätzliche Ermittlung der durchschnittlichen Werte für D_2 von Czyżak und Jaszkiewicz ergibt ein identisches Bild.

Tabelle 1. Durchschnittliche Ergebnisse für D_1

Instanz $n \times m$	Ziele	MOVNS	MOEA/ PMX	MOEA/ OBX	MOEA/ UOBX	MOEA/ TPOX
Ta 20 × 5 (1)	C_{max}, T_{sum}	0,0323	0,1318	0,0858	0,0825	0,0961
Ta 20 × 5 (2)	C_{max}, T_{sum}	0,1372	0,3049	0,3096	0,3236	0,3492
Ta 20 × 10 (1)	C_{max}, T_{sum}	0,0199	0,0933	0,0849	0,0806	0,0726
Ta 20 × 10 (2)	C_{max}, T_{sum}	0,0254	0,1433	0,1335	0,1349	0,1194
Ta 20 × 20	C_{max}, T_{sum}	0,0286	0,1559	0,1251	0,1316	0,1211
Ta 50 × 5	C_{max}, T_{sum}	0,0622	0,1712	0,1591	0,1930	0,1348
Ta 50 × 10	C_{max}, T_{sum}	0,3171	0,3857	0,3123	0,3605	0,3161
Ta 50 × 20	C_{max}, T_{sum}	0,3966	0,7980	0,4805	0,3974	1,2621
Ta 100 × 10	C_{max}, T_{sum}	0,3190	1,4758	0,6283	0,5603	2,8006
Ta 100 × 20	C_{max}, T_{sum}	0,2349	0,6948	0,2795	0,2645	1,0973
Ba 49 × 15	$C_{max}, \overline{F}, \overline{T}$	0,2440	0,6878	0,3604	0,2627	0,6376

Für Modellinstanz Ba 49 × 15 war es zudem möglich, bessere als in [1] berichtete Ergebnisse zu erzielen. Alle für diesen Datensatz bislang bekannten Alternativen wurden von den unter Einsatz des MOVNS Algorithmus ermittelten Lösungen dominiert.

3.2 Interaktive Entscheidungsunterstützung

Im Anschluss an die Erzeugung effizienter Alternativen wird unter Einbeziehung eines Entscheidungsträgers ein interaktiver Entscheidungsprozess modelliert. Dieser engt die Menge der Ergebnisse sukzessive eine, bis ein meistpräferierter Ablaufplan x^* identifiziert werden kann.

Grundlage des Verfahrens ist die Festlegung von Anspruchsniveaus $A = \{a_{g_1}, \ldots, a_{g_k}\}$, ein a_{g_i} je verwendeter Zielfunktion g_i. In einem ersten Schritt nehmen alle a_{g_i} maximale Werte an: $a_{g_i} = \max_{x \in P} g_i(x) \, \forall \, i = 1, \ldots, k$. Durch interaktive Verringerung der Werte wird es dem Entscheidungsträger dann möglich, die Menge P^{approx} schrittweise einzuengen und eine Teilmenge $P^{as} \subseteq P^{approx}$ zu erzeugen, welche alle Elemente enthält, die die jeweils festgelegten a_{g_i} erfüllen (unterschreiten). Aus Sicht des Entscheidungsträgers sind diese somit vorrangig von Interesse, und die Suche im Ergebnisraum terminiert naturgemäß mit der Bestimmung einer Menge A an Anspruchsniveaus so dass $|P^{as}| = 1$. Das System liefert das verbleibende Element in P^{as} als x^*.

Es versteht sich, dass Änderungen der Anspruchsniveaus frei möglich sind, so dass die potentielle Situationen in der $P^{as} = \emptyset$ auftritt nicht zu einem Abbruch mit anschließendem Neustart der Suche führen muss.

4 Zusammenfassung und Ausblick

Der vorliegende Artikel stellte ein System zur Lösung multikriterieller Probleme in der Ablaufplanung vor. Hierzu wurden heuristische Problemlösungskomponenten entwickelt, implementiert und erfolgreich auf Testprobleme angewendet. Ein lokales Suchverfahren auf der Grundlage wechselnder Nachbarschaftsdefinitionen wurde verschiedenen Evolutionären Algorithmen gegenübergestellt. Im Ergebnis erweist sich für die untersuchten Probleme die Konzeption des MOVNS den EA als überlegen.

Die Auswahl einer für einen Entscheidungsträger „optimalen“ Alternative wird durch die Integration eines interaktiven Verfahrens in das Gesamtsystem ermöglicht, welches direkt auf den Ergebnissen der Optimierungsläufe aufbaut. Der vorgestellte Lösungsansatz hat somit den Vorteil, ohne Präferenzinformationen des Entscheidungsträgers erste Ergebnisse erzielen zu können. Diese werden vielmehr in einem weiteren Schritt sukzessive in die Problemlösung integriert.

Denkbar und für weitere Untersuchungen geeignet ist in diesem Zusammenhang die Frage, wie etwaige Präferenzinformationen vorab in den Optimierungsprozess einfließen können. Dies ist insbesondere dann von Interesse, wenn diese nur partiell vorliegen oder nur unscharf beschrieben werden können.

Literaturverzeichnis

1. Tapan P. Bagchi. *Multiobjective scheduling by genetic algorithms.* Kluwer Academic Publishers, Boston, Dordrecht, London, 1999.
2. Matthieu Basseur, Franck Seynhaeve und El-ghazali Talbi. Design of multiobjective evolutionary algorithms: Application to the flow-shop scheduling problem. In *Congress on Evolutionary Computation (CEC'2002)*, Band 2, Seiten 1151–1156, Piscataway, NJ, 2002. IEEE Service Center.
3. Peter Brucker. *Scheduling Algorithms.* Springer Verlag, Berlin, 4. Auflage, 2004.
4. Piotr Czyżak und Andrzej Jaszkiewicz. Pareto simulated annealing - a metaheuristic technique for multiple-objective combinatorial optimization. *Journal of Multi-Criteria Decision Analysis*, 7:34–47, 1998.
5. V. Lotfi, T. J. Stewart und S. Zionts. An aspiration-level interactive model for multiple criteria decision making. *Computers & Operations Research*, 19(7):671–681, 1992.
6. Colin R. Reeves. Landscapes, operators and heuristic search. *Annals of Operations Research*, 86:473–490, 1999.
7. Vincent T'kindt und Jean-Charles Billaut. *Multicriteria Scheduling: Theory, Models and Algorithms.* Springer Verlag, Berlin, Heidelberg, New York, 2002.
8. E. L. Ulungu, J. Teghem, P. H. Fortemps und D. Tuyttens. MOSA method: A tool for solving multiobjective combinatorial optimization problems. *Journal of Multi-Criteria Decision Making*, 8:221–236, 1999.

On a Single Machine Due Date Assignment and Scheduling Problem with the Rate-Modifying Activity

Valery S. Gordon, Alexander A. Tarasevich

National Academy of Sciences of Belarus, United Institute of Informatics Problems, Minsk, Belarus
{gordon@newman.bas-net.by, Tarasevich.Alexander@gmail.com}

Abstract

In this paper we consider single machine common due date assignment and scheduling problem. The objective is to minimize the total weighted sum of earliness, tardiness and due date costs. There exists a possibility to perform some action (rate-modifying activity) to change processing times of the jobs following this activity. Thus, placing the rate-modifying activity to some position in the schedule can decrease the objective function value. We consider several properties of the problem which in some cases can reduce the complexity of the solution algorithm.

Keywords

machine scheduling, common due date assignment, earliness, tardiness, rate-modifying activity.

1 Introduction

Among the scheduling problems considered within the "Just-in-Time" concept there are problems in which all the jobs have to be completed as close as possible to a common due date. As examples of such problems one can consider an assembly line problem with common due date for processing the component parts, or a problem of minimizing the number of delayed requests to a server. Another example is the following problem arising in food industry. Fruits or vegetables from different farms are to be processed at a plant. If they are delivered too early then the plant has to pay for their storage. Otherwise, if they are delivered and processed too late, there would be the losses due to a large number of spoiled fruits or vegetables. The situation can be modeled by a single machine scheduling problem of minimizing earliness-tardiness and *assignable* common due date costs.

In some situations (repairing or upgrading the machine), a *rate-modifying activity* (RMA) can be applied to the machine to change (usually decrease) jobs processing times. The time p_j of processing job j changes after the RMA to $\delta_j p_j$.

Panwalkar, Smith and Seidmann [5] were the first to consider the common due date assignment in a single machine scheduling and to provide a polynomial time algorithm for solving the problem. Lee & Leon [2] consider several problems of scheduling on a single machine with the RMA (minimizing makespan, flow-time, weighted flow-time and maximum lateness). Mosheiov & Sidney [4] in their paper on new results with rate modifications study the problems of minimizing makespan with precedence relations, minimizing makespan with learning effect and minimizing the number of tardy jobs.

In a recent paper, Mosheiov & Oron [3] address the problem of Panwalkar et al. with RMA available, providing an $O(n^4)$ algorithm to solve the problem for any $\delta_j > 0$. In our paper we consider several useful properties of this problem.

The paper is organized as follows. Section 2 defines the problem. In Section 3 we describe the properties of the problem which allows us to reduce the runtime for solving some cases of the problem with $0 < \delta_j < 1$. Conclusions are placed into Section 4.

2 Problem formulation

Consider a scheduling situation when n jobs are to be processed on a single machine under common due date d, which is to be assigned. Each job $j, j = 1,2,\ldots,n$, becomes available at time zero and initially has processing time p_j. Let C_j be the completion time of job j in a certain schedule s. Let $E_j = \max\{0, d - C_j\}$ and $T_j = \max\{0, C_j - d\}$ be earliness and tardiness of job j. There exists a possibility to perform the RMA with duration t before some job. For each job j, let δ_j (with $0 < \delta_j \le 1$) be a processing time modification coefficient. If job j is scheduled after the RMA then its processing time changes to $\delta_j p_j$. The objective is to determine a job schedule with the position of the RMA in it and assign a common due date to minimize

$$f(d,s) = \sum_{j=1}^{n} (\alpha E_j + \beta T_j + \gamma d),$$

where $\alpha > 0$, $\beta > 0$ and $\gamma > 0$ are earliness, tardiness and due date per unit penalties, respectively.

So, we have to find an optimal position m^* of job before which the RMA is scheduled, an optimal schedule s^* of jobs and an optimal due date d^* to minimize total penalty $f(d,s)$.

Extending the standard scheme for scheduling notation [1], we refer to our problem as

$$1|\,RMA, d_j := d\,|\sum(\alpha E_j + \beta T_j + \gamma d)$$

3 Problem solving

Mosheiov & Oron [3] consider the common due date assignment and scheduling problem $1|\,RMA, d_j := d\,|\sum(\alpha E_j + \beta T_j + \gamma d)$ and proposed an $O(n^4)$ algorithm for solving this problem in case of $\delta_j > 0$. The idea of the algorithm is the following. The problem is modeled as bipartite matching problem which is solved for a given location of the RMA by Hungarian method. Placing the RMA at each of n possible positions and solving each of the corresponding problems in $O(n^3)$ time gives $O(n^4)$ runtime of the algorithm.

We restrict the value of coefficients δ_j by $0 < \delta_j \le 1$ that is natural for the maintenance operations and prove the properties which can reduce algorithm runtime significantly in some cases of the problem.

Panwalkar , Smith and Seidmann [5] provide a formula to obtain an optimal due date position K, which is valid also for the problem with the RMA [3]:

$$K := \lceil n(\beta - \gamma)/(\alpha + \beta) \rceil$$

Property 1

If the RMA is placed before the job in position m and $m > K$, then jobs are ordered in the LPT order in the optimal sequence before the position $K+1$ and in the SPT order in the positions j, where $K < j < m$ and $m \le j \le n$.

Here SPT and LPT stands for *shortest* and *longest processing time*, respectively. One can prove this property by interchanging adjacent jobs.

Property 2

There exists an optimal schedule where the RMA is scheduled either at time 0 or just before the job m, $K < m \le n$ (or is not scheduled at all).

Proof.

Suppose that the RMA is placed before some job m^*, $1 < m^* \leq K$, in an optimal schedule with cost function value F^*. If we place the RMA before the job in the (m^*-1) position, then the due date remains at the same position and the new schedule with the value F of the objective function will be not worse than the previous one. Indeed, the changing of the objective function is the following:

$$dF = F^* - F = \alpha(m^*-1)p_{[m^*-1]}(1-\delta_{[m^*-1]}) \geq 0,$$

where $[j]$ denotes the job in the position j. Repeating the moving of the RMA to the left, we can place the RMA at time zero, and the obtained schedule will be still optimal.

Corollary 1

If $K = n$ then the RMA is scheduled at time 0 or is not scheduled at all and the problem can be solved in $O(n \log n)$ time using the algorithm of Panwalkar et al.

Now consider the situation when the RMA is placed before some job m, $K < m \leq n$, and compare it to the schedule with the same sequence of jobs but without the RMA. Let $u_j = p_j(1-\delta_j)$. The difference between the objective function values F_{rma} (with the RMA) and F (without the RMA) is determined by (1):

$$F_{rma} - F = \beta(n-m+1)t - \beta \sum_{j=1}^{n-m+1} ju_{[n-j+1]}, \qquad (1)$$

If (1) is negative then placing the RMA before the position m may reduce the value of the objective function. It can be in case when

$$(n-m+1)t < A, \qquad (2)$$

where $A = \sum_{j=1}^{n-m+1} ju_{[n-j+1]}$

The following properties provide upper bounds for the value of A.

Property 3

The value of A obtains its maximum when $n-m+1$ jobs [m],[m+1],..., [n] are the jobs with the largest $p_j\delta_j$ ordered in descending order of $u_{[j]}$, i.e. $u_{[m]} \geq u_{[m+1]} \geq \ldots \geq u_{[n]}$.

Property 4

If $u_j \le u_i \Rightarrow p_j\delta_j \le p_i\delta_i$ *holds for all* $1 \le i, j \le n$, *then* A *obtains its maximum by placing* m *jobs with largest value* $p_j\delta_j$ *in the ascending order of* $u_{[j]}$.

One can prove this taking into account that, according to Property 1, the optimal order of jobs after the RMA is the SPT order (if the RMA is placed after the due date).

Properties 1-4 allow us to construct an optimal solution in $O(n^2)$ time in case when (2) does not hold for all m, $K < m \le n$. Thus, the complexity of Mosheiov and Oron algorithm can be reduced in this case from $O(n^4)$ to $O(n^2)$. Let (2) hold for l positions of the RMA. Then the complexity of Mosheiov and Oron algorithm will be $O(n^3 l)$. Note, that $l \le n - K$ according to Property 2.

4 Conclusions

In this paper we consider the scheduling problem of minimizing total weighted sum of earliness-tardiness and assignable common due date costs with the rate-modifying activity available during the jobs processing. For the case of $0 < \delta_j < 1$ we describe some properties which allow us to reduce significantly the runtime of the algorithm for solving the problem.

Similar approach to the problem analysis can be used for other due date assignment problems with the rate modifying activity and for scheduling problems with learning effect.

Acknowledgments

The research was partially supported by INTAS project 03-51-5501 and BRFFR project.

References

1. Lawler EL, Lenstra JK, Rinnooy Kan AHG., Shmoys DB (1993): Sequencing and scheduling: Algorithms and complexity. In: Graves SC, Zipkin PH, Rinnooy Kan AHG (eds.), Logistics of Production and Inventory. Handbooks in Operations Research and Management Science, vol. 4, North-Holland, Amsterdam : 445– 522.
2 Lee CL, Leon VJ (2001) Machine scheduling with a rate-modifying activity. European Journal of Operational Research 129 : 119-128.

3 Mosheiov, G., Oron, D (2005): Due-date assignment and maintenance activity scheduling problem. Computers & Operations Research (*to appear)*.
4 Mosheiov G, Sidney JB (2004) New results on sequencing with rate modifications. INFOR 41: 155-163.
5 Panwalkar SS, Smith ML, Seidmann A (1982) Common due date assignment to minimize total penalty for the one machine scheduling problem. Operations Research 30 : 391–399.

Primal-Dual Combined with Constraint Propagation for Solving RCPSPWET

András Kéri[1] and Tamás Kis[2]*

[1] Professur für BWL, insb. Verkehrsbetriebslehre und Logistik, Institut für Wirtschaft und Verkehr, Fakultät Verkehrswissenschaften "Friedrich List", Technische Universität Dresden, Andreas-Schubert-Str. 23, 01069 Dresden, andras.keri@mailbox.tu-dresden.de

[2] Computer and Automation Research Institute, 1111 Budapest, Kende u. 13-17, tamas.kis@sztaki.hu

Key words: Production Scheduling, Project Scheduling, Network Flows, Constraint Propagation

1 Introduction

In this paper we briefly describe a new exact method for solving the resource-constrained project scheduling problem with weighted earliness/tardiness penalty costs (RCPSPWET). The non-regular objective function makes the problem harder to solve to optimality than the thoroughly studied RCPSP with the (regular) makespan criterion. The input of the problem consists of a set of n activities and each activity $i \in N := \{1, \ldots, n\}$ has a processing time p_i, due date $\bar{d}_i$, earliness cost w_i^E and tardiness cost w_i^T (all numbers are integral and non-negative). If activity i completes at time C_i, the cost incurred is $f_i(C_i) = w_i^E \max\{\bar{d}_i - C_i, 0\} + w_i^T \max\{C_i - \bar{d}_i, 0\}$. Activities are connected by end-to-start precedence constraints, A, (no directed cycle is permitted). Finally, there are m renewable resources with constant capacities b_k, $1 \leq k \leq m$. The resource requirements of activity i are given by non-negative integer numbers r_{ik}. Once activity i is started, it cannot be interrupted and throughout its execution it requires r_{ik} units of resource k, $1 \leq k \leq m$. We have to find the completion time C_i of each activity i such that the demand for any resource does not exceed its capacity at any moment in time, the precedence constraints are respected and the total cost incurred, $\sum_i f_i(C_i)$, is minimized.

There are very few results on this problem. Vanhoucke et al. [6] modify and extend the branch-and-bound procedure of Demeulemeester and Herroelen [2]

* supported by Bolyai János research grant BO/00380/05, and OTKA T046509.

designed for RCPSP with the makespan objective. The branch-and-bound procedure uses minimal-delaying alternatives for branching and applies the subset dominance rule. For computing a lower bound in each node of the search-tree, the authors propose a very fast subroutine for solving the relaxation of the problem without resource constraints which can be expressed by the following linear program:

$$\min \sum_i w_i^E E_i + w_i^T T_i \tag{1}$$

$$-C_i + C_j \geq p_j, \quad (i,j) \in A \tag{2}$$

$$E_i + C_i \geq \bar{d}_i, \quad i \in N \tag{3}$$

$$T_i - C_i \geq -\bar{d}_i, \quad i \in N \tag{4}$$

$$E_i, T_i \geq 0, \quad i \in N \tag{5}$$

$$C_i \geq p_i, \quad i \in N \tag{6}$$

The decision variables C_i, E_i and T_i denote the completion time, the earliness and the tardiness of activity i, respectively.

Schwindt [5] studies the more general problem where there are minimum and maximum time lags between the project activities. He computes the lower bound by a steepest-descent procedure which exploits that the objective function is continuous, convex and differentiable from the left and from the right, but not in points C with $C_i = \bar{d}_i$.

The structure of the paper is as follows: In Section 2 we present (1) a new technique for computing a lower bound in each node of the search tree, and (2) a new method for computing time windows for activities, that we use in constraint propagation. Finally, in Section 3 we compare our method to results in the literature.

2 Our method

We use the same branch-and-bound procedure as Vanhoucke et al. [6], but compute lower bounds in a different way in each node of the search tree. In addition, we determine a time window for each activity in each node of the search tree and use them to derive new precedence arcs by standard constraint propagation techniques [3, 1]. To our best knowledge, constraint propagation has not been used before for solving RCPSPWET.

2.1 Lower bound computations

In order to compute lower bounds in each node of the search tree, we propose to consider the dual of the relaxation (1)-(6). In the following, we call (1)-(6) the *dual problem*, whereas its dual will be referred to as the *primal problem*. In fact, the primal problem is equivalent to a minimum cost circulation problem in a network with $n+1$ nodes. Our main idea is to use the primal-dual method in a clever way for solving the minimum cost circulation problem. Namely, suppose we have a pair of optimal primal-dual solutions in a node of the search-tree. Then, when branching using minimal delaying alternatives, each child is obtained by adding a set of new arcs to the network. As the network changes only a little bit, the dual solution at hand can easily be adjusted to be feasible for the child's network and the primal solution need to be only slightly changed so that complementary slackness conditions hold. ¿From this initial solution, the primal-dual method needs only very few iterations to obtain a pair of optimal primal-dual solutions for the network of the child. In contrast, both [6] and [5] compute the lower bound from scratch in each node of the search-tree. As a consequence, although the two methods may be faster than the primal-dual method when starting from scratch, with our method of initialization, primal-dual finds the lower bound much faster. Finally, notice that our approach is not restricted to the problem with precedence constraints, it works also with minimum and maximum time legs between project activities.

2.2 Time Windows

As a second enhancement, we suggest a method for computing a time window for each project activity and, using this, apply constraint propagation techniques to derive additional arcs between the project activities in each node of the search-tree. Given an upper bound ub on the optimal objective function value, we can determine a time window for each activity as follows. Let $g_i(t)$ be a function that for any t gives a lower bound on the least earliness-tardiness cost provided $C_i = t$. Moreover, suppose g_i is convex. If for all t, $g_i(t) \geq ub$ holds in some node of the search-tree, then the node can be fathomed. On the other hand, when there exists t_0 with $g_i(t_0) < ub$, then we try to find points $t_1 < t_0$ and $t_2 > t_0$ with $g_i(t_1) \geq ub$ and $g_i(t_2) \geq ub$, respectively. If t_1 exists, then let ef_i be the smallest $t > t_1$ with $g_i(t) < ub$. Similarly, let lf_i be the largest $t < t_2$ with $g_i(t) < ub$. Clearly, in any solution with earliness-tardiness cost smaller than ub, activity i cannot complete sooner than ef_i or later than lf_i. Notice that by exploiting the convexity of g_i, ef_i and lf_i can be computed in polynomial time in the size of the network.

Now we sketch our function g_i. First, we divide the set of activities into two subsets, N_{dep} and N_{indep}. N_{dep} consists of those activities that are connected to i by a directed path. In other words, these are the nodes that either precede or follow activity i in the network. N_{indep} contains all other activities. Clearly, for a given t, it is easy to compute the minimum earliness and tardiness of the activities in N_{dep}, if we omit the nodes in N_{indep}. On the other hand, for each node in N_{indep} we take the minimum possible earliness or tardiness cost. It is not necessarily 0, because the time window of the parent node is also valid for a child node in which we perform the computation.

Having computed the time windows, we use the so-called edge-finding procedure to deduce new arcs between project activities, see e.g., [1, 3].

3 Computational results

In order to evaluate and compare the methods, we have generated a large number of new test instances. The network and resource parameters are the same as described by Kolisch et al. in [4]. The due dates and the earliness-tardiness costs are generated the same way as Vanhoucke et al. [6]. We have also used a new additional control parameter, the *due date order*. It is 0 if for all $(i,j) \in A$ $d_i \leq d_j$, and it is 1 when for all $(i,j) \in A$ $d_i \geq d_j$. A value between 0 and 1 is the ratio of the arcs with reverse due-date relation.

We have divided the test instances into *three groups* based on the number of activities, N, where $N = 10, 20$ or 30. In each group we have generated five instances for each combination of the following parameter values:

- Network complexity: 0.25, 0.5, 0.75
- Resource Factor: 0.25, 0.5, 0.75, 1
- Resource Strength: 0, 0.25, 0.5
- Due Date Factor: 1, 1.25, 1.5
- Due Date Order: 0, 0.5, 1

The total number of combinations is 324, giving 1620 instances in each group.

We have implemented two algorithms: i) the simple primal-dual based algorithm, ii) the primal-dual algorithm with constraint propagation. We have also re-implemented Vanhoucke's algorithm [7], since his program can only solve instances with due date order 0. In the following, we refer to these three algorithms as MCF, MCF+CP and VA, respectively. We have run the tests on a PC with Pentium IV, 3.0 GHz CPU, and Linux operating system. The algorithms have been implemented in C++.

The test instances with 10 activities have been solved very quickly by all three solvers, so these instances are omitted from the following comparison.

Table 1 shows the number of instances solved to optimality in groups with $N = 20$ and 30 activities, respectively, by the algorithms in less than 1, 10, and 100 second, respectively.

Table 1. Number of instances solved within Time limit.

N	20			30		
Time limit	1s	10s	100s	1s	10s	100s
VA	1308	1486	1578	807	992	1158
MCF	1318	1507	1583	854	1020	1201
MCF+CP	1322	1507	1584	840	1022	1208

Table 2 depicts the average number of created nodes, the average number of branched nodes, and the average of the running time of instances which were solved in less than 100 seconds by all the three solvers. We can see that our two solvers outperform Vanhoucke's algorithm. The MCF+CP not only is the fastest algorithm, but it uses the least nodes to find the optimal solution.

Table 2. Results of instances solved in 100 seconds.

N	20			30		
	created nodes	branched nodes	running time	created nodes	branched nodes	running time
VA	41734	14505	2.16	51187	17598	6.07
MCF	41472	14389	1.73	52194	17779	4.21
MCF+CP	19774	6245	1.62	24777	7739	3.92

Table 3 shows the effect of the due date factor on the running time and the average number of the branched nodes. There is no strong correlation between this parameter and the difficulty of the instances.

Table 3. Effects of the due date factor on the algorithms' performance.

N	20			30		
DDF	1	1.25	1.5	1	1.25	1.5
VA	2.30s	2.11s	2.08s	5.96s	5.58s	6.63s
	17641	12607	13278	20305	16291	16354
MCF	1.97s	1.64s	1.59s	4.53s	4.34s	3.79s
	17272	12621	6847	20240	16609	10164
MCF+CP	1.67s	1.48s	1.71s	3.40s	3.69s	4.62s
	6245	5633	6847	6155	6689	10164

Table 4. Effects of the due date order on the algorithms' performance.

N	20			30		
DDO	0	0.5	1	0	0.5	1
VA	0.85s	2.14s	3.51s	5.17s	5.96s	7.33s
	8870	17186	17514	27611	11470	11929
MCF	1.10s	1.92s	2.18s	6.58s	3.06s	2.52s
	8859	16798	17564	27758	11840	11944
MCF+CP	1.67s	1.48s	1.71s	0.57s	1.65s	2.64s
	6245	5633	6847	1983	6126	10667

Finally, Table 4 shows the effect of the due date order. The larger is the due date order, the harder are the instances. Inspite of this, MCF solves easier the 30-activity instances with larger due date order than those with smaller due date order.

We can state that our two algorithms outperform that of Vanhoucke et al. in all cases, and in most cases MCF+CP is the best choice.

References

1. Baptiste P, Le Pape C and Nuijten W (2001), Constraint-Based Scheduling, Kluwer Academic Publishers, Massachusetts Dordrechts.
2. Demeulemeester EL, and Herroelen WS (1992), A branch-and-bound procedure for the multiple resource-constrained project scheduling problem, Management Science 38:1803-1818.
3. Dorndorf U (2002), Project Scheduling with Time Windows, Physica-Verlag, Heidelberg New York.
4. Kolisch R, Sprecher A, and Drexl A (1995): Characterization and Generation of a General Class of Resource-Constrained Project Scheduling Problem, Management Science 41, pp. 1693-1703.
5. Schwindt C (2000), Minimizing earliness-tardiness costs of resource-constrained projects, In: Inderfurth K, Schwoediauer G, Domschke W, Juhnke F, Kleinschmidt P and Waescher G (eds.), Operations Research Proceedings, Springer, Berlin, 1999, pp. 402-407.
6. Vanhoucke M, Demeulemeester EL, and Herroelen WS (2001), An exact procedure for the resource-constrained weighted earliness-tardiness project scheduling problem, Annals of Oper. Res. 102:179-196.
7. Vanhoucke M (2001), RCPSPWET executable program and benchmark instances, http://www.projectmanagement.ugent.be/RCPSPWET.htm

Ein Ameisenalgorithmus für die ressourcenbeschränkte Projektplanung mit Zeitfenstern und Kalendern

Thomas Knechtel und Jens Peter Kempkes

Decision Support and Operations Research Laboratory, Universität Paderborn, Warburger Str. 100, D-33100 Paderborn, Germany {knechtel;kempkes}@dsor.de

Viele praktische Projektplanungsprobleme erfordern die Berücksichtigung von Kalendern, die die Arbeitszeiten der gegebenen Ressourcen definieren. Während einige Aktivitäten während Pausen unterbrochen werden können, müssen andere Aktivitäten, beispielsweise aus technischen Gründen, ohne Unterbrechung ausgeführt werden. Minimale und maximale Vorrangbeziehungen können ebenfalls von Kalendern abhängen. Im folgenden Beitrag wird ein Ameisenalgorithmus für die Lösung des ressourcenbeschränkten Projektplanungsproblems mit Zeitfenstern und Kalendern (RCPSP/max-kal) vorgestellt. Anhand ausgewählter Testinstanzen wird die Lösungsgüte im Vergleich zu Prioritätsregelverfahren und einem Genetischen Algorithmus bestimmt.

1 Einleitung

Die ressourcenbeschränkte Projektplanung beschäftigt sich mit der Anordnung von Projektaktivitäten unter Beachtung von Zeit- und Ressourcenrestriktionen. Für das bekannte Resource-Constrained Project Scheduling Problem (RCPSP) wurde in den letzten Jahren eine Vielzahl von leistungsstarken Lösungsverfahren entwickelt. Die Untersuchungen in [5] zeigen, dass insbesondere Metaheuristiken sehr gute Ergebnisse liefern.

In der Praxis finden sich jedoch häufig Planungsprobleme, deren Anforderungen über die Modelleigenschaften des RCPSP hinausgehen. Durch den Einsatz minimaler und maximaler Vorrangbeziehungen entstehen einerseits Zeitfenster für die Durchführung der Projektaktivitäten. Andererseits definieren Kalender Zeitintervalle, in denen Ressourcen nicht zur Verfügung stehen und die Bearbeitung der Aktivitäten unterbrochen werden muss [3]. Während das Projektplanungsproblem mit Zeitfenstern (RCPSP/max) bereits ausführlich untersucht wurde und effiziente Lösungsverfahren vorgestellt wurden [7], werden lediglich in [2] Lösungsverfahren für das RCPSP/max-kal vorgestellt.

Im Rahmen dieses Beitrags wird am Beispiel eines in [6] vorgestellten Ameisenalgorithmus (AS) gezeigt, wie Metaheuristiken für das RCPSP auf das RCPSP/max-kal angepasst werden können. Dazu wird in Kapitel 2 zunächst die Problemstellung skizziert, bevor in Kapitel 3 der Aufbau des AS erläutert wird. In Kapitel 4 werden numerische Ergebnisse präsentiert und die Ergebnisse zusammengefasst.

2 Problemdefinition

Ein Projekt besteht aus einer Menge $\mathcal{V} = \{0, 1, \ldots, n, n+1\}$ von Aktivitäten. Die Bearbeitungsdauer einer Aktivität $i \in \mathcal{V}$ ist gegeben durch $p_i \in \mathbb{N}_0$. Bei den Aktivitäten 0 und $n+1$ handelt es sich um fiktive Aktivitäten mit $p_i = p_{n+1} = 0$, welche den Projektanfang bzw. das Projektende repräsentieren. Der Modellierung liegt ein Vorgangsknotennetzwerk zugrunde, in dem die Aktivitäten $i \in \mathcal{V}$ den Knoten und die Vorrangbeziehungen zwischen zwei Aktivitäten i und j den Kanten entsprechen.

Für die Bearbeitung der Aktivitäten wird die Menge $\mathcal{R}$ der Ressourcen benötigt. Die Kapazität einer Ressource $k \in \mathcal{R}$ ist gegeben durch R_k. Wie eingangs erwähnt, stehen die Ressourcen nicht durchgängig zur Verfügung. Für jede Ressource k wird ein *Ressourcenkalender* definiert:

$$A_k^{\mathcal{R}}(t) := \begin{cases} 0 & \text{, falls } R_k(t) = 0 \\ 1 & \text{, falls } R_k(t) > 0 \end{cases} \qquad \forall \quad t \in \mathbb{N}_0 \tag{1}$$

Der Bedarf r_{ik} einer Aktivität i an den Ressourcen $k \in \mathcal{R}$ ist konstant. Aus den Ressourcenkalendern der für die Bearbeitung der Aktivitäten $i \in \mathcal{V}$ benötigten Ressourcen leiten sich für jede Aktivität Arbeits- und Nichtarbeitsperioden ab. Mit Hilfe des Begriffs der Ressourcenmenge $\mathcal{R}_i := \{k \in \mathcal{R} | r_{ik} > 0\}$ der Aktivität i sei der *Aktivitätenkalender* von i wie folgt definiert:

$$A_i(t) := \begin{cases} \min\limits_{k \in \mathcal{R}_i} A_k^{\mathcal{R}}(t) & \text{, falls } \mathcal{R}_i \neq \emptyset \\ 1 & \text{, sonst.} \end{cases} \qquad \forall \quad t \in \mathbb{N}_0 \tag{2}$$

Die Menge der Aktivitäten $\mathcal{V}$ wird in zwei disjunkte Teilmengen geteilt: in die Menge der *unterbrechbaren* Aktivitäten $\mathcal{V}^p$ und die Menge der *nicht unterbrechbaren* Aktivitäten $\mathcal{V}^{np}$. Dabei heißt eine Aktivität $i \in \mathcal{V}$ nicht unterbrechbar, wenn $A_i(t)$ im Intervall $t \in [S_i, C_i]$ keine Nichtarbeitsperiode enthalten darf, wobei S_i den Start- und C_i den Endzeitpunkt der Aktivität i bezeichnet. Dagegen wird eine Aktivität i, deren Ausführung zum Zeitpunkt t mit $A_i(t) = 0$ unterbrochen werden darf, unterbrechbar genannt. Eine Unterbrechung der Aktivität $i \in \mathcal{V}^p$ ist jedoch nur dann zulässig, wenn ihre Bearbeitung zum nächsten Zeitpunkt $t' > t$ mit $A_i(t') = 1$ fortgesetzt wird.

Aufgrund der Unterbrechbarkeit der Vorgänge muss für jede Ressource $k \in \mathcal{R}$ defniert werden, ob sie während einer Unterbrechung einer Aktivität weiterhin beansprucht oder freigesetzt wird. Ob eine Ressource bei einer Vorgangsunterbrechung freigesetzt wird, ist durch $\rho_k^R \in \{0, 1\} (k \in \mathcal{R})$ anzugeben.

Eine Ressource k mit $\rho_k^R = 0$ steht bei Unterbrechung eines von ihr bearbeiteten Vorgangs bei dessen Unterbrechung für andere Vorgänge zur Verfügung.

Zwischen zwei Aktivitäten i und j können beliebige zeitliche Anordnungsbeziehungen definiert werden. Diesen werden unter Berücksichtigung der für sie relevanten Ressourcen $\mathcal{R}_{ij} \subseteq \mathcal{R}$ Anordnungskalender zugewiesen:

$$A_{ij}(t) := \begin{cases} \min\limits_{k \in \mathcal{R}_{ij}} A_k^{\mathcal{R}}(t) & \text{, falls } \mathcal{R}_{ij} \neq \emptyset \\ 1 & \text{, sonst.} \end{cases} \qquad \forall \quad t \in \mathbb{N}_0 \tag{3}$$

Für jeden zeitlichen Mindestabstand d_{ij}^{min} zwischen den Startzeitpunkten der Aktivitäten i und j wird eine gerichtete Kante $\langle i,j \rangle \in E$ mit einer Gewichtung $\delta_{ij} := d_{ij}^{min}$ eingeführt. Bei zeitlichen Höchstabständen d_{ij}^{max} zwischen den Startzeitpunkten der Aktivitäten i und j hingegen wird eine Rückwärtskante $\langle j,i \rangle$ mit einer Gewichtung $\delta_{ji} := -d_{ij}^{max}$ eingeführt. Zusätzlich erfolgt die Angabe, um welchen Anordnungstyp $e_ie_j \in \{SS, CS, SC, CC\}$ es sich handelt. Der Zeitpunkt des Ereignisses e_i von Vorgang i werde mit X_i beschrieben.

Die Menge $\mathcal{A}^{kal}(S,t) := \{i \in \mathcal{V} | S_i \leq t < C_i(S_i, \mathcal{R}_i)\}$ definiert die sich zum Zeitpunkt t in Ausführung befindlichen Aktivitäten. Die obere Schranke für die Dauer des Projektes ist gegeben durch $\overline{d}^{kal}$. Damit ergibt sich das Modell für das RCPSP/max-kal als:

$$\min S_{n+1} \tag{4}$$

$$\delta_{ij}^{e_ie_j} \leq \sum_{t=X_i}^{X_j-1} A_{ij}(t) - \sum_{t=X_j}^{X_i-1} A_{ij}(t) \qquad \forall \quad \langle i,j \rangle \in E \tag{5}$$

$$\sum_{t=S_i}^{S_i+p_i-1} A_i(t) = p_i \qquad \forall \quad i \in \mathcal{V}^{np} \tag{6}$$

$$C_i = S_i + p_i \qquad \forall \quad i \in \mathcal{V}^{np} \tag{7}$$

$$S_i = \max\{t \leq C_i - p_i | \sum_{\tau=t}^{C_i-1} A_i(\tau) = p_i\} \qquad \forall \quad i \in \mathcal{V}^p \tag{8}$$

$$C_i = \min\{t \geq S_i + p_i | \sum_{\tau=S_i}^{t-1} A_i(\tau) = p_i\} \qquad \forall \quad i \in \mathcal{V}^p \tag{9}$$

$$R_k(t) \geq \sum_{\substack{i \in \mathcal{A}^{kal}(S,t) \\ \wedge A_i(t)=1}} r_{ik} + \sum_{\substack{i \in \mathcal{A}^{kal}(S,t) \\ \wedge A_i(t)=0}} r_{ik}\rho_k^R \qquad \forall \quad k \in R, t \in \mathbb{N}_0 \tag{10}$$

$$S_{n+1} \leq \overline{d}^{kal} \tag{11}$$

$$S_i \in \mathbb{N}_0 \qquad \forall \quad i \in \mathcal{V} \tag{12}$$

Die Zielfunktion (4) minimiert den Startzeitpunkt der fiktiven Aktivität $n+1$ und damit die Projektdauer. Restriktionen (5) gewährleisten die Einhaltung der zeitlichen Anordnungsbeziehungen des Problems. Restriktionen (6) stellen für nicht unterbrechbare Aktivitäten sicher, dass ihre Bearbeitung nur dann begonnen wird, wenn ihr Aktivitätenkalender in dem gewählten Zeitraum keine Nicht-Arbeitsperioden enthält. Der Endzeitpunkt nicht unterbrechbarer Aktivitäten ergibt sich aus (7). Die Restriktionen (8) und (9) sorgen dafür, dass unterbrechbare Aktivitäten nur aufgrund einer Nicht-Arbeitsperiode ihres Aktivitätenkalenders unterbrochen werden und in der nächstfolgenden Arbeitsperiode wieder bearbeitet werden. Die Restriktionen (10) stellen sicher, dass die Ressourcenbeschränkungen nicht verletzt werden. Restriktion (11) fordert die Einhaltung des Planungshorizontes des Projektes.

3 Lösungsverfahren

Die Grundlage des AS-RCPSP/max-kal bilden ein serielles sowie ein paralleles Schedule Generation Scheme (sSGS, pSGS), die auf das RCPSP/max-kal angepasst wurden [2]. Beide Konstruktionsheuristiken fügen die Aktivitäten $i \in \mathcal{V}$ solange nacheinander einem partiellen Projektplan hinzu, bis alle Aktivitäten zeit- und ressourcenzulässig eingeplant sind. Die einzuplanende Aktivität i wird aus der Menge der aufgrund ihrer Vorrangbeziehungen einplanbaren Vorgänge ε ausgewählt. Da durch minimale und maximale Zeitabstände Zyklen im Vorgangsknotennetzwerk entstehen können, kann es im Verlauf der Einplanungen zu Verletzungen der zeitlichen Vorrangbeziehungen kommen. In diesem Fall wird durch Entfernen und Rechtsverschieben ausgewählter Vorgänge des partiellen Projektplans versucht, die Zeitzulässigkeit wieder herzustellen.

Die Idee des AS-RCPSP/max-kal ist es, die Einplanungsreihenfolge des sSGS und pSGS so zu steuern, dass gute Projektpläne entstehen [6]. Dazu konstruieren in jeder Generation m Ameisen mit Hilfe des sSGS oder pSGS eine Lösung. Für die Auswahl der nächsten einzuplanenden Aktivität berücksichtigt jede Ameise sowohl heuristische Informationen auf Basis einer Prioritätsregel (η_{ij}), als auch die Pheromoninformationen der Ameisen vorheriger Generationen (τ_{ij}). η_{ij} und τ_{ij} bewerten die Vorteilhaftigkeit, Aktivität j als $i$te Aktivität dem partiellen Projektplan hinzuzufügen. Nach [1] ergibt sich die Wahrscheinlichkeit, dass Aktivität j als $i$te Aktivität dem partiellen Projektplan hinzugefügt wird als:

$$p_{ij} = \frac{[\tau_{ij}]^\alpha \cdot [\eta_{ij}]^\beta}{\sum_{h \in \varepsilon} [\tau_{hj}]^\alpha \cdot [\eta_{hj}]^\beta} \tag{13}$$

Die Exponenten α und β gewichten dabei den Einfluss der Pheromon- sowie der heuristischen Informationen. In [6] wurde eine Erweiterung der in Formel (10) dargestellten lokalen Pheromonauswertung um eine gewichtete Summen-Pheromonauswertung vorgestellt. Über einen Faktor $c \in [0, 1]$ wird der relative Einfluss der lokalen und der gewichteten Summen-Pheromonauswertung bestimmt. In Formel (13) wird dazu τ_{ij} durch:

$$\tau'_{ij} := c \cdot x_i \cdot \tau_{ij} + (1 - c) \cdot y_i \cdot \sum_{k=1}^{i} \gamma^{i-k} \tau_{kj} \tag{14}$$

mit $x_i := \sum_{h \in \varepsilon} \sum_{k=1}^{i} \gamma^{i-k} \tau_{kh}$ und $y_i := \sum_{h \in \varepsilon} \tau_{ih}$ ersetzt. Für $c = 1$ liegt ausschließlich lokale, für $c = 0$ ausschließlich Summen-Pheromonauswertung vor. Der Parameter γ bestimmt die relative Gewichtung von Pheromonwerten aus früheren Einplanungsentscheidungen. Bei einem Wert $\gamma = 1$ wird jeder Pheromonwert $\tau_{kj}, k \leq i$ gleich gewichtet. Werte von $\gamma < 1$ ($\gamma > 1$) hingegen gewichten Pheromonwerte aus früheren Entscheidungen schwächer (stärker).

Auf Basis ihrer finalen Einplanungsreihenfolge aktualisieren die bisher beste Ameise sowie die beste Ameise der aktuellen Generation die Pheromoninformationen. Zuvor verdunsten jedoch die Pheromonspuren mit:

$$\tau_{ij} = (1 - \rho) \cdot \tau_{ij} \tag{15}$$

Die Aktualisierung der Pheromoninformationen erfolgt nach:

$$\tau_{ij} = \tau_{ij} + \rho \cdot \frac{1}{2T^\star} \tag{16}$$

Der Parameter $T^\star$ entspricht dabei der durch die bisher beste Ameise, bzw. beste Ameise der aktuellen Generation berechneten Projektdauer.

Die Berechnung der heuristischen Informationen erfolgt mit der Prioritätsregel Minimum Latest Starttime (*LSTmin)*. Eine Normalisierung der Regel führt dazu, dass der schlechteste Wert für η_{ij} 1 wird:

$$\eta_{ij} = \max_{h \in \varepsilon} LS_h - LS_j + 1 \tag{17}$$

Die Auswahl, welches *SGS* eine Ameise verwendet, wird ebenfalls über Pheromoninformationen beeinflusst. Neben der Aktualisierung von τ_{ij} wird ein Wert ν_{SGS} entsprechend den Formeln (11) und (12) aktualisiert. Die Wahrscheinlichkeit, dass eine Ameise ein *SGS* auswählt, ergibt sich als:

$$p_{SGS} = \frac{\nu_{SGS}}{\sum_{i \in SGS} \nu_i} \tag{18}$$

Durch dieses Vorgehen spezialisiert sich der Algorithmus im Verlauf der Optimierung auf die Erzeugung von Ameisen mit dem für die vorliegende Instanz geeigneteren SGS.

4 Numerische Ergebnisse

Der Test des AS-RCPSP/max-kal basiert auf dem Testset A^{cal} aus [2]. Es besteht aus 810 Instanzen mit je 10, 15 oder 20 Aktivitäten, 5 Ressourcen sowie zwei unterschiedlichen Kalendern. Bisher waren für 634 Instanzen zulässige Lösungen bekannt. Alle Testläufe wurden auf einem AMD Athlon mit 1,06 GHz und 256 MB Arbeitspeicher durchgeführt. Ein Durchlauf des *AS-RCPSP/max-kal* benötigte durchschnittlich 12 Sekunden pro Instanz.

Tabelle 1 zeigt die numerischen Ergebnisse des AS-RCPSP/max-kal im Vergleich zu einem angepassten Genetischen Algorithmus (GA) (vgl. [4]), sowie dem sSGS und pSGS in Kombination mit der Prioritätsregel LSTmin.

Tabelle 1. Vergleich des AS-RCPSP/max-kal mit ausgewählten Lösungsverfahren anhand des Testsets A^{cal}, 1000 berechnete Projektpläne pro Instanz

Lösungsverfahren	Abw. von LB0	gel. Instanzen
AS-RCPSP/max-kal ($c = 0,5; \gamma = 0,75$)	23,3 % (22,2 %)	644
GA	23,7 % (22,6 %)	644
sSGS (LSTmin)	27,6 % (27,2 %)	631
pSGS (LSTmin)	32,3 % (31,7 %)	634

Die Spalte „Abw. von LB0" enthält die Werte der durchschnittlichen relativen Abweichung von der unteren Schranke LB0. Die Werte in Klammern beziehen sich auf die Schnittmenge der durch die verschiedenen Verfahren gefundenen Lösungen. Die Spalte „gel. Instanzen" gibt die Anzahl der gelösten Instanzen für die Verfahren wieder. Als beste Parameterkombination für den AS-RCPSP/max-kal hat sich $c = 0,5$ und $\gamma = 0,75$ erwiesen. Die Einstellungen für die Parameter ρ, α und β wurden wie in [6] vorgeschlagen übernommen. Mit dieser Konfiguration erzielt der AS-RCPSP/max-kal die besten Ergebnisse im Vergleich der Verfahren.

Das in diesem Beitrag vorgestellte Lösungsverfahren verdeutlicht, dass die auf den SGS basierenden Metaheuristiken eine gute Möglichkeit darstellen, allgemeinere Projektplanungsprobleme zu lösen. Die bei den Berechnungen eingesetzten SGS sind bspw. in der Lage, weitere Aspekte wie beliebig unterbrechbare Aktivitäten oder ressourcenbeanspruchende Sammelvorgänge zu berücksichtigen und bieten damit einen Großteil an für den praktischen Einsatz notwendiger Flexibilität.

Literaturverzeichnis

1. Dorigo M (1992) Optimization, Learning and Natural Algorithms. Ph.D. Thesis, Dip. Elettronica e Informazione, Politecnico di Milano
2. Franck B (1999) Prioritätsregelverfahren für die ressourcenbeschränkte Projektplanung mit und ohne Kalender. Shaker, Aachen
3. Franck B, Neumann K, Schwindt C (2001) Project Scheduling with calendars. In: OR Spektrum 23:325-334
4. Hartmann S (2002) A Self-Adapting Genetic Algorithm for Project Scheduling under Resource Constraints. In: Naval Research Logistics 49: 433-448
5. Kolisch R, Hartmann S (2005) Experimental Investigation of Heuristics for Resource-Constrained Project Scheduling: An Update. To apper in European Journal of Operational Research
6. Merkle D, Middendorf M, Schmeck M (2002) Ant Colony Optimization for Resource-Constrained Project Scheduling. In: IEEE Transactions Vol. 6, No. 4: 333-346
7. Neumann K, Schwindt C, Zimmermann J (2003) Project Scheduling with Time Windows and Scarce Resources. Springer, Berlin Heidelberg New York

The Flow Shop Problem with Random Operation Processing Times *

Roman A. Koryakin[1] and Sergey V. Sevastyanov[2]

Sobolev Institute of Mathematics, prospekt Akademika Koptyuga 4
630090 Novosibirsk, Russia
[1]romank@mail.nsk.ru
[2]seva@math.nsc.ru

Summary. We consider the classical flow shop problem with m machines, n jobs and the minimum makespan objective. The problem is treated in stochastic formulation, where all operation processing times are random variables with distribution from a given class F. We present a polynomial time algorithm with absolute performance guarantee $C_{\max}(S) - L \leq 1.5(m-1)p + o(1)$ that holds *with high probability* (Frieze, 1998) for $n \to \infty$, where L is a trivial lower bound on the optimum (equal to the maximum machine load) and p is the maximum operation processing time. Class F includes distributions with regularly varying tails. The algorithm presented is based on a new algorithm for the compact vector summation problem and constructs a permutation schedule. The new absolute guarantee is superior to the best-known absolute guarantee for the considered problem ($C_{\max}(S) - L \leq (m-1)(m-2+1/(m-2))p$; Sevastyanov, 1995) that holds for all possible inputs of the flow shop problem.

1 Introduction

The flow shop problem is a classical multi-stage machine scheduling problem first time considered by Johnson [6]. We formulate it as follows. Jobs $J_1, \ldots, J_n$ are to be processed on machines $M_1, \ldots, M_m$. Each job J_j consists of m operations $o_{1j}, \ldots, o_{mj}$. Operation o_{ij} is processed on machine M_i and requires time $p_{ij} \geq 0$ for its processing. Operation processing must satisfy the following conditions:

(a) every job is processed by at most one machine at a time;
(b) every machine executes at most one job at a time;
(c) no operation preemption is allowed;
(d) every job is processed on machines in the same predefined order $M_1, \ldots, M_m$.

* This research was supported by the Russian Foundation for Basic Research (grant 05-01-00960-a)

Let s_{ij} denote the starting time of operation o_{ij} in a given schedule. The goal is to derive a schedule $S = \{s_{ij} \mid j = 1, \ldots, n; \quad i = 1, \ldots, m\}$ satisfying **(a)**–**(d)** (*feasible schedule*) and minimizing the maximum operation completion time (the *makespan*):

$$C_{\max}(S) \doteq \max_{j,i}(s_{ij} + p_{ij}) \longrightarrow \min_{S}. \tag{1}$$

Let $L_i = \sum_{j=1}^{n} p_{ij}$ be the load of machine M_i, $L_{\max} = \max_i L_i$ be the maximum machine load, $l_{\max} = \max_j \sum_{i=1}^{m} p_{ij}$ be the maximum job length and $p_{\max} = \max_{i,j} p_{ij}$ be the maximum operation length. Obviously, $\max\{L_{\max}, l_{\max}\} \leq C_{\max}(S)$ for any feasible S.

It is well-known that the two-machine flow shop problem is polynomially solvable [6] and the three-machine problem is strongly NP-hard [4] but possesses a PTAS [5]. Concerning the best known up-to-date results, Shmoys et al. constructed an approximation algorithm for the flow shop problem with ratio performance guarantee $C_{\max}(S)/C_{\max}(S_{\text{opt}}) \leq O(\log^2 m)$ [13], while there are algorithms with absolute guarantees: $C_{\max}(S) \leq L_{\max} + O(m^2)p_{\max}$ (Sevastyanov [11], algorithm $\mathcal{A}$) and $C_{\max}(S) \leq (1+\delta)L_{\max} + K_\delta m(\log m)p_{\max}$ for any $\delta > 0$, where K_δ is a function depending on δ only (Sviridenko [12]).

In our paper, we consider the flow shop problem in stochastic formulation, where operation processing times are random variables with distributions from class F, and present algorithm $\mathcal{B}$ that, with high probability, delivers a feasible schedule of length at most $L_{\max} + O(m)p_{\max}$. This is the best up-to-date approximation algorithm applicable to the considered stochastic flow shop problem.

2 Compact Vector Summation in Flow Shop

Our algorithm $\mathcal{B}$ has the same steps as the above mentioned algorithm $\mathcal{A}$ and is also based on a compact vector summation algorithm. The result of this paper is mainly derived due to a significant improvement of two steps of algorithm $\mathcal{A}$ while leaving all other steps unaltered. Each of the two algorithms delivers a permutation schedule, i.e., a schedule in which the jobs pass through each machine in the same order.

Firstly, a *correct levelling procedure* is needed, so as we could apply a compact vector summation algorithm afterwards. The procedure levels out machine loads L_i by increasing some p_{ij}, while leaving $L_{\max}$ and $p_{\max}$ the same. Thus, we define the correct levelling procedure as a function $Z : \{p_{ij}\} \to \{p^*_{ij}\}$ with the following three properties: **(e)** $p_{ij} \leq p^*_{ij}$; **(f)** $p^*_{ij} \leq p_{\max}$; **(g)** $L^*_i = L_{\max}$.

Secondly, the following *compact vector summation* problem is formulated. Given an instance with processing times $\{p^*_{ij}\}$ satisfying **(e)**–**(g)**, we introduce vectors $\mathbf{e}^*_\mathbf{j} = (p^*_{2j} - p^*_{1j}, \ldots, p^*_{mj} - p^*_{1j}) \in \mathbb{R}^{m-1}$ and a norm $\|\cdot\|$ defined for a given vector $\mathbf{x} = (x_1, \ldots, x_{m-1})$ by

$$\|\mathbf{x}\| = \max\left\{\max_i |x(i)|, \max_{i,j} |x(i) - x(j)|\right\}. \tag{2}$$

One needs to derive a permutation $\pi = \{\pi_1, \ldots, \pi_n\}$ of indices $\{1, \ldots, n\}$ delivering a solution to the problem

$$\mathcal{R}_\pi \doteq \max_{k=1,\ldots,n-1} \|\mathbf{E}_{\boldsymbol{\pi}\mathbf{k}}\| \longrightarrow \min_\pi, \tag{3}$$

where $\mathbf{E}_{\boldsymbol{\pi}\mathbf{k}} = \mathbf{e}^*_{\boldsymbol{\pi}_1} + \ldots + \mathbf{e}^*_{\boldsymbol{\pi}_\mathbf{k}}$. It is essential here, that $\sum_{j=1}^n \mathbf{e}^*_{\boldsymbol{\pi}_\mathbf{j}} = \mathbf{0}$ by property **(g)**. Since the length of any permutation schedule S_π is known [11] to be estimated as

$$C_{\max}(S_\pi) \le L_{\max} + (m-1)(\mathcal{R}_\pi + p_{\max}), \tag{4}$$

a better algorithm for problem (3) gives a better estimate for the length of permutation schedule S_π.

Finally, schedule S_π is built according to the permutation π found. By property **(e)**, this schedule preserves its feasibility for the original values of p_{ij}, while remaining makespan estimate (4) the same.

Algorithm $\mathcal{A}$ assumes that the correct levelling procedure is executed arbitrarily: it does not matter which exactly operation processing times are increased to satisfy properties **(e)**–**(g)**. At the stage of solving problem (3) it uses general algorithm $\mathcal{C}$ for the compact vector summation problem, which gives the estimate $\mathcal{R}_\pi(\mathcal{C}) \le O(m)p_{\max}$ [9] (see also survey [10]). Instead of this, algorithm $\mathcal{B}$ includes the concrete and precise correct levelling procedure where p_{ij} to be increased are chosen heuristically, in order to prepare a "convenient" input data for problem (3). Then, using this data, algorithm $\mathcal{B}$ solves problem (3) with estimate $\mathcal{R}_\pi(\mathcal{B}) \le O(1)p_{\max}$, which implies the makespan estimate to $C_{\max}(S_\pi) \le L_{\max} + O(m)p_{\max}$.

3 Stochastic Formulation and Results

We consider the flow shop problem in stochastic formulation. Let p_{ij} be random variables with distributions from the class defined by the following two properties:

- **(1)** all p_{ij} are independent identically distributed random variables with absolutely continuous non-degenerate distribution;
- **(2)** $\mathbb{P}(p_{11} > x) \sim 1/f(x)$ for $x \to \infty$, where $f(x)$ is a regularly varying function with index $k > 2$.

Function $f(x)$ is called *regularly varying* with index k, if $f(tx)/f(x) \to t^k$ holds for $x \to \infty$ and any $t > 0$ [1, 8]. Obvious examples of $f(x)$ are x^k, $x^k \ln x$, *etc.* The following property of a regularly varying function with index k makes property **(2)** much clearer: for any $\varepsilon > 0$, $f(x)/x^{k+\varepsilon} \to 0$ and $f(x)/x^{k-\varepsilon} \to \infty$

hold with $x \to \infty$. In particular, property **(2)** implies the existence of a finite variance ($\mathbb{D}p_{11} < \infty$).

With random variables on the input, problem (1) is adjusted by the following definition [3]: event $A(n)$ occurs *with high probability* if $\mathbb{P}\{A(n)\} \to 1$ with $n \to \infty$. Our goal is to find an approximate solution to problem (1) that satisfies an absolute guarantee of the form $C_{\max}(S) \le L_{\max} + O(m)p_{\max}$ with high probability. Now we formulate main results of the paper.

Theorem 1. *There exists an $O(m^2n^2)$-time algorithm for the flow shop problem with m machines, n jobs and random operation processing times satisfying* **(1)**–**(2)**, *that, with high probability, yields a schedule with length at most $L_{\max} + 1.5(m-1)p_{\max} + o(1)$.*

Corollary 1. *There exists an $O(m^2n^2)$-time algorithm for the flow shop problem with m machines, n jobs and random operation processing times satisfying* **(1)**–**(2)**, *that, with high probability, yields a schedule with length at most $L_{\max} + O(m)p_{\max}$.*

4 Proof Outline

In this section, we give an outline of algorithm $\mathcal{B}$ and some elements of the proof of theorem 1. A precise description of the algorithm and the complete proof of our statements can be found in [7].

As said above (in section 2), algorithm $\mathcal{B}$ consists of three steps: levelling procedure, compact vector summation, and schedule constructing. The last step is quite obvious and coincides with the one of algorithm $\mathcal{A}$, so we concentrate on the first two steps.

Levelling Procedure. Let $E_n = \{\mathbf{e_j}\}_{j=1}^n \subset \mathbb{R}^{m-1}$, where $\mathbf{e_j} = (p_{2j} - p_{1j}, \dots, p_{mj} - p_{1j})$, and let us define a real number y by equation $\mathbb{P}(\|\mathbf{e_1}\| \ge y) = n^{-1/2}\ln n$. We say that vector $\mathbf{e_j}$ is *big*, if $\|\mathbf{e_j}\| \ge y$; otherwise we call the vector *small.* Let $\mathbf{e_j}(i)$ be the ith component of vector $\mathbf{e_j}$. Big vector $\mathbf{e_j}$ is *positive* if it has at least one positive component greater than or equal to y, and *i-negative* if $\mathbf{e_j}(i) \le -y$. A big vector may be positive and i-negative for $m-2$ different i at a time.

The idea behind the levelling procedure within algorithm $\mathcal{B}$ is that only components of big vectors are changed. At the first step, we level the machine load L_1 up to $L_{\max}$ (in case of $L_1 < L_{\max}$) by means of big positive vectors.

Step 1. If $\mathbf{e_j}$ is a positive vector, then $p_{1j}^* := p_{1j} + \min\{y, L_{\max} - L_1\}$.

Clearly, property **(e)** of the correct levelling procedure holds. Let $\mathbf{e_j}(i_1 - 1)$ be the big positive component of $\mathbf{e_j}$. Then $\mathbf{e_j}(i_1 - 1) = p_{i_1j} - p_{1j} \ge y$ and $p_{1j}^* \le p_{1j} + y \le p_{i_1j} \le p_{\max}$. Thus, step 1 satisfies property **(f)** as well. At step k ($k = 2, \dots, m$), we level the machine load L_k up to $L_{\max}$ (in case of $L_k < L_{\max}$) in a similar way, but using big i-negative vectors instead.

Step k. If $\mathbf{e_j}$ is a $(k-1)$-negative vector, then $p^*_{kj} := p_{kj} + \min\{y, L_{\max} - L_k\}$.

Again, property **(e)** obviously holds, and $\mathbf{e_j}(k-1) = p_{kj} - p_{1j} \le -y$ implies $p^*_{kj} \le p_{kj} + y \le p_{1j} \le p_{\max}$ satisfying **(f)**.

Compact vector summation. At this stage, we have family $E^*_n = \{\mathbf{e_j}^*\}_{j=1}^n$ on the input. Let $\mathbf{S}$ be the sum of vectors already used in the summation process and H be the family of vectors not used so far. Clearly, $\mathbf{S} = 0$, $H = E^*_n$ in the beginning and $\mathbf{S} = 0$, $H = \emptyset$ in the end of the algorithm. Let $V = \{\mathbf{v_t}\}_{t=1}^{\nu}$ and $U = \{\mathbf{u_t}\}_{t=1}^{\mu}$ be non-overlapping families of small vectors from H. Denote $\mathbf{V_k} \doteq \sum_{t=1}^k \mathbf{v_t}$, $\mathbf{U_k} \doteq \sum_{t=1}^k \mathbf{u_t}$. Families V and U are called *α-compensating* for a big vector $\mathbf{e}^*_\mathbf{j}$, if:

- sequences $\mathcal{V}_k \doteq \|\mathbf{S} + \mathbf{e}^*_\mathbf{j}/2 + \mathbf{V_k}\|$ and $\mathcal{U}_k \doteq \|\mathbf{S} + \mathbf{V}_\nu + \mathbf{e}^*_\mathbf{j} + \mathbf{U_k}\|$ are strictly decreasing;
- $\mathcal{V}_\nu \le \alpha$, $\mathcal{U}_\mu \le \alpha$.

Thus, if $\|\mathbf{S}\| \le \alpha$ and we sum consecutively vectors from V, next vector $\mathbf{e}^*_\mathbf{j}$, and after that – vectors from U, then we get the middle of the vector $\mathbf{e}^*_\mathbf{j}$ being located within the ball of radius α with center in $\mathbf{0}$, and therefore, all these vectors are summed within the ball of radius $R \le \|\mathbf{e}^*_\mathbf{j}\|/2 + \alpha$. In the figure below, an example is given for $m = 3$ and the standard Euclidian norm in $\mathbb{R}^2$.

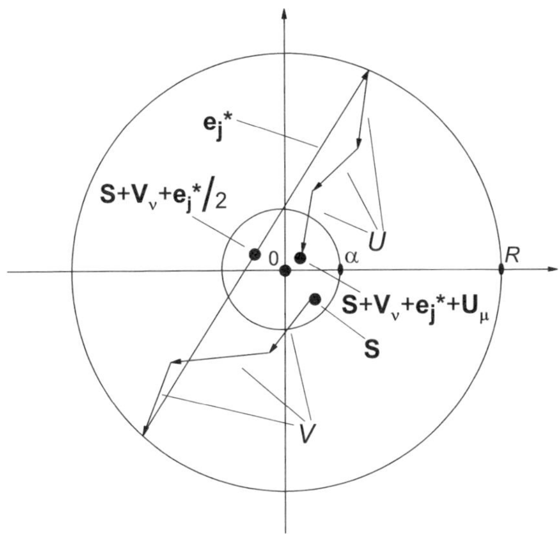

The idea of the compact vector summation within algorithm $\mathcal{B}$ is, firstly, to sum all big vectors along with their compensating families and, secondly, to sum all remaining small vectors from H using algorithm $\mathcal{C}$. Consider real number $y_0 > 0$ such that $y_0 = o(1)$ $(n \to \infty)$. The exact expression for this number is given in [7].

Step 1. For each big vector $\mathbf{e}_{\mathbf{j}}^*$, y_0-compensating families are found. Vectors are added to $\mathbf{S}$ in order $\mathbf{v}_1, \ldots, \mathbf{v}_\nu, \mathbf{e}_{\mathbf{j}}^*, \mathbf{v}_1, \ldots, \mathbf{v}_\nu$. Corresponding indices are written into π.

After step 1, we have $\|\mathbf{S}\| \le y_0$ and H containing small vectors only.

Step 2. All remaining vectors from H are summed by algorithm $\mathcal{C}$ and indices of result permutation $\tilde{\pi}$ are written into π.

Let $e_{\max}^* = \max_j \|\mathbf{e}_{\mathbf{j}}^*\|$. After step 2, we have all vectors from E_n^* summed within radius $\mathcal{R}_\pi(\mathcal{B}) = \max\{e_{\max}^*/2; \mathcal{R}\tilde{\pi}(\mathcal{C})\} + y_0$. To accomplish the proof of theorem 1, one still needs to show that, with high probability,

- big vectors from E_n suffice the levelling procedure to satisfy **(g)**;
- small vectors from E_n^* suffice for y_0-compensating families to be found for each big vector from E_n^*;
- $2\mathcal{R}\tilde{\pi}(\mathcal{C}) \le e_{\max} \le p_{\max}$.

The proof of these statements is based on using Chebyshev inequality [2] and can be found in [7].

References

1. Bingham NL, Goldie C, Teugels JL (1987) Regular Variation. Encyclopedia of Mathematics and Its Applications 27. Cambridge, Cambridge University Press
2. Borovkov AA (1998) Probabilities Theory. Amsterdam, Gordon and Beach
3. Frieze AM, Reed B (1998) Probabilistic analysis of algorithms. In: Habib M, McDiarmid C, Ramirez J, Reed B (eds) Probabilistic methods for algorithmic discrete mathematics. Springer, Berlin Heidelberg New York
4. Garey M, Johnson D, Sethi R (1976) The complexity of flowshop and jobshop scheduling. Math. Oper. Res. 1:117-129
5. Hofri M (1987) Probabilistic analisys of algorithms: on computing metodologies for computing algorithms performance evaluation. Springer, Berlin Heidelberg New York
6. Johnson SM (1954) Optimal two and three-stage production schedules with set-up times included. Nav. Res. Log. Quart. 1:61-68
7. Koryakin RA, Sevastyanov SV (2005) On the stochastic compact vector summation problem. Discr. Anal. and Oper. Res 12(1):71-100
8. Seneta E (1976) Regularly varying functions. Lecture Notes in Math. 508. Springer, Berlin Heidelberg New York
9. Sevastyanov S (1991) On a compact vector summation. Discretnaya Matematika 3:66-72 (in Russian)
10. Sevast'janov S (1994) On some geometric methods in scheduling theory: a survey. Discrete Applied Mathematics 55:59-82
11. Sevast'janov S (1995) Vector summation in Banach space and polynomial algorithms for flow shops and open shops. Math. Oper. Res. 20:90-103
12. Sviridenko M (2004) A note on permutation flow shop problem. Annals of Operations Research 129:247-252
13. Shmoys DB, Stein C, Wein J (1994) Improved approximation algorithms for shop scheduling problems. SIAM Journal on Computing 23:617-632

A Heuristic Solution for a Driver-Vehicle Scheduling Problem

Benoît Laurent[1,2], Valérie Guihaire[1,2], and Jin-Kao Hao[2]

[1] Perez Informatique, 41 avenue Jean Jaures, 67000 Strasbourg, France
`blaurent@perinfo.com`, `vguihaire@perinfo.com`

[2] LERIA, Université d'Angers, 2 Bd Lavoisier, 49045 Angers cedex 01, France
`jin-kao.hao@univ-angers.fr`

Summary. A driver-vehicle scheduling problem in a limousines rental company is studied. Given a set of trip demands to be covered, the goal is to find a driver-vehicle schedule that covers as many as possible of the required demands while satisfying a set of imperative constraints and optimizing several cost objectives. A formulation of the problem is given and a solution approach using local search is developed.

1 Introduction

This paper deals with a driver and vehicle scheduling problem in a limousines rental company. This application context induces specific operational constraints making the problem fairly distinct from other known crew and vehicle scheduling problems. However, some neighboring problems can be found in the literature dealing mainly with transport by bus. The most recent approaches related to this issue are based on a complete integration of drivers and vehicles during the scheduling process (e.g., see [2], [4] and [1]).

In our case, we are given daily sets of trips, drivers and vehicles, with the goal of scheduling resources in order to cover the maximum possible workload. The quality of service being a crucial issue, a schedule must comply with a set of imperative constraints, while optimizing some economic objectives.

2 Problem description

Only a subset of the constraints and objectives are presented here. We state:

- A set $\mathcal{T}$ of trips, each being defined by a time and a place for the departure and the destination, a number of passengers, required driver skills, etc.
- A set $\mathcal{D}$ of drivers, each being characterized by a daily allowed time spread, a set of skills, etc.
- A set $\mathcal{V}$ of vehicles, each being characterized among others by its capacity.

The problem is to find a daily assignment of driver-vehicle couples to the trips that satisfies a set $\mathcal{C}$ of constraints while optimizing a set $\mathcal{O}$ of objectives.

2.1 Constraints

We consider any trip $t \in \mathcal{T}$, a vehicle $v \in \mathcal{V}$ and a driver $d \in \mathcal{D}$ assigned to t. For quality requirements, a solution must satisfy the following constraints:

- v must be compatible with t, i.e. there must be enough seats in v to accommodate all the passengers.
- d must have all the skills required by t.
- The duration between the pick-up time of the first trip and the end of the last trip cannot be greater than the maximum spread allowed for d.
- There must be enough time between successive trips for d to move from one trip to the next one.

2.2 Objectives

Main objective: It is primordial to meet customers' trip demands. However, the available resources may not be sufficient to satisfy all of them. Therefore, the first goal is to cover as many as possible of the trip demands. From the point of view of the rental company, it is preferable to maximize the sum of the durations of the assigned trips since long trips are more profitable than short ones. Notice that this objective is equivalent to minimizing the sum of the durations of the trips to which no couple of resources is assigned. In addition, the trips starting in an imminent way must be favored.

Secondary objectives: For evident economic reasons, it is desirable to reduce the running costs, i.e. the number of working drivers and used vehicles. Furthermore, it is useful to minimize the drivers' waiting times between trips.

3 Problem formulation

3.1 Notations

Let T, D and V be the number of trips, drivers and vehicles respectively. Given $t \in \mathcal{T}$, $d \in \mathcal{D}$, $v \in \mathcal{V}$, we state:

- $capa(v)$, the capacity of v,
- $pas(t)$, the number of passengers for t,
- $st(t), et(t), sp(t)$ and $ep(t)$ are respectively the start time, the end time, the start place and the end place of trip t,
- $sd(d)$ and $ed(d)$ are respectively the start time and the end time for d,
- $S_{max}(d)$ is the maximum spread time allowed for d.

We also define a set of binary relations:

- $sk(d, s) \Leftrightarrow$ driver d owns skill $s \in \mathcal{S}$, the set of skills,
- $Sk(t, s) \Leftrightarrow$ trip t requires skill s,
- $compat(t_i, t_j) \Leftrightarrow$ trips t_i and t_j can be done by the same resources,
- $dh(t_1, t_2)$ is the deadhead between trips t_1 and t_2,
- $wt(t_1, t_2)$ is the waiting time between trips t_1 and t_2.

We define a total order $\prec$ on the set of trips $\mathcal{T}$ by:

$$\forall (t_i, t_j) \in \mathcal{T}^2, t_i \prec t_j \Leftrightarrow \begin{cases} st(t_i) < st(t_j) \\ \vee \\ st(t_i) = st(t_j) \wedge i < j \end{cases}$$

The following notations are used to handle "driver-vehicle to trip" assignments:

- $wd(d)$, (resp. $vu(v)$) $\Leftrightarrow d$ (resp. v) is assigned to at least one trip,
- $Seq(d)$ is the set of couples (t_1, t_2) that $d \in \mathcal{D}$ handles consecutively.

Eventually, we define $compat'(t_i, t_j), dh'(t_i, t_j), wt'(t_i, t_j)$ and $Seq'(d)$ that are similar to $compat(t_i, t_j), dh(t_i, t_j), wt(t_i, t_j)$ and $Seq(d)$ respectively except that they take into account a stop at the depot between trips.

3.2 Constraints

Capacity constraints The first type of constraints imposes that the vehicle is big enough to carry all the passengers. For each $t \in \mathcal{T}$ and each $v \in \mathcal{V}$:

$$CAPA(t, v) \Leftrightarrow pas(t) \leq capa(v)$$

Skills constraints Some trips require special skills from the driver, for instance spoken languages. For each $t \in \mathcal{T}$, each $d \in \mathcal{D}$ and each $s \in \mathcal{S}$:

$$SKILLS(t, d, s) \Leftrightarrow \neg Sk(t, s) \vee sk(d, s)$$

A similar type of constraints exists with vehicles features.

Maximum spread time constraints The third type of constraints imposes the maximum spread time for each driver. For each $d \in \mathcal{D}$ and each $(t_i, t_j) \in \mathcal{T}^2, t_i = (d, .), t_j = (d, .)$:

$$MAX\_SPREAD(t_i, t_j, d) \Leftrightarrow (et(t_j) - st(t_i)) \leq S_{max}(d)$$

Feasible sequences constraints This type of constraints imposes that the sequence of trips assigned to a driver is feasible, i.e. the driver has enough time to move from the end of a trip to the start of the following one. A possible change of vehicle, if needed, must take place at the depot. For each

$d \in \mathcal{D}$ and each $(t_i, t_j) \in \mathcal{T}^2, t_i = (d, v_k), t_j = (d, v_l), t_i \prec t_j$:

$$FEASIBLE\_D(t_i, t_j, d) \Leftrightarrow \begin{cases} ((v_k = v_l) \wedge compat(t_i, t_j)) \\ \vee \\ ((v_k \neq v_l) \wedge compat'(t_i, t_j)) \end{cases}$$

3.3 Objectives

Main objective Minimizing the total duration of the unassigned trips:

$$Min \sum_{t \in \mathcal{T}, t=(\epsilon,.) \vee t=(.,\epsilon)} (et(t) - st(t))$$

where $t = (\epsilon, .) \vee t = (., \epsilon)$ means the trip t is not covered.

Secondary objectives Minimizing the number of working drivers and vehicles in use:

$$Min \sum_{d \in \mathcal{D}} wd(d) + \sum_{v \in \mathcal{V}} vu(v)$$

Minimizing deadheads:

$$Min \sum_{d \in \mathcal{D}} \left(\sum_{(t_1,t_2) \in Seq(d)} dh(t_1, t_2) + \sum_{(t_1,t_2) \in Seq'(d)} dh'(t_1, t_2) \right)$$

Minimizing the total waiting time:

$$Min \sum_{d \in \mathcal{D}} \left(\sum_{(t_1,t_2) \in Seq(d)} wt(t_1, t_2) + \sum_{(t_1,t_2) \in Seq'(d)} wt'(t_1, t_2) \right)$$

3.4 Configuration and evaluation function

A configuration σ is a consistent assignment of "driver-vehicle" couples in $\mathcal{I} = (\mathcal{D} \cup \{\epsilon\}) \times (\mathcal{V} \cup \{\epsilon\})$ to trips in $\mathcal{T}$. The search space Ω is the set of all such assignments. A configuration is evaluated by a weighted aggregation of the objectives, augmented by a penalty function for broken constraints [3].

$$\forall \sigma \in \Omega, \quad eval(\sigma) = wbc \times \sum_{c \in C} f_c(\sigma) + \sum_{i \in \{1,\ldots,O\}} w_i \times f_i(\sigma)$$

with:

- $wbc > 0$ the weight associated to broken constraints,
- f_c the penalty for c. $f_c = 1$ if c is broken by σ, $f_c = 0$ otherwise,
- O the number of objectives,
- w_i the associated weight for i^{th} objective function,
- f_i the value of ith objective function.

3.5 Greedy algorithm for initial configuration

The first step is a pre-processing of the data. By examining the incompatibilities between drivers and vehicles, drivers and trips, and trips and drivers, we reduce the domains of the variables. A constructive greedy heuristic combined with constraint propagation techniques is then used to create an initial configuration σ. The stop criterion is that no additional assignment can be made without violating constraints.

3.6 Hill climbing and simulated annealing

In order to improve the initial configuration σ, both hill climbing and simulated annealing are experimented, based on a 1_change neighborhood. From the current configuration, we obtain a neighboring configuration by changing the driver and/or the vehicle assigned to one trip.

$$\forall \sigma \in \Omega, 1\_change(\sigma) = \{(t, (d, v)) \in \mathcal{T} \times \mathcal{I} | \exists \text{ a unique } t \text{ such that } \sigma(t) \neq (d, v)\}$$

4 Experimentations and results

Computational experiments were carried out on five real instances, representing different workloads. Table 1 shows the main characteristics of the instances and the results manually obtained in the limousines rental company for comparison purpose.

Our algorithms were programmed in C++, compiled with gcc 3.4.2, on a PC running Windows XP (256Mo RAM, 2.4Ghz). The program was run 10 times on each instance with different random seeds. The stop condition used is a maximum duration fixed to 10 minutes.

Table 1. Characteristics of the five instances and results of manual scheduling

Date	trips	total duration (hhh:mm)	Manual scheduling: broken constraints	drivers	vehicles	deadheads (hh:mm)	waiting time (hhh:mm)
08_05	91	133:46	27	44	52	27:01	119:04
10_05	126	238:58	48	65	70	39:32	200:32
18_05	153	453:06	47	74	74	56:30	196:20
19_05	163	504:47	48	79	77	55:26	165:35
23_05	202	457:59	79	84	81	63:53	248:07

Table 2 shows the results obtained with hill climbing and simulated annealing algorithms. The rules are slightly different between manual and computed scheduling: in the second case, broken constraints are strictly forbidden but unassigned trips are allowed. These are manually handled afterwards. Therefore, a column was added in Table 2 giving the percentage of unassigned work.

Table 2. Results using hill climbing and simulated annealing

Date	hill climbing					simulated annealing				
	Unassigned trips (%)	drivers	vehicles	deadheads (hh:mm)	waiting time (hh:mm)	Unassigned work (%)	drivers	vehicles	deadheads (hh:mm)	waiting time (hh:mm)
08_05	1.9	35	37	29:28	57:51	0.93	36	37	28:39	70:34
10_05	9.2	44	46	39:12	74:57	9.2	44	46	35:15	66:44
18_05	5.7	67	69	49:49	87:34	5.3	65	68	56:46	88:22
19_05	15	63	69	50:04	77:27	15	63	67	49:30	88:09
23_05	12	68	67	63:16	81:01	9.9	74	70	62:49	94:31

These results show a significant improvement regarding the actual practice. Without breaking any constraint, both algorithms assign most of the work. Furthermore, the number of required resources is substantially reduced. The total waiting time is divided by a factor of 2.36 in average. Actually, the workload for human schedulers is so high that costs reduction is only a secondary concern. This leads to the other major contribution of this work: the time needed to elaborate a planning is drastically decreased. Whereas people spend nearly 4 hours on this task, our program only takes a few minutes to get a much better quality schedule.

5 Conclusion

We tackled a practical driver and vehicle scheduling problem in an original context. The solution approach combines a pre-processing phase using constraint programming techniques and an optimization phase using local search. Results obtained on real data showed significant improvements compared with the actual practice in terms of solution quality and computing time.

Acknowledgments

This work was carried out within the framework of a CIFRE grant from "Agence Nationale de la Recherche Technique" (ANRT), which is acknowledged. Special thanks go to Joël Thibault for his role in fully supporting this work.

References

1. R. Borndörfer, A. Löbel, and S. Weider. A bundle method for integrated multi-depot vehicle and duty scheduling in public transit,CASPT 2004, Aug. 2004.
2. R. Freling, D. Huisman, and A.P.M. Wagelmans. Models and algorithms for integration of vehicle and crew scheduling. *Journal of Scheduling*, 6:63–85, 2003.
3. P. Galinier and J.K. Hao. A general approach for constraint solving by local search. *Journal of Mathematical Modelling and Algorithms*, 3(1):63–85, 2004.
4. D. Huisman. *Integrated and Dynamic Vehicle and Crew Scheduling.* PhD thesis, Erasmus University Rotterdam, 2004.

Scheduling Jobs with Uncertain Parameters: Analysis of Research Directions

Yakov Shafransky

United Institute of Informatics Problems of National Academy of Sciences of Belarus. Surganova 6, 220012 Minsk, Belarus. E-mail: *shafr@newman.bas-net.by*

The paper considers the scheduling of jobs under an assumption that values of some job parameters are unknown, instead, only lower and upper bounds are given for each such a parameter. The main aim of the paper is to outline perspective research directions for corresponding problems as well as to mark directions where a success is doubtful. Some results for single- and multi-stage systems are presented.

Introduction

Situations originating from practice are known, where constructing schedules for job processing is carried out when some parameters of the jobs are uncertain. For each such a parameter, everything known is an interval containing all its possible values. Uncertain parameters are uncontrollable, they can take any values from corresponding intervals irrespective of the will of the decision maker. The aim of the paper is not merely to present new results in this scanty investigated branch of scheduling, it is rather to outline perspective directions for a research, as well as to point out the directions, where the possibility of a success is doubtful.

The most of known optimization problems may be stated as follows. Given function $F(x, y, z)$ that depends on three groups of parameters. Here x are controllable parameters, y and z are uncontrollable parameters. The aim is to find a collection $x \in X$ of values of controllable parameters (an alternative) that minimizes function $F(x, y, z)$ provided that values of all parameters y are fixed and known. Depending on the nature of uncontrollable parameters z we obtain three types of problems.

Type 1. If the values of parameters z are assumed to be known in advance, then the problem under consideration is a deterministic optimization problem.

Type 2. Parameters z are random variables with known distribution functions. Then the problem under consideration is a problem of the stochastic optimization.

Type 3. Parameters z are not random or z are random variables with unknown distribution. Everything that is known is the range Z of possible values of z.

The problems of the third type (that are usually referred to as problems with uncertain parameters) are the subject for the discussion in this paper. For the simplicity of notations, we use $F(x, z)$ instead of $F(x, y, z)$ in what follows.

Possible research directions

The problems of the third type relate to the branch of the operations research known as the theory of the decision making under uncertainty. Within the frameworks of this theory, some approaches are developed to work with optimization problems in the presence of uncertain factors. These approaches actually determine possible directions for the research of corresponding scheduling problems.

It is usually senseless to speak about solving the third type problem in the traditional sense, i.e., about searching alternative $x^* \in X$ such that $F(x^*, z)$ is the minimum value of the function $F(x, z)$ over all possible choices $x \in X$ under any values $z \in Z$. If we do not deal with occasional degenerated situations, then for any alternative $x^* \in X$ there exists $z' \in Z$ such that $F(x^*, z') > F(x^0, z')$ for some $x^0 \in X$.

Whether the application of mathematical methods is possible when a researcher faces with problems of the third type? Whether the researcher can provide the decision maker with guidelines permitting to make a justified choice of this or that alternative as a solution? The operations research theory gives the positive answer to these questions. It is possible to recognize two directions for the development of guidelines for a decision maker operating in uncertain conditions.

In frames of the first (main) direction, some auxiliary criteria are introduced and the objective function of the initial problem is replaced by one of the new criteria. The auxiliary criterion is introduced so that to remove uncertainty. In the result, the initial problem with uncertain parameters is replaced by a deterministic optimization problem. Some auxiliary criteria of the mentioned type are as follows.

Principle of the guaranteed result (Wald criterion). The required alternative x^* should provide the minimum value to the objective function at worst circumstances (represented by uncertain parameters z). In other words, alternative $x^* \in X$ should provide the minimum value to the function $V(x)=\max\{F(x,z) | z \in Z\}$.

Hurwitz criterion. Alternative $x^* \in X$ should provide the minimum value to the function $\Phi(x) = \lambda \max\{F(x,z) \mid z \in Z\} + (1-\lambda)\min\{F(x,z) \mid z \in Z\}$, where $0 \le \lambda \le 1$ is a parameter to be chosen by the decision maker.

Savage criterion. Alternative $x^* \in X$ is to be found that provides the minimum value to the function $\Psi(x) = \max_{z \in Z}\{F(x,z) - \min_{x \in X} F(x,z)\}$.

Some other auxiliary criteria are known as well. The aim of the paper is not to analyze virtues and shortages of the mentioned criteria, neither to give their exhausting enumeration. The aim is to draw the attention of scheduling theory experts to the approach based on the use of such criteria. As a rule, the researchers try to get several alternatives, each of which being optimal for at least one of the auxiliary criteria. The providing the decision maker with such a collection of alternatives allows him to make a more justified decision.

The second research direction is aimed to reduce set X of alternatives.

Such an approach appears to be useful, if the reduced set X' is of a small capacity and besides, there are tools for the revealing properties of the alternatives. In this case, alternatives $x \in X'$ may be analyzed, and the researcher can provide the decision maker with information on possible consequences of acceptance this or that $x \in X'$ as a solution. In the absence of alternative $x^* \in X$ being optimal under any collection $z \in Z$, the reasonable reduction of set X may be considered only as an intermediate stage. Suppose, for example, that for each of the mentioned auxiliary criteria, the resulting set X' includes alternatives that are optimal with respect to this criterion. Then the reduction is useful, if the problem of the searching an alternative x^0 that is optimal with respect to an auxiliary criterion is of a high complexity. In such a case, a two-stage approach appears to be effective. At the first stage, we reduce the search area of x^0 using fast reduction procedures. At the second stage, we find alternative x^0 (or an alternative that is "close" to x^0) using some enumeration procedures. How to reduce X to set X' that satisfies the mentioned conditions? A possible approach is proposed in the next section.

Elimination of alternatives

The binary relation $\succ$ defined over set X is referred to as *the relation of H-dominance*, if it is transitive and for $x_1, x_2 \in X$ from $x_1 \succ x_2$ (alternative x_1 dominates alternative x_2) it follows that $F(x_1, z) \le F(x_2, z)$ holds for any collection $z \in Z$.

Alternatives $x_1, x_2 \in X$ are called *equivalent*, if $x_1 \succ x_2$ and $x_2 \succ x_1$.

The introduced concept of H-dominance describes a class of binary relations. Defining this class, the author just generalizes already existing approaches. Binary relations introduced in some papers belong to this class. It should be noted also that the structure of H-dominance relation and the way of its definition is similar to the concept of the preference relation from the multiple criteria decision making.

Set $X^H \subseteq X$ is called *H-effective*, if for any $x \in X \backslash X^H$ there exists alternative $x' \in X^H$ such that $x' \succ x$, and X^H does not contain alternatives x_1, x_2 such that $x_1 \succ x_2$ or $x_2 \succ x_1$.

Any H-effective set is minimal with respect to the inclusion. Any H-effective subsets $X_1^H, X_2^H \subset X$ coincide up to equivalent alternatives and $|X_1^H| = |X_2^H|$ if X is a finite set. Some different relations of H-dominance may exist over set X.

H-dominance relation $\succ$ defined over a finite set X is called *minimal*, if for any other H-dominance relation $\succ'$ we have $|X^H(\succ)| \le |X^H(\succ')|$. Here $X^H(\succ)$ is H-effective set defined by relation $\succ$.

Theorem 1. *H-dominance relation is minimal if and only if the existence alternatives $x_1, x_2 \in X$ such that $F(x_2, z) \le F(x_1, z)$ holds for any $z \in Z$ implies that $x_2 \succ x_1$ or $x_1 \succ x_2$, or there exists an alternative $x_3 \in X$ such that $x_3 \succ x_1$ or $x_3 \succ x_2$.*

Let $f_1(x) = \max_{z \in Z} \{F(x,z) - \min_{x \in X} F(x,z)\}$, $f_2(x) = \max_{z \in Z} F(x,z)$, $f_3(x) = \min_{z \in Z} F(x,z)$, $f_4(x) = F(x, z')$, $z' \in Z$, and $\theta(x) = \theta(f_1(x), f_2(x), f_3(x), f_4(x))$ be functions defined over X.

Theorem 2. *If θ ($f_1(x), f_2(x), f_3(x), f_4(x)$) is a non-decreasing function then for H-effective set X^H the inclusion $x^* \in X^H$ holds, where $x^* \in X$ is the alternative that minimizes function $\theta(x)$ over set X.*

Corollary 1. *H-effective set X^H includes alternatives that are optimal with respect to criteria of Wald, Hurwicz (for any $0 \le \lambda \le 1$) and Savage.*

Corollary 2. *For any $z \in Z$, H-effective set X^H includes an alternative $x(z)$ such that $F(x(z), z) = \min\{F(x,z) \mid x \in X\}$.*

A dominance relation may be introduced in a different "mild" way. Over set X, we define binary relation $\blacktriangleright_z$ for $z \in Z$.

Relation $\blacktriangleright_z$ is called *S-dominance relation*, if it is transitive and for $x_1, x_2 \in X$ from $x_1 \blacktriangleright_z x_2$ it follows that for $z \in Z$ the inequality $F(x_1, z) \le F(x_2, z)$ holds.

It is more correct to speak about a set of relations $\blacktriangleright_z$ defined over set X (each relation for each $z \in Z$), nevertheless, we use a "single" relation for the brevity.

Set $X' \subset X$ of alternatives is said to *S-dominate* alternative $x \in X$, if for each $z \in Z$ there exists alternative $x(z) \in X'$ such that $x(z) \blacktriangleright_z x$. In such a case we write $X' \blacktriangleright x$.

Set $X^S \subseteq X$ is called *S-effective*, if for any alternative $x \in X \backslash X^S$ the relation $X^S \blacktriangleright x$ holds, and for any $x' \in X^S$ there does not exist a subset $X' \subset X^S$ such that $X' \blacktriangleright x'$.

H-dominance relation is a special case of S-dominance relation, where for each dominated alternative there exists a single-element S-dominating set. S-dominance relation, being "milder" than H-dominance relation, does not inherit its major properties. It is easy to construct examples, where an analog of Corollary 1 is not valid for set X^S. The only valid analog is as follows.

Note. *For any $z \in Z$, S-effective set X^S contains alternative $x(z)$ such that $F(x(z), z) = \min\{F(x,z) \mid x \in X\}$.*

Scheduling problems

Now we consider some scheduling problems with uncertain parameters of jobs.

Elimination of schedules

Consider a class of problems, where the schedule may be presented by a permutation of jobs, and the objective function is priority-generating [1]. Precedence constraints $\rightarrow$ are defined over set $N=\{1, \ldots, n\}$ of jobs and are presented by precedence graph $G=(N, U)$. Denote by $B(j)$ and $A(j)$ the set of all predecessors and all successors of vertex $j \in N$ respectively, $B^0(j)$ and $A^0(j)$ are sets of direct predecessors and direct successors. We write $i \sim j$, if neither $i \rightarrow j$, nor $j \rightarrow i$ holds. For $v, h \in N$, denote $\overline{B}(h,v) = B(h) \backslash (B(v) \cup v)$, $\overline{A}(h,v) = A(v) \backslash (A(h) \cup h)$. Permutation π is a partial permutation if it includes not all the elements of set N. Denote by $\overline{\Pi}$ the set of all permutations of elements of set N (including partial and empty permutations). For $\Pi \subseteq \overline{\Pi}$, denote by $S[\Pi]$ the set of segments of permutations from Π, i.e., $\pi' \in S[\Pi]$ if there exist permutations $\sigma_1, \sigma_2 \in \overline{\Pi}$ such that $(\sigma_1, \pi', \sigma_2) \in \Pi$.

Function $F(\pi)$ defined over some set Π^0 is called *priority-generating over set* $\Pi \subseteq \Pi^0$, if a function $\omega(\pi)$ may be defined over set $S[\Pi]$ that has the following properties. For any $\alpha, \beta \in S[\Pi]$ and any permutations $\pi' = (\sigma_1, \alpha, \beta, \sigma_2)$ and $\pi'' =$

$(\sigma_1, \beta, \alpha, \sigma_2)$ such that $\pi', \pi'' \in \Pi$, the validity of inequality $\omega(\alpha) \geq \omega(\beta)$ implies that $F(\pi') \leq F(\pi'')$ holds. In such a case, $\omega(\pi)$ is called *priority function for* $F(\pi)$.

Denote by $\Pi(G)$ the set of full permutations of elements of set N that are feasible with respect to precedence constraints defined by graph G.

The problem is to find permutation $\pi^* \in \Pi(G)$ that minimizes the objective function $F(\pi)$, provided $F(\pi)$ is a priority-generating function over set $\Pi(G)$.

Theorem 3 [1]. *Let function* $F(\pi)$ *be priority-generating over set* $\Pi(G)$, $B^0(h) = v$ *and* $\omega(h) \geq \omega(j)$ *for all* $j \in \overline{A}(h,v) \cup v$. *Then for any permutation* $\pi = (\ldots, v, \sigma, h, \ldots) \in \Pi(G)$ *there exists permutation* $\pi^0 = (\ldots, v, h, \ldots) \in \Pi(G)$ *such that* $F(\pi^0) \leq F(\pi)$.

Theorem 4 [1]. *Let function* $F(\pi)$ *be priority-generating over set* $\Pi(G)$, $h \sim v$ *and* $\omega(j) \geq \omega(i)$ *for all* $j \in \overline{B}(h,v) \cup h$ *and* $i \in \overline{A}(h,v) \cup v$. *Then for any permutation* $\pi = (\ldots, v, \sigma, h, \ldots) \in \Pi(G)$ *there exists permutation* $\pi^0 = (\ldots, h, v, \ldots) \in \Pi(G)$ *such that* $F(\pi^0) \leq F(\pi)$.

Suppose that processing times p_j of jobs are uncertain parameters. For each job j we are given an interval $[P_j^{\min}, P_j^{\max}]$ such that $p_j \in [P_j^{\min}, P_j^{\max}]$. Then the value of the objective function depends on permutation π (controllable parameter) and on the collection of p_j values, $j=1, \ldots, n$, (collection of uncertain parameters). We write $F(\pi, p)$ for the uncertain situation instead of $F(\pi)$. Similarly, we use $\omega(j, p_j)$ instead of $\omega(j)$.

Introduce functions $\omega_{\min}(j) = \min\{\omega(j, p_j) \mid p_j \in [P_j^{\min}, P_j^{\max}]\}$ and $\omega_{\max}(j) = \max\{\omega(j, p_j) \mid p_j \in [P_j^{\min}, P_j^{\max}]\}$. Note that the functions $\omega_{\min}(j)$ and $\omega_{\max}(j)$ do not depend on the value p_j of the uncertain parameter.

Define binary relation $\triangleright$ over set N setting $i \triangleright j$ for $i, j \in N$, if $\omega_{\min}(i) \geq \omega_{\max}(j)$.

Theorem 5. *Let function* $F(\pi)$ *be priority-generating over set* $\Pi(G)$, *then relation* $\triangleright$ *defined over set N defines S-dominance relation over set* $\Pi(G)$.

Theorems 3 and 4 in the deterministic situation are used to reduce the range of location of an optimal permutation by transformations of precedence graph G. Under the conditions of Theorem 3, the pare v and h of the graph vertices is replaced by a single vertex with an associated permutation (v, h). Under the conditions of Theorem 4, a new arc (h, v) is added to set U. A new graph G' obtained from G in the result of a multiple application of the mentioned transformations is such that $\Pi(G') \subset \Pi(G)$ and $\Pi(G') \cap \Pi^*(G) \neq \phi$, where $\Pi^*(G)$ is the initial set of optimal permutations. Under the uncertainty, it follows from Theorem 5 that the similar technique for the reduction of set $\Pi(G)$ works if we replace inequality $\omega(h) \geq \omega(j)$ by $\omega_{\min}(h) \geq \omega_{\max}(j)$ in the condition of Theorem 3. In Theorem 4, we should replace $\omega(j) \geq \omega(i)$ by $\omega_{\min}(j) \geq \omega_{\max}(i)$. A grave disadvantage of such an approach is that the dominance relation we use is S-dominance. As the consequence, we have no guaranties that the resulting set $\Pi(G')$ includes permutations that are optimal with respect to criteria of Wald, Hurwicz and Savage.

Auxiliary criteria

Wald criterion. Schedule s is to be found that minimizes function $V(s)$= $\max\{F(s,p) \mid p \in \prod_{j=1}^{n}[P_j^{\min}, P_j^{\max}]\}$. Denote $p_{\max}$=($P_1^{\max}, P_2^{\max}, ..., P_n^{\max}$).

A scheduling problem is called regular if the corresponding objective function is regular (a non-decreasing function of the completion times $C_j(s)$ of jobs) and the following property holds. Let I_1 and I_2 be two instances of the problem that differ only in processing times of job operations and $p_{ij} \geq p_{ij}^{'}$, where p_{ij} and $p_{ij}^{'}$ are processing times of operations in I_1 and I_2 respectively. Then for any feasible schedule s for instance I_1 there exists a feasible schedule s' for I_2 such that $S_{ij}(s') \leq S_{ij}(s)$, where $S_{ij}(s)$ is the starting time of operation O_{ij} in schedule s.

Theorem 6. *If a scheduling problem with the objective function $F(s,p)$ is regular, then $V(s)=F(s,p_{\max})$.*

Thus, finding the optimal schedule for Wald criterion reduces to the deterministic version of the problem by giving maximum values to all processing times.

Hurwicz criterion. Schedule s is to be found that minimizes function $\Phi(s)$= $\lambda \max\{F(s,p) | p\in \prod_{j=1}^{n}[P_j^{\min}, P_j^{\max}]\} + (1-\lambda)\min\{F(s,p) \mid p \in \prod_{j=1}^{n}[P_j^{\min}, P_j^{\max}]\}$. Denote $p_{\min}$=($P_1^{\min}, P_2^{\min}, ..., P_n^{\min}$). Similarly to Wald criterion, for any regular problem we have $\min\{F(s,p) \mid p \in \prod_{j=1}^{n}[P_j^{\min}, P_j^{\max}]\}$ =$F(s, p_{\min})$. So, $\Phi(s)$= $\lambda F(s, p_{\max})+(1-\lambda)F(s, p_{\min})$. Consider some particular problems.

Jobs of set N are to be processed by a single machine under precedence constraints presented by graph G. For each job j, a function $f_j(t)$ gives the cost for the completion job j at time t, $p_j\in [P_j^{\min}, P_j^{\max}]$. The aim is to find a feasible sequence π of the jobs processing that minimizes the total cost $F(\pi, p)$= $\sum_{j=1}^{n} f_j(C_j(\pi, p_j))$. Consider two special cases of the problem, where (a) $f_j(t)$= w_jt+b_j and (b) $f_j(t)$= $w_j\exp(\gamma t)+b_j$, $\gamma\neq 0$. In case (a) $\Phi(\pi)= \sum_{j=1}^{n} w_j \left(\sum_{l=1}^{j}((\lambda P_l^{\max} + (1-\lambda)P_l^{\min})\right) + \sum_{j=1}^{n} b_j$. Thus, $\Phi(\pi)=F(\pi, \lambda P_l^{\max} + (1-\lambda)P_l^{\min})$ and the problem reduces to minimizing the initial objective function. In case (b) under $G=(N,\phi)$ and $w_j=w_i=w>0$ for all $j, i\in N$, it is sufficient to order the jobs in non-decreasing of values $\lambda\exp(\gamma P_j^{\max})+(1-\lambda)\exp(\gamma P_j^{\min})$.

The research is partly supported by ISTC project B-986.

References

1. Tanaev VS, Gordon VS, Shafransky YM (1994) Scheduling Theory. Single-Stage Systems. Kluwer Academic Publishers, Dordrecht-Boston-London

Job-Shop Scheduling by GA. A New Crossover Operator

Czesław Smutnicki and Adam Tyński

Wrocław University of Technology,
Institute of Computer Science, Automation and Robotics
Janiszewskiego 11-17, 50-372 Wrocław, Poland
czeslaw.smutnicki@pwr.wroc.pl, adam.tynski@pwr.wroc.pl

Summary. The new distance measure between job-shop solutions, based on Euclidean measure, has been proposed. The significant positive correlation of the proposed measure with its suitable version based on the Kendall's tau measure has been revealed. By applying this measure, a new, easy tunable, crossover quasi-operator for the genetic approach is designed. The genetic algorithm, equipped with the new operator, has been applied to the job-shop scheduling problem with the sum of job completion times criterion. Results provided by the algorithm, compared with the best results known in the literature, confirm superiority of the proposed method.

Key words: Genetic algorithms – Job-shop scheduling – Crossover operators

1 Introduction

In recent years, genetic algorithms were successfully adapted to many various scheduling problems. In this paper we concentrate our attention solely on the *job-shop* scheduling problem; the survey of achievements in this topic one can find, among others, in [1] and [2]. Although a lot of sophisticated theorems were proposed, a plenty of efficient heuristics were designed – this problem for some optimization criteria still remains hard. Particularly famous is the case of minimizing sum of job completion times for the sake of the lack of problem-specific properties and the absence of effective solving algorithms. Thus, application of genetic algorithms to such problems is fully justified due to its main advantage – minimum knowledge about the problem being solved.

The designer of a genetic algorithm is compelled to take several decisions concerning, among others, proper representation of solutions, suitable selection scheme, effective mating operators, etc. In particular, mating operators should be designed very carefully, since they are responsible, at least partially, for the proper balance between diversification and intensification of the solution space exploration. Certain new look on this subject provided phenomenon of *big valley* in the solution space, described and analyzed in recent years. The big valley means the significant positive correlation between the distance among solutions (in terms of a particular distance measure) and their goal function value. We refer here to [3] for the comprehensive discussion about the big valley for job-shop scheduling problems. One of sparse representatives of operators utilizing the big valley phenomenon is the quasi-operator MSXF proposed in [4]. It combines best features of a crossover operator and the local search algorithm which explores a path relinking between two mated solutions. In spite of good properties of MSXF, its sensitivity to numerous parameters frequently causes time-consuming process of algorithm tuning.

2 Problem formulation and representations of schedules

The job-shop scheduling problem can be described as follows. It is given a set of machines $M = \{1, 2, \ldots, m\}$ and a set of jobs $J = \{1, 2, \ldots, r\}$ where m and r is the number of machines and jobs, respectively. Every job $j \in J$ consists of a sequence of n_j operations $\tau_{j-1}+1, \tau_{j-1}+2, \ldots, \tau_{j-1}+n_j$ that have to be performed in that order, where $\tau_j = \sum_{i=1}^{j} n_i$ and $\tau_0 = 0$. The set of all $n = \sum_{j \in J} n_j$ operations will be denoted by $O = \{1, 2, \ldots, n\}$. Operation k has to be processed on a dedicated machine $m_k \in M$ during $p_k > 0$ time units, $k \in O$. It is assumed that: (i) every machine can process at most one operation at a time, (ii) at most one operation of every job can be processed at a time and (iii) no splitting of operations is allowed. Schedule $S = (S_1, \ldots, S_n)$ is the vector, where $S_k \geq 0$ is the starting time of operation k, $k \in O$. The schedule is feasible if all above constraints are satisfied. The problem is to determine the feasible schedule S that minimizes the goal function value $f(S)$; in our case we set $f(S) = \sum_{j \in J} C_{\tau_j}$, where $C_k = S_k + p_k$ denotes the completion time of operation k, $k \in O$.

There are known at least nine representations of schedules for the job-shop scheduling problem, see e.g. [1]. We refer here to the one basing on a set of preference lists. In this approach, the complete schedule is represented by a collection of processing orders defined for machines and introduced formally as follows. Set O can be naturally divided into m disjoint subsets $O_1, O_2, \ldots, O_m$, where $O_l = \{k \in O : m_k = l\}$ is the set of all operations that have to be processed on machine l. The processing order of operations from O_l can be defined by permutation $\pi_l = (\pi_l(1), \pi_l(2), \ldots, \pi_l(o_l))$, $o_l = |O_l|$, where $\pi_l(i)$, $1 \leq i \leq o_l$, denotes the operation from set O_l that is in i-th position in π_l. Let Π_l denote the set of all permutations of operations in set O_l. Thus, the processing order of operations on machines is m-tuple of permutations (for simplicity we will call it hereafter a permutation or a solution of the problem) $\pi = (\pi_1, \pi_2, \ldots, \pi_m)$, where $\pi \in \Pi = (\Pi_1 \times \Pi_2 \times \ldots \times \Pi_m)$.

It is already known that solution π determines a feasible schedule S if graph $G(\pi) = (O, R \cup E(\pi))$, where

The algorithm starts with processing order π and returns the feasible processing order σ.

1. Put $\sigma_l := \emptyset$, $l \in M$. Put $r_i := 0$, $i \in O$ and $U := \emptyset$.
2. Generate set $G = \{i \in O \setminus U : B_i \subset U\}$.
   Compute value $\Delta := \min_{i \in G}\{r_i + p_i\}$.
   Choose randomly machine l from set $MK = \{m_i : i \in G, r_i + p_i = \Delta\}$.
   Determine set $K = \{i \in G : m_i = l, r_i < \Delta\}$.
3. Choose operation k from set $\{j \in K : \pi^{-1}(j) = \min_{i \in K}\{\pi^{-1}(i)\}\}$.
4. Put $\sigma_l := \sigma_l, k$ and $U := U \cup \{k\}$.
   For every $j \in A_k \cup \{i \in O \setminus U : m_i = l\}$ put $r_j := \max\{r_j, r_k + p_k\}$.
   If $U = O$ then return σ and STOP. Otherwise go to step 2.

Fig. 1. Outline of algorithm P

$$R = \bigcup_{j \in J} \bigcup_{i=\tau_{j-1}+1}^{\tau_j - 1} (i, i+1), \quad E(\pi) = \bigcup_{l \in M} \bigcup_{j=1}^{o_l - 1} (\pi_l(j), \pi_l(j+1))$$

does not contain a cycle; S_k can be found as the longest path to node k in $G(\pi)$. Such solution we call feasible, and let Π^o denotes the set of all feasible solutions. Next, we denote by $f(\pi)$ the goal function value of schedule S determined by the feasible solution π, $\pi \in \Pi^o$. Obviously, if processing order π is infeasible, value $f(\pi)$ cannot be calculated. In such cases we propose to interpret each π_l as a preference list recommended for "repairing" procedure providing a feasible solution. To this purpose a priority-based algorithm P, outlined in Fig. 1, can be used. The algorithm constructs the feasible permutation σ from any (not necessarily infeasible) permutation π. Set B_k and set A_k denotes the set of all job predecessors and all job successors of operation $k \in O$, respectively. Symbol π^{-1} denotes the inverse permutation to permutation π, i.e. satisfies equality $\pi_{m_k}(\pi^{-1}(k)) = k$ for every $k \in O$.

3 Distance measurement

In order to introduce a new crossover operator we refer to a measure of the distance between solutions. At the beginning, we applied the natural generalization of commonly used Kendall's tau measure, recommended for permutation-like solutions, due to its link with local search methods. Indeed, the value of this measure between $\pi, \sigma \in \Pi$ is equal to the minimum number of adjacent swap moves that have to be performed in order to transform solution π into solution σ. For the stated problem needs, the Kendall's tau distance can be written as

$$h(\pi, \sigma) = \sum_{l \in M} |h_l(\pi_l, \sigma_l)|, \tag{1}$$

where

$$h_l(\pi_l, \sigma_l) = \{(\pi_l(j), \pi_l(k)) : \sigma^{-1}(\pi_l(j)) > \sigma^{-1}(\pi_l(k)), 1 \leq j < k \leq o_l\}. \quad (2)$$

After a few poor results with the use of this measure to our aim, we consider another measure basing on the inverse permutation π^{-1} which can be considered as a point in n-dimensional Euclidean space. For every permutation $\pi \in \Pi$ one can find the corresponding point $A = (a_1, a_2, \ldots, a_n) \in R^n$ such that $a_i = \pi^{-1}(i)$ for $i = 1, 2, \ldots, n$. Thus, we can define the distance measure between permutations that equals Euclidean distance between points in R^n

$$e(\pi, \sigma) = ||AB|| = \sqrt{(a_1 - b_1)^2 + (a_2 - b_2)^2 + \ldots + (a_n - b_n)^2} \quad (3)$$

where $A = (a_1, a_2, \ldots, a_n)$, $B = (b_1, b_2, \ldots, b_n)$ are points in R^n that correspond to permutations π and σ, respectively.

Although one can find correlation between Kendall's tau distance and inverse measure for separate π_l, $l \in M$, properties of the newly proposed measure for the job-shop problem still remain unknown. We found the significant positive correlation between these two measures, using 80 test benchmarks proposed by Taillard [5]. For every instance we generated $g = 2000$ (feasible) permutations π^i, $i = 1, 2, \ldots, g$. For all pairs $(h(\pi^i, \pi^{i+g/2}), e(\pi^i, \pi^{i+g/2}))$, $i = 1, 2, \ldots, g/2$, the correlation coefficient were calculated. The value of the correlation coefficient for every instance exceeded 0.99. Thus, most of results obtained for Kendall's tau measure (job-shop case) is also valid for the geometric measure, presented above.

4 Geometric crossover operator and the algorithm

Using proposed geometric measure we design the geometric crossover quasi-operator, called GX (see Fig. 2). The operator starts with two (not necessarily feasible) processing orders and returns one feasible child permutation. Due to correlation between Kendall's tau measure and geometric measure the child permutation is situated between parent permutations also in sense of Kendall's tau measure. Thus, operator GX utilizes the path relinking philosophy and is able to explore the big valley of the solution space of problems being solved.

The genetic algorithm utilizing operator GX is presented in Fig. 3. The selection and deletion scheme used in the algorithm bases on the concept presented in paper [4]. In step 5 simulated annealing with neighborhood $N(\pi) = \{(\pi)_v \in \Pi^o : v = (i, j), i, j \in O, i \neq j\}$ is used, where v denotes an insert move and $(\pi)_v$ is the processing order obtained by application of move v to π. We also used an exponential cooling scheme, according to which temperature c is periodically reduced by a constant factor λ, $0 < \lambda < 1$. In step 7 as stop condition we used a time limitation.

The operator starts with processing orders π, σ and parameter *maxiter*. The operator returns the feasible processing order π^* and its goal function value f^*.

1. Find points $A, B \in R^n$ that correspond to permutation π and σ, respectively. Put $A' := A$, $B' := B$, $\pi^* := \pi$, $f^* := f(\pi)$ and $iter := 0$.
2. Create set $W = \{W_1, W_2, \ldots, W_m\}$ where $W_l = \{1, 2, \ldots, o_l\}$, $l \in M$.
3. Generate list $\gamma = (\gamma(1), \gamma(2), \ldots, \gamma(n))$ such that $\gamma(i) = 0$, $i = 1, 2, \ldots, n$.
4. Generate random number r from range $(0, 1)$. Find point $C = (c_1, c_2, \ldots, c_n)$ so that $||AB|| = ||AC|| + ||CB||$ and $||AC|| = r\,||AB||$. For every $k = 1, 2, \ldots, n$ perform step 5.
5. Put $l := m_k$, Find $t = \min_{i \in W_l} |c_k - i|$. Choose randomly value z from set $\{i \in W_l : |c_k - i| = t\}$. Put $\gamma(k) := z$ and $W_l := W_l \setminus \{z\}$.
6. Create permutation δ so that $\delta^{-1} = \gamma$.
7. Create the feasible permutation ω by applying algorithm P to permutation δ. If $f(\omega) < f^*$ then set $\pi^* := \omega$, $f^* := f(\omega)$.
8. Find point $A \in R^n$ that corresponds to ω. Generate random number r from range $(0, 1)$. If $r \leq 0.5$ then put $B := A'$. Otherwise put $B := B'$.
9. Put $iter := iter + 1$. If $iter < maxiter$ then go to step 2. Otherwise return processing order π^* and goal function value f^* and STOP.

Fig. 2. Outline of operator GX

The algorithm starts with parameter s. The algorithm returns the best found feasible processing order and its goal function value.

1. Generate randomly s feasible processing orders and store them on list $L = (\pi^1, \pi^2, \ldots, \pi^s)$.
2. Sort processing orders on list L so that $f(\pi^i) \leq f(\pi^{i+1})$, $i = 1, 2, \ldots, s - 1$.
3. Select two processing orders π^i, π^j, $i \neq j$, with probability inversely proportional to their ranks i, j.
4. Apply operator GX to processing orders π^i, π^j and generate processing order γ.
5. Apply simulated annealing to processing order γ and generate processing order δ.
6. If $f(\delta) < f(\pi^s)$ and $f(\delta) \neq f(\pi^i)$, $i = 1, 2, \ldots, s$ put $\pi^s := \delta$.
7. Test stop condition. If stop condition is not satisfied then go to step 2. Otherwise return the best processing order on list L and its goal function value and STOP.

Fig. 3. Outline of the genetic algorithm

5 Computational results

We compared the efficiency of our algorithm with the best algorithms applied to job-shop problem with objective C_{sum} presented in the literature. In this purpose we used 80 test benchmarks provided by Taillard [5]. The best results for the considered problem was obtained using a genetic algorithm and presented in [3]. That algorithm was run once on a Pentium III (1 GHz) with time limit 1 000 seconds for every instance with $n \leq 500$ and 2 000 seconds for every larger instance.

Our algorithm was implemented in Delphi 6.0 and run on a computer with AMD Athlon XP 2500+ processor (1.84 GHz). For every instance the algorithm was run with the same parameters $s = 20$ and $maxiter = 100$. Since our processor is about 2.5 times faster than Pentium III, we decided to run the algorithm with comparable time limit: equal to 400 and 1200 seconds, respectively. We also calculated relative improvements of values C_{sum} presented in [3] by our algorithm. We found that our algorithm improved C_{sum} values (with respect to those from [3]) for 70 and 74 instances in first and second run, respectively. The average improvement for all instances is equal to 4.08% and 5.30% for the first and second run, respectively.

6 Conclusions

Job-shop problem with C_{sum} criterion is harder than that with the makespan criterion. While for the latter we are able to generate feasible solutions easily, for the former – a procedure that repairs numerous infeasible solutions is a necessity. The proposed operator GX shows that: proper selection of the measure between solutions plays significant role in the final quality of GA, GX is preferable for the use with GA. This research was partially supported by Grant 4T11A 01624 of the State Committee for Scientific Research.

References

1. Cheng R, Gen M, Tsujimura Y (1996) A tutorial survey of job-shop scheduling problems using genetic algorithms, part I: representation. International Journal of Computers and Industrial Engineering 30:983–997
2. Cheng R, Gen M, Tsujimura Y (1999) A tutorial survey of job-shop scheduling problems using genetic algorithms, part II: hybrid genetic search strategies. International Journal of Computers and Industrial Engineering 36:343–364
3. Hennig A (2002) Praktische job-shop scheduling-probleme. Dissertation, Friedrich-Schiller-Universität, Jena
4. Reeves C (1995) A genetic algorithm for flowshop sequencing. Computers and Operations Research 22:5–13
5. Taillard E (1993) Benchmarks for basic scheduling problems. European Journal of Operational Research 64:278–285

Robotic Cells: Configurations, Conjectures and Cycle Functions

Nadia Brauner[1] and Gerd Finke[1]

Laboratoire Leibniz-IMAG, 46, av. Félix Viallet 38031 GRENOBLE Cedex, France. {nadia.brauner, gerd.finke}@imag.fr

Summary. Robotic cells consist of a flow-shop with a circular layout and a single transporter, a robot, for the material handling. A single part is to be produced and the objective is to minimize the production rate. Different cell configurations have been studied, depending on the travel times of the empty robot: additive, constant or just triangular.

A k-cycle is a production cycle where exactly k parts enter and leave the system. Consider the set $S_{\mathcal{K}}$ of all k-cycles up to size $\mathcal{K}$ where $S_{\mathcal{K}}$ contains, for every instance, an optimal solution and $\mathcal{K}$ is minimal. The cycle function $\mathcal{K}$ depends on the cell configuration and the number of machines. Some of these functions are known and there are conjectures about others. We give new results invalidating in particular the so-called Agnetis' Conjecture for the classical robotic cell configuration.

1 Robotic cells

Robotic flow-shops consist of m machines arranged in a circular layout and served by a single central robot [Figure 1]. It is known that the robotic scheduling problem is already NP-hard for a flowshop with $m \geq 3$ machines and two or more different part types [9]. It remains the interesting case of the m-machine robotic cell in which one wants to produce identical parts. Then the problem reduces to finding the optimal strategy for the robot moves in order to obtain the maximal throughput rate for this unique part. A survey on general robotic cells can be found in [7].

The m machines of a robotic cell are denoted by M_1, $M_2 \dots M_m$ and we add two auxiliary machines, M_0 for the input station IN and M_{m+1} for the output station OUT. The raw material for the parts to be produced is available in unlimited quantity at M_0. The central robot can handle a single unit at a time. A part is picked up at M_0 and transferred in succession to M_1, $M_2 \dots M_m$, where it is machined in this order until it finally reaches the output station M_{m+1}. At M_{m+1}, the finished parts can be stored in unlimited amounts. We focus on the classical case as in [12], where the machines M_1,

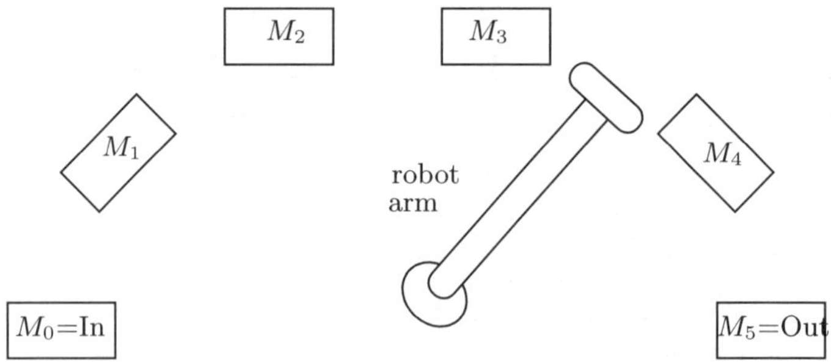

Fig. 1. Robotic cell with $m = 4$ machines

$M_2 \ldots M_m$ are without buffer facility. In this case, the robot has to be empty whenever it wants to pick up a part at M_h ($h = 0,1 \ldots m$).

Consider an instance I of an m-machine robotic cell. The *processing time* p_h represents the minimum time a part must remain on machine M_h ($h = 1, 2 \ldots m$). Once the part is finished, two policies may apply. In the *no-wait* case, the part must be removed immediately from the machine after p_h time units and transferred to the following machine. In the *classical* case, the part can remain on the machine waiting for the robot. Let ϵ be the time to load a part onto a machine from the robot or to unload a part from a machine onto the robot. Different cell configurations have been studied, depending mainly on the metrics for the travel times of the empty robot. These travel times may be additive, constant or just triangular. We shall concentrate on two classical metrics for the travel times of the robot. For *additive* travel times [12], let δ be the time for the robot to travel (idle or loaded) between two consecutive machines. The travel times are additive. Hence, the trip of the idle robot from M_h to $M_{h'}$ ($h \neq h'$) takes $\delta_{hh'} = |h - h'|\delta$. For *constant* travel times [8], δ is the time for the robot to travel between any two machines M_h and $M_{h'}$: $\delta_{hh'} = \delta$.

We consider cyclic robot moves for the production process of the parts and define a k-*cycle* as a production cycle of exactly k parts. It can be described as a sequence of robot moves where exactly k parts enter the system at M_0, k parts leave the system at M_{m+1} and each time the robot executes the k-cycle, the system returns to the same state, *i.e.* the same machines are loaded, the same machines are empty and the robot returns to the starting position. To describe k-cycles we use the concept of *activities* [5]. The activity A_h ($h = 0, 1 \ldots m$) consists of the following sequence:

- The idle robot takes a part from M_h.
- The robot travels with this part from M_h to M_{h+1}.
- The robot loads this part onto M_{h+1}.

In [5], the authors characterize the k-cycles as follows: A k-cycle C_k is a sequence of activities, in which each activity occurs exactly k times and between two consecutive (in a cyclic sense) occurrences of A_h ($h = 1, 2 \ldots m - 1$) there is exactly one occurrence of A_{h-1} and exactly one occurrence of A_{h+1}.

A ρ-cycle C_ρ is *optimal* if it minimizes $\frac{T(C_k)}{k}$ over all possible k-cycles $k = 1, 2\ldots$, where $T(C_k)$ denotes the cycle time of C_k. A set of cycles S is said to be *dominant* if, for any instance, there exists a cycle of S that is optimal.

2 Dominant sets of cycles

Ideally, one would like to determine, for a given instance, an optimal k-cycle. However, this is so far not possible, except for very particular cases, for instance for very slow or for very fast robots compared to the processing times. In [12] the authors conjectured that the 1-cycles are dominant. This conjecture is valid for classical two- and three-machine cells. However it is false for four-machine cells [3, 5, 8]. It has been replaced by the following conjecture:

Agnetis' Conjecture[1]: The set of k-cycles with $k \leq m - 1$ is dominant.

Note that this conjecture was originally formulated by Agnetis for additive no-wait cells. Let $S_{\mathcal{K}}$ be the set of all k-cycles with $1 \leq k \leq \mathcal{K}$. We are interested in the minimal dominant set $S_{\mathcal{K}}$, *i.e.*, $S_{\mathcal{K}}$ is dominant and no $S_{\mathcal{K}'}$ is dominant with $\mathcal{K}' < \mathcal{K}$. We can expect that $\mathcal{K} = \mathcal{K}(m)$ is a function of the number of machines m. We call $\mathcal{K}(m)$ the **cycle function**, denoted by $\mathcal{K}_{nw}(m)$ in the no-wait case and by $\mathcal{K}_c(m)$ in the classical case. Agnetis' Conjecture claims that $\mathcal{K}_{nw}(m) = m - 1$. In additive no-wait robotic cells, we know that

$$\mathcal{K}_{nw}(m) \begin{cases} = m-1 \text{ for } m = 2, 3 \text{ [1]} \\ \geq m-1 \text{ for } m \geq 4 \quad \text{[11]} \end{cases}$$

In classical robotic cells (with additive or constant travel times), one has

$$\mathcal{K}_c(m) \begin{cases} = 1 \ \text{ for } m = 2, 3 \\ \geq m \text{ for } m = 4 \end{cases}$$

For $m = 2$, and 3 this result was proven in [3, 5] for the additive case and in [8] in the constant case. It was shown in [3] that $\mathcal{K}_c(4) \geq 3$. We now strengthen this result to $\mathcal{K}_c(4) \geq 4$. We exhibit a cycle which proves that Agnetis' Conjecture is false in classical 4-machine robotic cells:

Proposition 1. *The 4-cycle*
$C_4 = (A_0A_1A_0A_3A_4A_2A_1A_0A_3A_2A_1A_4A_3A_2A_0A_1A_4A_3A_4A_2)$
strictly dominates all k-cycles for $k = 1, 2, 3$ for the following instance:
$m = 4$; $\delta = 1$; $\epsilon = 0$; $p_1 = 0$; $p_4 = 0$;
in the additive case: $p_2 = 10$; $p_3 = 10$;
in the constant case: $p_2 = 6$; $p_3 = 6$.

Proof for the additive case
One has $\frac{T(C_4)}{4} = 15$. We shall prove that for all k-cycles C_k $(k = 1, 2, 3)$, one has $\frac{T(C_k)}{k} > 15$.

$k = 1$: For the instance I, the best 1-cycle has cycle time 16.

$k = 2$: For the instance I, the 1-cycles dominate the 2-cycles (regular or equidistant case) [2, 3]. Therefore the best 2-cycle has cycle time greater or equal to 16.

$k = 3$: Let C_3 be a 3-cycle. Let m_i be the number of times the robot travels between machines M_i and M_{i+1} in both directions during one execution of C_3 and let $|S|$ be the number of occurrences of the sequence of activities S in C_3 and let $u_i = |A_{i-1}A_i|$. If the robot never makes any dummy moves, one has [3, 2]:

$m_0 = 2k \qquad m_2 \geq 4k - 2u_2 - 2|A_1A_0A_2|$
$m_m = 2k \qquad m_{m-2} \geq 4k - 2u_{m-1} - 2|A_{m-2}A_mA_{m-1}|$
$m_1 = 4k - 2u_1 \qquad m_{m-1} = 4k - 2u_m$

Moreover we know that the sequences $A_{i-1}A_i$ generate a waiting time of p_i and the sequence $A_1A_0A_2$ generates a waiting time of $\max(0, p_2 - 4\delta - 2\epsilon)$ and the sequence $A_{m-2}A_mA_{m-1}$ generates a waiting time of $\max(0, p_{m-1} - 4\delta - 2\epsilon)$. All those sequences do not interfere for the calculation of the waiting time. Therefore, one has

$$
\begin{aligned}
T(C_3) &= \sum_{i=0}^{4} m_i\delta + \text{waiting times} \\
&\geq 16k\delta + u_1(p_1 - 2\delta) + u_4(p_4 - 2\delta) + u_2(p_2 - \delta) + u_3(p_3 - \delta) \\
&\quad + |A_1A_0A_2|(\max(0, p_2 - 4\delta - 2\epsilon) - \delta) \\
&\quad + |A_2A_4A_3|(\max(0, p_3 - 4\delta - 2\epsilon) - \delta)
\end{aligned}
$$

and hence, for the instance I, the cycle time of C_3 satisfies

$$
T(C_3) \geq 48 - 2u_1 - 2u_4 + 5|A_1A_0A_2| + 5|A_2A_4A_3| + 9u_2 + 9u_3 \qquad (1)
$$

Suppose $T(C_3) \leq \frac{T(C_4)}{4} \times 3 = 15 \times 3 = 45$. In C_3, between two consecutive occurrences of A_2, there is exactly one occurrence of A_1 and one occurrence of A_3. If somewhere A_3 happens before A_1 between two consecutive occurrences of A_2, then C_3 is of the form described on Figure 2. In this case, the cycle time of C_3 satisfies $T(C_3) \geq 54$ (contradiction).

Therefore, C_3 is of the form

$$
C_3 = A_2 \cdots A_1 \cdots A_3 \cdots A_2 \cdots A_1 \cdots A_3 \cdots A_2 \cdots A_1 \cdots A_3 \ldots
$$

¿From inequality (1), $T(C_3) \leq 45$ implies $2(u_1 + u_4) \geq 3$, which means $u_1 + u_4 \geq 2$ since u_1 and u_4 are integers.

Consider now what happens between two consecutive activities A_2: there is exactly one occurrence of activity A_1 and between the end of A_1 and the following A_2 the time spent is at least $p_2 = 10$. Moreover, between the end of A_2 and the end of the following A_1, the robot has to travel from machine M_3 to machine M_1 and then from M_1 to M_2 (execution of A_1). Therefore, the time spent between two consecutive occurrences of A_2 is at least 14 plus

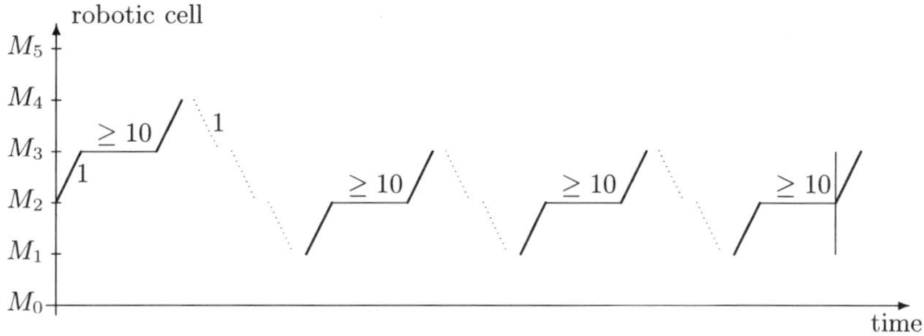

Fig. 2. Position of the robot in the cell for the case with A_3 before A_1

the time for the execution of the activities between A_2 and A_1 (at least 2 for each activity). The same reasoning applies to A_3: the time spent between two consecutive occurrences of A_2 is at least 14 plus the time for the execution of the activities between A_3 and A_2.

We know that the number u_1 of A_0A_1 plus the number u_4 of A_3A_4 is greater than two. If those two occurrences do not appear between the same consecutive A_2, then each generates an additive time of 2 and one has $T(C_3) \geq 14*3+2*2 = 46$ which leads to a contradiction. Therefore, C_3 is of the form:

$$C_3 = A_2A_1 \cdots A_3A_2A_0A_1 \underbrace{\cdots}_{S} A_3A_4A_2A_1 \cdots A_3$$

Indeed, no other activity can happen between A_2 and the following A_1 or between A_3 and the following A_2. Between two consecutive occurrences of A_1, there is exactly one occurrence of A_0. Therefore, the sequence S contains an occurrence of A_0. The same reasoning implies that S contains an occurrence of A_4. The sequences $A_2A_0A_1A_0A_4A_3A_4$ or $A_2A_0A_1A_4A_0A_3A_4$ imply a travel time of at least 20 before the next A_2. Since between the two other consecutive activities A_2 the total time is at least 14, one has $T(C_3) \geq 14*2+20 = 48$. Therefore, $T(C_3) > 45$ which is a contradiction. □

Proof in the constant case

One has $\frac{T(C_4)}{4} = 9.5$. The proof for the constant case is almost the same as for the additive case. It is a little bit simpler since, in the constant case, one has the following equality for the travel time $T_T(C_k)$ of the k-cycle C_k:

$$T_T(C_k) = 2k(m+1)\delta - \sum_{i=1}^{m} u_i\delta \tag{2}$$

The intuition for this equality is that between two activities, one has a time δ if and only if the two activities are not consecutive, *i.e.* they do not participate in a u_i. This equality is proven for $k = 1$ in [8]. The complete proof for this case can be found in [4]. □

3 Remaining challenging questions

Finding the best 1-cycle can be done in polynomial time ([6] for the additive classical case, [8] for the constant case and [10] for no-wait additive cells). Three challenging questions remain:

- Determine $\mathcal{K}_c(m)$ or find at least an upper finite bound for $\mathcal{K}_c(m)$;
- Settle Agnetis' conjecture in the no-wait case;
- Describe the complexity of finding the best cycle with degree smaller or equal to $\mathcal{K}_c(m)$ and/or $\mathcal{K}_{nw}(m)$.

References

1. Agnetis A. Scheduling no-wait robotic cells with two and three machines. *European Journal of Operational Research*, 123(2): 303-314, 2000.
2. N. Brauner. *Ordonnancement dans des cellules robotisées, analyse de la conjecture des un-cycles.* Thèse de doctorat, Université Joseph Fourier, Grenoble, France, 1999.
3. N. Brauner and G. Finke. Cycles and permutations in robotic cells. *Mathematical and Computer Modelling*, 34:565–591, 2001.
4. N. Brauner and G. Finke Cyclic scheduling in robotic cells: about Agnetis' conjecture for the classical case. Les cahiers du laboratoire Leibniz 120, Grenoble, France, 2005.
5. Y. Crama and J. van de Klundert. Cyclic scheduling in 3-machine robotic flow shops. *Journal of Scheduling*, 2:35–54, 1999.
6. Y. Crama and J. van de Klundert. Cyclic scheduling of identical parts in a robotic cell. *Operations Research*, 45(6):952–965, 1997.
7. Y. Crama, V. Kats, J. van de Klundert, and E. Levner. Cyclic scheduling in robotic flowshops. *Annals of Operations Research: Mathematics of Industrial Systems*, 96:97-124, 2000.
8. M. Dawande, C. Sriskandarajah, and S. Sethi. On throughput maximization in constant travel-time robotic cells. *Manufacturing and Service Operations Management*, 4(4):296-312, 2002.
9. N. G. Hall, H. Kamoun, and C. Sriskandarajah. Scheduling in robotic cells: Classification, two and three machine cells. *Operations Research*, 45(3):421–439, 1997.
10. E. Levner, V. Kats, and V.E. Levit. An improved algorithm for cyclic flowshop scheduling in a robotic cell. *European Journal of Operational Research*, 97:500–508, 1997.
11. F. Mangione, N. Brauner, and B. Penz. Optimal cycles for the robotic balanced no-wait flow shop. In *Proceedings IEPM'03, International Conference of Industrial Engineering and Production Management*, volume 2, pages 539-547, Porto, Portugal, 2003.
12. S. P. Sethi, C. Sriskandarajah, G. Sorger, J. Blazewicz, and W. Kubiak. Sequencing of parts and robot moves in a robotic cell. *International Journal of Flexible Manufacturing Systems*, 4:331–358, 1992.

Part XX

Technology and Innovation

Robot Task Planning for Laser Remote Welding

Jannis Stemmann[1] and Richard Zunke[2]

[1] Hamburg University of Technology (TUHH) j.stemmann@tuhh.de
[2] Hamburg University of Technology (TUHH) richard.zunke@tuhh.de

1 Introduction

Production cycle times are thought to be a key figure for industrial manufacturers, as they affect inventory, quality and order lead times. Laser welding offers primary processing times which are only fractions of those for conventional joining technologies. Moreover, laser remote welding (LRW) allows for drastic reductions of secondary processing times, owing to special optic tools called scanners [2]. It is a relatively new joining technology, envisaged in 1993 [9]. Remote welding is supposed to be a cornerstone of future automotive production [4]. Today, several major European and Asian car manufacturers are qualifying flexible robot-based systems with solid state lasers. These are intended to replace resistance spot welding robots in body-in-white manufacturing lines, starting in 2006.

Our article deals with task planning for such systems, which is demanding due to kinematic redundancy of the handling system. We propose to model robot-based remote welding jobs as instances of the Generalized Traveling Salesman Problem (GTSP), using the concept of open kinematic chains. Afterwards the GTSP instance can be tackled by appropriate algorithms, usually resulting in an approximate solution. That in turn can be decoded into a robot path and used as offline program or for simulation.

2 Generalized Traveling Salesman Problem

GTSP was introduced around 1970 [11]. It depicts combinatorial problems of simultaneous sequencing and selection [5]. In fact, a shortest tour through a set of clusters is sought, while each cluster is a set of vertices out of which (at least or exactly) one vertex has to be visited. As will be shown below, this makes it an attractive model for robot scheduling. The well-known Traveling Salesman Problem (TSP) can be considered as a special case of GTSP [7]. Obviously GTSP is NP-hard, and most of the few papers published on GTSP either treat only small problem sizes or suggest heuristics for approximation. The benchmark for exact solution methods is still the Branch & Cut routine developed by *Fischetti et al.* [3]. At the time of writing, the most powerful metaheuristics appear to be genetic algorithms. Though there is no publication in which GTSP is directly referred to as a model for robot task planning, TSP is very common for this purpose [1, 6].

3 Modeling robot-based remote welding tasks as GTSP instances

LRW systems under consideration are assumed to consist of a standard industrial robot (IR), a scanning unit mounted to the robot flange, and a workpiece. Industrial robots usually feature six degrees of freedom (DOF) that can be used to position the scanner, which in turn is equipped with 2 or 3 additional high-speed actuators. Two of these drives are used to deflect the laser beam in x- and y-directions, whereas the optinal third is integrated into a zoom device for axial focus shift. Thus the robot-based LRW system has up to 9 DOF for positioning and orientating the laser beam focus on the workpiece. However, only five DOF are needed to accomplish this task (presuming rotational symmetric laser beams). This results in kinematic redundancy and allows for changing of joint configurations while the laser beam focus keeps the identical pose (pose denotes position and orientation of a coordinate system). Furthermore, articulated robots generally show ambiguities in configuration space [10].

In order to cope with kinematic redundancy the whole system will be split up (s. Fig. 1). One subsystem includes the workpiece with the weld seams, the laser beam modeled as a rigid body and the scanner. Kinematic behaviour of the robot is described by a second subsystem. Sequential welding is assumed, i.e., the robot will not move during the joining process. Using homogeneous 4-by-4 transformation matrices (so called frames) $\{\mathrm{A}\}$, it is possible to model the kinematic chain by attributing coordinate systems to each revolute or prismatic joint and storing the poses in coordinates of a reference coordinate system, i.e., world frame $\{\mathrm{W}\}$. The first subsystem allows the determination of robot poses described by $\{\mathrm{F}\}$ in coordinates of $\{\mathrm{W}\}$. The according information in homogeneous coordinates is termed ${}^{W}_{F}\mathbf{A}$. By stepwise description of frame i in coordinates of system $i-1$, the whole chain is shaped. Consecutive frames are linked by multiplying. This coordinate transformation is called direct kinematics in terms of robotics. Given the values of all joint variables, it is possible to determine the pose of the last coordinate system. Solution of the inverse kinematic problem includes joint values for a given pose of the last frame in coordinates of a reference system. A special procedure by *Paul* allows the determination of analytical expressions out of the ensuing nonlinear equation system for anthropomorphic manipulators with spherical wrists, which is the basic structure of most articulated robots [8].

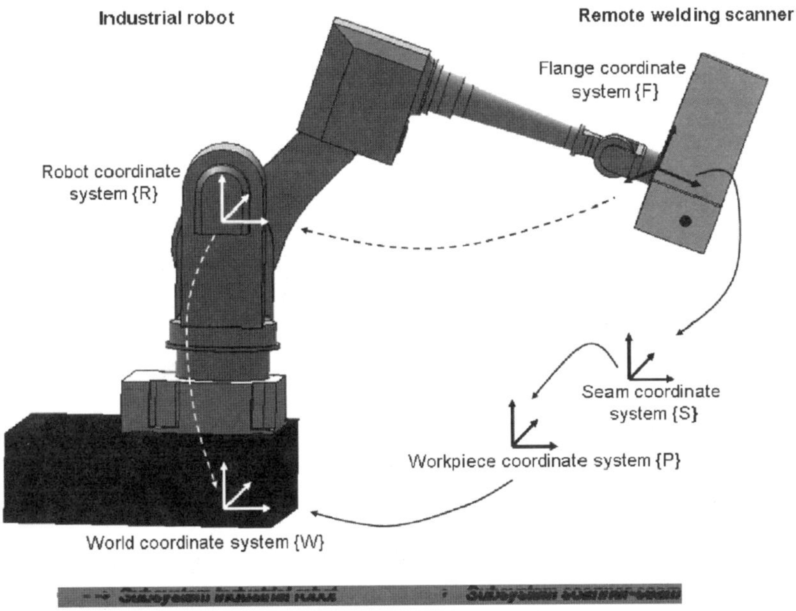

Fig. 1. Reduction of complexity by system splitting

The analogy permitting a formulation of robot-based LRW tasks as GTSP instances is as follows: Each cluster in GTSP terms is set equal to a scanning-feasible workspace. Scanning-feasible workspaces are portions of the overall robot workspace that allow for welding of seams (a task). Vertices are related to robot configurations. Edges between vertices correspond to robot motions. The cost of the tour that is to be minimized by GTSP solution algorithms depicts the entire motion time, or secondary processing time, for all welding tasks. Workpiece and handling system data are needed as input for the task planning approach. Computations can be divided into four rough steps which are explained below.

3.1 Determination of scanning-feasible workspaces

Scanning-feasible workspaces are represented by a discretized number of poses ${}^{W}_{F}\mathbf{A}$. The joint types of the first subsystem (starting with {W}) are: A prismatic joint with variable t describing the seam length, two perpendicular revolute joints to take the maximal incidence angles ϵ and δ into account, another revolute joint describing the possible rotation ξ around the rotation-symmetric laser beam, a prismatic joint for focus shift h and two consecutive revolute joints κ and ι describing the deflection of the laser beam (s. Fig. 2). The determination of viable welding poses of {F} is done in three steps. At first the pose of the robot flange coordinate system is calculated with regard to the initial point ($t = 0$) of every single seam for certain parameter values. This is repeated for seam end points. Afterwards each pose in a scanning-feasible workspace is checked for viability of both extreme points of its corresponding weld line at the same time, under the condition of not violating any constraints imposed by process and system.

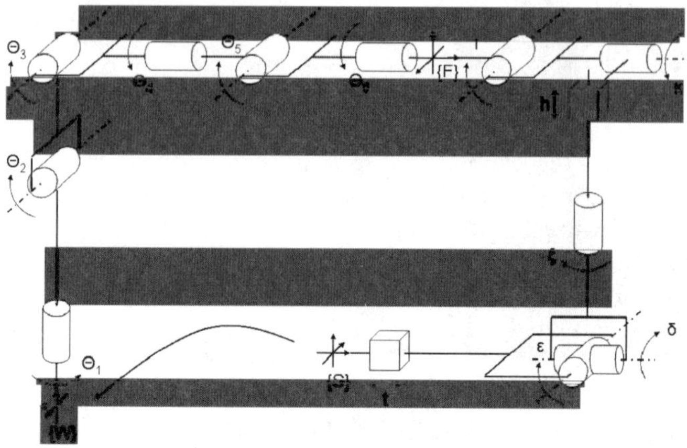

Fig. 2. Robot-based LRW system modeled as open kinematic chain

3.2 Validation of robot reachability

The reachability of determined poses is validated using the inverse kinematics solution for the IR subsystem. We propose a closed solution to allow a fast and precise calculation of joint values for the six revolute axes Θ_i with $i = 1 \ldots 6$ (s. Fig. 2). If one of the poses is reachable by multiple joint configurations, ambiguities can be taken into account at this stage.

3.3 Identification of overlapping scanning-feasible workspaces

The common GTSP formulation does not permit overlapping of clusters [7]. Mutually exclusive clusters therefore are generated by copying poses into superpositioned scanning-feasible workspaces, where necessary. The identification of poses that allow for welding of more than one seam without robot motion is of paramount importance to find a cost minimal tour. The actual identification routine is similar to the validation explained above, except that is has to be used more often. Time effort can be reduced by including sensible constraints (e.g., seams too far from each other do not have to be checked).

3.4 Calculation of the GTSP instance

A cost matrix containing data regarding the motion time between one joint configuration to another has to be calculated. Therefore a certain dynamic robot behaviour has to be assumed. The underlying model can either be obtained by the manufacturer or taken from literature and fitted to measurement values. Moreover, an affilitation list which assigns robot configurations (vertices) to clusters (weld seams) is needed.

4 Results

To demonstrate the effectiveness of the approach, a welding task of significant size was chosen as an example. In Fig. 3, a CAD drawing of a car sidepanel in a simplified robot workcell (on the left side) and seam frames (on the right side) are shown. The total number of seams is 207, and for each seam on

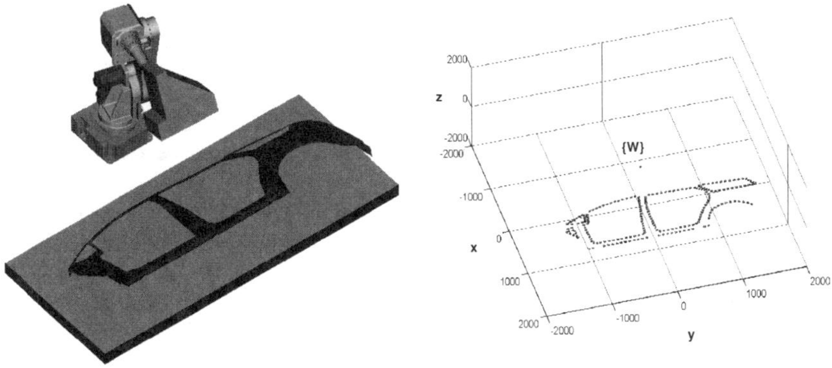

Fig. 3. Sidepanel remote welding task

average more than 20 feasible robot configurations were calculated. Note that the derived GTSP instance is about twice the size of the largest problem treated in literature up to date. The best tour found so far offers a total cost of 23.84 s. A simulation run in a proprietary offline programming environment revealed that the actual motion time for the generated robot path is only 22.65 s. This corresponds to a forecast error of 5.3% due to model imperfections. Relative to paths generated by teach-in programming without optimization, time savings of more than 20% could be realized.

5 Conclusion

Up to now, no commercial offline programming system for robot-assisted laser remote welding exists. The Generalized Traveling Salesman Problem seems to model sequential remote welding jobs with sufficient accuracy, if adequate constraints are taken into account. Robot kinematics based on homogeneous transforms was found to be a simple tool used throughout system abstraction. It could be proven that modeling and subsequent optimization results in shorter cycle times than for robot paths obtained by empirical knowledge and intuiton. Further research options are, for instance, multiple robot facilities and coupled axes systems.

References

1. S. Dubowski and T.D. Blubaugh. Planning time-optimal robotic manipulator motions and work places for point-to-point tasks. *IEEE Transactions on Robotics and Automation, Vol. 5*, (3):746–759, 1989.
2. C. Emmelmann. Laser remote welding - status and potential for innovations in industrial production. In A. Otto, editor, *Lasers in Manufacturing. Proceedings of the Third International WLT-Conference on Lasers in Manufacturing*, pages 1–6, Stuttgart, 2005. AT-Verlag.
3. M. Fischetti, J.J. Salazar Gonzalez, and P. Toth. A branch-and-cut algorithm for the symmetric generalized traveling salesman problem. *Operations Research, Vol. 45*, (3):378–394, 1997.
4. H.J. Herfurth and S. Heinemann. Robotic remote welding with state-of-the-art co_2 lasers. Automotive Laser Application Workshop (ALAW), 2005.
5. G. Laporte, A. Asef-Vaziri, and C. Sriskandarajah. Some applications of the generalized traveling salesman problem. *Journal of the Operational Research Society, Vol. 47*, (12):1461–1467, 1987.
6. O. Maimon, D. Braha, and V. Seth. A neural network approach for a robot task sequencing problem. *Artifial Intelligence in Engineering, Vol. 14*, (2):175–189, 2000.
7. Ch. E. Noon. *The generalized traveling salesman problem.* PhD thesis, The University of Michigan, 1988.
8. R.P. Paul. *Robot Manipulators: Mathematics, Programming, and Control.* MIT Press, Cambridge, Mass., 1981.
9. S. Ream. Laser scanner lap welding. Proceedings of the International Body Engineering Conference, Advanced Technologies and Processes, pp. 44-47, 1993.
10. L. Sciavicco and B. Siciliano. *Modeling and Control of Robot Manipulators, 2nd Edition.* Springer, London et al., 2000.
11. S.S. Srivastava, S. Kumar, R.C. Garg, and P. Sen. Generalized travelling salesman problem through n sets of nodes. *Journal of the Canadian Operational Research Society (CORS Journal), Vol. 7*, pages 97–101, 1969.

Technologischer Fortschritt in der deutschen Bankenwirtschaft

A. Varmaz[1] und Th. Poddig[2]

[1] Lehrstuhl für Finanzwirtschaft, Universität Bremen, Hochschulring 4, 28359 Bremen, varmaz@uni-bremen.de
[2] Lehrstuhl für Finanzwirtschaft, Universität Bremen, poddig@uni-bremen.de

1 Einleitung

Die Messung von Produktivität ist ein wichtiger Aspekt bei der Performancebeurteilung von Firmen. Ein wichtiges Konzept zur Produktivitätsmessung ist das Wachstum der totalen Faktorproduktivität (TFP), welches Änderungen von sowohl Ausbringungs- (Output) als auch Faktoreinsatzmengen (Input) berücksichtigt. Zur Messung von TFP werden oft Tornqvist- und Fisherproduktivitätsindizies herangezogen. Der Nachteil dieser Instrumente liegt in der Annahme einer effizienten Produktion von Firmen in den betrachteten Perioden. Erweitert man diese Ansätze durch die Möglichkeit ineffizienter Produktion, kann eine Produktivitätsänderung durch technologischen Fortschritt und eine Effizienzänderung bedingt werden. In diesem Beitrag wird ein auf dem Malmquistprduktivitätsindex basierender Ansatz vorgestellt, der eine Separierung der Produktivitätsänderung in Effizienzänderungen und technologische Änderung ermöglicht. Dazu wird der Malmquistindex auf Basis von Data Envelopment Analysis (DEA) modifiziert. DEA ist eine Methode zur relativen Effizienzmessung von unabhängigen Entscheidungseinheiten (EE) und wird im Kapitel 2 vorgestellt. Der modifizerte Malmquistindex wird in Kapitel 3 präsentiert. Eine empirische Studie zum Produktivitätswachstum innerhalb deutscher Kreditgenossenschaften wird in Kapitel 4 gegeben.

2 Data Envelopment Analysis

Die Grundidee einer Effizienzmessung mit DEA beruht auf der Anwendung des ökonomischen Prinzips. Danach ist eine EE effizient, wenn keine andere Alternative existiert, die mit einem gegebenen Mitteleinsatz einen höheren Zielerreichungsgrad erreicht (Maximalprinzip) oder einen gegebenen Zielerreichungsgrad mit geringerem Mitteleinsatz realisiert (Minimalprinzip). Im Folgenden wird die Operationalisierung des Minimalprinzips betrachtet.

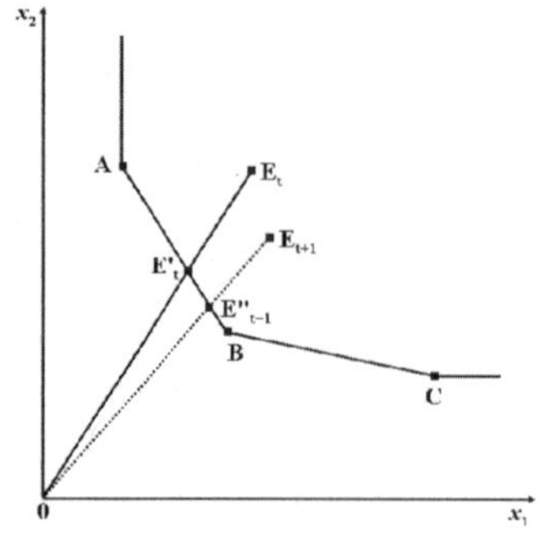

Abb. 1a.

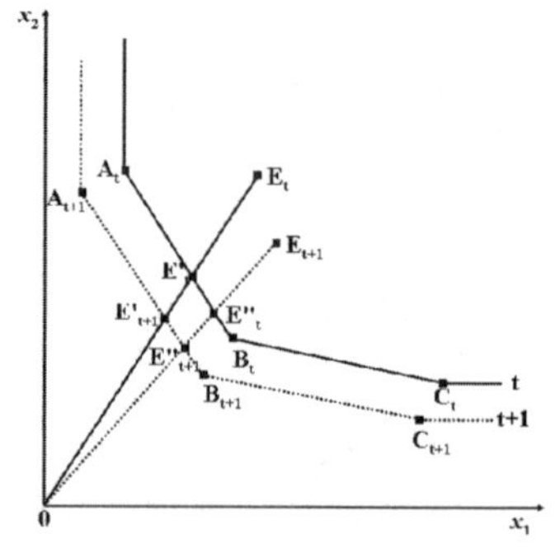

Abb. 1b.

Die Abbildung 1a veranschaulicht die Vorgehensweise einer DEA-Effizienzanalyse. In der Abbildung produzieren alle EE eine Einheit Output mit Hilfe von zwei unterschiedlichen Inputmengen. Nach der Anwendung des Effizienzkriteriums werden die EE A, B und C nicht dominiert und sind effizient. Die Linearkombination dieser EE bildet die Effizienzgrenze. EE E_t und E_{t+1} werden durch die Effizienzgrenze dominiert und sind ineffizient. Senkt E_t die Inputmengen proportional bis zum Punkt $E_t^{'}$ auf der Effizienzgrenze, definiert das Verhätnis der Strecken $\overline{0E_t^{'}}$ und $\overline{0E_t}$ ein Maß der Ineffizienz von E_t. Das Verhältnis gibt das Niveau an, auf welches E_t die Inputmengen proportional senken könnte, wenn es ihre Outputmenge effizient produzieren würde. EE D liegt zwar auf der konstruierten Effizienzgrenze, es ist jedoch offensichtlich, dass es von C dominiert wird, da C weniger vom Input x_1 einsetzen muss, um eine Einheit Output zu produzieren. DEA verallgemeinert die in Abbildung 1a dargestellte Vorgehensweise auf den Fall multipler Input- und multipler Outputmengen. Hierzu seien n EE angenommen. Jede EE_j $(j = 1, \ldots, n)$ stellt s verschiedene Produkte (Output) y_{rj} $(r = 1, \ldots, s)$ mit Hilfe von m verschiedenen Einsatzfaktoren (Input) x_{ij} $(i = 1, \ldots, m)$ her. Zur Berechnung der inputorientierten (= kostenminimierenden) technischen Effizienz einer beliebigen EE_0 wird das DEA–Optimierungsproblem in 1 gelöst (vgl. [1]),

$$
\begin{aligned}
& \min_{\theta_0, \lambda_j} \quad \theta_0 & & & (1)\\
s.t. \quad & \sum_{j=1}^{n} \lambda_j x_{ij} + s_i^- = \theta_0 x_{i0} & & i = 1, \ldots, m \\
& \sum_{j=1}^{n} \lambda_j y_{rj} - s_r^+ = y_{r0} & & r = 1, \ldots, s \\
& \theta_0, \lambda_j, s_i^-, s_r^+ \geq 0 & &
\end{aligned}
$$

wobei x_{i0} und y_{r0} der i-te Input bzw. r-te Output der gerade betrachteten EE_0 sind. Die Variablen s_i^- sowie s_r^+ repräsentieren Schlupfvariablen, die jeweils eine Verschwendung bzw. Nichterreichung von Input- bzw. Outputmengen anzeigen. θ_0 gibt die technische Effizienz von EE_0 unter der Annahme der konstanten Skalenerträge wieder, wobei $\theta_0 \in [0,1]$ gilt. Eine EE ist effizient, wenn $\theta_0 = 1$, $s_i^- = 0$ und $s_r^+ = 0$ gilt. Eine Effizienzanalyse der EE in der Abbildung 1a würde EE A, B und C als effizient klassifizieren. EE D hätte einen Effizienzwert von $\theta_0 = 1$, aber einen von Null verschiedenen Wert für $s_{x_1}^-$. EE E_t hätte einen Wert von $\theta_0 < 1$. Durch den Gewichtevektor λ_j wird der Abschnitt der Effizienzgrenze angegeben, von dem E_t dominiert wird.

3 DEA-Malmquist

Um Produktivitätsänderung zwischen zwei Perioden zu beobachten, kann auf den Malmquist(produktivitäts-)index zurückgegriffen werden. Werden die Ideen der DEA-Effizienzanalyse in den Kontext der Produktivitätsmessung einbezogen, kann der Malmquistindex zerlegt werden, um Effizienzänderung sowie technologischen Fortschritt zu beobachten. Die Grundidee wird in Abbildungen 1a und 1b veranschaulicht.

In der Abbildung produzieren EE mit zwei Inputmengen eine Einheit Output. Zunächst wird der Fall einer Produktivitätsänderung ohne eine Verschiebung der Effizienzgrenze betrachtet, die aus EE A, B und C gebildet wird (vgl. Abbildung 1a). Dies könnte als gleichbleibende Produktivität auf der Ebene der Gesamtbranche interpretiert werden. Im Beispiel wird die Produktivitätsänderung der EE E betrachtet, die in $t+1$ eine Einheit Output mit tatsächlich weniger Input herstellen konnte. Dabei kann E in $t+1$ die Produktivität alleine durch eine Effizienzverbesserung steigern. In der Abbildung 1b wird eine technologisch bedingte Produktivitätsänderung angenommen, die durch die Verschiebung der Effizienzgrenze zum Ursprung hin in $t+1$ angezeigt wird. Die effiziente EE A_t konnte die Produktivität in $t+1$ steigern. Da A bereits effizient ist, kann eine solche Verbesserung nicht durch Anwendung einer bekannten effizienten Produktionstechnologie, sondern durch Einsetzen neuer Technologien erklärt werden. Zur Berechnung der Produktivitätsänderung von E muss jetzt sowohl eine Änderung des Abstandes zur Effizienzgrenze als auch die Verschiebung der Effizienzgrenze berücksichtigt werden. Daher wird die Effizienz von E in den Perioden t und t+1 berechnet. Um die durchschnittliche Verschiebung der Effizienzgrenze zu berücksichtigen, muss zusätzlich die Effizienz von E in t zur Effizienzgrenze in $t+1$ sowie von E in $t+1$ zur Effizienzgrenze in t bestimmt werden. Die Berechnung des Produktivitätswachstum erfolgt dann durch die Bildung eines geometrischen Mittelwertes der Verhätnisse dieser Effizienzwerte. Nach [2] wird der Malmquistindex nach Gleichung 2 berechnet als (mit $\theta^{\text{Periode der Effizienzgrenze}}_{\text{Periode, in der sich } i \text{ befindet}}$):

$$M_0 = \sqrt{\frac{\theta^t_{i,t}}{\theta^{t+1}_{i,t+1}} \frac{\theta^{t+1}_{i,t}}{\theta^t_{i,t+1}}} \tag{2}$$

M_0 misst die Produktivitätsänderung bezüglich Inputmengen (inputorientierte Effizienz) zwischen t und $t+1$. Ein Wert von $M_0 > 1$ zeigt eine Verschlechterung der Produktivität und $M_0 < 1$ eine Verbesserung an. Bei $M_0 = 1$ bleibt sie unverändert. Besonders interessant hierbei ist jedoch die Möglichkeit einer Zerlegung des Index. Die Produktivitätsänderung von t zu $t+1$ kann durch eine managementbedingte Verschiebung der betrachteten EE relativ zur Benchmark (Effizienzverbesserung) und durch eine technologisch bedingte Verschiebung der Effizienzgrenze geschehen. Durch Separierung dieser Effekte können mehr Informationen über die Ursachen einer Produktivitätsänderung gewonnen werden. Dazu muss M_0 in zwei Komponenten entsprechend 3 zerlegt werden:

$$M_0 = \sqrt{\frac{\theta^t_{i,t}}{\theta^t_{i,t+1}} \frac{\theta^{t+1}_{i,t}}{\theta^{t+1}_{i,t+1}}} = \sqrt{\left(\frac{\theta^t_{i,t}}{\theta^{t+1}_{i,t+1}}\right)^2 \frac{\theta^{t+1}_{i,t}}{\theta^t_{i,t+1}} \frac{\theta^{t+1}_{i,t+1}}{\theta^t_{i,t}}} = \frac{\theta^t_{i,t}}{\theta^{t+1}_{i,t+1}} \cdot \sqrt{\frac{\theta^{t+1}_{i,t+1}}{\theta^t_{i,t+1}} \frac{\theta^{t+1}_{i,t}}{\theta^t_{i,t}}} \quad (3)$$

Der erste Term der Gleichung repräsentiert die Veränderung der Effizienz einer EE zwischen zwei Perioden. Der zweite Term gibt die Verlagerung der Effizienzgrenze wieder. Weitergehende Modifikationen und Zerlegungen von M_0 sind ebenfalls möglich (z.B. Beobachtung des Einflusses der Skaleneffizienz).

4 Empirische Untersuchung

Das vorgestellte DEA-Malmquist-Modell zur Messung von Produktivitätsänderungen wird im Rahmen einer empirischen Untersuchung eingesetzt. Da die Effizienz bzw. Produktivität von Banken bankgruppenspezifischen Einflüssen unterliegen könnte, konzentriert sich diese Analyse auf Genossenschaftsbanken (Geno). Dazu werden Jahresabschlussdaten von Geno für den Zeitraum 1998-2003 verwendet. Um eine Vergleichbarkeit der Ergebnisse für den gesamten Untersuchungszeitraum zu sichern, müssen für alle Banken in allen Jahren Jahresabschlussdaten verfügbar sein. Daher reduziert sich die Stichprobe auf 125 Banken pro Jahr. Zur Bestimmung der Effizienz von Banken müssen Input- und Outputfaktoren festgelegt werden. In Anlehnung an vorherige Arbeiten (vgl. [3]) werden folgende Inputdaten angenommen: Sachkapital, Anzahl der Mitarbeiter, Einlagen, Provisionsaufwendungen sowie Kreditrisikoprovisionen. Als Output gelten fest und nicht fest verzinsliche Anlagen, Kredite an Kunden, Kredite an Kreditinstitute sowie Provisionserträge.

Im ersten Schritt wurde technische Effizienz berechnet. Die Ergebnisse sind in Tabelle 1 zusammengefasst. Die durchschnittliche technische Effizienz von Geno sinkt im Zeitablauf. Gleichzeitig sinken die minimalen Werte der technischen Effizienz bei steigenden Standardabweichungen. Bei effizienter Produktion könnten Geno im Jahr 2003 im Mittel den gleichen Output mit 88% der Inputmenge herstellen.

Das Produktivitätswachstum wird mit Hilfe des Malmquistindex berechnet, mit dem auch eine Produktivitätsänderung auf eine Verschiebung der

Effizienz	1998	1999	2000	2001	2002	2003
Mittelwert	0,9262	0,9205	0,9047	0,8647	0,8706	0,8834
StdAbw	0,0742	0,0807	0,0856	0,1079	0,1105	0,1041
Min	0,7366	0,6880	0,6785	0,5612	0,5278	0,5621

Tabelle 1. Ergebnisse der Analyse der technischen Effizienz

Effizienzgrenze oder eine Effizienzverbesserungen zurückgeführt werden kann. Die Ergebnisse dieser Analyse sind in Tabelle 2 wiedergegeben. Danach konnten die Banken zwischen 98-99 und 99-00 die Produktivität leicht steigern. Diese Produktivitätssteigerung konnte durch technologischen Fortschritt realisiert werden. Zwischen 00-01 und 01-02 haben die Banken Produktivitätsrückgänge erfahren. In der letzten Betrachtungsperiode kann eine Produktivitätssteigerung beobachtet werden, die sowohl auf Effizienzverbesserungen als auch auf technologischen Fortschritt züruckgeführt werden kann.

	1998-1999	1999-2000	2000-2001	2001-2002	2002-2003
Malmquist	0,9915	0,9874	1,0557	1,0847	0,8971
Effizienzverb.	1,0069	1,0181	1,0499	0,9936	0,9844
Verschiebung	0,9848	0,9698	1,0055	1,0917	0,9113

Tabelle 2. Ergebnisse des DEA-Malmquist-Modells

Neben einer reinen Beobachtung von Produktivitätsänderung interessieren ebenfalls Faktoren, die den technologischen Fortschritt im Bankenbereich beeinflussen. Technologische Fortschritte in Bankbetrieben können durch Anwendung neuer Technologien, z.B. Informations- und Kommunikationstechnologien (IT), entstehen. Da bankinterne Daten fehlen, werden hier Zusammenhänge zwischen makroökonomischen Indikatoren zur IT-Nutzung und der Verschiebung der Effizienzgrenze für den Untersuchungsszeitraum mit Hilfe von Korrelationskoeffizienten geprüft. Als makroökonomische IT-Indikatoren werden die Zuwächse der Anzahl von Internetnutzern, von Geldausgabeautomaten, der Onlinetransaktionen sowie der Wertzuwächse von Onlinetransaktionen benutzt. Die durchschnittliche Korrelationen sind in der Tabelle 3 angegeben. Der Zusammenhang zwischen einer Änderung von Internetnutzerzahl und der Zahl von Geldautomaten ist nur sehr schwach. Dagegen korrelieren die Änderung der Zahl von Onlinetransaktionen sowie des Wertes dieser Transaktionen recht stark mit technologischem Fortschritt. Somit hat die IT-Nutzung eine produktivitätserhöhende Wirkung auf Geno.

Δ Internetnutzer	Δ Geldautomaten	Δ Onlinetrans.	Δ Wert Onlinetrans.
0,0463	-0,0719	0,5267	0,4895

Tabelle 3. Korrelation zwischen IT-Indikatoren und Effizienzgrenzeverschiebung

Effizienz	1998	1999	2000	2001	2002	2003
Mittelwert	1,2655	1,3519	1,6872	1,5602	1,8561	1,9824
StdAbw	0,2344	0,3591	0,1920	0,0972	0,1629	0,1873
Min	0,7891	0,8147	0,9256	0,9015	0,9562	0,9522

Tabelle 4. Effizienz von Online- und Direktbanken

Diese Beobachtung wird in einem abschließenden Untersuchungsschritt überprüft. Dazu wird die Effizienz von 10 Online-/Direktbanken gegenüber der Effizienzgrenze von Geno verglichen. Die Onlinebanken treten nur durch Nutzung von IT mit den Kunden in Kontakt und haben keine Filialen. Es ist zu erwarten, dass die Effizienzgrenze dieser Banken die Effizienzgrenze von Geno dominiert. In der Abbildung 1b wäre dann die Effizienzgrenze von Onlinebanken durch die gestrichelten Linienzug gekennzeichnet. Da die EE auf der Effizienzgrenze einen Effizienzwert von Eins haben, kann die Effizienz der neuen Banken über Eins werden, falls sie eine dominante Input-Output-Menge aufweist. Diese Art von Modellen wird als "variable benchmark models" bezeichnet (vgl. [4]). Die Ergebnisse dieser Untersuchung sind in der Tabelle 4 zusammengefasst. Die Onlinebanken dominieren im Durchschnitt die Effizienzgrenze von Geno ($\theta > 1$). Allerdings existieren auch Onlinebanken, die von herkömmlichen Banken dominiert werden (Zeile "Min").

5 Zusammenfassung

In dieser Arbeit wurde das Produktivitätswachstum von Genossenschaftsbanken mittels des Malmquistindex untersucht. Der besondere Beitrag dieser Arbeit liegt in einer Separierung des Malmquistindex auf eine managementbedingte Effizienzverbesserung und auf eine technologisch bedingte Verschiebung der Effizienzgrenze. Neben einer Messung der Produktivität wurde ebenfalls ein positiver Zusammenhang zwischen IT-Nutzung und dem technologischen Fortschritt festgestellt. Im letzten Untersuchungsschritt wurde die Effizienzgrenze von Onlinebanken gegenüber der Effizienzgrenze herkömmlicher Geno verglichen. Im Durchschnitt dominieren Onlinebanken die Geno.

Literaturverzeichnis

1. Cooper W, Seiford L (2000) Data Envelopment Analysis. Kluwer, Boston
2. Färe R, Grosskopf S, Lindgren B, Roos B (1989) Productivity developments in Swedish hospitals. Discussion Paper No. 89-3, Illionois University
3. Poddig Th, Varmaz A (2004) Effizienzprobleme bei Banken: Fusionen und Betriebswachstum als tragfähige Mittel? ZBB, 16:236–247
4. Zhu J (2003) Quantitative models for performance evaluation and benchmarking. Kluwer, Boston

Consistency Matrices Within Scenario Technique: An Empirical Investigation

Ewa Dönitz, Martin G. Möhrle, University of Bremen

1 Introduction

Scenario technique is a well known tool for strategic management. It helps to create alternative future scenarios based on quantitative and qualitative data and provides a systematic process.

An important step within this process is generating a consistency matrix, where columns and rows are both built by the influence factors of the future development and their projections, and where the fields contain consistency values between them. The consistency matrix is used for generating bundles of influence factors projections, which are the base for scenario writing. As it is very time consuming to fill in the consistency matrix manually, in a research project a semi-automatic way will be specified. As foundation, an empirical investigation of consistency matrices has been carried out, which results will be presented in this article.

Main answers are given to four aspects: (i) What is a consistency matrix and what are the challenges for research? (ii) What were the results of the empirical investigation? (iii) Are there differences between consistency matrices? (iv) What are the implications of the empirical investigation for semi-automatic filling in of a consistency matrix?

2 Using the Consistency Matrix: Basics and Challenges

There is a great number of approaches for scenario creating process. However, generally it is based on a qualitative sequence of steps (Geschka and Reibnitz 1981) or phases (Gausemeier et al. 1995). The goal is to generate future pictures regarding different topics, such as management, economics, environmental, science or policy science. Van Notten et al. (2003, 2005) show the diversity of these topics in the 50-year history of scenario planning and present an updated scenario typology. A wide overview of the most important scenario methods with references and practice applications worldwide is listed by Zinser (2000).

An important attribute of any scenario is its internal consistency. Especially by complex problems with a large number of influence factors the detailed analysis using the consistency matrix is recommended. This matrix contains rows and columns in which the future projections of uncertain, so-called critical influence factors, are inserted (see Fig 1). For each pair of projections of different influence factors experts estimate, how compatible the two projections are to each other:

- They could be strong consistent (value 5) or consistent (value 4). This is a strong recommendation for these two influence factors to appear in the same scenario.
- There could be no relationship between them (value 3).
- They could be partially (value 2) or totally (value 1) inconsistent. In the first case the consistency of the whole scenario could be reduced. In the second case the combination of the two influence factors projections should not be included in any scenario.

There are different methods for bundling projections to scenario frameworks based on the consistency matrix (Geschka 1999, Reibnitz 1991, Meyer-Schönherr 1991). Therefore the matrix is an important tool for scenario writing.

Influence factors	Projections		1A	1B	2A	2B	3A	3B	4A	4B	5A	5B	5C	6A	6B	7A	7B
Reliability of components	1A	cheaper			3	2	2	2	3	3	2	4	4	4	2	3	3
	1B	reliable			3	4	4	4	4	3	4	3	2	4	4	3	3
Worker moral	2A	civil service mentality					2	2	3	3	3	4	3	3	3	2	5
	2B	management thinking					4	2	3	3	4	4	4	4	4	5	2
Acceptance of new technologies	3A	high							4	4	5	4	5	2	4	3	3
	3B	oposition							3	3	2	4	4	4	2	3	3
Job satisfaction	4A	high									3	3	3	3	3	4	4
	4B	low satisfaction									3	3	3	3	3	4	4
R&D expences of industry	5A	+10%												2	5	3	3
	5B	+/- 0												4	4	3	3
	5C	- 10%												5	1	3	3
Horizon of investment	6A	low term														3	3
	6B	long term														3	3
Tasks on the job	7A	complex															
	7B	monotonous															

Figure 1: Consistency matrix (Geschka and Reibnitz 1981)

Until now the main weakness of using the consistency matrix is the long duration of the estimation process to get all consistency values of all pairs of uncertain projections. The challenge is to generate an algorithm for semi-automatic estimation of consistency values. Unfortunately, the relationships in the consistency matrix are stochastic, not deterministic. An example should illustrate this:

Consider three future projections P_{1A}, P_{3A} and P_{5A} (see Fig 1). Together they build a so-called triangular relationship. The index $_{i,j}$ indicates the influence factor i and its projection j. There is a partially inconsistency between P_{1A} and P_{3A}, which is described by the consistency value of $C_{1A3A}=2$. In contrary there is a strong consistency between P_{3A} and P_{5A} ($C_{3A5A}=5$). Which value should take the consistency C_{1A5A}, between P_{1A} and P_{5A}? The consistency value of 2 could be plausible (see

Fig. 2a), but there are also other values possible. Furthermore, in almost all cases there is not only one single triangular relationship, but several triangular relationships, which may be considered for estimating the needed consistency value (see Fig. 2b).

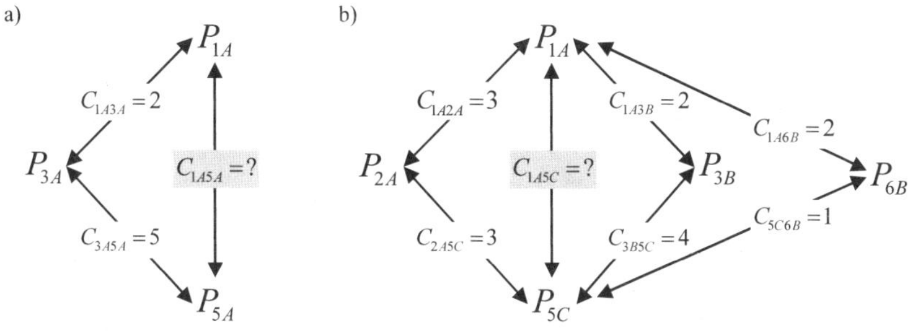

Figure 2: Relationships between influence factors projections in consistency matrix

For understanding these complex relationships among the projections in a consistency matrix, which can become difficult to manage even when only a few factors are involved, a great number of matrices has to be analysed empirically.

3 Design and Results of an Empirical Investigation

The aim of the empirical investigation was to analyse the relationships in consistency matrices. Therefore, an explorative study was carried out, which led to a reasonable number of cases (i.e. consistency matrices). As the cases differed in size, a standardisation was necessary. Based on this data a hierarchical cluster analysis was conducted, leading to three clusters of consistency matrices.

3.1 Data and Standardisation of Cases

The data for the investigation was extracted from several sources. Not only for this reason the consistency matrices differed in size, which was cause for standardisation.

Data: There were three ways to get a sufficient quantity of consistency matrices for further analysis: (i) an extensive investigation of literature and internet sources, (ii) an empirical study among students of the University of Bremen, and (iii) scenario workshops. In total 80 matrices were collected, based on different scales of consistency values. Besides the scale 1 to 5, which provides a basis for further statistical evaluation, there were other scales well known in the scenario practice. All different scaled matrices had to be adapted to the base scale before the evaluation.

Analysis of triangular relationships: The goal of the empirical investigation was to compare the consistency matrices with respect to the triangular relationships within them. This analysis may be illustrated: A triangular relationship between the projections P_{1A}, P_{3A} and P_{5A} is given (see Fig. 3). From this triangular relationship result the following three single relationships: Between P_{1A} and P_{3A} is a partially inconsistency and between P_{3A} and P_{5A} a strong consistency. There is third single relationship, a partially inconsistency between projections P_{1A} and P_{5A}. This result will included into a matrix for triangular relationships (see Fig 4). The same combination of projections will be considered from two other perspectives. So three values will be registered in the matrix for triangular relationships.

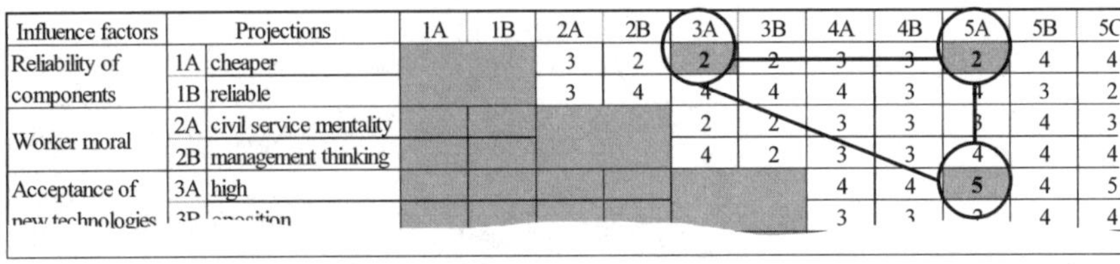

Figure 3: Triangular relationships as example

The results of the total evaluation of the whole consistency matrix (see Fig. 1) are 1.020 registered values, which are due to 340 different triangular relationships.

	1-1	1-2	1-3	1-4	1-5	2-2	2-3	2-4	2-5	3-3	3-4	3-5	4-4	4-5	5-5	Sum
1	0	0	0	0	0	0	0	3	0	5	0	0	1	1	0	10
2	0	0	0	3	0	9	16	12	6	45	23	6	26	3	1	150
3	0	0	10	0	0	8	90	23	6	42	206	40	28	16	0	469
4	0	3	0	2	1	6	23	52	3	103	56	16	45	18	1	329
5	0	0	0	1	0	3	6	3	2	20	16	0	9	2	0	62
Sum	0	3	10	6	1	26	135	93	17	215	301	62	109	40	2	1.020

Figure 4: Matrix for triangular relationships within a consistency matrix

Standardisation: The total results of the statistical evaluation of consistency matrices built the foundation for clustering of the matrices. The matrices for triangular relationships with the absolute values (see Fig. 4) are not comparable, because of different sizes. Therefore, relative values had to be calculated to enable the cluster analysis (all values within the matrix for triangular relationships were divided by the sum of the column).

3.2 Clustering of consistency matrices

An established procedure to classify cases is cluster analysis. Hierarchical cluster analysis is a statistical tool for solving classification problems by finding relatively homogeneous clusters of cases based on measured characteristics (Backhaus et al. 2003). In case of consistency matrices the cases are characterised by the number and kind of triangular relationships sets. The classification using cluster analysis can be verified by discriminant analysis, that is used for examination of group membership and understanding, which variables discriminate between two or more occurring groups (Backhaus et al. 2003).

The conducted cluster analysis consisted of two phases. In the first phase outliers should be sorted out. Squared Euclidean distance and Single linkage method were used and one consistency matrix was characterised as outlier. The main reason for the sorting out was a specific distribution of the evaluated values.

The aim of the second phase of cluster analysis was building of homogeneous groups using squared Euclidean distance and Ward linkage method. Generally, these methods are regarded as very efficient (Backhaus et al. 2003). The results of the clustering were three clusters with 31, 39 and 9 consistency matrices. The number of clusters was verified by Elbow criterion. The allocation of consistency matrices was also tested with discriminant analysis, which showed a robust solution.

A determining factor for this cluster solution are the particular variations between the distributions of the evaluated values (see Fig. 5). These results could be interpreted as follows: The consistency matrices in the first cluster comprised a broad spectrum of consistency values. In the second cluster there were mostly neutral and only a few other consistency values estimated. In the third cluster were no inconsistent values in matrices (nor consistency value 1 neither 2). The consistency value of 3 was most frequented in the second and the third cluster.

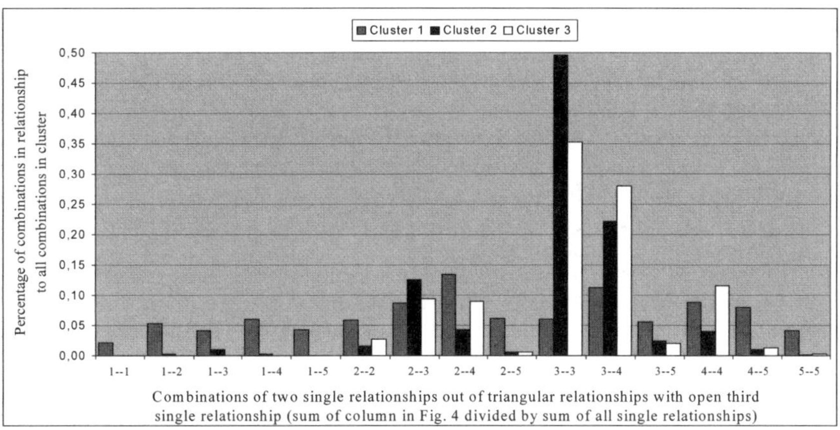

Figure 5: Frequency distributions in each identified cluster

Although the cluster solution is convincing, it is difficult to find a theory to explain these results. There are no differences between consistency matrices related to the domain of the application or to the size of consistency matrices. However, there are some indicators for the differences related to user origin and their qualifications: 22 consistency matrices in the first cluster were filled in by students, 33 matrices in the second cluster have been gotten by the investigation of literature and internet sources, the third cluster was mixed.

4 Conclusions

For this paper consistency matrices were analysed. Those consistency matrices were identified as being different, three clusters have been founded. These clusters differ in the distributions of the evaluated values in matrix for triangular relationships.

To come to a semi-automatic algorithm for estimation of consistency values two major implications may be drawn from this empirical investigation:

1) According to the three cluster solution three different algorithms (or one algorithm with three different parameter settings) are needed. The authors are working with fuzzy rules for that goal.
2) To start an algorithm for a semi-automatic matrix filling in it has to be decided to which of the three clusters the only partially filled input matrix should be assigned. Therefore, experiments with discriminant analysis and case-based reasoning are conducted.

References

Backhaus K, Erichson B, Plinke W, Weiber R (2003) Multivariate Analysemethoden: Eine anwendungsorientierte Einführung, 10th edn. Springer, Berlin et al.

Gausemeier J, Fink A, Schlake O (1995) Szenario-Management: Planen und Führen mit Szenarien. Hanser, München Wien

Geschka, Horst (1999) Die Szenariotechnik in der strategischen Unternehmensplanung. In Hahn D, Taylor B (eds) Strategische Unternehmensplanung, Strategische Unternehmensführung: Stand und Entwicklungstendenzen, 8th edn. Physica, pp 518-545

Geschka H, Reibnitz von U (1981) Die Szenario-Technik als Grundlage von Planungen. Battelle-Institut e.V., Frankfurt/Main

Meyer-Schönherr M (1992) Szenario-Technik als Instrument der strategischen Planung. Verlag Wissenschaft und Praxis, Ludwigsburg Berlin

Notten van PWF, Rotmans J, Asselt van MBA, Rothman DS (2003) An updated scenario typology. Futures 35 (5): 423-443

Notten van PWF, Sleegers AM, Asselt van MBA (2005) The future shocks: On discontinuity and scenario development. Technological Forecasting an Social Change 72: 175-194

Reibnitz von U (1991) Szenario-Technik: Instrumente für die unternehmerische und persönliche Erfolgsplanung. Gabler, Wiesbaden

Zinser S (2000) Eine Vorgehensweise zur szenariobasierten Frühnavigation im strategischen Technologiemanagement. Jost-Jetter, Heimsheim

Distributed Neurosimulation

Hans-Jörg v. Mettenheim[1] and Michael H. Breitner[2]

[1] Institut für Wirtschaftsinformatik, Universität Hannover, Germany, mettenheim@iwi.uni-hannover.de
[2] Institut für Wirtschaftsinformatik, Universität Hannover, Germany, breitner@iwi.uni-hannover.de

Summary. Distributed computing allows to combine the computing power of miscellaneous computers. The computers may be at different locations as long as they are connected via a network, e. g. the Internet or an intranet. In this paper the development of a distributed computing version of the neurosimulator FAUN (*F*ast *A*pproximation with *U*niversal Neural *N*etworks) is described. This offers the opportunity to use free computing resources, e. g. of a student and staff computer cluster. An easy to install client is part of the development work. The combined computation power is necessary for, e. g., fast forecasting or fast simulation problems to be solved with FAUN which would otherwise take hours or days on a single processor computer. Problems which computation time can be shortened significantly by distributed computing with FAUN include, but are not limited to, dynamic games, robust optimal reentry guidance of a space shuttle and currency forecasting.

1 Introduction

Often collected data have to be connected even if classical methods like linear or nonlinear regression don't work. In this case neurosimulators can help. A neurosimulator has the potential to discover relations in almost any data. This may sometimes cause the result to be difficult to interpret.

At the Institut für Wirtschaftsinformatik at the Universität Hannover the neurosimulator FAUN is used and under development. The first version of FAUN has been developed in 1996 by the second author. It has further on been upgraded and developed by him and other co-workers.

A neurosimulator trains so-called artificial neural networks for every problem. The networks can be thought of as a mathematical function with parameters. For detailed information on artificial neural networks and a list of more references see [3].

To get a good quality neural network it is necessary to compute a certain amount of neural networks. Calculation may take several hours depending on the problem using a single personal computer of type Pentium IV at 2.8 GHz.

The calculation time also increases linearly with the amount of training and validation data. This is the reason why a distributed and parallel version of FAUN is developed. Computing times of one day or more are not acceptable especially for real time applications. Miscellaneous personal computers are connected together to form one computation network. Even normal office computers can be used as the computation runs with the lowest priority. A user can still work without being slowed down by the computation.

The coarse-grained parallelization of FAUN offers a parallelization degree between 0.9 and 0.97. The meaning of this number is that if 100 computers work together, the computation will not be exactly 100times as fast as with a single computer. But it will still be 90 to 97 times as fast as with a single computer. The loss of computing power can be explained with the additional administration and communication overhead[3].

The scope of this paper is to give an introduction on the notions of distributed computing and grid computing and how these techniques can be used to obtain a high performance computer using standard hardware. It is not even indispensable to buy new computers. This is in contrast to very expensive massively-parallel supercomputers. The reader is encouraged to use the references on coding as source for self-development of distributed computing applications: See especially [8], [14] for an introduction to network functions and programming paradigms. See [10] for an overview of tools for distributed computing.

2 Distributed Computing and Grid Computing

Three types of high performance computing (HPC) techniques can be identified: Super computer systems, grid computing and distributed computing. While super computers, which are very expensive, allow very fast interprocessor communication the other techniques are interesting alternatives when communication speed isn't the bottleneck.

Distributed computing describes the fact that computation is done at various locations and on various types of computers. The challenge for distributed computing programs is to unify the computers to work together. See [2] or visit [13] for advanced links.

Grid computers can be thought of as a cluster with extra features. Especially, a grid computer is error tolerant and usually offers additional intelligent functionalities. Power saving is one of the important tasks a grid computer can perform. Idle computers are shutdown and restarted when they have to

[3] Imagine a bunch of potatoes that have to be peeled (famous "potato peeling problem"). A single person will work for a long time. If a second person helps the peeling will probably be twice as fast. But if more and more persons join in the peeling time will be spent on handing the potatoes from one to another and the entire process will be more and more inefficient. Finally the largest potato determines the minimum time achievable.

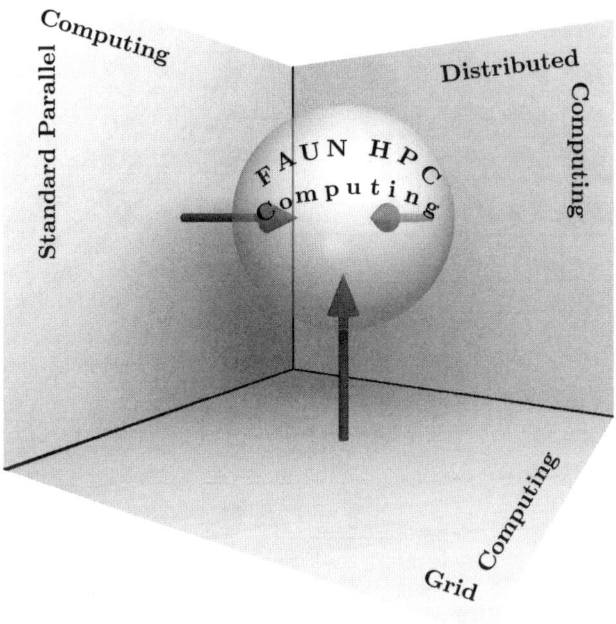

Fig. 1. The Enhanced Computing Sphere in the HPC Computing Cube

be used again. [1], [9] and [11] provide introductions to grid computing. The combination of distributed and grid computing used with parallel programming is the target of the FAUN client. These three aspects lead to enhanced computing and complement one another naturally to the enhanced computing sphere, see Figure 1. This sphere symbolizes that the advantages of distributed computing, grid computing and standard parallel computing are taken to do HPC computing in an enhanced way. Distributed computing offers the opportunity to use computers all over the world with an easy to install client. Grid computing leads to error tolerance and an easy configurable system with advanced features. Up to now usually only automatic wakeup and shutdown is concretely implemented. Parallel computing cuts a problem into small slices appropriate to every computer and coordinates the computation.

The FAUN client is developed to comply with the criteria for quality software, see Figure 3. The testbed will be the School of Economics with the hardware shown in Figure 2. Parallel computation is given by the fact that single neural networks are allocated to the computers. Distributed computing is necessary because the computers, e. g., are at different locations. Grid computer features are added as the computers of the cluster are started and shutdown remotely.

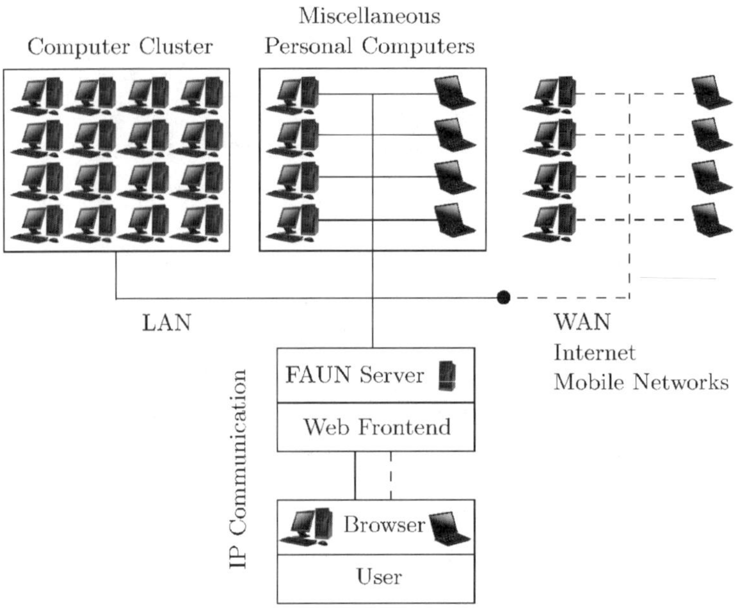

Fig. 2. FAUN HPC Computing configuration

3 Parallelization

For parallelization specialized programming packages can be used. For general information [10] and [7] are a good start. The most common packages are the Message Passing Interface (MPI) and the Parallel Virtual Machine (PVM). Both packages can realize similar tasks and keep the programmer away from low level network communication and allow, e. g., sending and receiving variables.

PVM is easy to implement while MPI offers additional functionalities and tends to be more efficient. Often vendor optimized versions of MPI exist for specific architectures. Nevertheless a first parallel version of FAUN uses PVM, see [6] and [5] for examples. An almost linear speedup is achieved on a cluster of homogeneous computers. The parallelization degree is about 0.97 for 200 computers.

A detailed description of FAUN can be found in [3], see also [4] and [12] for real life applications. For information on differences between a coarse-grained and fine-grained parallelization see [5].

4 Development of the Client

The restrictions of hardware environment (see Figure 2) in relation with the administrative privileges make it necessary to work with low level network

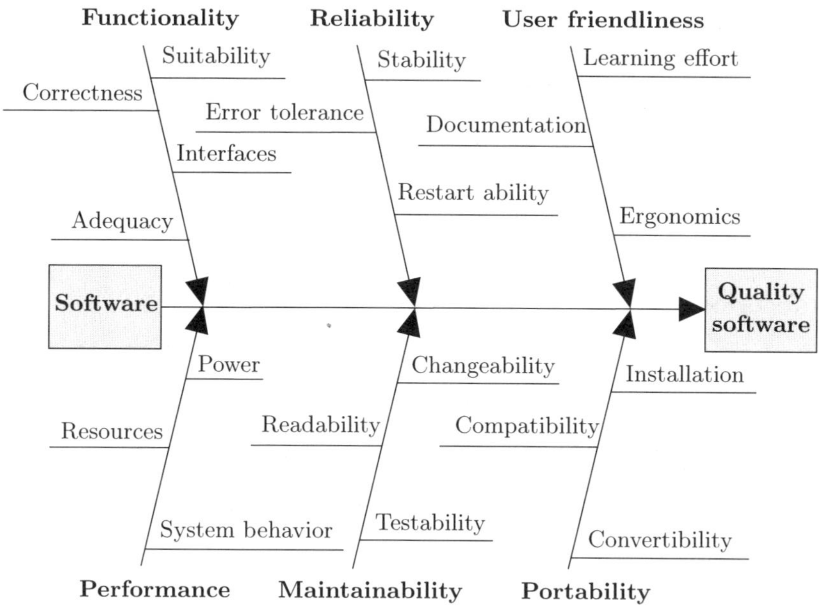

Fig. 3. Software quality criteria following different ISO and DIN standards

functions and develop an individual protocol for the communication between the different parts of FAUN. The criteria of quality software mentioned in Figure 3 have to be taken into account. Basic requirements for the client/server programs related to functionality and maintainability are defined:

- The client to be distributed on different computers is made of and runs from a single directory.
- The client runs without needing administrative privileges.
- The client itself is updateable via the network.

These basic requirements are realized by a simple self developed protocol on top of the reliable TCP/IP protocol which is often used in IP communication. PVM or MPI are not used as they would have to be installed first.

5 Conclusion and Outlook

Distributed and grid computing are techniques that have the potential to offer high performance computing at moderate costs or sometimes no additional costs at all. These techniques are especially interesting when standard hardware offers free computing power. This is often the case with public computer clusters at universities or personal computers at companies. The drawback is that the software requirements increase: Issues like security and integrity become important and also user friendliness and maintainability.

Artificial neural networks have the potential to be used as alternative method for several simulation or forecasting applications. Nevertheless the usage of neural networks is no black box schema as often assumed. The analysis and work necessitates deep knowledge. Additionally, knowledge about the special application is necessary. With the distributed computing version of the neurosimulator the user friendliness increases as usage is possible via a graphical web frontend and no manual editing of configuration files is necessary.

References

1. Abbas A (2004) Grid Computing: A Practical Guide to Technology and Applications. Charles River Media, Massachusetts
2. Attiya H, Welch J (2004) Distributed Computing: Fundamentals, Simulations, and Advanced Topics. Wiley-Interscience, New York
3. Breitner MH (2003) Nichtlineare, multivariate Approximation mit Perzeptrons und anderen Funktionen auf verschiedenen Hochleistungsrechnern (in German). Akademische Verlagsgesellschaft, Berlin
4. Breitner MH (2000) Robust optimal onboard reentry guidance of a space shuttle: Dynamic game approach and guidance synthesis via neural networks. Journal of Optimization Theory and Applications 107:484-505
5. Breitner MH, Mehmert P, Schnitter S (2000) Coarse- and fine-grained parallel computation of optimal strategies and feedback controls with multilayered feedforward neural networks. In: Nowak A et al. (eds) Proceedings of the Ninth International Symposium on Dynamic Games and Applications, Adelaide, Australia
6. Breitner MH, Mettenheim HJv (2005) Coarse-grained Parallelization of the Advanced Neurosimulator FAUN 1.0 with PVM and the Enhanced Cornered Rat Game revisited. To appear in: International Game Theory Review, World Scientific Publishing Company
7. Dongarra J, Foster I, Fox G, Gropp W, Kennedy K, Tonczon L, Whithe A (2003) Sourcebook of parallel computing. Morgan Kaufmann, Amsterdam
8. Hall B (2005) Programming guides (Guide to Network Programming 2001, Guide to C programming 2004). `http://www.ecst.csuchico.edu/~beej/guide/` last visited on 2005/03/08
9. Hey AJG, Fox G, Berman F (2003) Grid Computing. John Wiley & Sons, Chichester
10. Hughes C, Hughes T (2003) Parallel and Distributed Programming Using C++. Addison-Wesley, Boston
11. Joshy J, Fellenstein C (2003) Grid Computing. Prentice Hall PTR, London
12. Köller F, Breitner MH (2003) Efficient Synthesis of Strategies with the Advanced Neurosimulator FAUN 1.0. In: Breitner MH (ed) Proceedings of the Fourth International ISDG Workshop, Goslar
13. Pearson K (2005) Distributed Computing. `http://distributedcomputing.info` and `http://distributed computing.info/projects.html` last visited on 2005/03/08.
14. Stevens WR, Fenner B, Rudoff AM (2004) UNIX Network Programming - The Sockets Networking API - Volume 1. Addison-Wesley, Boston

Part XXI

Decision Theory

Multi-Criteria Decision Support and Uncertainty Handling, Propagation and Visualisation for Emergency and Remediation Management

Jutta Geldermann, Valentin Bertsch, Otto Rentz

Institute for Industrial Production (IIP), University of Karlsruhe,
Hertzstr. 16, 76187 Karlsruhe, Germany
jutta.geldermann@wiwi.uni-karlsruhe.de

Summary. The real-time online decision support system RODOS provides support throughout all phases of a nuclear or radiological emergency in Europe. The multi-criteria decision support tool Web-HIPRE has been integrated into RODOS for a transparent evaluation of long-term measures, taking into account the preferences of a decision making team. However, a decision making process in practice is subject to various sources of uncertainty. This paper describes a Monte Carlo approach to consistently model and propagate the uncertainties within the RODOS system, including Web-HIPRE, aiming at comprehensibly visualising and communicating the uncertainties associated with the results from the decision analysis.

1 Introduction

Since decision making in emergency management involves many parties with different views, responsibilities and interests, for which a consensus must be found, multi-criteria decision analysis (MCDA) is very important for ensuring a transparent resolution of such a complex decision situation [Belton and Stewart, 2002, Geldermann et al., 2006]. The handling of uncertainties is considered to be a fundamental part of good decision making [Basson, 2004], however, in practice it is also important not to overload the users of a decision support system with too much information about the uncertainties. This conflict points out the need for an understandable visualisation and communication of the uncertainties that arise in the decision making process.

According to their respective source, a distinction can be made between *"data uncertainties"* (uncertainties of the calculated potential consequences of the considered alternative actions), *"parameter uncertainties"* (uncertainties related to the parameters, such as the weighting factors, of a MCDA model) and *"model uncertainties"* (uncertainties resulting from the fact that models are ultimately only simplifications or approximations of reality [French and Niculae, 2005]). Since *model uncertainties* are difficult to quantify and can also be regarded as inherent to the nature of any model, they are not considered in this paper. *Parameter uncertainties* can in general be examined by means of "conventional" sensitivity analyses (investigating the impact of varying a model parameter, see [Bertsch et al., 2005, Geldermann et al., 2006] for the use within (nuclear) emergency management) or parametric analyses (performing the decision analysis for (all) combinations of available model parameters, see [Morgan and Henrion, 1990]). The focus in this paper is on *data uncertainties* and on the simultaneous consideration of data and parameter uncertainties.

2 Outline of the RODOS system

In the event of a nuclear or radiological emergency in Europe, the real-time online decision support system RODOS provides consistent and comprehensive decision support at all levels ranging from largely descriptive reports, such as maps of the contamination patterns and dose distributions, to a detailed evaluation of the benefits and disadvantages of various countermeasure or remediation strategies [Ehrhardt and Weiss, 2000, French et al., 2000, Raskob et al., 2005]. The conceptual structure of RODOS includes three subsystems:

- The Analysing Subsystem (ASY) processes incoming data and forecasts the location and quantity of contamination based upon monitoring and meteorological data and models including temporal variation.
- The modules of the Countermeasure Subsystem (CSY) simulate potential countermeasures, check them for feasibility and calculate their consequences.
- Web-HIPRE (see [Mustajoki and Hämäläinen, 2000]), a tool for MCDA, has recently been integrated into RODOS as Evaluation Subsystem (ESY) to support a team of decision makers in evaluating the overall efficacy of different countermeasure or remediation strategies according to their potential benefits/drawbacks (quantified by the CSY) and preference weights (provided by the decision makers) [Geldermann et al., 2005]. The tool is based on multi-attribute value theory (MAVT), a theory that develops and provides methods to structure and analyse complex decision problems by means of attribute trees and to elicit the relative importance of the criteria in such a tree.

3 Uncertainty handling, propagation and visualisation

On the basis of a hypothetical case study from nuclear emergency and remediation management, a Monte Carlo approach (see [Fishman, 1996]) is introduced to consistently model, propagate and visualise uncertainties within the RODOS system.

The hypothetical case study

It is assumed that a fictitious contamination situation was caused by a serious accident at a nuclear power plant which triggered the immediate shutdown of the reactor. Radioactive material was released into the atmosphere over a period of three hours. All necessary immediate and early countermeasures, including early food countermeasures, were taken in selected affected areas.

According to initial estimations by plant operators, approximately 50% of the plant inventory of radioactive noble gases and approximately 0.1% of the inventory of radioactive iodine and radioactive aerosols were released during the accident. Due to south-westerly winds the radioactive cloud from the nuclear power plant was blown over agricultural areas in a north-easterly direction and radioactive material from the cloud deposited on the ground.

Within a moderated decision making workshop in Germany, the problem structuring process of this case study resulted in an attribute tree showing the overall goal "total utility" (of a measure) as the top criterion which was further split into the criteria "radiological effectiveness", "resources", "impact" and "acceptance", each of which was split again. The considered countermeasure strategies included "No action", "Disp" (disposal), "Proc" (processing), "Stor" (storage), "Rmov, T=0" (removal of cows from contaminated feed at time T=0 (before the passage of the radioactive plume) and feeding with uncontaminated feed), "Rmov, T>0" (removal of cows from contaminated feed at time T>0 (after the passage of the plume) and feeding with uncontaminated feed), "Rduc, T=0" (feeding with uncontaminated feed) and "AddS+Proc" (adding of activity reducing concentrates to the food and subsequent processing). Further details of the case study and the processes of problem structuring and preference elicitation (inter alia the weighting of the criteria in the attribute tree) are described in [Bertsch et al., 2005, Geldermann et al., 2006].

Modelling and propagation of uncertainties

Since the forecasts of the ASY involve uncertainties, an adequate estimation of the uncertainties associated with modelling in the ASY is required and thus uncertainty assessment procedures were developed. The uncertainty modelling concerns the two variables *source term* and *wind direction*. A log-normal distribution is assigned to the source term, i.e. the quantity of released radioactive material, since a deviation of an order of magnitude is considered to be equiprobable in both directions. A normal distribution is assigned to the wind direction with a standard deviation of 30° (see [Gering, 2005]). Using a Monte Carlo approach the uncertainties can be consistently propagated from the ASY through the CSY to the ESY.

Uncertainty handling and visualisation in Web-HIPRE

As indicated above, a Monte Carlo approach is employed to propagate the uncertainties which means that, before the start of the ASY, multiple samples are drawn according to the (presumed) probability distributions of the variables *source term* and *wind direction.* These samples are then used as input data in multiple parallel runs of the system, leading to multiple results for the consequences of the countermeasures. Thus, the decision analysis in Web-HIPRE is not based on one (deterministic) decision table but on a set of decision tables where each table corresponds to one sample (realisation/scenario) which are simultaneously evaluated. The proposed visualisation aims at communicating the uncertainties associated with the results of the decision analysis without causing an information overload. In order to achieve this goal the simultaneously calculated results are not all visualised. However, in order to illustrate the *uncertainty ranges* (i.e. the ranges in which the results can vary due to the uncertainties of the input data), the results of the scenarios corresponding to the 5%- and 95%-quantiles (of the overall performance score) are shown alongside the results of the most probable scenario (cf. Figure 1). These scenarios will be referred to as *worst case* and *best case* scenarios respectively. For a given alternative this means, for instance, that the probability that the overall performance score of this alternative in a (randomly picked) scenario is smaller than the score in the scenario corresponding to the 95%-quantile (or *best case* scenario) is at least 95%.

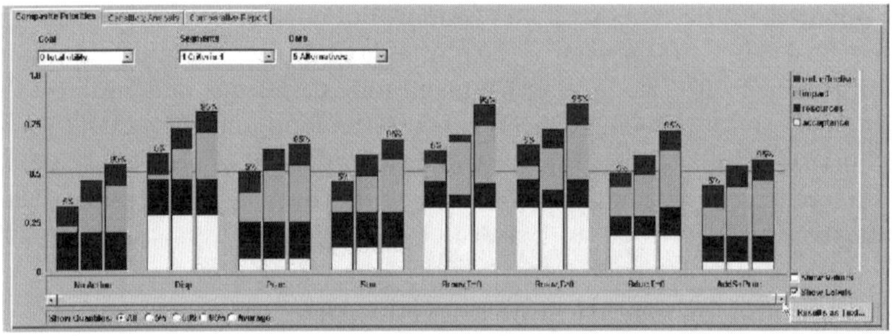

Fig. 1. Visualisation of results including uncertainties

Figure 1 contains important information for the decision makers. A stacked-bar chart has been chosen to visualise the results because it not only illustrates the *uncertainty ranges* of the overall goal but also indicates which of the considered criteria are subject to uncertainties and shows the *uncertainty ranges* of the individual criteria as well as their contribution to the uncertainties in the overall ranking. Furthermore, taking into account uncertainties in the decision making process allows to analyse whether or not the considered alternatives are distinguishable from each other (see [Basson, 2004]). In the case study for instance, it is hard to distinguish between the alternatives "Disp", "Rmov, T=0" and "Rmov, T>0" in consequence of their very similar performance scores.

The extended sensitivity analysis allows a simultaneous consideration of data uncertainties and the uncertainties associated with certain parameters of a MCDA model (e.g. the weights). Figure 2 shows a sensitivity analysis on the weight of acceptance where, for the alternative with the highest performance score for the current weight of acceptance, the performance scores of the *worst* and *best case* scenarios are shown alongside the performance score of the most probable scenario. Hence, such a figure also allows, for instance, an examination of the robustness of the results in the worst and best cases.

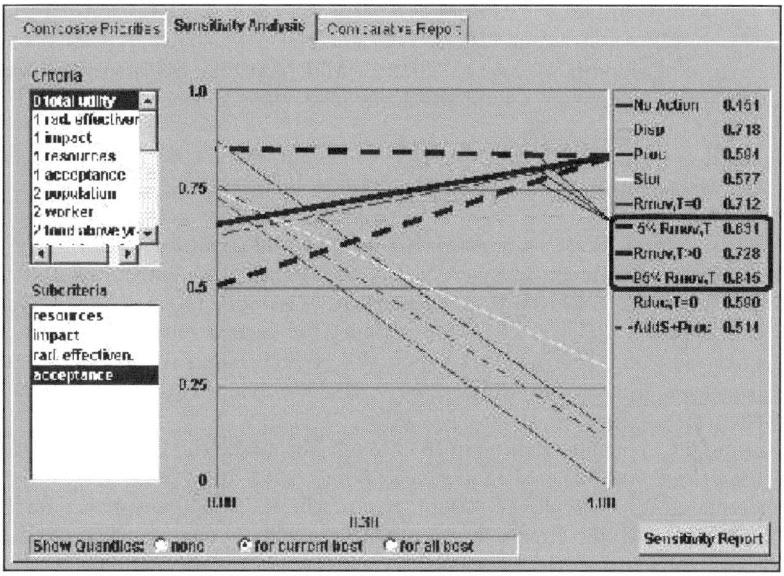

Fig. 2. Simultaneous consideration of parameter and data uncertainties

4 Conclusions

An approach has been introduced to consistently model, propagate and comprehensibly visualise uncertainties within the RODOS system, allowing an informative overview and deeper insight into the decision situation. This approach must now be tested and evaluated by potential users of RODOS and Web-HIPRE in further decision making workshops. However, it should be emphasised that the proposed approach is universally applicable and can thus be applied in any context where MCDA is used to support the resolution of a complex decision situation.

Acknowledgement. This research is closely related to work carried out within the EURANOS project (see http://www.euranos.fzk.de) funded by the European Commission. The authors wish to acknowledge the support by the Commission and all involved project partners and, in particular, wish to thank Florian Gering (Federal Office for Radiation Protection (BfS), Neuherberg, Germany) and Wolfgang Raskob (Forschungszentrum Karlsruhe (FZK), Karlsruhe, Germany) for many fruitful discussions.

References

L. Basson. *Context, Compensation and Uncertainty in Environmental Decision Making.* PhD thesis, Dep. of Chemical Engineering, University of Sydney, 2004.

V. Belton and T. Stewart. *Multiple Criteria Decision Analysis - An integrated approach.* Kluwer Academic Press, Boston, 2002.

V. Bertsch, J. Geldermann, and O. Rentz. Multi-Criteria Decision Support and Moderation Techniques for off-site Emergency Management. In *TIEMS 12th Annual Conference*, 2005.

J. Ehrhardt and A. Weiss. RODOS: Decision Support for Off-Site Nuclear Emergency Management in Europe. EUR19144EN. *Luxembourg, European Community*, 2000.

G. Fishman. *Monte Carlo - Concepts, Algorithms, and Application.* Springer Series in Operation Research. Springer, New York, Berlin, Heidelberg, 1996.

S. French, J. Bartzis, J. Ehrhardt, J. Lochard, M. Morrey, N. Papamichail, K. Sinkko, and A. Sohier. RODOS: Decision support for nuclear emergencies. In S. H. Zanakis, G. Doukidis, and G. Zopounidis, editors, *Recent Developments and Applications in Decision Making*, pages 379–394. Kluwer Academic Publishers, 2000.

S. French and C. Niculae. Believe in the Model: Mishandle the Emergency. *Journal of Homeland Security and Emergency Management*, 2(1), 2005.

J. Geldermann, V. Bertsch, M. Treitz, S. French, K. N. Papamichail, and R. P. Hämäläinen. Multi-criteria Decision Support and Evaluation of Strategies for Nuclear Remediation Management. *OMEGA - The International Journal of Management Science (submitted)*, 2006.

J. Geldermann, M. Treitz, V. Bertsch, and O. Rentz. Moderated Decision Support and Countermeasure Planning for off-site Emergency Management. In R. Loulou, J.-P. Waaub, and G. Zaccour, editors, *Energy and Environment: Modeling and Analysis.* Kluwer Academic Publishers, 2005.

F. Gering. *Data assimilation methods for improving the prognoses of radioecological models with measurements.* PhD thesis (in preparation), Leopold-Franzens-University Innsbruck, Austria, 2005.

M. G. Morgan and M. Henrion. *Uncertainty: A Guide to Dealing with Uncertainty in Quantitative Risk and Policy Analysis.* Cambridge University Press, New York, 1990.

J. Mustajoki and R. P. Hämäläinen. Web-HIPRE: Global Decision Support by Value Tree and AHP Analysis. *INFOR*, 38(3):208–220, 2000.

W. Raskob, V. Bertsch, J. Geldermann, S. Baig, and F. Gering. Demands to and experience with the Decision Support System RODOS for off-site emergency management in the decision making process in Germany. In B. van de Walle and B. Carlé, editors, *Proceedings of the Second International ISCRAM Conference*, 2005.

Interactive Decision Support Based on Multiobjective Evolutionary Algorithms

Thomas Hanne

Fraunhofer Institute for Industrial Mathematics (ITWM)
Department of Optimization
Fraunhofer-Platz 1, 67663 Kaiserslautern, Germany
hanne@itwm.fhg.de

Key words: multiple criteria decision making, multiobjective optimization, interactive methods, evolutionary algorithms, decision support systems.

1 Introduction

Both the development of evolutionary algorithms and interactive methods for multiobjective decision support originated at about the same time, in the late 1960s and early 70s (see [16] and [7]). Both of them were driven by the upcoming vision of using computers not just for 'pure' calculation and data processing but for a closer integration with the biological and human domain. Despite to this common cybernetics vision, both research fields developed mostly unaffected by the other.

The paradigm of interactive decision support is usually either that a set of solutions or alternatives is 'given' and a decision maker requires computer-support for finding a most appropriate or 'best' solution (by providing additional information on the DM's preferences). Or using feedback from the DM, new solutions are calculated iteratively, hoping to converge step by step to a best one (see, [15], [17]).

Evolutionary algorithms on the other hand were usually designed as non-interactive optimization methods aiming at the calculation of a best or almost best solution within an acceptable amount of time. With the upcoming of multiobjective evolutionary algorithms and the replacement of (scalar) optimality by Pareto-optimality (or efficiency or dominance), the methods did not any more result in a single solution but a solution set. The application scenario of multiobjective evolutionary algorithms (MOEAs) is usually as follows: The MOEA calculates a set of solutions which is provided to the DM (or a method supporting the DM) for further processing and, finally, for the selection of a most preferred method.

If evolutionary algorithms and traditional MCDM approaches are coupled, this is mostly done in an a priori approach (see [3]) which may impose several difficulties for a decision maker being uninformed about Pareto-optimal solutions when he/she is asked to specify preferences. For more details and exceptions a discussion of exceptions from that approach see, e.g., in [3], [5], [14], [12].

This clear distinction between the generation of appropriate solutions and the selection of a single solution has, however, some disadvantages which became more severe during the maturation of the research field and an increasing number of real-life applications of MOEAs. First, in many real-life situations of problem solving, there is a strong demand for an on-line computer support. This means that people require a real-time processing of their data or are willing to wait for a computer response at longest a few minutes. Requirements of just-in-time manufacturing and flexible and fast production processes and service provision requires an almost immediate processing of information.

On the other hand, many real-life optimization problems, either in combinatorial optimization (see, e.g., [6], [11], [9]) or in continuous optimization (see, e.g., [13]), impose a substantial amount of computation time. Multiobjective optimization problems may involve significantly higher requirements for computation than their single-criterion counterparts as well as other difficulties (see [10]).

Considering the fact that only a rather small part of the objective space and also of the parameter space of a multiobjective optimization problem may be 'interesting' for the decision maker in the end, it should be sensible not to waste computational effort for the 'uninteresting' regions. User input should allow an MOEA to focus on accelerating its search towards the most interesting regions.

2 What Kind of Interaction?

During the history of multiple criteria decision making (MCDM) (see, e.g., [7]) a vast range of methods have been developed, many of them being interactive. Besides the specific type of optimization problem and optimization process, interactive methods can be distinguished by the nature of information they provide to the decision maker and require from him/her. For instance, many methods deal with preference information interpreted as 'weights'. Some methods based on utility or value functions frequently require judgments on trade-offs between criteria (see, e.g., [20]). Reference point approaches allow the DM to formulate most desired (or most undesired) solutions (see [21]), etc.

Experiences with real DMs using such methods as well as psychological results in general suggest that human beings are often overwhelmed with specifying the required information (see, e.g., [2] and [1] for comparative studies on interactive MCDM methods). Limitations in human information processing impose obstructions on rational decision making. For these reasons, the success of computer-based decision support mostly depends on how information are presented to a DM and what information is elicited from him/her.

In [19] we have discussed such limitations on human information processing. Our experiences with various applications in multicriteria decision support led to the development of a graphical user interface based (among other concepts) on a spider-web visualization of solutions in the objective space and various means of navigation in the solutions space, e.g. by locking specific objective values or setting bounds on objective values. According to our experiences, such an interface can easily be utilized for multiobjective decision support also by non-expert DMs.

For hard-to-solve decision problems it is usually not possible to provide all data behind the navigation interface in an on-line fashion. In some cases, the interface is fed by a database of solutions calculated in beforehand. In other cases (or combined with database usage) solutions are interpolated for a fast assessment. Considering this issue, such an interface provides an approach for integrations with an MOEA as discussed in the following section.

3 Interactivity for an MOEA

A significant disadvantage of some MCDM methods is not only that they require too much or too inappropriate information from the decision maker. They also require information too frequently, i.e. typically after each single processing step (from one solution to another one). Since the calculation time is often rather small compared with the DM's effort in understanding the methods results and providing new preference information, the effort is unbalanced between machine and human user. For a considerable time, the computer is just waiting for a DM's input. This problem can be solved by an asynchronous communication concept as discussed below.

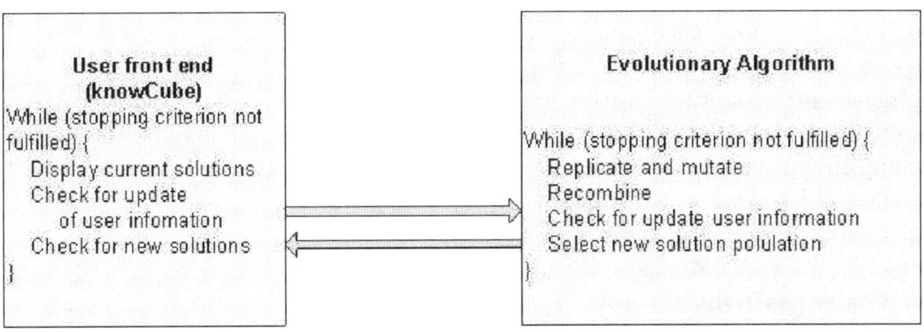

Fig. 1. Interactive user interface and evolutionary optimization

Fig. 1 shows the basic algorithmic loops of user interface and evolutionary algorithm and their communication. While the user interface may update its information on currently best solutions (i.e., the current population of the evolutionary algorithm), the evolutionary algorithm may proceed without an updated user information in each iteration. (For more details on multiobjective evolutionary algorithms see, e.g., [4], [18].) Thus, there is an essentially asynchronous communication between the two components.

As worked out in [19], an effective means for interactive solution navigation can be based on aspiration levels by locking specific objective values. Ideal or utopia values for specific objectives may be an easy-to-understand means to guide the further exploration of the solution space into preferred directions. Such information used by the graphical user interface can easily be processed by the multiobjective evolutionary algorithm: Aspiration values are used as hard bounds for feasible solutions when deciding on the next population on solutions (selection step). A softening of these bounds may also be possible when a decision maker is not really sure about aspiration values or when too many solutions are excluded by using them as hard bounds. A decision maker may also be informed during the process that currently the population consists of too many solutions such that stricter aspiration levels may be applied and are, indeed, welcome by the evolutionary algorithm because of a faster processing during each iteration.

Ideal or utopia values on the other hand are basically soft information on the DM's preferences. This information may be used as additional selection criteria for forming a subsequent population. Since many multiobjective evolutionary algorithms use dominance-related information only during the selection step (see [4]) there is few discrimination of solutions during the progress of the algorithm (see [8]). The distance to a reference point (or ideal solution) may thus be used for punishing solutions being Pareto-optimal (within a current population) but being to far from a desired location in the objective space.

Apart from the selection step, the user input also modifies the stopping criterion applied by the evolutionary algorithm. While this criterion is usually defined by conditions of the EA, for instance a maximum no. of generations or some measure of progress during the iterations, it is now basically the DM's business to stop the solution navigation (because of having found a 'best solution' or because of time restrictions or motivation). As long as the user is using the interface, the evolutionary algorithm should continue for improving the considered set of solutions.

4 Conclusions

Although, we do not yet have practical experiences with using the new approach for multicriteria decision support and optimization, it is appealing because the interaction between navigation interface and MOEA solves two problems at the same time. First, the MOEA provides a new and effective means for feeding the navigtion interface with data on solutions to a hard-to-solve multiobjective optimization problem in an on-line fashion. The usage of solution interpolation or a database resulting from a computationally expensive apriori calculation of solutions can be avoided, at least in part.

Secondly, information from the navigation interface allows the MOEA to concentrate on a preferable subset of solutions which is frequently much smaller than the whole set of Pareto-optimal solutions. Thus, the computational effort of the MOEA may decrease drastically.

The asynchronous concept of the communication between user interface and MOEA avoids waiting times both for the human user and for the computer. A user has still the possibility to let the MOEA run autonomously for a longer time (or until a stopping criterion for the MOEA is reached) and then navigate through the final result of the algorithm. Thus, complexity of user input and frequency may be kept to an acceptable amount.

References

1. Buchanan JT (1994) An experimental evaluation of interactive MCDM methods and the decision making process. J. Oper. Res. Soc. 45, 9, 1050–1059
2. Buchanan JT, Daellenbach HG (1987) A comparative evaluation of interactive solution methods for multiple objective decision models. European Journal of Operations Research 29:353–359.
3. Coello Coello CA (2000) Handling preferences in evolutionary multiobjective optimization: A survey. In: 2000 Congress Evolutionary Computation Proc. IEEE
4. Deb K (2001) Multi-objective optimization using evolutionary algorithms. Wiley
5. Cvetkovic D, Parmee IC (2002) Preferences and their application in evolutionary multiobjective optimization. IEEE Trans. Evolutionary Computation 6(1):42–57
6. Ehrgott M, Gandibleux X (eds) (2002) Multiple criteria Optimization: State of the art annotated bibliographic surveys. Kluwer, Boston
7. Gal T, Hanne T (1997) On the development and future aspects of vector optimization and MCDM. A tutorial. In: Climaco J (ed) Multicriteria analysis. Springer Berlin, 130–145
8. Hanne T (2001) Selection and mutation strategies in evolutionary algorithms for global multiobjective optimization. Evolutionary Optimization 3(1):27–40
9. Hanne T (2005) On the Scheduling of Construction Sites Using Single- and Multiobjective Evolutionary Algorithms. In: Proceedings of MIC2005: The Sixth Metaheuristics International Conference. Vienna.
10. Hanne T (2004) Five open issues in solving MOCO problems. Discussion of the article Approximative solution methods by for multiobjective combinatorial optimization by M. Ehrgott and X. Gandibleux. TOP 12(1): 70–76
11. Hanne T, Nickel S. (2005) A multi-objective evolutionary algorithm for scheduling and inspection planning in software development projects. European Journal of Operational Research 167:663–678

12. Jaskiewicz A (2005) The use of pairwise comparisons in interactive hybrid evolutionary algorithms. Multiple objective knapsack problem case study. In: Proceedings of MIC2005: The Sixth Metaheuristics International Conference. Vienna
13. Küfer K-H, Monz M, Scherrer A, Süss P, Alonso F, Sultan ASA, Bortfeld T, Craft D, Thieke C (2005) Multicriteria optimization in intensity modulated radiotherapy planning. Report of the Fraunhofer ITWM 77
14. Phelps SP, Köksalan M (2003) An Interactive Evolutionary Metaheuristic for Multiobjective Combinatorial Optimization Management Science 49, 12, 2003, 1726-1738
15. Stewart T (1999) Concepts of interactive programming. In: Gal T, Stewart T, Hanne T (eds): Multicriteria decision making. Advances in MCDM models, algorithms, theory, and applications. Kluwer, Boston
16. Schwefel H-P (1994) On the evolution of evolutionary computation. In: Zurada JM, Marks II RJ, Robinson CJ (eds): Computational intelligence - Imitating life. IEEE Press, Piscataway NJ, 116–124
17. Steuer RE (1986) Multiple criteria optimization. John Wiley and Sons, New York
18. Tan KC, Khor EF, Lee TH (2005) Multiobjective evolutionary algorithms and applications. Springer, Berlin
19. Trinkaus HL, Hanne T (2005) knowCube: a visual and interactive support for multicriteria decision making. Computers & Operations Research 32:1289–1309
20. Vincke P (1992) Multicriteria decision-aid. Wiley, Chichester
21. Wierzbicki AP (1999) Reference point approaches. In: Gal T, Stewart T, Hanne T (eds): Multicriteria decision making. Advances in MCDM models, algorithms, theory, and applications. Kluwer, Boston

Using a Combination of Weighting Methods in Multiattribute Decision-Making

Antonio Jiménez, Sixto Ríos-Insua and Alfonso Mateos

Technical University of Madrid, School of Computer Science, Campus de Montegancedo s/n, 28660 Boadilla del Monte, Madrid, Spain

Within the framework of multiattribute decision-making, it is well-known that weights derived from different weighting methods differ widely, even if are based on the same theoretical assumptions. There is not a best weighting method, and each one has advantages and drawbacks. We introduce the use of a combination of weighting methods in the GMAA system, a user-friendly decision support system that makes provision for all the stages in the Decision Analysis cycle. It includes several features aimed at overcoming biases and inconsistencies associated with weighting methods. Specifically, weights are elicited hierarchically, and different weighting methods can be used at different levels. Also incomplete information is allowed in the stakeholders' responses for the different methods, consistency is checked, attribute ranges are emphasized to avoid the so-called *range effect*, and the system provides thorough documentation of the methodologies and possible biases and inconsistencies.

1 Introduction

Multi-Attribute Utility Theory (MAUT) has become a widely accepted and frequently applied tool for assisting decision makers (DM) in making choices among complex alternatives that vary on multiple conflicting objectives, see, e.g., [3, 10]. In this paper we focus on one stage in the decision-making process, the quantification of DM preferences, specifically weight elicitation.

There are many weighting methods that use different questioning procedures to elicit weights, see [8, 11] for a review. Several experimental studies demonstrate that weights derived from different weighting methods differ widely, see, e.g., [1].

A number of authors have given explanations for the above paradox. First, the formulation of elicitation questions and the way they introduce the meaning of an attribute weight to the DM play an important role. For example, weighting methods differ in the way the attribute ranges are shown, which could lead to the so-called *range effect* [9].

On the other hand, the numerical scale that is explicitly or implicitly used restricts the resulting weights [4, 5, 7]. Another source of change in the attribute weights against theoretical expectations is the structure of the value tree. For example, the division of attributes in value trees can either increase or decrease the weight of an attribute (*splitting bias*) [6].

Several improvements have been proposed to overcome the above sources of biases [7] and the following conclusions have been reached by a number of authors:

- Any value tree weighting method is equally aceptable.
- Being aware of the biases is the first and probably most important step towards overcoming them. Consequently, a full description of the methodology and its assumptions should be provided.
- Attribute ranges should be stressed during the elicitation process to prevent the so-called range effect.
- On the other hand, imprecision concerning the DM responses should be allowed. This is less stressful for DMs and also makes the method suitable for group decision support. Moreover, consistency checks should be performed throughout the elicitation process.
- Splitting biases are more frequent when attribute weights are elicited non hierarchically. However, they may also appear locally with hierarchical weighting.
- Interactive evaluation of the results during weight elicitation would improve the process.

In this paper we propose a preliminary approach for combining weighting methods on the basis of the above ideas to improve the Generic Multi-Attribute Analysis[1] (GMAA) System, a PC-based decision support system (DSS) founded on an additive multi-attribute utility model that is intended to allay many of the operational difficulties involved in decision analysis (DA), see [2].

2 An approach for combining weighting methods

Suppose we have a generic complex decision-making problem whose value tree is shown in Fig. 1. As a starting point we also assume that weights can be hierarchically elicited, i.e., local weights, w_i, representing the relative importance of nodes at each branch and level of the value tree can be elicited, and the attribute weights, k_i, are then obtained by multiplying the above local weights in the path from the *overall objective* to each attribute.

Moreover, as imprecision concerning the DM responses will be allowed in the different weighting methods, an average normalized weight w_i and a normalized weight interval $[w_i^L, w_i^U]$ rather than a precise weight will be calculated for each node under consideration, where *L* means *L*ower and U means *U*pper.

[1] http://www.dia.fi.upm.es/· ajimenez/GMAA

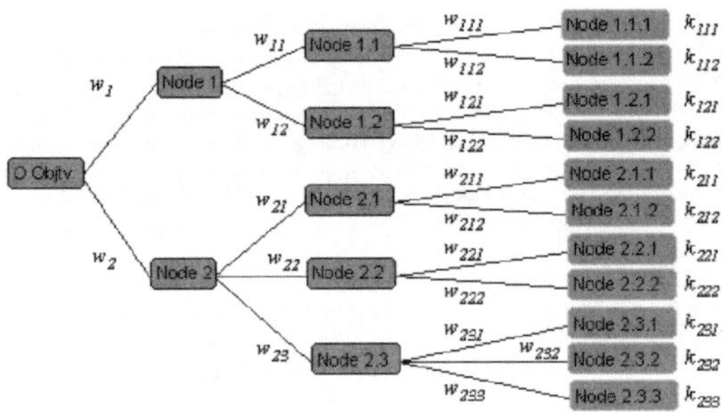

Fig. 1. Generic value tree

Consequently, attribute weights will be assessed as follows:

$$k_{111}^L = w_1^L \times w_{11}^L \times w_{111}^L \qquad k_{111} = w_1 \times w_{11} \times w_{111} \qquad k_{111}^U = w_1^U \times w_{11}^U \times w_{111}^U \tag{1}$$
$$\ldots \qquad \ldots \qquad \ldots$$
$$k_{233}^L = w_2^L \times w_{23}^L \times w_{233}^L \qquad k_{233} = w_2 \times w_{23} \times w_{233} \qquad k_{233}^U = w_2^U \times w_{23}^U \times w_{233}^U$$

We consider precise and equally weighted nodes throughout the value tree as default local weights, i.e., all nodes are equally important at any branch and level. Then attribute weights are assessed on the basis of these local weights.

The basic idea of the approach we propose is that if we are able to assess local weights representing the relative importance of nodes throughout the hierarchy from the attribute weights, then we could apply *SWING weighting* [10] to elicit attribute weights and use local weights at intermediate levels to run consistency checks. Moreover, local weights could be updated or reelicited using *TRADE-OFFS weighting* [3] or *Direct point allocation*, as appropriate, and attribute weights would be automatically reassessed.

Let us show how to propagate attribute weights in ascending order throughout the value tree to obtain local weights. This is straightforward for average normalized weights, as shown in Fig. 2 for nodes stemming from Node 1.

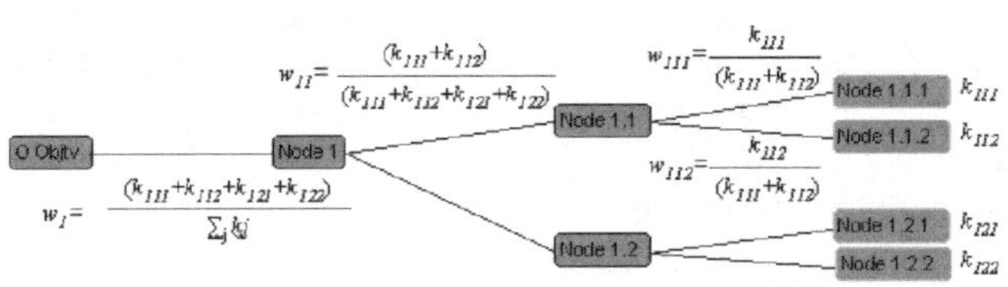

Fig. 2. Propagating average normalized weights

Let us explain the process for propagating normalized weights intervals for nodes stemming from Node 1.1. Note that it is equivalent for the remaining nodes. We know that

$$\begin{aligned} k_{111}^L &= w_1^L \times w_{11}^L \times w_{111}^L, \quad k_{111}^U = w_1^U \times w_{11}^U \times w_{111}^U, \\ k_{112}^L &= w_1^L \times w_{11}^L \times w_{112}^L, \quad k_{112}^U = w_1^U \times w_{11}^U \times w_{112}^U, \end{aligned} \tag{2}$$

so we can state that

$$
\begin{aligned}
c^L &= w_1^L \times w_{11}^L = k_{111}^L / w_{111}^L = k_{112}^L / w_{112}^L \Rightarrow w_{111}^L = k_{111}^L \times w_{112}^L / k_{112}^L, \\
c^U &= w_1^U \times w_{11}^U = k_{111}^U / w_{111}^U = k_{112}^U / w_{112}^U \Rightarrow w_{112}^U = k_{112}^U \times w_{111}^U / k_{111}^U,
\end{aligned} \tag{3}
$$

where w_1^L and w_{11}^L are unknown at this moment, being c^L and c^U constants.

Assuming that average normalized weights are the midpoints of the normalized weight intervals:

$$
w_{111} = (w_{111}^L + w_{111}^U)/2 \quad \text{and} \quad w_{112} = (w_{112}^L + w_{112}^U)/2. \tag{4}
$$

From Eqs. (3, 4), we find that

$$
w_{111}^L = \frac{k_{111}^L \times k_{112}^L \times 2\left(w_{112} - w_{111} \times k_{112}^U\right)}{\left(1 - \dfrac{k_{111}^L \times k_{112}^U}{k_{112}^L \times k_{111}^U}\right)}
$$

and the remaining values are obtained by substituting w_{11}^L in Eqs. (3,4).

Let us illustrate the weight elicitation process with an example. First, the DM applies *SWING weighting* to elicit attribute weights, see Fig. 3.

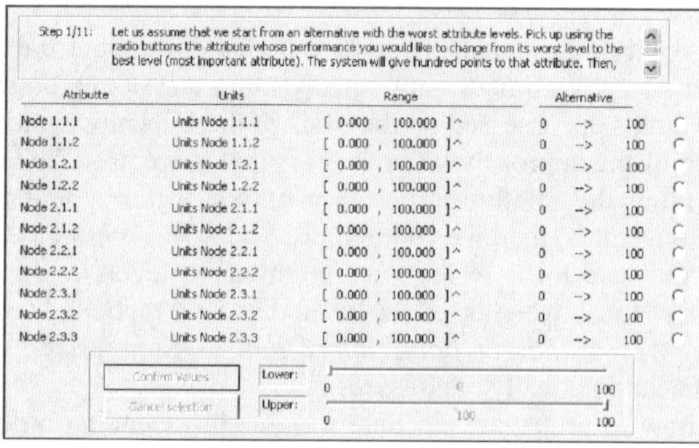

Fig. 3. SWING weighting

The DM starts with an alternative with the worst attribute levels and picks the attribute whose performance he/she would like to change from its worst to the best level. This attribute is given hundred points. Next, the attribute whose performance the DM would like to change from its worst to the best level is selected again, and an imprecise value between 0 and 100 representing its relative importance regarding the most important attribute is provided. Finally, average normalized weights and normalized weight intervals are obtained by means of a normalization process [3], see Fig. 4.

We benefit from the attribute weight propagation throughout the value tree and run consistency checks in which the DM is informed about the relative importance of intermediate nodes in ascending order, see Fig. 5.

If the DM disagrees with the local weights, he/she can modify them using *TRADE-OFFS weighting* or *Direct point allocation*, and the attribute weights stemming from the updated nodes will be automatically reassessed, see Eq. (1).

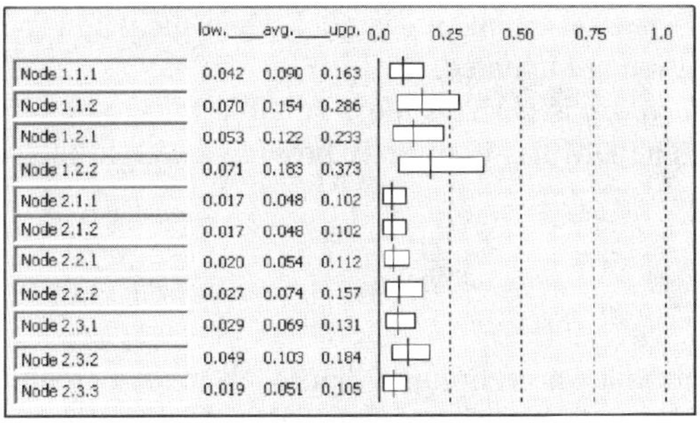

Fig. 4. Attribute weights

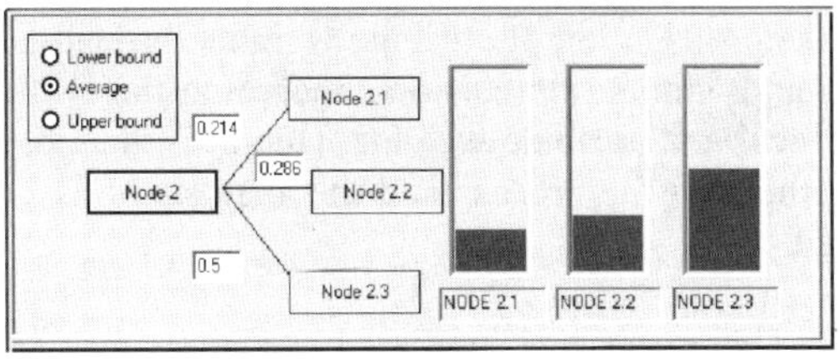

Fig. 5. Consistency checks

TRADE-OFFS weighting is perhaps more suitable for the low-level nodes in the value tree, because it involves a more specific area of knowledge. It is based on trade-offs among the respective attributes of the lowest-level objectives stemming from the same objective, [3]. For example, Fig. 6 shows the probability question for eliciting weights for Node 1.1 and Node 1.2.

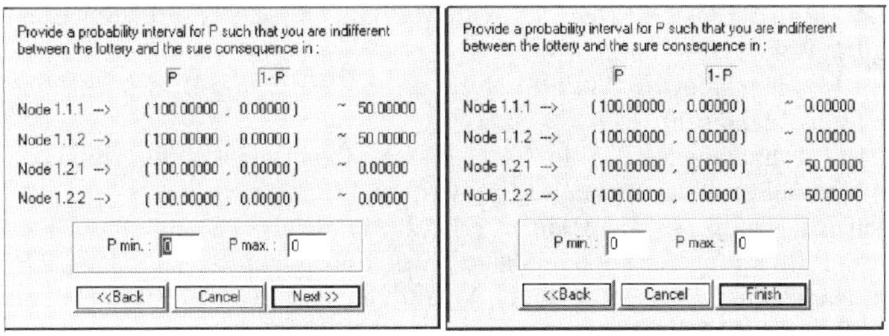

Fig. 6. *TRADE-OFFS weighting*

The DM is asked to give an interval of probabilities such that he/she is indifferent with respect to a gamble and sure consequences. Average normalized weights and normalized weight intervals are output by taking expected utilities from gambles and following a normalization process.

Note that for *TRADE-OFFS weighting* attributes have to be measured on a continuous scale, and the shape of the utility function affects the attribute weights. In this sense, the original method has been modified to admit imprecision either in the DM responses or component utilities. Moreover, *TRADE-OFFS weighting* requires a continuous reassessment of weights when local lower-level weights involved in the elicitation are updated.

Direct point allocation is perhaps more suitable for the possibly more political upper-level nodes. The DM has to directly provide a weight interval for each node under consideration.

3 Conclusions

In this paper we introduce a preliminary approach for a weighting method that is intended to overcome possible biases detected in traditional weighting methods. The procedure is based on a combined application of several methods, *SWING weighting*, *TRADE-OFFS weighting* or *Direct point allocation*. It benefits from the value tree and the propagation of attribute weights through the tree to perform consistency checks and admits imprecision concerning the DM responses, whcih leads to imprecise local and attribute weights. However, the possible inclusion of other weighting methods in the process and the analysis of associated biases still needs to be analyzed.

Acknowledgments. This paper was supported by the Spanish Ministry of Education and Science TSI2004-06801-C04-04.

References

1. Borcherding K, Eppel T, von Winterfeldt D (1991) Comparison in Weighting Judgements in Multiattribute Utility Measurement. Manage Sci 36: 1603-1619.
2. Jiménez A, Ríos-Insua S, Mateos S (2003) A Decision Support System for Multiattribute Utility Evaluation based on Imprecise Assignments. Decis Support Syst 36: 65-79.
3. Keeney RL, Raiffa H (1976) Decision with Multiple Objectives: Preferences and Value-Tradeoffs. Wiley, New York.
4. Pöyhönen M, Hämäläinen RP (1998) Notes on the Weighting Biases in Value Trees, J Behav Decis Making 11: 139-150.
5. Pöyhönen M, Hämäläinen RP, Salo AA (1997) An Experiment on the Numerical Modelling of Verbal Ratio Statements, J Multi-Criteria Decis Analysis 6: 1-10.
6. Pöyhönen M, Vrolijk H, Hämäläinen RP (2001) Behavioral and Procedural Consequences of Structural Variation in Value Trees, Eur J Oper Res 134: 216-227.
7. Salo AA, Hämäläinen RP (1997) On the Measurement of Preferences in the Analytic Hierarchy Process. J Multi-Criteria Decis Analysis 6, 309-343.
8. Stewart T (1992) A Critical Survey on the Status of Multiple Criteria Decision Making Theory and Practice. OMEGA 20: 569-586.
9. von Nitzsch R, Weber M (1993) The Effect of Attribute Ranges on Weights in Multiattribute Utility Measurements. Manage Sci 39: 937-943.
10. von Winterfeldt D. Edwards W (1986) Decision Analysis and Behavioral Research. Cambridge Universisty Press, Cambridge.
11. Weber M, Borchering K (1993) Behavioral Influences on Weight Judgements in Multiattribute Decision Making. Eur J Oper Res 67: 1-12.

Gremienentscheidungen bei partiellen Präferenzordnungen

Eva Ponick

Technische Universität Clausthal, Institut für Wirtschaftswissenschaft,
Julius-Albert-Str. 2, D-38678 Clausthal-Zellerfeld
eva.ponick@tu-clausthal.de

Zusammenfassung. Im Rahmen der individuellen Entscheidungstheorie finden partielle Präferenzordnungen immer stärker Beachtung. Fehlt die Angabe, ob ein Entscheider eine Alternative strikt präferiert oder indifferent ist, so mag dies schlicht daran liegen, dass kein Vergleich stattgefunden hat. Es ist aber auch möglich, dass nach Meinung des Entscheiders mit den zur Verfügung stehenden Informationen oder unter den gegebenen Rahmenbedingungen kein Vergleich durchgeführt werden kann. Letztere Auffassung sollte ein Entscheider auch bei Entscheidungen, die von einem Gremium getragen werden, einbringen können. Regeln zur paarweisen Aggregation individueller Präferenzrelationen erlauben die Interpretation fehlender Präferenzen. Wie im Falle vollständiger Präferenzordnungen lässt sich auch hier die Transitivität der kollektiven Präferenzordnung durch ein speziell erweitertes Eingipfligkeitskriterium gewährleisten.

1 Einleitung

Schon vor der konsequenten Ausbildung demokratischer Strukturen war die Übertragung von Entscheidungen auf eine ausgewählte Gruppe von Personen eine feste Größe im Leben der Menschen. Die Entscheidungskompetenz von Gremien hat sich bis heute in vielen Unternehmensbereichen stets aufs Neue bewährt. Zur geregelten Durchführung einer Abstimmung haben sich im Laufe der Zeit unzählige Verfahren herausgebildet, die mit dem Ziel Anwendung finden, strategisches Verhalten der Gremienmitglieder einzudämmen und ein gerechtes Abstimmungsergebnis zu erhalten. Beide Ziele sind nicht zu erfüllen, wie von Arrow, Gibbard und Satterthwaite eindrucksvoll bewiesen wurde [1, 8, 12]. Eines haben alle diese Verfahren gemein, die einzelnen Mitglieder des Gremiums müssen ihre Präferenzen bezüglich der zur Auswahl stehenden Alternativen kennen und gemäß dieser Präferenzen am Abstimmungsverlauf teilnehmen. Der Großteil der Literatur, welche sich mit Regeln zur Ermittlung kollektiver Präferenzaussagen beschäftigt, geht davon aus, dass alle Gremienmitglieder in der Lage sind, die zu bewertenden Alternativen in eine vollständige Rangordnung zu bringen. Das tägliche Leben lehrt, dass eine vollständige Rangordnung der vorliegenden Alternativen nicht immer möglich ist. Auch Forschungsergebnisse aus dem Bereich der individuellen Entscheidungstheorie lassen diesen Schluss zu. So ist bekannt, dass unter der Annahme bestimmter Verhaltensaxiome anhand von Auswahlentscheidungen eine vollständige individuelle Präferenzordnung abgeleitet werden kann [13, 5]. Handlungen der Entscheider stehen jedoch häufig im Widerspruch zu der so ermittelten Präferenzordnung. Als Erklärung werden unter anderem auch partielle Präferenzordnungen herangezogen [14, 6]. Die Relevanz der Bewertung von Alternativen als unvergleichbar wird ebenfalls durch Untersuchungen im Bereich der multikriteriellen Entscheidungstheorie gestützt [11, 4].

Ein Gremium besteht aus einzelnen Personen, deren Alternativenbewertungen zu einer Aussage aggregiert werden. Sind eventuell auch Mitglieder ohne vollständige Rangordnung der Alternativen vertreten, so muss dies im Abstimmungsverfahren zur Geltung kommen. In der vorliegenden Arbeit wird eine Regel verwendet, die durch paarweise Vergleiche von Alternativen innerhalb eines Gremiums die kollektive Präferenzordnung ermittelt. Den Gremienmitgliedern wird die Möglichkeit gegeben, auch dann mit ihren tatsächlichen Alternativenbewertungen an der Entscheidung teilzunehmen, wenn sie nicht in der Lage sind, eine vollständige Rangordnung zu bilden. Handelt es sich um eine Expertenrunde, so kann Unvergleichbarkeit von Alternativen als bewusste Aussage verstanden werden, dass mit den zur Zeit verfügbaren Informationen oder unter den geltenden Umständen kein Vergleich erfolgen kann, darf oder soll. Ein Problem, das dem des bekannten paarweisen Vergleichs bei vollständigen Präferenzordnungen gleicht, ist ein eventueller Verstoß der kollektiven Präferenzordnung gegen Transitivität. Daher wird ein erweitertes Eingipfligkeitskriterium eingeführt, das auch auf Gremien angewendet werden kann, deren Mitglieder eventuell partielle Präferenzordnungen besitzen. Dieses Kriterium baut auf dem von Black [2] definierten Kriterium auf und schließt analog bei Gültigkeit Intransitivität der kollektiven Präferenzordnung aus. Zur Anwendung des Kriteriums wird eine grafische Darstellungsmöglichkeit partieller Präferenzordnungen aufgezeigt.

2 Begriffe und Definitionen

In diesem Beitrag wird stets eine endliche Menge A von Alternativen vorausgesetzt. Die Anzahl der Gremienmitglieder $i \in G$ ist ebenfalls endlich. Präferenzen werden durch eine Relation $\succeq \subseteq A \times A$ beschrieben. Mit $a_i \succeq a_j$ wird ausgedrückt, dass die Alternative a_i mindestens so gut eingeschätzt wird wie die Alternative a_j. Strikte Präferenz $a_i \succ a_j$ bedeutet, dass zwar $a_i \succeq a_j$ jedoch nicht $a_j \succeq a_i$ gilt, Indifferenz $a_i \sim a_j$, dass sowohl $a_i \succeq a_j$ als auch $a_j \succeq a_i$ und Unvergleichbarkeit $a_i ? a_j$, dass weder $a_i \succeq a_j$ noch $a_j \succeq a_i$ gilt [7]. Mit $\succeq_i$ wird die Präferenzrelation eines Gremienmitglieds i und mit $\succeq_G$ die kollektive Präferenzrelation bezeichnet. Dies gilt analog für strikte Präferenz, Indifferenz und Unvergleichbarkeit. Die Relation $\succeq$ ist eine partielle

Präferenzordnung, wenn sie reflexiv, transitiv und antisymmetrisch ist. Ist sie ebenfalls vollständig, so ist sie eine vollständige Präferenzordnung. Gilt Transitivität, müssen folgende drei Bedingungen erfüllt sein: gilt $a_i \succ a_j$ und $a_j \succ a_k$ so folgt $a_i \succ a_k$, gilt $a_i \sim a_j$ und $a_j \sim a_k$ so folgt $a_i \sim a_k$ und gilt $a_i \succ a_j$ und $a_j \sim a_k$ so folgt $a_i \succ a_k$ (vgl. [9], S. 32). Die Präferenzordnungen der Gremienmitglieder können in einem n-Tupel zusammengefasst werden. Dieses wird als Präferenzordnungsprofil bezeichnet (vgl. [9], S. 440).

3 Paarweiser Vergleich von Alternativen

Bei dem klassischen paarweisen Vergleich, auch Mehrheitsregel genannt, werden den Gremienmitgliedern in einer vorher bestimmten Reihenfolge stets zwei Alternativen vorgelegt. Jedes Mitglied besitzt eine Stimme und die Alternative mit den meisten Stimmen wird beibehalten, die andere verworfen. Bei Gleichstand wird kollektive Indifferenz vorausgesetzt und durch eine festgelegte Regel ermittelt, welche Alternative beibehalten wird (vgl. [9], S. 421/422). Allerdings kann bei Verwendung dieses Verfahrens eine intransitive kollektive Präferenzordnung ermittelt werden. Besteht ein Gremium aus drei Mitgliedern mit den Präferenzordnungen $a_1 \succ_1 a_2 \succ_1 a_3$, $a_2 \succ_2 a_3 \succ_2 a_1$ und $a_3 \succ_3 a_1 \succ_3 a_2$, so ergeben sich die kollektiven Präferenzen $a_1 \succ_G a_2$, $a_2 \succ_G a_3$, $a_3 \succ_G a_1$ (vgl. [2], S. 47). Ein solcher Ringschluss macht eine sinnvolle Entscheidung unmöglich.

Bewertet ein Gremienmitglied zwei Alternativen als unvergleichbar, besitzt es keine Möglichkeit, diese Einschätzung in die Entscheidung einfließen zu lassen. Tatsächlich bietet jedoch der paarweise Vergleich von Alternativen zur Ermittlung der kollektiven Präferenzordnung vielfältige Möglichkeiten auch partielle Präferenzordnungen zu beachten. Eine Variante ist in Tabelle 1 aufgeführt. Dabei steht (P) für die Ermittlung kollektiver strikter Präferenz. Hier werden nur die Gremienmitglieder mit einbezogen, die eine der Alternativen strikt präferieren. Dieser Ansatz findet sich auch in [10]. Um den Bedenken eines Gremienmitglieds, das Alternativen als unvergleichbar ansieht, mehr Gewicht zu verleihen, könnten in (P) nicht nur die Gegenstimmen sondern auch die Stimmen derjenigen, die Unvergleichbarkeit angeben, als Schwelle für strikte Präferenz gesetzt werden. Die Bedingung (I) steht für kollektive Indifferenz und (U) dafür, dass die Gremienmitglieder gemeinsam nicht in der Lage sind, die Alternativen zu vergleichen. Erhalten zwei Alternativen die gleiche Anzahl Stimmen, so werden sie nicht wie in der klassischen Regel kollektiv als indifferent, sondern als unvergleichbar eingestuft. Auf notwendige Änderungen im Abstimmungsverlauf für den Fall, dass Alternativen kollektiv als unvergleichbar angesehen werden, kann an dieser Stelle nicht eingegangen werden. Der entscheidende Nachteil des klassischen paarweisen Vergleichs, die Existenz von intransitiven kollektiven Präferenzordnungen, bleibt erhalten.

Tabelle 1. Erweiterter paarweiser Vergleich

(P)	$a_i \succ_k a_j$ für alle $k \in D \subseteq G, D \neq \emptyset$ und $a_j \prec_l a_i$ für alle $l \in E \subseteq G, \mid D \mid > \mid E \mid$	$\Rightarrow a_i \succ_G a_j$
(I)	$a_i \sim_k a_j$ für alle $k \in D \subseteq G, D \neq \emptyset$ und $a_j ?_l a_i$ für alle $l \in E \subseteq G, E \cup D = G, \mid D \mid > \mid E \mid$	$\Rightarrow a_i \sim_G a_j$
(U)	alle anderen Kombinationen	$\Rightarrow a_i ?_G a_j$.

4 Eingipfligkeit bei partiellen Präferenzordnungen

Das Eingipfligkeitskriterium von Black [2] stellt die wohl bekannteste Eigenschaft eines Präferenzordnungsprofil dar, die gewährleistet, dass die kollektive Präferenzordnung transitiv ist. In Abbildung 1 ist die Präferenzordnung eines Entscheiders eingipflig dargestellt. Dabei wird für jede Alternative die Höhe der Präferenz im Vergleich zu den anderen Alternativen aufgetragen und die ermittelten Punkte miteinander verbunden. Es entsteht ein Polygonzug, in dessen Verlauf genau ein Gipfel zu erkennen ist (vgl. [2], Kapitel 2). Ein Präferenzordnungsprofil heißt eingipflig, wenn eine Anordnung der Alternativen existiert, so dass alle Präferenzordnungen der Gremienmitglieder eingipflig dargestellt werden können.

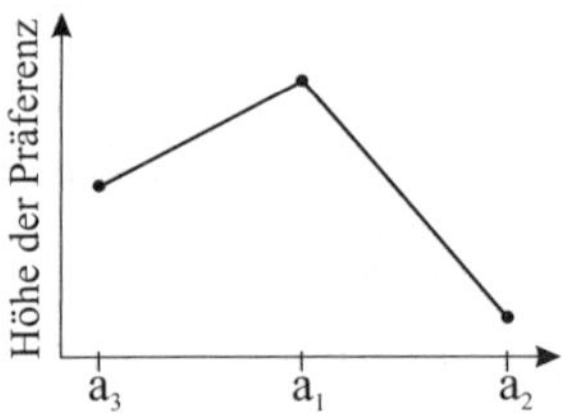

Abb. 1. $a_1 \succ_i a_3 \succ_i a_2$

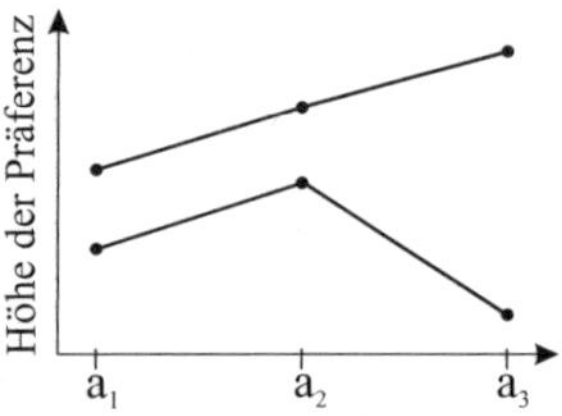

Abb. 2. $a_1 \prec_i a_2$, $a_1 ?_i a_3$, $a_2 ?_i a_3$

Bevor eine Definition der Eingipfligkeit für Präferenzordnungsprofile mit nicht notwendig vollständigen Präferenzordnungen aufgestellt werden kann, muss eine grafische Darstellung gefunden werden, die auch Unvergleichbarkeit ausdrückt. Bewertet ein Entscheider zwei Alternativen als unvergleichbar, kann diese Aussage nicht durch die Verwendung nur eines Eintrags der Höhe der Präferenzen je Alternative erzielt werden. Vielmehr wird jede Alternative nunmehr mit zwei Einträgen dargestellt. Diese können als Bewertung der Alternativen bezüglich zweier Kriterien interpretiert werden. Im weiteren Sinne ist dieses Vorgehen vergleichbar mit der Ermittlung einer Präferenzordnung durch Vergleich der Ausgangs- und Eingangsflüsse bei dem Outrankingverfahren PROMETHEE [4]. Jeder Punkt einer Alternative wird eindeutig einem Polygonzug zugeordnet. Es entstehen zwei Polygonzüge, wie in Abbildung 2 beispielhaft gezeigt, die als oberer und unterer Polygonzug angeordnet werden können. Durch Vergleich der Höhe der Präferenzen kann für die einzelnen Polygonzüge strikte Präferenz bzw. Indifferenz definiert werden. Es bezeichne

$a_k \succ_i^o a_j$ strikte Präferenz bezüglich des oberen und $a_k \succ_i^u a_j$ bezüglich des unteren Polygonzuges, bestimmt jeweils durch einen höheren Eintrag für a_k im Vergleich zu a_j. Entsprechend ist bei gleich hohem Eintrag $a_k \sim_i^o a_j$ und $a_k \sim_i^u a_j$ definiert. Gilt $a_k \succ_i a_j$, so ist sowohl $a_k \succ_i^o a_j$ als auch $a_k \succ_i^u a_j$. Gilt $a_k \sim_i a_j$, so ist sowohl $a_k \sim_i^o a_j$ als auch $a_k \sim_i^u a_j$. Bei $a_k ?_i a_j$ gilt entweder $a_k \succ_i^o a_j$ und $a_j \succ_i^u a_k$ oder $a_j \succ_i^o a_k$ und $a_k \succ_i^u a_j$.

Definition 1. *Ein nicht notwendig vollständiges Präferenzordnungsprofil des Gremiums G heißt genau dann* **erweitert eingipflig** *auf* $A = \{a_1, a_2, a_3\}$, *wenn eine Anordnung* (a_j, a_k, a_l) *der Alternativen aus A existiert, so dass für alle* $i \in G$ *gilt:* $a_j \succeq_i^o a_k \Rightarrow a_k \succ_i^o a_l$ *und* $a_j \succeq_i^u a_k \Rightarrow a_k \succ_i^u a_l$.

Die Eingipfligkeitsbedingung wird sowohl für den oberen als auch für den unteren Polygonzug verlangt. Dabei wird eine modifizierte Version des ursprünglichen Kriteriums verwendet. Siehe [1], S. 77, und [3], S. 134/135, für das urspüngliche Kriterium für vollständige Präferenzordnungsprofile.

Theorem 1. *Gilt Definition 1 bezüglich aller Teilmengen von drei Alternativen aus A, so ist die kollektive Präferenzordnung auf A, die gemäß des Mechanismus aus Tabelle 1 ermittelt wurde, transitiv.*

Die klassische Variante von Theorem 1 gilt nur für Gremien mit ungerader Anzahl Mitglieder (vgl. [1], S. 78). Das ist in der Erweiterung nicht mehr der Fall. Allerdings wird die Eingipfligkeitsbedingung nun für alle Gremienmitglieder gefordert. Es sind somit keine gleichgültigen Entscheider zugelassen.

5 Zusammenfassung und Ausblick

Die Integration von Gremienmitgliedern mit partiellen Präferenzordnungen ist gerade bei Verwendung eines paarweisen Vergleichs möglich. Das Ergebnis einer Gruppenentscheidung kann somit mehr Informationen beinhalten und auch Bedenken bezüglich der Realisierbarkeit von Alternativen ausdrücken. Die Ermittlung einer transitiven kollektiven Präferenzordnung verursacht weiterhin Probleme. Allerdings kann darüber hinaus aus bestimmten Eigenschaften eines Präferenzordnungsprofils wiederum auf Transitivität geschlossen werden. Diese Aussage bleibt auch bei bestimmten Variationen des paarweisen Vergleichs aus Tabelle 1 erhalten.

A Beweisskizze zu Theorem 1

Bei zwei Alternativen ist die kollektive Präferenzordnung stets transitiv. Gilt für jede Teilmenge von drei Alternativen aus A, $|A| \geq 3$, Transitivität, so kann gezeigt werden, dass die kollektive Präferenzordnung auf A insgesamt transitiv ist. Siehe [13], S. 492, für den Fall einer vollständigen Präferenzordnung.

Bei Verwendung des paarweisen Vergleichs gemäß Tabelle 1 können unter Ausschluss gleichgültiger Entscheider folgende Verstöße gegen Transitivität auftreten: $[a_1 \succ_G a_2, a_2 \succ_G a_3, a_3 \succ_G a_1]$, $[a_1 \succ_G a_2, a_2 \succ_G a_3, a_1 \sim_G a_3]$, $[a_1 \succ_G a_2, a_2 \succ_G a_3, a_1 ?_G a_3]$, $[a_1 \succ_G a_2, a_2 \sim_G a_3, a_1 ?_G a_3]$ und $[a_1 \succ_G a_2, a_2 ?_G a_3, a_1 \sim_G a_3]$.

Es existieren sechs Möglichkeiten, drei Alternativen anzuordnen. Untersucht werden müssen nur (a_1, a_2, a_3), (a_2, a_3, a_1) und (a_3, a_1, a_2), da sich die anderen Varianten durch eine spiegelverkehrte Darstellung herleiten lassen.

Werden nun die drei Anordnungen der Alternativen nacheinander vorausgesetzt und für jede Anordnung die bei Gültigkeit von Definition 1 möglichen Präferenzordnungen der Gremienmitglieder ermittelt, so kann gezeigt werden, dass keiner der oben aufgeführten Verstöße gegen Transitivität existiert.

Literaturverzeichnis

1. Arrow KJ (1966) Social choice and individual values. Wiley, New York, 3. Auflage
2. Black D (1998) The theory of committees and elections. In: McLean I, McMillan A, Monroe BL (eds) "The theory of committees and elections" by Duncan Black and "Committee decisions with complementary valuation" by Duncan Black and R.A. Newing. Kluwer, Boston Dordrecht London, 2. Auflage
3. Bossert W, Stehling F (1990) Theorie kollektiver Entscheidungen. Springer, Berlin Heidelberg New York
4. Brans JP, Vincke Ph, Mareschal B (1986) How to select and how to rank projects: the PROMETHEE method. European Journal of Operational Research 24:228-238
5. Chernoff H (1954) Rational selection of decision functions. Econometrica 22:422-443
6. Eliaz K, Ok EA (2004) Indifference or indecisiveness? Choice-theoretic foundations of incomplete preferences. Mimeo, New York University
7. Esser J (2001) Vollständigkeit, Konsistenz und Kompatibilität von Präferenzrelationen. OR Spektrum 23:183-201
8. Gibbard A (1973) Manipulation of voting schemes: a general result. Econometrica 41:587-601
9. Laux H (2003) Entscheidungstheorie. Springer, Berlin Heidelberg New York, 5. Auflage
10. Regenwetter M, Marley AAJ, Grofman B (2002) A general concept of majority rule. Mathematical Social Sciences 43: 405-428
11. Roy B (1991) The outranking approach and the foundations of ELECTRE methods. Theory and Decision 31:49-73
12. Satterthwaite MA (1975) Strategy-proofness and Arrow's conditions: Existence and correspondence theorems for voting procedures and social welfare functions. Journal of Economic Theory 10:187-217
13. Sen AK (1971) Choice function and revealed preference. Review of Economic Studies 38:307-317
14. Sippel R (1997) An experiment on the pure theory on consumer's behaviour. Economic Journal 107:1431-1444

The Impact of Preference Structures in Multi-Issue Negotiations - an Empirical Analysis

Rudolf Vetschera

Department of Business Studies, University of Vienna, Bruenner Strasse 72, A-1210 Vienna, Austria rudolf.vetschera@univie.ac.at

1 Introduction

Several Negotiation Support Systems like Inspire [3], Negotiation Assistant [8] or Joint Gains [1] use a utility-based representation of negotiator preferences to evaluate offers, determine the efficiency of proposed compromise solutions or suggest a fair compromise. All these systems implicitly assume that preferences as encoded in the utility functions can be used to guide a negotiator's behavior during the negotiation.

Despite of its importance for analytical negotiation support, the question whether the utility functions elicited from negotiators are actually reflected in their behavior during negotiations, and consequently in the outcomes of negotiations, has not yet been subject to much empirical analysis. Mumpower [4] gives a theoretical overview of how structural properties of negotiator preferences, like attribute weights, convexity of single-attribute utility functions, or the method used to aggregate preferences across attributes, influence the complexity of negotiations and thus negotiation outcomes. Empirical studies were performed by Stuhlmacher [9], who studied the impact of attribute weights, and Northcraft [6], who analyzed the impact of convexity of single attribute utilities on negotiator behavior.

In this paper, we present the results of an exploratory analysis of experiments conducted with the NSS Inspire [3]. Inspire elicits additive multi-attribute utility functions from negotiators using an extended conjoint measurement method. Section two of the paper develops the hypotheses how structural properties of the negotiators' utility functions will affect their behavior during negotiations and the outcomes. Section three presents the empirical results and section four concludes the paper with a short discussion.

2 Hypotheses

Three structural properties of additive utility functions are analyzed in this paper: The attribute weights, the monotonicity of marginal utility functions, and their shape. We expect these properties to influence the behavior of negotiators, and consequently the outcome of negotiations. Behavior during negotiations is characterized by two variables for each attribute: the value in the first offer made by a negotiator (e.g. the price of goods initially asked for by a seller), and the total amount of concessions made. The final outcome is characterized by the compromise value in each attribute.

Attribute weights reflect the importance which an attribute has for a negotiator. We expect that negotiators, who assign high weights to an attribute, will negotiate more toughly concerning this attribute. Referring to the process variables we analyze in this paper, this means they will start with higher initial offers and make less concessions. This behavior should lead to a better outcome in that attributes.

The second structural property is the monotonicity of marginal utility functions. The case description used in the experiments explicitly mentioned the direction of improvement for each attribute (e.g. that sellers should prefer a higher price over a lower price). But since Inspire leaves the subjects considerable freedom in the specification of their utility functions, it is possible that users specify non-monotonic utility functions or even utility functions which are monotonic in the opposite direction than specified in the case description. Depending on the attribute, the data used in this study contains between 8.7% and 16.8% of non-monotonic utility functions and between 2.4% and 8.0% of monotonic utility functions with incorrect directions.

In our analysis, we interpret attribute values in a way which is consistent with the case description. Thus we consider an initial offer from a seller, who demands a higher price, as a "tougher" initial offer than one demanding a lower price. Therefore we expect subjects who have non-monotonic marginal utility functions or monotonic utility functions contradicting the case description to make "weaker" initial offers. During bargaining, these subjects might also move in the wrong direction, so we expect them to make less concessions. But since they intend to move an attribute value in the wrong direction, we also expect them to have worse outcomes than subjects with correct monotonic utility functions.

As a final property of utility functions, we consider the shape of marginal utility functions. The usual assumption of decreasing marginal utility would lead to a concave shape of the marginal utility functions. However, it is sometimes argued that this assumption is not valid in a multi-attribute context [2]. Our data contains between 21.7% and 35.8% of convex marginal utility functions in the different attributes.

A negotiator who has a convex utility function will incur a significant loss of utility when deviating from the best possible value in an attribute, while for a negotiator with concave marginal utility, this loss is much smaller

[4, 5, 6]. Therefore, we expect negotiators having convex utility functions to make higher initial offers and less concessions. This should also result in better outcomes for them.

Concerning the relationship between the process variables, we have argued that tougher negotiation behavior is reflected both in higher initial demands and less concessions. But these two variables might also be seen as substitutes. Therefore, we finally formulate the hypothesis that negotiators who make more demanding higher offers are then willing to give up more and make more concessions than negotiators who already start with a weak position.

3 Results

To test our hypotheses, we analyze data from 1,448 negotiation experiments conducted with Inspire in the years 1996 to 2004. All experiments are based on a buyer-seller negotiation involving four attributes: price, delivery time, payment terms and conditions for the return of defective goods. Only experiments in which an agreement was reached are used in this study.

All hypotheses were tested using linear regression models. For these models, monotonicity was coded on a five-point scale ranging from -2 for strictly decreasing monotonic to +2 for strictly increasing monotonic. To test for convexity of utility functions, a generalized exponential function was fitted to the utility values in each attribute [7] and a dummy variable for convex utility functions was set to one when the parameter of this function indicated a convex shape. The weights were directly entered into the regression. An additional dummy variable was used to indicate the role of buyer in the negotiation.

To test the hypotheses concerning initial offers, the first offers of all negotiators in each attribute were scaled between zero and one according to the direction of improvement specified in the case description. These transformed values were then regressed on the explanatory variables. Table 1 shows the results of these regressions.

Most of the hypotheses are confirmed by this analysis. Negotiators who have a convex marginal utility function make significantly tougher initial offers in the attributes price and returns. The monotonicity effect is significant in all attributes. Sellers who have increasing monotonic marginal utility functions in the attributes price, delivery and returns (which is correct for their role) make higher initial demands in those attributes. For the attribute payment, this effect is reversed, since here sellers should have decreasing utility. Higher weights also lead to higher demands in the initial round.

Table 2 shows the results for concessions, which were measured as the differences between a negotiator's initial offer and the compromise value, both rescaled to the (0,1)-interval. Again, most of the hypotheses are confirmed, the only exception is the lack of a significant effect of convexity on concessions in the attribute payment. There is also a significant substitution effect between initial offers and concessions: negotiators who demand more in their initial

Table 1. Initial offers

Property		Price	Delivery	Payment	Returns
Convexity	β	0.0150	0.0116	0.0135	0.0279
	t	2.1200	1.1200	1.3500	3.2100
	p	0.0338	0.2607	0.1781	0.0014
Monotonicity	β	0.0865	0.1365	-0.1335	0.1368
	t	21.3800	29.0300	-28.4100	27.1000
	p	< .0001	< .0001	< .0001	< .0001
Weight	β	0.1071	0.2140	0.4398	0.3806
	t	4.3100	4.4000	7.7400	9.1800
	p	< .0001	< .0001	< .0001	< .0001
Role	β	0.0425	0.1739	-0.0473	0.0276
	t	6.4200	16.4200	-4.5700	3.1600
	p	< .0001	< .0001	< .0001	0.0016
	N	2896	2896	2896	2896
	R^2	0.1651	0.3829	0.2547	0.2463

Table 2. Concessions

Property		Price	Delivery	Payment	Returns
Convexity	β	-0.0536	-0.0279	-0.0049	-0.0820
	t	-6.2000	-2.5900	-0.3400	-5.7600
	p	< .0001	0.0097	0.7321	< .0001
Monotonicity	β	-0.0108	-0.0373	0.0515	-0.0521
	t	-2.0400	-6.6800	8.4200	-6.3800
	p	0.0419	< .0001	< .0001	< .0001
Weight	β	-0.5657	-0.8978	-1.0984	-1.1766
	t	-18.5600	-17.6200	-16.8000	-19.5500
	p	< .0001	< .0001	< .0001	< .0001
Role	β	0.0310	-0.3445	0.1707	-0.1242
	t	3.8000	-29.9000	14.3300	-9.7900
	p	0.0001	< .0001	< .0001	< .0001
Initial Offer	β	0.7556	0.8571	0.8660	0.8716
	t	33.2500	44.1300	40.5200	32.5700
	p	< .0001	< .0001	< .0001	< .0001
	N	2896	2896	2896	2896
	R^2	0.3561	0.5002	0.4202	0.3354

offers are then willing to make larger concessions. However, the coefficients of initial offers on concessions are significantly smaller than one for all attributes. This indicates that although most of the initially higher demands are later on given up again during the negotiations, negotiators were still able to retain a part of their additional initial demands.

Table 3. Outcomes

Property	Role		Price	Delivery	Payment	Returns
Weight	Buyer	β	-0.5068	-35.9192	69.6970	-13.2903
		t	-14.4100	-13.9700	13.5200	-18.4800
		p	< .0001	< .0001	< .0001	< .0001
	Seller	β	0.3782	26.1269	-52.1411	0.4249
		t	10.7400	9.1300	-10.6800	2.0500
		p	< .0001	< .0001	< .0001	0.0409
Convexity	Buyer	β	-0.0565	-1.5374	-2.8419	-0.8879
		t	-5.6000	-3.1100	-2.5400	-5.5500
		p	< .0001	0.0019	0.0112	< .0001
	Seller	β	0.0406	-0.0998	-4.8174	0.4249
		t	4.0500	-0.1400	-4.4100	2.0500
		p	< .0001	0.8905	< .0001	0.0409
Monotonicity	Buyer	β	0.0196	0.9466	3.3827	0.5942
		t	2.9500	2.6600	8.9300	5.3200
		p	0.0033	0.0080	< .0001	< .0001
	Seller	β	0.0256	2.3610	4.2119	0.6550
		t	5.0200	10.7400	8.9000	8.2700
		p	< .0001	< .0001	< .0001	< .0001
		N	1448	1448	1448	1448
		R^2	0.2467	0.2385	0.2708	0.3326

Table 3 shows the impact of structural properties of the utility functions on negotiation outcomes measured in terms of attribute values. Since outcomes depend on both sides of the negotiation, parameters describing the utility functions of both parties were used as explanatory variables. Again, most of the hypotheses are confirmed: when buyers put a higher weight on an attribute, they obtain a lower price, shorter delivery times, longer payment terms or better conditions for returns. For sellers, we observe the opposite effect. When negotiation parties have monotonically increasing marginal utility functions, the compromise value in that attribute increases, too. In most cases, negotiators with a convex utility function achieve better outcomes for their role. The only two results which contradict our initial hypotheses concern the lack of impact of convexity of the sellers' utility functions on delivery time, and the unexpected sign of the parameter for convexity of the buyers' utility functions on payment terms.

4 Conclusions

Most of the hypotheses we have tested seem to be rather straightforward: when negotiators consider an attribute to be more important, it is quite natural that they negotiate tougher regarding this attribute and obtain better outcomes than negotiators who do not really care about an attribute.

The important conclusion of this paper is that the multi-attribute utility functions elicited in Inspire really reflect these attitudes of negotiators. Even in cases where the utility functions contradict the case descriptions, e.g. when a buyer has a monotonically increasing utility function for price and thus assigns higher utility to a higher price, the actual behavior reflects the utility function, and not the more natural case description. Thus we can conclude that multi-attribute utility functions as used in Inspire can really capture the preferences of negotiators, even when they seem to be quite odd.

This is an important and positive message concerning the possibility to use analytical tools for negotiation support. When the preferences which really guide a negotiator's behavior can be captured with sufficient precision in a utility function, such a utility function can also be used to provide substantial and useful advise to a negotiator.

References

1. Hämäläinen R (2003) Decisionarium - aiding decisions, negotiating and collecting opinions on the web. Journal of Multi-Criteria Decision Analysis 12:101–110
2. Keeney R, Raiffa H (1976) Decisions with Multiple Objectives: Preferences and Value Tradeoffs. J. Wiley & Sons, New York
3. Kersten G, Noronha S (1999) WWW-based negotiation support: Design, implementation. Decision Support Systems 25:135–154
4. Mumpower J (1991) The judgement policies of negotiators and the structure of negotiation problems. Management Science 37:1304–1324
5. Northcraft G, Brodt S, Neale M. (1995) Negotiating with nonlinear subjective utilities: Why some concessions are more equal than others. Organizational Behavior and Human Decision Processes 63:298–310
6. Northcraft G, Preston J, Neale M, Kim P, Thomas-Hunt M (1998) Non-linear preference functions and negotiated outcomes. Organizational Behavior and Human Decision Processes 73:54–75
7. Pennings J, Smidts A (2003) The shape of utility functions and organizational behavior. Management Science 49:1251–1263
8. Rangaswamy A, Shell G (1997) Using computers to realize joint gains in negotiations: Toward an "Electronic bargaining table". Management Science, 43:1147–1163
9. Stuhlmacher A, Stevenson M (1997) Using policy modeling to describe the negotiation exchange. Group Decision and Negotiation 6:317–337

Part XXII

Applied Probability

Stochastic Analysis of the Traffic Confluence at the Crossing of a Major and a Minor Road

Frank Recker[1]

University of Hagen, Department of Mathematics,
Lützowstraße 125, D-58084 Hagen frank.recker@fernuni-hagen.de

Summary. We present a stochastic model for the traffic situation at a crossing of a major and a minor road. Afterwards, we give an overview of two recent results for this model. First, the distribution of the waiting time for a large gap is given. This might also have applications in other settings, where one has to wait for a large gap to occur in a Poisson process. The second result shows, that the queue length on the minor road is either a V-ergodic Markov chain or it is transient. This justifies the estimation of the stationary distribution by the time mean of the queue length in a computer simulation.
Key Words: Renewal process; V-ergodic Markov chain; Poisson process; Stopping time; Queuing theory; Traffic problems.
AMS Subject classification: 60G40, 60K30, 90B20.

1 Introduction

Assume, that there is a crossing of a major and a minor road. On both roads, cars arrive according to a Poisson process. Let the rate of the Poisson processes be $\lambda > 0$ on the major road and $\mu > 0$ on the minor road. The cars on the minor road can drive, whenever two conditions are fulfilled:

1. The next car on the major is far enough away (say at least time $\tau_1 > 0$).
2. The previous car on the minor road left some time ago (say at least time $\tau_2 > 0$ with $\tau_2 < \tau_1$).

Let X_t be the number of cars on the minor road at time $t \in \mathbb{R}_+$. Then $(X_t)_{t\in\mathbb{R}_+}$ is a stochastic process with state space $\mathbb{N}_0$ and we can analyze the stochastic properties of it. Figure 1 shows a typical situation. On the minor road, some cars a waiting at the crossing. New cars a driving towards the crossing on both roads with exponentially distributed distance.

In this article we will present two recent results from us for this confluence model. First, we will analyze the distribution of the service time (the time, that a car has to wait at the first position of the minor road). Our result

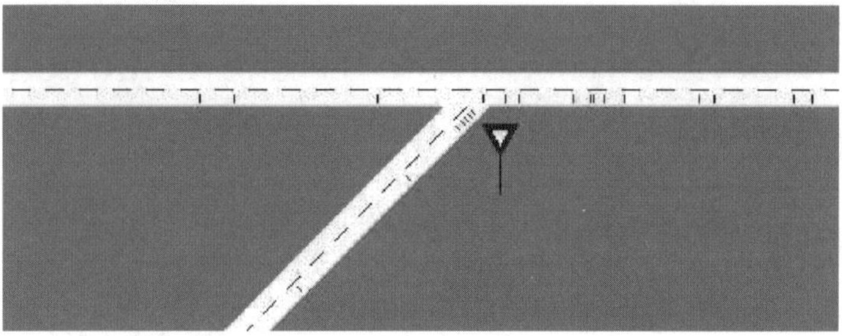

Fig. 1. A snapshot of the crossing

is a step towards the analysis of the model as a queuing system. There are however unsolved technical problems, which we will present.

Second, we will define an embedded Markov chain which is V-ergodic. Hence, the distribution of the queue length at these time points converges V-uniformly towards the stationary distribution and this stationary distribution can be estimated quite easily.

The stochastic properties of gaps in the traffic flow where already analyzed by Cohen. In [1, 2] Cohen analyzed the question, how long a pedestrian has to wait for a gap that allows crossing the street. But since the pedestrians are not queuing up, the results are of course different. Moreover, Cohen uses a different model for the traffic on the road. In [3], the vehicles on the minor road can theoretically all leave the minor road at once. Each vehicle has an acceptance probability for the gap. Contrary, in our model the number of leaving cars depends on the size of the gap.

Our first result gives a recursive formula for the law of the random sum of exponentially distributed random variables. Such sums should always occur, when one has given a Poisson process and one has to wait for a large gap.

The second result solves practical problems for traffic simulations on a computer. We have an exact criterion, whether we should expect a traffic collapse, that is, the queue length goes beyond any bound almost surly. Moreover, the theorem gives the justification for estimating the stationary distribution of the queue length by the time-mean, which can be implemented on a computer. The stationary distribution can be used to answer questions of the form: How likely is it, that there queue up more then n cars (which might be a problem, if the end of the queue reaches another crossing in the road network).

The organization of the paper is as follows: In Section 2 we will define the underlying stochastic process and we will show the result on the waiting time for a gap. In Section 3, we will present the ergodic theorem and in Section 4 we will estimate the stationary distribution for two parameter sets.

2 The Distribution of the Service Time

The above defined stochastic process $(X_t)_{t\in\mathbb{R}_+}$ is a $M/G/1$-queue, that is, new cars enter the queue according to a Poisson process (the M) and the service time has a general law G. There exists a lot of theory about such queues, e.g. [4]. It is however necessary, to know something about G. A step in this direction is the analysis of the distribution of the time between two crossings of cars on the major road.

Assume, that a car on the major road just passed the crossing. The distance between two consecutive cars is Exp(λ)-distributed. We have to sum these distances, until the first distance is greater or equal to a constant $\tau > 0$.

Definition 1. *Let be $\lambda, \tau \in \mathbb{R}$, $\lambda, \tau > 0$, and let $(Y_n)_{n\in\mathbb{N}}$ be a sequence of i.i.d.* Exp(λ)*-distributed random variables. Further let T be the stopping time*

$$T := \min\{n \in \mathbb{N} \,|\, Y_n \geq \tau\}.$$

Then the **waiting time for intensity λ and minimal gap τ** *is the random variable*

$$Z := \sum_{n=1}^{T-1} Y_n.$$

In [5] we proved a closed form for the distribution of Z. For this we write F_Z for the distribution function of Z, that is $F_Z(x) = P(\{Z \leq x\})$ for all $x \in \mathbb{R}$. Then we have the following theorems:

Theorem 1. *Let Z be the waiting time for the parameters $\lambda, \tau > 0$. Then $F_Z(x) = 0$ for all $x < 0$. Further for each $n \in \mathbb{N}$ the restriction of F_Z on the interval $[(n-1)\tau, n\tau]$ is a polynomial of degree n. Finally F_Z has only one jump point at 0 (and hence F_Z is continuous in all other points of $\mathbb{R}$).*

Theorem 2. *Let λ, τ, Z, and F_Z be as in Theorem 1. For each $n \in \mathbb{N}$ define the polynomial p_n as*

$$p_n(x) = \sum_{k=0}^{n} a_{n,k} x^k,$$

where the coefficients are defined as follows:

$$a_{1,0} = e^{-\lambda\tau},$$

$$a_{1,1} = \lambda e^{-\lambda\tau},$$

and for each $n \geq 2$:

$$a_{n,0} = \sum_{k=0}^{n-2} \left((-1)^k \frac{1}{k!} (\lambda\tau)^k e^{-k\lambda\tau} a_{n-1-k,0} \right) - \sum_{k=1}^{n-1} \left((-1)^k \frac{1}{k!} (\lambda\tau)^k e^{-k\lambda\tau} \right),$$

$$a_{n,k} = (-1)^{k+1} \frac{1}{k!} \lambda^k e^{-k\lambda\tau} (1 - a_{n-k,0})$$

for each $k = 1, \ldots, n-1$*, and*

$$a_{n,n} = (-1)^{n+1} \frac{1}{n!} \lambda^n e^{-n\lambda\tau}.$$

Then for each $n \in \mathbb{N}$ *and each* $x \in [(n-1)\tau, n\tau]$*:*

$$F_Z(x) = p_n(x - (n-1)\tau).$$

For each $n \geq 2$ and each $k = 0, \ldots, n$, the definition of $a_{n,k}$ uses only $a_{n',k'}$ with $n' < n$. Hence the definition is indeed recursive.

The distribution of Z is a step towards determining the distribution of G but there is a subtle difficulty: A car can reach the first position in the queue in two ways. The preceding car has just left the queue or the car enters an empty queue. The service time is different in both cases. The distribution of the service time in the first case can be computed with the distribution of Z. The second case, however, is far more complicated and up to now, we do not have an explicit result. Thus, we cannot analyse the queue with the standard techniques. Instead, we will estimate the stationary distribution. The justification for the used estimator is given in the following section.

3 The distribution of the queue length

We define the stochastic processes in a semi-formal way. An exact mathematical definition can be found in [6]. As above, we assume, that $(X_t)_{t \in \mathbb{R}_+}$ is a Poisson process with rate $\lambda > 0$. The jump points of X_t are the time points, when a new car on the main road passes the crossing. Let $(Y_t)_{t \in \mathbb{R}_+}$ be a Poisson process with rate $\mu > 0$. The jump points are the time points, when a new car on the minor road enters the queue.

Assume, that at time 0 there are $x \in \mathbb{N}_0$ cars in the queue on the minor road and that a car on the major road just passed the crossing. The cars on the minor road can drive according to the rules in Section 1. Let $(U_{x,t})_{t \in \mathbb{R}_+}$ be the number of cars, that left the minor road. Then $Q_{x,t} := Y_t - U_{x,t}$ are the number of cars at time t.

We analyse the (continuous time) stochastic process with the usual technique, that is, we define an embedded Markov chain. For this we define the random time point T_i as the i-th jump point of X_t for all $i \in \mathbb{N}$ and $T_0 := 0$. The distribution of $Y_{T_{i+1}} - Y_{T_i}$ (that is the number of occurring cars on the minor road in the time interval from T_i to T_{i+1}) is folklore. It is just a geometrical distribution with Parameter $\frac{\mu}{\lambda+\mu}$. However, we do not have an expression for the distribution of $U_{T_{i+1}} - U_{T_i}$. Hence, we have to estimate the stationary distribution. The following theorem states, that such an estimation is indeed possible. We define

$$(\Phi_{x,n})_{n\in\mathbb{N}_0} = (Q_{x,T_n})_{n\in\mathbb{N}_0}.$$

From the strong Markov property of the Poisson processes follows, that $(\Phi_{x,n})_{n\in\mathbb{N}_0}$ is indeed a Markov chain. As is proved in [6] we have the following theorem (for a definition of V-ergodic, we refer to [7]):

Theorem 3. *Let* $(\Phi_{x,n})_{n\in\mathbb{N}_0}$ *be as above for the parameters* $\lambda, \mu, \tau_1, \tau_2 > 0$, $x \in \mathbb{N}_0$. *Define*

$$\mu^* = \lambda \frac{e^{-\lambda\tau_1}}{1 - e^{-\lambda\tau_2}}.$$

If $\mu < \mu^*$, *then there exist an* $s > 0$ *such that the Markov chain* $(\Phi_{x,n})_{n\in\mathbb{N}_0}$ *is V-ergodic with* $V(x) = e^{sx}$.

If $\mu > \mu^*$, *then* $(\Phi_{x,n})_{n\in\mathbb{N}_0}$ *is transient.*

In the case of a V-uniform ergodic Markov chain, the Markov chain is especially positive recurrent and therefore the queue vanishes from time to time. Furthermore, we can estimate the stationary distribution (c.f. [8] for recent results on the convergence rate). In contrast, a transient Markov chain on $\mathbb{N}_0$ has the property, that it converges almost surely towards infinity. We hence get a traffic collapse.

4 Estimating the Distribution of the Queue-Length

As an application of Theorem 3, we have estimated the stationary distribution for two parameter sets. We define $\tau_1 = 3$ (the gap has to be at least 3 seconds), $\tau_2 = 1.1$ (every 1.1 seconds, a new car can drive), and $\lambda = 0.2$ (the average time between two cars on the main road is 5 seconds).

In Figure 2, we chose $\mu = 0.2$. Looking at the queue at a random time point, we can be quite confident, that there are at most 10 cars in the queue.

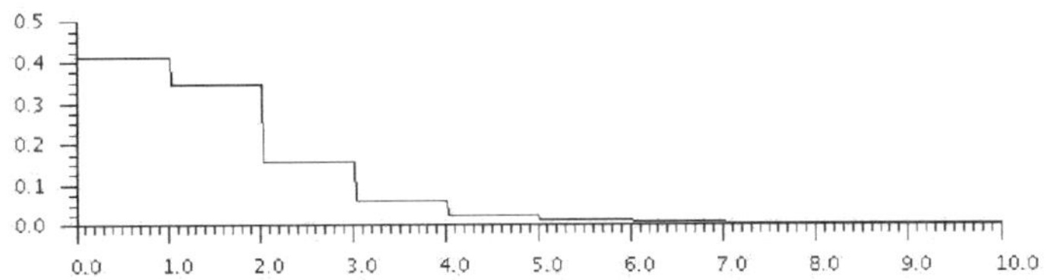

Fig. 2. The stationary distribution for $\mu = 0.2$

In Figure 3, we chose $\mu = 0.5$. We see, that ergodic distribution has a heavier tail. The queue can be quite large.

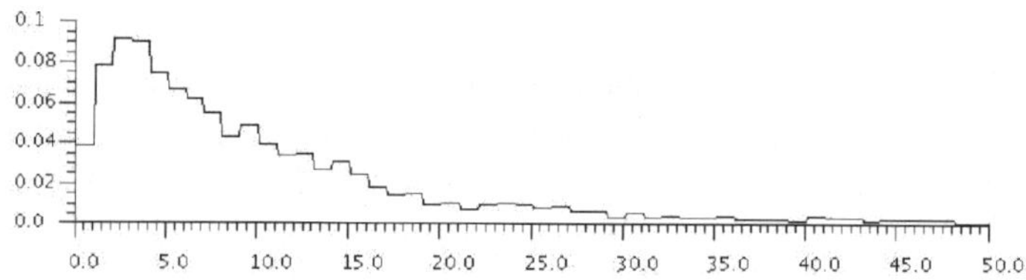

Fig. 3. The stationary distribution for $\mu = 0.5$

References

1. Cohen J W (1963) A note on the delay problem in crossing a traffic stream; Statistica Neerlandica 17, No. 1: 3–11
2. Cohen J W (1965) Some Comments on: "Another note on the delay problem in crossing a traffic stream"; Statistica Neerlandica 19, No. 1: 1–2
3. Cohen J W, Lange S J de (1967) Numerical Results for Queueing for Gaps in a Traffic Stream; In: Edie L C, Herman R, Rothery R (eds) Vehicular traffic science. Proceedings of the third international symposium on the theory of traffic flow, New York, June 1965. American Elsevier Publishing Company: 306–307
4. Asmussen S (1987) Applied Probability and Queues. Wiley, New York
5. Recker F (2005) On the distribution of the Waiting Time in a Confluence Process; submitted manuscript
6. Recker F (2005) On the asymptotical queue length in vehicular traffic confluence; submitted manuscript
7. Meyn S P, Tweedie R L (1993) Markov Chains and Stochastic Stability, 2nd. Ed. Springer, London
8. Baxendale P H (2005) Renewal theory and computable convergence rates for geometrically ergodic Markov chains; Ann. Appl. Probab. 15, No.1B: 700–738

Decomposition in Multistage Stochastic Programs with Individual Probability Constraints

Vlasta Kaňková

Institute of Information Theory and Automation
Academy of Sciences of the Czech Republic
Pod Vodárenskou věží 4, 182 08 Praha 8, Czech Republic
kankova@utia.cas.cz

1 Introduction

Multistage stochastic programming problems correspond to applications that are reasonable to consider with respect to some finite "discrete" time interval and that can be (simultaneously) decomposed into individual time points. Multistage stochastic programming problems depend on a random factor and the control of the corresponding activity is considered with respect to the mathematical expectation of the objective function. In the case of "classical" multistage stochastic programming problems the constraints sets are given by a system of algebraic inequalities. However, the constraints set depending on an "underlying" system of the probability measures appears (in the stochastic programming literature) also (see e.g. [9]). The aim of the contribution is to analyze the case of the constraints sets given by individual probability constraints and of the random element following a (generally) nonlinear autoregressive sequence. Namely, we can see that under these assumptions the multistage problems (with constraints depending on the probability measure) have a pleasant properties and, moreover, they are (from the mathematical point of view) very "near" to the "classical" case.

To introduce M-stage ($M \geq 2$), generally nonlinear, stochastic programming problem let $\xi^j (:= \xi^j(\omega))$, $j = 1, \ldots, M$ denote an s-dimensional random vector defined on a probability space (Ω, S, P); $F^{\xi^j} (:= F^{\xi^j}(z^j))$, $z^j \in R^s$, $j = 1, 2 \ldots, M$ denote the distribution function of the ξ^j and $F^{\xi^k | \bar{\xi}^{k-1}} (:= F^{\xi^k|\bar{\xi}^{k-1}}(z^k|\bar{z}^{k-1}))$, $z^k \in R^s$, $\bar{z}^{k-1} \in R^{(k-1)s}$, $k = 2, \ldots, M$ denote the conditional distribution function (ξ^k conditioned by $\bar{\xi}^{k-1}$); $Z_F^j \subset R^s$, $j = 1, 2, \ldots, M$ denote the support of the probability measure corresponding to F^{ξ^j}.

Let, moreover, $g_0^M(\bar{x}^M, \bar{z}^M)$ be a continuous function defined on $R^{nM} \times R^{sM}$; $\mathcal{X}, X^k \subset R^n$, $k = 1, \ldots, M$ be nonempty sets; $\mathcal{K}_{\mathcal{F}}^{k+1}(\bar{x}^k, \bar{z}^k)(:=$

$\mathcal{K}^{k+1}_{F^{\xi^{k+1}|\bar{\xi}^k}}(\bar{x}^k, \bar{z}^k))$, $k = 1, \ldots, M-1$ be a multifunction mapping $R^{nk} \times R^{sk}$ into the space of subsets of R^n. $\bar{\xi}^k (= \bar{\xi}^k(\omega)) = [\xi^1, \ldots, \xi^k]$; $\bar{z}^k = [z^1, \ldots, z^k]$, $z^j \in R^s$; $\bar{x}^k = [x^1, \ldots, x^k]$, $x^j \in R^n$; $\bar{X}^k = X^1 \times \ldots \times X^k$, $\bar{Z}^k_F = Z^1_F \times Z^2_F \ldots \times Z^k_F$, $j = 1, \ldots, k$, $k = 1, 2, \ldots, M$. (R^n, $n \geq 1$ denotes an n-dimensional Euclidean space.)

A general M-stage stochastic programming problem ($M \geq 2$) can be introduced as an optimization problem considered with respect to some general abstract (say $\mathcal{L}_p$) mathematical space (see e.g. [1]).

Find
$$\inf_{\bar{x}^M = (\bar{x}^M(\bar{\xi}^M))} E_F g_0^M(\bar{x}^M, \bar{\xi}^M) \tag{1}$$

where
$$\begin{aligned} &\bar{x}^M(\bar{\xi}^M) = (x^1, x^2(\bar{\xi}^1), \ldots, x^M(\bar{\xi}^{M-1})), \quad x^1 \in \mathcal{K}^1, \\ &x^k(\bar{\xi}^{k-1}) \in \mathcal{K}^k_{\mathcal{F}}(\bar{x}^{k-1}(\bar{\xi}^{k-2}), \bar{\xi}^{k-1}) \quad \text{a.s.}, \\ &x^k(\bar{\xi}^{k-1}) \in \mathcal{L}_p, \quad k = 2, \ldots, M, \quad \mathcal{K}^1 = X^1. \end{aligned} \tag{2}$$

Evidently, the optimization problems defined by the relations (1), (2) are complicated and mostly suitable only for a theoretical investigation. However, in the literature, there exists also another (perhaps for practical treatment more suitable) definition (see e.g. [1] or [5]). We can recall it as the problem:

Find
$$\varphi_{\mathcal{F}}(M) = \inf \{\mathsf{E}_{F^{\xi^1}} g^1_{\mathcal{F}}(x^1, \xi^1) |\ x^1 \in \mathcal{K}^1\}, \tag{3}$$

where the function $g^1_{\mathcal{F}}(x^1, z^1)$ is defined recursively

$$\begin{aligned} &g^k_{\mathcal{F}}(\bar{x}^k, \bar{z}^k) = \inf\{\mathsf{E}_{F^{\xi^{k+1}|\bar{\xi}^k = \bar{z}^k}}\ g^{k+1}_{\mathcal{F}}(\bar{x}^{k+1}, \bar{\xi}^{k+1}) \,|x^{k+1} \in \mathcal{K}^{k+1}_{\mathcal{F}}(\bar{x}^k, \bar{z}^k)\}, \\ &k = 1, \ldots, M-1, \\ &g^M_{\mathcal{F}}(\bar{x}^M, \bar{z}^M) := g^M_0(\bar{x}^M, \bar{z}^M), \quad \mathcal{K}^1 = X^1. \end{aligned} \tag{4}$$

$\mathsf{E}_{F^{\xi^1}}$, $\mathsf{E}_{F^{\xi^{k+1}|\bar{\xi}^k = \bar{z}^k}}$, $k = 1, \ldots, M-1$ denote the operators of mathematical expectation corresponding to F^{ξ^1}, $F^{\xi^{k+1}|\bar{\xi}^k}$.

Generally these two definitions are not equivalent. The assumptions under which the both problems are solvable and, moreover, the optimal values coincide are introduced in [7]. The crucial assumptions for this are the following:

1. $\{\xi^k\}_{k=-\infty}^{\infty}$ follows (generally) a nonlinear autoregressive sequence, $\mathcal{L}_p = \mathcal{L}_\infty$,
2. $\mathcal{K}^{k+1}_{\mathcal{F}}(\bar{x}^k, \bar{z}^k) = \mathcal{K}^{k+1}(\bar{x}^k, \bar{z}^k)$, $k = 1, \ldots, M-1$ is given by the system of algebraic inequalities independently on the probability measure.

In this contribution we assume that the random element follows an autoregressive random sequence too, however moreover, we admit the dependence of the constraints set $\mathcal{K}^{k+1}_{\mathcal{F}}(\bar{x}^k, \bar{z}^k)$ on the probability measure. We assume:

A.1 $\{\xi^k\}_{k=-\infty}^{\infty}$ follows (generally) nonlinear autoregressive sequence

$$\xi^k = H(\xi^{k-1}) + \varepsilon^k, \; k = \ldots, -1, 0, 1, \ldots, \tag{5}$$

where $\xi^1, \varepsilon^k, k = 2, \ldots$ are stochastically independent and, moreover, $\varepsilon^k, k = 1, 2, \ldots$ are identically distributed; $H(z)$ is a continuous function defined on R^s. (We denote the distribution function and the support corresponding to $\varepsilon^k, k = 1, \ldots$ by the symbols F^{ε}, $Z_{F^{\varepsilon}}$.)

A.2 there exist functions $f_i^{k+1}(\bar{x}^{k+1}), \; i = 1, \ldots, s, \quad k = 1, \ldots, M-1$ defined on $R^{n(k+1)}$ such that

$$\mathcal{K}_{\mathcal{F}}^{k+1}(\bar{x}^k, \bar{z}^k) = \bigcap_{i=1}^{s} \{x^{k+1} \in X^{k+1} : P_{F^{\xi^{k+1}|\xi^k}}\{f_i^{k+1}(\bar{x}^{k+1}) \le \xi_i^{k+1}\} \ge \alpha_i\},$$
$$\alpha_i \in \langle 0, 1\rangle, i = 1, 2 \ldots, s, \quad \xi^{k+1} = (\xi_1^{k+1}, \ldots, \xi_s^{k+1}). \tag{6}$$

2 Problem Analysis

Evidently, generally, a complete information on the system

$$\mathcal{F} = \{F^{\xi^1}(z^1), \quad F^{\xi^k|\bar{\xi}^{k-1}}(z^k|\bar{z}^{k-1}), \; k = 2, \ldots, M\} \tag{7}$$

is a necessary condition to solve the both above defined types of the problems. However in applications very often one of the following cases happen:

- the system (7) must be replaced by its statistical estimates,
- the system (7) must be (for numerical difficulties) replaced by simpler one,
- the actual system (7) is a little modified (e.g. by a contamination).

Consequently, to apply responsibly the multistage problems, the stability (w.r.t. the system (7)) and statistical estimates must be investigated. Employing the relations (3), (4) we can see that results achieved for one-stage problems can be employed. The approach of a decomposition has been already employed e.g. in [5], [6] when the Markov dependence has been considered or in [8] when, moreover, probability measure dependence on a decision is admitted. The "deterministic" constraints set has been (mostly) assumed there.

Evidently, under A.1 and A.2, the system $\mathcal{F}$ is determined by the distribution functions F^{ξ^1}, F^{ε}. Consequently, if we define for $\alpha_i \in (0, 1), \; i = 1, \ldots, s, \quad k = 2, \ldots, M,$

$$k_{F_i^{\xi^k|\bar{\xi}^{k-1}=\bar{z}^{k-1}}}(\alpha_i) = \sup_{z_i^k}\{z_i^k : P_{F^{\xi^k|\bar{\xi}^{k-1}}}\{z_i^k \le \xi_i^k\} \ge \alpha_i\}, \quad z^k = (z_1^k, \ldots, z_s^k),$$
$$k_{F_i^{\varepsilon}}(\alpha_i) = \sup_{z_i^k}\{z_i^k : P_{F^{\varepsilon}}\{z_i^k \le \varepsilon_i^k\} \ge \alpha_i\}, \quad \varepsilon^k = (\varepsilon_1^k, \ldots, \varepsilon_s^k),$$

then $k_{F_i^{\varepsilon}}(\alpha_i) = k_{F_i^{\xi^k|\bar{\xi}^{k-1}=\bar{z}^{k-1}}}(\alpha_i) - H_i(z^{k-1})$ and, moreover, for $k = 1, 2, \ldots, M-1,$

$$\mathcal{K}_{\mathcal{F}}^{k+1}(\bar{x}^k, \bar{z}^k) = \bigcap_{i=1}^{s} \{x^{k+1} \in X^{k+1} : f_i^{k+1}(\bar{x}^{k+1}) \le k_{F_i^{\varepsilon}}(\alpha_i) + H_i(z^k)\}. \quad (8)$$

According to (3), (4) the problem can be decomposed (for details see e.g. [5]) into M one-stage parametric optimization problems in which constraints sets depend on the former random element realizations and the former decisions.

Of course if F^{ε} is replaced by another G^{ε}, then we obtain another system

$$\mathcal{G} = \{G^{\xi^1}(z^1), G^{\xi^k|\bar{\xi}^{k-1}}(z^k|\bar{z}^{k-1}),\ k = 2, \dots, M\} \quad (9)$$

and a new multistage stochastic programming problem. Let the symbol $\varphi_{\mathcal{G}}(M)$ denote its optimal value. It follows from the analysis presented e.g. in [5] that

$$|\varphi_{\mathcal{F}}(M) - \varphi_{\mathcal{G}}(M)| \quad (10)$$

depends on the relationships between $g_{\mathcal{F}}^k(\bar{x}^k, \bar{z}^k)$, $g_{\mathcal{G}}^k(\bar{x}^k, \bar{z}^k)$, $k = 1, \dots, M$ and $\mathcal{K}_{\mathcal{F}}^{k+1}(\bar{x}^k, \bar{z}^k)$, $\mathcal{K}_{\mathcal{G}}^{k+1}(\bar{x}^k, \bar{z}^k)$, $k = 1, \dots, M-1$. Furthermore, it follows from the relation (8) that if the system $\mathcal{F}$ is known, then the multifunctions $\mathcal{K}_{\mathcal{F}}^{k+1}$, $k = 1, \dots M-1$ are given exactly; consequently to construct deterministic approximate schemes, the technique of [6], [11] can be employed. However, in the general case the relationship between $\mathcal{K}_{\mathcal{F}}^{k+1}(\bar{x}^k, \bar{z}^k)$, $\mathcal{K}_{\mathcal{G}}^{k+1}(\bar{x}^k, \bar{z}^k)$, $k = 1, \dots, M-1$ must be also included in the evaluation of (10).

3 Main Results

First, we recall definitions of the Kolmogorov d_K^s and the Wasserstein metrics $d_{W_1}^s$. To this end, let F^{ξ}, G^{ξ} be s-dimensional distribution functions and $P_{F^{\xi}}$, $P_{G^{\xi}}$ the corresponding probability measures.

$$\begin{aligned}
&d_K^s(F^{\xi}, G^{\xi}) = d_K^s(P_{F^{\xi}}, P_{G^{\xi}}) = \sup_{z \in R^s} |F^{\xi}(z) - G^{\xi}(z)|,\\
&d_{W_1}^s(F^{\xi}, G^{\xi}) = d_W^s(P_{F^{\xi}}, P_{G^{\xi}}) =\\
&\inf\{ \int_{R^s \times R^s} \|z - \bar{z}\| \kappa(dz \times d\bar{z}) : \kappa \in \mathcal{D}(P_{F^{\xi}}, P_{G^{\xi}})\},\ P_{F^{\xi}}, P_{G^{\xi}} \in \mathcal{M}_1(R^s)\},\\
&\mathcal{M}_1(R^s) = \{\nu \in \mathcal{P}(R^s) : \int_{R^s} \|z\| \nu(dz) < \infty\},
\end{aligned} \quad (11)$$

where $\mathcal{D}(P_{F^{\xi}}, P_{G^{\xi}})$ is the set of those measures in $\mathcal{P}(R^s \times R^s)$ whose marginal distributions are $P_{F^{\xi}}$ and $P_{G^{\xi}}$ and $\mathcal{P}(R^s)$ denotes the set of all (Borel) probability measures on R^s. The symbol $\|\cdot\|$ denotes a suitable norm in R^s.

Taking the technique employed for one-stage stochastic programming problems (for details see e.g. [2]) it can be shown that, under some additional rather general assumptions, for given $\alpha_i \in (0, 1)$, $i = 1, \dots, s$ there can be found constants $\bar{K}$, $\delta > 0$ such that

$$\Delta[\mathcal{K}_{\mathcal{F}}^{k+1}(\bar{x}^k, \bar{z}^k), \mathcal{K}_{\mathcal{G}}^{k+1}(\bar{x}^k, \bar{z}^k),] \le \bar{K} \max_{i\in\{1,\dots,s\}} d_K^1[P_{F_i^\varepsilon}, P_{G_i^\varepsilon}],$$
$$k = 1, \dots, M-1 \quad \text{whenever when} \quad \max_{i\in\{1,\dots,s\}} d_K^1[P_{F_i^\varepsilon}, P_{G_i^\varepsilon}] < \delta. \tag{12}$$

($F_i^\varepsilon, G_i^\varepsilon$, $i = 1, 2, \dots, s$ denote one-dimensional marginal distribution functions corresponding to F^ε and G^ε, Δ denotes the Hausdorff distance in the space of closed subsets of R^n; for definition see e.g. [10].)

Furthermore, under A.1 with the Lipschitz function H on R^s, if

1. P_{F^ε} is absolutely continuous w.r.t. the Lebesgue measure on R^s,
2. for every $\bar{x}^k \in \bar{X}^k$, $\bar{z}^{k-2} \in R^{s(k-2)}$ the function $g_{\mathcal{F}}^k(\bar{x}^k, \bar{z}^k)$ is Lipschitz on $Z_{F^{\xi^{k-1}}} \times Z_{F^{\xi^k}}$,
3. there exists a finite $\mathsf{E}_{F^{\xi^k|\bar{\xi}^{k-1}}} g_{\mathcal{F}}^k(\bar{x}^k, \bar{\xi}^k)$,

then $\mathsf{E}_{F^{\xi^k|\bar{\xi}^{k-1}}} g_{\mathcal{F}}^k(\bar{x}^k, \bar{\xi}^k)$ is for every $\bar{x}^k \in \bar{X}^k$, $\bar{z}^{k-2} \in Z_{F^{\bar{\xi}^{k-2}}}$ a Lipschitz function on $Z_{F^{\xi^{k-1}}}$, $k = 1, \dots, M$.

It follows from [3], [4] that if A.1 is not valid, then a stronger assumptions must be fulfilled to obtain the Lipschitz property of $\mathsf{E}_{F^{\xi^k|\bar{\xi}^{k-1}}}\, g_{\mathcal{F}}^k(\bar{x}^k, \bar{\xi}^k)$.

Theorem.

Let $\mathcal{K}^1$ be a nonempty, compact set, $\alpha_i \in (0, 1)$, $i = 1, \dots, s$, $P_{F^\varepsilon} \in \mathcal{M}_1(R^s)$. Let, moreover the assumptions A.1 and A.2 be fulfilled. If

1. $\mathcal{K}_{\mathcal{F}}^{k+1}(\bar{x}^k, \bar{z}^k) \subset \mathcal{X}$, $\bar{x}^k \in \bar{X}^k$, $\bar{z}^k \in \bar{Z}_F^k$, $k = 1, \dots, M-1$ and , moreover, $\mathcal{K}^1$, $\mathcal{K}_{\mathcal{F}}^{k+1}(\bar{x}^k, \bar{z}^k)$ are compact sets,
2. a. G^ε is an arbitrary s-dimensional distribution function, $P_{G^\varepsilon} \in \mathcal{M}_1(R^s)$,
   b. G^ε determines the system $\mathcal{G}$ by the relation (9),
3. a. $g_F^1(x^1, z^1)$ is a uniformly continuous function on $\mathcal{K}^1 \times R^s$ and, moreover, for every $x^1 \in \mathcal{K}^1$ a Lipschitz function on R^s with the Lipschitz constant not depending on $x^1 \in \mathcal{K}^1$,
   b. for every $k \in \{2, \dots, M\}$ and every $\bar{x}^{k-1} \in \bar{X}^{k-1}$, $\bar{z}^{k-1} \in \bar{Z}_F^{k-1}$, $g_F^k(\bar{x}^k, \bar{z}^k)$ is a a Lipschitz function on $X^k \times R^s$ with the Lipschitz constant not depending on $\bar{x}^{k-1} \in \bar{X}^{k-1}$, $\bar{z}^{k-1} \in \bar{Z}_F^{k-1}$,
4. the relation (12) is fulfilled,

then there exists a constant C_K, $C_{W_1}^i$, $i = 1, \dots, s$, $C_{W_1} \ge 0$ such that

a. $|\varphi_{\mathcal{F}}(M) - \varphi_{\mathcal{G}}(M)| \le C_{W_1} d_{W_1}^s(F^\varepsilon, G^\varepsilon) + C_K \max_{i\in\{1,\dots,s\}} d_K^1[P_{F_i^\varepsilon}, P_{G_i^\varepsilon}]$,

b. $|\varphi_{\mathcal{F}}(M) - \varphi_{\mathcal{G}}(M)| \le \sum_{i=1}^{s} C_{W_1}^i d_{W_1}^1(F_i^\varepsilon, G_i^\varepsilon) + C_K \max_{i\in\{1,\dots,s\}} d_K^1[P_{F_i^\varepsilon}, P_{G_i^\varepsilon}]$.

Sketch of the Proof. The proof of Theorem can be obtained (as a special case) by the technique employed in [5]. To achieve the result a. the $\mathcal{L}^2$ norm (classical Euclidean norm) has been employed in the definition of the Wasserstein metric. To obtain the result b. the $\mathcal{L}^1$ norm has been employed. The idea of employing the $\mathcal{L}^1$ norm appears in [11]. This approach is very suitable for a construction of approximative schemes.

4 Conclusion

Evidently the approach of this contribution can be employed for the construction of approximative solution schemes. Of course, to this end the detail analysis have to be done. Especially, great attention must be paid to a verification of the inequality (12); (the results of [2]) can be employed for it). The actual probability of the constrains fulfilling is reasonable also to investigate in the case when the F^{ε} is replaced by G^{ε}.

If an empirical distribution function F_N^{ε} replaces F^{ε}, then an empirical estimate of the value $\varphi_{\mathcal{F}}(M)$ can be obtained. The rate of the convergence is the same as under the Markov dependence assumption (for details see e.g. [4]). However, to to this end (under the assumption A.1) a rather weaker assumptions can be assumed. Of course, all above mentioned investigations are over the possibilities of this contribution.

Acknowledgement. The research was supported by the Grant Agency of the Czech Republic under Grants 402/04/1294 and 402/05/0115.

References

1. Dupačová J (1995) Multistage stochastic programs: the state-of-the-art and selected bibliography. Kybernetika 31:151–174
2. Kaňková V (1997) On the stability in stochastic programming: the case of individual probability constraints. Kybernetika 33:525–624
3. Kaňková V (1998) A note on multistage stochastic programming. In: Proceedings of 11th joint Czech–Germany–Slovak Conference: Mathematical Methods in Economy and Industry. University of Technology, Liberec (Czech Republic) 1998, pp 45–52
4. Kaňková V (2000) A note on multistage stochastic programming with individual probability constraints. In: Fleischmann B, Lasch R, Derigs U, Domschke W, Rieder U (eds) Operations Research Proceedings 2000. Springer, Berlin, pp 91–96
5. Kaňková V (2002) A remark on the analysis of multistage stochastic programs, Markov dependence. Z. angew. Math. Mech. 82:781–793
6. Kaňková V, Šmíd M (2004) On approximation in multistage stochastic programs: Markov dependence. Kybernetika 40:625–638
7. Kuhn D (2005) Generalized bound for convex multistage stochastic programs. Lectures Notes in Economics and Mathematical Systems. Springer, Berlin
8. Mänz A, Vogel S (2005) On stability of multistage stochastic decision problem. In: Seeger (ed.) Recent Advances in Optimization. Lecture Notes in Economics and Mathematical Systems, to appear
9. Prékopa A (1998) Stochastic programming. Akadémiai Kiadó and Kluwer Publisher
10. Salinetti G (1983) Approximations for chance constrained programming problems. Stochastics 10:57–179.
11. Šmíd M (2004) On approximation of stochastic programming problems. Doctoral Thesis, Charles University, Prague

Algorithmic Procedures for Mean Variance Optimality in Markov Decision Chains

Karel Sladký and Milan Sitař

Institute of Information Theory and Automation
Academy of Sciences of the Czech Republic
Pod Vodárenskou věží 4, 182 08 Praha 8, Czech Republic
sladky@utia.cas.cz, milan.sitar@tiscali.cz

1 Introduction

In this note we discuss some algorithmic procedures for finding optimal policies of Markov decision chains with respect to various mean variance optimality criteria. To this end, we present formulas for the growth rate and asymptotic behavior of the variance of total cumulative reward. Finally, algorithmic procedures of policy iteration type for finding efficient policies with respect to various mean variance optimality criteria along with computational experience are discussed.

We shall consider a Markov decision chain $X = \{X_n, n = 0, 1, \ldots\}$ with finite state space $\mathcal{I} = \{1, 2, \ldots, N\}$, finite set $\mathcal{A}_i = \{1, 2, \ldots, K_i\}$ of possible decisions (actions) in state $i \in \mathcal{I}$ and the following transition and reward structure (we assume that in state $i \in \mathcal{I}$ action $k \in \mathcal{A}_i$ is selected):

$$\begin{aligned} p_{ij}^k &: \text{ transition probability from } i \to j \ \ (i, j \in \mathcal{I}), \\ r_{ij} &: \text{ one-stage reward for a transition from } i \to j, \\ r_i^k &: \text{ expected value of the one-stage rewards incurred in state } i, \\ r_i^{(2),k} &: \text{ second moment of the one-stage rewards incurred in state } i. \end{aligned}$$

Obviously, $r_i^k = \sum_{j \in \mathcal{I}} p_{ij}^k \cdot r_{ij}$, $r_i^{(2),k} = \sum_{j \in \mathcal{I}} p_{ij}^k \cdot [r_{ij}]^2$ and hence the corresponding one-stage reward variance $\sigma_i^k = r_i^{(2),k} - [r_i^k]^2$.

A policy controlling the chain is a rule how to select actions in each state. In this note, we restrict on stationary policies, i.e. the rules selecting actions only with respect to the current state of the Markov chain X. Then a policy, say π, is fully identified by some decision vector f whose ith element $f_i \in \mathcal{A}_i$ identifies the action taken if the chain is in state X. Stationary policy $\pi \sim (f)$ then completely identifies the transition probability matrix $\boldsymbol{P}(f)$. Observe that the ith row of $\boldsymbol{P}(f)$ has elements $p_{i1}^{f_i}, \ldots, p_{iN}^{f_i}$ and that $\boldsymbol{P}^*(f) = \lim_{n\to\infty} n^{-1} \sum_{k=0}^{n-1} [\boldsymbol{P}(f)]^k$ exists.

Let the vectors $\boldsymbol{R}^{\pi}(n)$, $\boldsymbol{S}^{\pi}(n)$, and $\boldsymbol{V}^{\pi}(n)$ denote the first moment, the second moment and the variance of the (random) total reward ξ_n respectively received in the n next transitions of the considered Markov chain X if policy $\pi \sim (f)$ is followed, given the initial state $X_0 = i$. More precisely, for the elements of $\boldsymbol{R}^{\pi}(n)$, $\boldsymbol{S}^{\pi}(n)$, and $\boldsymbol{V}^{\pi}(n)$ we have

$$R_i^{\pi}(n) = \mathsf{E}_i^{\pi}[\xi_n], \quad S_i^{\pi}(n) = \mathsf{E}_i^{\pi}[\xi_n^2], \quad V_i^{\pi}(n) = \boldsymbol{\sigma}_i^{2,\pi}[\xi_n]$$

where $\xi_n = \sum_{k=0}^{n-1} r_{X_k,X_{k+1}}$ and E_i^{π}, $\boldsymbol{\sigma}_i^{2,\pi}$ are standard symbols for expectation and variance if policy π is selected and $X_0 = i$. Recall that

$$R_i^{\pi}(n+1) = r_i^{f_i} + \sum_{j \in \mathcal{I}} p_{ij}^{f_i} \cdot R_j^{\pi}(n) \tag{1}$$

Throughout the paper make the following assumption:

AS 1. For any stationary policy $\pi \sim (f)$, the transition probability matrix $\boldsymbol{P}(f)$ has a single class of recurrent states and is aperiodic.

2 Reward Variance of Markov Chains

Since for any integers $m < n$ $[\xi_n]^2 = [\xi_m]^2 + 2 \cdot \xi_m \cdot \xi^{(m,n)} + [\xi^{(m,n)}]^2$ (where $\xi^{(m,n)}$ is reserved for the reward obtained from the mth up to the nth transition) we get

$$\mathsf{E}_i^{\pi}[\xi_n]^2 = \mathsf{E}_i^{\pi}[\xi_m]^2 + 2 \cdot \mathsf{E}_i^{\pi}[\xi_m \cdot \xi^{(m,n)}] + \mathsf{E}_i^{\pi}[\xi^{(m,n)}]^2. \tag{2}$$

In particular, for $m = 1$, $n := n+1$ if policy $\pi \sim (f)$ is followed we get:

$$S_i^{\pi}(n+1) = \sum_{j \in \mathcal{I}} p_{ij}^{f_i} \cdot \{[r_{ij}]^2 + 2 \cdot r_{ij} \cdot R_j^{\pi}(n)\} + \sum_{j \in \mathcal{I}} p_{ij}^{f_i} \cdot S_j^{\pi}(n). \tag{3}$$

Since $V_i(\cdot) = S_i(\cdot) - [R_i(\cdot)]^2$ by (1), (3) we arrive after some algebra at

$$V_i^{\pi}(n+1) = \sum_{j \in \mathcal{I}} p_{ij}^{f_i} \cdot \{[r_{ij} + R_j(n)]^2\} - [R_i(n+1)]^2 + \sum_{j \in \mathcal{I}} p_{ij}^{f_i} \cdot V_j^{\pi}(n). \tag{4}$$

From the literature (see e.g. [1]) it is well known that under AS 1 there exist vector $\boldsymbol{w}^{\pi}$, constant vector $\boldsymbol{g}^{\pi}$ and vector $\boldsymbol{\varepsilon}(n)$ (where all elements of $\boldsymbol{\varepsilon}(n)$ converge to zero geometrically) such that

$$\boldsymbol{R}^{\pi}(n) = \boldsymbol{g}^{\pi} \cdot n + \boldsymbol{w}^{\pi} + \boldsymbol{\varepsilon}(n) \Rightarrow \lim_{n \to \infty} n^{-1} \boldsymbol{R}^{\pi}(n) = \boldsymbol{g}^{\pi} = \boldsymbol{P}^*(\pi) \cdot \boldsymbol{r}(f). \tag{5}$$

The constant vector $\boldsymbol{g}^{\pi}$ along with vectors $\boldsymbol{w}^{\pi}$ are uniquely determined by

$$\boldsymbol{w}^{\pi} + \boldsymbol{g}^{\pi} = \boldsymbol{r}(f) + \boldsymbol{P}(f) \cdot \boldsymbol{w}^{\pi}, \quad \boldsymbol{P}^*(\pi) \cdot \boldsymbol{w}^{\pi} = 0. \tag{6}$$

By using the relations (1), (4) and (5) in a number of steps we arrive at (for details see [2], [3]):

$$\boldsymbol{V}^{\pi}(n+1) = \boldsymbol{s}(\pi) + \boldsymbol{P}(f) \cdot \boldsymbol{V}^{\pi}(n) + \boldsymbol{\varepsilon}^{(1)}(n) \tag{7}$$

where for elements of the vector $\boldsymbol{s}(\pi)$ we have

$$s_i(\pi) = \sum_{j \in \mathcal{I}} p_{ij}^{f_i} \cdot \{[r_{ij} + w_j^{\pi}]^2\} - [g^{\pi} + w_i^{\pi}]^2 \tag{8}$$

$$= \sum_{j \in \mathcal{I}} p_{ij}^{f_i} \cdot \{[r_{ij} - g^{\pi} + w_j^{\pi}]^2\} - [w_i^{\pi}]^2 \tag{9}$$

and elements of the vector $\boldsymbol{\varepsilon}^{(1)}(n)$ converge to zero geometrically.

In analogy with (5), (6) we can conclude that there exists vector $\boldsymbol{w}^{(2),\pi}$ along with a constant vector $\boldsymbol{g}^{(2),\pi}$ uniquely determined by

$$\boldsymbol{w}^{(2),\pi} + \boldsymbol{g}^{(2),\pi} = \boldsymbol{s}(\pi) + \boldsymbol{P}(f) \cdot \boldsymbol{w}^{(2),\pi}, \quad \boldsymbol{P}^*(\pi) \cdot \boldsymbol{w}^{(2),\pi} = 0 \tag{10}$$

such that

$$\boldsymbol{V}^{\pi}(n) = \boldsymbol{g}^{(2),\pi} \cdot n + \boldsymbol{w}^{(2),\pi} + \boldsymbol{\varepsilon}(n) \Rightarrow \boldsymbol{g}^{(2),\pi} = \lim_{n \to \infty} \frac{\boldsymbol{V}^{\pi}(n)}{n} = \boldsymbol{P}^*(\pi) \cdot \boldsymbol{s}(\pi). \tag{11}$$

Moreover, for the vector $\tilde{\boldsymbol{s}}(\pi)$ with elements

$$\tilde{s}_i(\pi) = \sum_{j \in \mathcal{I}} p_{ij}^{f_i} \cdot \{[r_{ij} - g^{\pi}]^2 + 2 \cdot r_{ij} \cdot w_j^{\pi}\} \tag{12}$$

$$= r_i^{(2),f_i} - [g^{\pi}]^2 + 2 \cdot \sum_{j \in \mathcal{I}} p_{ij}^{f_i} \cdot r_{ij} \cdot w_j^{\pi} \tag{13}$$

we get ($[\cdot]_{\text{sq}}$ denotes that elements of the vector are squared)

$$\boldsymbol{g}^{(2),\pi} = \boldsymbol{P}^*(\pi) \cdot \tilde{\boldsymbol{s}}(\pi) = \boldsymbol{P}^*(\pi) \cdot \{\boldsymbol{r}^{(2)}(f) - [\boldsymbol{g}^{\pi}]_{\text{sq}} + 2 \cdot \tilde{\boldsymbol{r}}(f,\pi)\} \tag{14}$$

$$= \tilde{\boldsymbol{g}}^{(2),\pi} - [\boldsymbol{g}^{\pi}]_{\text{sq}} + 2 \cdot \boldsymbol{P}^*(\pi) \cdot \tilde{\boldsymbol{r}}(f,\pi) = \bar{\boldsymbol{g}}^{(2),\pi} + 2 \cdot \boldsymbol{P}^*(\pi) \cdot \tilde{\boldsymbol{r}}(f,\pi) \tag{15}$$

where

$\boldsymbol{r}^{(2)}(f)$ is a column vector with elements $r_i^{(2),f_i} = \sum_{j \in \mathcal{I}} p_{ij}^{f_i} \cdot [r_{ij}]^2$,
$\tilde{\boldsymbol{r}}(f,\pi)$ is a column vector with elements $\sum_{j \in \mathcal{I}} p_{ij}^{f_i} \cdot r_{ij} \cdot w_j^{\pi}$,
$[\boldsymbol{g}^{\pi}]_{\text{sq}}$ is a constant vector with elements $[g^{\pi}]^2$,
$\tilde{\boldsymbol{g}}^{(2),\pi} = \boldsymbol{P}^*(\pi) \cdot \boldsymbol{r}^{(2)}(f)$ is a constant vector with elements $\tilde{g}^{(2),\pi}$, and
$\bar{\boldsymbol{g}}^{(2),\pi} = \tilde{\boldsymbol{g}}^{(2),\pi} - [\boldsymbol{g}^{\pi}]_{\text{sq}}$ is a constant vector with elements $\bar{g}^{(2),\pi}$.

Obviously, $\tilde{g}^{(2),\pi}$ averages expected values of the second moments of one-stage rewards, $\bar{g}^{(2),\pi}$ denotes the average "one-stage reward variance" considered with respect to the mean reward g^{π} instead of the one-stage expected reward $r_i^{f_i}$ in state $i \in \mathcal{I}$, and the last term in (15) expresses the Markov dependence that occurs if the total variance of cumulative rewards is considered.

3 Mean Variance Selection Rules

The following definitions slightly extends various mean variance selection rules originally introduced in [4]. Stationary policy $\hat{\pi} \sim (\hat{f})$ is called *δ-mean variance optimal* (with $\delta \in \langle 0,1 \rangle$) if for every stationary policy $\pi \sim (f)$

$$h^{\delta}(\hat{\pi}) = \delta \cdot \frac{g^{(2),\hat{\pi}}}{g^{\hat{\pi}}} - (1-\delta) \cdot g^{\hat{\pi}} \le h^{\delta}(\pi) = \delta \cdot \frac{g^{(2),\pi}}{g^{\pi}} - (1-\delta) \cdot g^{\pi}. \qquad (16)$$

In particular, 1-mean variance optimal policy is the mean variance optimal policy, i.e. it minimizes the ratio $g^{(2),\pi}/g^{\pi}$, 0-mean optimal policy maximizes the mean (average) reward.

Similarly, the *square mean variance optimal policy* $\pi^\star$ minimizes the ratio of the mean variance to the squared mean reward in the class of stationary policies, i.e. stationary policy $\pi^\star \sim (f^\star)$ is called *square mean variance optimal* if for every policy $\pi \sim (f)$

$$h^{(2)}(\pi) = g^{(2),\pi}/[g^{\pi}]^2 \ge g^{(2),\pi^\star}/[g^{\pi^\star}]^2 = h^{(2)}(\pi^\star). \qquad (17)$$

Using formulas (5), resp. (6), and (14), (15), resp. (10), for a given stationary policy $\pi \sim (f)$ we can easily calculate the values of $h^{\delta}(\pi)$ and $h^{(2)}(\pi)$ to evaluate the policy according to δ-mean variance and square mean variance optimality criteria.

However, this approach is not possible when we are seeking δ-mean variance optimal and square mean variance optimal policies; some algorithmic procedures must be employed not to evaluate all stationary policies. Unfortunately, there is the following substantial difference: in (5), (6) the ith element of the vector $\boldsymbol{r}(f)$ depends only on the action selected in state i, however in (11), (10) the ith element of the vector $\boldsymbol{s}(\pi)$ depends on the decisions taken in all states through the term $\sum_{j\in\mathcal{I}} p_{ij}^{f_i} \cdot r_{ij} \cdot w_j^{\pi}$. The same is also true for (14) where $\boldsymbol{s}(\pi)$ is replaced by a simplified $\tilde{\boldsymbol{s}}(\pi)$ with elements given by (12), (13).

For this reason for the mean variance tradeoff in the literature only "one-stage reward variance" (called also the myopic variance) instead of the variance of total (cumulative) rewards is considered, i.e., we ignore the last term on the RHS in (14), (15) and employ the fact that $\boldsymbol{g}^{\pi}$ is a constant vector. This "one-stage reward variance" simplification enables to employ some techniques of stochastic dynamic programming, however, considering "one-stage reward variance" simplification may select policies not optimal with respect to the above criteria.

If we characterize every stationary policy $\pi \sim (f)$ by the pair of values g^{π}, $\tilde{g}^{(2),\pi}$ (mean rewards calculated from one-stage rewards $r_i^{f_i} = \sum_{j\in\mathcal{I}} p_{ij}^{f_i} r_{ij}$, $r_i^{(2),f_i} = \sum_{j\in\mathcal{I}} p_{ij}^{f_i}[r_{ij}]^2$) in [2] a policy iteration type algorithm was suggested for finding the convex hull of all stationary policies (see Figure 1) and it was shown that if the "one-stage reward variance" (myopic variance) is considered the δ-mean variance optimal and square mean optimal policies can be selected in the vertices of the south-east boundary of this convex hull. It it important that the stationary policies corresponding to adjacent vertices of the convex hull select different actions only in one state.

In what follows present computational experience with the above method on a family of numerical examples.

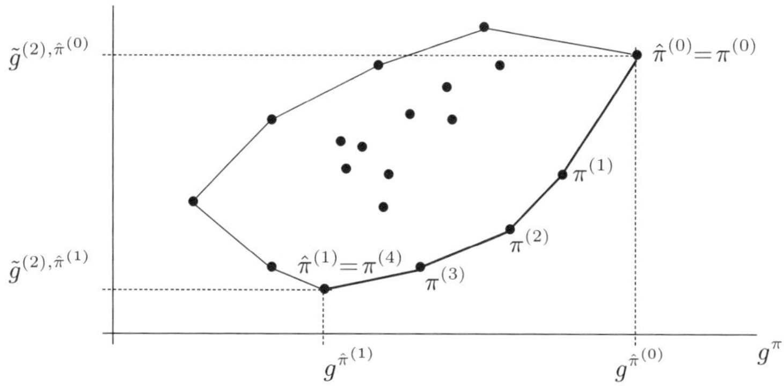

Fig. 1. Convex hull of the set of stationary policies

The above algorithm was tested on many large scale numerical examples (with dimensions of the state space $\mathcal{I}$ up to 100 states and each action set $\mathcal{A}_i$ up to 100 actions). The obtained results are depicted in Figures 2 and 3.

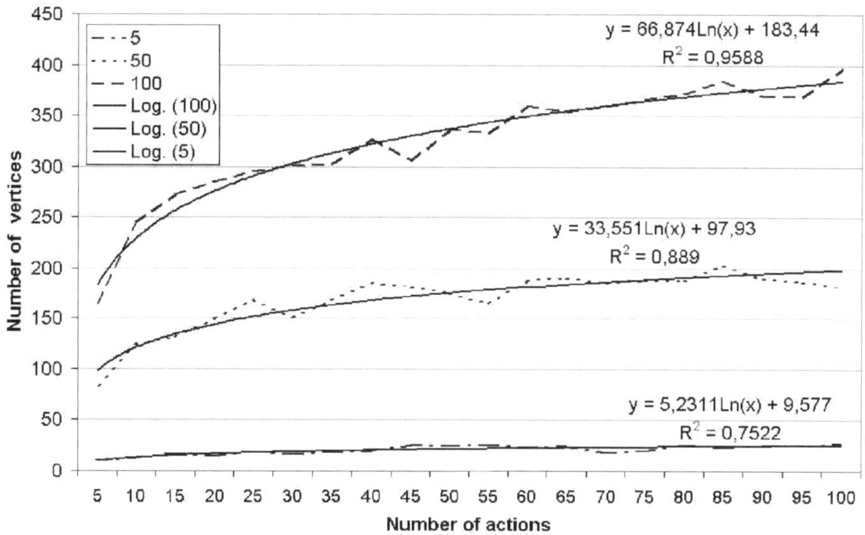

Fig. 2. Number of vertices on the south-east boundary of the policy set as a function of the number of actions (for 5, 50 and 100 states of the Markov decision chain)

Using a standard Pentium computer finding an optimal mean-variance stationary policy takes less than 10 minutes (observe that there exists some $100^{100} = 10^{200}$ stationary policies). However, the number of efficient policies (i.e. vertices on the south-boundary of the policy set, cf. Figure 1) for which the mean and variance must be calculated explicitly is usually less than 400. As it is shown in Figure 2 (some 300 examples with randomly generated parameters were tested) the number of efficient policies grows linearly in the number of states and logarithmically in the number of actions (see Figure 3).

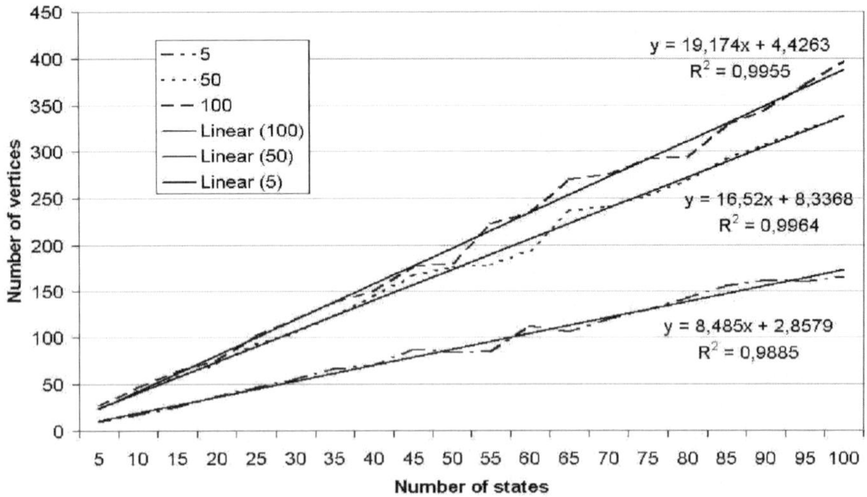

Fig. 3. Number of vertices on the south-east boundary of the policy set as a function of the number of states (for 5, 50 and 100 admissible actions in each state)

Acknowledgement. The research was supported by the Grant Agency of the Czech Republic under Grants 402/05/0115 and 402/04/1294.

References

1. Puterman ML (1994) Markov decision processes – discrete stochastic dynamic programming. Wiley, New York
2. Sladký K, Sitař M (2004) Optimal solutions for undiscounted variance penalized Markov decision chains. In: Marti K, Ermoliev Y, Pflug G (eds) Dynamic stochastic optimization. Springer, Berlin Heidelberg New York, pp 43–66
3. Sladký K (2005) On mean reward variance in semi-Markov processes. Math Methods Oper Res 62:387–397
4. Sobel MJ (1985) Maximal mean/standard deviation ratio in an undiscounted MDP. Oper Res Lett 4:157–159

On State Space Truncation of Finite Jackson Networks

Nico M. van Dijk

University of Amsterdam, Department of Economics
Roetersstraat 11, 1018 WB, Amsterdam, Netherlands

Abstract: An error bound result is provided for the truncation of a finite Jackson network from which both a computational and analytic error bound can be concluded. The results are illustrated for a special application of a cellular mobile communication network.

1. Introduction

Jackson type networks are well-known as a powerful modeling tool in a variety of application fields most notably among which manufacturing, reliability, computer communications, the service industry and last but not least telecommunications with present day applications such as for internet and mobile networks.

Unfortunately, closed form expressions, more precisely product form solutions, are available only under special conditions, generally referred to as separability conditions. These conditions are typically violated by interactions between separate service stations such as most commonly due to finite capacity constraints (buffers).

Numerical or approximate computations then become required. However, the state spaces involved might be (prohibitively) large. An approach that to some extent justifies a state space truncation is thus of interest. In fact, error bounds for a performance effect of a state space truncation could be useful in a twofold manner:

i) *to justify a numerical reduction*
ii) *for comparison with an infinite solvable (product form) model*

First results in this direction were established in [1] as applied to a tandem queue and an overflow model. Recently, in [2] these results were generalized and tailored to the truncation of finite Jackson networks with its associated technical complications. The present paper can be regarded as a restricted version of this reference. Nevertheless, it differs in that it contains a slightly different and more direct proof for the main result when relying upon results from this reference. The present paper is thereby kept self-contained and can be read independently.

For a brief discussion on somewhat related references, such as for general Markov chains and on stochastic monotonicity results by the reader is referred to [2].

The present paper extends [1] in two essential ways:

1. It provides an explicit computational error bound for the general case of a FJN directly expressed in the steady state distribution of the truncated FJN (This bound in turn will also lead to an analytical error bound for specific applications))
2. In a technical way in that no growth condition is required, in that the truncation is allowed to be transition dependent and in allowing a randomized routing. These extensions will be used for a specific application of interest.

2. Model and truncation

Model

Consider an open Jackson network with S service stations, numbered 1, ..., S and Poisson arrival rates λ_i at station i. Assume that $\lambda = \sum_i \lambda_i > 0$ and let $p_{0i} = \lambda_i / \lambda$. Upon service completion at station i a job routes to station j with probability p_{ij} or leaves the system with probability p_{io}. Station i has an exponential service rate $\mu_i(n_i)$ when n_i jobs are present.

Furthermore, station i has a capacity constraint for no more than N_i jobs. When station i is saturated ($n_i = N_i$), a job requesting service at station i (arriving from outside or from another node) is lost (i.e. it clears the system) (loss protocol). This assumption of a loss protocol (instead of a production type protocol in which blocked jobs are delayed) is more in line with communication applications natural for the special application of a cellular mobile communication network, as will be dealt with in the next section.

Truncation

Let us consider the truncation by restricting the queue lengths to $L_i \leq N_i$ for each station i. We assume that the truncated model also operates under the loss protocol. We aim to investigate the consequence of this truncation for the total throughput.

Let the queue length vector $\boldsymbol{n} = (n_1, n_2, ..., n_S)$ denote the number of jobs n_i at each station i=1, 2, ..., S. By $\boldsymbol{e}_i$ we denote the unit vector with the i-th component equal to 1, i.e.: $\boldsymbol{e}_i = (0, ..., 0, 1, 0, ..., 0)$. Hence by $\boldsymbol{n} - \boldsymbol{e}_i + \boldsymbol{e}_j$ we denote the state with one job more at station j and one less at station i. Similarly we use the notation $\boldsymbol{n} + \boldsymbol{e}_i$ and $\boldsymbol{n} - \boldsymbol{e}_i$. Furthermore we denote the state spaces of the original and truncated model by $\boldsymbol{S} = \boldsymbol{S}_N$ and $\overline{\boldsymbol{S}} = \boldsymbol{S}_L$.

Truncation result

We will be interested in the throughput $\boldsymbol{H}_N$ of the system. To this end, let $h=B^{-1}$ with $B \geq [\lambda + \max_{n\in S} \sum_i \mu_i(\boldsymbol{n})]$ and define the operator $\boldsymbol{T}$ by

$$\begin{aligned} &\boldsymbol{T} f(\boldsymbol{n}) = h \sum_j \lambda_j p_{0j} \mathbf{1}\{n_j < N_j\} f(\boldsymbol{n}+e_j) + \\ &\sum_i \mu_i(n_i) \sum_j p_{ij} \mathbf{1}\{n_j < N_j\} f(\boldsymbol{n}-e_i+e_j) + [1 - h\lambda - h\sum_i \mu_i(n_i)] f(\boldsymbol{n}). \end{aligned} \tag{1}$$

The throughput $\boldsymbol{H}_N$ of the original system is then determined by

$$\begin{aligned} &\boldsymbol{H}_N = \lim_{k\to\infty} \frac{B}{k}\left[\boldsymbol{V}^k(\boldsymbol{n})\right] \quad \text{for arbitrary } \boldsymbol{n} \in S \text{ with for } k = 0, 1, 2, \ldots: \\ &\boldsymbol{V}^k(\boldsymbol{n}) = h\, r(\boldsymbol{n}) + \boldsymbol{T}\boldsymbol{V}^{k-1}(\boldsymbol{n}) \quad \text{where} \\ &r(\boldsymbol{n}) = \sum_p \lambda_p \mathbf{1}\{n_p < N_p\} \quad \text{and} \\ &\boldsymbol{V}^0(\boldsymbol{n}) = 0 \quad \text{for all } \boldsymbol{n}. \end{aligned} \tag{2}$$

Result 2.1

Consider a FJN with capacity constraints N_i at station i, (i=1, ..., S) and its truncation with capacity constraints $L_i \leq N_i$ at station i, (i=1, ..., S). Let $\boldsymbol{H}_N$ and $\boldsymbol{H}_L$ be the corresponding system throughputs and $\{\pi_L(\boldsymbol{n}) | \boldsymbol{n}\in\boldsymbol{S}_L\}$ the steady state distribution of the truncated FJN. Then:

$$\left|\boldsymbol{H}_N - \boldsymbol{H}_L\right| \leq \sum_{\boldsymbol{n}} \pi_L(\boldsymbol{n}) \sum_{j=1}^{S} \mathbf{1}\left\{n_j = L_j\right\} \max\left[\lambda_j, \sum_k \mu_k(n_k)\, p_{kj}\right] \tag{3}$$

Proof

Let $\overline{\boldsymbol{V}}^k$ and $\overline{\boldsymbol{T}}^k$ be defined similarly to $\boldsymbol{V}^k$ and $\boldsymbol{T}^k$ for the truncated network restricted to $\overline{S}$ instead of S and note that $\overline{S} \subseteq S$. Since for any $\boldsymbol{n} \in \overline{S}$:

$$r(\boldsymbol{n}) = \sum_p \lambda_p \mathbf{1}\{n_p < N_p\} = \overline{r}(\boldsymbol{n}) = \sum_p \lambda_p \mathbf{1}\{n_p < L_p\} \tag{4}$$

and as the transitions for the operator $\overline{\boldsymbol{T}}$ remain restricted to $\overline{S} \subseteq S$, for arbitrary $\boldsymbol{n} \in \overline{S}$ by (2):

$$\begin{aligned} (\boldsymbol{V}^k - \overline{\boldsymbol{V}}^k)(\boldsymbol{n}) &= (\boldsymbol{T}\boldsymbol{V}^{k-1} - \overline{\boldsymbol{T}}\overline{\boldsymbol{V}}^{k-1})(\boldsymbol{n}) \\ &= (\boldsymbol{T} - \overline{\boldsymbol{T}})\boldsymbol{V}^{k-1}(\boldsymbol{n}) + \overline{\boldsymbol{T}}(\boldsymbol{V}^{k-1} - \overline{\boldsymbol{V}}^{k-1})(\boldsymbol{n}) \end{aligned} \tag{5}$$

Now by virtue of the truncation for any t we have:

$$\begin{aligned} &(\boldsymbol{T} - \overline{\boldsymbol{T}})\, \boldsymbol{V}^t(\boldsymbol{n}) = \sum_j \mathbf{1}\{n_j = L_j\} + \\ &\sum_{j=1,\ldots,S} \lambda_j \mathbf{1}\{n_j = L_j\}\left[\boldsymbol{V}^t(\boldsymbol{n}+\boldsymbol{e}_j) - \boldsymbol{V}^t(\boldsymbol{n})\right] + \\ &\sum_{k=1,\ldots,S} \mu_k(n_k)\, p_{kj}\, \mathbf{1}\{n_j = L_j\}\left[\boldsymbol{V}^t(\boldsymbol{n}-\boldsymbol{e}_k+\boldsymbol{e}_j) - \boldsymbol{V}^t(\boldsymbol{n}-\boldsymbol{e}_k)\right] \end{aligned} \tag{6}$$

By lemma 2.2 below, for any t we may thus conclude:

$$\left|(\boldsymbol{T}-\bar{\boldsymbol{T}})\,\boldsymbol{V}^{t}(\boldsymbol{n})\right| \leq \gamma(\boldsymbol{n}) \equiv \sum_{j=1}^{S}\mathbf{1}\{n_j = L_j\}\max\left[\lambda_j, \sum_k \mu_k(n_k)\,p_{kj}\right] \tag{7}$$

Furthermore, since π_L represents the steady state distribution and thus $\sum_n \pi_L(\boldsymbol{n})\left(\bar{T}\,f\right)(\boldsymbol{n}) = \sum_n \pi_L(\boldsymbol{n}) f(\boldsymbol{n})$, by combining (5) and (7) and by pre-multiplication (weighing) by the steady state distribution π_L:

$$\begin{aligned}\sum_{\boldsymbol{n}} \pi_L(\boldsymbol{n})\left|\left(\boldsymbol{V}^k - \bar{\boldsymbol{V}}^k\right)(\boldsymbol{n})\right| &\leq h\sum_{\boldsymbol{n}} \pi_L(\boldsymbol{n})\gamma(\boldsymbol{n}) + \sum_{\boldsymbol{n}} \pi_L(\boldsymbol{n})\left|\left(\boldsymbol{V}^{k-1} - \bar{\boldsymbol{V}}^{k-1}\right)(\boldsymbol{n})\right| \\ &\leq k\left[h\sum_{\boldsymbol{n}} \pi_L(\boldsymbol{n})\gamma(\boldsymbol{n})\right]\end{aligned} \tag{8}$$

where the latter inequality follows by iteration and $\boldsymbol{V}^0(\boldsymbol{n}) = \bar{\boldsymbol{V}}^0(\boldsymbol{n}) = 0$ for all $\boldsymbol{n}$. Dividing the left and the right hand side of (8) by k, substituting $h = B^{-1}$ and recalling (2) the limit expressions from (2) for the original and truncated chain to be independent of the initial state, the proof is completed by using for $\boldsymbol{H}_L$ and $\boldsymbol{H}_N$. □

Lemma 2.2

For all states $\boldsymbol{n} \in \boldsymbol{S}$, any station l such that $\boldsymbol{n} + \boldsymbol{e}_l \in \boldsymbol{S}$, for all $k \geq 0$:

$$0 \geq \Delta_l \boldsymbol{V}^k(\boldsymbol{n}) = \boldsymbol{V}^k(\boldsymbol{n}+\boldsymbol{e}_l) - \boldsymbol{V}^k(\boldsymbol{n}) \geq -1 \tag{9}$$

Proof

The proof follows by induction in k, writing out the dynamic reward relation (2) with the transitions for $\boldsymbol{T}$ substituted, and by appropriately (re)arranging difference terms in a pair-wise manner. The details are technical and can be found in [2].

Remark 2.3 By inductive monotonicity arguments, again based on the dynamic reward relation from (2) in [2] it was also shown that π_L in (3) could be replaced by π_∞ with π_∞ exhibiting a closed product form expression by assuming infinite capacities and that with $\gamma(\boldsymbol{n})$ as defined in (7), (3) can also be replaced by

$$\begin{aligned}\boldsymbol{H}_L \leq \boldsymbol{H}_N &\leq \boldsymbol{H}_L + \sum_{\boldsymbol{n}} \pi_L(\boldsymbol{n})\gamma(\boldsymbol{n}) \\ &\leq \boldsymbol{H}_L + \sum_{\boldsymbol{n}} \pi_\infty(\boldsymbol{n})\gamma(\boldsymbol{n})\end{aligned} \tag{10}$$

3. Application: A Mobile Communication Network

A special application of practical interest that actually motivated this research is that of a mobile communication network. In such a network calls arrive in a cell j at some arrival rate λ_j (fresh calls). A call residing in cell j will move to (another) neighbouring cell k at a rate λ_{jk} (a handover call). Within a cell a call requires a frequency channel that is not used by another call within that cell (a free channel). Each cell has a finite number of frequency channels. Neighbouring cells cannot have the same frequency channel. When a fresh call cannot find a free channel it is lost. When a handover call cannot find a free new channel in the cell that it is moving to, it is broken off and also lost. Let N_i be the number of channels in cell i, $i = 1, \ldots, S$. Under the assumption of exponential call durations this mobile communications network can be parameterized as a Finite Jackson Network by:

$$\begin{cases} \mu_i = \mu + \sum_j \lambda_{ij} & \text{(holding / service rate in cell } i) \\ p_{ij} = \lambda_{ij} / \mu_i & \text{(handover probability from } i \text{ to } j) \\ p_{i0} = \mu / \mu_i & \text{(call completion probability in cell } i) \end{cases}$$

By result 2.1 in combination with remark 2.3, the following result can then be proven (see [2]).

Result 3.1 *With* $\rho_j = \left[v_j / \mu_j \right]$ and $B_j(N_j) = e^{-\rho_j} \sum_{k=N_j}^{\infty} \frac{1}{k!} \left[\rho_j \right]^k$ *(the loss probability of a multi-server (* $M \mid M \mid N_j \mid N_j$*) loss system):*

$$\Delta = \left[\frac{H_N - H_L}{H_N} \right] \leq \frac{\sum_j v_j B_j(N_j)}{\sum_j \lambda_j \left[1 - B_j(N_j) \right]} \tag{11}$$

Numerical examples

Some numerical examples will be provided in table 1 below. Herein, the number of channels is chosen identical in each cell (i.e. $N = N_i$ for all $i = 1, \ldots, S$). The number N is chosen such that a given loss probability maximum $\boldsymbol{B}$ is guaranteed, as according to a standard $M \mid M \mid \boldsymbol{N} \mid \boldsymbol{N}$-queue. (For interest, also in the reduced case these loss probabilities are listed). The relative error bound (in percentage) Δ is computed (estimated from above) by expression (11).

For the completion and handover rates, the assumption is made that 2/3 of all calls will be completed within a cell, i.e. $\mu/\mu_i = 2/3$, while handovers are equally (randomly) spread over the neighbouring cells.

S	N	$\boldsymbol{B}$ *(in %)*	M	$\boldsymbol{R}$ (%)	Δ (%)
7	10	0.5	19.5×10^6	100%	
	8	2.9	4.78×10^6	24.5%	5.2%
	7	6.1	2.10×10^6	10.7%	9.3%
	6	11.4	8.24×10^3	4.2%	17.6%
10	5	0.5	60.5×10^6	100%	
	4	2.0	9.77×10^6	16.6%	2.57%
	3	7.5	1.05×10^6	1.6%	9.95%
50	3	0.5	1.27×10^{30}	100%	
	2	4.3	7.18×10^{21}	$5{\times}10^{-7}$%	4.8%

Legend

S: Number of cells	$\boldsymbol{B}$: Standard loss percentage
N: Number of channels in a cell	$\boldsymbol{R}$: reduction factor (in %)
M: Number of states	Δ: Relative error bound by (11)

Table 1: Numerical results of state space reduction and analytic relative error bound (11)

Some remarks:

1. The results show that the relative error bound is substantially less than the reduction in number of states and thus also the computational effort for an approximate computation of the throughput with given accuracy!
2. The relative error bound, as based upon the analytic results, appears to remain in the order (roughly by at most a factor 1.5) of the loss probabilities as if the cells are in isolation (which they are obviously not). Such a result could have been expected intuitively but no formal support seemed to be available.
3. In line with remark 2, the results also provide formal support for approximating finite cells by infinite cells. (Think of $N_j = \infty$ for all j.)

References

[1] N.M. van Dijk, Truncation of Markov chains with applications to queueing, *Operations Research,* Vol. 39, (1991), 1018-1026.

[2] N.M. van Dijk, Error Bounds for state space truncation of continuous time Markov chains and finite Jackson Networks. Research report. University of Amsterdam, 2005

Numerical Method for the Single-Server Bulk-Service Queuing System with Variable Service Capacity, $M/G^y/1$, with Discretized Service Time Probability Distribution

Mejía-Téllez, Juan., Ph.D.

Sección de Estudios de Posgrado e Investigación, Escuela Superior de Ingeniería Mecánica y Eléctrica, Instituto Politécnico Nacional
Unidad Profesional Adolfo López Mateos, Col. Lindavista, 07738, D.F. México

Departamento de Sistemas, División de Ciencias Básicas e Ingeniería, Universidad Autónoma Metropolitana
Av. San Pablo no. 180, Col. Reynosa Tamaulipas, 02200, D.F, México
jmt@correo.azc.uam.mx

1 Introduction

In the queueing theory when the server is occupied by more than 1 units demanding service such a case is named as bulk-service, if the service capacity available is not the same at every service, it is then said of variable capacity (Y). The time the server spends attending customers is a random variable commonly, and when in the model it is not declare specifically the type of the service time distribution then is declared as generic and the simbol G is used. When the arrival stream is of Markovian type, the system describe here is symbolized as $M/G^y/1$, and it is considered in the literature as of transportation type.
Transportation type queueing systems have received considerable attention in the past, since the original work of Bailey (1954). Bailey analysed the system with constant service capacity giving analytical solution for the queue length, applying the imbedded Markov chain technique (IMCT) due to Kendall (1951, 1953). Jaiswal (1961) extended Bailey's problem to the case in which the maximum number of units to be taken for service is not constant, but depends upon the number of units already present with the server as well as upon capacity. He applied two techniques to solve the problem, the IMCT and the phases technique. Singh (1971) analysed Jaiswal's paper considering the waiting room of (N+1) fixed capacity (including those in service). An arrival finding the waiting room full baulks and an arrival after joining the system does not renege. Alfa (1982) studied Jaiswal's model in discrete time, and extended the problem by allowing time-dependent bulk arrivals. Brière and Chaudhry (1989) took over the work done by Jaiswal, proposing two methods for the numerical inversion of Jaiswal's analytical expression. They compute steady-state probabilities and moments of the number of customers in the system at three different epochs, post-departure, random, and pre-arrival. Numerical results are illustrated, given a service time distribution,

such as deterministic, Erlang, hyperexponential, and uniform distribution. They claim that with these algorithms it is possible to find solution for the system with maximum capacity of up to 200 units and high traffic intensity. Chaudhry et al (1991) discussed the Singh's model, with the feature that the waiting room includes those in the queue with maximum value N and B the maximum capacity of the server, i.e., the M/G(0,y)/1:(N+B)) system. In this paper comparative analysis of computational aspects is carried out using the Jacobi iterative method and a root finding method. Steady state probabilities and moments of the number of customers in system at post-departure epochs have been obtained. A variety of numerical results have been obtained for Erlang, deterministic and hyper-exponential distributions. In all analytical solutions given in the work mentioned above, the solution depends upon a number of roots of an equation, whose form is determined by the type of service time distribution. The number of roots to be determined depends either upon the size of the batch which is served at each service epoch, or upon the type of service time distribution. If the batch size is large the determination of these roots is difficult. Mejia (1993) and Mejia and Worthington (1994) developed practical algorithms to give approximate solution to Bailey's model, at transient epochs and at steady state for homogeneous and inhomogeneous stream arrivals. A scaling approximation was proposed for high capacity system. The approaches applied were: imbedded Markov chain, discrete time and phases technique,

This briefly review about the work done concerning the single server bulk service system, was done to emphasize the techniques applied, the form of results and their practicability.

2 Proposed Model

Let be the service capacity a random variable with a maximum value, and following the imbedded Markov chain approach, the number of units waiting in the system is observed just before the server arrival and just before the server departure. Let $E[T^r]$ be the r-th moment about the origin of the empirical or theoretical service time random variable, now, our goal is to find a discrete service time random variable that match the first moments of the continuous one. To do this we adopt the method proposed by Miller and Rice (1983) which is reproduced here since it is essential for our model development:

Let us assume that the service time takes the value t_i with probability p_i and the number of them are finite i.e. i = 1,2,...,m, such that

$$\sum_{i=1}^{m} t_i^r\, p_i = E\left[T^r\right]. \qquad (2.1)$$

Defining a constant length of time say v such that t_i = iv the equations above take the form

$$v^r \sum_{i=1}^{m} i^r\, p_i = E\left[T^r\right] \quad r = 0,1,2,\ldots \qquad 2.2()$$

Having a discretized probability mass distribution { p_i }, i=1, 2, 3, ..., m and $p_i \geq 0$, with a unit service time of length v, where the product i times v is the service time with occurrence probability p_i, then, the probability of k arrivals occurring in a time period of length v is Poisson, i.e.

$$\Lambda_k = \frac{e^{-\lambda_v}}{k!} \lambda_v^k \qquad k= 0,1,2,\ldots.. \tag{2.3}$$

$$\lambda_v = \rho v E[Y]/E[T] \tag{2.4}$$

Where ρ is the traffic intensity, E [Y] is the mean service capacity, E[T] is the mean service time: When the service time is of length i times v , the probability of k arrivals is again Poisson with mean $i\lambda_v$. The probability of k arrivals while service is taking place is then

$$\Lambda_k = \sum_{i=1}^{m} p_i e^{-i\lambda_v} (i\lambda_v)^k / k! \tag{2.5}$$

So that, Λ is the vector holding the probabilities of the number of units arriving while the service is in progress. Let: Q be a vector holding the probabilities of the number of customers in the queue just before the server arrives., U be a vector holding the probabilities of the number of customers left behind by the server.., C be a vector holding the probabilities of the number of capacity units available for service, and be b its maximum value. Then at the service number n (n = 1,2,..)

$$Q^n = U^{n-1} * \Lambda \tag{2.6}$$

the symbol * denotes convolution, i.e.

$$Q_k^n = \sum_{j=0}^{k} U_j^{n-1} \Lambda_{k-j} \tag{2.7}$$

The elements of the vector U at epoch n are given by

$$U_0^n = \sum_{j=0}^{b} Q_j^n \sum_{i=j}^{b} C_i \tag{2.8}$$

$$U_k^n = \sum_{j=k}^{b+k} Q_j^n C_{j-k} \qquad k= 1,2,3,\ldots \tag{2.9}$$

The model is completely defined if the initial condition is given. For example if the system starts empty $U_0^0 = 1.0$, with k units waiting $U_k^0 = 1.0$, etc.

Implementing the algorithm in a computer program stopping rules for vectors expansion and reaching the steady state must be given, Usually an error of 10^{-10} for cumulative probability reaching the unit, and 10^{-6} error reaching steady state is common for practical purposes

3 Validation

Results given by the application of the model were contrasted with those presented by Brière and Chaudhry for the queueing system $M / E_r^Y / 1$, for different values of traffic intensity, and for the following service capacity probability distribution: $Y_{50} = 0.1$, $Y_{60} = 0.2$, $Y_{70} = 0.4$, $Y_{80} = 0.4$. As an example of discretized distribution of an Erlang-3 used in this contrasting procedure was: v = 0.8763950317, p_2 = 0.4784945418, p_3 = 0.296103352, p_7 = 0.2254021062. It was found that no errors occur.

4 Application

Customers arrive randomly to a train station. Train service capacity is assumed to present a truncated Poisson distribution with parameter λ, (λ=127.5), and maximum capacity b = 150. So that

$$c_k = \frac{e^{-\lambda}\lambda^k / k!}{\sum_{j=0}^{b} e^{-\lambda}\lambda^j / j!} \qquad k = 0,1,2,\ldots,b = 150 \tag{4.1}$$

Train inter-departure times in the rush hour are illustrated in figure 4.1

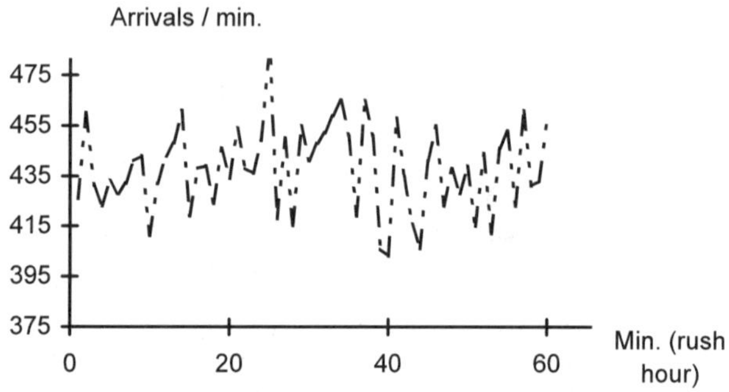

Fig. 4.1. Passenger arrival rates during the rush hour

Moments about the origin: 1st 2.2277, 2nd 5.5916, 3rd 15.5848
The station manager wants to know the behaviour of the system concerning the queue length and left behind customers, in order to have information for decision making

Applying the discretization method, it was found a probability distribution for the number of units to be taken of length v, matching the first two moments and the third one was found underestimated in about 1%: v = 1.257408, p_1 = 0.343188, p_2 = 0.542521, p_3 = 0.114291.

Table 4.1. . Steady state system behaviour given several values of traffic intensity

Traffic intensity	Average queue length	Average left behind
0.5	6305	0.08
0.6	77.1	0.94
0.7	92.9	4.10
0.8	115.3	13.8
0.9	168.1	54.0

In table 4.1 results are given for a series of scenarios varying traffic intensity, so that the station manager can get a wider information of the system behaviour. An interesting feature of results is observed, when the traffic intensity (percentage occupancy average) is less than 70%, there are practically no customers left behind, but when traffic intensity is larger than 70%, the manager be worried about the number of customers waiting for service on platform and of the number left behind,

The following graphs illustrate the system behaviour when traffic intensity is about 80%.

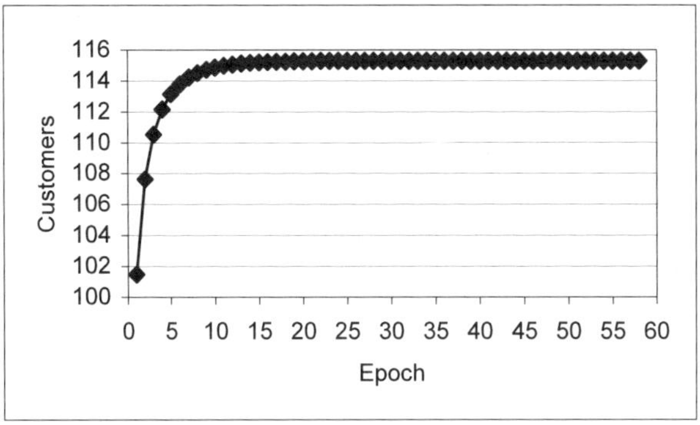

Fig. 4.2. Queue length just before service start, $\rho = 0.8$

From figure 4.2 it is observed that the system gets steady state at say epoch (service) 10, for traffic intensity 80%. For a higher level of traffic intensity, it is expected that the number of epochs to reach steady state will increase dramatically.

The running time of a C program on a PC for the case of $\rho = 90\%$ was about of three seconds.

5 Conclusions

Some of the most relevant papers tackling bulk service queueing systems with random service capacity were briefly described, through the analysis of the results given it was found that other authors solutions are difficult to apply. The solution method presented here does not require to fit a theoretical probability distribution for service time, but there is a need to find a discrete distribution matching at least the first two moments of the empirical service time distribution. Recursive numerical convolution of probability vectors must be carried out on a computer program, which is easy to implement. Numerical results are accurate enough for practical purposes.

An important advantage this model has among other approximation queueing model, such as those analysing the system at phase epochs, is that the system is observed at service epochs, having in consequence a substantial saving in computer running time.

References

Alfa AS (1982) Time-inhomogeneous bulk server queue in discrete time: A transportation type problem. Operations Research 30: 650-658

Bailey NTJ (1954) On queueing processes with bulk service. Journal of the Royal Statistical Society Series B 16: 80-87

Brière G, Chaudhry ML (1989) Computational analysis of single-server bulk-service queues $M/G^Y/1$. Advances in Applied. Probability 21: 207-225

Chaudhry ML, Gupta UC, Madill BR (1991) Computational aspects of bulk-service queueing system with variable capacity and finite waiting space $M/G^Y/1/N+B$. Journal of the Operations Research Society of Japan 34(4): 404-421

Jaiswal NK (1961) A bulk service queueing problem with variable capacity. Journal of the Royal Statistical Society Series B 23: 143-148

Kendall DG (1951) Some problems in the theory of queues. Journal of the Royal Statistical Society 13: 151-185

Kendall DG (1953) Stochastic processes occurring in the theory of queues and their analysis by the method of the imbedded Markov chain. Annals of Mathematical Statistics 24: 338-354

Mejía-Téllez J (1993) Models for single-server bulk-service queueing system of transportation type. Ph.D. thesis, Lancaster University U.K

Mejía-Téllez J, Worthington D (1994) Practical methods for queue length behaviour for bulk service queues of the form $M/G^{0,C}/1$ and $M(t)./G^{0,C}/1$. European Journal of Operational Research 74: 103-113

Miller AC, Rice TR (1983) Discrete approximations of probability distributions. Management Science 29(3): 352-362

Singh VP (1971) Finite waiting space bulk service system. Journal of Engineering Mathematics 5: 241-248

Worst-case VaR and CVaR

Jana Čerbáková

Charles university, Prague, Faculty of Mathematics and Physics, Department of Probability and Mathematical Statistics, Sokolovská 83, Praha 8, 186 75
janacerb@karlin.mff.cuni.cz*

Summary. The main goal of this paper is to derive and compare values of worst-case VaR and CVaR under different type of information on distribution of random parameter. To this purpose we exploit results from moment problem theory and apply upper bound of loss probability of univariate random variable with special properties, given expected value and variance. Subsequently, we suppose that except the first two moments of the distributions, we know further characteristics of the class of distributions. We assume symmetry and/or unimodality. The bounds are also illustrated on the case of interbank exchange rate.

Key words: exchange rate, moment problem, worst-case CVaR, worst-case VaR
MSC2000 Subject Classification: 62P20, 90C47, 91B30

1 Introduction

The main goal of this paper is to compare values of risk measures such as Value at Risk and Conditional Value at Risk under different type of information on distribution of random variable. To overcome the incomplete information about the distribution we apply the worst-case strategy, i.e. we define the worst-case VaR and CVaR.

We deal with univariate random parameter X defined on $(\Omega, \mathcal{F})$, where $\mathcal{F}$ is the Borel σ-algebra of a nonempty set $\Omega \subseteq \mathbb{R}$. The set of considered probability distributions of X is denoted by $\mathcal{P}$. Existence of all mentioned expected values is assumed.

In this paper the basic choice of $\mathcal{P}$ corresponds to specification of the distributions by its first and second order moments. Problem of moments is already discussed in [3] including results for unimodal distributions. The most important results of this paper come from dual formulation of moment problem. Essential outcomes in duality are also derived in [5]. As the next

* This work was supported by the Czech Grant Agency (grant 201/05/H007).

step we suppose besides of the knowledge of expected value and variance other properties of considered distributions - symmetry and unimodality.

The worst-case VaR for general, symmetric and/or unimodal distributions is derived in [2]. In this paper we extend these results for two examples of the worst-case CVaR. We also show the equality of worst-case VaR and CVaR in general and symmetric case. We illustrate the results obtained by applications of the discussed worst-case risk measures on a currency market, specially we study increments in the interbank exchange rate between EUR and HRK.

2 Var and CVaR

For loss random variable $-X$ with left continuous distribution function F_{-X}, and for $\alpha \in (0,1]$ we define Value at Risk by

$$VaR_\alpha(-X) := \min_{\gamma\in\mathbb{R}} \gamma \qquad \text{s.t.} \quad P(-X \geq \gamma) \leq \alpha. \tag{1}$$

For perfectly known distribution of $-X$ we obtain the optimal value $VaR_\alpha(-X) = F_{-X}^{-1}(1-\alpha)$, where F_{-X}^{-1} is the inverse of F_{-X}, i.e. $F_{-X}^{-1}(\alpha) := \inf\{x : F_{-X}(x) \geq \alpha\}$. Value at Risk is a quantile and hence it holds

$$VaR_\alpha(-X) = F_{-X}^{-1}(1-\alpha) = -F_X^{-1}(\alpha) = -VaR_{1-\alpha}(X).$$

Another popular risk measure is Conditional Value at Risk. For $\alpha \in (0,1)$ is define by

$$CVaR_\alpha(-X) := \inf_{a\in\mathbb{R}}\{a + \frac{1}{\alpha}E[(X+a)^-]\}, \tag{2}$$

where $(c)^- := \max\{0,-c\}$. CVaR may be also defined as a conditional tail expectation $CVaR_\alpha(-X) := -E[X|X \leq -VaR_\alpha(-X)]$. Both definitions are equivalent, see [1].

Specially if $-X$ is normally distributed with expected value $-\mu_X$ and variance σ_X^2 we obtain the results (Φ is distribution function of $N(0,1)$)

$$VaR_\alpha^N(-X) = -\mu_X + \Phi^{-1}(1-\alpha)\sigma_X,$$

$$CVaR_\alpha^N(-X) = -\mu_X + \frac{\sigma_X}{\sqrt{2\pi}\alpha}\exp\left\{-\frac{[\Phi^{-1}(1-\alpha)]^2}{2}\right\}.$$

2.1 Worst-case VaR and CVaR for general distributions

In most cases we do not exactly know the distribution of random variable X. We are only able to specify the set $\mathcal{P}$ of feasible probability distributions, e.g. $\mathcal{P}$ is the set of all distribution that fulfil certain moment conditions, or the set of symmetric or unimodal distributions. In that case we apply the min-max strategy and define the worst-case VaR, resp. CVaR, with respect to the set of probability distributions $\mathcal{P}$ as

$$wc\,VaR_\alpha(-X) := \min_{\gamma\in\mathbb{R}} \gamma \qquad \text{s.t.} \quad \sup_{\mu\in\mathcal{P}} P_\mu(-X \geq \gamma) \leq \alpha, \tag{3}$$

$$wc\,CVaR_\alpha(-X) := \sup_{\mu\in\mathcal{P}}\inf_{a\in\mathbb{R}}\{a + \frac{1}{\alpha}E_\mu[(X+a)^-]\}. \tag{4}$$

For given expected value $EX = \mu_X$ and variance $E[X-\mu_X]^2 = \sigma_X^2$ we can rewrite expression (4) in successive steps. First, due to finite expected value and convexity of the inner objective function in (4) we can interchange supremum and infimum. From [3] we obtain

$$\max_{\mu\in\mathcal{P}} E[(X+a)^-] = \max_{\mu\in\mathcal{P}} E[(-a-X)^+] = \frac{1}{2}[-a-\mu_X+\sqrt{\sigma_X^2+(\mu_X+a)^2}].$$

Then by solving $\min\limits_{a\in\mathbb{R}}\{a+\frac{1}{2\alpha}[-a-\mu_X+\sqrt{\sigma_X^2+(\mu_X+a)^2}]\}$ we obtain the optimal solution $a^* = -\mu_X + \frac{1-2\alpha}{2\sqrt{\alpha(1-\alpha)}}\sigma_X$ and formula

$$wcCVaR_\alpha(-X) = -\mu_X + \sqrt{\frac{1-\alpha}{\alpha}}\sigma_X. \tag{5}$$

For distribution identified by its first two moments and for $\gamma > -\mu_X$ the expression (5) coincides with that for $wcVaR_\alpha(-X)$. This results from one-sided Chebyshev bound, see [4]. If $\gamma \le -\mu_X$ $wcVaR_\alpha(-X)$ is equal to 1.

2.2 Worst-case VaR and CVaR for symmetric distributions

Definition 1. *Let $\Omega = I \subseteq \mathbb{R}$ be either a compact interval, or $I = \mathbb{R}$. We say that a distribution μ of random variable X defined on $(\Omega, \mathcal{F})$ is μ_X-symmetric if $\mu[\mu_X - x, \mu_X] = \mu[\mu_X, \mu_X + x]$ $\forall x \in I_{\mu_X}$, where $I_{\mu_X} := \{x \ge 0 : \mu_X - x \in I$ and $\mu_X + x \in I\}$.*

The set $\mathcal{P}^S_{\mu_X}$ of all μ_X-symmetric probability distributions is convex and closed. We can also write $\mathcal{P}^S_{\mu_X} = \text{cl}(\text{conv}(\mathcal{T}^S_{\mu_X}))$, where $\mathcal{T}^S_{\mu_X} = \{\mu = \frac{1}{2}\delta_{\mu_X+x} + \frac{1}{2}\delta_{\mu_X-x}, x\in I_{\mu_X}\}$ is the set of μ_X-symmetric Dirac distributions.

Lemma 1. *Let X be a random variable with symmetric distribution and given finite expected value μ_X and variance σ_X^2 then*

$$\sup_{\mu\in\mathcal{P}^S_{\mu_X}} E_\mu[(X+a)^-] = \begin{cases} \frac{\sigma_X^2}{8(a-\mu_X)} & \text{for } a > \mu_X + \frac{\sigma_X}{2}, \\ \mu_X - a + \frac{\sigma_X^2}{8(\mu_X - a)} & \text{for } a < \mu_X - \frac{\sigma_X}{2}, \\ \frac{\sigma_X - a + \mu_X}{2} & \text{for } \mu_X - \frac{\sigma_X}{2} \le a \le \mu_X + \frac{\sigma_X}{2}, \end{cases} \tag{6}$$

$$\sup_{\mu\in\mathcal{P}^S_{\mu_X}} P_\mu[X\ge\gamma] = \begin{cases} \frac{1}{2}\min\{1, \frac{\sigma_X^2}{(\gamma-\mu_X)^2}\} & \text{for } \gamma > \mu_X, \\ 1 & \text{for } \gamma \le \mu_X. \end{cases} \tag{7}$$

If the bounds (6) and (7) are achievable, then they are attained for a discrete distribution concentrated on a finite support, see [5].

Worst-case CVaR for symmetric distribution identified by its first two moments can be calculated as

$$wcCVaR^S_\alpha(-X) := \inf_{a\in\mathbb{R}}\{a + \frac{1}{\alpha}\sup_{\mu\in\mathcal{P}^S_{-\mu_X}} E_\mu[(X+a)^-]\},$$

where we insert the formula (6). We obtain the solution

$$wcCVaR^S_\alpha(-X) = \begin{cases} -\mu_X + \frac{\sigma_X}{\sqrt{2\alpha}} & \text{for } \alpha < \frac{1}{2}, \\ -\mu_X + \sqrt{\frac{1-\alpha}{2}}\frac{\sigma_X}{\alpha} & \text{for } \alpha > \frac{1}{2}, \\ -\mu_X + \sigma_X & \text{for } \alpha = \frac{1}{2}. \end{cases} \tag{8}$$

Worst-case VaR for loss random variable $-X$ with symmetric distribution and given expected value $-\mu_X$ and variance σ_X^2 results from

$$wcVaR_\alpha^S(-X) := \min_{\gamma > -\mu_X} \gamma \qquad \text{s.t.} \quad \frac{1}{2}\min\{1, \frac{\sigma_X^2}{(-\mu_X - \gamma)^2}\} \le \alpha.$$

For $\alpha < \frac{1}{2}$ the optimal value is $wcVaR_\alpha^S(-X) = -\mu_X + \frac{\sigma_X}{\sqrt{2\alpha}}$. If $\alpha \ge \frac{1}{2}$ the minimum is not achieved. There only exists infimum and its value is $-\mu_X$. If $\gamma \le -\mu_X$ the solution exists only for $\alpha = 1$.

To summarize: For general and symmetric distributions and for $\alpha < \frac{1}{2}$ *the worst-case VaR and CVaR are identical.*

2.3 Worst-case VaR for symmetric and unimodal distributions

Definition 2. *Let* $\Omega = I \subseteq \mathbb{R}$ *be either a compact interval, or* $I = \mathbb{R}$. *Distribution* μ *of random variable* X *defined on* $(\Omega, \mathcal{F})$ *is said to be* μ_X*-unimodal on* $I \ni \mu_X$ *if the corresponding distribution function is convex on the left of* μ_X *and concave on the right of* μ_X.

Let $\mathcal{P}_{\mu_X}^U$ denote the set of continuous μ_X-unimodal probability distributions. Then $\mathcal{P}_{\mu_X}^U$ is convex and $\text{cl}(\mathcal{P}_{\mu_X}^U)$ is the set of all μ_X-unimodal distributions. The set of all μ_X-symmetric and μ_X-unimodal distributions $\text{cl}(\mathcal{P}_{\mu_X}^{SU}) = \mathcal{P}_{\mu_X}^S \cap \text{cl}(\mathcal{P}_{\mu_X}^U) = \text{cl}(\text{conv}(\mathcal{T}_{\mu_X}^{SU}))$ is convex, where $\mathcal{T}_{\mu_X}^{SU} = \{\delta_{[\mu_X - x, \mu_X + x]} : x \in I_{\mu_X}, x \ne 0\}$. By $\delta_{[a,b]}$ we denote probability distribution with uniform density on $[a, b]$. For details see [2], [3], [5].

Lemma 2. *Let* X *be a random variable with symmetric unimodal distribution and given finite expected value* μ_X *and variance* σ_X^2 *then*

$$\sup_{\mu \in \text{cl}(\mathcal{P}_{\mu_X}^{SU})} P_\mu[X \ge \gamma] = \begin{cases} \frac{1}{2}\min\{1, \frac{4}{9}\frac{\sigma_X^2}{(\gamma - \mu_X)^2}\} & \text{for } \gamma > \mu_X, \\ 1 & \text{for } \gamma \le \mu_X. \end{cases} \tag{9}$$

The corrected proof can be found in [2]. By analogy with symmetric case, the discrete optimal distribution exists if the bound is achievable.

Now we can handle the univariate worst-case VaR for symmetric unimodal distribution of loss random variable $-X$ with known expected value $-\mu_X$ and variance σ_X^2. The optimal value is solution of the problem

$$wcVaR_\alpha^{SU}(-X) := \min_{\gamma > -\mu_X} \gamma \quad \text{s.t.} \quad \frac{1}{2}\min\{1, \frac{4}{9}\frac{\sigma_X^2}{(-\mu_X - \gamma)^2}\} \le \alpha. \tag{10}$$

For $\alpha < \frac{1}{2}$ the minimum is $wcVaR_\alpha^{SU}(-X) = -\mu_X + \sqrt{\frac{2}{9\alpha}}\sigma_X$. If $\alpha \ge \frac{1}{2}$, the minimum is not achieved. There only exists infimum and its value is $-\mu_X$. If $\gamma \le -\mu_X$ the solution exists only for $\alpha = 1$.

3 Numerical results

The assumption of normality is frequently used in practice. When it is not fulfilled, the maximal error in VaR, CVaR caused by holding on the fixed normal distribution can be evaluated as the difference between the worst-case VaR, resp. worst-case CVaR, and the VaR, resp. CVaR, value provided by a specific distribution.

In our example we study interbank exchange rate increments between Euro EUR and Croatian Kuna HRK during the time period 1.6.2004 – 30.6.2005. Our data represent increments of daily closing values to insure data independence. We apply the last 30 observations to estimate expected value and variance for a calculation of the next worst-case level. The results are presented in Figure 1. The confrontation illustrates the fact that a lower information level concerning the class of distributions causes an increase of gap between worst-case levels.

We also test symmetry, unimodality and normality by standard tests. We reject normality on the level $\alpha = 0.05$ in 37% of all cases. Therefore, it makes sense to deal with the relaxation of normality assumption. We do not reject the supposition of unimodality in one third of all cases. Symmetry cannot be rejected.

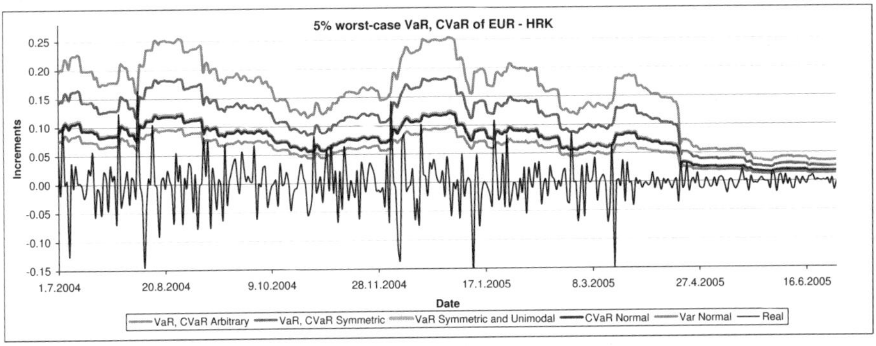

Fig. 1. Worst-case VaR, CVaR for increments of EUR - HRK interbank exchange rate.

The numerical experiments and tests of symmetry, unimodality and normality have been computed in Maple 9, R 1.7.0 using packages base and diptest, see [6], and in Microsoft Excel 2000 under Windows 2000.

Appendix

Proof of Lemma 1.
The proof of equation (7) can be found in [5].

The corresponding dual formulation of the problem (6) is, see [3], [5],

$$\begin{aligned} &\min_{y \in \mathbb{R}^3} y_0 + y_1\mu_X + y_2(\sigma^2 + \mu_X^2) \\ &\text{s.t.} \quad 2y_0 + 2\mu_X y_1 + 2y_2(\mu_X^2 + x^2) \geq (\mu_X - x + a)^- + (\mu_X + x + a)^- \quad \forall x \geq 0. \end{aligned} \tag{11}$$

In order to simplify let $\mu_X = 0$. (The results for $\mu_X \neq 0$ can be derived by substituting $a - \mu_X$ for a.) We solve following problem

$$\begin{aligned} &\min_{y \in \mathbb{R}^2} y_0 + y_2\sigma^2 \\ &\text{s.t.} \quad 2y_0 + 2y_2x^2 \geq (-x + a)^- + (x + a)^- \quad \forall x \geq 0. \end{aligned} \tag{12}$$

Evidently $y_2 > 0$. We distinguish three cases:

CASE 1: $a > 0$. Constraints of the problem (12) imply

$$2y_0 + 2y_2x^2 \geq \begin{cases} 0 & \text{for } 0 \leq x \leq a, \\ x - a & \text{for } x > a. \end{cases} \tag{13}$$

First suppose that the parabola tangents the ray $x-a$, i.e. $2y_0+2y_2x^2 = x-a$. We obtain $y_0 = \frac{1}{16y_2} - \frac{a}{2} > -\frac{a}{2}$. In that case parabola cannot have more than one common point with x-axe, therefore $2y_0 + 2y_2x^2 \geq 0$. The discriminant cannot be greater than zero. We realize the condition $y_0 \geq 0$ and thereout $y_2 \leq \frac{1}{8a}$. Problem (12) reduces to

$$\min \frac{1}{16y_2} - \frac{a}{2} + y_2\sigma_X^2 \quad \text{s.t.} \quad 0 < y_2 \leq \frac{1}{8a}.$$

In case when Lagrangian multiplier corresponding to the upper constrain is equal to zero we obtain solution $y_2^* = \frac{1}{4\sigma_X}$, $a \leq \frac{\sigma_X}{2}$ and optimal value is $\frac{\sigma_X - a}{2}$. Otherwise for nonzero multiplier is $y_2^* = \frac{1}{8a}$, $a > \frac{\sigma_X}{2}$ and optimal value is equal to $\frac{\sigma_X^2}{8a}$. If we assume that parabola $2y_0 + 2y_2x^2$ tangents the x-axe, we obtain the same results.

CASE 2: $a < 0$. Constraints of the problem (12) are reduced to

$$2y_0 + 2y_2x^2 \geq \begin{cases} -2a & \text{for } 0 \leq x \leq -a, \\ x - a & \text{for } x > -a. \end{cases} \tag{14}$$

At first let the parabola $2y_0 + 2y_2x^2$ tangent the ray $x - a$. We realize the condition $y_0 = \frac{1}{16y_2} - \frac{a}{2} > -\frac{a}{2}$. The parabola cannot have more than one common point with the ray $-2a$. By analogy with case 1 we get $y_0 \geq -a$ and $y_2 \leq -\frac{1}{8a}$. We obtain the optimal value $\frac{\sigma_X - a}{2}$ for $-a \leq \frac{\sigma_X}{2}$ and $-a - \frac{\sigma_X^2}{8a}$ for $-a > \frac{\sigma_X}{2}$. The same result is obtained when assuming that parabola tangents the ray $-2a$.

CASE 3: $a = 0$. We solve problem (12) under condition $2y_0 + 2y_2x^2 \geq x$ for $x \geq 0$. The parabola cannot have more than one common point with the axe of the first quadrant, form here it follows $y_0 \geq \frac{1}{16y_2}$. The problem (12) reduces to $\min\limits_{y_2>0} \frac{1}{16y_2} + y_2\sigma_X^2$. The minimum is achieved for $y_2^* = \frac{1}{4\sigma_X}$. The optimal value is equal to $\frac{\sigma_X}{2}$.

References

1. Acerbi C, Tasche D (2002) On the coherence of expected shortfall. J. of Bank. Fin. 26: 1487–1503
2. Čerbáková J (2005) Moment Problem and worst-case Value-at-Risk. In: Skanská H (ed) MME 2005 Proceedings: 33-38
3. Dupačová J (1977) Minimaxová úloha stochastického lineárního programování a momentový problém. Ekonomicko matematický obzor 13: 279–307
4. Isii K (1963) On the sharpness of Chebyshev-type inequalities. Ann. Inst. Stat. Math. 14: 185–197
5. Popescu I (2005) A semidef. program. approach to optimal moment bounds for convex classes of distribution. forthcomming in Math. of Oper. Research
6. R Develop. Core Team (2003) R: A language and environment for stat. comput.. R Found. for Stat. Comp., Vienna, Austria, URL http://www.R-project.org

Printing: Krips bv, Meppel
Binding: Stürtz, Würzburg